CRC Mathematical Modeling Series

Series Editor
Nicola Bellomo
Politecnico di Torino, Italy

Titles included in the series:

Nicola Bellomo/Luigi Preziosi: **Modelling Mathematical Methods and Scientific Computation**

Paolo Brandimarte/Agostino Villa: **Advanced Models for Manufacturing Systems Management**

Fabio Casciati/Brian Roberts: **Mathematical Models for Structural Reliability Analysis**

Jean-Pierre Kernévez: **The Sentinel Method and Its Application to Environmental Pollution Problems**

Wolfgang Klieman/N. Sri Namachchivaya: **Nonlinear Dynamics and Stochastic Mechanics**

Andrew Kusiak/Maurizio Bielli: **Designing Innovations in Industrial Logistics Modelling**

Alexander B. Movchan/Nataliya V. Movchan: **Mathematical Modelling of Solids witb Nonregular Boundaries**

Mathematical Analysis for Modeling

Judah Rosenblatt
Director of Biomathematics
University of Texas Medical Branch

with the collaboration of

Stoughton Bell
Professor Emeritus of Computer Science
University of New Mexico

CRC Press
Taylor & Francis Group
Boca Raton London New York

CRC Press is an imprint of the
Taylor & Francis Group, an **informa** business

CRC Press
Taylor & Francis Group
6000 Broken Sound Parkway NW, Suite 300
Boca Raton, FL 33487-2742

First issued in paperback 2019

ISBN-13: 978-0-8493-8337-3 (hbk)
ISBN-13: 978-0-367-40010-1 (pbk)
Library of Congress Card Number 98-40054

Library of Congress Cataloging-in-Publication Data

Rosenblatt, Judah I. (Judah Isser), 1931–.
Mathematical analysis for modeling / Judah Rosenblatt ; with the collaboration of Stoughton Bell.
p. cm. -- (CRC mathematical modelling series)
Includes bibliographical references and index.
ISBN 0-8493-8337-4 (alk. paper)
I. Mathematical analysis. 2. Mathematical models I. Bell, Stoughton. III. Title. II. Series.
QA3OO.R626 1998
515—dc21 98-40054
CIP

Visit the Taylor & Francis Web site at
http://www.taylorandfrancis.com

and the CRC Press Web site at
http://www.crcpress.com

Preface

Overview

Science and technology of all kinds are concerned with describing the behavior of, and relationships among observations. What distinguishes the various fields are the quantities being described and related. In medicine we are interested in measurements such as blood pressure, metabolic rate, heart rate, body temperature, LDL[1] level and recovery times — as well as treatment measurements such as drug dosage and measurements describing surgical procedures. In mechanical engineering, the measurements we deal with are positions and velocities of mechanical system components such as various types of engines. In chemistry we focus on the rates of formation and breakdown of different chemicals.

In all of these systems we need to describe the quantities of interest and find appropriate descriptions of how they are related. The function of the mathematics developed in this work is to provide the most useful framework that we can for accomplishing these aims.

Our emphasis is not on presenting a large grab-bag of techniques for solving most standard problems, but rather, to increase understanding and provide a natural approach to problem formulation and solution. Specifically, we examine how to produce simple descriptions of systems we are likely to encounter, and then develop the tools for extracting information and making predictions from these descriptions.

Our approach is based on the development of a few fundamental ideas. Whenever possible we start with a problem, then develop as simple a description as we can which adequately encompasses this problem; finally, we try to extract the information we want in as natural a fashion as possible.

Since relationships between various quantities are described by functions, this concept is fundamental. We introduce functions starting with the simplest ones, sequences — progressing to real valued functions of a real variable, then real valued functions of several real variables, and finally to generalized functions. Our aim is to get comfortable with each new concept before progressing further.

In studying the various classes of functions, we initially concentrate on local, usually linear, descriptions because these are usually the easiest to formulate from basic principles. This simple unifying idea relies on the concept that it's easiest for most of us to describe a system by examining a typical small piece (examining the system *locally*). After all, it should be

---

1. *Bad cholesterol.*

easier to comprehend and describe accurately what goes on in such a local area, or in a short time, than what's happening overall.

Most local descriptions appear *linear* — in the sense that short pieces of most curves we encounter look like segments of straight lines, and most small pieces of surfaces look like parts of a plane, etc., **and** lines and planes are relatively easy to deal with.

Once we have our initial description, we are faced with the problem of extracting useful information. Developing tools for this purpose leads us very naturally to the concepts of appropriate coordinate systems and change of coordinates. A coordinate system is a means of describing objects numerically; e.g., the positions of points can be described by Cartesian coordinates (x,y) or alternatively by polar coordinates (r, ϑ) which are different descriptions of the same objects. One coordinate system may be useful for the initial system description and another for extracting certain types of information.

Keeping in mind our ultimate goals, it's essential to become familiar with how to change descriptions as we progress toward the problem's solution. So we examine various coordinate systems whose structures are motivated by the purposes we have in mind. These ideas lead us to developing integrals and to examining ordinary differential equations (our local description of functions of one variable), and to the elementary transcendental functions, *sin, cos, ln, exp* and to Taylor's theorem. In our study of descriptions related to functions of several variables, we cover linear algebra and change of coordinates in n dimensions, as well as matrices, which simplify the handling of systems of differential equations.

The search for more appropriate coordinate systems for dealing with functions of several variables leads to the study of eigenvalues and eigenvectors, and to their obvious extensions, Fourier series and other transforms. Finally, to deal with some of the problems which this theory doesn't seem to handle adequately, we are led to construct generalized functions, in much the same way that the real numbers were developed to handle issues that could not be dealt with properly by the rational numbers.

The material developed here is the foundation for understanding specialized subject areas, such as partial differential equations, probability and statistics, used in many sciences.

Our aim is a *unified* approach, cemented by the basic concepts of *local linear descriptions*, *functions* and *appropriate coordinate systems.*

s. harris
"PUTTING A BOX AROUND IT, I'M AFRAID, DOES NOT MAKE IT A UNIFIED THEORY."

Also, we've tried to avoid *proof by intimidation.*

s. harris
"YOU WANT PROOF? I'LL GIVE YOU PROOF!"

Finally, we've tried to supply sufficient detail and motivating ideas so that readers won't become frustrated or discouraged by obstacles that are too formidable.

"I THINK YOU SHOULD BE MORE EXPLICIT HERE IN STEP TWO."

A few comments are in order about a reasonable way of approaching mathematics.

Many people dislike proofs — quite often for good reason, when they've been forced to memorize them without understanding. But ***a proof is just a convincing argument.*** So in a work like this one, if a result is really obvious it's best omitted. A fine example of a worthwhile proof is the one below for the Pythagorean theorem. A proof is needed because this result is certainly not obvious to most of us.

> Draw a square whose sides are of length $a + b$, as shown below, and draw the inner square with side length c as indicated.

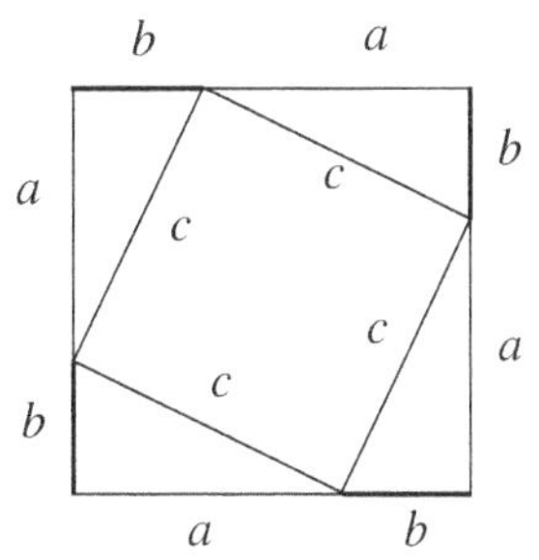

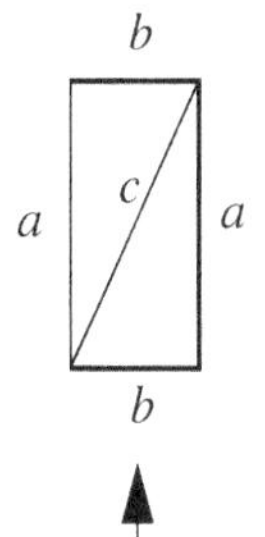

Now compute the area of the outer square in two ways:

$$A = (a+b)^2$$

and as the sum of the inner square area and that of the four triangles $\qquad A = c^2 + 2ab$

inner square area (↑ c^2) — area of the two rectangles gotten by fitting the triangles as shown at the right (↑ $2ab$)

Since $(a+b)^2 = a^2 + 2ab + b^2$, we find

$$a^2 + 2ab + b^2 = c^2 + 2ab$$

or

$$a^2 + b^2 = c^2,$$

which is the Pythagorean theorem.

It's actually hard to forget this proof.

Definitions seem to be a source of confusion for many students. We think of a ***definition*** simply as an ***abbreviation***, to be introduced when needed. And a definition is needed when a phrase, sentence or paragraph is being repeated too often. Definitions tend to arise at a fairly steady rate, and they should only be introduced when they will serve to make reading and understanding easier.

The logic on which mathematics is based has been presented somewhat informally, with more detailed explanation (as in proof by contradiction, and proof by induction) when it seemed needed. Time will tell whether our efforts here were adequate.

We hope we've succeeded in making the presentation natural and understandable, so that readers can continue their development and have the capability of comprehending the mathematics being used in their own fields. We'd be grateful to readers who communicate with us with suggestions for improvement. Our e-mail addresses are: judah.rosenblatt@utmb.edu and sto@www.cs.unm.edu. We will post any corrections and useful supplementary explanations on the world wide web at http://judah.rosenblatt.utmb.edu or http://www.cs.unm.edu/~sto.

Intended Audience

This work is intended for those who need a solid understanding of the mathematics used in most fields of application. It is self-contained, in that the only required background is some familiarity with pre-college mathematics. However, previous exposure to more advanced material does make the going a lot easier, because it means that less effort has to be spent on becoming familiar with new notation. So, the more comfortable the reader is with various branches of mathematics, the easier it will be to concentrate on the less familiar ideas. Typically, students who have been exposed to calculus, but have either forgotten much of what was presented or who felt that they didn't have an adequate understanding of that material, can benefit substantially from this book. Those with less background can also benefit, provided they are willing to put forth the additional effort that is needed. For them, the book requires a more patient approach.

Recommendations for Use

In most cases the theory is built up naturally, with the direction supplied by some type of problem. The important results which this process produces are stated formally (usually as theorems) for future reference. For readers who may not recall pertinent earlier material (especially in the theorem statements) we frequently refer to previous relevant results and definitions. These references are usually in parentheses such as (see Definition ... on page ...) or (see Example ... on page ...).

Material starting off with the symbol ⌈, in small font, and concluding with the symbol ⌊ is optional — possibly rather advanced, or of interest to a limited number readers.

Omitting this material should cause no difficulties.

There are a small number of exercises to help readers check whether they can use the material that is presented. Now that computer packages such as *Mathematica*, *Mlab*, *Maple*, *Matlab*, *Mathcad,* and *Splus* are available, it should be easy for readers who need more practice to make up their own exercises and verify their answers with the aid of such programs. There are a substantial number of exercises and problems extending the theory, and most are supplied with detailed hints. We recommend that the reader try doing these before looking at the hints — because even if they don't succeed, the effort put in and the obstacles encountered should help provide an understanding of the approaches used in the hints.

Chapter Summaries

Chapter 1 presents many of the fundamental ideas of calculus in the most understandable setting possible. This setting consists of

- the conceptually simplest functions (sequences)
- the simplest *local description operator* (the forward difference operator, Δ)
- the simplest *integrator* (summation).

In addition to these important concepts, the fundamental summation theorem

- serves as an excellent precursor to the *fundamental theorem of calculus* of Chapter 4, taking the mystery out of this latter result.
- provides a systematic approach to finding formulas for sums such as $1^n + 2^n + \cdots + k^n$ and $1 + x + x^2 + \cdots + x^n$.

A local model for population growth via a difference equation is presented and analyzed, to prepare the reader for differential equation descriptions.

Chapter 2 introduces the foundations of differential calculus, whose purpose is convenient local linear approximation and description, using derivatives. The main differentiation rules (rules for calculating the slope of the tangent line approximation) are developed using geometric motivation for the algebraic results that are derived. The fundamental idea of continuity as reflecting well-behaved systems — i.e., systems in which small input changes lead only to small output changes — is introduced.

In this chapter the student is first exposed to the *Mathematica* program to check answers.

Chapter 3 further develops the use of derivatives as a tool for determining the behavior of functions — for graphing them and for solving equations. The differential equations describing a radioactive decay process are introduced and properties of this process are deduced, along with illustration of the need for care in such analysis. The mean value theorem is introduced and its importance in deriving practical results is illustrated. *Mathematica* is used to help implement the solution of equations, and for graphing. Newton's method and bisection are introduced, and their basic properties are studied with emphasis on the geometric basis of their behavior.

Chapter 4 is motivated by the need to develop useful information about mechanical and biological systems from the local linear descriptions we have set up for them. This leads up to the Riemann integral, and its analytic and numerical evaluation. A section is devoted to the Lebesgue integral as a more suitable tool for theoretical development However, the advanced nature of this subject does not permit the detailed presentation required for its practical use here.

Chapter 5 introduces the exponential, logarithmic and trigonometric functions from the viewpoint of describing various systems — the exponential arising as a solution to growth, decay and elementary enzyme kinetic problems — the trigonometric functions introduced as the solution to problems involving mechanical systems in which the effect of friction is small enough to be neglected. All of the development here proceeds along natural lines, in a way that is an especially compatible one for those using mathematics in research.

Use of computers for investigating these systems is an integral part of this approach. Resonance is studied both from a numerical and an analytic viewpoint. The shift rule for a systematic theoretical solution of constant coefficient linear differential equations is used extensively.

Chapter 6 further extends the capabilities of local descriptions by showing how they can often be used to obtain the function values of interest, to arbitrary accuracy (Taylor's theorem with error term). Taylor's theorem appears in its main role here as one of the most essential tools for investigating the behavior of functions in more advanced work. The Taylor coefficients may be viewed as the first of many coordinate changes we will be examining — i.e., alternate descriptions better adapted to investigating certain problems.

Chapter 7 develops the ideas of infinite series as a natural generalization of Newton's method and Taylor's theorem — and is used to find theoretical solutions of more general differential equations.

In **Chapter 8** we begin the study of functions of several variables — to examine systems with multiple inputs. The development starts with the extension of the idea of derivative to that of *directional derivative*. The object here is to view a function of several variables as a family of functions of one variable, so as to use the previously developed methods. Partial derivatives, the gradient vector and dot products are an immediate consequence of finding out how to represent directional derivatives. The vector notation introduced here is carefully chosen so that it hardly changes from what has already become familiar. This makes the additional concepts required to study functions of several variables — *projections, orthogonality* and the meaning of the *gradient vector*, stand out as the needed extensions to be concentrated on, with the previous structures remaining essentially intact.

In **Chapter 9** the algebra needed to handle the concepts and properties of coordinate systems appropriate to functions of several variables (in particular, *orthogonal coordinate systems*) is developed. The Gram-Schmidt procedure for construction of orthogonal coordinate systems is presented. This is then used to solve the problem of fitting curves to experimental data (*least squares*) and solving systems of linear equations which arise in finding maxima and minima of functions of several variables.

Chapter 10 continues the development of linear algebra with the introduction of matrices. The work needed to get the relevant matrix machinery developed really begins to pay off with the extension of matrix infinite series (in particular, the matrix geometric series) and matrix norms. These concepts are useful for improving accuracy in solution of systems of equations. They lead to a geometric understanding of the normal equations used in curve fitting, and to an improved Gram-Schmidt procedure. Matrices allow the generalization of Taylor's theorem to n dimensions with little notational change. With the availability of matrix tools, both necessary and sufficient conditions for maxima and minima are easy to derive. Also, Newton's method (including proof of convergence) is extendable with little notation or conceptual change, for application to the nonlinear fitting that is needed in kinetic study.

Orthogonal complements are introduced in **Chapter 11** to obtain a much fuller understanding of the nature of the solution of systems of n linear equations in m unknowns — together with a geometric interpretation of the inverse of a matrix. Finally, orthogonal complements are used to provide a very understandable geometrically oriented solution to the problem of finding maxima/minima under constraints — problems such as finding when the earth is closest to the sun, and determining the significance of the gradient vector as giving the direction of steepest descent.

Chapter 12 extends the concept of integral to functions of several variables — to multiple integrals, line and surface integrals, motivated by applications from physics. We conclude this chapter with a brief outline of complex function theory essential for the Fourier series and transform theory needed to better understand vibrating systems.

Chapter 13 investigates the concept of preferred coordinate systems, used in the study of linear operators. Eigenvalues and their associated eigenvector coordinate systems are examined. Here we see how the appropriate choice of coordinate system can make complicated looking problems turn into simple ones. Newton's method and maxima/minima theory both become more comprehensible in this framework.

Chapter 14 extends the concept of preferred coordinate systems from matrices (finite dimensional linear operators) to linear differential operators (infinite dimensional operators) which enables a much simpler understanding of the nature of solutions to linear differential equations.

The concluding chapter, **Chapter 15**, on generalized functions, extends the concept of function to handle certain important limiting cases not very amenable to conventional theory. It replaces a number of clumsy attempts at solution with an approach that is more powerful, simpler, and even more in accord with physical concepts than its predecessors.

Table of Contents

Chapter 1

Finite Differences

1 Introduction

As a tool of science and technology, mathematics is initially used to furnish quantitative descriptions of systems; systems such as a scientific instrument, or a human organ system (for instance the cardiovascular system). Then from these descriptions we try to derive some of their interesting and important properties. Since we rarely have much understanding of a system when we first examine it, we usually try to develop an initial description which is as simple as possible. Most of the time this means keeping the description **local** and **linear**.

A **local** description is one in which we describe the system by specifying the relations between parts which are *close* to each other — physically close, or close in time. A local description often simplifies matters, because **locally** "most" curves and surfaces look **linear**. That is, *small portions of a curve look like straight line segments, and small portions of a surface look like pieces of a plane*, as illustrated in Figure 1-1.

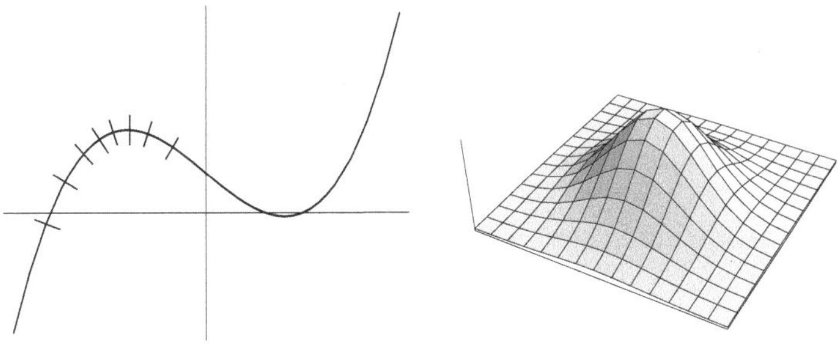

Figure 1-1

Because of this local linearity, we are able to make use of our knowledge of lines and planes to simplify analyzing the system.

In the next chapter we'll begin a systematic development of the analytical techniques based on local linearity. In this chapter we'll try to get a basic understanding of how we make use of local descriptions in their simplest setting.

2 Sequences — The Simplest Functions

In science, engineering and medicine we often want to describe a system by specifying its responses to its possible stimuli or inputs. For instance, we might want to determine

- the effect of specified amounts of ultraviolet radiation (input) on the amount of DNA damage (output)

or

- the effect of a car's speed (input) on its mileage (output)

or

- the effect of specified amounts of alcohol (stimuli) on reaction time (response).

All input-output descriptions are examples of functional relations — or (more simply) functions.

Definition 1.1: Function

A **function** or **functional relation** consists of a set of allowable inputs[1] and a rule of association which links a unique output to each allowable input. We illustrate the concept of a function as an input-output black box in Figure 1-2. The set of allowable inputs to a function is called its **domain**. The **range** of a function consists of all of its possible outputs. Members of the range of a function are referred to as its values.

❑

If we let the symbol f denote some function, and denote a possible input by x, then the associated output will be denoted as $f(x)$, as shown in Figure 1-2

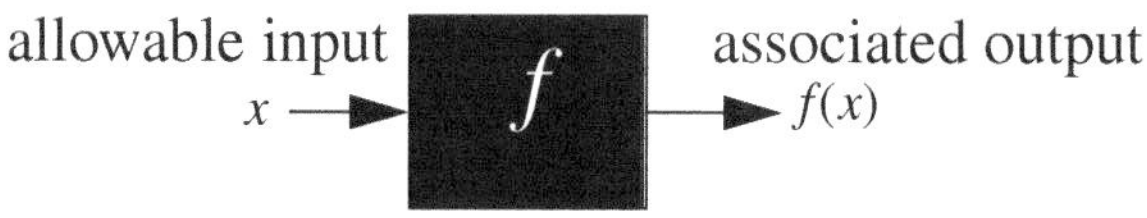

Figure 1-2

1. It is a subtlety whose significance will be clearer as we proceed, that we get distinct functions if the sets of allowable inputs are distinct, even if the rules of association are the same. This will certainly be evident when we represent a function by its graph — as can be seen in Figure 13-1 on page 584. You might also compare the graphs of $f(x) = x^2$ for x real and $g(x) = x^2$ for $x \geq 0$.

We'll be spending much of the remainder of this book on methods of describing and determining the properties of functions as they arise in commonly encountered situations. Now the possible inputs and outputs of functions can be quite complicated. In order to avoid being overwhelmed, we'll start off by taking a careful look at the functions with the simplest useful inputs and outputs; namely, we'll first examine **sequences of real numbers. We will see that such sequences can frequently substitute for more complicated functions. In fact, that is essentially what we're doing when we make use of a table of some function when the function is called for.**

In everyday usage a sequence is just a collection of objects listed in some order; for example, a sequence of 24 systolic blood pressure measurements taken at the beginning of each hour on some given day. One natural order here is the one dictated by the times at which the measurements were taken. The various blood pressures are referred to as the *elements* of this (blood pressure) sequence. We will soon see how the following definition embodies the essential idea of what a sequence is.

Definition 1.2: Sequence, real valued sequence

A **sequence** is a function (Definition 1.1 on page 2), whose possible inputs consist of counting numbers[1],[2] (the numbers 0,1,2, ...). A **real valued sequence** is a sequence whose possible outputs are real numbers (numbers represented as infinite decimals, such as $\pi = 3.14159...$ etc.). A sequence is called **finite** if its domain (Definition 1.1 on page 2) consists of a finite number of counting numbers (a finite number of **elements**).

❑

If F stands for some given sequence (like the blood pressure sequence mentioned above), then $F(j)$ or F_j would represent the j-th element of this sequence; e.g., the j-th blood pressure, $j = 1,2,...,24$.

There are a number of alternate ways to describe a sequence

- One way is just to *list* its elements as what we call a *vector*, say (132, 84, 70.5, 38.5), which might represent a sequence of medical measurements; the first element maybe standing for systolic blood pressure, the second one diastolic blood pressure (both in mm of mercury), the third one weight in pounds, and the last one

---

1. If f denotes a sequence, it is fairly common to use the notation f_k in place of $f(k)$
2. Unless we specifically say otherwise, we will only be dealing with sequences that have some collection of successive integers (k, k+1,...,m, or all counting numbers $\geq k$) as their domain (possible inputs).

temperature in degrees Celsius. If we want to emphasize the functional nature of this sequence, we could denote it by f, e.g., writing $f(1) = 132$, $f(2) = 84$, $f(3) = 70.5$, $f(4) = 38.5$.

Sometimes it is possible and useful to describe a sequence, F, by a formula; e.g., $F(n) = n^2$ *for n any counting number in the interval from* 1 *to* 4.

The two sequences just presented are called **finite sequences** because they have only a finite number of possible inputs (four to be precise).

The vector

$$(1, 4, 9, 16) \quad 1.1$$

can also be specified as a *table, F, of ordered pairs* as shown in Table 1-3

n	$F(n)$
1	1
2	4
3	9
4	16

Table 1-3

or *as a set* consisting of the ordered pairs (1,1), (2,4), (3,9), and (4,16); or *graphically* as shown in Figure 1-4.

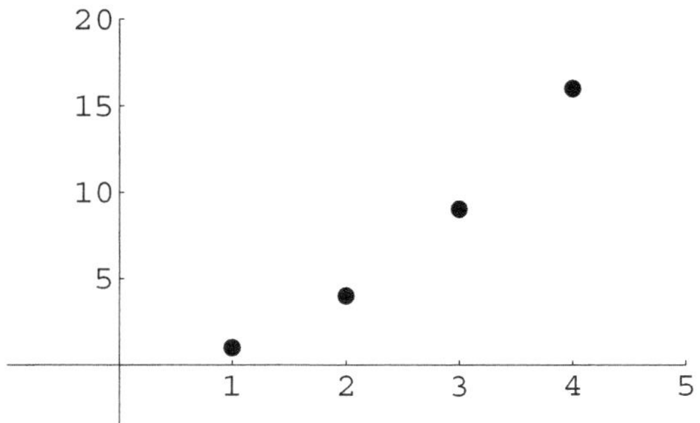

Figure 1-4

Further on we'll be dealing with functions that are specified by a precise set of computational rules (called an **algorithm**), suitable for implementation on a digital computer.

How we choose to describe a function depends on what is most convenient for the purpose we have in mind. For the present we'll concentrate on what we call local descriptions of real valued sequences, seeing how helpful they can be. The local description we will use in this chapter, rather than just using the heights of the points of a real valued sequence, as shown in Figure 1-4, makes use of the **changes in height** as we move from one of the points to the adjacent one, as illustrated in Figure 1-5, for $f(n) = n^2$.,

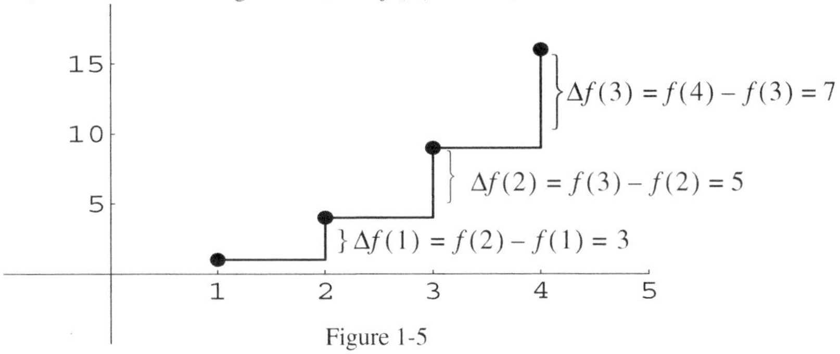

Figure 1-5

Using the changes in function values to describe the function's behavior is a process we're familiar with, as for example, when we describe the state of the economy by giving its yearly percentage inflation rate.

For easy symbolic reference to the changes in height between adjacent points as in Figure 1-5, we give the following algebraic definition:

Definition 1.3: Forward difference sequence Δf

For any real valued sequence f, (Definition 1.2 on page 3), we let

$$(\Delta f)(n) = f(n+1) - f(n)$$

whenever the right hand side is well defined for the counting number n.

❑

Note that Δf is a sequence, whose value at n is $(\Delta f)(n)$. Thus, the *forward difference operator,* Δ, is itself a function - whose inputs and outputs are functions, namely, sequences. So strictly speaking we should write $(\Delta(f))(n)$ rather than $(\Delta f)(n)$. ***We will nonetheless go in the opposite direction for easier reading;*** we will generally omit the parentheses around Δf in evaluating Δf at n, writing $\Delta f(n)$ rather than $(\Delta f)(n)$ [1]or

1. Actually, since parentheses are used for grouping, some other symbol should be used to indicate the input to a function. The *Mathematica* program actually does such distinguishing, using square brackets, [ ], to enclose arguments (inputs) to functions, and parentheses only for grouping.

$(\Delta(f))(n)$. This abuse of notation allows another convenient abuse, namely, preceding the formula for $f(n)$ by Δ to represent $\Delta f(n)$. This avoids having to write out the formula for $f(n)$ prior to writing $\Delta f(n)$.

Although it may not be apparent, the use of the height changes, $\Delta f(n)$, to describe f, may greatly simplify the job of obtaining important properties of f.

Notice that **knowledge of the difference sequence Δf alone isn't enough to determine the sequence f completely; for this we must also be given the value of f at some specific input.**

Here are a few exercises to familiarize yourself with the notation just introduced. As hints to help those with trouble getting started, note that if you know $f(1)$ and $\Delta f(1)$, you can find $f(2)$. Now knowing $f(2)$ and $\Delta f(2)$ allows computation of $f(3)$, and so forth. Careful bookkeeping does the rest.

Exercises 1.1:

Compute $f(5)$ given

1. $\Delta f(n) = n$ for all n and $f(1) = 1$
2. $\Delta f(n) = 1/n$ for all n and $f(1) = 2$
3. $\Delta f(n) = 1$ for all n and $f(1) = -1$
4. $\Delta f(n) = a_{n+1}$ for all n and $f(1) = a_1$, where $a_1, a_2, ...$ is some given sequence of real numbers.

Now that you have some familiarity with the difference notation, we provide a simple example of a situation which lends itself to a local description, and see how this information can be used.

Example 1.2: Local description of population growth

A population of cancer cells is often described by saying that its rate of growth is proportional to its size. We can interpret this to mean that there is some positive constant, K, such that the local change in population size

$$\Delta P(t) = KP(t), \quad t = 0, 1, 2, ... \qquad \textit{1.2}$$

This description, (called a **difference equation**), may be adequate as long as there are sufficiently many cells to make a stable increase likely, but where the population is not so large as to alter its surroundings and create inhibiting factors (such as lack of nourishment).

If now we are given the initial population size

$$P(0) = p_0,$$

then with some effort we can calculate the population size for later times, namely,

$$\Delta P(0) = P(1) - P(0) = KP(0),$$

so $$P(1) = (1+K)P(0) = (1+K)p_0,$$

and $$\Delta P(1) = P(2) - P(1) = KP(1),$$

so $$P(2) = (1+K)P(1) = (1+K)^2 p_0, \text{ etc.}$$

Generally we may rewrite equations 1.2 as

$$P(t+1) = (1+K)P(t). \qquad \textbf{\textit{1.3}}$$

Each additional application of equation 1.3 raises the argument to P by 1 and raises the exponent of the factor $(1+K)$ by 1; so we see that the argument to P and the exponent of $(1+K)$ are always the same, yielding

$$P(j) = (1+K)^j p_0 \qquad j = 0, 1, 2, \ldots . \qquad \textbf{\textit{1.4}}$$

■

Thus our local description, equations 1.2, has led to the *exponential growth* that is exhibited in equations 1.4. In this case we were able to derive an explicit formula for the population size at time j. In contrast to this situation, it is often unfeasible to derive a simple formula for the locally defined sequence using the kind of direct approach we just carried out; although, a digital computer, which is well suited to repetitive computation would allow numerical determination of as many specific values as we might want. But ***in order to provide more insight than we can get from numerical answers alone, we will also be developing analytical methods to investigate the properties of such functions.***

Let's now take a look at how the local description we have introduced can help us obtain certain properties *more easily.* Notice that if the differences $\Delta f(n)$ are positive for all values of n in some interval, then the *heights, $f(n)$,* increase as n increases within this interval, as you can see from Figure 1-5 on page 5. Because we will frequently be interested in whether or not functions that we are examining are increasing as their inputs increase, it is reasonable to make the following formal definitions.

Definition 1.4: Positive sequences, increasing sequences

We say that a sequence (Definition 1.2 on page 3), g, is **positive** [on the interval $[m; n]$, $m \le n$][1], and write $g > 0$ if for all counting numbers, k in its domain [satisfying $m \le k \le n$]

$$g(k) > 0.$$

1. In many definitions and theorems it is convenient to present a general case alongside restricted versions. We do this using square brackets, [...], to indicate the restriction(s) being introduced. So, above, we defined both a positive sequence, and a sequence which is positive on the interval $[m; n]$.

We say that g is **increasing** [on the interval $[m; n]$, $m < n$], if for all values j and k in its domain [and with $m \le j < k \le n$] with $j < k$, we have

$$g(j) < g(k) .$$

❑

When the context is clear we may omit specific mention of $[m; n]$.

We now state formally the obvious result relating the increase of a sequence f (function f with integer domain, Definition 1.1 on page 2), to the positivity of Δf.

Theorem 1.5: If $\Delta f > 0$ [on $[m; n]$], then f is increasing (Definition 1.4 on page 7) [on $[m; n]$].

Frequently it's easier to examine Δf rather than examining f directly to determine whether or not f is increasing, as we will see next.

Also we will see that it is very useful to realize that since Δf is a sequence, we can form *its* difference sequence, $\Delta(\Delta f)$, called the **second forward difference** of f, and usually written as $\Delta^2 f$. From Theorem 1.5, we obtain

Corollary to Theorem 1.5: If $\Delta^2 f > 0$ [on $[m; n]$], then Δf is increasing [on $[m; n]$].

Geometrically $\Delta^2 f > 0$ means that f is *convex* ("concave up"), as illustrated in Figure 1-6.

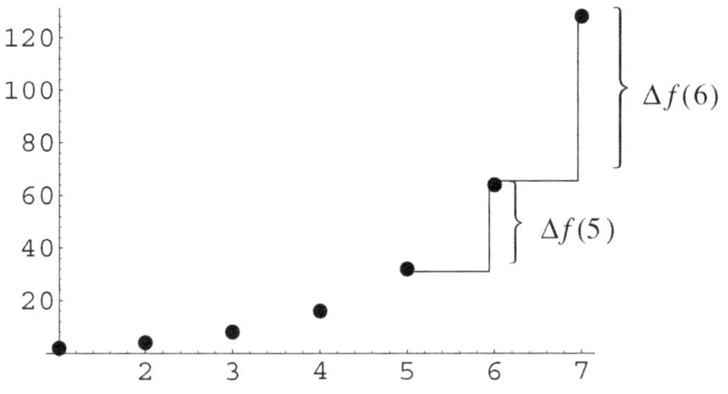

Figure 1-6

Between them Theorem 1.5 and the Corollary to Theorem 1.5 can often simplify graphing sequences considerably. In fact, to determine the shape of the graph of f, it is almost always simpler to examine $\Delta^2 f$ rather than the formula for f itself, Here are some illustrative examples.

Example 1.3

Let $f(n) = n^3$ for $n = 0,1,2,\ldots$. Then $\Delta f(n) = 3n^2 + 3n + 1$ and $\Delta^2 f(n) = 6n + 6$. Hence f is increasing (since $\Delta f > 0$) and Δf is increasing (because $\Delta^2 f > 0$). So the graph of f rises as you move to the right, and because Δf is increasing, this graph rises faster as we move to the right. That is, f is increasing and convex (concave up), as illustrated in Figure 1-7.

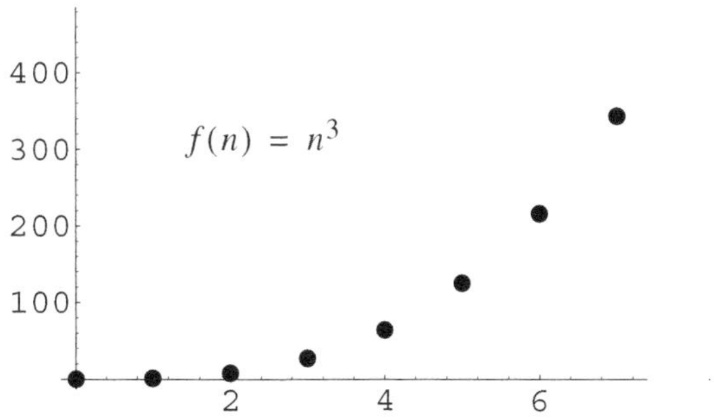

Increasing sequence with increasing differences

Figure 1-7

■

In Figure 1-8 we see the complementary situation, illustrated in Example 1.4 which follows.

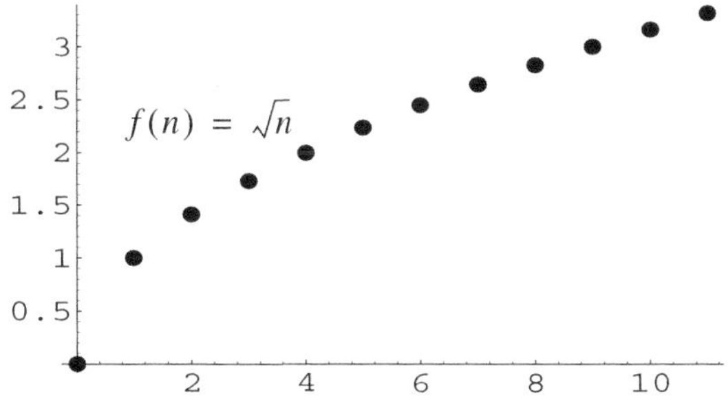

Increasing sequence with decreasing differences

Figure 1-8

Example 1.4

Let $f(n) = \sqrt{n}$ for $n = 0,1,2,\ldots$. Then,

$$\Delta f(n) = \sqrt{n+1} - \sqrt{n} = (\sqrt{n+1} - \sqrt{n})\frac{\sqrt{n+1} + \sqrt{n}}{\sqrt{n+1} + \sqrt{n}}$$
$$= \frac{1}{\sqrt{n+1} + \sqrt{n}}.$$

From this computation we see immediately that $\Delta f > 0$, and also by direct examination of this last expression that Δf is **decreasing,** so that the graph of f rises more and more slowly as we move to the right. That is, f is increasing and concave down, as illustrated in Figure 1-8.

■

Exercises 1.5

1. Compute a formula for $\Delta f(n), \Delta^2 f(n)$, for the following:
   a. $f(n) = n$
   b. $f(n) = n^3$
   c. $f(n) = 2^n$
   d. $f(n) = x^n$ x a fixed constant
   e. $f(n) = 1/n$ $n = 1,2,...$
   f. $f(n) = 1/n^2$ $n = 1,2,...$
2. Graph the sequences of the previous exercise.
3. Graph the sequences given by
   a. $f(n) = n + 10/n$ $n = 1,2,...$
   b. $f(n) = n/1.2^n$
4. Verify directly from the local description of the sequence P of Example 1.2 on page 6 that P is convex. Confirm this from the derived solution, equation 1.4 on page 7.

If you went through Exercises 1.1 on page 6, you may have noticed that given the local description specified by Δf, together with an initial value, such as $f(0)$, any value $f(k)$ is obtainable by adding to $f(0)$ all of the successive height changes (differences) to get us to $f(k)$. That is,

$$f(k) = f(0) + \Delta f(0) + \Delta f(1) + \cdots + \Delta f(k-1) \qquad \textit{1.5}$$

as we see in Figure 1-9. That is, the **total change in height,** $f(k) - f(0)$ **is the sum of the individual successive height changes** $\Delta f(j)$, $j = 0,...,k-1$ a result which is geometrically obvious. From an algebraic viewpoint, equation 1.5 is also easy to verify, as we can see from the following manipulations:

$$f(0) + \Delta f(1) + \cdots + \Delta f(k-1)$$

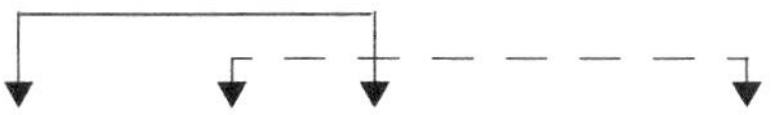

$$= f(0) + [f(1) - f(0)] + [f(2) - f(1)] + \cdots + [f(k) - f(k-1)]$$
$$= f(k),$$

since all of the remaining terms are canceled out by terms close by.

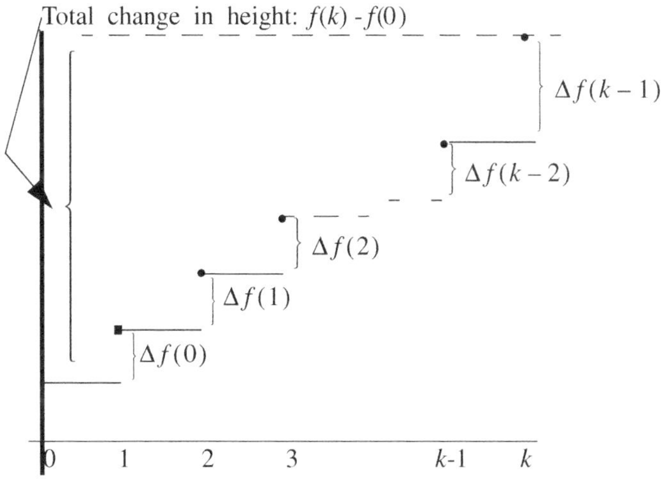

Figure 1-9

At this point in the development, it is advisable to introduce the standard abbreviated notation for sums - both to save space and to make it easier to read through mathematical discussions involving them.

Definition 1.6: Summation Notation: $\sum_{k=m}^{n} a_k$

If $a_m, a_{m+1}, \ldots$ is any given sequence of numbers, we denote the **sum**, $a_m + a_{m+1} + \cdots + a_n$, by $\sum_{k=m}^{n} a_k$.

❑

The *three dot notation,* in $a_m + a_{m+1} + \cdots + a_n$, which we used above to define the sum, $\sum_{k=1}^{n} a_k$, is an imprecise notation, requiring mind-reading to figure out just what the three dots really stand for. Furthermore it is not a notation that is very well suited to proofs, and it is inappropriate for any computer work. A precise definition of $\sum_{k=1}^{n} a_k$ is the following:

Definition 1.7: Rigorous definition of $\sum_{k=m}^{n} a_k$

$$\sum_{k=m}^{m} a_k = a_m \text{, and, for each counting number, } n \geq m,$$

$$\sum_{k=m}^{n+1} a_k = \left(\sum_{k=m}^{n} a_k\right) + a_{n+1}.$$

❑

This *inductive* definition is easy to implement on a computer, since it only involves adding two numbers at a time, and it is more suitable in proofs involving sums. This definition simply says that for each counting number n, the sum of *one more* element of the sequence a is obtained by adding the next element to the previously computed sum.

In terms of the summation notation, our result, equation 1.5 on page 10, may be rewritten very economically as

$$\sum_{j=0}^{k-1} \Delta f(j) = f(k) - f(0).$$

For reference purposes, we will state the useful, slightly more general result, which is established in the identical fashion.

Theorem 1.8: Fundamental summation theorem

If f is a real valued sequence (Definition 1.2 on page 3), whose domain includes the interval $[m; k]$, $k > m$, then

$$\sum_{j=m}^{k-1} \Delta f(j) = f(k) - f(m).$$

❑

As we stated before, this result allows determination of $f(k)$ from any initial value, say $f(0)$, and the differences $\Delta f(j)$. It is also extremely useful for finding simple formulas for many sums, and such formulas often make it much easier to determine important properties of the sums being investigated. Here's how we go about obtaining such formulas.

Suppose $a_0, a_1, a_2, \ldots$ is some given sequence of numbers, and ***we want to find a formula for***

$$\sum_{j=0}^{k-1} a_j.$$

We hunt for any sequence, f, whose *first difference* is the sequence $a_0, a_1, a_2, \ldots$. That is, we hunt for any sequence, f, satisfying

$$\Delta f(j) = a_j \quad \text{for each } j. \qquad \textbf{\textit{1.6}}$$

Then,

$$\sum_{j=0}^{k-1} a_j = \sum_{j=0}^{k-1} \Delta f(j)$$

$$= f(k) - f(0), \quad \text{by Theorem 1.8.}$$

So all that is needed to find a formula for $\sum_{j=0}^{k-1} a_j$ **is to find a sequence** f **satisfying equation 1.6, and then write**

$$\sum_{j=0}^{k-1} a_j = f(k) - f(0). \qquad \textbf{\textit{1.7}}$$

Sometimes, *albeit rarely*, this is not difficult, as we see in the next examples. Nevertheless, these rare cases can be useful, see Exercise 1.10.2 on page 19.

Example 1.6: Sum of the first k – 1 integers

This sum, $\sum_{j=0}^{k-1} j$ (together with the sum, $\sum_{j=1}^{k-1} j^2$, of the squares of the first k – 1 integers), can be used in determining the efficiency of *interchange* (also called *bubble sort*), computer alphabetizing programs. The simple formulas for these sums make it easy to determine when any of these programs will perform rapidly enough, and when they need improvement.

This particular problem will be solved if we can find a sequence, f, satisfying the condition

$$\Delta f(j) = j.$$

Some experience with specific sequences is helpful here. In particular,

$$\Delta j^2 = (j+1)^2 - j^2 = 2j + 1 \qquad \textbf{\textit{1.8}}$$

$$\Delta j = (j+1) - j = 1. \qquad \textbf{\textit{1.9}}$$

(**Note** that here we are using Δj^2 as an abbreviation for $\Delta g(j)$ where $g(j) = j^2$.)

Looking at equation 1.8, it's natural to try $j^2/2$, that is, to see that $\Delta(j^2/2) = j + 1/2$. Then looking at equation 1.9, we see that we can *knock out the 1/2 by subtracting j/2.* That is $\Delta(j^2/2 - j/2) = j$, or

$$\Delta\left(\frac{j^2 - j}{2}\right) = j. \qquad \textbf{\textit{1.10}}$$

Hence,

$$\sum_{j=0}^{k-1} j = \sum_{j=0}^{k-1} \Delta\left(\frac{j^2 - j}{2}\right)$$

$$= \frac{k^2 - k}{2}$$

(the last step following from Theorem 1.8 on page 12, the fundamental summation theorem)

or

$$\sum_{j=0}^{k-1} j = \frac{k(k-1)}{2}. \qquad \textbf{\textit{1.11}}$$

■

Exercises 1.7

1. Extend the reasoning of Example 1.6 to determine a formula for $\sum_{j=0}^{k-1} a_j$ when

   a. $a_j = j^2$

   b. $a_j = j^3$.

   Hints: $\Delta j^3 = (j+1)^3 - j^3 = 3j^2 + 3j + 1$

   $\Delta j^4 = (j+1)^4 - j^4 = 4j^3 + 6j^2 + 4j + 1.$

2. Provide a geometric explanation of formula 1.11 above. We suggest looking at the number of dots on the diagonal and the number of dots above the diagonal, in the diagram to follow. Relate these values to the total number of dots and to the sum being examined.

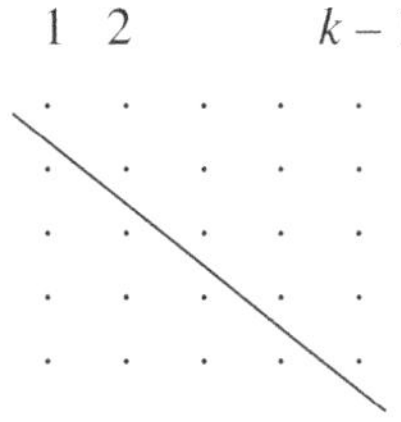

3. *If* $f(n)$ is the number of molecules of some substance, such as glucose in the *blood compartment* of the body at time n, one commonly used description is that the amount of this substance *leaving* this compartment per unit time is proportional to (i.e., a fixed positive multiple, k, of), the number of molecules of this substance in the compartment at that time. What can you determine about the behavior of $f(n)$ from this description?

The next example illustrates one of the many situations leading to a *geometric series*, one of the most important sums in mathematics.

Example 1.8: Enzyme kinetics recycling

Enzyme kinetics is concerned with the behavior of chemicals inside the bodies of animals, especially people. There is a great deal of interest in determining the total amount of certain chemicals produced by various organs, for example, the amount of cholesterol produced by the liver, in some given time period.

To accomplish this aim, a determination may be made, of just how much of this chemical passes through some specified arterial cross section just downstream of where this chemical is fed from the organ into the bloodstream. If the chemical was immediately used and completely broken down soon after its introduction into the bloodstream, then the amount of chemical passing by the specified cross section in the given time frame would represent a reasonable estimate of the amount the organ produces in the given time period. However, many a chemical doesn't just get completely eliminated without any of it returning to previously visited sites.

The simplest model of such *recycling* is one in which a proportion, p, of the chemical passing the specified cross section will rather quickly return to pass this cross section yet again. If A represents the amount of chemical produced by the organ in the time period being considered, then during this time period not only would the amount A be observed to pass the specified cross section, but also most of the recycled amount pA and most of the recycled part of the already recycled amount pA, that is, $p(pA) = p^2A$ and so on.

So the total amount, T, passing the given site, would be nearly

$$T = A + pA + p^2A + p^3A + \cdots + p^nA = \sum_{j=0}^{n} p^jA$$

for some n, assuming (as we shall soon show) that any remaining recycling would contribute only negligibly to the amount passing by the specified cross section during the chosen time period.

■

This leads naturally to the problem of finding a simple formula for $\sum_{j=0}^{n} p^jA$. To help answer this question notice that, considering p as fixed,

$$\Delta p^j = p^{j+1} - p^j$$
$$= p^j(p-1).$$

Hence

$$\Delta\left(\frac{p^j}{p-1}\right) = p^j. \qquad \textbf{\textit{1.12}}$$

Thus from the fundamental summation theorem, (Theorem 1.8 on page 12), with $k - 1 = n$,

$$\sum_{j=0}^{n} p^j = \sum_{j=0}^{n} \Delta\frac{p^j}{p-1} = \frac{p^{n+1}}{p-1} - \frac{p^0}{p-1} = \frac{1-p^{n+1}}{1-p}.$$

We summarize this result as

Theorem 1.9: Geometric series

$$\sum_{j=0}^{n} p^j = \frac{1-p^{n+1}}{1-p}, \quad \text{for all real } p \text{ different from 1.}$$

❑

So we see that if n is large enough so that p^{n+1} is negligible compared to 1, and at the same time the recycling time (time from the first appearance of some batch to the first reappearance of some of this batch) multiplied by n is much smaller than the time period of interest, the measured amount passing the given cross-section over the time period is approximately $A/(1-p)$, which can be a significant overestimate of A, the value desired.

If p is a number ***very close to*** 0, then from Theorem 1.9 above, we obtain the most useful approximation:

$$\frac{1}{1-p} \cong 1 + p. \qquad \textit{1.13}$$

Because

$$\frac{1}{1-p} - (1+p) = \frac{1-(1-p)-p(1-p)}{1-p} = \frac{p^2}{1-p},$$

the error in approximation 1.13 equals

$$\frac{p^2}{1-p}, \qquad \textit{1.14}$$

which is very small when p is close to 0.

Results 1.13 and 1.14 are useful for quick estimation of reciprocals, as we see in

Example 1.9: Decimal evaluation of $\frac{1}{24}$

$$\frac{1}{24} = \frac{1}{25-1} = \frac{1}{25(1-1/25)} \cong 0.04(1+0.04) = 0.0416$$

with an error E *satisfying*

$$E \le 0.04\frac{0.0016}{0.96} \cong 0.000064.$$

■

Results 1.13 and 1.14 will also be seen useful in making certain complicated expressions more tractable.

Exercises 1.10

1. Without any *pencil and paper arithmetic*, write down a reasonable decimal approximation to
   a. $\frac{1}{0.99}$
   b. $\frac{1}{1.02}$
   c. $\frac{3}{2.04}$
   d. $\frac{5}{1.96}$

2. With almost no *pencil and paper arithmetic*, estimate the error in your answers to the previous exercise.

Problems 1.11

1.. a. In conducting independent trials each of which has probability p of success, the probability that the first failure will occur at the n-th trial is usually taken to be $p^{n-1}(1-p)$, because the first $n-1$ trials must all be successes, and the n-th trial must be a failure. Knowing that the probability of a failure prior to trial k is the sum of the probabilities of failure at trials prior to the k-th one, find a simple formula for the probability of failure prior to trial k.

b. Suppose an experiment has probability of .02 of success. How many times would you run this experiment if you wanted to be about 90% sure of at least one successful run.

2. At what time between 3 and 4 o'clock are the minute and hour hands pointed in exactly the same direction? Hint: When the minute hand reaches 3:15, where is the hour hand ? Where is the hour hand when the minute hand **is** where the hour hand **was** at 3:15, and so on.

3. A simple amplifier is said to have amplification factor A if the output is A times as loud as the input. A simple amplifier can be modified so that a proportion, say p, of the output is *fed back* as an input. (This is what happens at concerts, when the sound of the orchestra coming out of the loudspeakers gets back to the microphone. If the proportion is too great, the amplifier starts to squeal because with too much input actual amplifiers no longer behave as simple amplifiers the way they were defined above.) What is the effective amplification factor of the modified simple amplifier above, assuming that proportion fed back doesn't drive the amplifier over the edge?

4. Use the geometric series result, Theorem 1.9 on page 16, to derive the formula

$$a^n - b^n = (a-b)\sum_{j=0}^{n-1} a^{n-1-j} b^j .$$

Hint: $a^n - b^n = a^n\left(1 - \left[\frac{b}{a}\right]^n\right)$.

As already hinted just prior to Example 1.6 on page 13, in most situations it's impossible to obtain a simple formula for a sum that arises, but it is often possible to find a pair of sums each of which has a simple formula, which are close to each other and bracket the recalcitrant sum of interest. The next two exercises illustrate this case

Exercises 1.12

1. Derive a formula for $\sum_{k=m}^{n} \frac{1}{k(k+1)}$, where $0 < m < n$.

   Hint: Examine $\Delta \frac{1}{k}$.

2. Use the result of the previous question to find good upper and lower bounds for $\sum_{k=20}^{40} \frac{1}{k^2}$, i.e., two values, U and L, close to each other in a *relative error* sense (L estimating U with small relative error, See Definition 1.10, below, for the meaning of relative error), and satisfying

$$L \le \sum_{k=20}^{40} \frac{1}{k^2} \le U.$$

   Hint: $\frac{1}{k(k+1)} \le \frac{1}{k^2} \le \frac{1}{(k-1)k}$.

3 As in the previous exercise, derive good upper and lower bounds for

$$\sum_{k=20}^{40} \frac{1}{k^3}.$$

Sums of the form $\sum_{k=m}^{n} p_k$ arise all the time in probability, representing the probability of an outcome falling between m and n. The technique just illustrated can be very helpful in getting a *handle* on the behavior of these probabilities.

Definition 1.10: Relative error

The **relative error** E, in approximating a quantity q by the value Q is defined by the formula

$$E = \frac{q - Q}{q}.$$

❑

Relative error is often more meaningful than the actual error, $q - Q$, because it remains unchanged even when the units of measurement are changed. So, for example, a change from the English to the metric system would require change from the statement that the length of a tube was measured at 500 inches with a maximum error of 1 inch. However, the relative error would be about 1/500 which would remain the same no matter which of the measuring units was used.

With the definition of relative error available, we append one more exercise to the set presented just prior to this interjection.

4. What is the relative error using
   a. $1 + p$ as an approximation to $\frac{1}{1-p}$?
   b. $\sum_{j=0}^{k-1} p^j$ as an approximation to $\frac{1}{1-p}$?
   c. What conclusions can you draw from the answer to part b?

Chapter 2

Local Linear Description

1 Introduction

In the previous chapter it was implied that just about all functions of interest in practical applications could be adequately represented by sequences. Furthermore, sequences are conceptually simpler than almost any other types of functions used in practical applications. Despite this simplicity, restricting consideration solely to sequences would make it overly difficult to develop some of the most important and useful results in applied mathematics. This shouldn't be too surprising. A hammer, saw, and nails may be all the tools needed to construct a building. And their proper use can be learned in less time than it takes to learn how to use more sophisticated tools. Nonetheless, very few would advocate requiring all construction to be done with these tools alone. The more sophisticated class of functions that we need at this stage of development are functions whose inputs and outputs are real numbers, (which we represent as infinite decimals).

Definition 2.1: Real valued function of a real variable

A function, f, whose domain and range (recall Definition 1.1 on page 2), both consist of real numbers[1] is called a **real valued function of a real variable**.

❑

We will see that many of the results which we obtain using these conceptually more complicated functions are quite a bit simpler than the corresponding sequence results.

Our object in this chapter is to set up the basic framework needed for local description and analysis of real valued functions of a real variable.

2 Tangent Lines: Convenient Linear Approximations

We start off by noticing that most real valued functions of a real variable that we have met previously — functions such as parabolas ($y = x^2$), cubics

1. A summary of the properties of real numbers is presented in Appendix 1 on page 757.

$(y = x^3)$, hyperbolas $(y = 1/x,\ x \neq 0)$ are usually graphed as smooth curves, as illustrated in Figure 2-1.

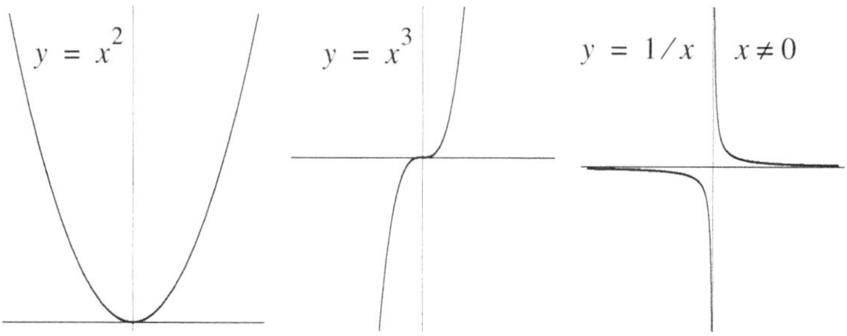

Figure 2-1

As already mentioned in the introduction to Chapter 1, we expect that we can locally approximate most of these functions by straight lines. So, if we call a typical such function f, we will begin by **hunting for the *most convenient* line which approximates f well near** $(x_0, f(x_0))$. An obvious candidate is the line through the points $(x_0, f(x_0))$ and $(x_0+h, f(x_0+h))$, as shown in Figure 2-2.

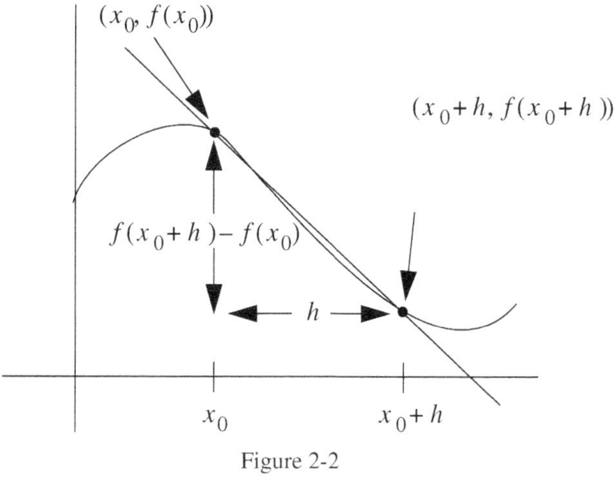

Figure 2-2

Here, h is supposed to be small enough so that the graph of f looks pretty straight between these points. While any such line is a suitable approximation to f near $(x_0, f(x_0))$, we'll have to go through some algebra to show that there is an approximating line which, for reasons of simplicity, is more suitable than any of these.

Even though it is assumed that most readers can deal adequately with lines, due to their critical importance in our development, the following optional material is included.

Lines

Each nonvertical line in the plane consists of all points (x,y) satisfying the relation

$$y = sx + b, \qquad \textbf{2.1}$$

where s and b are fixed numbers whose values determine the specific line

To be fully convinced that this is so, you must be convinced that

a. For each choice of s and b, the set of all points satisfying equation 2.1 really forms a nonvertical line,

b. For any nonvertical line specified, say, by two distinct points, there is only one choice of s and b such that the coordinates of these points satisfy equation 2.1.

To help in this process we need to introduce a small amount of analytic geometry. First we present two interpretations of any ordered pair, (a,b), of numbers. This ordered pair can be thought of as a point in the plane, as illustrated in Figure 2-3;

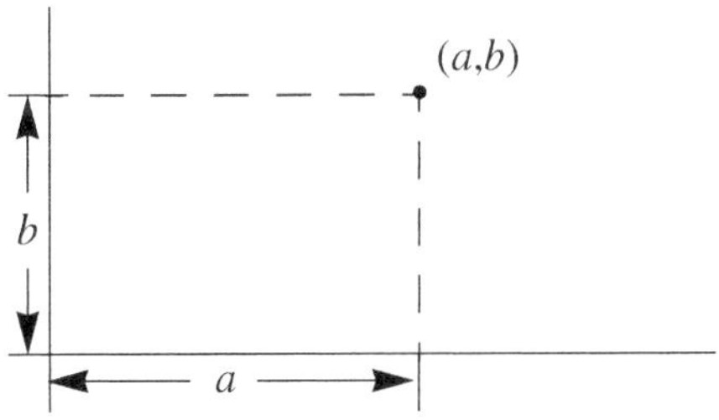

Figure 2-3

or it may also be thought of as a vector (arrow, from the origin (0,0) to the point (a,b)), as shown in Figure 2-4.

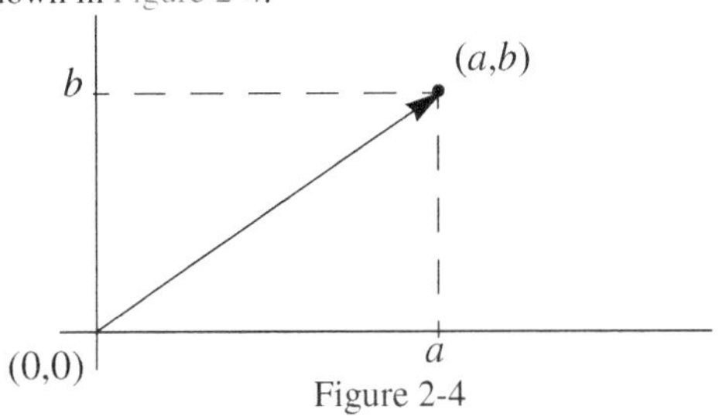

Figure 2-4

Next we introduce the operation of multiplying a vector by some real number, c. We often refer to c as a *scalar* and this operation as *scalar multiplication,* which is given by

Definition 2.2: $c(a, b) = (ca, cb)$

For $c > 0$, the geometric interpretation of this operation is that the vector $c(a,b)$ can be viewed as an arrow along the same direction as (a,b), but c *times as long* - since both coordinates have been multiplied by c, see Figure 2-5. Here c is about 1.5.

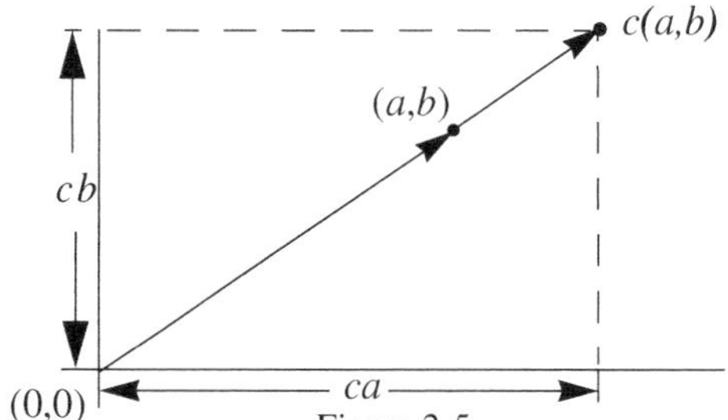

Figure 2-5

We illustrate the case c negative, and close to -1.5 in Figure 2-6.

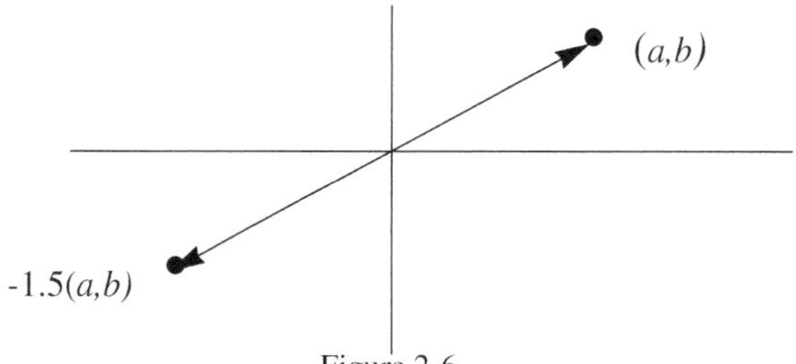

Figure 2-6

Finally, we introduce the operation of *adding two vectors*, providing

Definition 2.3: $(a, b) + (A, B) = (a + A, b + B)$

This is geometrically illustrated in Figure 2-7.

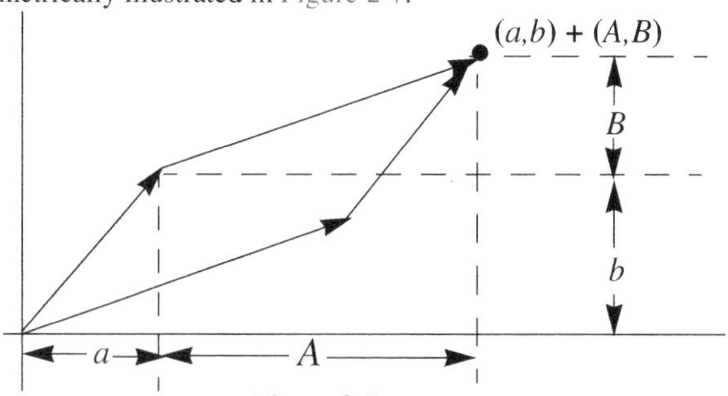

Figure 2-7

In Figure 2-7 the end of the $(a,b) + (A,B)$ arrow is found by sliding the vector (A,B) without altering its direction, until its initial point is located at the end of the vector (a,b).

We can now show that the set of all (x,y) values for which $y = sx + b$ does yield a line - because the set of all such (x,y) is the set of all points of the form $(x, sx+b)$. But from Definitions 2.2 and 2.3 just preceding we may write

$$(x, sx+b) = (0,b) + x\,(1,s), \qquad 2.2$$

that is, the set of points we obtain by adding to the vector $(0,b)$ all scalar multiples of the vector $(1,s)$. The geometrical interpretations that we have just introduced show that this set of points consists of the line which includes the point $(0,b)$ and which lies in the direction of the vector $(1,s)$. This establishes assertion a. following equation 2.1 on page 23 and is illustrated in Figure 2-8, where here s is taken to be negative.

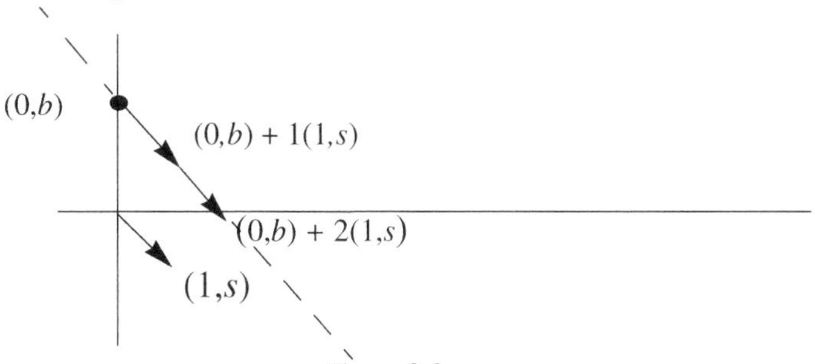

Figure 2-8

Now suppose we are given two points (x_1, y_1) and (x_2, y_2) with $x_1 \neq x_2$. (We exclude the possibility $x_1 = x_2$ to avoid the possibility of a vertical line.) To determine the formula of the form 2.1 on page 23 for the line through these two points, we see if we can uniquely determine s and b satisfying

$$\begin{aligned} y_1 &= sx_1 + b \\ y_2 &= sx_2 + b \end{aligned} \quad \textit{or} \quad \begin{aligned} y_1 - sx_1 &= b \\ y_2 - sx_2 &= b \end{aligned} \qquad 2.3$$

But these two equations are valid if and only if

$$y_1 - sx_1 = y_2 - sx_2,$$

i.e., if and only if

$$y_1 - y_2 = s(x_1 - x_2).$$

Since we excluded the possibility that $x_1 = x_2$, we find that

$$s = \frac{y_1 - y_2}{x_1 - x_2} = \frac{y_2 - y_1}{x_2 - x_1}.$$

Substituting this expression for s in any of the equations in 2.3, say, the top right one, yields

$$
\begin{aligned}
b &= y_1 - sx_1 \\
&= y_1 - x_1\frac{y_1 - y_2}{x_1 - x_2} \\
&= \frac{y_1x_1 - y_1x_2 - x_1y_1 + x_1y_2}{x_1 - x_2} \\
&= \frac{x_1y_2 - x_2y_1}{x_1 - x_2}.
\end{aligned}
$$

This establishes assertion b following equation 2.1 on page 23. The value s is called the ***slope*** of this line, and is geometrically the *rise*, $y_2 - y_1$, over the *run*, $x_2 + -x_1$, as seen in Figure 2-9

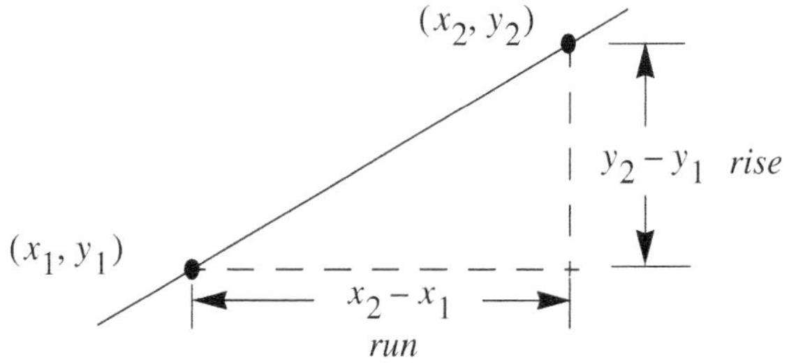

Figure 2-9

If the line's height decreases with motion to the right, then s is negative, while $s = 0$ corresponds to a horizontal line.

Now any line through $(x_0, f(x_0))$ consists of points (x,y) whose coordinates are related by the equation

$$y = s(x - x_0) + f(x_0), \qquad \textbf{2.4}$$

It's easy enough to see that the point $(x_0, f(x_0))$ is on this line no matter what the value of the *slope s*, and since all nonvertical lines are completely determined by specifying one point on the line and specifying its slope, this formula specifies all nonvertical lines which include the point $(x_0, f(x_0))$. Further, if (x_1, y_1) and (x_2, y_2) are any two *distinct* points on this line, then because

$$y_1 = s(x_1 - x_0) + f(x_0)$$
$$y_2 = s(x_2 - x_0) + f(x_0)$$

we have $$y_2 - y_1 = s(x_2 - x_0) - s(x_1 - x_0) = s(x_2 - x_1).$$

It follows that

$$s = \frac{y_2 - y_1}{x_2 - x_1}. \qquad \textbf{2.5}$$

From this, by letting (x_1, y_1) be the point $(x_0, f(x_0))$, and (x_2, y_2) be the point $(x_0 + h, f(x_0 + h))$, we can see that the slope of the line connecting the points $(x_0, f(x_0))$ and $(x_0+h, f(x_0+h))$, is given by

$$\frac{f(x_0 + h) - f(x_0)}{h}. \qquad \textbf{2.6}$$

We illustrate this in Figure 2-10.

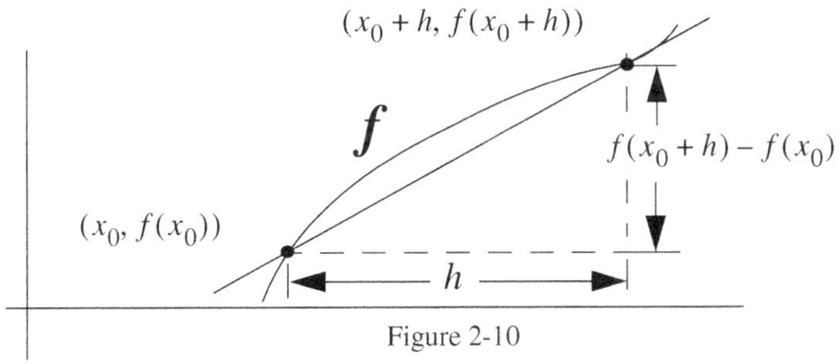

Figure 2-10

To help decide which line to choose, let us examine the formulas that we get for this slope for several different *arithmetically computable*[1] functions *f*, and for the square root function (which is not arithmetically computable, but can be easily approximated) to see if some sort of pattern emerges.

Example 2.1: Slope for $f(x) = x^2$

Here, since we are dealing with distinct points, we assume $h \neq 0$, and find that under this assumption

For small values of h, all of these slopes are close to the simple value $2x_0$.

So as the linear approximation to this function *f* near the point $(x_0, f(x_0))$,

1. These are functions with formulas such as 1, *x*, 1/*x*, and any functions that can be constructed from them using arithmetic operations of addition, subtraction, multiplication and division. These functions are used because they are the building blocks on which all computations are based.

$$\frac{f(x_0+h)-f(x_0)}{h} = \frac{(x_0+h)^2 - x_0^2}{h} = \frac{2x_0h+h^2}{h}$$
$$= 2x_0 + h$$

it's reasonable to use the line going through the point $(x_0, f(x_0))$ whose slope is $2x_0$, i.e., the line whose formula is

$$y = 2x_0(x-x_0) + x_0^2.$$

This line does include the point $(x_0, f(x_0)) = (x_0, x_0^2)$, and certainly seems simpler than its competitors. We illustrate this *tangent line to* $f(x) = x^2$ at the point $(x_0, x_0^2) = (1.5, 2.25)$ in Figure 2-11.

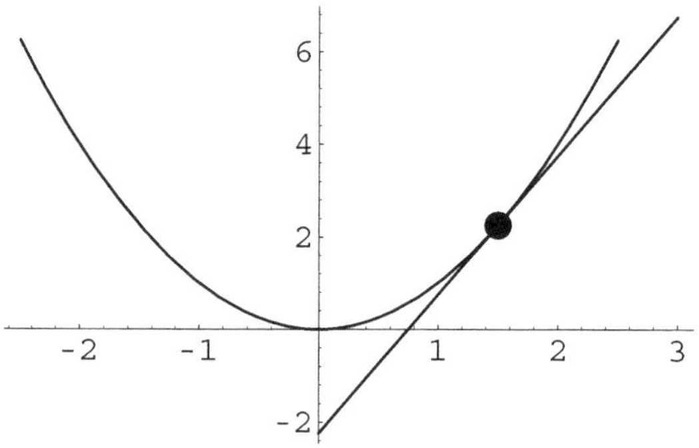

Figure 2-11

■

Example 2.2: Slope for $f(x) = 1/x$, $x \neq 0$

For $h \neq 0$ we have

$$\frac{f(x_0+h)-f(x_0)}{h} = \frac{\frac{1}{x_0+h} - \frac{1}{x_0}}{h} = \frac{x_0-(x_0+h)}{h(x_0+h)x_0}$$
$$= \frac{-h}{h(x_0+h)x_0}$$
$$= \frac{-1}{(x_0+h)x_0}.$$

Here we can see that for all nonzero values of h close to 0, these slopes are close to the simpler value $-1/x_0^2$. So here it makes sense to use the line through the point $(x_0, f(x_0))$ with slope $-1/x_0^2$ to approximate the function f near $(x_0, f(x_0))$. Notice that this slope is negative, as you would expect, since $f(x)$ is decreasing as x increases. (It's usually a wise idea to examine the sign of the slope you obtain in such calculations, to see if it makes sense.)

■

Example 2.3: Slope for $f(x) = \sqrt{x}$, $x > 0$

For nonzero h sufficiently close to 0, we calculate

$$\frac{f(x_0+h)-f(x_0)}{h} = \frac{\sqrt{x_0+h}-\sqrt{x_0}}{h}$$

$$= \frac{\sqrt{x_0+h}-\sqrt{x_0}}{h}\;\frac{\sqrt{x_0+h}+\sqrt{x_0}}{\sqrt{x_0+h}+\sqrt{x_0}}$$

$$= \frac{h}{h(\sqrt{x_0+h}+\sqrt{x_0})}$$

$$= \frac{1}{\sqrt{x_0+h}+\sqrt{x_0}}.$$

For small h, that is, h close enough to 0, this slope is close to $1/(2\sqrt{x_0})$. *This* should be the slope of the approximating line to the square root function at $(x_0, \sqrt{x_0})$.

■

In the previous examples and from here on, we take for granted that any line used for approximating a function f near x_0 should include the point $(x_0, f(x_0))$. The last three examples suggest that as the slope of the desired approximating line, we should use the simpler value that all of the ratios

$$\frac{f(x_0+h)-f(x_0)}{h}$$

are close to for small nonzero h (nonzero values of h which are close to 0). This leads to the following definitions.

Definition 2.4: Limit

If the real valued function F, of a real variable (Definition 2.1 on page 21) is defined for all points in some interval which includes the point x_0 strictly inside, and if there is a value which $F(x_0 + h)$ is close to for ***all sufficiently small nonzero values of h***, then this value is denoted by

$$\lim_{h \to 0} F(x_0 + h)$$

and is called the **limit of $F(x_0 + h)$ as h approaches 0**. This limit is also called the **limit of $F(x)$ as x approaches x_0** and is written

$$\lim_{x \to x_0} F(x) .$$

❑

Definition 2.5: Derivative, f', Df, derivative operator D, tangent, differentiable function, $f^{(n)}$, D^n

If the limit

$$\lim_{h \to 0} \frac{f(x_0 + h) - f(x_0)}{h}$$

exists, it is called the **derivative** of f at x_0 and will be denoted by the symbol

$$f'(x_0) \quad \text{or by the symbol[1]} \quad Df(x_0)$$

The line through $(x_0, f(x_0))$ whose slope is $f'(x_0)$ is called the **tangent line to f at x_0**, or the **tangent line to f at the point (x_0, y_0)**, where $y_0 = f(x_0)$. When f has a derivative at every point in its domain [at x_0] it is called **differentiable** [at x_0]. The ***n*-th derivative** $f^{(n)}(x_0) = D^n f(x_0)$ is defined inductively, via $D^{n+1} f = D(D^n f)$, with $D^0 f = f$.

Note that $f' = Df$ is a function — the slope function, whose value at x_0 is $f'(x_0) = Df(x_0)$. The **derivative operator, D,** is the natural extension of the **difference operator, Δ**, (see Definition 1.3 on page 5 ff.), to real valued functions of a real variable. D is a function whose inputs and outputs are themselves functions. Use of the derivative operator notation will turn out to simplify certain investigations considerably; it is introduced here so that when the time comes to make use of it, you can concentrate on how it is used, without the distraction of using an unfamiliar notation.

❑

---

1. We will avoid the commonly used symbol $\frac{df}{dx}(x_0)$ to denote the derivative.

In Figure 2-12, we plot the function f, and its derivative f' for $f(x) = x^2$ and $f(x) = 1/x$.

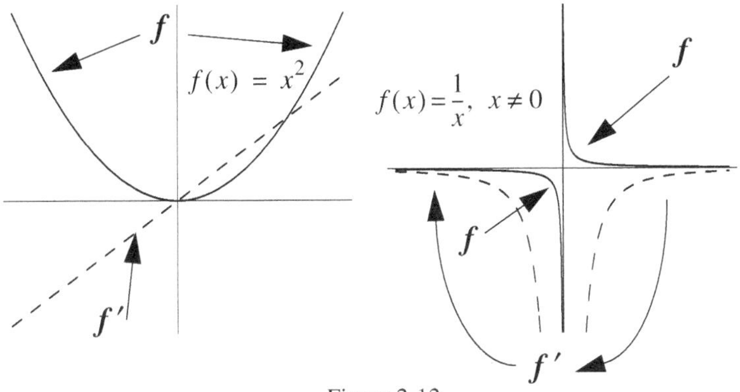

Figure 2-12

Exercises 2.4

1. Find the derivative, $f'(2)$ for
   a. $f(x) = x^2 + 3x$
   b. $f(x) = 1/x^2$
   c. $f(x) = 1/\sqrt{x}$
   d. $f(x) = 1/(1-x)$
2. Find the equation of the tangent line to the functions in question 1 above at the point $x = 2$.

3 The Fundamental Linear Approximation

One of the most important uses of derivatives, (Definition 2.5 on page 30), is for *quick and dirty* approximations. The reasoning is pretty simple; the idea behind the derivative is that $f'(x)$ is the value that the slope

$$\frac{f(x+h) - f(x)}{h}$$

is *close to*, for small nonzero h. Hence we may write the following.

Fundamental approximation

$$f(x_0 + h) \cong f(x_0) + f'(x_0)h, \quad \text{valid for } \textit{small } h. \qquad \textbf{2.7}$$

Approximation 2.7 is of practical use for approximate computation of $f(x_0 + h)$ in terms of $f(x_0)$ and $f'(x_0)$. It will have widespread application when we develop simple formulas for computing derivatives, in the next section. As we extend the scope of our methods, this approximation will be extended repeatedly to more complicated types of functions, where it will help us to gain insight into their behavior and enable us to derive many of their properties in a very simple fashion.

We conclude this section with a few of the simplest applications of this most important result.

Example 2.5: Approximation of $\sqrt{1+h}$

In approximation take $x_0 = 1$ and $f(x) = \sqrt{x}$. We know from Example 2.3 on page 29 that $f'(x_0) = 1/(2\sqrt{x_0})$. So $f'(1) = 1/2$. Substituting, we find

$$\sqrt{1+h} \cong \sqrt{1} + \frac{1}{2\sqrt{1}}h$$

or

$$\sqrt{1+h} \cong 1 + \frac{h}{2} \quad \text{for small } h.$$

This approximation is more useful than it might appear at first glance because it can help to improve most initial guesses for square roots. For instance, if you wanted to compute $\sqrt{26}$, you could write

$$\sqrt{26} = \sqrt{25+1} = \sqrt{25}\sqrt{1+\frac{1}{25}} \cong 5(1+1/50) = 5.1.$$

■

Later on we will investigate the accuracy of this type of approximation. But with the knowledge of the square root function gained in Example 1.4 on page 9, we would guess that the square root function is concave down, so that the tangent line lies above the graph of the function. Thus we expect that this approximation overestimates $\sqrt{26}$. This is easily verified by noting that $5.1^2 = 25.01$.

Example 2.6: Approximation of $\frac{1}{1-h}$

There are several ways of handling this expression.[1] The one we will use here relies on the derivative of the function given by $f(x) = 1/x$. We know from Example 2.2 on page 28 that $f'(x) = -1/x^2$. So we take $x_0 = 1$, getting $f(x_0) = 1$, $f'(x_0) = -1$; replacing h by $-h$, we obtain

$$f(x_0 - h) = \frac{1}{1-h} \cong 1 - (-h) = 1 + h\ .$$

---

1. Previously we used the geometric series, see approximation 1.13 on page 17.

As we already know from the answer to Exercise 1.5.1e on page 10, the sequence whose n-th term is $1/n$, is concave up. So here we would expect f to be concave up, and hence, its tangent lines to lie below its graph; so it would appear that this approximation underestimates the desired value. This result should come as no surprise, in light of the geometric series result, Theorem 1.9 on page 16.

■

Exercises 2.7

1. Approximate 3.1^3.
2. Derive the result of example Example 2.6 above by examination of the function given by $f(x) = 1/(1-x)$.
3. Approximate $\sqrt[3]{28}$.
   Hint:

   $$(\sqrt[3]{x+h} - \sqrt[3]{x})(\sqrt[3]{(x+h)^2} + \sqrt[3]{x+h}\cdot\sqrt[3]{x} + \sqrt[3]{x^2}) = h$$

4. Approximate $1/\sqrt{24}$.

4 Continuity and Calculating Derivatives

At this stage it's worthwhile to examine what we've done to compute derivatives, so that we can develop a systematic, dependable approach to this task. We are using the value

$$\lim_{h \to 0} \frac{f(x_0 + h) - f(x_0)}{h},$$

which $[f(x_0 + h) - f(x_0)]/h$ gets close to for small nonzero h, as the slope of the line approximating f locally near the point $(x_0, f(x_0))$.

It would be convenient if we could find this limit by simply plugging $h = 0$ into the ratio $([(f(x_0 + h) - f(x_0))]/h)$. **This won't work**, because it takes two distinct points to determine a straight line, and the pairs $(x_0, f(x_0))$, $(x_0 + h, f(x_0 + h))$ yield only one point when $h = 0$.

The way that we overcame this obstacle in Examples 2.1, 2.2 and 2.3 was to use algebraic manipulation to find an expression, which will be denoted by $C(h)$, that had the following properties:

$$\text{For } h \neq 0 \quad C(h) = \frac{f(x_0 + h) - f(x_0)}{h}, \qquad \textbf{\textit{2.8}}$$

$$C(0) \text{ is meaningful and } \quad C(0) = \lim_{h \to 0} \frac{f(x_0 + h) - f(x_0)}{h}. \qquad \textbf{2.9}$$

That is, we *play around (legitimately)* with the ratio $[(f(x_0 + h) - f(x_0))]/h$ until we find an expression $C(h)$ equal to it for nonzero h, into which we can meaningfully plug in $h = 0$. The correct answer will be obtained if the function C is *continuous at* $h = 0$. By this we mean that $C(h)$ is close to $C(0)$ when h is near 0. More generally we define continuity at h_0 as follows.

Definition 2.6: Continuity

The real valued function C, *of a real variable* (Definition 2.1 on page 21), is said to be **continuous** at h_0 if

- $C(h_0)$ is meaningful

  and, for h in the domain of C,

- $C(h)$ is close to $C(h_0)$ when h is close to h_0.

C is said to be **continuous** if it is continuous at each point in its domain (i.e., at each possible input).

❑

If the function C is thought of as an *input/output* box, then it is continuous at h_0 if a small change in the input, h_0, yields only a small change in the output, $C(h_0)$.

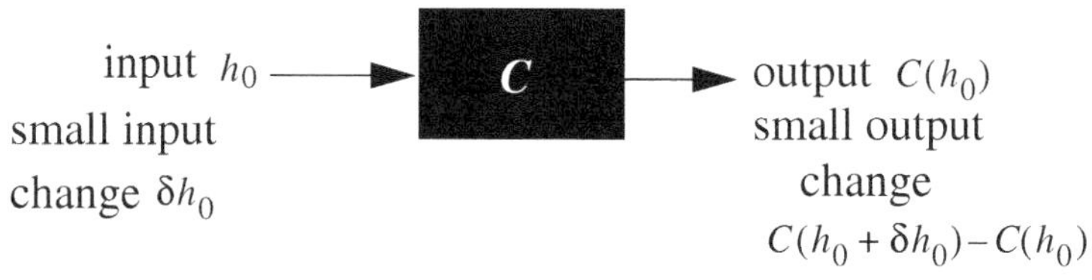

Behavior of function C which is continuous at h_0

Figure 2-13

Continuity is an extremely important concept in mathematical modeling, especially when making use of imperfect measuring instruments. For example, if the orbit of a space vehicle was not a continuous function of various critical parameters such as the energy expended in its launch, the temperatures of the medium through which it is traveling, and so forth, there would be no hope of controlling/correcting its orbit. Or, in medicine, if the results of medication were not continuous functions of dosage, treatment with drugs would be unfeasible. What is so dangerous about plutonium might be thought of as akin to a lack of continuity at zero dosage — any amount of ingested plutonium is considered to be likely to induce cancer.

Our immediate need for continuity is to allow calculation of derivatives from the definition. But we don't even have a precise operational definition of continuity. Later, (in Chapter 7, Section 2), we will present such a definition and develop some familiarity with it[1]. Lacking these resources here, we will make do with the results summarized in Theorem 2.10 on page 38, which will allow us to ***recognize continuity*** in most of the cases that will arise in the material being considered here. The following definitions are needed in order to provide a solid foundation for the terms used in Theorem 2.10.

Definition 2.7: cf, $f + g$, $f - g$, fg, $1/f$, f/g, $H(f)$

Let c be a given real number (a *constant* or *scalar*), and let *f, g, and H* be real valued functions of a real variable.

- cf is the function whose value at each allowable input, x, to f is $cf(x)$.
- $f+g$ is the function whose value at each allowable input, x, is $f(x) + g(x)$, provided f and g have the same *domain* — i.e., the same allowable inputs, which is indicated by writing $dmn(f) = dmn(g)$. We let $f - g$ be defined as $f + (-1)g$, so that $(f-g)(x) = f(x) - g(x)$.
- fg is the function whose value at each allowable input, x, is the product $f(x)g(x)$ (provided f and g have the same domain, see Definition 1.1 on page 2).
- $1/f$ is the function whose value at input x is $1/f(x)$, provided that $f(x)$ is never 0. Like $(f - g)(x)$, let $(f/g)(x) = f(x)/g(x)$.
- $H(f)$ is the function whose value at x is $H(f(x))$, provided that all values $f(x)$ are in the domain of H (allowable inputs to H). $H(f)$ is called the **composition** of H with f.

❑

The ***product*** fg can be thought of as the function whose value at x is the ***area*** of a rectangle whose side lengths are $f(x)$ and $g(x)$, as shown in Figure 2-14. This interpretation turns out to be surprisingly useful.

---

1. A thorough treatment of continuity is presented in *Modern University Calculus*, by Bell, Blum, Lewis and Rosenblatt [1].

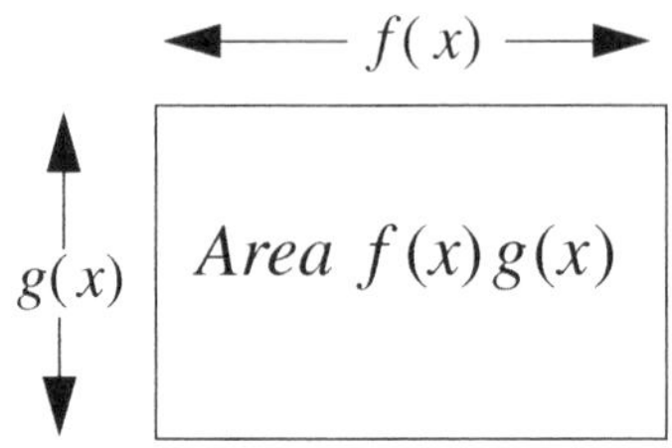

Figure 2-14

The ***composition*** $\boldsymbol{H(f)}$ can be visualized as a pair of *input/output black boxes,* as illustrated in Figure 2-15.

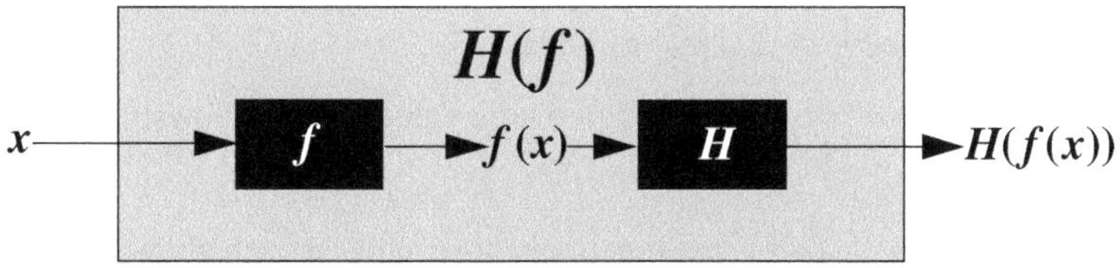

Figure 2-15

Two more definitions, that of the *inverse function* and that of a *strictly monotone function,* are needed. The inverse function of some function *f*, is the function which ***reverses*** the action of *f*, ***provided that such a function exists.*** For example, the *cube root function, cr,* whose formula is

$$cr(y) = \sqrt[3]{y} \qquad \textbf{2.10}$$

is the inverse of the *cube* function, *cu*, whose formula is

$$cu(x) = x^3. \qquad \textbf{2.11}$$

The function, *f*, given by

$$f(x) = x^2 \quad \text{for } x \text{ a real number,} \qquad \textbf{2.12}$$

does not possess an inverse function, because every positive real number has two square roots; hence knowing that the value of this function at x was 4 does not allow us to reverse the action of f to determine which input, $x = 2$, or $x = -2$, gave rise to the output 4. However, the function given by

$$f(x) = x^2 \quad \text{for } x \text{ real,} \quad x \geq 0 \qquad \textbf{2.13}$$

does have an inverse function, often denoted by *sqrt; if we let* $\sqrt{y}$ denote the nonnegative square root of the nonnegative number y. *sqrt* is given by

$$sqrt(y) = \sqrt{y}, \quad y \geq 0. \qquad \textbf{2.14}$$

Definition 2.8: Inverse function

If f is some function (not necessarily a real valued function of a real variable) and there is a function, g, whose domain is the range of f (see Definition 1.1 on page 2), and which satisfies the equation $g(f(x)) = x$ for all x in the domain of f, then g is called the **inverse function of** f, and is denoted by the symbol f^{-1} or sometimes by *invf.*

❑

In Figure 2-16 we illustrate the concept of functions and their inverse functions just presented in equations 2.10, 2.11, 2.13, and 2.14.

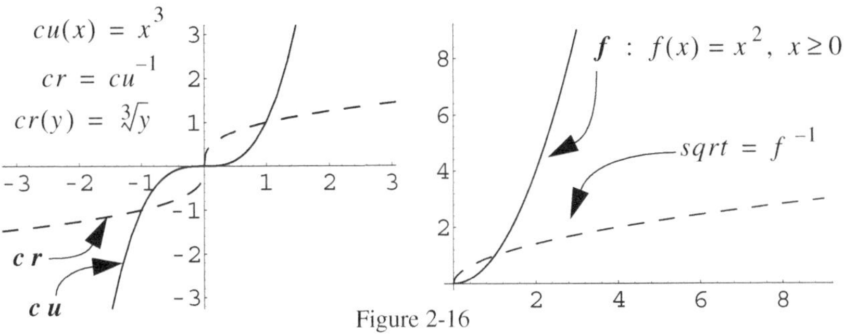

Figure 2-16

If we consider the graph of the function f as a set of points (x,y), then the graph of *inv f* consists of the points (y,x); so it is obtained from the graph of f as its mirror image about the 45 degree line, as seen in Figure 2-16.

The other definition that is needed, of a *strictly monotone function,* is used to provide easily verifiable conditions which ensure that a function is *invertible* (has an *inverse function*).

Definition 2.9: Monotone and strictly monotone functions

The real valued function f, of a real variable is said to be **monotone increasing [nondecreasing, decreasing, nonincreasing]** if for each x and y in the domain of f, with $x < y$, we have $f(x) < f(y)$ [$f((x) \leq f(y))$, $f(x) > f(y), f(x) \geq f(y)$]. The function f is said to be **strictly monotone** if it is either monotone increasing or monotone decreasing. Often we refer to monotone increasing functions simply as **increasing** functions, etc.

❑

The graph of a monotone increasing function rises with movement to the right, while that of a monotone decreasing function falls with such movement, as illustrated in Figure 2-17.

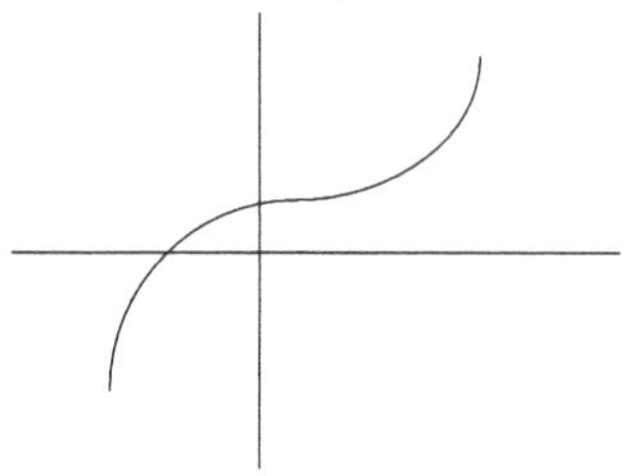

Graph of monotone increasing function

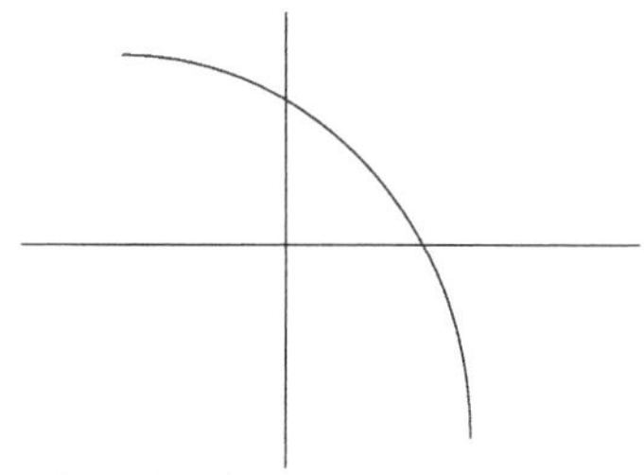

Graph of monotone decreasing function

Figure 2-17

Theorem 2.10: Continuity theorems

Let f and g be real valued functions of a real variable (Definition 2.1 on page 21), with the same domain (inputs) and the domain of the real valued function H includes the range (outputs) of the function f, and c is a given constant. If f and g are continuous (Definition 2.6 on page 34), at x_0, then

- cf is continuous at x_0;
- $f+g$ is continuous at x_0;
- fg is continuous at x_0;
- $1/f$ is continuous at x_0, (provided $f(x_0) \neq 0$).
- If H is continuous at $f(x_0)$, then $H(f)$ is continuous at x_0;
- If the domain of f is an interval, and f is continuous at every point in its domain, and is also strictly monotone (Definition 2.9 on page 37), then f is invertible (has an inverse function, Definition 2.8 on page 37), which is continuous at every point in its domain (the range of f).

❑

The first four of these results are needed to ensure continuity of functions created by arithmetic operations on continuous functions. Since many functions are created by composition (see Definition 2.7 on page 35), the next to last result is sure to be useful. The final result is what is used here to establish the continuity of n-th root functions.

The optional section which follows is an attempt to provide some insight as to the reasonableness of parts of Theorem 2.10. Those more interested in seeing how the results of this theorem are applied may omit or postpone looking through it.

Let us see how small h must be in order to make the quantity $f(x_0+h)g(x_0+h)-f(x_0)g(x_0)$ as small as we might desire, to establish the continuity of the ***product***, fg, at x_0. We rewrite this expression in terms of the quantities that we can control. Namely let

$$\delta f(h) = f(x_0+h)-f(x_0)$$

$$\delta g(h) = g(x_0+h)-g(x_0)$$

and substitute $\quad f(x_0)+\delta f(h)$ for $f(x_0+h)$

and $\quad g(x_0)+\delta g(h)$ for $g(x_0+h)$

to obtain

$$f(x_0+h)g(x_0+h)-f(x_0)g(x_0)=f(x_0)\delta g(h)+g(x_0)\delta f(h)+\delta f(h)\delta g(h). \quad \textbf{\textit{2.15}}$$

We illustrate equation 2.15 in Figure 2-18.

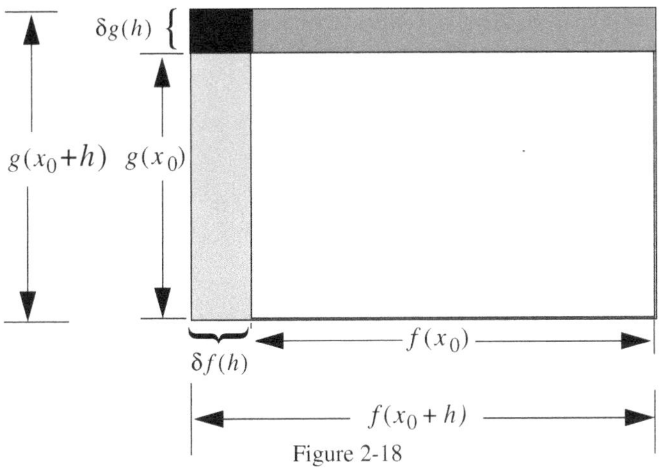

Figure 2-18

Here the left side of equation 2.15 above is the total area of the darker shaded three strips. Each of these strips represents one of the terms on the right side of equation 2.15. From this figure it is evident that if we keep h sufficiently small, these three regions have a small total area.

Algebraically we see this because continuity of f at x_0 means that $\delta f(h)$ can be made as small as desired by taking h close enough to 0, and similarly for $\delta g(h)$. So, if we wanted the change in the product fg to have magnitude smaller than, say, 10^{-6}, just choose h close enough to 0 so that the following four conditions are satisfied:

- $|\delta f(h)| \le \frac{1}{|g(x_0)|}\frac{10^{-6}}{3}$, disregarding this if $g(x_0) = 0$,
- $|\delta g(h)| \le \frac{1}{|f(x_0)|}\frac{10^{-6}}{3}$, disregarding this if $f(x_0) = 0$,
- $|\delta f(h)| \le \frac{10^{-3}}{\sqrt{3}}$,
- $|\delta g(h)| \le \frac{10^{-3}}{\sqrt{3}}$.

Then each of the terms on the right of equation 2.15 will have magnitude no larger than $10^{-6}/3$, so that the right side of equation 2.15 cannot have magnitude larger than 10^{-6}, as desired.

The continuity of the ***composition***, $H(f)$, can be understood by examining Figure 2-15 on page 36. Because of the continuity of f, a small change in the input x causes only a small change in the output $f(x)$. But the change in output of f is the input change to H. So the input change to H can be made small — and thus, using the continuity of H, its output change can therefore be made small. But this output of H happens to be the output of the composition $H(f)$ when its input is x.

So, by keeping the change in x small enough, we can keep the change in $H(f(x))$ as small as is desired.

To establish continuity of the ***reciprocal*** $1/f$, at x_0, we want to show that we can make the magnitude of

$$\frac{1}{f(x_0+h)} - \frac{1}{f(x_0)}$$

as small as desired, by keeping h close enough to 0. We see how to accomplish this as follows:

Since

$$\begin{aligned}\frac{1}{f(x_0+h)} - \frac{1}{f(x_0)} &= \frac{f(x_0) - f(x_0+h)}{f(x_0+h)f(x_0)} \\ &= \frac{-\delta f(h)}{[f(x_0) + \delta f(h)]f(x_0)} \\ &= \frac{1}{(f(x_0))^2}\frac{-\delta f(h)}{1 + \frac{\delta f(h)}{f(x_0)}}\end{aligned}$$

if we wanted to keep this smaller than, say 10^{-9}, we need only choose h close enough to 0 so that the following conditions hold:

- $|\delta f(h)| \leq 0.5|f(x_0)|$

 so that the second denominator, $1 + \dfrac{\delta f(h)}{f(x_0)} \geq 0.5$,

and

- $|\delta f(h)| \leq 0.5(f(x_0))^2 10^{-9}$.

We next look at the result on continuity of inverse functions. This result is not of major importance here, because it turns out to be much easier to establish the differentiability of inverses of strictly monotone differentiable functions using purely geometrical arguments. Its only use in our context was for direct derivation of the derivative of square roots and cube roots.

Continuity of ***inverses*** is somewhat trickier to establish, basically because the more continuous that f is, the *less continuous* that *inv f* seems to be. That is, the less $f(x)$ is changing for x near x_0, the more $f^{-1}(y)$ changes for y near $y_0 = f(x_0)$. In such a situation an indirect proof *by **contradiction*** is called for. We prove that f^{-1} is continuous by ***showing that it is impossible for it to fail to be continuous***. So we start off by saying "*Suppose f^{-1} failed to be continuous at some value y_0.*" We then logically reason our way along until we arrive at an assertion which is palpably false. Assuming our reasoning was correct, the only possible conclusion that we can come to is that the assumption of lack of continuity is untenable. Proof by contradiction has an aura of unreality — basically because we are always dealing with a correct derivation of ***false*** statements. As such, the details of such a proof may not seem very enlightening — and hence we will omit them. In outline, here's what needs to be done:

- Establish that if f is strictly monotone, then f has an inverse function (not very difficult; see Appendix 2 on page 761).).

- Determine what would occur if f^{-1} failed to be continuous at some point y_0 in its domain (the range of f). That is, determine how to ***phrase a denial of continuity***; i.e., rephrase the statement "*It is false that $g = f^{-1}$ is continuous at* y_0", which means rephrasing the statement "*It is false that as h gets close to* 0, $g(y_0 + h)$ *gets close to* $g(y_0)$." The essence of this rephrasal is that if g were not continuous at y_0, there would be some value $c > 0$ such that no matter what the (nonzero) distance, *d,* to 0, there are values of h even closer to 0 than d, for which

$$|g(y_0 + h) - g(y_0)| = |g(y_0 + h) - x_0| \geq c.$$

 The missing details now involve establishing that $f(x)$ fails to be defined for the x values satisfying $x_0 - c < x < x_0$ or for those x values satisfying $x_0 < x < x_0 + c$, but is defined for x_0 and some x values further than c units from x_0.

- But this last assertion is false for the type of functions, *f,* that we are examining, since it violates the given conditions that we are considering only functions, f, whose domains are intervals.

So, for each function, f, which is strictly monotone and continuous with an interval domain, a lack of continuity of the inverse function, $g = f^{-1}$ would lead to the conclusion that the domain (inputs) of f would not be an interval — i.e., leads to a clearly false statement. The only way around this problem is to admit that a lack of continuity of f^{-1} cannot hold true — i.e., to admit that $g = f^{-1}$ must be continuous.

The final result that we need concerning continuity (Definition 2.6 on page 34) is the following innocent looking theorem.

Theorem 2.11: Continuity of constant and identity functions

Each real valued function of a real variable whose formula is

$$K(x) = c,$$

(a function whose value is always the constant c for all x in its domain), or

$$I(x) = x$$

(called an ***identity function*** on whatever domain is specified) is continuous.

❑

These conclusions are easily established; since for *constant* functions the quantity

$$K(x+h) - K(x) = c - c = 0$$

is always small, while for the *identity* function, I,

$$I(x+h) - I(x) = (x+h) - x = h$$

can surely be made small by making h small.

Now, from successive applications of the continuity of products of continuous functions (see Theorem 2.10 on page 38), we can establish the continuity of functions given by $f(x) = x^n$, n a positive integer,[1] where x

---

1. A strict proof requires induction: we already know it holds for $n = 1$. Next show that whenever the assertion holds for a particular value of n, then it also must hold for the *next value*, $n + 1$. That is, show that if $f(x) = x^n$ is continuous, then $x^{n+1} = x \cdot x^n = xf(x)$ is continuous. But this follows from the third result of Theorem 2.10 because both factors are continuous.

is restricted to some specified domain.

Then, from the continuity result on inverse functions (again Theorem 2.10), it follows that any reasonable n-th root function is continuous.

In fact, just about any function that can be constructed algebraically, such as polynomials, rational functions (ratios of polynomials), and so forth, are continuous at each point in their domains.

What we now have is the ability to recognize continuity of most of the functions that will be encountered just by examining how they were constructed.

Let us see how these results can be applied to put some of the previous derivative computations on a firmer foundation.

Example 2.8: Cleaning up Examples 2.1 and 2.2

In Example 2.1 on page 27 (derivative of $f(x) = x^2$), we found

$$C(h) = 2x_0 + h.$$

The first term to the right of the = sign is constant, hence continuous, while the second term is the value at h of the identity function, $I(h) = h$, so that C, being the sum of two continuous functions, is continuous. So, from the definition of continuity, we have the result that

$$\lim_{h \to 0} C(h) = C(0) = 2x_0,$$

really is the desired derivative of f at x_0.

In Example 2.2 on page 28 (derivative of $f(x) = 1/x$), we found

$$C(h) = \frac{-1}{(x_0 + h)x_0}.$$

The denominator is the product of values of two continuous functions, x_0 representing a constant function, which we know is continuous, and $x_0 + h$, which is the sum of values of two continuous functions, (the constant function whose value is x_0 and the identity function), which is therefore continuous. Hence the denominator represents the value of a continuous function, and thus from the fourth result in Theorem 2.10 on page 38 it follows that C is continuous. From this continuity we have that

$$\lim_{h \to 0} C(h) = C(0) = \frac{-1}{x_0^2}$$

is the derivative of $f(x) = 1/x$ at $x = x_0 \neq 0$.

■

Exercise 2.9

Redo Exercise 2.4.1 on page 31 properly, for arbitrary values of x.

5 Rules for Computing Derivatives

Up to this point we have calculated derivatives directly from the definitions. As a general method for derivative computation this is too cumbersome for all but the simplest functions. Nevertheless, it is essential to understand the direct method for use in cases that aren't covered by the rules we are about to derive. In particular, a thorough understanding of the definition of the derivative is often necessary when new problems must be formulated and analyzed.

Now we'll turn our attention to the rules which simplify computing of derivatives. By making computation of derivatives so easy, they make local description convenient as a standard technique, rather than a method solely for expert specialists.

Our discussion here will be fairly informal, and we will summarize all of the results for future reference in Theorem 2.13 on page 53.

The main *algebraic* tool for deriving these rules is the fundamental approximation, 2.7 on page 31:

$$f(x_0 + h) \cong f(x_0) + f'(x_0)h.$$

Let us first determine the derivative of the sum, $F + G$, of differentiable functions, F and G.

For nonzero h we may write

$$\begin{aligned}\frac{(F+G)(x_0+h)-(F+G)(x_0)}{h} &= \frac{F(x_0+h)+G(x_0+h)-F(x_0)-G(x_0)}{h}\\ &\cong \frac{F(x_0)+F'(x_0)h+G(x_0)+G'(x_0)h-F(x_0)-G(x_0)}{h}\\ &\qquad \text{by applying approximation 2.7 to } F \text{ and to } G\\ &= \frac{[F'(x_0)+G'(x_0)]h}{h}\\ &= F'(x_0)+G'(x_0)\end{aligned}$$

So, it appears that $(F + G)' = F' + G'$. It isn't hard to extend this result to *weighted sums*, $rF + sG$ — namely,

$$(rF + sG)' = rF' + sG' \qquad \textbf{2.16}$$

where r and s are fixed constants, which holds provided that $rF + sG$ is defined, and both F and G are differentiable.

A more *water-tight* argument for the simple version of this result is the following:

$$\frac{(F+G)(x_0+h)-(F+G)(x_0)}{h} = \frac{F(x_0+h)-F(x_0)}{h} + \frac{G(x_0+h)-G(x_0)}{h}.$$

Define

$$f(h) = \begin{cases} \dfrac{F(x_0+h)-F(x_0)}{h} & \text{for} \quad h \neq 0, \\ F'(x_0) \quad \text{when } h = 0. \end{cases} \qquad \textbf{2.17}$$

and

$$g(h) = \begin{cases} \dfrac{G(x_0+h)-G(x_0)}{h} & \text{for} \quad h \neq 0, \\ G'(x_0) \quad \text{when } h = 0. \end{cases} \qquad \textbf{2.18}$$

Then, because of the differentiability of F and G at x_0, f and g are continuous at $h = 0$. So, by Theorem 2.10 on page 38, $f + g$ is continuous at $h = 0$. Hence,

$$\frac{(F+G)(x_0+h)-(F+G)(x_0)}{h} = f(h) + g(h) \quad \text{for } h \neq 0$$

has a limit as $h \to 0$, and this limit, by the continuity of $f + g$ is

$$f(0) + g(0) = F'(x_0) + G'(x_0),$$

proving, in a somewhat more precise fashion, that

$$(F+G)'(x_0) = F'(x_0) + G'(x_0).$$

The next expression whose derivative we turn to is the product, FG, of two differentiable functions. In order to handle this task in the simplest way, we think of the product, $F(x)G(x)$, as a rectangular area, as depicted in Figure 2-14 on page 36. Since we are interested in the behavior of the expression

$$\frac{F(x_0+h)\,G(x_0+h) - F(x_0)\,G(x_0)}{h}$$

as h gets small, we notice that its numerator is represented by the area of the three darker shaded regions in Figure 2-19.

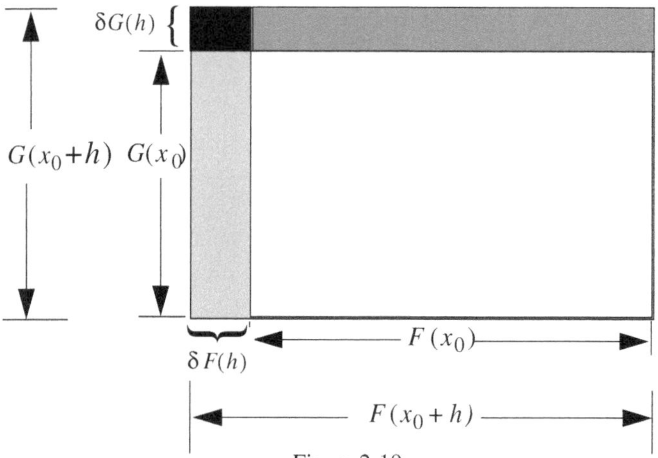

Figure 2-19

From this illustration it is natural to introduce the abbreviation

$$\delta F(h) = F(x_0+h) - F(x_0)$$
$$\delta G(h) = G(x_0+h) - G(x_0)$$

and

$$\delta FG(h) = (FG)(x_0+h) - (FG)(x_0)$$

and then rewrite the ratio of interest as

$$\frac{\delta FG(h)}{h} = \frac{\delta F(h)G(x_0) + F(x_0)\delta G(h) + \delta F(h)\delta G(h)}{h}. \qquad \textbf{\textit{2.19}}$$

Now use the fundamental approximation 2.7 on page 31, to write

$$\delta F(h) \cong F'(x_0)h \quad \text{and} \quad \delta G(h) \cong G'(x_0)h,$$

and substitute these approximations into equation 2.19, to obtain

$$\frac{\delta FG(h)}{h} \cong \frac{F'(x_0)hG(x_0) + F(x_0)G'(x_0)h + F'(x_0)hG'(x_0)h}{h}$$

or $$\frac{\delta FG(h)}{h} \cong F'(x_0)G(x_0) + F(x_0)G'(x_0) + F'(x_0)G'(x_0)h$$

for $h \neq 0$.

It seems, therefore, that the ratio

$$\frac{\delta FG(h)}{h} = \frac{(FG)(x_0 + h) - (FG)(x_0)}{h}$$

gets close to

$$F'(x_0)G(x_0) + F(x_0)G'(x_0)$$

as h gets small. That is, that

$$(FG)'(x_0) = F(x_0)G'(x_0) + F'(x_0)\,G(x_0), \qquad \textbf{2.20}$$

which is, in fact, the case. This may alternately be written

$$D(FG) = FD(G) + GD(F), \qquad \textbf{2.21}$$

which is often easiest to remember in the form

$$D(hi\ ho) = hiDho + hoDhi\,.$$

A more precise proof still starts out with equation 2.19

$$\frac{\delta FG(h)}{h} = \frac{\delta F(h)\,G(x_0) + F(x_0)\,\delta G(h) + \delta F(h)\,\delta G(h)}{h}$$

However we then proceed by using abbreviations 2.17 and 2.18 on page 45, and then using the appropriate continuity theorems on page 38 to determine a representation of the above ratio which is defined and continuous at $h = 0$, just as was done in the more precise proof of the derivative of a sum. Those interested might try

Exercises 2.10

1. Supply the details in establishing the *product rule* (equation 2.20 above for the derivative of a product of two differentiable functions).
2. Supply the details for establishing the more water- tight version of the weighted *sum rule* (derivative of $rF+sG$, equation 2.16 on page 44).

Example 2.11 : $D(x^n)$

The function I^n, the n-th power of the identity has the formula $I^n(x) = x^n$, and is often symbolized by its formula — i.e. referred to simply as x^n, an abuse of notation which can be justified (see Chapter 5, Section 4). We already know that for $n = 2$, $D(x^n) = nx^{n-1}$.. This formula is easily verified for $n = 1$. To show that it holds for all of the other positive integers, examine any n for which this rule is valid; so far only $n = 1$ and $n = 2$ qualify. We may write $x^{n+1} = x\,x^n$. Now just apply the product rule, as expressed in equation 2.21, to this expression, to obtain

$$\begin{aligned} D(x^{n+1}) &= D(x)x^n + xD(x^n) \\ &= 1 \cdot x^n + x \cdot n\,x^{n-1} \\ &= (n+1)x^n. \end{aligned}$$

So we see that whenever we have

$$D(x^n) = n\,x^{n-1}$$

it follows that

$$D(x^{n+1}) = (n+1)x^n.$$

Thus, starting with $n = 2$, using this we find the rule $D(x^n) = nx^{n-1}$ holds for $n = 3$, from which it now follows that it holds for $n = 4$, and so forth. This should be convincing enough evidence that

$$D(x^n) = nx^{n-1}, \qquad \textbf{2.22}$$

for all positive integers, n.

■

The ***local change magnification interpretation of the derivative*** that we are about to introduce provides further insight into the meaning of the derivative, which will enable us to derive many results in a simple fashion (the first application being a simple, understandable derivation of the rule for differentiating a composition of two differentiable functions [the *chain rule*]). We begin by rewriting the fundamental approximation, given by 2.7 on page 31, as

$$f(x+h) - f(x) \cong f'(x)h. \qquad \textbf{2.23}$$

Now interpreting the function f as an input-output *black box*, we note that h is the ***input change*** to the function f, while the right side of approximation 2.23 is the corresponding ***approximate output change of*** f, as indicated in Figure 2-20.

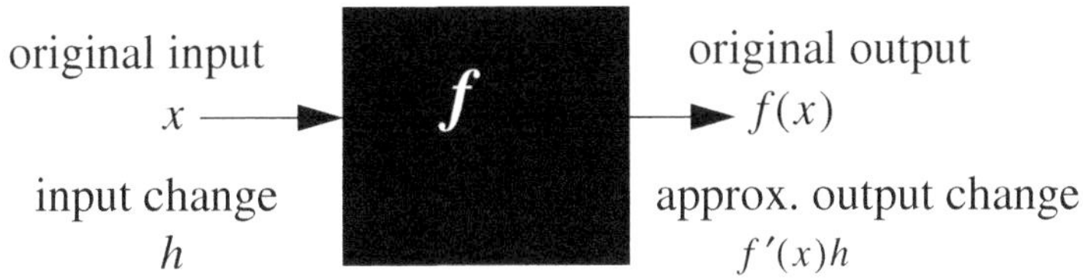

Figure 2-20

We therefore see that we can introduce

Assertion 2.12: Derivative interpretation

The derivative, $f'(x)$ (Definition 2.5 on page 30), may be thought of as the ***local change magnification factor of the function f near x,*** i.e., the value that must *multiply* the input change to obtain the (local - approximate) output change.

❑

Let us now see how we can use this interpretation to determine the derivative of the composition $H(F)$. In Figure 2-21, the initial input to the $H(F)$-box is denoted by x. Let this be changed by an amount h, to the value $x + h$. This causes the output of the F-box to change approximately by an amount $F'(x)h$, which is also the approximate input change to the H-box. The approximate corresponding H-box output change is the product of the (approximate) input change, $F'(x)h$, with the H-box *(local) change magnification factor,* $H'(F(x))$; ***note*** that H' must be evaluated at the input, $F(x)$, to the H-box.

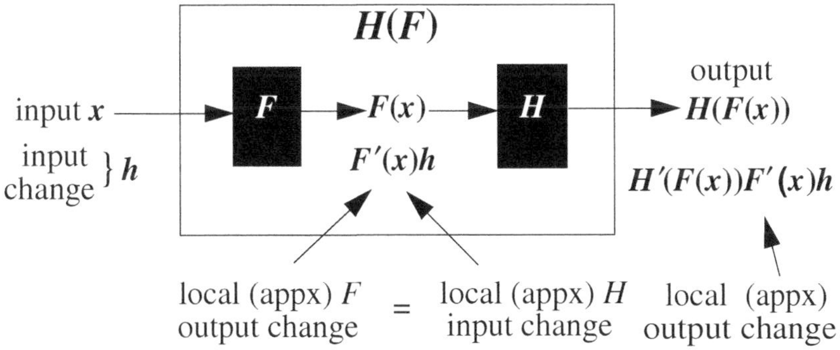

Figure 2-21

So when the input change to the $H(F)$ box is h, as shown in Figure 2-21, the (local) output change of the $H(F)$ box is $H'(F(x))F'(x)h$. From this we see that the *local change magnification factor of the function H(F) near x is the factor above preceding the* h - i.e., $H'(F(x))F'(x)$, providing these derivatives exist, i.e.,

$$(H(F))'(x) = H'(F(x))F'(x) . \qquad \textbf{2.24}$$

Exercises 2.12

Supply a more precise justification of equation 2.24.

Example 2.13: Derivative of $(x^2 + 1)^4$

Here we want to find $f'(x)$ for $f(x) = (x^2 + 1)^4$. We recognize f as the composition of two functions, H and F, whose derivatives we know, namely, $F(x) = x^2 + 1$ and $H(F) = F^4$.

Then $F'(x) = 2x$ (see Example 2.11 on page 47) and $H'(F) = 4F^3$, (also Example 2.11), so that

$$H'(F(x)) = 4(F(x))^3 = 4(x^2 + 1)^3$$

and hence from equation 2.24 above

$$(H(F))'(x) = H'(F(x))\,F'(x) = 4(x^2 + 1)^3\, 2x .$$

■

The final rule that we need yields the derivative of the inverse function of a strictly monotone differentiable function. From a geometric viewpoint this may be the most obvious result of all. This is because the graph of the function inverse to f consists of the ordered pairs on the graph of f ***but with the coordinates reversed.*** So if (x, y) is a point on the graph of f, then (y, x) is a point on the graph of f^{-1}. If (X, Y) is another point on the tangent to f at (x, y), then

$$f'(x) = \frac{Y - y}{X - x} \quad \text{and} \quad (f^{-1})'(y) = \frac{X - x}{Y - y},$$

as illustrated in Figure 2-22 (provided the original tangent line has nonzero slope).

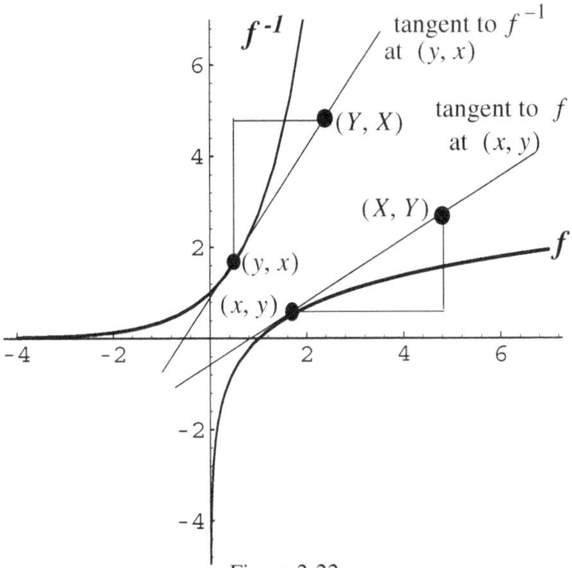

Figure 2-22

So the derivative of f^{-1} at y is the *reciprocal of the derivative of* f *at the* x *for which* $y = f(x)$, i.e., at $x = f^{-1}(y)$. To be concise

$$(f^{-1})'(y) = \frac{1}{f'(f^{-1}(y))} \qquad \textbf{2.25}$$

as long as $f'(f^{-1}(y)) = f'(x) \neq 0$.

We may also write this result as

$$Df^{-1}(y) = \frac{1}{Df(f^{-1}(y))}. \qquad \textbf{2.26}$$

Here $Df^{-1}(y)$ stands for the derivative of f^{-1} at y, and $Df(f^{-1}(y))$ stands for the derivative of f at $f^{-1}(y)$.[1]

Another way of getting this result on the derivative of the inverse is to notice that from the definition of inverse function, f^{-1},

$$f^{-1}(f(x)) = x. \qquad \textbf{2.27}$$

1. The extra parentheses that would make this notation unambiguous, namely, $(D(f^{-1}))(y)$ for $Df^{-1}(y)$ and $(D(f))(f^{-1}(y))$ for $Df(f^{-1}(y))$, would also make it just about impossible to read.

So if f^{-1} has a derivative at the point y, and f has a derivative at the x for which $y = f(x)$, then by the *chain rule* (see equation 2.24 on page 50, or equation 2.30 on page 53), for the derivative of a composition of two differentiable functions, taking the derivative of both sides of equation 2.27, we have

$$Df^{-1}(f(x))Df(x) = 1$$

Now substituting $y = f(x)$ for the first $f(x)$ on the left, and $x = f^{-1}(y)$ for the rightmost x above, we again obtain equation 2.26. ***A word of caution*** however, just differentiating both sides of equation 2.27 does not, by itself establish the validity of equation 2.26; we must know that f^{-1} has a derivative. For this, the geometric approach we first used should be fairly convincing.

Example 2.14: Derivative of $\sqrt[3]{y}$

Suppose $f(x) = x^3$ for x real. Then $f^{-1}(y) = \sqrt[3]{y}$. Since $f'(x) = 3x^2$, (Example 2.11 on page 47), it follows from equation 2.25 that

$$f'(f^{-1}(y)) = 3(f^{-1}(y))^2 = 3(\sqrt[3]{y})^2 = 3y^{2/3}$$

Hence, for $y \neq 0$ $(f^{-1})'(y) = \dfrac{1}{3y^{2/3}} = \dfrac{1}{3}y^{-2/3}$;

i.e. the derivative of $g(y) = \sqrt[3]{y} = y^{1/3}$ is $g'(y) = \frac{1}{3}y^{-2/3}$, for $y \neq 0$.

This is a situation in which f has a derivative for all x, but f^{-1} does not have a derivative, (Definition 2.5 on page 30), at $0 = f(0)$. However, f^{-1} does have a geometrical tangent line at $y = 0$, a vertical line (the second coordinate axis), as seen in Figure 2-23.

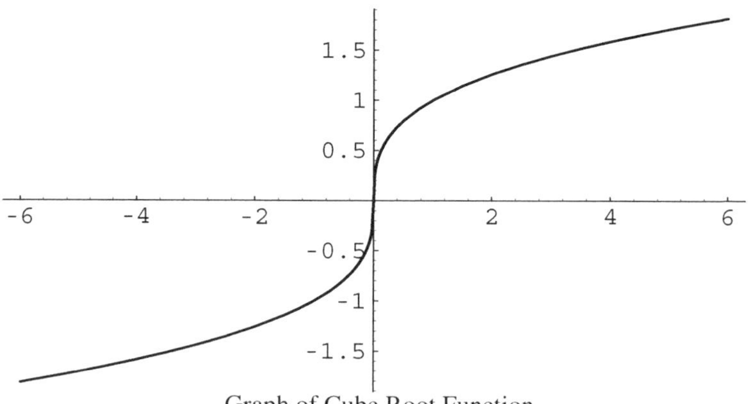

Graph of Cube Root Function

Figure 2-23

But a vertical line does not represent a real valued function of a real variable; sometimes this situation is described by saying that $\sqrt[3]{y}$ has an *infinite derivative* at $y = 0$.

■

Here is a precise statement of the rules we have just developed.

Theorem 2.13: Differentiation rules

Let F and G be real valued functions of a real variable (Definition 2.1 on page 21), with the same domain (inputs) and having a derivative (Definition 2.5 on page 30), at the point x. Let H be a real valued function whose domain includes the range (outputs) of F, and having a derivative at the point $F(x)$. Finally, let r and s denote fixed real numbers (constants). Then $rF + sG$, FG, and $H(F)$ are differentiable at the point x with

$$(rF + sG)'(x) = rF'(x) + sG'(x), \quad \text{([}\textit{weighted}\text{] } \textit{sum rule}\text{)} \qquad \textbf{2.28}$$

$$(FG)'(x) = F'(x)G(x) + F(x)G'(x), \quad (\textit{product rule}) \qquad \textbf{2.29}$$

$$(H(F))'(x) = H'(F(x))F'(x), \quad (\textit{chain rule}). \qquad \textbf{2.30}$$

If f is a strictly monotone real valued differentiable function of a real variable, (Definition 2.9 on page 37), it has an inverse function, f^{-1} (Definition 2.8 on page 37), with derivative

$$(f^{-1})'(y) = \frac{1}{f'(f^{-1}(y))}, \qquad \textbf{2.31}$$

at all points, y, for which the denominator above is nonzero.

❑

A few observations about the differentiation rules are in order here.

- First, the sum, product, and composition of functions can be differentiable even when the hypotheses listed in Theorem 2.13 are not satisfied. For the *sum rule*, clearly if $F = -G$ and $r = s = 1$, then $rF + sG$ is differentiable with derivative 0. If F and G are functions not identically 0, it is certainly possible that their product, FG, is 0, and hence differentiable (with derivative 0). If either of the functions in a composition is constant, then the composition is differentiable with derivative 0.
- Second, applying any rule when the hypotheses are not satisfied can lead to incorrect results. A classic example of this is the function F defined as

$$F(x) = \begin{cases} 1 & \text{if } x \text{ is a rational number} \\ -1 & \text{if x is irrational,} \end{cases}$$

- which is not differentiable anywhere. However, $F^2(x) = 1$. If you (mis)applied the product rule to $F^2(x) = F(x)F(x)$,
- you would obtain

  $$2\,F(x)F'(x) = 0.$$

- Since $F(x)$ is never 0, it would follow that $F'(x) = 0$. So the product rule can produce a meaningful, but incorrect, result when it is applied in a situation where its hypotheses are not satisfied.
- When equation 2.28 (the weighted sum rule on page 53) is written as

  $$D(rF + sG) = rDF + sDG; \qquad \textbf{\textit{2.32}}$$

  it is said to illustrate the *linear operator property* of the derivative operator, D.

The following are routine exercises for those who want to develop some facility with the differentiation rules above. *Mathematica* can be used to check the results, e.g., for Exercise 2.15.7 below, D[1/x^n,x].

Exercises 2.15

Using Theorem 2.13, find f' for

1. $f(x) = 2x - 5$
2. $f(x) = 3x - 7$
3. $f(x) = (2x - 5)(3x - 7)$
4. $f(x) = 1/(2x - 5)$, $x \neq 2.5$
5. $f(x) = 1/\{(2x - 5)(3x - 7)\}$, $x \neq 2.5$ and $x \neq 7/3$
6. $f(x) = (2x - 5)(3x^2 + x - 1)$
7. $f(x) = x^{-n} = 1/x^n$, $x \neq 0$
8. $f(x) = 1/(2x - 5)^4$, $x \neq 2.5$
9. $f(x) = \sqrt{\sqrt[3]{x^2} + 1 + 2/x}$, $x \neq 0$ and the argument to $\sqrt{\ }$ is > 0
10. $f(x) = x^{1/n} = \sqrt[n]{x}$ n even, $x > 0$
11. $f(x) = x^{1/n} = \sqrt[n]{x}$ n odd, $x \neq 0$
12. $f(x) = x^{m/n} = \sqrt[n]{x^m}$ n odd, $x \neq 0$
13. $f(x) = \sqrt[3]{(x^2 + 2x - 7)}$, $f(x) = \sqrt[3]{(x^2 + 2x - 7)^2}$.

Chapter 3

Graphing and Solution of Equations

1 Introduction

This chapter is devoted to seeing how local description via derivatives can be applied to graphing real valued differentiable functions of a real variable, and to solving algebraic equations involving a real variable. In optional sections, some of the results that are stated without proof are discussed in greater detail.

2 An Intuitive Approach to Graphing

Just as we used the difference operator, Δ, to help graph sequences (see Theorem 1.5 on page 8), we can use the derivative (Definition 2.5 on page 30) for graphing real valued differentiable functions of a real variable (see Definition 2.1 on page 21). Intuitively we expect a function f to be

- monotone increasing (see Theorem 1.5 on page 8), if $f' > 0$;
- convex (concave up, see Chapter 1, Corollary to Theorem 1.5 on page 8), if the second derivative, $f'' = D^2 f = D(Df)$ satisfies the condition $f'' > 0$.

We will provide a formal, exact statement of these results in Theorem 3.3 on page 66, Theorem 3.4 on page 68, Problem 3.5.1 on page 67, and Problem 3.5.4 on page 67. Before doing that we will present a few examples illustrating how useful derivatives can be for graphing functions whose behavior may not be completely obvious from their definitions.

Example 3.1: $f(x) = 3x + 1/\sqrt{x}, \quad x > 0$

You may notice that f is the sum of an increasing function and a decreasing function, (Definition 2.9 on page 37). As we will see, this information alone is of no use in predicting the behavior of f. However, using the various differentiation rules (see Theorem 2.13 on page 53), we find

$$f'(x) = 3 - \frac{1}{2x^{3/2}} \quad \text{and} \quad f''(x) = \frac{3}{4x^{5/2}}.$$

We see that $f'(x) > 0$ for $x^{3/2} > 1/6$ and $f''(x) > 0$ for all $x > 0$. So, assuming our intuition about the geometric meaning of derivatives is valid, we conclude that the graph of f is increasing for

$x^{3/2} > 1/6$, (i.e., $x > .3029$), decreasing for all other positive values of x, and concave up, as illustrated in Figure 3-1.

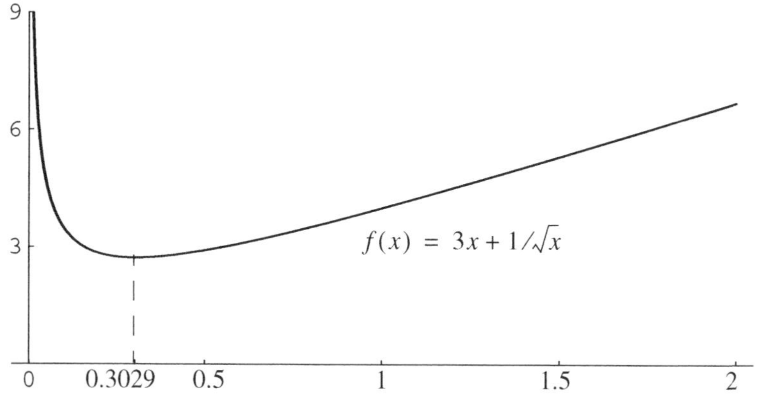

Figure 3-1

■

The next example is one in which we don't supply a formula for the function of interest, but instead describe it by specifying its derivative and a point on the graph of the function.

Example 3.2: Function whose derivative is 1/x

If there is a function, f, for which $f'(x) = 1/x$, for all $x > 0$, and satisfying the condition $f(1) = 0$, what would be the shape of this function?

Since $1/x > 0$ for positive x, this function would be ***increasing***. Furthermore $f''(x) = -1/x^2$, so we expect the function to be ***concave down.*** Just from this information it would seem that this function would have the shape illustrated in Figure 3-2.

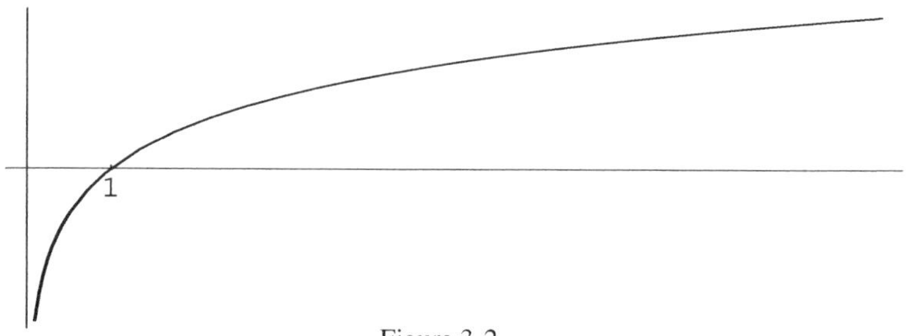

Figure 3-2

In fact, it appears as if you could draw a reasonably accurate graph of this function by first plotting the point (1,0). Then notice that the slope at $x = 1$ is $1/x = 1$, so you can draw a small segment going through the point (1,0) with slope 1. This gets the first segment of our approximation to this function. At the end points of this segment you compute new slopes and add two more small segments, as illustrated in Figure 3-3.

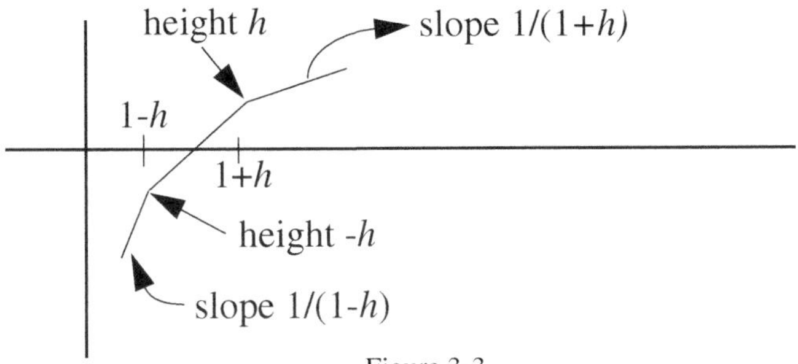

Figure 3-3

This process may be continued — of course, the further you extend it, the greater the error you expect — because the errors tend to accumulate. In most commonly encountered situations, this error accumulation can be controlled by reducing the size of the increment, h, that you are using.

■

Yet another example arises in describing a radioactive decay process.

Example 3.3: Radioactive decay

Suppose we are examining a homogeneous sample containing radioactive atoms of a single species, maybe carbon 14 which comes up in determining the time of death of some plant or animal, or maybe tritium (radioactive hydrogen), which is used in tracer studies to determine the behavior of a specified chemical containing hydrogen in some animal's body.

We let $A(t)$ denote the amount of this radioactive material in the sample at time t. It is of interest to determine the behavior of $A(t)$ as t increases, to allow prediction of radioactivity levels, for radioactive dating. As time passes, some of the atoms of this material are splitting apart to form different elements, and the amount of radioactive material decreases. The amount of radioactive material lost in the time interval from time t to time $t+h$ is $A(t)-A(t+h)$. When atoms decay independently of each other (which would be expected if the atoms are not packed very tightly, so that the decay products have very little chance of causing other decays), a change in the amount of radioactive material by a factor of M would lead to losing M times as many radioactive atoms per unit time. So it seems reasonable to assume that the quantity of this radioactive material lost in the time interval $[t;t + h]$ is roughly proportional to

- the amount of radioactive material, $A(t)$, at time t.

In a breakup of any ***small time interval*** into equal length nonoverlapping subintervals, the decay mechanisms on these subintervals (as determined by the number of atoms *available* for decay) would be essentially identical. This would lead us to expect the same number of decays on each such subinterval. So, provided h is sufficiently small, the number of atoms lost in the time interval $[t; t+h]$ should also be proportional to

- the duration, h, of this time interval.

We embody the above assumptions in the following approximation:

$$A(t) - A(t+h) \cong khA(t) \qquad \text{for small } h. \qquad \textbf{\textit{3.1}}$$

The proportionality constant, k, depends on which radioactive elements constitutes the sample; its value is usually available from previous studies.

Divide both sides of this approximation by h, to obtain

$$\frac{A(t) - A(t+h)}{h} \cong kA(t). \qquad \textbf{\textit{3.2}}$$

The closer h is to 0, the more accurate this approximation. So we let h approach 0, to obtain in the limit the *differential equation*

$$-A'(t) = kA(t),$$

which we rewrite in the standard form

$$A'(t) = -kA(t). \qquad \textbf{\textit{3.3}}$$

The value, $k > 0$, as already mentioned, depends on the particular radioactive element in the sample. We would like to use this description to determine the behavior of A.

The left side of approximation 3.2 is referred to as the **average rate of change of amount of radioactive material over the time interval** $[t; t+h]$, and the quantity $A'(t)$ is called the **[time] rate of change of A at time t**, or sometimes the **instantaneous rate of change of A at time t.**

Suppose we are told that the amount of radioactive material, $A(0)$, at time 0 is the specified value a_0. It is often very important to determine how much radioactive material will remain after some time has passed. This is true, for instance, when we want to investigate the amount of radioactivity still present in water contaminated by nuclear waste.

Specifically, suppose we want to determine the amount of radioactive material, $A(t_0)$, still remaining at time t_0. For simplicity here, we will now only consider the case $k = 1$ and $a_0 = 1$. Reduced to a mathematical formulation, we want to determine $A(t_0)$ where the function A must satisfy the conditions

$$A'(t) = -A(t) \text{ for all } t \quad \text{and} \quad A(0) = 1. \qquad \textbf{3.4}$$

This problem can be handled in almost the same way as the previous example, Example 3.2. First, if this model is any good at all, $A(t)$ is always positive. In that case, $A'(t)$ is always negative, as it should be, since the amount of radioactive material is presumably decreasing. Furthermore, differentiating both sides of the left equation in 3.4, and then replacing $A'(t)$ in the resulting equation by $-A(t)$ (justified by equation 3.4), we find $A''(t) = -A'(t) = A(t)$, so that A'' is positive. From this we conclude that the graph of A is concave up.

So we expect the graph of A to be decreasing and concave up, as shown in Figure 3-4.

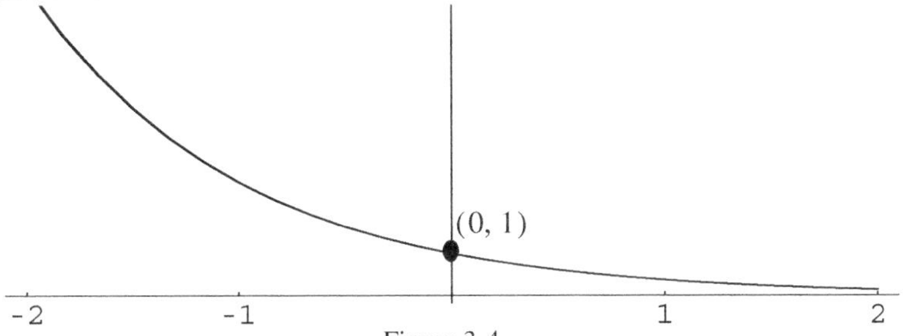

Figure 3-4

We also expect that we can draw a reasonably accurate graph of A, in almost the same way as in Example 3.2. Since $A(0) = 1$, using the left equation in 3.4, we find $A'(0) = -1$. So we may draw a short -45 degree segment through the point (0,1). As seen below in Figure 3-5, this allows approximate computation of the values $A(h)$ and $A(-h)$. Then, using equation 3.4 to obtain $A'(h) = -A(h)$, we apply this new slope on the interval from h to $2h$. We continue on in this same fashion, until we have as much of the approximate graph of A as we want.

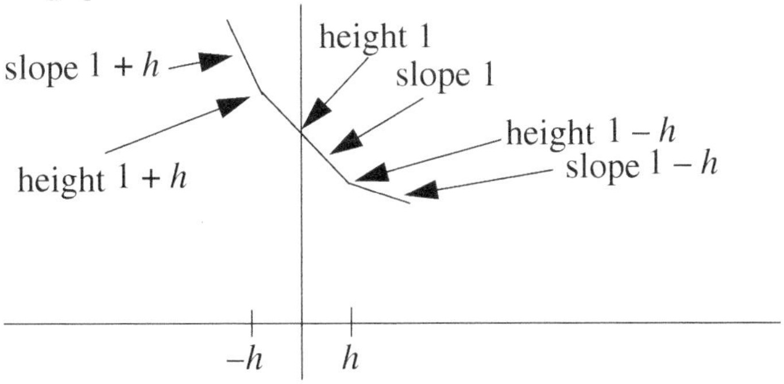

Figure 3-5

The smaller the value of h, for fixed t_0, the more accurate you expect the computation of $A(t_0)$ to be.

■

Equations 3.4 on page 59 constitute an example of a ***differential equation***, describing a function of interest — ***not by a formula*** *which can be evaluated for any given x to yield f(x),* ***but rather by conditions that this function must satisfy***. Very similar differential equations are used to describe such phenomena as

- the flow of chemicals in human and animal bodies (substrate and tracer kinetics and compartmental modeling),
- the behavior of electrical/electronic systems, such as amplifiers and radio and TV receivers,
- the behavior of mechanical vibrating systems, such as portions of automobiles, and airplanes.

In general a very large proportion of systems arising in engineering and scientific applications are described by differential equations. The approach used in this chapter for determining the behavior of functions from their differential equation local descriptions seems reasonable and plausible. But even for the very elementary situations that we have examined here, and certainly for more advanced problems, this discussion is nowhere near exact enough to ensure correct results. Here's an example to point out the need for some dependable tools for exploiting local descriptions.

Example 3.4: $f'(x) = -.5\,(f(x))^3, \quad f(1) = 1$

If we try to calculate $f(0)$ in the same way as in Example 3.3, for any choice of h, we would obtain some finite numerical value. Let us rewrite the given differential equation in the functional form

$$\frac{-2}{f^3} f' = 1.$$

However because the derivative of the function g given by $g = 1/f^2$ is, using the chain rule (Theorem 2.13 on page 53) $g' = \frac{-2}{f^3} f'$, we can rewrite the given differential equation in the form

$$\left(\frac{1}{f^2}\right)' = 1.$$

It shouldn't be hard to convince yourself that if a function has derivative always equal to 1, this function has to be some 45 degree line. That is,

$$\frac{1}{(f(x))^2} = x + c.$$

where c is some fixed constant.[1] The condition $f(1) = 1$ shows that $c = 0$.

So we conclude that for all positive x,

$$f(x) = \frac{1}{\sqrt{x}}.$$

But this shows that the type of computation we were doing[2] does not properly describe the behavior of f for small positive values of x.

■

So we really do need a more trustworthy approach for proper interpretation of local descriptions. In order to have at our disposal such an approach, in the section to follow we now present a few precise tools, and show how they can be used effectively.

3 Using the Mean Value Theorem

The most fundamental tool for relating a *derivative* local description to the behavior of the function itself is the *Mean Value Theorem.* In spite of the fact that it's such a simple and obvious result, it is one of the most important tools of applied mathematics.

Geometrically the mean value theorem states that if f is a well-behaved function whose domain includes the closed interval[3] from a to b, then at least one tangent line to f somewhere inside this interval is parallel to the *secant* line connecting the point $(a, f(a))$ with the point $(b, f(b))$, as illustrated in Figure 3-6.

1. This is easy to establish precisely via the mean value theorem, Theorem 3.1, which is introduced in the next section. See Problems 3.5.3 on page 67.
2. This example was constructed by **starting** with a function which behaves *badly* near 0, and then differentiating it, to find a differential equation it satisfied.
3. The *closed interval from* a *to* b, denoted by $[a; b]$, includes the end points, a and b, while the *open interval,* $(a; b)$, *from* a *to* b, does not include them.

In this figure two such tangents are illustrated. Notice that the slope of the

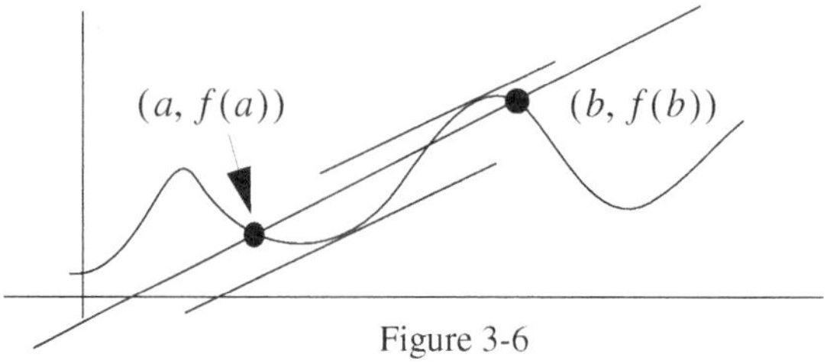

Figure 3-6

secant line connecting $(a, f(a))$ with $(b, f(b))$ is $(f(b) - f(a))/(b - a)$. Keeping this in mind, here is the simplest version of this theorem.

Theorem 3.1: Mean value theorem

Suppose f is a real valued function of a real variable which is differentiable for each x in the closed interval $[a; b]$ (see Footnote 3 on page 61) of points x satisfying $a \le x \le b$, where $a < b$. Then there is at least one value, c, strictly inside this interval (i.e., in the *open interval* $(a; b)$) such that

$$f'(c) = \frac{f(b) - f(a)}{b - a}. \qquad \textbf{3.5}$$

❑

There are several observations which should be stressed. First, if the condition of differentiability fails at even one point ***strictly inside*** the closed interval $[a; b]$ (i.e., at even one point in $(a; b)$), then the conclusion need not hold, as we illustrate in Figure 3-7. Second, the location of the value, c, is not specified, and is usually unknown. As we will see, this omission does not seem to create any difficulties in applying the mean value theorem.

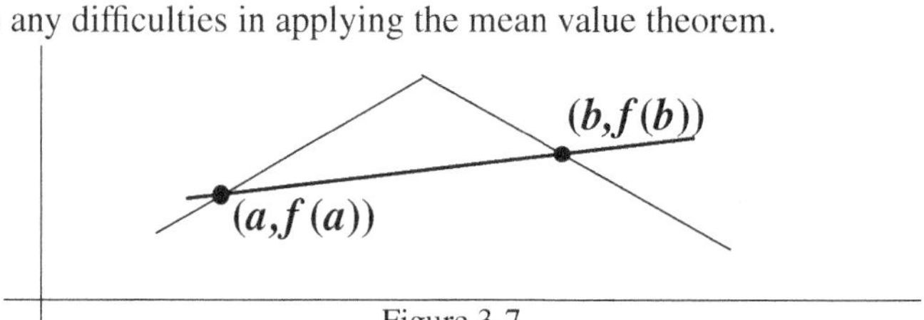

Figure 3-7

It's worth keeping in mind that the mean value theorem is the natural tool when you have some knowledge of the behavior of f' (e.g., $f' > 0$) and want to relate it to properties of f, such as whether $f(b) > f(a)$. ***So whenever you want to know something about, say, $f(X) - f(x)$, and know something about f', the mean value theorem should come to mind,*** as we'll see in its applications.

For future reference it is important to know the following extension, which allows application of equation 3.5 to such functions as the square root at $a = 0$.

Theorem 3.2: Relaxed mean value theorem conditions

The conclusion, equation 3.5, of the mean value theorem holds if at a and b the function f is only required to be *continuous*, rather than differentiable.

❑

The proof of the mean value theorem is not particularly difficult. However since this result is so plausible, not knowing the proof is unlikely to handicap its proper application. So those not interested in learning the main ideas in its establishment should feel free to skip over the optional section which follows.

In order to understand the proof of the mean value theorem it is necessary to look at the real number system a bit beyond the definition of a real number as an infinite decimal.

Suppose a positive real number, r, is represented as the infinite decimal

$$0_{\bullet}a_1a_2... \times 10^b$$

where b represents an integer (positive, negative, or zero), and each of the a_i stands for one of the digits 0 through 9, with $a_1 \neq 0$. What this means is that for each positive integer, k, the value r lies in the closed interval

$$[0_{\bullet}a_1a_2...a_k;\ 0_{\bullet}a_1a_2...a_k + 10^{-k}] \times 10^b.$$

This method of describing real numbers is really all that we must have for any theoretical development. However, it is a bit cumbersome — because at each stage of pinning down the value of interest, we must look at 10 possible subintervals (corresponding to the digits 0 through 9). So to simplify the presentation, we are better off with a binary representation — which makes use of only the digits 0 and 1. This would be too cumbersome for hand computation, but is the way digital computers really do arithmetic. With binary, we *zero in* on any value by the process of bisection — i.e., successively cutting in half the uncertainty in describing the location of a number. Now the binary system starts off with an initial interval with integer length. This is an unnecessary restriction; we can start off with any interval, and if we successively bisect and replace the bisected interval by one of its two halves, we will generate a real number (Appendix 1 on page 757).

There are several stages to establishing the mean value theorem:

a. Show that any function f which is differentiable at each point of some closed interval $[a;b]$ is continuous at each point of this interval, (a very simple task).

b. Show that if a function f is continuous at each point of a closed interval $[a;b]$, then f is *bounded* on this interval - i.e., there are values m, M such that $m \leq f(x) \leq M$ for all x in $[a;b]$.

c. Show that for f continuous there is a smallest value of M, denoted by Max, and a largest value of m denoted by min satisfying $min \le f(x) \le Max$ for all x in $[a;b]$.

d. Show that then there are values x_{min} and x_{Max} in $[a;b]$, satisfying

$$f(x_{Max}) = Max \quad \text{and} \quad f(x_{min}) = min.$$

e. Using these results, the Mean Value Theorem is first established for the case $f(a) = f(b)$ (**Rolle's theorem**) in which case, referring to equation 3.5, the theorem asserts the existence of a value c strictly inside $[a;b]$ at which $f'(c) = 0$. If the function f is constant, then the theorem clearly holds, because the derivative f' would be zero at all values. If f is not constant, then at least one of the values x_{min} and x_{Max} must lie strictly inside $[a;b]$— i.e., in the open interval $(a;b)$, and at this value it can be seen that f' is 0, from the following argument: suppose that x_{min} is the value strictly inside $[a;b]$, and that $f'(x_{min}) \ne 0$. From the definition of derivative we know that for all nonzero sufficiently small h, the ratio

$$\frac{f(x_{min}+h) - f(x_{min})}{h}$$

is as close as we want to $f'(x_{min})$. So choose h small enough so that

$$\left| \frac{f(x_{min}+h) - f(x_{min})}{h} - f'(x_{min}) \right| \le \frac{|f'(x_{min})|}{2}$$

Then for all such h

$$\frac{-|f'(x_{min})|}{2} \le \frac{f(x_{min}+h) - f(x_{min})}{h} - f'(x_{min}) \le \frac{|f'(x_{min})|}{2}$$

or

$$f'(x_{min}) - \frac{|f'(x_{min})|}{2} \le \frac{f(x_{min}+h) - f(x_{min})}{h} \le f'(x_{min}) + \frac{|f'(x_{min})|}{2}$$

The two outer expressions have the same algebraic sign as $f'(x_{min})$. So, for ***all*** sufficiently small h, the middle expression cannot change sign when the sign of h is reversed. But when h changes sign, the denominator of this expression changes sign, so the numerator must also change sign. This means that for fixed sufficiently small h, at least one of the values, $f(x_{min}+h)$, $f(x_{min}-h)$ must be less than $f(x_{min})$, so that $f(x_{min})$ cannot be the smallest value of $f(x)$ (which we know it must be.) Since our reasoning is correct, this contradictory (false) result can only have arisen from the assumption that $f'(x_{min}) \ne 0$. To eliminate this false result we must conclude that $f'(x_{min}) = 0$. The reasoning is unchanged if it is x_{Max} that is strictly inside $[a, b]$.

f. Finally, if the function f does not satisfy the condition

$f(a) = f(b)$, modify f by subtracting from it the linear function which will make it satisfy this condition — i.e., replace f by F where $F(x) = f(x) - qx$ and q is chosen so that $F(a) = F(b)$. When this is done, you find that

$$q = \frac{f(b) - f(a)}{b - a}$$

g. Now just apply the already proved version of the mean value theorem to the function F, to obtain the usual version of this result.

Next, we fill in a few of the less obvious details.

For part a, let

$$S(h) = \begin{cases} \dfrac{f(x+h) - f(x)}{h} & \text{for } h \neq 0 \\ f'(x) & \text{for } h = 0. \end{cases}$$

S is continuous at $h = 0$ because *of the assumption that* f has a derivative at x. But $f(x+h) - f(x) = hS(h)$ for all h, and the term on the right side of this expression is a continuous function of h, whose value at $h = 0$ is 0. Thus, $\lim_{h \to 0} (f(x+h) - f(x)) = 0$, showing that f is continuous at x.

For part b. we observe that if f is not bounded on $[a;b]$, it must be unbounded (not bounded) on at least one of the two halves of this interval. Repeatedly applying this argument to the half on which it is known that f is not bounded, yields a number, call it u, belonging to the interval $[a;b]$; in every one of the sequence of intervals narrowing down on u (each of which is wholly contained in $[a;b]$), f is unbounded. But f is continuous at u, so when we are *far enough along* in the sequence of intervals defining u, all values in these intervals are close to $f(u)$. Since it is not possible for a set of values of $f(x)$ to be unbounded, and simultaneously all close to $f(u)$, the assumption that f is unbounded on $[a;b]$ must be rejected. Notice how crucial it is that the function f be continuous at the end points, a and b. If we only insisted that f be continuous at each point of the open interval $(a;b)$, we could not prove that f is bounded on this interval, as can be seen by noticing that the function given by $f(x) = 1/x$ is continuous at each point in the open interval $(0;1)$, but is not bounded on this interval, since $1/x$ can be made arbitrarily large for positive values of x sufficiently close to 0. The result fails because no matter what value is chosen for $f(0)$, f cannot be continuous at 0. The proof would fail for $f(x) = 1/x$ because the process used in the proof would drive us to the number 0, at which we just saw, f cannot be continuous.

For part c, construct *Max* via bisection of the interval $[B;M]$, where M is a strict upper bound for f, (i.e., $f(x) < M$ for all x in $[a;b]$) and B satisfies the condition that $f(x) \geq B$ for at least one x in $[a;b]$. The midpoint, call it *MID,* of the *current interval* replaces M if all values $f(x) < MID$, and replaces B otherwise. It is not very difficult to then show that $f(x) \leq Max$ for all x in $[a, b]$, and that *Max* is the smallest number with this property. (We say that *Max* is the ***least upper bound*** of the set of values of $f(x)$ for x in $[a; b]$.)

A similar argument produces the value *min.*

For part d. we find a value x_{Max} satisfying $f(x_{Max}) = Max$ as follows: On at least one of the two halves of $[a, b]$, Max is still the *least upper bound* of $f(x)$ values on that half; If this were not so, the two least upper bounds, $LMax$ and $RMax$ would satisfy $LMax < Max$ and $RMax < Max$, so that the larger of the two numbers, $LMax$, $RMax$, which is smaller than Max, would replace Max as the least upper bound of the set of $f(x)$ values on $[a; b]$, and Max would not be the least upper bound, which we know it is. Choose the half of $[a;b]$ on which Max is still the least upper bound of the $f(x)$ values, and continue the process, generating a number x_{Max} in $[a; b]$. Using the continuity, it is evident that $f(x_{Max}) = Max$. A similar argument is used to generate x_{min}.

As the first illustration of application of the mean value theorem to establish important results firmly, we have

Theorem 3.3: If $f' > 0[f' < 0]$, then f is increasing [decreasing]

Suppose that the real valued function, f, of a real variable has a positive [negative] derivative at every point of an interval. Then f is increasing [decreasing] on this interval (see Definition 2.9 on page 37).

❑

Proof: What we need to show is that if x and X are any two values in the interval, with $x < X$, then

$$f(x) < f(X). \qquad \textbf{3.6}$$

By the Mean Value Theorem, Theorem 3.1, page 62, there is a value, c, satisfying $x < c < X$, such that

$$f'(c) = \frac{f(X) - f(x)}{X - x}.$$

Rewrite this as

$$f(X) - f(x) = (X - x)f'(c). \qquad \textbf{3.7}$$

Even though we don't know the exact location of c, we do know from the hypotheses of this theorem that $f'(c) > 0$, and by construction $X - x > 0$, so that the right hand side of equation 3.7, and hence its left-hand side are both positive. That is, $f(X) - f(x) > 0$, so that inequality 3.6 is established, proving Theorem 3.3.

Notice how natural the use of the mean value theorem is, for determining if $f(x) < f(X)$, i.e., if $f(X) - f(x) > 0$, knowing only that $f' > 0$.

Problems 3.5

In these problems assume $a < b$. (The closed interval $[a;b]$ includes its end points.)

1. Show that if $f'(x) \geq 0$ for all x in $[a;b]$, then f is nondecreasing on $[a;b]$, i.e. if $x < X$ then $f(x) \leq f(X)$.
2. Show that if the real valued function, f, of a real variable has derivative $f'(x) = 0$ for all x in $[a;b]$, then there is a constant, c, such that $f(x) = c$ for all x in $[a;b]$.
3. Show that if the real valued function, f, of a real variable has derivative $f'(x) = c$ for all x in $[a;b]$, then there is a constant, d, such that $f(x) = cx + d$ for all x in $[a;b]$. So the only functions with constant slope are straight lines.
4. Show that the conclusion of Theorem 3.3, that f is increasing, still holds if $f'(x) > 0$ for all x in the open interval $(a;b)$, (see and f is continuous at a and b.
5. What conclusion can be drawn if for all x in $[a;b]$ $f'(x) < g'(x)$ and $f(a) = g(a)$?
6. What conclusion can be drawn if for all x in $[a,b]$ $f'(x) < g'(x)$ and $f(b) = g(b)$?

Exercises 3.6:

Plot the functions whose formulas are given below, using a program such as *Mathematica* as a check on your results.

1. $\sqrt{x}\,\sqrt[3]{x-5}$, $0.001 \leq x \leq 4$
2. $(x-1)^3(x-3)$, $0 \leq x \leq 10$
3. $\dfrac{x}{1+x^2}$, $-100 \leq x \leq 100$
4. $\sqrt{\dfrac{x^2}{1+x^4}}$, $-4 \leq x \leq 4$

Just as in the Corollary to Theorem 1.5 on page 8, where the condition $\Delta^2 f > 0$ meant that the sequence f had a *concave-up* shape, the condition $D^2 f > 0$ results in the graph of the real valued function, f, of a real variable having a *concave-up* shape, as illustrated in Figure 3-8. In this figure, we indicate that the graph of the concave-up function, f, lies *above its tangent lines,* and *below each secant line segment* connecting any two distinct points on this function's graph. Although this claim is intuitively plausible, it's worth establishing in a reliable manner, to serve as an introduction to proper application of the mean value theorem in more complicated situations.

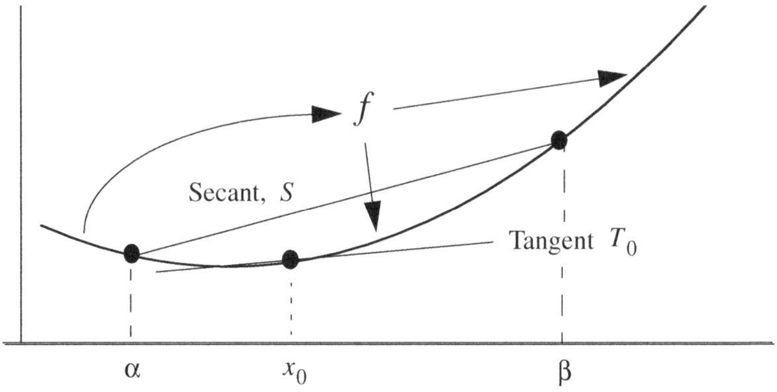

Figure 3-8

Theorem 3.4: Consequences of positive f''

Suppose that for each x in the closed interval $[\alpha;\beta]$ (which includes its endpoints) where $\alpha < \beta$, the condition $f''(x) > 0$ is satisfied. Then for all x satisfying $\alpha < x < \beta$,

$$f(x) < S(x), \tag{3.8}$$

where the *secant,* S is the line segment connecting the points $(\alpha, f(\alpha))$ and $(\beta, f(\beta))$ and for all x satisfying $\alpha \le x \le \beta$ except for x_0,

$$f(x) > T_0(x), \tag{3.9}$$

where T_0 is the tangent to f at $(x_0, f(x_0))$.

❑

As a start to establishing these results, we need to write the equations for S and T_0. To determine these formulas we use the result that any line of slope s through the point $(z_0, f(z_0))$ consists of the points (x, y) satisfying the condition

$$y = s(x - z_0) + f(z_0),$$

(see equation 2.4 on page 26). The tangent line to f at $(x_0, f(x_0))$ has slope $f'(x_0)$ (Definition 2.5 on page 30). So we see that the formula for the tangent line to f at x_0 is given by

$$T_0(x) = f'(x_0)(x - x_0) + f(x_0). \tag{3.10}$$

The secant line through the points $(\alpha, f(\alpha))$ and $(\beta, f(\beta))$ has slope

$$\frac{f(\beta)-f(\alpha)}{\beta-\alpha},$$

(again using Definition 2.5.) So the formula for this secant line is

$$S(x)=\frac{f(\beta)-f(\alpha)}{\beta-\alpha}(x-\alpha)+f(\alpha). \qquad \textit{3.11}$$

To establish inequality 3.8 above, it is convenient to work with the function $S-f$, (recall Definition 2.7 on page 35), because the mean value theorem, and Theorem 3.3 on page 66, which we will be using, apply only to *individual* functions. A look at Figure 3-8 shows that the function $S-f$ should have the shape given in Figure 3-9.

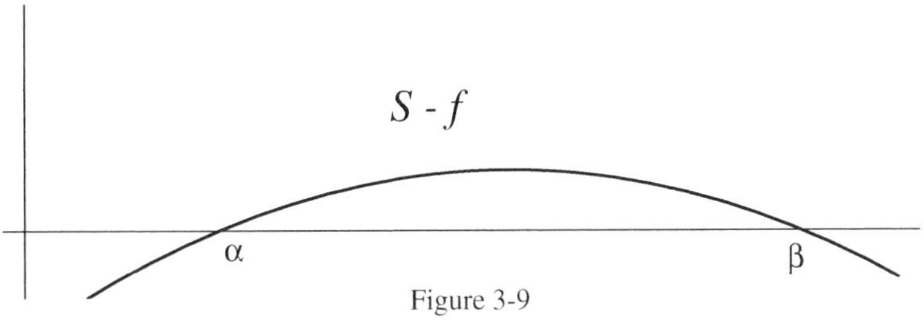

Figure 3-9

That is, it looks as if $(S-f)(x)$ is 0 at $x = \alpha$, then increases for a while, then starts decreasing until it reaches 0 at $x = \beta$. To verify the first part, we can see directly from equation 3.11 that

$$(S-f)(x)= 0 \text{ at } x = \alpha \text{ and } x = \beta. \qquad \textit{3.12}$$

To verify our intuition about the behavior of $(S-f)(x)$ as x increases from α to β we need only examine its derivative, $(S-f)'$. From equation 3.7 on page 66, and the mean value theorem on page 62, we know that there is a value, c, somewhere (strictly) between α and β such that

$$(S-f)'(c)=0.$$

Also $(S-f)''=-f''$, and by hypothesis $f''>0$. Hence,

$$(S-f)''<0.$$

So from Theorem 3.3 on page 66 (applied to $(S-f)'$ rather than to f) we see that $(S-f)'(x)$ is decreasing as x increases. Since it is 0 at $x=c$, we see that

- $(S - f)'(x)$ must *start out positive* (at $x = \alpha$), decrease to 0 (at $x = c$), continue decreasing (and hence be negative) for $x > c$ until $x = \beta$.

Let's summarize what we have shown so far:

$$(S - f)(x) = 0 \text{ at } x = \alpha \text{ and } x = \beta$$

and

$$(S - f)'(x) > 0 \text{ for } \alpha \le x < c;$$

hence

$$(S - f)(x) > 0 \quad \text{for } \alpha < x \le c.$$

We can really stop here, because the same arguments apply if we start at $x = \beta$ and decrease x until it reaches c to show that

$$(S - f)(x) > 0 \quad \text{for } c \le x < \beta.$$

This establishes inequality 3.8 on page 68.

The proof of inequality 3.9 of Theorem 3.4 is essentially the same — we simply examine $(f - T_0)(x)$ in the same way we examined $(S - f)(x)$, but instead of starting at the end points, we begin the reasoning at the point of tangency, x_0. In this case, $(f - T_0)(x_0) = 0$, and $(f - T_0)(x)$ is seen to increase as x moves away from x_0.

So we have now established Theorem 3.4. It should be evident how to modify the conclusions of this theorem when $f'' < 0$.

Here is an example of the kind of results we can get from Theorem 3.4.

Example 3.7: Simple bounds for $\sqrt{x}$

If we're doing a lot of work with the square roots of numbers in some small range, say 1 to 1.21, we might very well prefer to deal with simple linear functions, rather than with the less tractable square root function. It's easy to verify that $D^2\sqrt{x} < 0$, so the square root function is concave down — lying below its tangent lines and above its secant lines. The formulas for the tangent line at $x_0 = 1$ and the secant line through the points $(1, 1)$ and $(1.21, 1.1)$ are, respectively,

$$T_0(x) = \frac{1}{2}(x - 1) + 1 \quad = .5\,x + .5$$

and

$$S(x) = \frac{\sqrt{1.21} - \sqrt{1}}{0.21}(x - 1) + 1 \quad = 0.476x + 0.524\,.$$

The average of these two linear functions has the formula $(T_0(x) + S(x))/2 = .488\,x + .512$. Since these functions all have the same value at $x = 1$, and diverge from each other as x increases, as seen in the greatly exaggerated Figure 3-10,

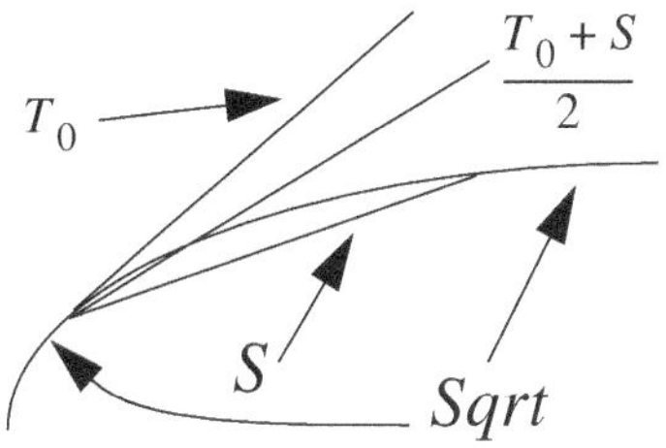

Figure 3-10

at any x between 1 and 1.21 this average, $.488\,x + .512$, is not more than

$$\frac{T_0(x) - S(x)}{2} = 0.012x - 0.012$$

from $\sqrt{x}$. For example, at $x = 1.16$ the approximation to $\sqrt{x}$ is 1.07808 with error magnitude not exceeding .00192, and hence with percentage error not exceeding .1783 %.

■

Exercises 3.8

Determine simple linear upper and lower bounds for the following, and find upper bounds for the errors made using the average of these linear bounds.

1. $\frac{1}{x}$ for $10 \le x \le 12$
2. $\sqrt[3]{x}$ for $26 \le x \le 27$
3. $\sqrt{x}$ for $24 \le x \le 26$

4 Solving $g(x) = 0$: Bisection and Newton's Method

The final topic treated in this chapter is the solution of equations of the form

$$g(x) = 0, \qquad \textbf{3.13}$$

where g is a real valued function of a real variable, whose value, $g(x)$, can be computed for any specified value of x.

Our interest in solving such equations arises in a variety of situations. Here are a few of them.

Determining where f is increasing and decreasing

Most of the functions, f, that we will meet will be increasing on a finite number of intervals inside of which the derivative, f', is positive, and decreasing on a finite number of other intervals inside of which the derivative is negative, with the end points of most of these intervals being x values for which $f'(x) = 0$. So to find the intervals of interest here, we need to solve the equation $g(x) = 0$ where $g = f'$.

Determining the maxima and minima of some function

Of particular interest in engineering are the maximum and minimum values of a differentiable function, f, where $f(x)$ might represent

- mileage, (miles per gallon), at a speed of x miles per hour,

or

- profit (dollars), where x is the price of the item being sold,

or

- yield of some process at a temperature of x degrees C.

In most cases that we will meet, f will be differentiable with domain being a single interval which itself consists of a finite number of subintervals, inside each of which, the derivative is of one algebraic sign (positive or negative), as illustrated in Figure 3-11.

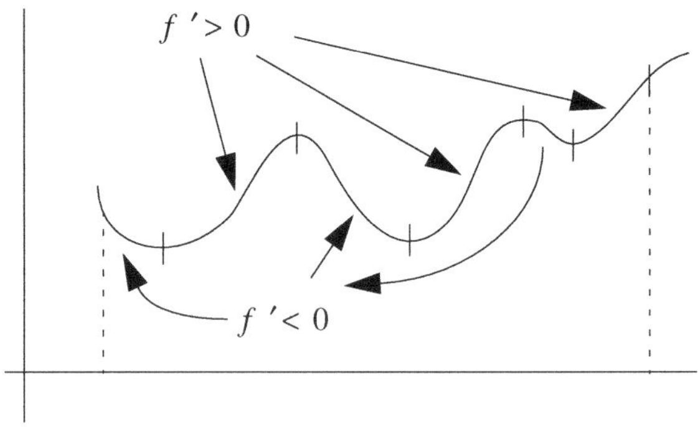

Figure 3-11

Because of Theorem 3.3 (page 66), and its extension given in Problem 3.5.4, (page 67), on any such interval f is either increasing or decreasing. In either case, when x is strictly inside one of these intervals, $f(x)$ cannot be either the largest or the smallest value of *f*; because if $f'(x) > 0$, then by Theorem 3.3, for sufficiently small $h > 0$ we have

$$f(x+h) > f(x) \quad \text{and} \quad f(x-h) < f(x),$$

and similarly if $f'(x) < 0$. This argument shows that except for the end points of the domain of *f*, if $f'(x)$ is not 0, then $f(x)$ cannot be either the maximum or minimum of the function *f*. So the only points, *x*, at which $f(x)$ can possibly be the largest or smallest values of $f(x)$ are the end points, *a* and *b*, and *x* values at which $f'(x) = 0$. So here again we want to solve $g(x) = 0$, where $g = f'$.

Determining where *f* is convex

Suppose we are interested in the shape of the graph of some twice-differentiable[1] function, *f*. We know from Theorem 3.3 (page 66) applied to f', that

- $f'(x)$ increases as x increases if $f'' > 0$ (and hence looks concave up),

and

- $f'(x)$ decreases as x increases if $f'' < 0$ (concave down),

so that it's of interest to determine where second derivatives are positive and where they are negative. In most cases the easiest way to solve this problem is to find the values of x at which $f''(x) = 0$. For most commonly encountered functions there will only be a finite number of such values. So for this problem $g = f''$.

Determining $\sqrt[n]{x}$

Here we attack this problem by solving the equation $y^n = x$ for y where x is specified. We convert this to the form $g(x) = 0$ by letting

$$g(x) = y^n - x.$$

Finding upper $1-\alpha$ points on distribution functions

In carrying out statistical test procedures, it is often necessary to determine the value of x for which the test statistic distribution function, *F*, satisfies the equation $F(x) = 1 - \alpha$, where α represents the risk of a *false positive* that the researcher is willing to accept (i.e., the probability of claiming a new treatment is beneficial, when, in fact, it is not beneficial).

1. Twice-differentiable means that the second derivative, f'', exists.

Determining the proper amount of caloric intake or medication

Given an individual's weight, activity level and possibly other measured values, there are formulas available which specify the equilibrium weight the individual will attain as a function of the person's daily caloric intake. For a specific person, the weight goal of the person is chosen, and it is desired to determine how many calories per day the person should ingest to achieve this goal. Similar questions are asked for the necessary amount of medication — e.g., insulin dosage for diabetics, where the desired blood sugar level is specified, and a formula is available relating blood sugar level to the person's weight, age, severity of the diabetes, etc.

In spite of the impression left by many elementary algebra and calculus textbooks, ***for most cases of scientific application of mathematics, there is no simple formula for the solution of equations of the form*** $\boldsymbol{g(x) = 0}$, even for fifth degree polynomial functions, or for fairly simple functions such as the one given by $g(x) = x + ln(x) - 4$, (where ln is the *natural logarithm* function, which we will be developing shortly). Lacking useful formulas for solving such equations, and needing these solutions in many problems forces us to look for *numerical solutions* - i.e., for ***decimal approximations*** to x values for which $g(x) = 0$. The ***only tools available*** for such problems ***are the abilities to calculate g or its derivatives*** at any finite number of domain values.

Our main interest is in using local methods to determine these approximations. However, local methods may not work very well, or may not even work at all if we do not have available a value of x reasonably close to the desired value, r, for which $g(r) = 0$.

So the first topic we will look into is how to get close enough to the desired solution, so that local methods can then be applied. For real valued functions of a real variable the most reliable method for finding numerical solutions of $g(x) = 0$ uses ***bisection***. Interestingly enough, bisection is also essential for the theoretical investigation of such solutions. The applicability of the bisection method depends on the property of continuous functions which is illustrated in Figure 3-12 , which shows that a continuous function which

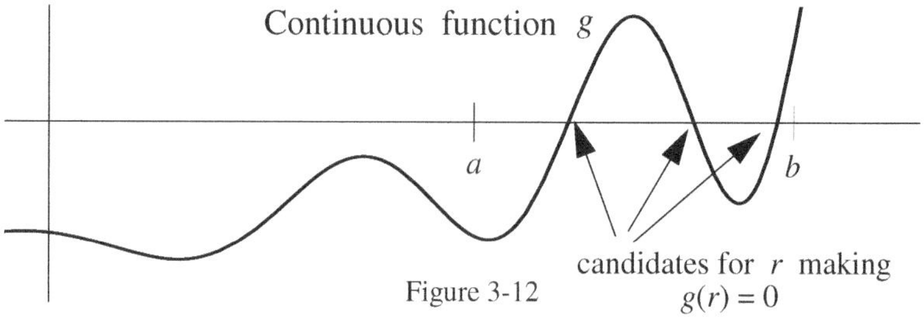

Figure 3-12

which assumes both positive and negative values, must somewhere assume the value 0. The precise version is given in

Theorem 3.5: The intermediate value theorem for continuous functions

Suppose g is a real valued continuous function of a real variable (Definition 2.1 on page 21), defined for all values of x in the closed interval $[a; b]$, (see Footnote 3 on page 61), with $g(a)$ and $g(b)$ having opposite algebraic signs.

Then there is a value, r, inside of $[a; b]$ satisfying

$$g(r) = 0. \qquad \textit{3.14}$$

❑

Notice, as illustrated in Figure 3-12, that there may be more than one value of x for which $g(x) = 0$.

We could establish this result by directly constructing r as an infinite decimal satisfying equation 3.14. However such a construction is somewhat clumsy; it's much simpler to make use of *bisection* (dividing each chosen interval in half, and choosing the half in which the desired number is located) rather than *decisection* (dividing the interval in which the number lies into 10 parts, and choosing the particular one of these 10 in which the desired number is located). Bisection, in essence, constructs r as a binary number,[1] in a manner which is easy to implement (approximately), on a computer. We leave the direct decimal number construction as an exercise.

To find a value, r, satisfying equation 3.14, we first notice that the condition that $g(a)$ and $g(b)$ be of opposite signs, is neatly expressed by the inequality

$$g(a)g(b) < 0. \qquad \textit{3.15}$$

It's not difficult to program a computer to check such an inequality, but in certain computations, computers can attempt to generate numbers which they cannot handle properly, (numbers out of the computer's range). In such circumstances, computer checking of inequalities like the one above can give incorrect, and possibly very misleading answers. For the present we will not be concerned with practical computational difficulties which arise in trying to implement our constructions.

---

1. Here is where we first make significant use of the properties of the system of real numbers. Appendix 1 on page 757, on the real numbers, is meant to provide the necessary understanding of their main properties.

Bisection to construct *r*

It's easy to see that ***if*** the value of g at the midpoint, m, of the interval $[a;b]$ has the opposite sign from $g(a)$, then the value we are seeking lies in the *left* half of $[a;b]$ as shown in Figure 3-13.

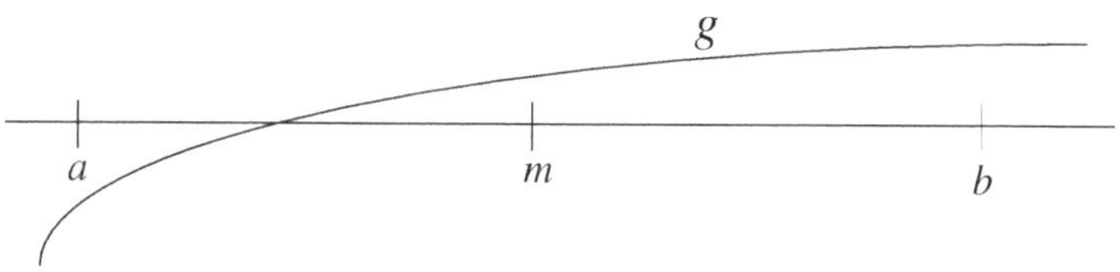

Figure 3-13

Similarly, if this value has the *same* sign as $g(a)$, then the desired number is in the *right* half of $[a;b]$. If the value of g at the midpoint of $[a;b]$ is 0, then this midpoint is the number r that we want.

To be precise, let m be the midpoint of $[a;b]$; i.e.,

$$m = \frac{a+b}{2}. \qquad \textbf{\textit{3.16}}$$

If $g(m)g(b) < 0$, then replace a by m, and repeat the process,

If $g(m)g(b) > 0$, then replace b by m, and repeat the process.

If $g(m) = 0$, stop, since m is the exact value you want for r.

We illustrate all but the last of these possibilities in Figure 3-14.

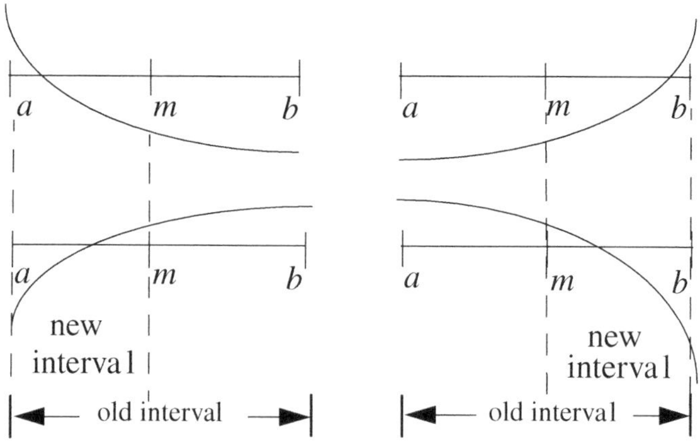

Figure 3-14

This process generates a number, r (essentially in binary). It still must be demonstrated that $g(r) = 0$. But (except for the case where $g(m) = 0$, where we already know that $g(r) = 0$ because we chose $r = m$) in each of the intervals defining r, the function g took on both positive and negative values. Now $g(r)$ has some value. When the intervals defining r get small enough (and, of course, they will, because each interval is half the length of the preceding one), all values $g(x)$ for all x in these intervals become close to the value $g(r)$ **because of the continuity** (Definition 2.6 on page 34), **of g at r.** But $g(r)$ cannot be any specific positive number, because the negative numbers in the successive intervals cannot possibly be coming arbitrarily close to any fixed positive number. Similarly, $g(r)$ cannot be negative. The only possibility remaining is the one we must accept, namely, $g(r) = 0$. This establishes Theorem 3.5.

Bisection has the advantage of providing an upper bound for the error in determining r, because the length of the n-th interval is $(b-a)/2^n$, and the midpoint of this interval cannot be further from r than

$$\frac{b-a}{2^{n+1}}. \qquad \textit{3.17}$$

So, assuming that the function value, $g(x)$ can be computed for each x, when the hypotheses of Theorem 3.5 on page 75 are satisfied, we proceed with the bisection process until the *error*, 3.17, is as small as we desire, and use as our approximation to r (the midpoint of the final interval that was constructed). The bisection process, being very repetitive, has the desirable property of being easy to implement on a digital computer.

Example 3.9: Computing $\sqrt{2}$ using bisection

We take $g(x) = x^2 - 2$ and $a = 1, b = 2$, because $1^2 < 2$ and $2^2 > 2$.

So, the first bisection interval is $[1,2]$.

The next bisection interval is $[1, 1.5]$ because $1.5^2 > 2$.

The next bisection interval is $[1.25, 1.5]$ because $1.25^2 < 2, \ldots$.

Notice that this process is zeroing in on $\sqrt{2} = 1.41421356273095\ldots$.

■

Bisection is reliable, but unfortunately, as shown here, it is pretty slow. As part of a larger problem in which the bisection procedure must be carried out a great many times, its inefficiency can be a serious handicap - on even the fastest computers. Fortunately, once bisection has provided a moderately good estimate of r, we can switch to the local method (*Newton's method*) that will be presented next for greater efficiency. The reason that we often do not start out with Newton's method is that it may work poorly, or fail to work at all, unless it is initiated with a decent estimate of r.

Exercises 3.10

Use bisection to compute the following to 3 decimals

1. $\sqrt{3}$
2. $\sqrt[3]{5}$
3. The solution of $x^3 + x = 3$ that lies between 1 and 1.5.

Newton's method

Newton's method for solving $g(x) = 0$ for a differentiable function (Definition 2.5 on page 30), g, starts off with an initial guess, x_0. As the next approximation it uses the easily found value where the current local approximation to g (the tangent line) crosses the x-axis. That is, the next approximation to the solution of $g(x) = 0$ is chosen to be the value x_1, at which the current tangent line function is 0. This process then continues always replacing g by its tangent line at the latest approximation to the value r, as illustrated in Figure 3-15.

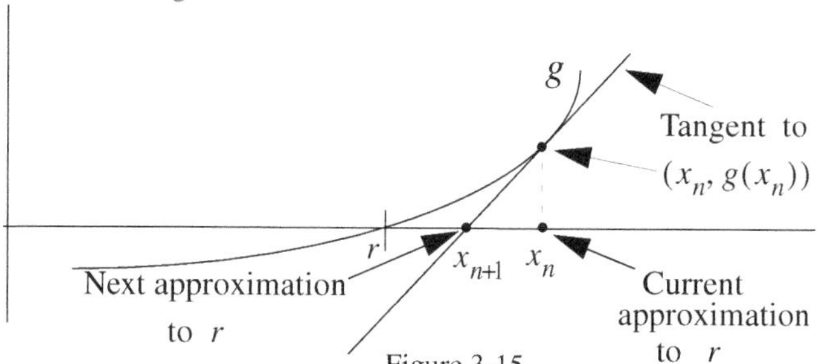

Figure 3-15

From this picture we would expect great performance when the *current point* on the graph of g lies on the *straight-looking* portion of the graph of g through $(r,0)$.

In order to be able to carry out Newton's method we have to translate it into algebraic terms. We do this with the aid of Figure 3-16.

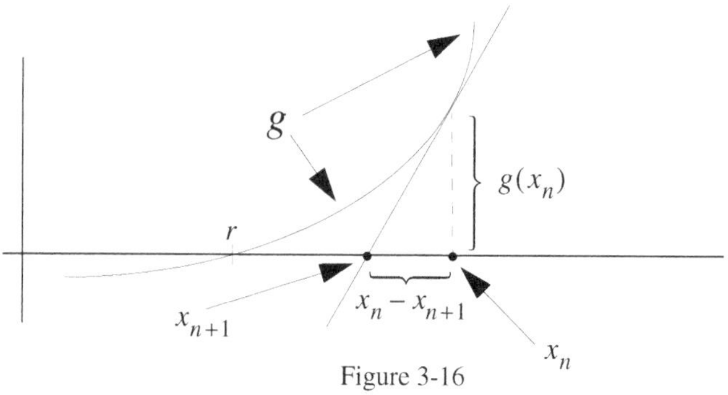

Figure 3-16

We want to express our *next* approximation, x_{n+1}, in terms of the current approximation, x_n. From Figure 3-16 we see that the slope, $g'(x_n)$, of the indicated tangent line must equal $g(x_n)/(x_n - x_{n+1})$; i.e.,

$$g'(x_n) = \frac{g(x_n)}{x_n - x_{n+1}}.$$

Solving for x_{n+1} we obtain

Newton's method formula

$$x_{n+1} = x_n - \frac{g(x_n)}{g'(x_n)}. \qquad \textit{3.18}$$

This is the computational scheme for getting the sequence, $x_1, x_2, \ldots$ of *Newton iterates.* We start out with an initial guess, x_0, possibly obtained from the bisection method. If we know that the solution lies in between two given values (like $\sqrt{3}$ lies between 1 and 2) we usually modify equation 3.18 to take this into account.

Example 3.11: Computing $\sqrt{3}$ by Newton's method

Here we take $g(x) = x^2 - 3$, so that $g'(x_n) = 2x_n$. Substituting into equation 3.18, we obtain

$$x_{n+1} = x_n - \frac{x_n^2 - 3}{2x_n} = \frac{x_n}{2} + \frac{3}{2x_n}.$$

Starting out with $x_0 = 2$, we find $x_1 = \frac{x_0}{2} + \frac{3}{2x_0} = 1 + \frac{3}{4} = 1.75$.

Continuing on in this manner, we find $x_2 = 1.732143$, and $x_3 = 1.732051$ (to 7 significant digits). This is pretty accurate already, since $x_3^2 = 3.0000007$ to 9 decimal places.

■

Newton's method seems to work so well when the initial guess appears to be on a very linear-looking piece of the graph of g which includes the point $(r,0)$, that it should be easy to establish some good quantitative results about its accuracy. However, before we do that, knowing that it is a *local* method should suggest caution. In fact it's very easy to draw a picture, to indicate some circumstances in which Newton's method is likely to do badly, or even fail completely. In Figure 3-17, we illustrate a situation in which Newton's method goes charging off in the wrong direction

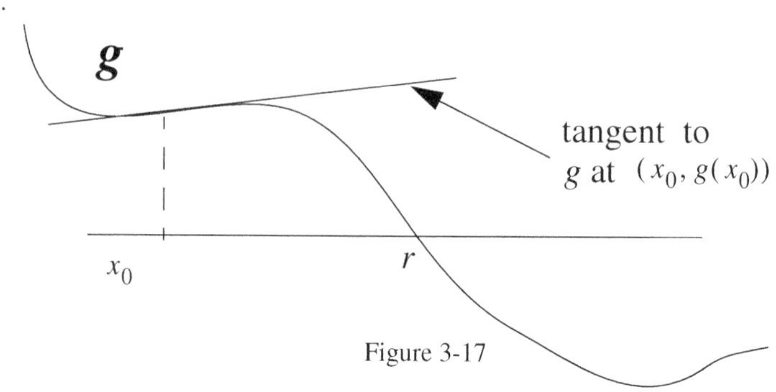

Figure 3-17

If, however, x_0 was shifted just a *little* to the left, the tangent line would be horizontal, and Newton's method would fail. Even worse, however, are cases in which, for a while Newton's method seems to be in the process of generating an answer, when in fact, there is none. This will be explored in Problem 3.14 on page 84. The failure of Newton's method when a poor starting value is chosen arises in a number of commercially written computer programs which make use of Newton's method. This is one area in which you cannot correctly assume that a program will work properly just because it is available commercially.

We now address the problem of determining some useful quantitative results concerning Newton's method. So suppose we want to solve the equation $g(x) = 0$, and we have what seems to be a *reasonable* initial guess x_0. We recall equation 3.18 which specifies Newton's method:

$$x_{n+1} = x_n - \frac{g(x_n)}{g'(x_n)}.$$

Since we are interested in the error at any given stage, it's reasonable to subtract the sought-after solution, *r*, from both sides of this equation to obtain

$$x_{n+1} - r = x_n - r - \frac{g(x_n)}{g'(x_n)}.$$

Notice that if the left side, being $x_{n+1} - r$, is thought of as the error at the current stage, the right side involves the error, $x_n - r$, at the previous stage. We would like to get some simple relationship between these two errors. So in the numerator of the last term on the right we would like to somehow obtain $x_n - r$. Since $g(r) = 0$, we subtract it from this numerator, to obtain

$$x_{n+1} - r = x_n - r - \frac{g(x_n) - g(r)}{g'(x_n)}.$$

Now is the time to use the mean value theorem, Theorem 3.1 on page 62, to replace the numerator, obtaining

$$x_{n+1} - r = x_n - r - \frac{(x_n - r)g'(x_n^*)}{g'(x_n)}, \tag{3.19}$$

where x_n^* is some (unknown) value between r and x_n. We rewrite equation as

$$x_{n+1} - r = (x_n - r)\left(1 - \frac{g'(x_n^*)}{g'(x_n)}\right). \tag{3.20}$$

We could stop here and investigate conditions on g' alone which would guarantee that the magnitude, $|x_{n+1} - r|$, is much smaller than $|x_n - r|$. That will be left as an exercise for anyone sufficiently interested, and we will proceed in another, natural fashion. Namely, we will rewrite equation 3.20 with a common denominator in the second factor, to obtain

$$x_{n+1} - r = (x_n - r)\frac{g'(x_n) - g'(x_n^*)}{g'(x_n)}. \tag{3.21}$$

Once again, the expression we have obtained suggests use of the mean value theorem, to yield

$$x_{n+1} - r = (x_n - r)\frac{(x_n - x_n^*)g''(x_n^{**})}{g'(x_n)} \tag{3.22}$$

where x_n^{**} is somewhere between x_n and x_n^* as illustrated in Figure 3-18.

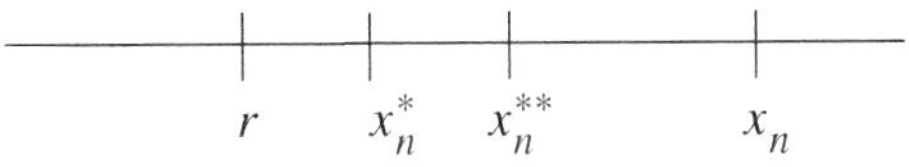

Figure 3-18

Even though we don't know the precise locations of x_n^* and x_n^{**}, we can see from Figure 3-18 that

$$|x_n - x_n^*| \le |x_n - r|$$

and hence,

$$|x_{n+1} - r| \le |x_n - r|^2 \frac{|g''(x_n^{**})|}{|g'(x_n)|}. \tag{3.23}$$

Since we don't know the exact location of x_n^{**} and the values $g'(x_n)$ are changing with n, we will replace the right-hand ratio with a simpler one; namely, let M_2 be an *upper bound* for $|g''(x)|$ on the interval in which r is known to lie. That is,

$$|g''(x)| \le M_2 \quad \text{for all } x \text{ of interest.} \tag{3.24}$$

And, let m_1 be a nonzero lower bound for $|g'(x)|$, i.e.,

$$|g'(x)| \geq m_1 > 0 \quad \text{for all } x \text{ of interest.} \qquad \mathbf{3.25}$$

At least in the simple cases m_1 and M_2 are easy to find. (We'll present an example shortly.) With these definitions, inequality 3.23 yields the following result:

Theorem 3.6: Newton's method error estimates

If on the x-interval, I, which we know contains the solution, r, that we want of the equation $g(x) = 0$, m_1 and M_2 satisfy inequalities 3.24 and 3.25 respectively, then Newton's method, modified to keep all x_n within I, relates the error magnitude, $|x_{n+1} - r|$, at the $n+1$st stage to the error magnitude, $|x_n - r|$, at the previous stage by the inequality

$$|x_{n+1} - r| \leq |x_n - r|^2 \frac{M_2}{m_1}. \qquad \mathbf{3.26}$$

❑

Examining inequality 3.26, we see that once x_n is near r and also close enough to r so that $|x_n - r|(M_2/m_1)$ is much smaller than 1, it will always be the case that x_{n+1} is much closer to r than x_n is. In fact, once x_n is close enough to r, each successive iteration just about doubles the number of decimal places of accuracy in the Newton method approximation of r. This is considerably better than bisection, which picks up one additional place of guaranteed accuracy in somewhere between 3 and 4 iterations (on the average).

It's not surprising that the smaller M_2 is, the better the results - since the smaller M_2, the *more linear* g *looks.* Also it's not surprising that the larger m_1 is, the better the results - since a sharp slicing of the axis near r is easier to spot than a *flatter crossing.*

Example 3.12: Continuation of Example 3.11 ($\sqrt{3}$)

The function given by $g(x) = x^2 - 3$ satisfies the equations

$$g'(x) = 2x \text{ and } g''(x) = 2.$$

Knowing that $\sqrt{3}$ is between 1 and 2, we see that $g'(x) \geq 2$, which allows the choices $m_1 = 2$ and $M_2 = 2$. Hence, from inequality 3.26 we find

$$|x_{n+1} - r| \leq |x_n - r|^2. \qquad \mathbf{3.27}$$

If we choose as the initial guess the value $x_0 = 1.5$, we know that $|x_0 - \sqrt{2}| \leq 0.5$, and we can use inequality 3.27 to see the rapid convergence of

x_n to $\sqrt{3}$. Letting $e_n = |x_n - r|$, we find

$$e_2 \le 0.25,\ e_3 \le 0.0625,\ e_4 \le 0.00391,\ e_5 \le 0.00001529,$$

and so forth.

■

The type of results given above are theoretical, and only apply until the limits of computer accuracy, usually about 16 or so decimal places, are reached. Some computer packages, such as *Mathematica*, permit *arbitrary precision* arithmetic, which can overcome this common computer limitation.

Exercises 3.13

1. Find $\sqrt{5}$ to 3 decimal accuracy

   a. As the solution of $x^2 - 5 = 0$, starting with $x_0 = 1.5$.

   b. As the solution of $1/x^2 - 0.2 = 0$, starting with $x_0 = 1.5$.

   c. For each of the above try various other positive starting values.

   d. For each of the above, try some negative starting values.

   e. Explain the observed behavior of parts c and d for starting values near 0.

   f. Determine theoretically the behavior of the iterates of parts a and b, using Theorem 3.6.

For those with access to a mathematical computing package, such as *Mathematica, Maple* or *Splus*:

   g. Use the package to implement the computations of parts a and b, printing out the successive results. Use of such packages can greatly reduce the computational tedium that such computations may entail.

   h. Find the positive value at which the graphs of $v(x) = x^6$ and $w(x) = 8 - 6x$ intersect, to an accuracy of .000001 Hint: let $g = v - w$.

   i. Determine to an accuracy of .00000001 the negative value closest to 0 for which $x^4 - 10x = 10 - 3x^3$.

   j. To an accuracy of .000001, find the positive value at which the derivative of $1/x^3 + x^5 + 4x - 1$ is 0. Plot this function for positive values of x, to see what this value represents.

2. Use Newton's method to try to solve ($g(x) = 0$), with starting value $x = 1$, explaining the behavior that you observe, for

   a. $g(x) = x^2$

   b. $g(x) = 1/x$

   c. $g(x) = 1/x - 2$

   d. $g(x) = x^2 + 1$

Problem 3.14

Construct a function g for which the sequence of Newton iterates seems, for a while, to be generating a solution of $g(x) = 0$, but for which there is, in fact, no solution at all.

Hints:

- Start off with a function, F, for which there is a solution to $F(x) = 0$.
- Choose a starting value, and for, say, the first 6 or so iterates, retain on the graph that part including the associated points.
- Replace the remaining portion of the graph by a curve which is still differentiable, but does not intersect the x-axis. The only problem you might run into is that, if you're not careful, the newly constructed function may fail to be differentiable at the *patch point.*

There is one situation in which the behavior of the iterates from Newton's method is particularly simple — namely, when the function g

- is strictly monotone,
- satisfies the condition $g(a)g(b) < 0$,
- is either concave up or concave down on $[a; b]$.

We illustrate the four possible graph types in Figure 3-19.

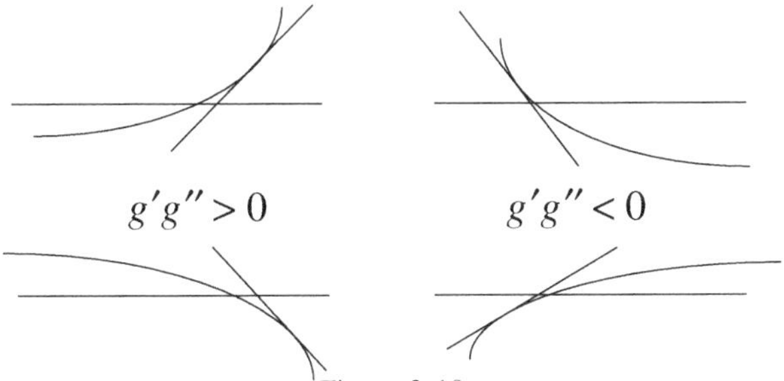

Figure 3-19

We will examine only one of these cases, the others being essentially the same, in Figure 3-20.

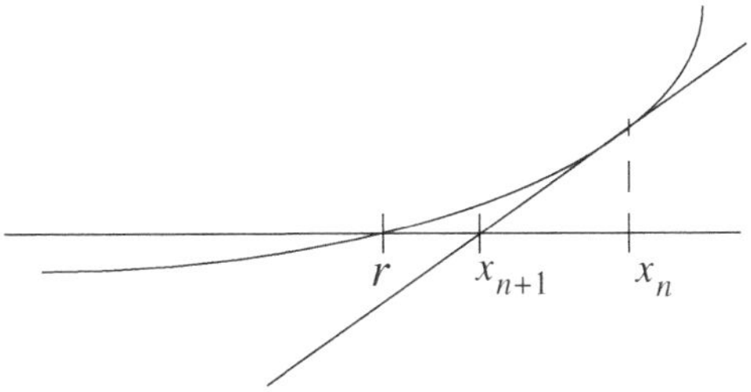

Figure 3-20

In this case, with $g'' > 0$, because the tangent line always lies below the graph of the function (See Theorem 3.4 equation 3.9 on page 68), we see that the sequence $x_1, x_2, x_3, \ldots$ is decreasing, with each value exceeding r. (Note that even if x_0 were to the left of r, it would follow from equation 3.9 that the next Newton iterate, x_1, would lie to the right of r, so that the above illustration is valid so long as $n > 0$.) What we have here is a particular case of a *bounded, monotone* (here, decreasing) *sequence. Bounded* means that all the values lie in some finite interval, say $[L; U]$. It can be shown that for bounded monotone sequences, as n grows large, the values x_n, get close to some specific value. This should bring to mind the definition of *limit*, given in Definition 2.4 on page 30. So at this point it seems desirable to extend this definition to include such sequences.

Definition 3.7: Limit of a sequence

A sequence of real numbers, $x_1, x_2, x_3, \ldots$ is said to have *limit* L (or *converge to* L), written

$$\lim_{n \to \infty} x_n = L$$

if ***all values*** $x_n, x_{n+1}, x_{n+2}\ldots$ can be made as close to L as desired, by choosing n sufficiently large.

❑

For reference purposes we state the following result for sequences of real numbers:

Theorem 3.8: Bounded monotone sequence has limit

To establish this result, we simply construct the claimed limit using bisection. Suppose the interval containing all values of this sequence is $[a;b]$, and the sequence is increasing. Bisect the interval $[a;\ b]$ and choose the right half if it contains at least one point of the sequence. Otherwise choose the left half. We illustrate this in Figure 3-21. Now repeat this process. Notice that the first time you must choose the right half, and it *looks like* you choose the left half for the second and third times, and so forth. The number defined by this

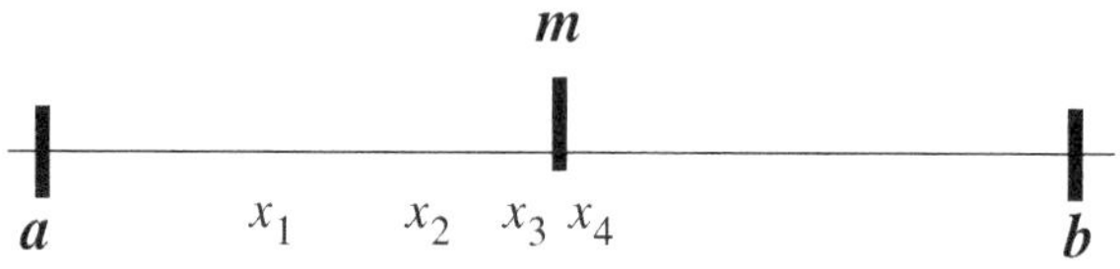

Figure 3-21

process is seen (see Problem 3.15 which follows shortly) to be the limit of the sequence, $x_1, x_2, \ldots$.

Now we can apply this result to the Newton iterates generated in accordance with Figure 3-20. This sequence is defined by equation 3.18, which we repeat here as

$$x_{n+1} = x_n - \frac{g(x_n)}{g'(x_n)}. \qquad \textbf{\textit{3.28}}$$

At this point, if we knew that g and g' were continuous, (see Definition 2.6 on page 34), using the properties of continuous functions, (Theorem 2.10 on page 38), if we let $R = \lim_{n\to\infty} x_n$, we know that $\lim_{n\to\infty} g(x_n) = g(R)$, and $\lim_{n\to\infty} g'(x_n) = g'(R)$.

So we could take the limit of both sides of equation 3.28 to obtain

$$\lim_{n\to\infty} x_{n+1} = \lim_{n\to\infty} x_n - \frac{\lim_{n\to\infty} g(x_n)}{\lim_{n\to\infty} g'(x_n)}.$$

$$\text{or } R = R - \frac{g(R)}{g'(R)},$$

which yields $g(R)/g'(R) = 0$. Since $g'(R)$ is never 0, we finally obtain

$$g(R) = 0, \qquad \textbf{\textit{3.29}}$$

showing that the sequence $x_1, x_2, x_3, \ldots$ approaches a value R satisfying equation 3.29, as we were hoping. It can be shown that if a function is differentiable, then it is continuous.[1] So if we assume that g'' exists for all x in $[a; b]$, the continuity of g' and g follow. Putting all of this together yields

Theorem 3.9: Convex function convergence of Newton's Method

If the real valued function, g, of a real variable

- is strictly monotone (Definition 2.9 on page 37) on the interval $[a; b]$ (see Footnote 3 on page 61),
- satisfies the condition $g(a)g(b) < 0$,
- is twice differentiable and either concave up or concave down on $[a; b]$ ($g'' > 0$ or $g'' < 0$ will ensure this),

then the Newton iterates, defined by equation 3.28, and modified to ensure that all iterates are in the interval $[a; b]$, have a limit, R, satisfying equation 3.29.

❑

A **convex function** is one whose graph is *bowl-shaped facing up.* The strict definition is that on each given interval, it never goes above its secant line — if the interval is $[a; b]$ and the function is f, the secant line is the segment connecting the points $(a, f(a))$ and $(b, f(b))$.

Problem 3.15

Show that the number constructed by the bisection process used in Theorem 3.8 on page 86 is $\lim_{n \to \infty} x_n$.

Hint: Each interval chosen in this process includes all but a finite number of values of the sequence, and these intervals get short.

---

1. We actually established this in the optional section on page 65.

For later reference, we conclude this chapter with some formal definition and theorem statements, most of which were established in the optional section outlining the proof of the mean value theorem.

Definition 3.10: Boundedness of a set and of a function

Let A be a set of real numbers (see Appendix 1 on page 757).

A is said to be bounded if there is a real number $B > 0$ such that for all x in A

$$-B < x < B.$$

B is called an **upper bound**, and **$-B$** a **lower bound for A.**

Let f be a real valued function of a real variable (Definition 2.1 on page 21). **The function f is said to be bounded** if there is a real number, $M > 0$ such that for all x in the domain of f (Definition 1.1 on page 2)

$$-M \leq f(x) \leq M.$$

M is call an **upper bound** and **$-M$** a **lower bound** for **f.**

❑

Definition 3.11: Closed and open intervals

The symbol $[a;b]$, where a and b are real numbers with $a \leq b$ stands for the set of all real numbers, x, satisfying $a \leq x$ and $x \leq b$. This latter is frequently written $a \leq x \leq b$. $[a;b]$ is called the *closed bounded*[1] *interval from a to b*. The *open interval* $(a;b)$, where $a < b$, consists of those real x satisfying $a < x < b$.

❑

Theorem 3.12: Boundedness of continuous function

If f is a real valued continuous function (see Definition 2.6 on page 34) whose domain (see Definition 1.1 on page 2) is a closed and bounded interval $[a; b]$, see Definition 3.11 above, then f is bounded. Furthermore, f achieves its maximum and minimum — i.e., there are numbers, min, Max, x_{min}, and x_{Max} such that for all x in the domain of f

$$min \leq f(x) \leq Max \quad \text{and} \quad f(x_{min}) = min, \qquad f(x_{Max}) = Max.$$

❑

---

1. Sometimes the word *bounded* is omitted.

Definition 3.13: Piecewise continuous and piecewise smooth functions

A real valued function, f, of a real variable is called piecewise continuous on the closed bounded interval $[a\,;b]$ (see Definition 3.11) if

- f is continuous at all points of $[a\,;b]$ with the exception of a finite number of points, $t_1,...,t_n$ (see Definition 2.6 on page 34)
- the limits (see Definition 2.4 on page 30)

$$\lim_{\substack{h \to 0 \\ h > 0}} f(t_i + h), \quad \lim_{\substack{h \to 0 \\ h < 0}} f(t_i + h)$$

exist for all $i = 1,...,n$ (except for the two end points, where only one of these limits is defined.)

The function f is called piecewise smooth if it has a derivative at all points of $[a\,;b]$, except for a finite number of values $t_1,...,t_n$, and this derivative is piecewise continuous.

❑

Theorem 3.14: Continuity of differentiable functions

If f is a real valued function of a real variable (Definition 2.1 on page 21), which is differentiable at some point (Definition 2.5 on page 30) then f is continuous at that point (see Definition 2.6 on page 34 as well as Definition 7.2 on page 213 for a more precise version).

❑

Chapter 4

Recovering Global Information, Integration

1 Introduction

In a great many situations we have a local description of a function available, and want to learn something about this function, e.g., calculate its values with some specified accuracy. Here are a few examples.

Example 4.1: Acceleration

Let $x(t)$ denote the position at time t of some object, such as a railroad car, which is constrained to move along a straight line. That is, $x(t)$ is the *signed distance* of this object from some specified location (the origin) on this line.[1] Then $Dx(t) = x'(t)$ is this object's velocity at time t. In terms of the *rate of change* concept, see page 58, velocity is the *(time) rate of change of position;* velocity at time t measures how fast position is changing near time t. The time rate of change of velocity, $D^2x(t) = x''(t)$, is called the *acceleration* at time t; it measures how fast the velocity is changing near time t. Devices for measuring acceleration are essential for navigation, because, as we will see, knowing *initial* position and *initial* velocity of an object, we can determine the object's position at all times t for which we have available all accelerations from time 0 to time t. So here we want to calculate values $x(t)$ from $x(0), x'(0)$, and $x''(\tau)$ all τ in $[0;t]$.

■

Example 4.2: Accelerometer

It is of some interest to present a mathematical model of a *spring-mass* system coupled to a pen and moving paper, and see how such a model allows the determination of acceleration. This system is illustrated in Figure 4-1. One end of the spring is firmly attached to the flatcar, the other end to an object of mass m, to which is attached a pen. The object is assumed free to move only along the direction of the track, and its motion is recorded on the paper which is moving from left to right.

---

1. Alternatively, $x(t)$ may be one of the coordinates of the position vector of some object, such as a ship, airplane, or spacecraft at time t. For the present we will restrict consideration to the one-dimensional situation. It will not be very difficult to extend the one-dimensional results to objects requiring more coordinates to describe their position.

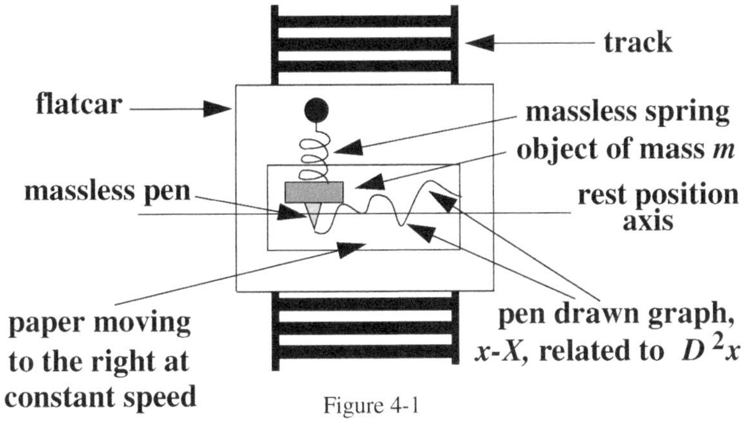

Figure 4-1

The reasoning which shows why the graph produced by this device should allow determination of the acceleration vs. time function is the following:

First we take note of the simplest version of

Newton's second law[1] (of motion)

$$F(t) = m\,x''(t). \qquad \textit{4.1}$$

Here $F(t)$ represents the force (push or pull) on some object at time t, measured in some agreed upon fashion,[2] $x''(t)$ is the acceleration[3] this object undergoes at time t in response to the force, and m is the *mass* of this object. What this law states is simply that the acceleration at time t is proportional to the applied force at time t. The positive quantity $1/m$ is the *constant of proportionality.*[4]

Next we introduce *Hooke's law*, which describes the force exerted by a spring to extending/compressing it from its *rest configuration* — its length when no external forces are applied to it.

Hooke's law

$$f = -k\,x, \qquad \textit{4.2}$$

where f is the amount of force exerted by the spring in response to extension or compression, x, of the spring from its length when no other force

1. A *law* is just a description relating measured quantities.
2. Possibly the extension a calibrated standard spring undergoes in applying this force to some object, or the required number of objects of some standard specification so that the force just neutralizes the force of gravity on these objects, i.e., keeps the objects in place, neither rising nor falling.
3. In the usage here, acceleration can be positive or negative.
4. The units for force and acceleration determine the units for m.

is acting on it. This law states that the extension/compression of the spring is proportional to the applied force. The constant of proportionality, $-k$, is negative, indicating that this response is a *restoring force,* since the spring opposes being extended or compressed. As anyone who has ever stretched a spring too far knows, Hooke's law is a reasonable description only in a limited range. The quantity k is called the **spring constant**.

Now we can apply Newton's and Hooke's laws to the situation illustrated in Figure 4-1. If the acceleration of the flatcar at time t is $x''(t)$, this acceleration is transmitted to the firmly fastened spring, which then attempts to transmit it to the indicated object of mass m. The spring can only transmit this acceleration to the mass if it is expanded (or compressed if the acceleration is negative). This will happen if the flat car is accelerating, since we are assuming that the only force exerted on the object due to the flatcar motion is transmitted through the spring.

Let P denote the rest position[1] of the object in the flatcar; let $x(t)$ denote the displacement of P from some fixed point, Q, on the track at time t; finally, let $X(t)$ be the displacement of the object from the point Q at time t, the positive direction being *up.* We assume that $x(t) - X(t)$, the displacement of the object from its rest position at time t, is available to us from the graph produced by the pen attached to the object, but that neither $x(t)$ nor $X(t)$ is directly available. We would like to use the available $x - X$ data to determine the function x. So we have to determine how the functions x and $x - X$ are related.

To do this note that the force at time t exerted by the spring on the object is

$$k(x(t) - X(t)).$$

From Newton's second law of motion, equation 4.1 on page 92, the object, which has mass m responds to this force with an acceleration $X''(t)$ determined by

$$m X''(t) = k(x(t) - X(t)).$$

Notice here that when $x(t) - X(t)$ is positive, if the position where the spring is attached to the flatcar has *moved up* on the page being viewed, the spring is stretched, and the force on the object is in the positive (upward) direction. So we have the correct algebraic signs in the above equation.

Since we want to examine the quantity $x - X$, which is essentially the graph recorded by the pen, we rewrite this equation as

$$m(x''(t) - X''(t)) + k(x(t) - X(t)) = m x''(t)$$

and, dividing by the positive value, m, and converting to functional notation,

1. The object's flatcar location where the spring exerts zero force on it.

$$(x - X)'' + \frac{k}{m}(x - X) = x''.$$

Letting $Z = x - X$, the recorded graph, we therefore find that for given position function, x, of the flatcar, under the assumptions made thus far, the recorded graph, Z, and x are related by the differential equation

$$Z'' + \frac{k}{m} Z = x''. \qquad \textbf{4.3}$$

It is assumed that we know the values of k and m. Since the graph of Z is available to us, we can generate approximations to Z', and Z'' (just using $(Z(t + h) - Z(t))/h$ to approximate $Z'(t)$, and repeating the process to approximate $Z''(t)$ from the approximations to $Z'(t)$). Then, using equation 4.3, we can approximate the acceleration function, x'', that we wanted.

Before leaving this example it should be noted that for the actual system being considered, this model is a very simplistic one, totally ignoring the effect of friction of the pen against the paper. A model of this system which properly takes account of friction would be difficult to deal with. However, if the pen, and paper were replaced by a device such as a piezoelectric crystal, which produces an electrical voltage when it is compressed, this model for the voltage produced can be a very good description. If this voltage were fed into a computer, the above computations could be carried out quickly and effectively.

■

Example 4.3: Radioactive dating

In Example 3.3 on page 57, the simplest model of radioactive decay was presented, in the form of equation 3.3, which we now display.

$$A'(t) = -k A(t).$$

Here $A(t)$ is the amount[1] of radioactive material still present at time t, and k is some positive constant depending on which radioactive substance is under consideration.

---

1. It should be realized that this differential equation model treats the amounts of radioactive material as real numbers, and that the set of values, $A(t)$, as t varies, has infinitely many elements. This contradicts the atomic nature of radioactive substances, which seems a more reasonable description. From a practical point of view, this discrepancy is not serious. However, to avoid this paradox, a statistical model should really be used, and in the statistical model quantities such as $A(t)$ represent, not the actual amounts of radioactive material at time t, but the *mathematical expectations* of these quantities (which, despite what it seems, is not the number of such atoms that one expects, but the long-run average of these quantities over many experimental repetitions).

One particularly interesting application of this model is to the determination of the time of death of some *once-living* object — animal, insect or plant, from a tissue sample of this organism. Such information can be important to historians, archaeologists and geologists.

Legitimate application of this model relies on the assumptions[1] that when they are alive, the carbon in these living beings is in equilibrium with atmospheric carbon, and that radioactive carbon-14 is a constant percentage of all atmospheric carbon (being steadily created from constant cosmic ray bombardment of our atmosphere, and steadily disappearing via the process of radioactive decay governed by equation 3.3 reproduced above). Because of the equilibrium just mentioned, in this model the amount of carbon-14 in any living tissue, as a percent of the total carbon in this tissue, is the same as the atmospheric percentage.

We now suppose that we have a sample from some organism (plant, animal or insect), that has been dead for some time, t_d, and we want to determine t_d, from this sample. For our analysis we let $R_{org}(t)$ denote the sample's amount of radioactive carbon-14 at time t, and let C_{org} denote the sample's amount of nonradioactive carbon. Present time is taken to be $t = 0$. Then, applying equation 3.3 to R_{org}, for all $t > -t_d$, we have

$$R'_{org}(t) = -k\,R_{org}(t). \qquad \textit{4.4}$$

We assume that C_{org} is known from some accepted chemical procedure.

■

In Chapter 5 we will develop important properties of the solution of equation 4.4, and show how to use it to determine $R_{org}(t)$ accurately from k, C_{org}, and any specified value $R_{org}(\tau)$. These results will be applied to illustrate the use of scintillation counter measurements to estimate the carbon-14 decay constant, k, and the time of death, t_d.

Example 4.4: Probability density

If $F(x)$ is the probability of observing an experimental value less than x, in many cases there is no simple formula for $F(x)$, but there is one for its derivative, $F'(x)$. F is the *(cumulative) distribution function* and F' is called a *probability density*, usually denoted by f. Knowing f, there is frequent call for accurate calculation of $F(b) - F(a)$, which is the probability that an observed experimental value lies in the interval from a to b.

■

1. If these assumptions are badly in error, many scholarly results would go down in flames. Fortunately, many such conclusions seem in agreement with results determined by other methods.

Example 4.5: Compartmental models

In Example 1.8 on page 15, the subject of enzyme kinetics was introduced. A reasonable approach to this subject is to represent the body as a set of interacting compartments, with the rate of flow of chemicals from any specified compartment to another one being proportional to the amount of this chemical in the specified compartment. Although at first it may seem unnatural, flow in both directions is assumed, but the (rate) constants of proportionality for the two directions of flow are often assumed unequal.

What follows describes a simple two-compartment model for the behavior of a drug infused into the *blood compartment*, from which it can migrate to and from another compartment (possibly representing the rest of the body). We introduce

- $X_B(t)$: the amount of this drug in the blood compartment at time t,
- $X_O(t)$: the amount of this drug in the other compartment at time t,
- $R_a(t)$: the infusion rate of the drug into the blood compartment at time t, assumed known.

The equations which we write to describe this system are the following *mass-balance* differential equations:

$$\begin{aligned} X_B' &= -k_{OB}\,X_B - k_{EB}\,X_B + k_{BO}\,X_O + R_a \\ X_O' &= k_{OB}\,X_B - k_{BO}\,X_O\,. \end{aligned} \qquad \textbf{4.5}$$

In the first of these equations, the flow ***out*** of the blood compartment is assumed to be either to the *Other* compartment (first term to the right of the = sign) or to be irretrievably lost to the *External* environment (second term); the flow ***into*** the blood compartment is either from the *Other* compartment (third term), or via intravenous infusion (last term). The *Other* compartment receives the drug from the blood compartment (the first term to right of the = sign in the second equation) or loses drug to the blood compartment (last term). The system is usually represented schematically as show in Figure 4-2.

The drug infusion function, R_a, is here assumed known. It is sometimes reasonable to assume that at the start of infusion, ($t = 0$),

$$X_B(0) = 0 \quad \text{and} \quad X_O(0) = 0. \qquad \textbf{4.6}$$

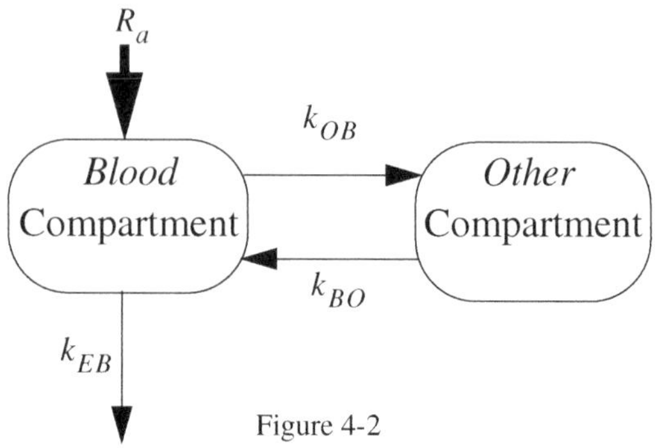

Figure 4-2

If the infusion is very rapid, only lasts for a very short time, and consists of injecting a fixed amount, d, of the drug (a so-called *Bolus of size d*), the R_a term is omitted from equations 4.5, and the initial conditions, equations 4.6, are replaced by

$$X_B(0) = d \quad \text{and} \quad X_O(0) = 0. \qquad \textbf{4.7}$$

If the values of the rate constants, k_{OB}, k_{BO} and k_{EB} are known, solution of the system (i.e., determination of $X_B(t), X_O(t)$ for specified value(s) of t), would indicate how much of the drug remains in each compartment at time t. From this type of information the amount of drug remaining in the system at various times can be determined. Other quantities, such as the total system exposure to the drug, especially important for computing *total radiation dosage* from a radioactive drug, can also be determined; it should be pointed out that if this model overlooks important other compartments which store the drug, the computed value of total radiation dosage could seriously underestimate the actual total radiation dosage, with serious consequences.

Solution of equations 4.5, subject to appropriate initial conditions, has to be carried out many times when the rate parameters, k_{OB}, k_{BO} and k_{EB} are unknown, and we want to use observed data to estimate them. The most common criterion for such parameter estimation is that of *least squares*. The least squares criterion attempts to choose the parameters so that, in a reasonable sense, the solution of equations 4.5 with the specified initial conditions (either equations 4.6 or 4.7) using these parameter values furnishes the *best* agreement with the observed data. Here is a bit more detail:

For an arbitrary choice of k_{OB}, k_{BO} and k_{EB}, let

$$L(k_{OB}, k_{BO}, k_{EB}) = \sum_i \left\{ [X_B(t_i) - x_{B,i}]^2 + [X_O(t_i) - x_{O,i}]^2 \right\},$$

where $x_{B,i}$ and $x_{O,i}$ are the measured (observed) values of drug amounts at time t_i in the *Blood* compartment and the *Other* compartment, respectively; the values $X_B(t_i)$, $X_O(t_i)$ are the computed solution values of equations 4.5 at t_i subject to the appropriate initial conditions, such as provided in equations 4.6 or 4.7, obtained for the specified choice of k_{OB}, k_{BO} and k_{EB}. The *least squares* criterion chooses k_{OB}, k_{BO} and k_{EB} to minimize $L(k_{OB}, k_{BO}, k_{EB})$. Notice that if the chosen values of k_{OB}, k_{BO} and k_{EB} happened to yield the result

$$L(k_{OB}, k_{BO}, k_{EB}) = 0$$

this model would furnish a perfect description of the data. Furthermore, if this choice only made $L(k_{OB}, k_{BO}, k_{EB})$ quite small, it would still appear that the model associated with this choice provided a pretty good description of the data. Actual choice of k_{OB}, k_{BO} and k_{EB} to achieve the objective of minimizing $L(k_{OB}, k_{BO}, k_{EB})$ usually involves some sort of *search procedure*, which starts off with an initial estimate (often just an educated guess) $k_{OB,0}, k_{BO,0}, k_{EB,0}$ for k_{OB}, k_{BO} and k_{EB}. Then successive *improvements*, $k_{OB,j}, k_{BO,j}, k_{EB,j}$ $j = 1,2,...$are determined, usually by examining the local behavior of the function L near the three-dimensional point $k_{OB,j-1}, k_{BO,j-1}, k_{EB,j-1}$. It is these examinations that require repeated solution of equations 4.5. An extension of Newton's method for finding maxima and minima (see Section 4 of Chapter 3) is usually used once the estimates are close enough to the desired value. The techniques for getting close enough to be able to apply the extended Newton's method tend to be much trickier in higher dimensions (i.e., when several parameters must be determined), than they are in one dimension. At this stage of our development there isn't much more that can be stated regarding experimental determination of rate parameters. We will take up the subject of extending Newton's method to higher dimensions in Chapter 10.

■

2 Integration: Calculating *f* from *Df*

In this section we introduce the simplest useful procedure for precise determination of a function from its local description. We will investigate the determination of the value $f(b)$, as precisely as we want, from the values $f(a)$ and $f'(x)$ for all x with $a \le x \le b$.

To get a handle on precise determination of f from an *initial value*, $f(a)$, and values of f', we will use the fundamental approximation, formula 2.7 on page 31, which we reproduce here as

$$f(x_0 + h) \cong f(x_0) + f'(x_0)h, \quad \text{valid for small } h. \qquad \textbf{4.8}$$

To compute an approximation to $f(b)$,

- choose the integer, n, large enough so that $h = \frac{b-a}{n}$ is *small*,
- then successively in approximation 4.8 substitute
  $x_0 = a,\ a+h,\ a+2h, \ldots,\ b-h$.

Proceeding according to this plan, here is what results:

1. $f(a+h) \cong f(a) + f'(a)h \qquad (x_0 = a)$

Next $f(a+2h) \cong f(a+h) + f'(a+h)h \qquad (x_0 = a+h)$
and substituting from formula 1 $f(a+h) \cong f(a) + f'(a)h$
which we rewrite as

2. $f(a+2h) \cong f(a) + [f'(a) + f'(a+h)]h$

Next $f(a+3h) \cong f(a+2h) + f'(a+2h)h \qquad (x_0 = a+2h)$
and substituting from the just determined approximation 2
$f(a+2h) \cong f(a) + [f'(a) + f'(a+h)]h$ which we rewrite as

3. $f(a+3h) \cong f(a) + [f'(a) + f'(a+h)]h + f'(a+2h)h$

Proceeding inductively, we obtain the n-th result

n. $f(a+nh) \cong f(a) + [f'(a) + f'(a+h) + \cdots + f'(a+\{n-1\}h)]\,h.$

Since $nh = b-a$ and $h = (b-a)/n$, we rewrite this last approximation as

$$f(b) \cong f(a) + \sum_{k=0}^{n-1} f'\left(a + k\frac{b-a}{n}\right)\frac{b-a}{n}. \qquad \textbf{\textit{4.9}}$$

We expect this approximation to be as accurate as we desire, provided n is chosen large enough. So we expect the quantity

$$\sum_{k=0}^{n-1} f'\left(a + k\frac{b-a}{n}\right)\frac{b-a}{n} \qquad \textbf{\textit{4.10}}$$

to have a limiting value (see Definition 3.7 on page 85), as n gets large. Before presenting the precise results that can be established, it's worthwhile making a few observations:

Expression 4.10 has a meaningful *geometric interpretation* related to the graph of f' as shown in Figure 4-3 (which was simplified by choosing $a = 0$).

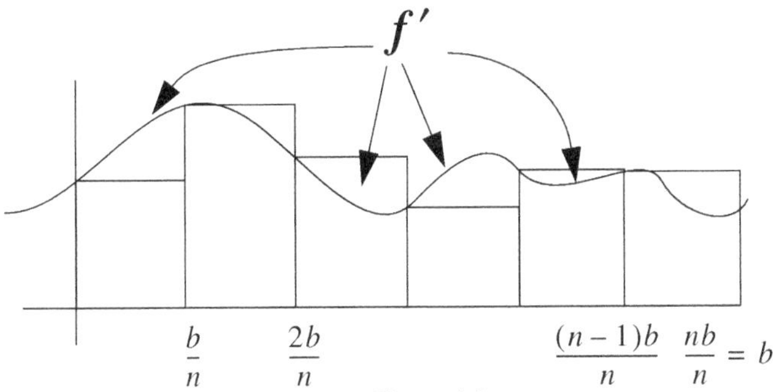

Figure 4-3

We see that this sum is just an approximation to the *signed area* between the graph of f' and the *first coordinate axis* on the interval $[a, b]$. So, we would expect the limit of this sum as n gets large to be the signed area between the graph of f' and the interval from a to b on the first coordinate axis. Here it is the *signed area*, in which the area corresponding to negative function values is negative, that the limit of expression 4.10 would represent.

It would seem natural to take the limit of the expression in 4.10 as the definition of the object (the *integral*), which will yield $f(b)$ from f' and $f(a)$. This approach won't work, because if we defined the integral in this fashion, sometimes, as we will see in Example 4.12 on page 111, we'd get values we wouldn't really want to get as the integrals of some functions. The so-called *Riemann sum* definition that we will adopt below will help avoid this difficulty, and is still reasonable and relatively easy to understand.

Before proceeding with a systematic development of integrals, it should be pointed out that integrals (limits of sums such as in expression 4.10), arise in important contexts other than the one we are currently considering. Here are a couple of examples.

Example 4.6: Work

It's common knowledge that it takes energy to pump water from a low to a high elevation, or to lift an elevator from the bottom to the top floor of a building. The amount of energy required for such tasks is of great interest, since it translates directly into fuel cost. This amount of energy is referred to as the *work* required for the given task. In the simplest elevator situation, in order to raise the elevator, a fixed force is applied upward to the elevator cable until the task is completed; the work, W, required, turns out to be the product of the force, F, applied, and the distance, d, that the elevator is raised, i.e.,

$$W = Fd.$$

But applying a constant force to the elevator cable would result in an uncomfortable ride, and cause a great deal of wear on parts of the elevator as well. So the applied force is set to start out small, then gradually increase to attain a decent speed, and then diminish to avoid too sudden a stop. The work done can be approximated by a sum

$$\sum_i F(t_i)d(t_i), \tag{4.11}$$

where the t_i are time points spanning the duration of the lifting process, whose separation is taken small enough so that from time t_i to t_{i+1} the force varies little from its value, $F(t_i)$ at t_i, and $d(t_i)$ is the distance the elevator is raised during this time interval. The exact work done in raising the elevator is taken to be the limit of the sum in expression 4.11 as the duration of the largest of the time intervals approaches 0.

■

Example 4.7: Density functions

The distribution of electrical charge along an insulator, or the distribution of mass along a thin cylinder is often specified by a density function, which may have been arrived at from theoretical considerations. When we say that there is a charge (or mass) *density* function, g, we mean that the amount of charge (or mass) in a short interval from position x to position $x + h$ is approximately $g(x)h$. What this is taken to mean is that if the x_i form an increasing sequence spanning the interval $[a;b]$, and the largest of the values $x_{i+1} - x_i$ becomes small, then the sum

$$\sum_i g(x_i)(x_{i+1} - x_i)$$

gets close to the total charge (or mass) in the cylinder from position a to position b.

■

Because integrals arise in so many diverse situations, rather than defining integrals solely for functions f' which are derivatives, we will define them more generally.

As a preliminary we introduce the notion of a *Riemann sum*, which is seen to be a somewhat less restricted approximation to the signed area between the graph of a function, g, and the *first coordinate* (x) axis, lying over the interval $[a;b]$ than that given above in Example 4.7.

Definition 4.1: Partition, Riemann sum

Let g be a real valued function of a real variable whose domain (input values) is the interval $[a;b]$, where $a < b$. Let $x_0, x_1, ..., x_n$ be an increasing sequence of numbers from this interval (recall Definition 1.4 on page 7), with

$x_0 = a$ and $x_n = b$. This sequence defines a *partition of the interval* $[a;b]$. Let X_i stand for an arbitrary number in the interval from x_i to x_{i+1}.

The quantity

$$\sum_{i=0}^{n-1} g(X_i)(x_{i+1} - x_i) \qquad \textbf{\textit{4.12}}$$

is called a *Riemann sum for g on* $[a;b]$. The largest of the values $x_{i+1} - x_i$ is called the *norm* of the partition (defined by) $x_0, x_1, ..., x_n$.

❑

Notice that a Riemann sum for g is an approximation to the signed area between the graph of g and the x-axis interval from a to b, as illustrated in Figure 4-4. To get the area, multiply base length, $x_{i+1} - x_i$, by height, $g(X_i)$.

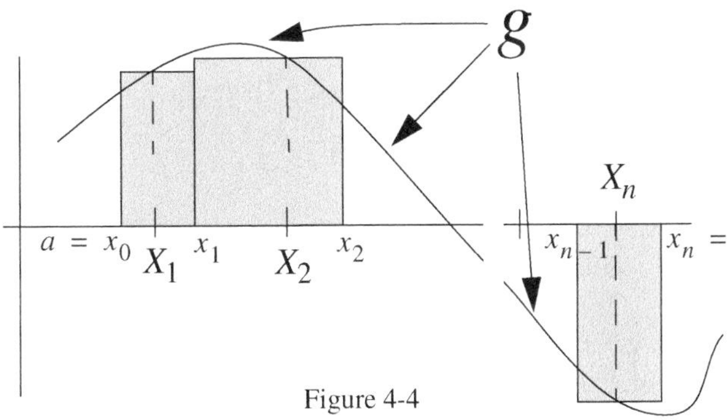

Figure 4-4

Definition 4.2: Integral $\int_a^b g$

If g is a real valued function of a real variable, and there is a single value, say, I, which all Riemann sums, (expressions 4.12), are close to, provided only that their partition norms are small enough, and independent of what *typical point*, X_i, is chosen for evaluation of g on the i-th interval, then we say that the function g is (*Riemann*) *integrable on* $[a;b]$, with (*Riemann*) *integral* (or ***definite** integral*), I.

The integral, I, is denoted by the symbols

$$\int_a^b g \quad \text{or} \quad \int_a^b g(x)dx.$$

In the latter notation, the symbol x may be replaced by any symbol which does not conflict with symbols already in use. The integral sign is an *OLDE ENGLISH S*, as in the word **sum.**

The process of determining integrals is called **integration**. The function being integrated is called the **integrand**.

❑

The alternate notation above is indicative of the way in which the integral is formed, the integral sign, $\int$ (being an *Old English S*) as essentially a *Sum*, of terms of the form $g(x)$ times dx (dx representing a small change in x). It is especially convenient in dealing with functions specified by their formulas.

From the discussion leading up to approximation 4.9 (page 99), it is reasonable to believe that

$$\int_a^b f' = f(b) - f(a). \qquad \textit{4.13}$$

This result, which is valid under fairly general conditions, is important, because it allows evaluation of many (but by no means *most*) integrals arising all over the field of applied mathematics (see Theorem 4.3 below).

Example 4.8: Calculation of $\int_a^b x^2 dx$

Here, if we write $g(x) = x^2$, we know that $g = f'$ where $f(x) = x^3/3$. Thus, assuming that equation 4.13 holds in this situation, we see that

$$\int_a^b x^2 dx = \int_a^b f' = f(b) - f(a) = b^3/3 - a^3/3.$$

■

We conclude the development portion of this section with a statement of the main result on existence and evaluation of integrals via formula.

Theorem 4.3: Existence and formula evaluation of integrals

If the real valued function, g, of a real variable, is continuous at each x from the closed bounded interval $[a;b]$, then the integral

$$\int_a^b g(x)\,dx$$

exists. That is, all of the associated Riemann sums have a limit as the norms of their partitions become small.

Moreover, g is also the derivative of some function f, i.e., $g = f'$, and

$$\int_a^b g = \int_a^b f' = f(b) - f(a). \qquad \textit{4.14}$$

This will be referred to as the **fundamental (integral) evaluation formula.**

❑

In the optional section to follow, we will discuss some of the mathematical issues that arise in establishing this result (whose first part is proved in Appendix 3 Corollary to Theorem A3.5 on page 771). In particular, we will see why a satisfactory theoretical development would be impossible were we to adopt the limit of the expression in 4.10 as the *definition* of the integral of f' from a to b. Equation 4.14, the fundamental integral evaluation formula is established just following the two versions of the fundamental theorem of calculus, on page 127)

If g has an integral over $[a;b]$, then for suitably chosen partitions, **all Riemann sums** are **close** to the integral of f and hence, **to each other**. It follows in Appendix 3, from Theorem A3.1 on page 766 that the converse holds too — i.e.,

Theorem 4.4: Existence and numerical evaluation of integrals

Let g be a real valued function with domain the closed bounded interval $[a; b]$, $a < b$ (see Definition 3.11 on page 88). If we can choose a sequence of partitions, $P_1, P_2, \ldots$ of $[a, b]$ and numbers $d_1, d_2, \ldots$ decreasing to 0, such that for **all** pairs of Riemann sums (Definition 4.1 on page 101) R_n, R_n^*, associated with the partition P_n, we have

$$\left|R_n - R_n^*\right| \le d_n$$

then the integral (Definition 4.2 on page 102)

- $$\int_a^b g \quad \text{exists}$$

and

- $$\left|\int_a^b g - R_n\right| \le d_n .$$

❑

This result will lead to practical results in Section 6 of this chapter..

Sufficient material has already been covered in this section to make it worthwhile to go over the main ideas in a brief:

Survey and Summary of Section 2 of this Chapter

- Starting out with the desire to determine $f(b)$ from $f(a)$ and f', we were led to the sum, expression 4.10 on page 99. We cannot stress strongly enough that the basic result embodied in equation 4.14 on page 103 is that ***the value $f(b)$ can essentially be recovered from $f(a)$ and f' via the summation process exhibited in expression* 4.10.**

- Note the similarity of equation 4.14 to the fundamental summation theorem of sequence calculus, Theorem 1.8 on page 12 This is more than just a formal similarity, since the k-th term of the sum in approximation 4.9 on page 99 is roughly the change

$$f\left(a + (k+1)\frac{b-a}{n}\right) - f\left(a + k\frac{b-a}{n}\right),$$

as can be seen from the fundamental linear approximation, $f(x_0 + h) \cong (f(x_0) + f'(x_0))h$, formula 2.7 on page 31, valid for small h; thus, the right equality of equation 4.14 on page 103 is asserting that:

the total height change of $f(x)$ as x goes from a to b is approximately the sum of the successive approximate height changes.[1]

We see that equation 4.13 on page 103 is plausible. But there are some nontrivial obstacles to providing a completely convincing proof of what is being claimed, and they will be discussed in Appendix 3 on page 763.

- Finally, formula 4.14 on page 103 is usable, (just like the fundamental evaluation theorem for sums), for evaluating integrals, when we can find a simple formula for a function whose derivative is g. Finding such formulas will be explored in Section 7 of this chapter. It should be pointed out here and further on, that most integrals which come up in real applications, just like most such sums, cannot be evaluated exactly by a simple formula (despite the impression given in many elementary calculus courses.) Nonetheless, as we will see in Section 3 of this chapter, such difficult integrals can often be well approximated by *easy-to-evaluate* ones.

The discussion following expression 4.10, illustrated in Figure 4-3 (see page 99 ff) leads to the following.

---

1. The use of the fundamental linear approximation formula 2.7 on page 31 in re-examining equation 4.14 from scratch is just a slight variation on its original derivation, which also used the fundamental linear approximation.

Area Interpretation of $\int_a^b g$

The integral, $\int_a^b g$, may be interpreted as the *signed area* between the graph of the function g and the first coordinate axis, (the x-axis), as shown in Figure 4-5.

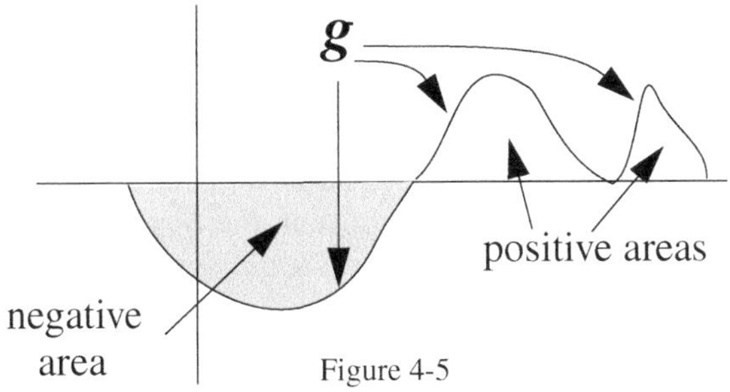

Figure 4-5

Portions of the graph of g lying below the x-axis contribute negatively to the area. The ***significance of this interpretation*** is that *it makes it easier to anticipate and remember many of the properties of integrals*, since most people already have a reasonable intuitive understanding of the concept of area.

Exercises 4.9

Evaluate the following integrals from your knowledge of derivatives, and compute Riemann sums for these integrals, based on a partition consisting of four equal-length intervals, with the points X_i being the center of the intervals in the partition. For those with access to a programming environment like *Splus* or *Mathematica*, try investigating how rapidly the Riemann sums seem to be approaching the integral as the norm of their partition becomes small.

1. $\int_1^3 \frac{1}{x^2} dx.$
2. $\int_0^4 x^3 dx$
3. $\int_0^3 x\sqrt{x^2 + 7}\, dx$ Hint: Recall the chain rule for differentiation of a product, formula 2.30 on page 53.

3 Some Elementary Aspects of Integration

It's convenient to extend the definition of the integral $\int_a^b g$ to the situation $a > b$, via the general convention

$$\int_a^b g = -\int_b^a g. \qquad \textit{4.15}$$

Notice that under this convention, the fundamental evaluation formula (equation 4.14 on page 103) is still valid.

By thinking of the definite integral as essentially a ***sum***, (expression 4.12 followed by Definition 4.2 on page 102), or, when more convenient, as a ***signed area*** (Figure 4-5 on page 106) the following results should be easy to accept.

Theorem 4.5: Properties of definite integrals

If the real valued function, g, of a real variable has an integral from a to b, and c lies between a and b, then the three integrals below are meaningful and have the indicated *interval additivity* relation:

$$\int_a^b g = \int_a^c g + \int_c^b g. \qquad \textit{4.16}$$

If the functions f and g both have an integral from a to b, and r and s are any real numbers, then the three integrals below are meaningful, and have the indicated relation.

Linear operator property

$$\int_a^b (rf + sg) = r\int_a^b f + s\int_a^b g \qquad \textit{4.17}$$

If, in addition, $f \le g$ on the interval $[a, b]$ (i.e., $f(x) \le g(x)$ for all x in the closed interval from a to b, $a \le b$), then

$$\int_a^b f \le \int_a^b g. \qquad \textit{4.18}$$

❑

Inequality 4.18 is just the extension of the algebraic rule for *adding inequalities*:

If $A \le B$ and $C \le D$, then $A + C \le B + D$.

It can be very useful for obtaining bounds for integrals, where it may be difficult to evaluate $\int_a^b g$, but we may be able to find functions, f and F satisfying the condition $f \le g \le F$ whose integrals $\int_a^b f$ and $\int_a^b F$ are easy to evaluate. If these two integrals are close enough to each other, then we have effectively evaluated the difficult integral $\int_a^b g$, using, say, the average of $\int_a^b f$ and $\int_a^b F$ to estimate $\int_a^b g$, with an error not exceeding

$\frac{1}{2}\left(\int_a^b F - \int_a^b f\right)$. Here is an illustration of such a case.

Example 4.10: Reasonably accurate, simple estimate of

$$\int_{15.4}^{35.1} \frac{1}{x^3+1} dx$$

We know from the continuity Theorem 2.10 on page 38 that the functions (whose formulas are) $1/x^3$ and $1/(x^3+1)$ are continuous for $x > 0$, and it is evident that

$$0 < \frac{1}{x^3} - \frac{1}{x^3+1} = \frac{1}{x^3(x^3+1)} \le \frac{1}{x^6} \quad \text{for} \quad x > 0.$$

Since $1/x^6$ also represents a continuous function, it follows from Theorem 4.3 on page 103, that these functions are *integrable* over the interval [15.4, 35.1]. From inequality 4.18 of Theorem 4.5 above, it follows that

$$0 = \int_{15.4}^{35.1} 0\, dx \le \int_{15.4}^{35.1} \left[\frac{1}{x^3} - \frac{1}{x^3+1}\right] dx \le \int_{15.4}^{35.1} \frac{1}{x^6} dx.$$

Hence from the linear operator property of integrals, equation 4.17 on page 107, we find

$$0 \le \int_{15.4}^{35.1} \frac{1}{x^3} dx - \int_{15.4}^{35.1} \frac{1}{x^3+1} dx \le \int_{15.4}^{35.1} \frac{1}{x^6} dx,$$

or, rearranging

$$\int_{15.4}^{35.1} \frac{1}{x^3} dx - \int_{15.4}^{35.1} \frac{1}{x^6} dx \le \int_{15.4}^{35.1} \frac{1}{x^3+1} dx \le \int_{15.4}^{35.1} \frac{1}{x^3} dx \qquad \textit{4.19}$$

It is easy to verify that if $f(x) = -1/(2x^2)$ then $f'(x) = \frac{1/x^3}{f}$.

So, by the fundamental integral evaluation formula, 4.14 on page 103,

$$\int_{15.4}^{35.1} \frac{1}{x^3} dx = \int_{15.4}^{35.1} f'(x)\, dx = f(35.1) - f(15.4) = -\frac{1}{2(3.1)^2} + \frac{1}{2(15.4)^2}.$$

Similarly, letting $\varphi(x) = -1/(5x^5)$, we find $\varphi'(x) = 1/x^6$, so that

$$\int_{15.4}^{35.1} \frac{1}{x^6} dx = -\frac{1}{5(35.1)^2} + \frac{1}{5(15.4)^2}.$$

Substituting into inequalities 4.19, and evaluating each of the integrals as a decimal, we find

$$0.0017024 \le \int_{15.4}^{35.1} \frac{1}{x^3+1}\,dx \le 0.0017024 + 0.0000006. \qquad \textbf{4.20}$$

So as our estimate of $\int_{15.4}^{35.1} \frac{1}{x^3+1}\,dx,$ we use 0.0017027, the average of the left and right sides of inequalities 4.20. An upper bound for the magnitude of the error of this estimate is half the difference of these two sides, 0.0000003. The magnitude of the *relative error* of our estimate (see Exercise 1.10 on page 19) does not exceed 0.0000003/0.0017027 = 0.000176, showing our estimate to be very accurate indeed for most purposes.

■

Exercises 4.11

Using Example 4.10 as a guide, find reasonably accurate estimates of the following integrals:

1. $\int_{15}^{50} \frac{1}{x^3+x}\,dx.$

2. $\int_{18}^{74.1} \frac{1}{x^3-1}\,dx.$

3. $\int_{15}^{50} \frac{1}{x^3-x}\,dx.$

4. $\int_{16}^{30} \sqrt{x^3+1}\,dx$ Hint: for $x > 0$

$$0 \le \sqrt{x^3+1} - \sqrt{x^3} = (\sqrt{x^3+1} - \sqrt{x^3})\,\frac{\sqrt{x^3+1}+\sqrt{x^3}}{\sqrt{x^3+1}+\sqrt{x^3}} \le \frac{1}{2\sqrt{x^3}} = \frac{1}{2x^{3/2}}.$$

5. $\int_{17}^{30} \frac{1}{\sqrt{x^3+1}}\,dx$ Hint: $\left|\frac{1}{\sqrt{x^3+1}} - \frac{1}{\sqrt{x^3}}\right| = \left|\frac{\sqrt{x^3} - \sqrt{x^3+1}}{\sqrt{x^3+1}\sqrt{x^3}}\right|.$

Now see hint from Exercise 4, above.

4 Overview of Proper Integral Development

Our first approach to defining the integral, the one suggested by expression 4.10 on page 99, is deficient in several major aspects:

- First, there are functions, f, for which the limit,

  $$\sum_{k=0}^{n-1} f\left(a + k\frac{b-a}{n}\right)\frac{b-a}{n},$$

  exists, where you wouldn't want the integral, $\int_a^b f$ to exist. Such a function will be exhibited in the optional part to follow. A proper definition of integral should exclude such possibilities.
- Second, expression 4.10 doesn't lend itself to establishing useful practical rules for numerical evaluation of integrals.
- Both expression 4.10 and even Definition 4.2 on page 102, which overcomes the first two objections, aren't very good for simple justified theoretical manipulation of integrals. Actually, even the *upper-sum lower-sum* approach that we use for our theoretical development in Appendix 3 to this chapter, is far from being best for theoretical uses. The ***Lebesgue integral*** — in contrast to the Riemann integral presented here — is much simpler for theoretical purposes. We'll get a glimpse of why this is so in Section 5 of this chapter.

The approach to developing the results that are needed to allow adequate use of the Riemann integral in applications is based on the concept of *upper and lower sums.* For a continuous bounded function, see Theorem 3.12 on page 88, the upper sum corresponding to a given partition is the largest possible Riemann sum associated with that partition (see Definition 4.1 on page 101), and the *lower sum* is the smallest Riemann sum. We illustrate upper and lower sums in Figure 4-6. The total signed area indicated by

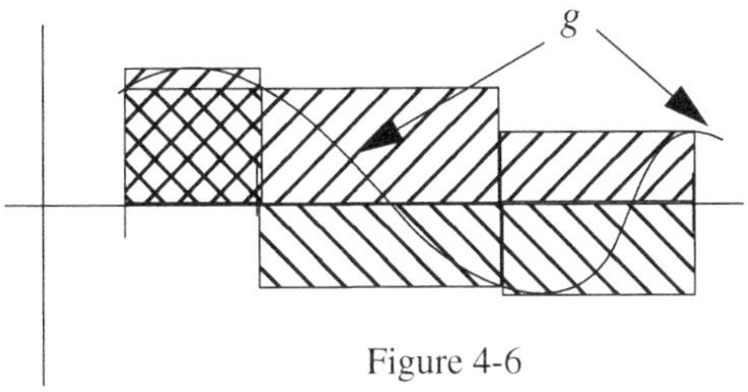

Figure 4-6

any  marking corresponds to the upper sum, while the signed area indicated by any  marking corresponds to the lower sum associated with the given partition. The Riemann integral can be shown to exist if and only if the difference between the upper and lower sum can be made arbitrarily small by suitable choice of partition. From a geometrical sense this claim is easy to accept. The advantage of using lower and upper sums for theoretical work with the integral is simply that there are only two numbers that need to be examined when dealing with any partition of the interval of integration (rather than all possible Riemann sums associated with the partition). However, the details of its establishment are not easy to follow, which is why they are relegated to Appendix 3.

In the optional part which concludes this section we justify the claim that the limit of expression 4.10 on page 99 would not be suitable as the definition of the integral, by means of an example.

The function *f,* to be defined in Example 4.12 below, is not one that you'd expect to meet in practical applications. However, theoretical structures, such as the real numbers, are not really met in real-life computation, and therefore, we must know how to handle such *unpractical* cases, if we are to get the benefit of these tools for easy establishment of useful results.

To get some idea of the significance of such theoretical tools, notice that digital computers don't use any numbers but rationals, and only a finite number of them. So, you might think it would be better to restrict all mathematical discussions to a finite set of rationals, maybe all decimals from $-10^{1,000,000,000}$ to $10^{1,000,000,000}$, having, say, $10^{1,000,000}$ digits to the right of the decimal point. This should take care of all foreseeable practical computational needs; but it would make it incredibly difficult to establish many useful numerical results; in fact, with such restrictions, we couldn't even establish the Pythagorean theorem, because there would be available values a, b for which $a^2 + b^2$ wouldn't be available. Furthermore, there are moderate size numbers, a, b for which $a^2 + b^2$ doesn't have a rational square root,[1] e.g., $a = 1, b = 1$. The lack of rational square roots of some rationals, as well as the absence of quantities such as π, e, etc., makes restriction to the rationals too inconvenient for practical development of results — i.e., too inconvenient for useful theory.

Example 4.12: Case where $\lim_{n \to \infty} \sum_{k=1}^{n} f\left(a + k\frac{b-a}{n}\right)\frac{b-a}{n}$ ***shouldn't*** $= \int_a^b f$

For $0 \leq x \leq 1$, let

$$f(x) = \begin{cases} 0 & \text{for } x \text{ rational,} \\ 1 & \text{otherwise.} \end{cases} \qquad \textbf{\textit{4.21}}$$

This function isn't one you're likely to meet in practical computations, but if, as already stated, you want the advantages of simple establishment of results that the real numbers provide, you must be able to handle such weird functions adequately.

Were it to exist, with the limit of expression 4.10 motivating its definition, the integral $\int_0^1 f$ would be the limit as $n \to \infty$ of

$$\sum_{k=0}^{n-1} f(a + k((b-a)/n))\frac{b-a}{n} \quad \text{with } a = 0 \text{ and } b = 1.$$

But $a + k\dfrac{b-a}{n} = a + k\dfrac{1-0}{n} = \dfrac{k}{n}$, a rational number. Hence,

$$\sum_{k=0}^{n-1} f(a + k((b-a)/n))\frac{b-a}{n} = \sum_{k=0}^{n} 0 = 0.$$

Thus, if $\int_0^1 f$ were to exist, using $\lim_{n \to \infty} \sum_{k=1}^{n} f(a + k(b-a)/n) \cdot \dfrac{b-a}{n}$ as its definition would lead to the result that $\int_0^1 f = 0$.

So, what's so bad about this? Mainly, if we calculate $\int_0^{1/\sqrt{2}} f$ and $\int_{1/\sqrt{2}}^1 f$ since $1/\sqrt{2}$ is irrational, for $k > 0, n > k$ we have

$$0 + k\frac{1 - 1/\sqrt{2}}{n} \quad \text{and} \quad \frac{1}{\sqrt{2}} + k\frac{1 - 1/\sqrt{2}}{n}$$

are irrational (see Exercise 4.13.1), and hence,

---

1. A short proof of this is as follows: suppose m and n were integers for which $(m/n)^2 = 2$. It can be shown (and it's not hard to convince yourself), that we can factor m into a unique product $m_1 m_2 \cdots m_k$, of integers, m_i, where

   - each $m_i > 1$;
   - except for the order in which they are written, the m_i are unique (e.g., $6 = 3*2$, $18 = 3*2*3$.) Similarly, $n = n_1 n_2 \cdots n_j$.

   Writing $(m/n)^2 = 2$ as $m^2 = 2n^2$ and substituting for m and n, we find

   $$(*) \qquad m_1 m_2 \cdots m_k m_1 m_2 \cdots m_k = 2 n_1 n_2 \cdots n_j n_1 n_2 \cdots n_j$$

   The left and right sides of (*) are two ***distinct*** ways of factoring the same integer — distinct, because the left side has an even number of 2's and the right side has an odd number of 2's. But because of the unique factorization property of integers, this cannot be; what led to this was the assumption that $(m/n)^2 = 2$. Hence, this assumption cannot be valid, which means that ***2 cannot have a rational square root.*** It has a ***real square root,*** represented by the infinite decimal 1.414..., whose digits can be found, one at a time via *decisection* (see page 75); this number can also be found via Newton's method (see Section 4 of Chapter 3).

$$\int_0^{1/\sqrt{2}} f = \frac{1}{\sqrt{2}} \quad \text{and} \quad \int_{1/\sqrt{2}}^{1} f = 1 - \frac{1}{\sqrt{2}}.$$

Thus we would have

$$\int_0^{1/\sqrt{2}} f + \int_{1/\sqrt{2}}^{1} f = 1 \quad \text{but} \quad \int_0^1 f = 0$$

i.e., the *area property,*

$$\int_a^b g = \int_a^c g + \int_c^b g$$

would not hold.

To summarize, with $\lim_{n \to \infty} \sum_{k=1}^{n} f(a + k(b-a)/n) \cdot \frac{b-a}{n}$ defining $\int_a^b f$, we could not satisfy the desirable *area property*, embodied in equation 4.16. This would lose us too much, and hence is not acceptable. For this reason we use the Riemann sum definition of integral embodied in expression 4.12 which is used in Definition 4.2 on page 102. ***With the Riemann sum definition, the function f defined by equation 4.21 on page 111, given in Example 4.12, does not have an integral*** (see Exercise 4.13.2). The ***Lebesgue integral***, which we will briefly discuss in Section 5 of this chapter, assigns integral 1 to f from 0 to 1, basically because there *many more* irrationals than there are rationals.

■

Exercises 4.13

1 Convince yourself that $0 + k\frac{1/\sqrt{2} - 0}{n}$ and $\frac{1}{\sqrt{2}} + k\frac{1 - 1/\sqrt{2}}{n}$ are irrational (not rational) numbers for k a positive integer, and n a positive integer exceeding k.

Hint: Were either of these numbers rational, you should show that it would follow that $\sqrt{2}$ is rational.

2. Show that the function f given in equation 4.21 does not have a Riemann integral, since choosing the numbers X_i rational always leads to a Riemann sum, equal to 0, while choosing them irrational always leads to a Riemann sum of 1.

3. Show that an unbounded function with domain $[a;b]$, $a < b$ (i.e., a function which is not bounded (see Definition 3.10 on page 88) cannot have a Riemann integral.

Hint: No matter what the partition, X_i can be chosen in expression 4.12 on page 102 so that this expression has arbitrarily large magnitude.

5 The Lebesgue Integral

Although the level of mathematical depth in this work only permits use of the Riemann integral, it is worth discussing the advantages and ideas behind a preferable one for theory, namely, the (abstract) Lebesgue integral. Basically the Lebesgue integral is, like the Riemann integral, a limit of sums. It is somewhat more general, but its big advantage is the ease of its theoretical manipulation.[1] The kind of manipulation we're referring to involves *reversing the order of limiting operations.* The particular type of reversal we have in mind is of the form

$$\lim_{h} \int_a^b C_h(x)dx = \int_a^b \lim_{h} C_h(x)dx.$$

There are good reasons to expect this type of result to hold, and here are a few of them. Recall the *sum rule for differentiation* (see Theorem 2.13, equation 2.28 on page 53)

$$(f+q)' = f' + q'$$

From the definition of derivative, see Definition 2.5 on page 30, we can write this equation as

$$\lim_{h\to 0}\left[\frac{f(x+h)-f(x)}{h} + \frac{q(x+h)-q(x)}{h}\right] = \lim_{h\to 0}\frac{f(x+h)-f(x)}{h} + \lim_{h\to 0}\frac{q(x+h)-q(x)}{h}$$

or, abbreviating, as

$$\lim_{h\to 0}[F_h(x) + Q_h(x)] = \lim_{h\to 0} F_h(x) + \lim_{h\to 0} Q_h(x)$$

An integral is essentially a sum, so if for each x in some interval $[a;b]$ $\lim_{h\to 0} C_h(x)$ exists (together with all the other objects we're about to write), we should expect, or at least not be surprised to find, in light of the above equation, that

$$\lim_{h\to 0} \int_a^b C_h(x)\,dx = \int_a^b \lim_{h\to 0} C_h(x)dx \qquad \textbf{4.22}$$

This is even more intuitively plausible using the area interpretation of integrals (see Figure 4-5 on page 106). Geometrically the existence of $\lim_{h\to 0} C_h(x)$, which we will denote by $C(x)$, means that for small h the

---

1. Just as the real numbers are better for theoretical purposes than the rationals.

graph of the function C_h should, in some sense, be close to the graph of the function C, as illustrated in Figure 4-7.

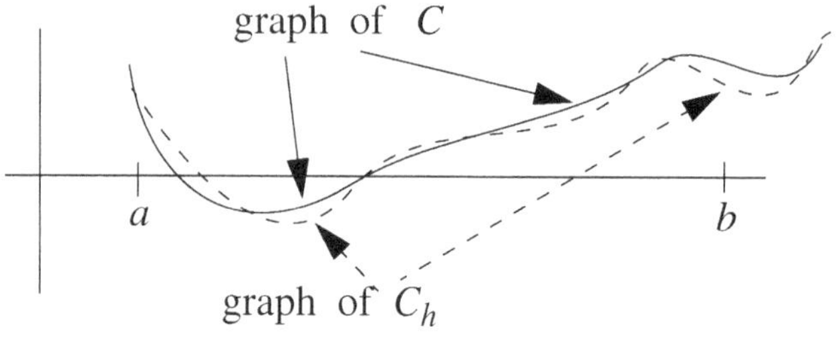

Figure 4-7

But then you would expect the signed area between the graph of C_h and the x axis to be close to the signed area between the graph of C and the x- axis for small h; i.e., you would expect

$$\int_a^b C_h \text{ to be close to } \int_a^b C \text{ for small } h.$$

That is, from geometrical considerations you would expect equation 4.22 to hold.

The importance of a result like equation 4.22 comes about in situations where a function is defined as an integral of the sort given in Example 4.7 on page 101. In particular, suppose $g(t,x)$ represents the mass density of some fluid in a blood vessel at a point x, at time t. Then $G(t)$ defined by the formula

$$G(t) = \int_a^b g(t, x)dx$$

is the total fluid mass between points a and b at time t. To get a handle on how this mass behaves as time passes, we might want to determine $G'(t)$. Now from the definition of derivative, and the *linear operator property* of integrals, equation on page 107

$$G'(t) = \lim_{h \to 0} \frac{G(t+h)-G(t)}{h} = \lim_{h \to 0} \int_a^b \frac{g(t+h, x)-g(t, x)}{h} dx.$$

It would often be convenient if, justified by equation 4.22, we could simply claim that for each t, identifying $C_h(x)$ with $[g(t+h, s)-g(t, x)]/h$, the above limit is

$$\int_a^b \lim_{h \to 0} \frac{g(t+h, x)-g(t, x)}{h} dx,$$

because, as you might notice, the integrand computation here is that of computing a derivative — acting as if the variable x is just a constant.

In essence, the Lebesgue integral builds this capability into its definition. Later, in Example 12.14 on page 478 we'll see how satisfaction of equation 4.22 will greatly simplify evaluation of certain integrals. For this we will also need to establish conditions under which equation 4.22 holds for Riemann integrals — conditions that are more restrictive than those for Lebesgue integrals (see Theorem 12.26 on page 477).

To reiterate, the Lebesgue integral is a limit of sums, in which useful properties, such as equality 4.22 are easier to come by, because they are essentially built into the definition.

It turns out that the Lebesgue integral can be naturally developed from a probability viewpoint, with equation 4.22 as its defining property.

6 Elementary Numerical Integration

In the material presented thus far, we still have not shown exactly how to obtain precise numerical estimates for most integrals. This section will overcome this omission, although not in the most efficient manner.

The easiest functions amenable to simple derivation of a numerical integration formula are the *bounded monotone* functions. Theorem 4.4 on page 104 will be used to obtain the desired formula. So, suppose that g is monotone increasing on $[a, b]$. Divide $[a, b]$ into n equal-length subintervals as shown in Figure 4-8.

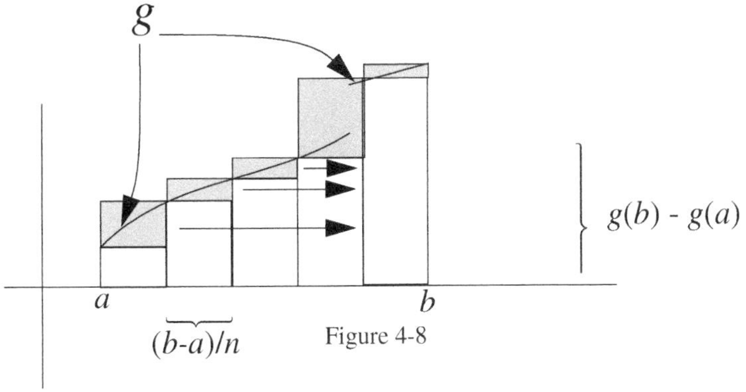

Figure 4-8

It is evident from this figure that the difference between any two Riemann sums cannot exceed the heavily shaded areas ▭. By sliding these areas to all lie over the last interval, we see that the total heavily shaded areas total up to $\frac{b-a}{n}(g(b) - g(a))$. So we see that if R_n and R_n^* are any two Riemann sums associated with the partition we have chosen,

$$\left|R_n - R_n^*\right| \le \frac{b-a}{n}[g(b) - g(a)] \equiv d_n.$$

Thus from Theorem 4.4 on page 104 we have shown that if we partition the interval $[a. b]$ into n equal length subintervals of length $(b-a)/n$, then the monotone increasing function g has an integral, and furthermore

$$\left|R_n - \int_a^b g\right| \le \frac{b-a}{n}[g(b) - g(a)].$$

The general result for monotone functions is the following.

Theorem 4.6: Integral of monotone function

If g is a real valued monotone function with the closed bounded interval $[a; b]$ as its domain, then g is integrable on $[a; b]$. If R_n is any Riemann sum associated with a partition of $[a; b]$ into n equal length subintervals, then

$$\left|R_n - \int_a^b g\right| \le \frac{b-a}{n} \mid g(b) - g(a)\mid \qquad \textit{4.23}$$

❑

We can now choose n to make d_n as small as desired, and any associated Riemann sum, R_n will be within d_n of $\int_a^b g$.

Example 4.14: Determination of $\int_1^2 \frac{1}{x}dx$ to within .01

Here $g(x) = 1/x$, and we find $\frac{b-a}{n}\mid g(b) - g(a)\mid = \frac{1}{2n} = d_n$. Choosing $n = 50$ yields $d_n = .01$. Thus the Riemann sum

$$R_n = \frac{1}{50}\sum_{i=1}^{50}\frac{1}{X_i} = \frac{1}{50}\sum_{i=1}^{50}\frac{50}{50+i}$$

is satisfactory for our purposes.[1]

■

---

1. If we use the *Mathematica* command 1/50 Sum[50/(50 + i),{i,1,50}] to try to carry out this summation operation, we obtain the fraction

$$\frac{479796225641557869184786090396628981226 17}{697203752297124771645338089353123035568 00},$$

which is not exactly what was desired. The *Mathematica* command

1/50 N[Sum[50/(50+i),{i,1,50}]]

yields the result .688172 (The N function in *Mathematica* yields decimal approximations to its arguments.) To six digits the desired integral is .693147, as we will see when the natural logarithm is treated (see Definition 5.2 on page 137).

Exercises 4.15

1. Approximate $\int_1^3 \frac{dx}{x^3+1}$ to within .01 using Theorem 4.6.[1]

2. Find an n adequate to obtain accuracy .00001 for exercise 1 above.

The majority of functions that we will want to integrate numerically will be differentiable with a bounded derivative. So, suppose g is differentiable with

$$|g'(x)| \leq B \qquad \textbf{4.24}$$

for all x in $[a; b]$. Let $[x_i ; x_i + h_i]$ be a typical interval in some partition of $[a; b]$. As illustrated in Figure 4-9, on $[x_i ; x_i + h_i]$ you would expect the graph of g to lie between the two indicated line segments L_l and L_u, of slope

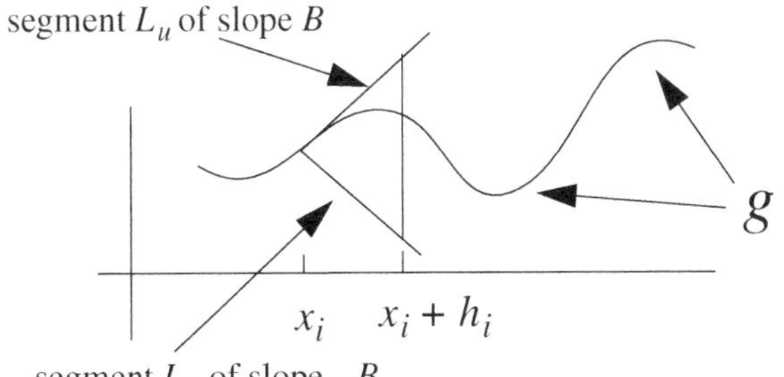

Figure 4-9

$-B$ and B respectively, because L_u rises no slower than g while L_l rises no faster.[2] So the contribution of this interval to any Riemann sum must lie between the areas under the two line segments, L_l, L_u. If all intervals of the partition are of length $(b-a)/n$, it's easy to see from Figure 4-10 that the

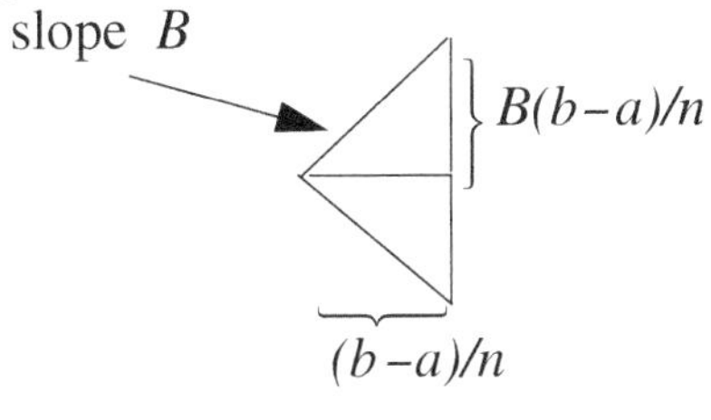

Figure 4-10

---

1. checking your results with *Mathematica*'s NIntegrate or the equivalent
2. A precise argument is simply that $(L_u - g)'(x) = L_u'(x) - g'(x) = B - g(x) \geq 0$; so $f = L_u - g$ is non-decreasing, see Theorem 3.3 on page 66. Since we've set things up so that $L_u(x_i) - g(x_i) = 0$, it follows that $L_u(x) \geq g(x)$ for $x > x_i$, i.e., $L_u(x) - g(x) \geq 0$.

difference between these two areas on any one interval cannot exceed the indicated triangular area, which is

$$\frac{b-a}{n}B\frac{b-a}{n} = B\frac{(b-a)^2}{n^2}.$$

The difference between any two Riemann sums cannot exceed the sum of the areas of the triangles formed by such segments, which is

$$\sum_{j=1}^{n} B\frac{(b-a)^2}{n^2} = B\frac{(b-a)^2}{n} = d_n.$$

So from Theorem 4.4 on page 104, we see the following.

Theorem 4.7: Integration of differentiable function

If g is a real valued differentiable function of one real variable on the closed bounded interval $[a; b]$, with $|g'(x)| \leq B$ for all x in $[a, b]$, then g is integrable on $[a, b]$; if R_n is any Riemann sum associated with a partition of $[a, b]$ into n equal length subintervals, then

$$\left|R_n - \int_a^b g\right| \leq B\frac{(b-a)^2}{n}. \qquad 4.25$$

❑

The art of using this type of result, as well as others which will be introduced, is determining good upper bounds for (in this case) $|g'|$. We will provide only one example here, waiting to develop such tools when the problems get more interesting.

Example 4.16: Application to the function g of Example 4.14

In this example, $g(x) = 1/x$ and $[a, b] = [1, 2]$. Since $g'(x) = -1/x^2$, we see that $|g'(x)| \leq 1$ on $[a, b]$. So from inequality 4.25 of Theorem 4.7 just preceding we find

$$\left|R_n - \int_1^2 dx/x\right| \leq \frac{1}{n}$$

(not as good an answer as with using the monotone function result, *but not all functions are monotone*).

So, to guarantee an estimate of this integral to an accuracy of .01 using Theorem 4.7 would require a partition into $n = 100$ (equal-length) subintervals.

■

Exercises 4.17

Redo Exercises 4.15, but using Theorem 4.7 in place of Theorem 4.6.

7 Integration via Antidifferentiation

Section 2 of this chapter emphasized use of the summation process (*integration*, see the discussion near Definition 4.2 on page 102) to find an antiderivative, f, of f' — i.e., calculating f whose derivative is the specified function designated as f'. In this section we reverse that approach, and investigate the determination of integrals (i.e., functions produced by the summation process), by antidifferentiation, which consists of using our knowledge of derivatives to determine $\int_a^b g(x)dx$, by finding a function G whose derivative $G' = g$. The technique we are developing is an exact parallel to the one used for evaluation of sums,[1] (see Theorem 1.8 on page 12 ff.). It depends on equality 4.14 on page 103 as follows.

In our desire to evaluate $\int_a^b g$, using our knowledge of derivatives (and, in particular, the differentiation rules developed in Chapter 2, summarized in Theorem 2.13 on page 53) we try for direct determination of f satisfying $f' = g$. If we are successful, then by equation 4.14

$$\int_a^b g = \int_a^b f' = f(b) - f(a).$$

To cut down on the writing when $f(x)$ or $f(x, y)$ is given by a complicated formula, we introduce

Definition 4.8: $f(x)\Big|_a^{x=b}$, $f(x)\Big|_a^{b}$, $f(x, y)\Big|_a^{x=b}$

The first two of these symbols stand for $f(b) - f(a)$. The last one stands for $f(b, y) - f(a, y)$.

❑

Now to proceed to apply the differentiation rules in reverse to determine integrals. The first of the differentiation rules, Theorem 2.13 equation 2.28 on page 53, is

$$(rF + sG)'(x) = rF'(x) + sG'(x) \quad .$$

Using equality 4.14 on page 103 with $f = rF + sG$ yields

$$\int_a^b (rF+sG)'(x)dx = [rF(x)+sG(x)]\Big|_a^{x=b} = rF(x)\Big|_a^{x=b} + sG(x)\Big|_a^{x=b} = r\int_a^b F' + s\int_a^b G'$$

---

1. And as in the case of sums, there are very few functions whose integral can be evaluated by a simple formula. Nonetheless, as with sums, the relatively few integrals which can be so evaluated can prove quite useful; recall Example 4.10 on page 108.

(the latter steps justified by again using equality 4.14, first with $f = F$, and then with $f = G$). So we have

$$\int_a^b rF + sG' = r\int_a^b F' + s\int_a^b G'.$$

Now we get rid of the unnecessary derivatives, by substituting $f = F'$ and $g = G'$ to obtain

$$\int_a^b rf + sg = r\int_a^b f + s\int_a^b g.$$

But this is the same as the *linear operator property,* on page 107 introduced earlier. So equation 2.28 doesn't generate anything new.

On the other hand, Theorem 2.13 equation 2.29 on page 53 (the product rule) yields

$$\int_a^b (fg)' = \int_a^b f'g + \int_a^b fg'.$$

By the fundamental evaluation formula, equation 4.14 on page 103 (with fg in place of f) this becomes $f(b)\,g(b) - f(a)\,g(a) = \int_a^b f'g + \int_a^b fg'.$

We rewrite this, using the abbreviation of Definition 4.8 above, as

Theorem 4.9: Integration by parts formula

$$\int_a^b f'g = f(x)g(x)\Big|_a^{x=b} - \int_a^b fg',$$

valid when f' and g' are continuous.

❑

This formula is useful when we have difficulty evaluating $\int_a^b f'g$ but easily recognize the value of $\int_a^b fg'$.

Example 4.18: Evaluation of $\int_0^2 x\sqrt{1+x}\,dx$ using integration by parts

Here we have the choice of which factor to call g and which to call f'. Examination of the integrand indicates that if we want the right side of the integration by parts formula to be *simpler* than the left side (that we started with), it's reasonable to choose g so that g' is *simpler* than g, hoping that f is no more complicated than f'. Our choice is indicated below, where we also systematically carry out the necessary computation.

Apply the integration by parts formula, Theorem 4.9 just established, to the integrand above, choosing $g(x) = x$, $f'(x) = \sqrt{1+x}$, $a = 0$, $b = 2$, first noting that

$$g'(x) = 1, \quad f(x) = \frac{2}{3}(1+x)^{3/2}$$

(using Definition 4.8 on page 120 again), to obtain

$$\int_0^2 x\sqrt{1+x}\,dx = \frac{2}{3}(1+x)^{3/2}x\Big|_0^{x=2} - \int_0^2 \frac{2}{3}(1+x)^{3/2}\cdot 1\,dx$$

To evaluate the last integral on the right, note that because the derivative of $\frac{2}{3}\frac{2}{5}(1+x)^{5/2}$ is $\frac{2}{3}(1+x)^{3/2}$, from the fundamental evaluation formula, equation 4.14 on page 103, again using Definition 4.8,

$$\int_0^2 \frac{2}{3}(1+x)^{3/2}dx = \frac{2}{3}\frac{2}{5}(1+x)^{5/2}\Big|_0^{x=2}.$$

Hence

$$\int_0^2 x\sqrt{1+x}\,dx = \left[\frac{2}{3}(1+x)^{3/2}x - \frac{2}{3}\frac{2}{5}(1+x)^{5/2}\right]\Bigg|_0^{x=2}. \qquad \textbf{4.26}$$

Carrying out the evaluations indicated above, we find

$$\int_0^2 x\sqrt{1+x}\,dx = \frac{8}{15}3^{3/2} + \frac{4}{15}.$$

If we let $f(x)$ denote the formula on the right of equation 4.26, i.e.,

$$f(x) = \frac{2}{3}(1+x)^{3/2}x - \frac{2}{3}\frac{2}{5}(1+x)^{5/2}, \qquad \textbf{4.27}$$

then we know from the fundamental evaluation formula, 4.14 on page 103 that $f'(x) = x\sqrt{1+x}$ (which you can verify directly). So using the product rule for derivatives in reverse enabled computation of an antiderivative[1] of $f'(x) = x\sqrt{1+x}$, namely expression 4.27.

■

The final technique of antidifferentiation uses the *chain rule*, Theorem 2.13 equation 2.30 on page 53. Together with two applications of the fundamental evaluation formula for integrals, equation 4.14 on page 103, this leads to the formula:

---

1. This really gets us all of the antiderivatives of $x\sqrt{1+x}$, since we know from Problem 3.5.2 on page 67 that the difference between any two functions having the same derivative function must be constant. Thus each antiderivative is the above plus some constant.

$$\int_a^b H'(F)F' = \int_a^b [H(F)]' = H(F(x))\Big|_a^{x=b} = \int_{F(a)}^{F(b)} H' \qquad 4.28$$

We can get from the first integral to the last (at which point it's easier in many cases to recognize the desired antiderivative) via the following formal manipulation, which can often be carried out successfully, even when you don't have the faintest notion of what's going on. Namely, you *substitute*

$$t = F(x), \qquad dt = F'(x)dx \qquad 4.29$$

and make the appropriate substitution of limits of integration for t, namely, $F(a)$ and $F(b)$. With luck you may be able to evaluate what you get more easily than what you started with. This method is referred to as ***integration by substitution***. Let's look at an example.

Example 4.19: Evaluation of $\int_0^2 x\sqrt{1+x}\,dx$ via *substitution*

Try substituting $t = 1 + x = F(x)$. Then, since

$$x = t - 1, \quad dt = dx = F'(x)dx,$$

while

$$f(a) = F(0) = 1 + 0 = 1, \quad F(b) = F(2) = 1 + 2 = 3,$$

we find

$$\int_0^2 x\sqrt{1+x}\,dx = \int_1^3 (t-1)\sqrt{t}\,dt = \int_1^3 (t^{3/2} - t^{1/2})dt = \left[\frac{2}{5}t^{5/2} - \frac{2}{3}t^{3/2}\right]\Bigg|_1^{t=3},$$

the last step coming from equation 4.14 on page 103 (the ubiquitous fundamental integral evaluation theorem).

Since $t = 1 + x$, we also see that

$$\int_0^2 x\sqrt{1+x}\,dx = \left[\frac{2}{5}(1+x)^{5/2} - \frac{2}{3}(1+x)^{3/2}\right]\Bigg|_0^{x=2},$$

from which we conclude that

$$\frac{2}{5}(1+x)^{5/2} - \frac{2}{3}(1+x)^{3/2} \qquad 4.30$$

is an antiderivative of $\quad f'(x) = x\sqrt{1+x}$.

■

Now expression 4.30 does not look the same as expression 4.27, even though these expressions are both antiderivatives of $f'(x) = x\sqrt{1+x}$. But we know that their difference must be constant (recall the footnote on page

122). If we rewrite expression 4.27, we can verify that it is equal to expression 4.30, as follows:

$$\frac{2}{3}x(1+x)^{3/2} - \frac{2}{3}\frac{2}{5}(1+x)^{5/2} = \frac{2}{3}(1+x-1)(1+x)^{3/2} - \frac{2}{3}\frac{2}{5}(1+x)^{5/2}$$

$$= \quad \frac{2}{3}(1+x)^{5/2} - \frac{2}{3}(1+x)^{3/2} - \frac{2}{3}\frac{2}{5}(1+x)^{5/2}$$

$$= \quad \frac{2}{5}(1+x)^{5/2} - \frac{2}{3}(1+x)^{3/2}$$

The *Mathematica* program can be used both to carry out antidifferentiation and to compare symbolic answers to see if they are equal, see exercises 4.20.5 on page 125 and 4.22.4 on page 129.

In using substitution, notice that you often need to *solve* the equation $t = F(x)$ for x in terms of t. If you are unable to do this, you may have to give up this particular substitution.

Before leaving this section some comments are in order.

- Most functions you encounter in real applications don't have antiderivatives whose formulas can be determined by either of the preceding methods. So don't be misled into thinking these techniques are going to solve most of your integration problems. Nonetheless, they are worth learning for the reasons presented next. Even when they work, there are often easier ways to obtain the same results, as we will see in Example 12.14 on page 478.
- One use of these methods is to develop exact formulas to use as test cases for various computer packaged numerical integration methods. (*Mathematica* actually cautions against uncritical use of its own such packages.)
- These methods are of importance in transforming integrals to forms which may be more suited to theoretical or numerical investigations.
- It doesn't seem worth spending much effort practicing integration techniques in a vacuum. A great deal of time is spent doing just that in many elementary calculus courses, time which could be spent more profitably on other topics. Techniques of integration by antidifferentiation are learned best by being used when they are needed in real problems. The necessity for

having great facility in this area has been diminished substantially by the existence of packaged programs, such as *Mathematica*, *Maple*, and *Splus* that are now widely available.

- Having used either of the previous techniques (integration by parts, and integration by substitution), it is worth writing the result as

$$f(x)\Big|_a^{x=b}$$

and differentiating f to make sure it yields the *integrand* (function being integrated).

A few exercises are supplied to help those who desire to verify that they can properly apply these methods.

Exercises 4.20.

Evaluate the following integrals and verify by differentiation that the function obtained for this evaluation is an antiderivative of the integrand.

1. $\int_a^b x\sqrt{1-x}\,dx$.
2. $\int_a^b x\sqrt{1+x^2}\,dx$.
3. $\int_a^b x^2\sqrt{1+x^2}\,dx$.
4. $\int_a^b x\sqrt{1+x}\,dx$ using the substitution $t = \sqrt{1+x}$.
5. For those with access to the *Mathematica* computer program (or *Maple*, or any similar program having antidifferentiation capabilities) find an antiderivative of the integrands treated in this section (including the exercises), and verify the correctness of the answers. The command to do this in *Mathematica* is

   Integrate[f(x),x] <CR>,

   where <CR> represents the *Enter key* and f(x) is the formula for the integrand of interest — e.g.,

   x^2 + 2 x Sqrt[1+x] represents the formula $x^2 + 2x\sqrt{1+x}$.

   The uppercase letters and square brackets are mandatory for any function, such as Integrate and Sqrt, *built in* to *Mathematica*. You can also use *Mathematica* to check that two answers, say, ans1 and ans2, agree, via typing

   Simplify[ans1 - ans2],

   which should evaluate to 0 if the answers are the same.

8 The Fundamental Theorem of Calculus

In many situations involving functions initially introduced by a local description, **we will be led** by a very natural development **to functions defined as integrals.** This will be the case for the exponential function (describing radioactive decay) and the trigonometric functions (which, together with the exponential functions, describe many vibrating systems and metabolic processes).

When a function, G, is defined by

$$G(x) = \int_a^x g \qquad \textit{4.31}$$

one of the first questions of interest is the determination of its derivative, G'. The answer is available almost immediately from the fundamental integral evaluation formula, 4.14 on page 103; namely, if $g = f'$, then from 4.14

$$G(x) = \int_a^x g = \int_a^x f' = f(x) - f(a),$$

and hence, since f is differentiable, looking first at the outer terms, and then using $g = f'$, we find

$$G'(x) = f'(x) = g(x).$$

The *reasoning* here may be a bit misleading, because it depends on g being the derivative of something — and if this should fail to be the case, then the conclusion that $G' = g$ could not legitimately be drawn. All of this can be put right with the help of results from Appendix 3 on page 763, where we show the existence of integrals of continuous functions whose domains are closed and bounded intervals and from the integrability of monotone functions (see Theorem 4.6 on page 117). If now g is a function which is continuous at each point in the closed bounded interval $[a;b]$, and x is strictly between a and b — i.e., $a < x < b$, then we may legitimately write (justified by Theorem 4.5 equation 4.16 on page 107) for h sufficiently close to (but not equal to) 0 so that $x + h$ is in $[a;b]$

$$\frac{G(x+h) - G(x)}{h} = \frac{1}{h}\int_x^{x+h} g(t)dt.$$

But for all t in $[x, x+h]$, by the continuity of g at x, $g(t)$ is *close* to $g(x)$, say, $g(t) = g(x) + s(t)$, where $s(t)$ is near 0 for all t in $[x, x+h]$. So

$$\frac{G(x+h) - G(x)}{h} = \frac{1}{h}\int_x^{x+h} g(x)dt + \frac{1}{h}\int_x^{x+h} s(t)dt = g(x) + \frac{1}{h}\int_x^{x+h} s(t)dt$$

Now if $L \le s(t) \le U$, where L and U both approach 0 as t is restricted to approach x, then by inequality 4.18 on page 107

$$Lh \le \int_x^{x+h} s(t)dt \le Uh$$

yielding

$$g(x) + L \le \frac{G(x+h) - G(x)}{h} \le g(x) + U \quad \text{for} \quad h \neq 0.$$

Since L and U approach 0 as h approaches 0, we see from the definition of derivative (Definition 2.5 on page 30) that $G'(x) = g(x)$. We state this result formally as

Theorem 4.10: The fundamental theorem of calculus

If g is a real valued function of a real variable, which is continuous at each point of the closed bounded interval $[a; b]$, with $a < b$, and $a < x < b$, then the function G *defined by*

$$G(x) = \int_a^x g \qquad \textbf{4.32}$$

is differentiable with

$$G'(x) = g(x) \qquad \textbf{4.33}$$

❑

In a similar manner it's not very hard to establish

Theorem 4.11: Monotone version of fundamental theorem

If g is monotone on the closed bounded interval $[a; b]$ and g is continuous at the value x_0 where $a < x_0 < b$, then G defined by equation 4.32 is differentiable at x_0 with

$$G'(x_0) = g(x_0). \qquad \textbf{4.34}$$

❑

The above fundamental theorem is readily understandable using the area interpretation of integrals, see Figure 4-5 on page 106. Namely, the rate of change, $G'(x)$, of area from a to x under g is simply $g(x)$, the height of g at x. Not too surprising, since the larger $g(x)$, the faster the area changes when x changes by a given small amount (provided g is continuous at x).

We are now in a position to supply a simple proof of the fundamental integral evaluation formula, Theorem 4.3, equation 4.14 on page 103.

From the results of Problem 3.5.2 on page 67, we know that if two differentiable functions have the same derivative, they differ by some constant. In the case being considered, from Theorem 4.10, it is seen that if $g = f'$ is continuous, then both the function f, and the function G given by

$$G(x) = \int_a^x f'(t)dt \qquad \mathbf{4.35}$$

ave the same derivative, namely f'. So it must be that they differ by some constant, say c — i.e.,

$$G - f = c. \qquad \mathbf{4.36}$$

But $G(a) = 0$, from which it follows that $-f(a) = c$. Hence, substituting into equation 4.36 we find

$$G = f - f(a),$$

i.e., for all x

$$\int_a^x f'(t)dt = G(x) = f(x) - f(a),$$

which is the conclusion we wanted to establish, proving the last part of Theorem 4.10.

We provide one computational example applying Theorem 4.10.

Example 4.21: $Arcsin(x) = \int_0^x \frac{1}{\sqrt{1-t^2}} dt, \quad -1 < x < 1$

Later we'll introduce this function in a natural manner. Here we are simply using the Arcsin function solely as an illustration of a function defined as an integral. It follows from the fundamental theorem, Theorem 4.10, that its derivative

$$Arcsin'(x) = \frac{1}{\sqrt{1-x^2}}.$$

Furthermore, if $H(x) = \int_0^{x^3} \frac{1}{\sqrt{1-t^2}} dt$, then because

$$H(x) = Arcsin(x^3)$$

it follows from the chain rule (Theorem 2.13, equation 2.30 on page 53) that

$$H'(z) = Arcsin'(x^3)3x^2 = \frac{3x^2}{\sqrt{1-x^6}}.$$

■

Exercises 4.22

1. Let $Ln(x) = \int_1^x \frac{1}{t}\,dt, \quad x > 0$

   a. Find $Ln'(x)$.

   b. For $G(x) = Ln(x^2)$ find $G'(x)$.

   c. What does this tell you about the function Ln?

2. Let $G(x) = \int_0^x t\sqrt{1+t}\,dt, \quad x > -1$.

   a. Find $G'(x)$.

   b. Let $H(x) = G(x^3)$ Find $H'(x)$ in two ways - first using the chain rule and the results of part a. above, and second by direct differentiation using the results of Example 4.18 on page 121, or Example 4.19 on page 123. (Be careful for changes in notation which may have arisen.)

3. Establish Theorem 4.11 on page 127.

4. Create your own drill exercises to practice use of the fundamental theorem of calculus, page 127, if you need to. For those with access to *Mathematica*, use it to help verify your results. Derivatives there are obtained via the D function, e.g., D[(x+1)^2,x] which yields the output 2(1 + x).

Chapter 5

Elementary Transcendental Functions

1 Introduction

From the viewpoint of modeling of physical processes, one of the most natural ways to introduce the elementary transcendental functions[1] is as solutions of the differential equations describing such processes. Recall that in Example 3.3 starting on page 57 we begin the investigation of the phenomenon of radioactive decay, resulting in equation 3.3 on page 58, which we reproduce here as

$$A'(t) = -kA(t). \qquad \textit{5.1}$$

Here the function value $A(t)$ stands for the amount of radioactive material at time t; in this model of the radioactivity process, every such function must satisfy the differential equation above. In Example 3.3 we also started intuitively investigating how this equation could furnish qualitative information about the shape of such functions, and quantitative information about their values.

Along similar lines, we also introduced differential equations used to describe enzyme kinetics — the behavior of chemicals flowing between compartments in living systems — in Example 4.5 on page 96, and in Example 4.2 on page 91 concerning accelerometers, we were led to the differential equation 4.3 on page 94 relating actual acceleration to the graph produced by a connected mechanical device, and indicated how these equations could be analyzed. Equations like these are fundamental to determining the behavior of vibrating systems, ranging from airplane wings, bridges (like the Tacoma Narrows bridge, which collapsed due to uncontrolled vibration) to quartz crystals running our digital watches.

Thus far the reasoning we used to examine such equations seems plausible, but neither quantitatively precise, nor completely trustworthy. The purpose of this chapter is to provide the solid foundation adequate to draw reliable conclusions from such differential equation descriptions. In the next section we will provide a precise development of solutions of differential equations like the radioactive decay equation above, leading to the logarithm and exponential functions. In the succeeding section a very similar development will be presented for functions arising in systems described by Hooke's law for springs (equation 4.2 on page 92) and Newton's second law of motion

---

1. These are the exponential, logarithm and trigonometric functions.

(equation 4.1 on page 92). In the next chapter, on Taylor's theorem, we will be able to connect these results in a very natural fashion, and also see how they are related to solutions of the compartmental modeling equations.

In all of this development, every attempt is made to see how the results develop naturally, from reasonable attempts to investigate the problems under consideration.

2 The Logarithm and Exponential Functions: Intuitive Development

We will initiate our study of equations related to the radioactive decay equation, 5.1, by looking at the simplest version, namely the equation and initial condition

$$x' = x, \quad x(0) = 1. \qquad \textit{5.2}$$

Our object is to determine what this equation tells us about the function x. Paralleling the discussion in Example 3.3 which began on page 57, we see that from this description we can immediately draw a reasonable picture of the function x, since the first equation in 5.2 above states that *the slope equals the height.* Starting out with a slope of 1 at 0 (from the second equation above), the graph of x rises as we move a bit to the right, and from the first equation above, the slope must increase, so that the graph of x increases at an even faster rate as we move to the right, as seen in Figure 5-1.

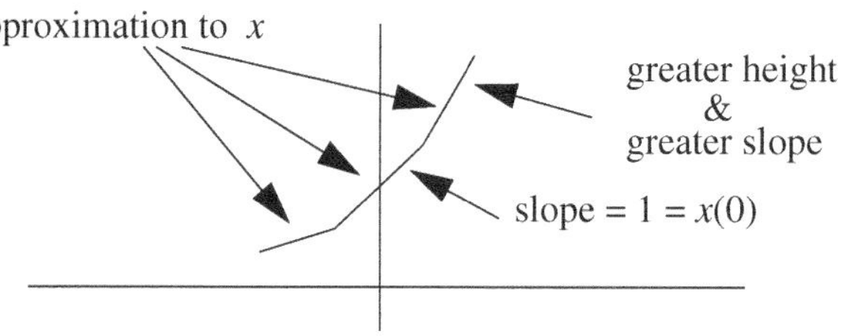

Figure 5-1

So, the slope increases as the height rises, leading to the type of picture we have drawn. We will be referring to the function described by equation 5.2 as the exponential function, denoted by *exp*. But so far we don't even know for certain that equation 5.2 really does determine a function, and even if it does, we don't know its domain — i.e., for which t the value $x(t)$ is defined. (We know from Example 3.4 on page 60, that differential equations can lead to functions which *blow up* at some values; so it should be evident that we need to develop a trustworthy approach to investigating such problems.) To get a better handle on the exponential function, we'll start out by trying for an algebraic approximate solution of equations 5.2.

Since $x'(t)$ is approximately equal to $[x(t+h)-x(t)]/h$ for small nonzero h (from the definition of *derivative*, Definition 2.5 on page 30) the equation $x' = x$ leads to the approximation $[x(t+h)-x(t)]/h \cong x(t)$, which we write as

$$x(t+h) \cong (1+h)x(t).$$

Letting $t = 0, h, 2h, \ldots$, and using $x(0) = 1$, we find that since every time we replace $t = kh$ by $t = (k+1)h$ a factor of $1 + h$ is tacked on, for n a counting number

$$x(nh) \cong (1+h)^n.$$

Now, for fixed t (positive or negative) choose n and h so that $nh = t$, i.e., $h = t/n$. Then substituting $h = t/n$ into the above approximation, we find

$$x(t) \cong (1+t/n)^n. \qquad \textbf{5.3}$$

The larger n (i.e., *the smaller* the magnitude of h), the more accurate we expect this approximation to be. Now that we have some idea about its structure, it's reasonable to start using the name *exp* (the *exponential function,* a name suggested by the exponent in approximation 5.3).

So we see that we expect

$$exp(t) \cong (1+t/n)^n, \qquad \textbf{5.4}$$

the approximation becoming arbitrarily accurate as n becomes large. Approximation 5.4 suggests that

$$exp(2t) = exp(t)\, exp(t) \qquad \textbf{5.5}$$

because

$$\begin{aligned} exp(t)exp(t) &\cong (1+t/n)^n(1+t/n)^n \\ &= [(1+t/n)^2]^n \\ &= [1+2(t/n)+t^2/n^2]^n \\ &\cong exp(2t) \end{aligned}$$

since t^2/n^2 is very small compared to $2(t/n)$ for large n.

In fact, for all u,v, positive, negative or 0, we expect

$$exp(u+v) = exp(u)exp(v), \qquad \textbf{5.6}$$

because, if $u = nh$ and $v = mh$

$$
\begin{aligned}
exp(u+v) &= exp(nh+mh)\\
&= exp[(m+n)h]\\
&\cong (1+h)^{m+n}\\
&= (1+h)^m(1+h)^n\\
&\cong exp(mh)exp(nh)\\
&= exp(u)exp(v).
\end{aligned}
$$

So far we've been deducing the properties of the exponential function using rather loose reasoning. In order to put matters on a firmer basis, we'll go one step further, loosely reasoning our way to the inverse of the exponential function, and starting our precise development from that point.

Recall Definition 2.8 on page 37 of *inverse function* and Definition 2.9 on page 37 of a *strictly monotone function.* It is geometrically evident that a real valued monotone increasing [decreasing] function, f, of a real variable has an inverse function, f^{-1} which is monotone increasing [decreasing], and that f is also the inverse function of f^{-1}. Because we will be making very specific use of this result, we will state it formally, but relegate its proof and a discussion of some of the subtleties which require care to Appendix 2 on page 761, the appendix on inverse functions.

Theorem 5.1: Inverses of strictly monotone functions

If f is a real valued strictly monotone function,(Definition 2.9 on page 37), of a real variable, it has an inverse function, g, (Definition 2.8 on page 37), whose domain[1] is the range[2] of f, which is strictly monotone of the same type as f; furthermore,

$$g(f(x)) = x \quad \text{for all } x \text{ in the domain of } f, \quad (g = f^{-1}) \qquad 5.7$$

$$f(g(y)) = y \quad \text{for all } y \text{ in the range of } f, \quad (f = g^{-1}). \qquad 5.8$$

❑

We expect the exponential function to take on only positive values, and to be defined for all real arguments, because

- for positive x, $exp(x)$ seems to be increasing, from its value, 1, at $x = 0$, and therefore positive for all $x \geq 0$ for which it is defined,
- from equations 5.2 on page 132 it certainly seems reasonable for *exp* is defined for a small interval whose

1. Collection of allowable inputs
2. Set of possible outputs

left end is the 0. Then, from equation 5.5 it will be defined for all positive reals,

- from equation 5.6, which seems to hold for setting $u = x$, $v = -x$ and using the condition $exp(0) = 1$ which arises from the second equation in equations 5.2 we expect $exp(-x) = 1/(exp(x))$, and hence the exponential function at negative values of its argument should also be defined and positive.

Also, because of the expected positivity of *exp* and the first of the equations in 5.2, we expect the exponential function to be strictly increasing and to have a positive derivative. So, using Theorem 5.1 above, we expect *exp* to have a strictly increasing inverse function, whose domain consists of the positive real numbers, and whose range consists of the set of real numbers. Now making use of Theorem 2.13 on page 53, we expect this inverse function to be differentiable. Before proceeding further, it is first convenient to supply a name[1] for the *inverse* function that we expect the *exponential* function to have; we will call this function the *natural logarithm,* and denote it by the symbol *ln*. We're now in position to determine a formula for the derivative of the natural logarithm.

From Theorem 5.1 above, with $f = exp$ and $g = ln$, using equation 5.8 of this theorem it seems proper to write

$$exp(ln(y)) = y \qquad \textbf{5.9}$$

Now applying formula 2.31 on page 53 from Theorem 2.13, we would obtain

$$exp'(ln(y))ln'(y) = 1.$$

But from the first of the equations in display 5.2 on page 132 we would have

$$exp' = exp.$$

Hence the previous equation would become

$$exp(ln(y))ln'(y) = 1.$$

Substituting from equation 5.9 would yield

$$y\,ln'(y) = 1,$$

which would finally yield the formula

$$ln'(y) = \frac{1}{y} \qquad \textbf{5.10}$$

that we've been hunting for.

1. We will do all of this formally following this intuitive development.

Again using Theorem 5.1, but this time using equation 5.7, from this theorem, we find

$$ln(exp(x)) = x.$$

But from the second of the equations in 5.2 on page 132 we know that

$$exp(0) = 1.$$

Replacing x in the preceding equation by 0 then leads to the result

$$ln(1) = 0. \qquad \textbf{\textit{5.11}}$$

Equations 5.10 and 5.11 provide the legitimate starting point for our investigation of the *exponential* and *natural logarithm* functions. You might be wondering why we have gone to all of this trouble. The reason is that, together with the fundamental theorem of calculus, Theorem 4.10 on page 127, equations 5.10 and 5.11 provide an absolutely solid foundation for further development, tools much more reliable than the intuition used so far. Using these tools, we will define the natural logarithm as an integral — a definition which is completely dependable and which allows us to compute the *natural logarithm* values as accurately as we want. We will then define the *exponential function* as the inverse of the *natural logarithm function*, again absolutely legitimately, derive its properties and be able to calculate its values with whatever precision is desired — all of this starting from a problem of interest, and investigated in a way that is a good model for investigation of many other real problems.

Problems 5.1

1. Check numerically the behavior of the expression $(1 + t/n)^n$ as the counting number n increases, for both positive and negative t, and try to explain this behavior.
2. Assuming the intuitively derived results of this section are correct, determine both lower and upper bounds for $exp(t)$ based on the results of problem 1 above. Hint: Examine $1/(1 - t/n)^n$.
3. Intuitively explain why we really lose no generality investigating the equation $x' = x$ rather than the more general looking $x' = kx$. Hint: Time, t, can be measured in any units desired.

3 Precise Development of *ln* and *exp*

In line with the approach generated in Section 2, in order to properly develop the *exponential* function, motivated by its satisfaction of equations 5.2 on page 132, which we rewrite as

$$exp' = exp \quad \text{and} \quad exp(0) = 1,$$

we are led to define the inverse of the *exponential* function — the natural logarithm function, *ln*, as the function satisfying equations 5.10 and 5.11, i.e.,

$$ln'(y) = \frac{1}{y} \quad \text{and} \quad ln(1) = 0\,, \qquad \textbf{5.12}$$

where y is restricted to being a positive real number.

The fundamental theorem of calculus, on page 127, shows that for every positive real number, a, the function G given by

$$G(y) = \int_a^y \frac{1}{t}\,dt \quad \text{for } x > 0$$

satisfies the condition

$$G'(y) = \frac{1}{y} \quad \text{for } y > 0.$$

Since

$$G(1) = \int_a^1 \frac{1}{t}\,dt,$$

we see that choosing $a = 1$ yields $G(1) = 0$. Hence, we give

Definition 5.2: The *natural logarithm* function, *ln*

$$ln(y) = \int_1^y \frac{1}{t}\,dt \quad \text{for } y > 0.$$

❑

We leave it to Problem 5.3 on page 143 to show that ln is the unique solution of equations 5.12, i.e., that if $g'(y) = 1/y$ for all positive y, and $g(1) = 0$, then $g(y) = ln(y)$ for all such y.

What remains now is tying up loose ends. We have to work our way back to verifying all of the intuitively derived results concerning the inverse, *exp*, of *ln*, as well as first phrasing some of these results in terms of the (*natural*) *logarithm* function. We now proceed to this task.

The first result that we will establish is the *logarithm* version of equation 5.6 on page 133, that $exp(u + v) = exp(u)exp(v)$. Solely to see what we want to establish, take *natural logarithms* of both sides of this equation to obtain

$$ln(exp(u + v)) = ln(exp(u)exp(v));$$

now act as if ln and exp are inverses of each other, to obtain

$$u + v = ln(exp(u)exp(v)). \qquad \textbf{5.13}$$

Next let

$$U = exp(u) \quad \text{and} \quad V = exp(v), \qquad \textbf{5.14}$$

which also yields

$$u = ln(U) \quad \text{and} \quad v = ln(V). \qquad \textbf{5.15}$$

Substitute from equations 5.15 into the left side of equation 5.13 and from equations 5.14 into the right-hand side of equation 5.13, to obtain

$$ln(U) + ln(V) = ln(UV). \qquad \textbf{5.16}$$

This is the equation we want to establish properly for all positive real numbers, U and V. Please realize that the reasoning we have just given doesn't prove anything — it just points the way to what we should be able to prove. But it's not hard to establish equation 5.16 properly from the definition of ln by letting

$$G_V(U) = ln(UV) = \int_1^{UV} \frac{1}{t} dt \qquad \textbf{5.17}$$

and

$$H_V(U) = ln(U) + ln(V) = \int_1^{U} \frac{1}{t} dt + \int_1^{V} \frac{1}{t} dt. \qquad \textbf{5.18}$$

The desired result will be established if we can show that for each fixed positive real number V, the functions H_V and G_V are equal (i.e., are the same function).

Because we often want to draw conclusions about the relationship between functions whose derivatives are equal, we will formally establish the following result for future reference.

Theorem 5.3: Functions with identical derivatives

If G and H are differentiable real valued functions of a real variable having the same interval domain, with

$$G' = H',$$

then there is some constant, c, such that

$$G = H + c.$$

From this, it follows that if $G(w) = H(w)$ for some number, w, in the above common domain, then

$$G = H.$$

❑

The proof of Theorem 5.3 follows immediately from the result of Problem 3.5.2 on page 67 by choosing $f = G - H$.

So all we need do to establish equation 5.16 is show for each positive real number, V, that $G_V' = H_V'$ and $G_V(1) = H_V(1)$. This latter equality is immediately recognized by substituting $U = 1$ in equations 5.17 and 5.18.

To show that $G_V' = H_V'$, recognizing that V just stands for a fixed constant, apply the fundamental theorem of calculus, on page 127, and the chain rule for differentiating compositions, see Theorem 2.13 on page 53, to equation 5.17 and the fundamental theorem of calculus and the sum rule for derivatives to equation 5.18. Doing this then establishes the following.

Theorem 5.4: $ln(U\,V) = ln(U) + ln(V)$ for all $U, V > 0$.

Because the *natural logarithm* function is defined for all positive real x, and, from the fundamental theorem of calculus has derivative

$$ln'(x) = \frac{1}{x} > 0,$$

it follows that ln is increasing, and hence, from Theorem 5.1 on page 134, it has an inverse function, prompting the introduction of the following notation.

Definition 5.5: The *exponential function, exp*

The inverse function of the *natural logarithm* function is called the *exponential* function and denoted by *exp*.

❑

The properties that we expect of the *exponential* function are now easily established. From the rule for the derivative of the inverse function of a strictly monotone differentiable function, Theorem 2.13, equation 2.31 on page 53, with $f = ln$, $f^{-1} = exp$, we find that *exp* is differentiable with

$$\begin{aligned} exp'(y) &= (ln^{-1})'(y) \\ &= \frac{1}{ln'(\,ln^{-1}\,)(y)} \\ &= \frac{1}{1/ln^{-1}(y)} \\ &= ln^{-1}(y) \\ &= exp(y). \end{aligned}$$

Furthermore, since it is evident from its definition, Definition 5.2 on page 137, that $ln(1) = 0$, applying the *exponential* function to both sides of this equation, using the fact that *exp* is the inverse function of ln, so that $exp(ln(x)) = x$, yields the result that

$$exp(0) = 1.$$

To establish equation 5.6 on page 133, that

$$exp(u+v) = exp(u)exp(v)$$

we just reverse what we did to obtain 5.16 on page 138; namely, first take *exponentials*[1] of both sides of the equation in Theorem 5.4 above, to obtain

$$exp(ln(UV)) = exp(ln(U) + ln(V)).$$

Now using $exp(ln(x)) = x$, justified by equation 5.7 of Theorem 5.1 on page 134, applied to $f = ln$, we obtain from the equation just derived that

$$UV = exp(ln(U) + ln(V))$$

Next make the substitution $U = exp(u)$, $V = exp(v)$ in this equation, that is, let $u = ln(U)$, $v = ln(V)$, to obtain

$$exp(u)exp(v) = exp(ln(exp(u)) + ln(exp(v))).$$

Finally we apply equation 5.8 of Theorem 5.1 on page 134 twice to the right side, to obtain the desired result that

$$exp(u)exp(v) = exp(u+v).$$

We summarize what has been derived in

Theorem 5.6: Properties of the *exponential* function

The *exponential* function, *exp,* defined by Definition 5.5 on page 139 satisfies the following equations:

$$exp' = exp \qquad \textbf{5.19}$$

For all real numbers, u, v

$$exp(u+v) = exp(u)exp(v), \qquad \textbf{5.20}$$

$$exp(0) = 1. \qquad \textbf{5.21}$$

We are now in a position to define what is meant by a^r for $a > 0$, and r any real number. To see how to do this in the easiest way, note first that from equation 5.20 it follows (strictly speaking using proof by *induction*[2]) that for each integer m and all reals $a > 0$,

$$ln(a^m) = m\, ln(a). \qquad \textbf{5.22}$$

❑

---

1. Here what we mean precisely is that if r and s are reals with $r = s$, then $exp(r) = exp(s)$.
2. That is, establish the result for $m = 1$, and then establish that whenever the result is true for $m = k$, it must hold for $m = k + 1$. This is the reasoning used to show that the result holds for all positive integers. Recall Example 2.11 on page 47.

Recall that the *n-th root* function f given by $f(x) = x^{1/n}$ for $x > 0$, n a nonzero integer is the inverse function to the function g given by $g(y) = y^n$, $y > 0$. So, for $a > 0$, and n a nonzero integer, we find, using equation 5.22 on the numerator in the exponent which follows, that

$$ln(a) = ln(a^{n/n}) = n\,ln(a^{1/n})$$

or

$$ln(a^{1/n}) = \frac{1}{n} ln(a). \qquad \textit{5.23}$$

The usual definition of a positive real to a rational power is given by

Definition 5.7: $a^{m/n}$

For integers $n \neq 0$ and m, and real $a > 0$, we define

$$a^{m/n} = (a^{1/n})^m.$$

❑

Now combining equations 5.22 and 5.23, we obtain

$$ln(a^{m/n}) = \frac{m}{n}\ ln(a)$$

for integers $n \neq 0$ and m, and real $a > 0$. So, for $a > 0$ and r rational, we have

$$ln(a^r) = r\ ln(a). \qquad \textit{5.24}$$

For fixed $a > 0$, the right-hand side of the above equation represents a continuous function of r; where r can stand for any ***real*** number. By taking *exp* of both sides of equation 5.24, and using the fact that *exp* and *ln* are inverse functions, we finally arrive at the definition we want, namely,

Definition 5.8: a^r

For fixed real $a > 0$ and r **any real number** we let

$$a^r = exp(r\ ln(a)).$$

❑

Thus the natural *logarithm* function allowed the definition of a^r without having to work very hard to establish either the continuity or the differentiability of $f(r) = a^r$. If $f(r) = a^r$, this definition makes it easy to find f'.

Exercises 5.2

In the first five exercises find $f'(x)$ for

1. $f(x) = a^x$ for $a > 0$. Be careful to ***avoid the trap*** of assuming that the answer is $x a^{x-1}$.
2. $f(x) = x^a$ for $x > 0$. Hint: $x^a = exp(a\, ln(x))$.
3. $f(x) = exp(x^3)$.
4. $f(x) = ln(1-x)$, , $x < 1$.
5. $f(x) = \int_0^x \frac{1}{1-t} dt$, $x < 1$.
6. Combining the results of Exercises 4 and 5 above, show that

   $$ln(1-x) = \int_0^x \frac{-1}{1-t} dt \quad \text{for } x < 1.$$

7. Using the geometric series result $1 + t + t^2 + \cdots + t^{n-1} = \frac{1-t^n}{1-t}$

   (see Theorem 1.9 on page 16) to show that for $|x| < 1$

   $$ln(1-x) = -x - x^2/2 - \cdots - \frac{x^n}{n} + \varepsilon \text{ where } |\varepsilon| \le \frac{|x|^{n+1}}{1-|x|}.$$

8. Show that the results of the previous exercise allow computation of $ln(1-x)$ to any desired accuracy for $-1 < x < 1$. This result is usually written as

   $$ln(1-x) = \sum_{k=1}^{\infty} \frac{-x^k}{k} \quad \text{for } -1 < x < 1.$$

9. Show how to calculate $ln(x)$ to any desired accuracy for all $x > 0$ using the result of the previous exercise. Hints: see Theorem 5.4 on page 139, and show that $ln(1/x) = -ln(x)$ for $0 < x \le 1$.
10. Evaluate $\int_a^b c^t dt$ where $c > 0$.
11. Evaluate $\int_a^b ln(t) dt$ for positive a, b. Hint: Use integration by parts.
12. Evaluate $\int_a^b t\, ln(t) dt$ for positive a,b.
13. Evaluate $\int_a^b t\, exp(t^2) dt$. Hint: Try substituting $u = t^2$.
14. Show that exp and ln are increasing functions.
15. Show that for $a > 0$ and r real, $ln(a^r) = r\, ln(a)$.

Problem 5.3

Show that the natural logarithm given in Definition 5.2 on page 137 is the only function, g, satisfying $g'(y) = 1/y$ for all positive y, and $g(1) = 0$. Hint: Use Theorem 5.3 on page 138.

The first logarithm that you were probably introduced to was log_{10}. For $y > 0$ the relation $r = log_{10}(y)$ meant that $y = 10^r$, and in general, for positive a different from 1,

$$r = log_a(y) \quad \textit{means that} \quad y = a^r. \qquad \textbf{5.25}$$

Taking natural logarithms of the right-hand equation above, using equation 5.24 on page 141 and then substituting for r from the first equation in 5.25 yields the formula

$$log_a(y) = \frac{ln(y)}{ln(a)} \qquad \textbf{5.26}$$

The positive value a different from 1 is called the *base* of the log_a function. From equation 5.26 we see that if we denote by the symbol[1] e the value for which

$$ln(e) = 1, \qquad \textbf{5.27}$$

then

$$log_e = ln. \qquad \textbf{5.28}$$

That is, the *natural logarithm* function is the logarithm to the base e. But from equation 5.25, if $t = log_e(y)$ then $y = e^t$, whereas from equation 5.28 it also follows that $t = ln(y)$, and hence, because *ln* and *exp* are inverse functions, $y = exp(t)$. Thus we find

$$exp(t) = e^t.$$

For reference purposes we summarize these results as follows:

Definition 5.9: log_a, e

For positive reals a,y with $a \neq 1$, $r = log_a(y)$ means $y = a^r$, see Definition 5.8 on page 141. The value $e = exp(1)$; see Problems 5.4.2 and Problems 5.4.3 on page 144 for numerical determination of e.

□

---

1. This symbol e was chosen by the mathematician Euler, who, based on this choice, would not seem noted for his humility.

Theorem 5.10: Formulas for log_a, exp

For positive reals $a \neq 1$ and y

$$log_a(y) = \frac{ln(y)}{ln(a)} \quad \text{and} \quad exp(t) = e^t.$$

❑

Problems 5.4

1. Show that $\lim_{n \to \infty} (1 + t/n)^n = exp(t) = e^t$.

   Hints:
   $$(1 + t/n)^n = exp(ln[(1 + t/n)^n]) = exp(n\, ln(1 + t/n))$$
   But by the mean value theorem, Theorem 3.1 on page 62,

   $$ln(1+t/n) = ln(1+t/n) - ln(1) = \frac{t}{n} ln'(1+\vartheta\, t/n) = \frac{t}{n} \frac{1}{1+\vartheta\, t/n}$$

   where $0 \leq \vartheta \leq 1$.

2. Show how to determine $e = exp(1)$ via bisection and the ability to compute $ln(x)$ for any $x > 0$ — i.e., to solve $ln(x) = 1$. You need y_L and y_U satisfying $ln(y_L) < 1$ and $ln(y_U) > 1$. You may let $y_L = 1$. If you don't already have y_U, you may use Theorem 5.4 on page 139 to obtain it.

3. Generate $e = exp(1)$ via Newton's method (Theorem 3.9 on page 87) and the ability to compute $ln(x)$ for any $x > 0$ (see Exercise 5.2.9 on page 142).

4 Formulas as Functions

We have previously stated that it is a common abuse of mathematical notation to let the formula for a function stand for the function itself. Actually, the use of its formula to represent a function can be legitimately justified by a very simple device. Namely, the formula $f(t)$ will stand for the function f provided that the symbol t is taken to be the *identity function* with the same domain as that of the function f. Because, the symbol $f(t)$ is the composition of f with t, and then because $t(z) = z$, we have

$$f(t(z)) = f(z)$$

for all z in the domain of f. Thus,

$$f(t) = f$$

Use of this device will greatly simplify the *shift rule* manipulations that we will introduce in the next section.

Example 5.5: $D(f\ exp)$

Suppose that f is differentiable, and we want to find the derivative of the product of the functions f and *exp*. Using the derivative operator, D (see the material in and following Definition 2.5 on page 30, noting that $(D + a)f$ means $Df + af$, and also remembering that $De^t = Dexp = exp = e^t$) we can write the desired derivative as

$$\begin{aligned} D(fe^t) &= (Df)e^t + fD(e^t) \\ &= e^t(D+1)f. \end{aligned}$$

We rewrite this result in the form

$$(D+1)f = e^{-t}D(e^t f). \qquad \textbf{5.29}$$

■

Equation 5.29 is a special case of the *shift rule* which will be very important to us, both in establishing properties of differential equations, and in solving them as well. Some extensions of this example will be dealt with in the exercises to follow, and then made use of in Section 5 on applications of the exponential function.

Exercises 5.6

1. Let k be a fixed real number, and let x be a differentiable function. With t being the appropriate identity function, show that

$$(D+k)x = e^{-kt}D(e^{kt}x).$$

2. Let q and f be differentiable functions with the same domain. Show that

$$(D+q')f = e^{-q}D(e^q f).$$

5 Applications of *exp*

It should not surprise you that the solution of the radioactive decay equation

$$x' = -kx \qquad \textbf{5.30}$$

will turn out to be some form of the exponential function, but you should realize that we don't even know yet that *exp* is the only function satisfying the equations

$$x' = x, \ x(0) = 1.$$

It is tempting to try the following to establish what we want. If $x' = x$, then

$$x'/x = 1. \qquad \textbf{5.31}$$

But by the chain rule

$$[ln(x)]' = \frac{x'}{x}.$$

Hence equation 5.31 can be written

$$[ln(x)]' = 1 ,$$

from which it would follow (see Problem 3.5.3 on page 67) that there is a real number, c, for which

$$ln(x(t)) = t + c.$$

and since $x(0) = 1$, $ln(1) = 0 + c$, which yields

$$c = 0.$$

Therefore, we would have

$$ln(x(t)) = t,$$

and taking exponentials of both sides would yield the desired result that

$$x(t) = exp(t).$$

The flaw here is that we didn't know that we were allowed to *divide* both sides of the equation $x' = x$ by x, since x could *conceivably* take on the value 0. So a more airtight argument must be supplied.

The key to a trustworthy approach to solution of equations of the form,

$$x' = -kx, \quad x(0) = x_0$$

lies in the result of Exercise 5.6.1 on page 145, which we state as

Theorem 5.11: The *shift rule*

Using the notation introduced in Section 4 of this chapter, for x a differentiable function, and D the derivative operator, we have

$$(D + k)x = e^{-kt}D(e^{kt}x).$$

❑

First we rewrite the equation $x' = -kx$ as

$$(D + k)x = 0.$$

Using the *shift rule*, this equation can be written

$$e^{-kt}D(e^{kt}x) = 0.$$

Multiplying both sides of this equation by the nonzero value e^{kt} yields

$$D(e^{kt}x) = 0. \qquad \textbf{5.32}$$

It's worth mentioning here that the reasoning we've just been through merely shows that

$$x' = -kx \text{ if and only if } D(e^{kt}x) = 0;$$

this latter equation is much easier to deal with, as we will now show. From the result of Problem 3.5.2 on page 67 we know that if $D(e^{kt}x) = 0$, then for some constant, c we have $e^{kt}x = c$, or $x = ce^{-kt}$. If we require that $x(0)$ be the given value x_0, then we see that $c = x_0$. What we have established is the following.

Theorem 5.12: Solution of $x' = -kx,\ x(0) = x_0$

The unique solution of the differential equation $x' = -kx$ subject to the *initial condition* $x(0) = x_0$ is

$$x(t) = x_0 e^{-kt}.$$

❑

We will now investigate how this result can be applied to the radioactive dating problem. The first issue that we will look into is that of determining the radioactive decay constant, k, for radioactive carbon-14. A Geiger counter is a device for counting the number of radioactive decays emitted by some object. Actual Geiger counters only count decays which impinge on their receptors, and only if they do not occur at times too close to each other. For our purposes we will model a Geiger counter rather crudely, as a device which registers the quantity

$$q\int_0^t A'(\tau)\,d\tau \qquad \textit{5.33}$$

when the counter operates for t time units starting at time 0, and where $A(\tau)$ is the amount of radioactive carbon-14 in the object at time τ, and q is a (most likely unknown) constant determined by the design of the counter. From Theorem 5.12 above, we see that under the model we are using, a piece of wood freshly cut at time 0 has in it an amount of radioactive carbon-14

$$A_{new}(\tau) = x_0 e^{-k_{14}\tau}$$

at time τ. The Geiger counter, running from time 0 to time t in some fixed position relative to the piece of wood will register an amount

$$\begin{aligned} R_{new} &= q\int_0^t A'_{new}(\tau)d\tau \\ &= q[A_{new}(t) - A_{new}(0)] && \text{using equation 4.14 on page 103} \\ &= qx_0(e^{-k_{14}t} - 1). \end{aligned}$$

A piece of wood of the same shape,[1] which had been cut T time units earlier than the new piece will have in it an amount

$$A_{old}(\tau) = x_0 e^{-k_{14}(T+\tau)}$$

of carbon 14 at time τ. If the Geiger counter is exposed to it in the same way as just described, it will register

$$R_{old} = q[A_{old}(t) - A_{old}(0)] = qx_0 e^{-k_{14}T}(e^{-k_{14}t} + -1).$$

Hence

$$R_{old}/R_{new} = e^{-k_{14}t},$$

and thus, taking natural logarithms of both sides of this equality, and solving for k_{14}, we find

$$k_{14} = -\frac{1}{T} ln(R_{old}/R_{new}). \qquad \mathbf{5.34}$$

Determination of k_{14}:

Assuming a Geiger counter runs for t time units when applied to given objects, and is applied to a freshly harvested piece of organic material to yield a reading of R_{new}, and to an identical piece of material which was freshly harvested T time units earlier, yielding a reading of R_{old}, under the given model the decay constant, k_{14} is given by equation 5.34. This model completely ignores the question of how the values of t and T relate to the accuracy of determination of k_{14}. A statistical analysis beyond the scope of our work here is required to resolve this question. For practical purposes we would run the Geiger counter until the value on the right side of equation 5.34 appears stable — e.g., changing very little, percentagewise, when we increase the observation time, *t,* by a substantial percentage.

❑

Since k_{14} has been determined very accurately[2] to be about .000121/year, we need not further concern ourselves with its determination.

---

1. What is essential in Geiger counter comparisons for dating purposes is that we know what the Geiger counter reading for the piece of material being dated should have been when it possessed its full amount of carbon 14.
2. Actually, the *half-life* is what is usually tabled — i.e., the value $\tau_{.5}$ satisfying the equation $e^{-k_{14}\tau_{.5}} = .5$. To four digits $\tau_{.5} = 5730$ years, which explains why carbon-14 is so well suited to archaeological application.

To use the known value of k_{14} (or, equivalently, the known half-life of carbon-14) to determine the date of death of some organic object, we take a sample from this object, and a comparable sample from a living object of the same type. Similar to equation 5.34, we obtain

$$R_{old-sample}/R_{new-sample} = e^{-k_{14}T}$$

from which we find

$$T = -\frac{1}{k_{14}} ln(R_{old-sample}/R_{new-sample}) \qquad 5.35$$

Estimating the date of death of an organic object

Knowing the carbon-14 radioactive decay constant, k_{14}, to be about .000121/yr, identical Geiger counter application to two identically shaped samples, one just harvested from a living object, yielding a reading denoted by $R_{new-sample}$, and a reading $R_{old-sample}$, from a comparable object which ceased living an unknown T time units ago, the quantity T can be estimated using equation 5.35.

❑

Exercises 5.7

1. An object has $R_{old-sample}/R_{new-sample} = .046$. How long ago does the model being used calculate as its time of death?
2. Under the model being considered, what value would the ratio $R_{old-sample}/R_{new-sample}$ have for an object which died in the year 1492?
3. Under the model introduced above, what value would the ratio $R_{old-sample}/R_{new-sample}$ have for an object which died 30 million years ago? Would you trust carbon 14 dating for such a situation? (Explain your answer.)
4. If you had two comparable pieces of wood, one known to have been cut 500 years ago, and the other 102.4 years ago, how would you use them to investigate how large the Geiger counter observation time, *t*, should be to achieve reasonable accuracy?

To finish this section we will examine a simple, but nonetheless useful, enzyme kinetics model.

Example 5.8: Single compartment kinetics

Suppose that a drug not previously present in the body is infused at a constant rate into the bloodstream. If it is the case that this drug *washes out* of the bloodstream at a rate proportional to the amount in the blood, then we may write

$$A'(t) = -kA(t) + I \quad \text{for} \quad t > 0, \quad A(0) = 0, \qquad \mathbf{5.36}$$

where $A(t)$ is the amount of drug in the bloodstream at time t and I is the constant rate of infusion of this drug.

The first problem of interest is that of determining the rate parameter, k, from observed data. One approach to this problem involves first determining the form of the solution to equations 5.36. To do this we rewrite the first of these equations as

$$(D + k)A = I$$

and once again apply the *shift rule*, Theorem 5.11 on page 146, to obtain

$$e^{-kt}D(e^{kt}A) = I,$$

from which we see that A must satisfy the equation

$$D(e^{kt}A) = Ie^{kt}.$$

Since $D\left(\frac{I}{k}e^{kt}\right) = Ie^{kt}$, it follows from Theorem 5.3 on page 138 that

there is some constant, d, such that $e^{kt}A = \frac{I}{k}e^{kt} + d$, or

$$A(t) = \frac{I}{k} + de^{-kt}.$$

Utilizing the second equation of 5.36 above, shows that

$$A(t) = \frac{I}{k} - \frac{I}{k}e^{-kt}. \qquad \mathbf{5.37}$$

From this equation we see that

$$\lim_{t \to \infty} A(t) = \frac{I}{k},$$

by which we mean that *for all* t *sufficiently large,* $A(t)$ *is close to* I/k.

So if we continue the infusion until the value, $A(t)$, of A, seems to have reached a plateau, $A_{plateau} = \lim_{t \to \infty} A(t)$, then, since the infusion rate, I, *is* known, being under the experimenters control, we may estimate k via

$$k \cong \frac{I}{A_{plateau}}. \qquad \mathbf{5.38}$$

There are more sophisticated and reliable methods for estimating k, but this approach is commonly used when a one-compartment model is assumed.

As already mentioned in the discussion of compartmental modeling, Example 4.5 on page 96, a model such as this one, while it may reasonably well describe the behavior of the drug in the bloodstream, may lead to serious errors if it is assumed that the amount of drug remaining in the body is well represented by the amount of drug in the bloodstream. It is now believed that there is great danger to the body if even a small amount of a chemical such as dioxin, while washing out of the bloodstream, accumulates in fatty tissues. So great care should be exercised in applying such a model to draw conclusions about *whole- body behavior* of infused drugs or other chemicals.

■

Exercises 5.9

1. Another commonly used means for estimating the washout rate parameter, k, of a single-compartment kinetic model of the type discussed in Example 5.5 on page 145 is to observe the amount of drug, $A(t)$, in the bloodstream at various times t, following a bolus of size B, of this drug. We may assume that this system is governed by the equation
$$A'(t) = -kA(t) \quad \text{for} \quad t > 0, \quad A(0) = B.$$
Determine how to estimate k graphically, from data
$$(t_i, \; ln(A(t_i))) \quad i = 1, 2, ..., n..$$
This method is also commonly used to estimate k.

2. Given the situation in exercise 1 above, suppose we have determined that $k = .014$/minute. How long does it take for 99% of the bolus to wash out of the bloodstream?

3. Assume that in the situation described in exercise 1 above, we have determined that $k = .02$/minute. Suppose that pills containing the drug act as boluses (one shot instant doses) of 30 mg. of drug, entering the system 20 minutes after ingestion.

   a. A patient comes in suffering an overdose, having taken a full bottle of 10 pills all at once. If the blood is determined to contain 110 mg. of the drug (by measuring concentration and estimating blood volume), at what time, relative to the time of this latest measurement, were the pills ingested? (From this information, we may be able to determine how much damage was done.)

   b. Suppose the number of pills in part a is unknown, but they were ingested 55 minutes prior to the blood level measurement. How many pills were taken?

4. Money in a saving account pays *simple annual interest at a rate of p percent* returns an amount $\$(1 + p/100)M$ at the end of one year, when $\$M$ was the amount of money initially deposited. We say the interest is *compounded* if the interest earned in a given time period is added to the principal which earned the interest, (so that subsequently, the previously earned interest itself

earns interest). We say that interest is *compounded* k *times per year* if the process of compounding takes place at every time of the form j/k years, $j = 1,2,...,k$.

a. What amount of money is on deposit at year's end if interest is compounded k times per year, $k = 1,2,...,$ when the initial principal is $\$M$?

b. How does this amount behave as the number of compounding times grows arbitrarily large? This situation is referred to as *continuous compounding.*

c. What simple annual rate of interest is equivalent to continuous compounding at a rate of $p\%$ of interest at the end of one year? Which of the two methods — simple interest at rate $p\%$ or the equivalent compound interest — is preferable for the depositor? (Explain your answer.)

Problems 5.10

1. Suppose a drug is infused at a rate $I(t) = at + b$ starting a time 0. Given that the system is governed by the equation
$$A' = -kA, \quad A(0) = 0,$$
what is $A(t)$ for $t > 0$? Hint: Use the shift rule, Theorem 5.11 on page 146 and the appropriate rules of integration in Section 7 of Chapter 4.

2. Referring to Example 5.8 on page 149, in an effort to shorten the time it takes to reach a plateau, a *priming* bolus of the drug of interest is often given at the outset of the infusion. To investigate the effect of such a bolus, solve the following equations: $(D + k)A = I$, $A(0) = B$, *where* B is the amount of drug in the bolus. From this, show that the closer the bolus amount is to I/k, the better the result. What is the effect of B being much greater than I/k?

3. Pills are often prescribed to be taken every H hours, with each pill having an amount B of drug in it — each pill assumed to constitute a bolus of size B, entering the bloodstream some fixed time T after ingestion. For simplicity, restrict the analysis below to refer to times relative to when the drug entered the bloodstream. Our object in this problem is to determine the behavior of the peak blood levels — those levels reach when the drug in each pill enters the bloodstream. It is important to know these levels in trying to avoid drug toxicity, for example, too great a drop in blood pressure if the blood concentration of some blood pressure medication is too high. Let x_n represent the blood drug level at the time the drug of the n-th pill entered the bloodstream.

a. Derive the equations $x_{n+1} = B + dx_n,\ x_0 = 0,$ where $d = e^{-kH}$, k being the washout rate parameter for the model of Example 5.8 on page 149 that we will use to describe this situation. Notice that, assuming this model is valid up to the concentration at which this drug is toxic, determination of k does not require observation of plateau values anywhere near toxic concentration. But there should be good evidence that this model is reasonable over this range.

b. Now solve this system. Hints:
(i) First write the system as $(\Delta + (1-d))x_n = B$ (recall the definition of the forward difference operator Δ,

c. Definition 1.3 on page 5.
(ii) Next derive the discrete shift rule:

$$(\Delta + q)x_n = (1-q)^{n+1}\Delta((1-q)^{-n}x_n)$$

by examining and experimenting with $\Delta(r^n x_n)$

(iii) Observe that if $\Delta a = \Delta b$ then there is a constant, c, such that $a = b + c$ (i.e. $a_n = b_n + c$ for all n).

(iv) Use the results just derived to derive the solution of the equation determined in part a, by observing that

$$\Delta\left(\frac{d^{-(n+1)}}{1/d - 1}\right) = d^{-(n+1)}$$

The solution is

$$x_n = \frac{B}{1-d} + cd^n$$

where c is a constant to be determined by the condition $x_0 = 0$.

d. Suppose each pill has 300 mg. of drug and the half life of the drug in the bloodstream is 8 hours (see Footnote 2 on page 148). What level would the peak amounts approach if the pills were taken every 3 hours?

e. If the maximum safe blood level were 800 mg., and you wanted the pills taken every 4 hours, assuming the same half-life as in part d, what is the largest safe amount of drug per pill?

4. Solve the differential equation $x' + tx = t^2,\ x(0) = 1$.
Hint: See Exercise 5.6.2 on page 145

5. See if you can derive equation 5.38 on page 150 directly from equation 5.36 on page 150 without solving equation 5.36.

6 Trigonometric Functions, Intuitive Development

The development of the sine and cosine functions parallels that of the exponential function. We start out with a local description in the form of a differential equation describing the motion of a frictionless spring-mass system.[1] We next examine the algebraic solution of finite difference approximations to this differential equation, using them to get an idea of the behavior of the solution of the differential equation. From this we are able to plausibly conjecture the form of the derivative of the inverse function of the solution being examined. It is here that the formal development begins.

The prototype spring-mass system being presented here makes use of Hooke's law, which was introduced in equation 4.2 on page 92. Hooke's law states that the force, F, exerted by a spring on a mass which displaces the spring's free end to a position, x, from its rest position is given by the formula

$$F = -kx, \qquad \textbf{\textit{5.39}}$$

where k is a positive constant determined by the stiffness of the spring. The negative sign indicates that the force is in the direction opposite to the displacement. So the spring is opposing its displacement. Hooke's law starts to fail as an adequate description if the spring's vibrations generate too much heat, or if the spring is extended too far. Newton's second law (of motion) relates the force on an object to the rate of change (derivative with respect to time) of momentum (product of its mass and velocity). Assuming the object has constant mass (i.e., neglecting the increase in mass accompanying very high speeds, as described by Einstein's *Special Theory of Relativity*), Newton's second law may be written as

$$F = mx'', \qquad \textbf{\textit{5.40}}$$

where $x(t)$ is the spring's displacement from its rest position at time t, and m is the mass of the object hooked to the spring, as shown in Figure 5-2.

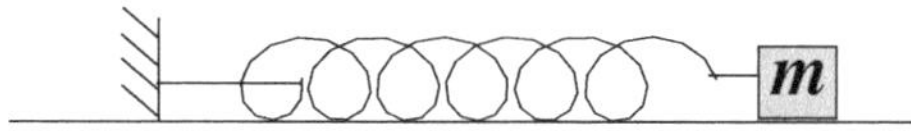

Figure 5-2

If an external *driving force*, $f(t)$, also acts on the mass at time t, we obtain the equation

$$mx'' = -kx + f.$$

1. Although this furnishes only the simplest model of a vibrating system, it exhibits many of the characteristics common to such systems, and forms the basis for analysis of most of them.

This is the basic equation governing the position versus time function, x, of a frictionless spring-mass system subjected to an external driving force f. In an attempt to account for frictional forces, the usual first approach is to assume a force proportional to and opposing the velocity, in which case, assuming a nonnegative constant, c, of proportionality, we obtain the equation

$$x'' + \frac{c}{m}x' + \frac{k}{m}x = \frac{f}{m},$$

which we rewrite in the form

$$x'' + qx' + rx = g, \qquad \textbf{5.41}$$

where q and r are constants, g is a known function, and $x(t)$ is the position of the object of mass m at time t. This differential equation provides an *implicit* description of functions, x, by specifying an equation they must satisfy — an equation which does not itself provide an explicit formula for x. This is in contrast to the very similar looking differential equation, 4.3 on page 94,

$$Z'' + \frac{k}{m} Z = x'',$$

which provides an explicit formula for the quantity, x'', in terms of the observed function Z and its approximately computable second derivative, Z''.

Just as we often do, we begin by studying a very simple form of equation 5.41, namely,

$$x'' + x = 0. \qquad \textbf{5.42}$$

To see how to obtain information from this equation, we will approximate it by an algebraic equation, obtained from the approximations

$$x'(t) \cong \frac{x(t+h) - x(t)}{h}, \qquad \textbf{5.43}$$

$$x''(t) \cong \frac{x'(t+h) - x'(t)}{h}, \qquad \textbf{5.44}$$

which are suggested from the definition of derivative. Substituting from approximation 5.43 into approximation 5.44 yields

$$x''(t) \cong \frac{x(t+2h) - 2x(t+h) + x(t)}{h^2} \qquad \textbf{5.45}$$

Substituting this last approximation into differential equation 5.42 yields the following approximation:

$$x(t+2h) \cong 2x(t+h) - (1+h^2)x(t). \qquad \textbf{5.46}$$

From this it seems evident that in order to be able to compute values of x for values of t that might be of interest, we must know two *starting* or *initial* values of x, so that we can start evaluating the right side of approximation 5.46. For now, let's assume that we know the two values, $x(0)$ and $x(h)$. Then putting $t = 0$, and using approximation 5.46 we can calculate an approximation to $x(2h)$. Next, setting $t = h,$ from the known value of $x(h)$ and the just computed approximation to $x(2h)$ we can again use approximation 5.46, setting $t = h,$ to compute an approximation to $x(3h)$. Then set $t = 2h$ again using approximation 5.46 and the computed approximations to $x(2h)$ and $x(3h)$ to approximate $x(4h)$. We continue on in this fashion, successively computing approximations to $x(5h), x(6h),...$. Let x_k denote our approximation to $x(kh)$. In the procedure we have been describing, approximation 5.46 translates into the equations

$$x_{k+2} = 2x_{k+1} - (1 + h^2)x_k. \qquad \textbf{5.47}$$

To this we must append initial conditions specifying the values of x_0 and x_1. If we want to reflect initial conditions on $x(0)$ and $x'(0)$, using the Fundamental Linear Approximation, 2.7 on page 31, we require

$$x_0 = x(0), \quad x_1 = x(0) + hx'(0) \qquad \textbf{5.48}$$

Equations 5.47 and 5.48 are easy to program in *Mathematica.* For the case $x(0) = 0, x'(0) = 1,$ with h chosen to be .01, and k running from 0 to 999, we obtain[1] the graph of the points (k, x_k) shown in Figure 5-3. The graph in

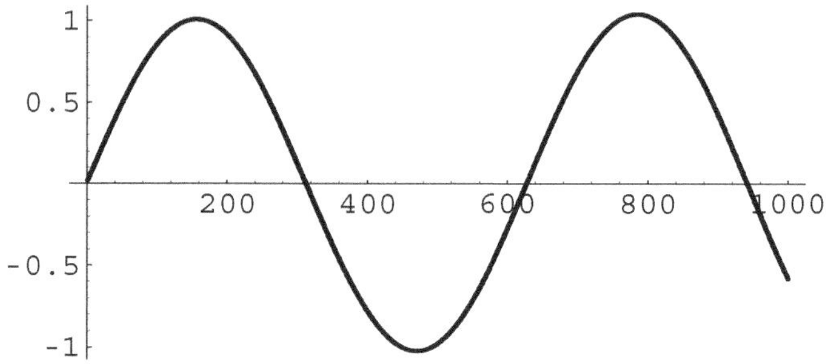

Figure 5-3

Figure 5-3 is extremely close to the portion of the sine function with domain [0,9.99] (because the scale corresponds to the multiple of $h = .01$).

---

1. The following *Mathematica* commands constitute the program which produced the graph in Figure 5-3.

```
For[a=0;b=.01;k=0,k<1000,k++,c = 2 b -1.0001a;a = b;b = c; c >>>
"output"]
x=ReadList["output"]
ListPlot[x]
```

The *For* loop computes the successive x_{k+2} values, storing x_{k+1} and x_k in the (*Mathematica*) variables a and b respectively, and putting x_{k+2} in *the variable c*, which value is then appended to the file called *output.* This file is next read into the variable x as a list and then plotted.

This numerical approach to the trigonometric functions is useful, but somewhat less illuminating about the nature of what is going on than a more direct qualitative examination of equation 5.42 itself, that we now present.

The general shape of x is not difficult to establish from equation 5.42 as follows:

> If $x(t) > 0$ and $x'(r) > 0$, then since $x''(t) < 0$, as seen from equation 5.42, x' must be decreasing, as illustrated in Figure 5-4.

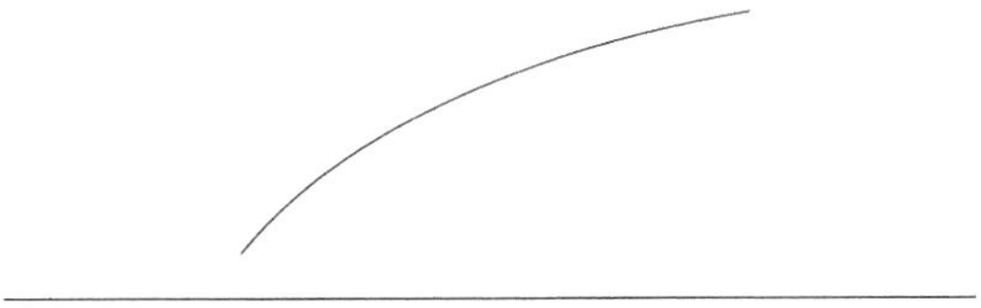

Figure 5-4

> In fact, so long as $x > 0$, its slope must decrease, while if $x < 0$, its slope must increase, yielding a picture that might well resemble the one in Figure 5-3.

Let us now begin a more detailed investigation of the solution of equation 5.42 satisfying the initial conditions $x(0) = 0, x'(0) = 1$, i.e., of the equations

$$x'' + x = 0, \quad x(0) = 0, x'(0) = 1. \qquad \textbf{5.49}$$

From here on we will refer to the (conceivably mythical) function satisfying equations 5.49 as the ***sine*** function, abbreviated *sin* — i.e.,

$$sin'' + sin = 0, \; sin(0) = 0, \; sin'(0) = 1. \qquad \textbf{5.50}$$

Throughout the remainder of this section we will omit any more mention of the hypothetical nature of the entire discussion. In the section to follow, and Appendix 4, the results conjectured here will all be stated and established properly.

We let the derivative of the sine function be called the *cosine*, denoted by *cos* — i.e.,

$$cos = sin'. \qquad \textbf{5.51}$$

Now multiplying the equation $sin'' + sin = 0$ by $2 sin'$ we find

$$2 sin' sin'' + 2 sin' sin = 0;$$

using the *chain rule*, equation 2.30 on page 53, we find

$$[(sin')^2]' + [sin^2]' = 0$$

or

$$[cos^2]' + [sin^2]' = 0$$

from which we obtain, by means of the result of Problem 3.5.2 on page 67, that $cos^2 + sin^2$ is constant.[1] But from equations 5.50 and 5.51 we know that $sin(0) = 0$ and $cos(0) = 1$, so that we finally obtain

$$cos^2 + sin^2 = 1 \qquad \textbf{5.52}$$

(which should remind you of the Pythagorean theorem).

We are now ready to look at the inverse of the sine, as promised — but if Figure 5-3 on page 156 is to be believed, we had better restrict ourselves to a short enough piece of the sine function so that there is an inverse. We will refer to the longest portion of the sine function which includes the value (0,0) and which is monotone increasing with a positive derivative as *SIN*. The *SIN* function should look as shown in Figure 5-5.

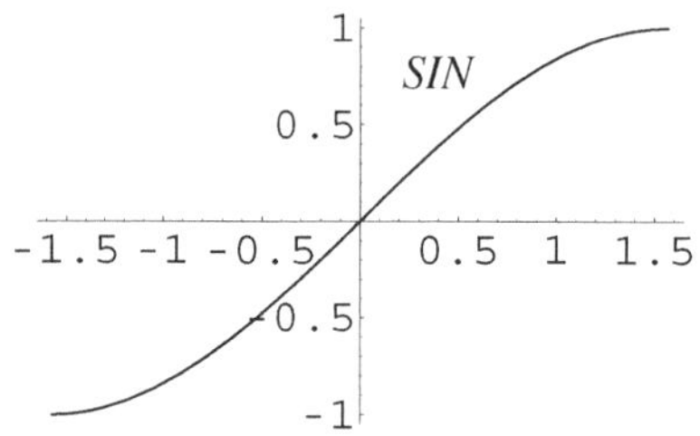

Figure 5-5

From equation 2.31 on page 53 (the rule for differentiation of inverse functions), we find, letting $COS = SIN'$ that

$$(SIN^{-1})'(t) = \frac{1}{SIN'(SIN^{-1}(t))} = \frac{1}{COS(SIN^{-1}(t))}.$$

But equation 5.52 above surely applies with *SIN* and *COS* replacing *sin* and *cos*. Because *COS* is always positive, we see that

$$COS(SIN^{-1}(t)) = \sqrt{1 - SIN^2(SIN^{-1}(t))} = \sqrt{1-t^2}.$$

Substituting into the previous equation then yields

$$(SIN^{-1})'(t) = \frac{1}{\sqrt{1-t^2}}. \qquad \textbf{5.53}$$

---

1. The manipulations we have just gone through here can hardly be labeled as anything but a trick, for which apologies are in order. This trick will also be used here to establish uniqueness of the solution of equations 5.49. In the chapter to follow we will provide a more natural proof of uniqueness using an extension of the shift rule involving complex numbers.

Furthermore, since by equation 5.50 on page 157, $sin(0) = 0$, we have

$$SIN^{-1}(0) = 0. \qquad \textbf{5.54}$$

Equations 5.53 and 5.54 constitute the start of our formal development in the next section.

Problems 5.11

1. Provide a reasonable argument to show that every solution of $x'' + x = 0,\ x(0) = a,\ x'(0) = b$ is bounded, i.e., that for all t there is a quantity c (determined by a and b), such that $|x(t)| \le c$. Hint: Recall the derivation of equation 5.52.

2. Provide a reasonable argument to show that the solution of $x'' + x = 0,\ x(0) = a,\ x'(0) = b$ cannot approach a limit for large t unless a and b are 0. Hint: Examine the differential equation to see what it indicates is happening to the graph of x; consider several cases, e.g.,

$$x(t) > 0 \text{ and } x'(t) > 0$$

$$x(t) > 0 \text{ and } x'(t) = 0$$

etc.

3. Make a convincing argument showing that the equations $x'' + qx = 0,\ x(0) = a,\ x'(0) = b$ have at most one solution (here q,a,b are constants). Hint: Suppose there were two solutions, X and x. Then if $z = X - x$,

$$z'' + qz = 0,\ z(0) = 0,\ z'(0) = 0.$$

Now recall the derivation of equation 5.52.

4. See if you can convince yourself that

$$(f(t) = sin(2t) = 2\,sin(t)cos(t) = g(t)).$$

Suggested approach: Show that both f and g satisfy $x'' + 4x = 0,\ x(0) = 0,\ x'(0) = 2$ and use the result of problem 3 above.

5. Show that $cos(2t) = 2cos^2(t) - 1$. Hint: Use the result of problem 4 above.

6. Provide a convincing intuitive explanation for why no generality is really lost by studying the equation $x'' + x = 0$, rather than the more physically meaningful equation $x'' + \frac{k}{m}x = 0$. Hint: The units of measurement for mass and time may be chosen arbitrarily.

7 Precise Development of *sin, cos*: Overview

A precise development of the *sine* and *cosine* functions starts out by defining the *inverse* function, called *arcsin*, of the *principal part, SIN,* of the *sine* function. The reasoning which leads to the definition we are about to give is the same as that which led from equations 5.12 on page 137 to Definition 5.2 on page 137, equations 5.53 and 5.54 on page 159.

Definition 5.13: *arcsin*

For $-1 < t < 1$, recalling Definition 4.2 on page 102, of integral,

$$arcsin(t) = \int_0^t \frac{1}{\sqrt{1-\tau^2}} d\tau$$

❑

This starting point has the merit of allowing use of our previous knowledge about accurate computation of integrals (Section 6 of Chapter 4) to obtain arbitrarily accurate values of this function. From the fundamental theorem of calculus, Theorem 4.10 on page 127, ***arcsin* is differentiable, with derivative**

$$arcsin'(t) = 1/(\sqrt{1-t^2}.) \qquad \textbf{5.55}$$

With this knowledge, Newton's method (Section 4 of Chapter 3) can be used to calculate the values of *SIN* with prescribed accuracy.

Equation 5.55 together with Theorem 3.3 on page 66 shows that *arcsin* is monotone increasing. So from Theorem 5.1 on page 134 ***arcsin* has an inverse *function defined in the following.***

Definition 5.14: *SIN*

$$SIN = arcsin^{-1}$$

(see Definition 5.13 above and Definition 2.8 on page 37, of inverse.)

❑

Since

$$arcsin''(t) = \frac{2t}{(1-t^2)^{3/2}}$$

arcsin is concave down for $-1 < t < 0$ and concave up for $0 < t < 1$ (see the discussion surrounding Figure 3-8 on page 68.) Hence the graph of *arcsin* has the shape illustrated in Figure 5-6, and therefore that of *SIN* has the shape illustrated in Figure 5-5 (page 158).

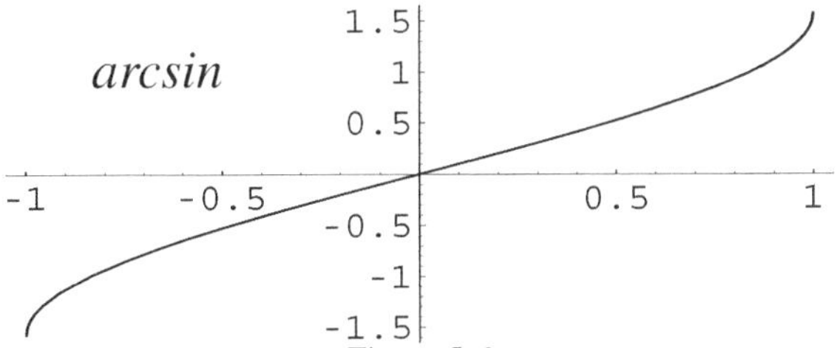

Figure 5-6

From Theorem 2.13, equation 2.31 on page 53, on the derivative of the inverse of a strictly monotone function, we find

$$SIN'(\vartheta) = \frac{1}{arc\sin'(SIN(\vartheta))} = \sqrt{1 - SIN^2(\vartheta)}. \qquad \textbf{5.56}$$

We ***define***

$$SIN' = COS, \qquad \textbf{5.57}$$

and note that

$$SIN^2 + COS^2 = 1. \qquad \textbf{5.58}$$

To see how we should now proceed, we introduce the geometric meaning of *SIN* and *COS*. Since $SIN^2 + COS^2 = 1$, the point $(COS(\vartheta), SIN(\vartheta))$ lies on the *unit circle* (circle of radius 1), centered at the origin. As t increases

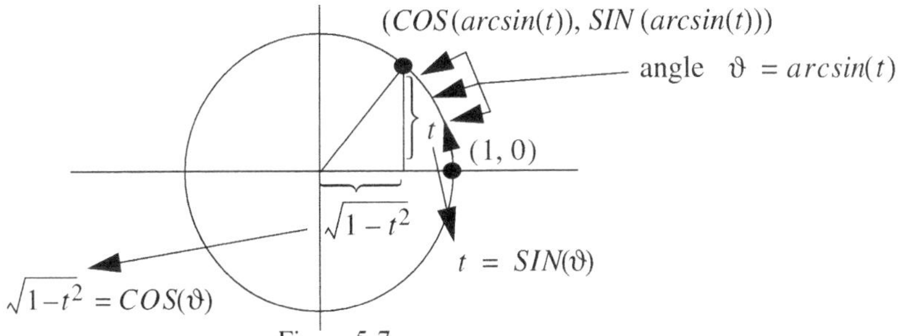

Figure 5-7

from 0 toward 1, we see below that the point $(COS(arcsin(t)), SIN(arcsin(t)))$ travels along this unit circle. In fact, as we will show in Appendix 4, $\vartheta = arcsin(t)$ represents the radian measure of the *angle* (signed length of the arc of the unit circle) going from (1,0) to the point $(COS(arc\sin(t)), SIN(arc\sin(t)))$.

The quantity $t = SIN(\vartheta) = SIN(arc\sin(t))$, being the *signed length* of the vertical leg of the *right triangle* in Figure 5-7 is the familiar ratio *opposite/hypotenuse* since the hypotenuse is of unit length.

The quantity $COS(\vartheta) = SIN'(\vartheta) = \sqrt{1 - SIN^2(\vartheta)} = \sqrt{1-t^2}$ is the familiar *adjacent/hypotenuse.*

From Figure 5-7 it is seen that the natural way to extend the definition of *SIN* and *COS* to get the *sine* and *cosine* functions is to let $sin(\vartheta)$ always be the second coordinate, *t,* of the point a (signed) radian measure (arclength) ϑ from the point (1,0), and $cos(\vartheta)$ be the first coordinate of this point. To make this definition precise, we need a symbol for the arclength of the quarter-circle of unit radius, and a formula for this value. The number π is defined as the ratio of the perimeter of a circle to its diameter, so that the arclength of a quarter-circle of unit radius is $\pi/2$. We can't use the formula for $arcsin(t)$ directly for $t = 1$, because the integrand on the right side of the definition of $arcsin(t)$ takes on arbitrarily large values for $0 < t < 1$, and thus the Riemann integral is not defined at $t = 1$. In Appendix 4 we will show that this integral does approach a limit as t approaches 1 from below. However, this does not furnish a really practical way of determining $\pi/2$. Rather we will make use of Figure 5-7 to note that at one eighth of a full traversal of the unit circle starting at (1,0), $SIN(\vartheta) = COS(\vartheta)$. Together with 5.58, that $SIN^2 + COS^2 = 1$ yields that the radian measure of the angle representing one eighth of a circle, denoted by $\pi/4$ must satisfy

$$SIN(\pi/4) = 1/\sqrt{2}$$

from which it follows that

$$\pi = 4\, arcsin(1/\sqrt{2}) = 4\int_0^{1/\sqrt{2}} \frac{1}{\sqrt{1-\tau^2}}\, d\tau. \qquad \textit{5.59}$$

We now have the necessary machinery to define the *sine* function.

Definition 5.15: *sin*

For *SIN* as defined in Definition 5.14 on page 160, $sin(\vartheta)$ is first defined for ϑ in the closed interval $[-\pi, \pi]$ as follows:

$$sin(\vartheta) = \begin{cases} SIN(\vartheta) \text{ for } -\pi/2 < \vartheta < \pi/2 \\ 1 \text{ for } \vartheta = \pi/2 \\ -1 \text{ for } \vartheta = -\pi/2 \\ SIN(\pi - \vartheta) \text{ for other } \vartheta \text{ with } -\pi \le \vartheta \le \pi \end{cases}$$

Now that we have defined $sin(\vartheta)$ for one full circle of ϑ values, we notice that the *sine line* in Figure 5-7 repeats its behavior with each succeeding revolution (clockwise or counterclockwise). To reflect this in the definition of the *sine function*, let k be any nonzero integer (positive or negative), and define

$$sin(\vartheta + 2k\pi) = sin(\vartheta),$$

The graph of *sin* is shown in Figure 5-8 below.

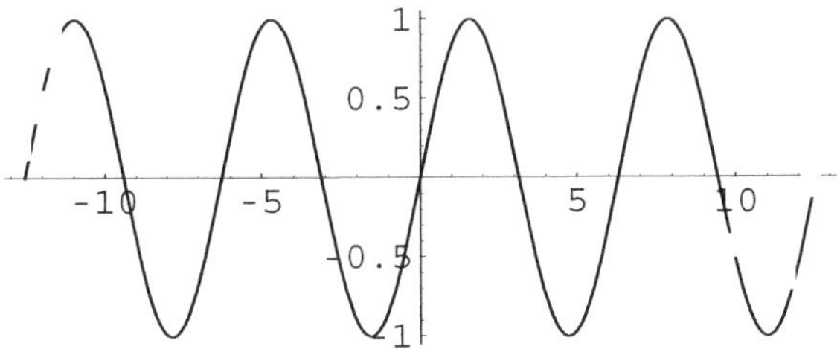

Figure 5-8

❑

The repetitive nature of the sine function is reflected in

Definition 5.16: Periodic Function, Period

A function, *f*, whose domain consists of all of the real numbers is called **periodic with period** $p > 0$, if for all real x and all integers k (positive or negative)

$$f(x + kp) = f(x).$$

❑

When referring to the period of a function, we usually mean the smallest period, provided one exists. Note that *sin* is periodic with (smallest) period 2π. Periodic functions dominate the study of electromagnetic radiation, (including light waves), as well as that of mechanical and electrical vibrating systems such as musical instruments.

Determination of *sin′*, *sin″*

We already know from equation 5.56 on page 161 and equation 5.57 that

$$SIN'(\vartheta) = \sqrt{1 - SIN^2(\vartheta)} = COS(\vartheta) \quad \text{for } -\pi/2 < \vartheta < \pi/2.$$

From the chain rule (Theorem 2.13 on page 53) and the rule for differentiating the square root function, Example 2.3 on page 29, it follows that

$$\begin{aligned} SIN''(\vartheta) &= COS'(\vartheta) \\ &= \frac{1}{2\sqrt{1 - SIN^2(\vartheta)}}(-2SIN(\vartheta)COS(\vartheta)) \\ &= -SIN(\vartheta) \end{aligned}$$

for $-\pi/2 < \vartheta < \pi/2$.

Except at points ϑ which are odd multiples of $\pi/2$ it's a straightforward, albeit tedious, task to show from the definition that *sin* is differentiable at all other points in its domain, and determine the formula for $sin'(\vartheta)$. To take care of the just mentioned exceptions, we need only handle the case $\vartheta = \pi/2$; this can be accomplished without too much difficulty by applying the mean value theorem. All of these details are left for Problem 5.12.1 on page 165. Assuming these details are disposed of, we can now present the theorem summarizing what has been established above.

Theorem 5.17: Properties of *sin* and *cos*

The *sine* function, *sin,* given by Definition 5.15, is

- differentiable at all points in its domain, with derivative *cos,* which is computationally specified by the algorithm presented below
- periodic, with smallest period 2π

The *cosine* function, whose formula for $-\pi/2 \le \vartheta \le 3\pi/2$ is given by

$$cos(\vartheta) = \begin{cases} \sqrt{1 - [sin(\vartheta)]^2} & \text{for } -\pi/2 \le \vartheta \le \pi/2 \\ -\sqrt{1 - [sin(\vartheta)]^2} & \text{for } \pi/2 \le \vartheta \le 3\pi/2 \end{cases}$$

- is periodic with smallest period 2π
- is differentiable everywhere, with derivative $-sin$

❑

Note that the above enables computation of $cos(\vartheta)$ for all real ϑ.

The formulas for $cos(\vartheta)$ are easy to remember from Figure 5-9, which shows clearly the algebraic signs of $sin(\vartheta)$ and $cos(\vartheta)$ as $cos(\vartheta), sin(\vartheta)$ travels around the unit circle, with

$$(cos(0), sin(0)) = (1, 0)$$
$$(cos(\pi/2), sin(\pi/2)) = (0, 1)$$
$$(cos(\pi), sin(\pi)) = (-1, 0)$$
$$(cos((3\pi)/2), sin((3\pi)/2)) = (0, -1)$$

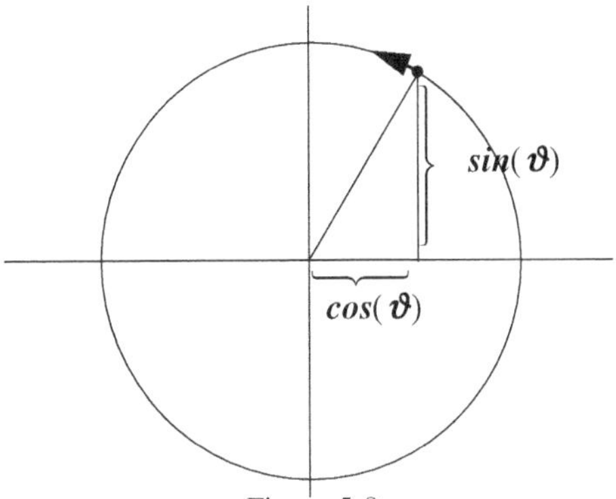

Figure 5-9

A graph of the *cosine* function is shown in Figure 5-10.

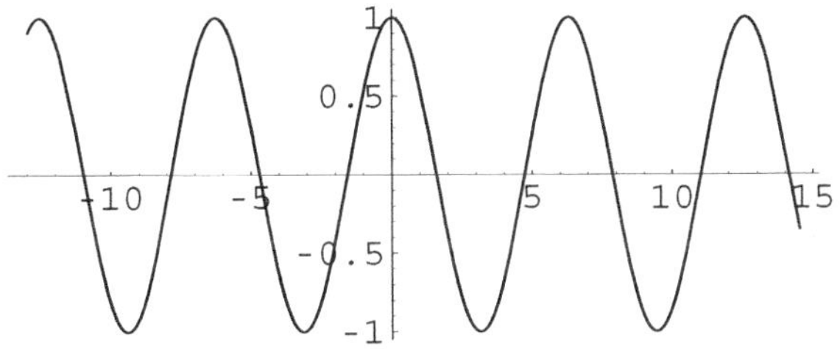

Figure 5-10

This section, along with the details determined in Appendix 2 provides the required basic material for proper development of the sine and cosine functions, motivated by examination of simple vibrating systems.

In the next section we will investigate the simplest interesting results about such systems which these functions provide. After Taylor's theorem and complex numbers have been developed in the next chapter, we can carry out some theoretical investigation of multicompartmental kinetic models and vibrating systems with friction.

Problems 5.12

1. Prove that *sin* is differentiable at all reals by
   a. establishing its derivative at all points not of the form $(2k + 1)\pi/2$.
   b. showing that $sin'(\pi/2) = 0 = sin'(-\pi/2)$. Hint: Use the relaxed conditions mean value theorem, Theorem 3.2 on page 63.
   c. extending part b to show $sin'((2k + 1)\pi/2) = 0$.

2. Establish that $cos' = -sin$.

3. Show that

   a. $sin'' + sin = 0,\ sin(0) = 0,\ sin'(0) = 1.$

   b. $cos'' + cos = 0,\ cos(0) = 1,\ cos'(0) = 0.$

4. Suppose we define y by the equation $y(t) = x(qt)$, for all nonnegative t, where $x'' + x = 0$, and q is a positive real number. Show that for all nonnegative t, $y''(t) + q^2 y(t) = 0$. Hint: From the chain rule, (see equation 2.30 on page 53)

$$y''(t) = q^2 x''(qt).$$

8 First Applications of *sin, cos*

Example 5.13: $x'' + x = 0,\ x(0) = a,\ x'(0) = b$

This example examines the behavior of the simplest frictionless spring-mass system, in which the ratio k/m (spring constant/mass) equals 1.
Since

$$cos(0) = 1,\ cos'(0) = -sin(0) = 0,$$

$$sin(0) = 0,\ sin'(0) = cos(0) = 1,$$

and $cos'' + cos = 0$, $sin'' + sin = 0$ (see Problem 5.12.3), we see that $x = a\,cos + b\,sin$ is one solution of the above differential equation satisfying the given initial conditions. This is, in fact, the only solution. To see this, suppose y is another solution — by which we mean that suppose $y'' + y = 0,\ y(0) = a,\ y'(0) = b.$ Letting $z = x - y$ we see that

$$z'' + z = 0,\ z(0) = 0,\ z'(0) = 0.$$

Now we use the same trick that led to equation 5.52 on page 158. That is, we multiply the differential equation above by z, obtaining $z''z' + zz' = 0$, which is equivalent to

$$[(z')^2 + z^2]' = 0.$$

This last equation implies that $(z')^2 + z^2$ is constant (see Problem 3.5.2 on page 67). But since $z(0) = 0$ and $z'(0) = 0$, this constant is 0, so that the function $z = 0$, i.e., $x - y = 0$. So we must have $y = x$ as asserted.

If we specified, say $a = 2$, $b = -3$, a graph of the solution with domain [0, 20] is shown[1] below in Figure 5-11. This graph suggests that

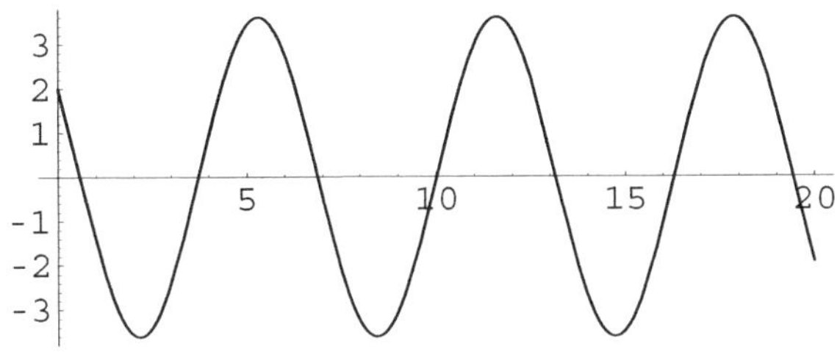

Figure 5-11

a cos + *b sin* is some sort of *sine wave* — a topic we will treat properly in the next chapter.

Just note here that the system being considered is essentially a general frictionless spring-mass system subject to no external forces, because the time and mass units can always be chosen so that the ratio $k/m = 1$. So we see that in such a system the vibrations continue indefinitely, and periodically, neither growing nor diminishing. As a description of an actual system, this is often adequate for substantial time periods, as long as friction takes a long time to have a significant effect.

■

In the next chapter we will carry out a theoretical investigation to take account of friction which is assumed proportional to velocity. In this study, we will determine how long it takes until the solution from the description above becomes inadequate due to friction. In the next example we will do some preliminary numerical computations to get an idea of the effect of this type of friction.

Problem 5.14

Find a solution of the differential equation

$$x'' + \frac{k}{m}x = 0, \; x(0) = a, \; x'(0) = b$$

and show that this solution is unique. Hints: See Problem 5.12.4 on page 166, and Example 5.13.

To give some idea of what happens when we try to account for friction in a spring-mass system, we present

---

1. The *Mathematica* command to obtain this graph is: precisely the following. Plot[2 cos[x] - 3 sin[x],{x,0,20}]

Example 5.15: Friction proportional to velocity

We have already mentioned that the position of a mass in a spring-mass system with friction proportional to velocity may be described by a differential equation of the form of equation 5.41 on page 155

$$x'' + qx' + rx = g,$$

where q and r are positive constants, with $q = c/m$ representing the frictional component, and $r = k/m$ the spring-mass component. For the usual reasons of simplicity, choose $r = 1$; and let $q = .1$, to represent a moderately small frictional component. Also, assume the forcing function, $f = mg = 0$, indicating the absence of any other external forces on the mass.[1] So we are investigating numerical solution of the differential equation

$$x'' + 0.1x' + x = 0.$$

In order to obtain a numerical solution, we must specify some initial conditions, in this case choosing $x(0) = 0,\ x'(0) = 1$. Now using exactly the same approach which led to equation 5.47 on page 156, if we let x_k denote the numerical approximation to $x(kh)$, we are led to the equations

$$x_{k+2} = (2 - 0.1h)x_{k+1} + (-1 + 0.1h - h^2)x_k.$$

To these we append the initial conditions $x_0 = 0,\ x_1 = h$. As the domain of interest we choose [0,50], yielding k values going from 0 to about $50/h$. For a choice of $h = .05$, this leads to the equations

$$x_{k+2} = 1.995x_{k+1} - 0.9975x_k,\quad k = 0, 1, ..., 1000,\quad x_0 = 0, x_1 = 0.05.$$

We solve these equations and plot the graph of the solution in Figure 5-12, using *Mathematica*, employing the same approach which led to Figure 5-3 on page 156:

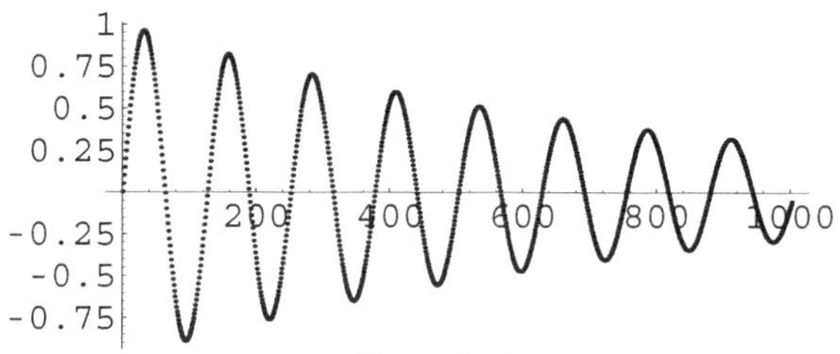

Figure 5-12

1. The reason for the stress placed on the solution of differential equations in the absence of external forces, is that this solution characterizes the behavior of the system described by these equations, as is explained in the remark on page 410, where the general structure of the solution of linear algebraic equations, (which can be used to approximate the differential equations), is investigated.

Here is the graph we obtain had we chosen $h = .01$, rather than $h = .05$. Note that this requires $x_1 = 0.01$.

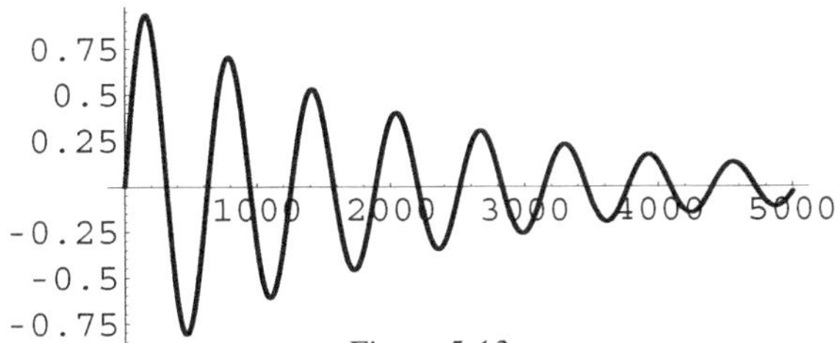

Figure 5-13

There are two points worth noticing here. First, while the graphs are similar, the peaks on the first one (presumably the less accurate result) are substantially larger than those on the second graph. Second, the computation time to obtain the assumed more accurate graph was substantially greater than that needed for the other graph. So, we need to develop results to determine how accurate our numerical computations are, and it would appear that we need to develop more efficient ways of numerically determining solutions of such differential equations. The tools for both of these tasks will be initiated in the chapter to follow. The results of our numerical investigation are in line with what we expect — namely, that friction damps out the vibrations.

■

Exercises 5.16

1. The equation $x'' - 0.1x' + x = 0$ includes a *negative friction* term. While this may not have a standard mechanical interpretation, it could be realized via electronic circuitry, where $x(t)$ is interpreted as electrical current at time t.

   a. For those with access to *Mathematica* or similar programs, numerically approximate the solution of this equation for the same time interval and subject to the same initial conditions as those used in Example 5.15 just preceding.

   b. Provide a theoretical explanation for the behavior exhibited by the graph of the solution of part a. Hint: For the function y defined by the formula $y(t) = x(-t)$, we have $y''(t) = x''(-t)$, $y'(t) = -x'(-t)$.

2. Numerically approximate the solutions of the following equations, subject to the same initial conditions as given in Example 5.15; graph these solutions, and comment on them.

   a. $x'' + \sqrt{3}x' + x = 0$.

   b. $x'' + 2x' + x = 0$.

c. $x'' + 8x' + x = 0$.

3. Solve the equations $x'' + 2x = 0,\ x(0) = 0,\ x'(0) = 1$
Hints: To obtain a 2 from the x'' term, you need to try $sin(\sqrt{2}\,t)$. To satisfy the initial conditions choose the proper constant to multiply $sin(\sqrt{2}\,t)$. Examining just the differential equation itself, and assuming there is a solution, draw some conclusions about the shape of the solution — i.e., its concavity.

4. Numerically solve $x'' + 2x' + 2x = 0$ subject to initial conditions $x(0) = 0,\ x'(0) = 1$, and graph the solution. Examining just the differential equation itself, and assuming there is a solution, draw conclusions about the shape of the solution — contrasting its shape with the shape of the solution of the previous question.

If you went through exercises 3 and 4 above, you should have noticed that the effect of friction was not only to damp out the vibrations, but to reduce their frequency as well — i.e., to increase the length of time between successive values of time at which the solution takes on the value 0. This can be verified analytically by showing that the solution of $x'' + 2x' + 2x = 0$ subject to initial conditions $x(0) = 0,\ x'(0) = 1$ is $x(t) = exp(-t)sin(t)$. (You can use this solution to verify your conclusions in the answer to exercise 4.) In the next chapter we shall see how this result can be systematically derived.

We now turn our attention to solutions of a few simple but important equations of the form

$$x'' + x = g \qquad \textbf{5.60}$$

where g is a known function. This describes a frictionless system in which the mass is acted on by a known forcing function. This force may be applied to the wing of an airplane by an engine, or may represent the force of winds acting on a bridge, such as the Tacoma Narrows bridge, which collapsed as a result of these forces.

The first situation we will look into concerns the response to a sinusoidal forcing function of the form

$$g(t) = sin(ft),$$

where f is some constant, which, for reasons that will soon be clear, is chosen to be different from 1 and -1. In order to solve

$$x'' + x = sin(ft),\ x(0) = 0,\ x'(0) = 0$$

we notice that if $y(t) = sin(ft)$, then $y''(t) = -f^2 y(t)$, so that $(y''(t) + y(t) = (1 - f^2)sin(ft))$. From this it is seen that one solution of the equation $x'' + x = sin(ft)$ is given by

$$x(t) = x_p(t) = \frac{1}{1-f^2} sin(ft) .$$

(This explains why we excluded the case $f = \pm 1$.) This also satisfies the initial condition $x(0) = 0$, but not the condition $x'(0) = 0$. But now we observe that if to the solution x_p we add any solution, $x_h = a\ cos + b\ sin$ of the *homogeneous* equation $x'' + x = 0$ (see Example 5.13 on page 166) the function $x = x_p + x_h$ is also a solution of $x''(t) + x(t) = sin(ft)$. Moreover we can choose the constants a and b so that $x = x_p + x_h$ satisfies any specified initial conditions on $x(0)$ and $x'(0)$. The details of this computation are left to the reader. Using the same reasoning as in Example 5.13, the solution satisfying such initial conditions is unique. We summarize the results concerning this case in

Theorem 5.18: Solution of $x'' + x = sin(ft)$, $x(0) = 0$, $x'(0) = 0$

For $f \neq \pm 1$ the unique solution of the equations being considered is

$$x(t) = \frac{1}{1-f^2} sin(ft) - \frac{f}{1-f^2} sin(t).$$

❑

Exercises 5.17

1. Show that there is at most one solution of the equations
$$x'' + x = g,\ x(0) = \alpha,\ x'(0) = \beta,$$
where g is a given function. Hint: See Example 5.13 on page 166.
2. Determine the unique solution of the equations
$$x'' + x = sin(ft),\ \ x(0) = \alpha,\ x'(0) = \beta.$$
Hint: See the discussion immediately preceding Theorem 5.18.

We have seen from Theorem 5.18 that the unique solution of

$$x'' + x = sin(ft),\ x(0) = 0,\ x'(0) = 0$$

for $x \neq \pm 1$ is

$$x_f(t) = \frac{1}{1-f^2} sin(ft) - \frac{f}{1-f^2} sin(t).$$

We would like to see what's going on as f gets close to the excluded value 1 ($sin(1 \cdot t)$ being the *natural* vibration of this system). That is, we want to see what happens if we supply a forcing vibration whose frequency is close to the natural vibration of the system.

To facilitate the computation, let $f = 1 + h$, and note that f is near 1 precisely when h is near 0. Then for nonzero h,

$$\begin{aligned} x_f(t) &= \frac{1}{1-(1-h)^2}[sin((1+h)t) - (1+h)sin(t)] \\ &= \frac{1}{-2h-h^2}[sin((1+h)t) - sin(t) - h\ sin(t)] \\ &= \frac{1}{-(2+h)h}[sin((1+h)t) - sin(t)] + \frac{sin(t)}{2+h}. \end{aligned}$$

i.e.,

$$x_f(t) = \frac{1}{-(2+h)}\left[\frac{sin((1+h)t) - sin(t)}{h}\right] + \frac{sin(t)}{2+h}. \qquad \textbf{\textit{5.61}}$$

The last term is a simple sine wave. To see the behavior of the first term, choose a small value for h, say $h = .001$, and graph $x_f(t)$ for t in $[0, 8\pi]$. In *Mathematica* we would just type

Plot[1000 (Sin[1.001 t] - Sin[t]),{t,0,8 Pi}]

to obtain the graph shown in Figure 5-14.

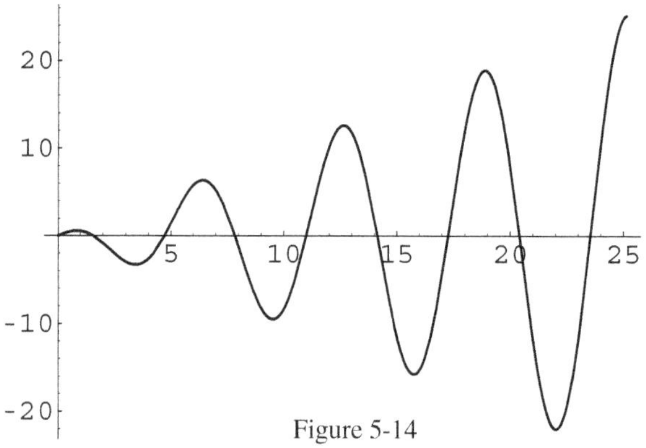

Figure 5-14

For any fixed $f = 1 + h$, the magnitude of $x_f(t)$ can be seen to be bounded by the expression

$$\frac{1}{(2-|h|)|h|}2 + \frac{1}{2-|h|} .$$

What seems to be happening is that for small h a sinusoidal forcing function whose frequency is close to the system's natural frequency has an effect which grows very large as time passes. Such growth can be very dangerous if it happens to a bridge, airplane wing, or automobile axle. To get a more quantitative idea of what happens, let us see if we can analytically determine the solution to

$$x'' + x = sin, \quad x(0) = 0 = x'(0),$$

by letting h approach 0 in equation 5.61. The mean value theorem, Theorem 3.1 on page 62, shows that

$$\frac{sin((1+h)t) - sin(t)}{h} = t\, cos((1+\vartheta h)t)$$

where $0 < \vartheta < 1$, so that

$$x_f(t) = \frac{t}{-(2+h)} cos((1+\vartheta h)t) + \frac{sin(t)}{2+h}.$$

The limit of this as $f \to 1$ (i.e ., as $h \to 0$) appears to be

$$x_1(r) = -\frac{t}{2} cos(t) + \frac{1}{2} sin(t). \qquad \textbf{5.62}$$

Evidently, $x_1(0) = 0,\ x_1'(0) = 0,$ and furthermore, since

$x_1'(t) = \frac{t}{2} sin(t)$, so that $x_1''(t) = \frac{1}{2} sin(t) + \frac{t}{2} cos(t)$, we find

$$x_1''(t) + x(t) = sin(t).$$

Thus, using Exercises 5.17.1 on page 171, we obtain the following result.

Theorem 5.19: Resonance, the solution of $x'' + x = sin$

The unique solution of the equations

$$x'' + x = sin, \quad x(0) = 0 = x'(0)$$

is

$$x(t) = -\frac{t}{2} cos(t) + \frac{1}{2} sin(t).$$

❑

This shows that applying a time varying force which is a sinusoid with the same frequency as the natural vibration of the system, generates a response with the same frequency, whose peaks rise linearly with time, as we see illustrated in Figure 5-15.

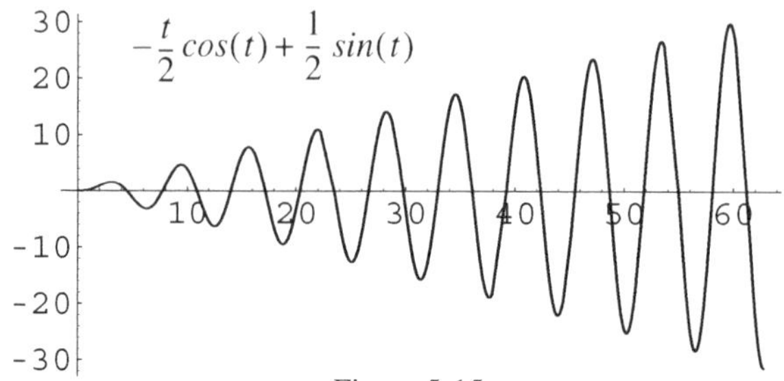

Figure 5-15

This phenomenon is called *resonance*, and as already stated, can be very dangerous. In the design stage of vibrating systems, such as airplanes and bridges, care must be taken to exclude energy input at frequencies near the natural vibration frequency(s) of the system. Most of us have had experience with resonance in trying to gain height on a swing. We supply energy in small amounts at the top of the swing's trajectory. This input is supplied at the same frequency as the natural frequency of the swing (with you on it). This effect is precisely what you don't want to get from the engines on an aircraft. Avoidance of the possibility of resonance is the reason behind a platoon breaking formation when crossing a bridge.

Even in the presence of friction, resonance is important. To see this try the following exercises.

Exercises 5.18

Using the appropriate modification which led to the equations and graphs in Example 5.15 on page 168 numerically solve and graph the following:

1. $y''(t)+0.1y'(t)+y(t) = sin(t),\ y(0) = 0 = y'(0),$ for $0 \le t \le 10\pi$.

2. $y''(t)+0.1y'(t)+y(t) = sin((2t),\ y, e, p(0)) = 0 = y'(0),$ for $0 \le t \le 10\pi$.

In the next chapter we will investigate analytic solution of these equations.

Chapter 6

Taylor's Theorem

1 Introduction

It is difficult to imagine being able to do much applied mathematics without Taylor's theorem. Its importance stems from its incredibly widespread use as a means of representing functions in a simple form which is nonetheless ideal for much powerful theoretical manipulation. Taylor's theorem approximates a function by a polynomial whose local behavior (value and first n derivatives) matches that of the function of interest. This approximation is useful because, in addition to its simplicity there is available a simple estimate of its accuracy.

After we establish the simplest version of Taylor's theorem, and present the version in most common use, we will conclude this chapter with a variety of applications.

2 Simplest Version of Taylor's Theorem

We'll introduce this subject from a geometric viewpoint — letting our picture lead us to the algebraic approach. Continuing on with the algebraic approach, it is then very easy to establish the general result.

Start off by supposing that we know the value $f(0)$ of some real valued function, f, of a real variable, and bounds

$$L \leq f'(x) \leq U, \qquad \textbf{\textit{6.1}}$$

valid for $0 \leq x \leq b$. Under these conditions it seems evident that the graph of f on the interval [0, b] lies between two lines through the point $(0, f(0))$, with slopes L and U, as indicated in Figure 6-1.

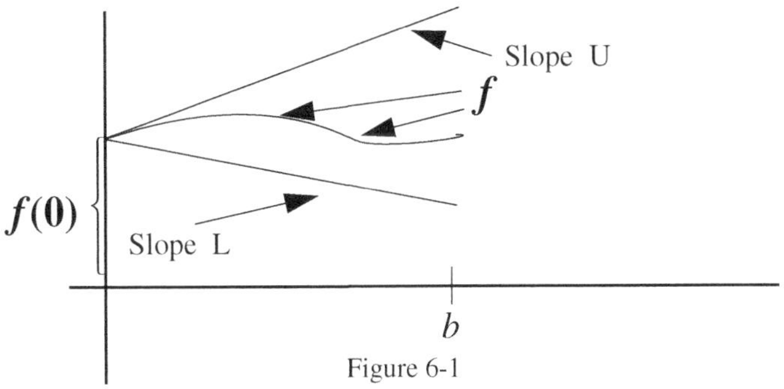

Figure 6-1

This geometrical assertion translates into the following algebraic claim:

$$Lx + f(0) \le f(x) \le Ux + f(0) \quad \text{for} \quad 0 \le x \le b. \qquad \textbf{6.2}$$

It is now reasonable to ask how we might arrive at inequalities 6.2 algebraically, starting from inequalities 6.1. The answer is simply to integrate (*add up*) inequalities 6.1 — that is, use inequality 4.18 of Theorem 4.5 on page 107 to obtain

$$\int_0^x L \le \int_0^x f' \le \int_0^x U \quad \text{for all} \quad x \quad \text{in} \quad a, b. \qquad \textbf{6.3}$$

Evaluating the three integrals of these inequalities using equation 4.14 of Theorem 4.3 on page 103 yields

$$Lx \le f(x) - f(0) \le Ux,$$

which is equivalent to inequalities 6.2.

It is now pretty evident how to extend our result when we have more knowledge about the local behavior of f near 0, and *global* bounds for some higher derivative — i.e., bounds for a higher derivative over the entire interval $[0, b]$. Specifically suppose we know the values

$$f(0), f'(0), f''(0), \ldots, f^{(n)}(0) \qquad \textbf{6.4}$$

and values L_{n+1}, U_{n+1} such that $f^{(n+1)}$ is continuous with

$$L_{n+1} \le f^{(n+1)}(x) \le U_{n+1} \qquad \textbf{6.5}$$

at all x with $0 \le x \le b$.

All we need do is integrate inequalities 6.5 $n + 1$ times to get the general result that we want; but in order to keep these manipulations simple and at the same time proper, it's best to use the functional notation of Section 5.4 (starting on page 144). Using this notation, we let t stand for the identity function with domain [0, b] — i.e., for each x with $0 \le x \le b$, $t(x) = x$. Then $\int_0^t f^{(k)}$ is the function whose value at the number x is

$$\int_0^x f^{(k)} = f^{(k-1)}(x) - f^{(k-1)}(0)$$

(this latter step justified by equation 4.14 of Theorem 4.3 on page 103), or

$$\int_0^t f^{(k)} = f^{(k-1)}(t) - f^{(k-1)}(0)$$

(the last term being the constant function[1] with value $f^{(k-1)}(0)$).

1. In standard abuse of notation we fail to distinguish between a real number, c, and the constant function whose value at all x in its domain is c.

Then we may legitimately write

$$\int_0^t t^k = \frac{t^{k+1}}{k+1}, \qquad \textbf{\textit{6.6}}$$

where t stands for the identity function whose domain is the set of real numbers.

At this point it is easy to repeatedly integrate (apply $\int_0^t$) to the inequalities

$$L_{n+1} \le f^{(n+1)} \le U_{n+1}$$

to first obtain

$$\int_0^t L_{n+1} \le \int_0^t f^{(n+1)} \le \int_0^t U_{n+1}$$

(justified by inequality 4.18 of Theorem 4.5 on page 107). Now making use of the *fundamental integral evaluation formula,* equation 4.14 of Theorem 4.3 on page 103, we obtain

$$L_{n+1} t \le f^{(n)} - f^{(n)}(0) \le U_{n+1}\, t$$

or

$$f^{(n)}(0) + L_{n+1} t \le f^{(n)} \le f^{(n)}(0) + U_{n+1} t.$$

Integrating again, we obtain

$$f^{(n)}(0)t + L_{n+1}\frac{t^2}{2} \le f^{(n-1)} - f^{(n-1)}(0) \le f^{(n)}(0)t + U_{n+1}\frac{t^2}{2},$$

or

$$f^{(n-1)}(0) + f^{(n)}(0)\,t + L_{n+1}\frac{t^2}{2} \le f^{(n-1)} \le f^{(n-1)}(0) + f^{(n)}(0)\,t + U_{n+1}\frac{t^2}{2}.$$

Continuing on in this fashion, strictly speaking using mathematical induction, but where informally we define $k!$ to be $1 \cdot 2 \cdot 3 \cdots k$, we obtain

$$f(0) + f'(0)t + f''(0)\frac{t^2}{2} + \cdots + f^{(n)}(0)\frac{t^n}{n!} + L_{n+1}\frac{t^{n+1}}{(n+1)!} \le f$$

$$\le f(0) + f'(0)\,t + f''(0)\frac{t^2}{2} + \cdots + f^{(n)}(0)\frac{t^n}{n!} + U_{n+1}\frac{t^{n+1}}{(n+1)!}.$$

To cut down the burden of writing, we define the *n-th* degree Taylor polynomial, T_n, associated with the n times differentiable real valued function f, of a real variable, by the following formula.

Definition 6.1: n-th degree Taylor polynomial about 0

If f is a real valued function of a real variable, (Definition 2.1 on page 21), with existing n-th derivative, $f^{(n)}(0)$ (Definition 2.5 on page 30) then its n-th degree Taylor polynomial, denoted by T_n is defined by the formula

$$T_n(t) = \sum_{k=0}^{n} f^{(k)}(0)\frac{t^k}{k!}, \tag{6.7}$$

where $f^{(k)}$ is the k-th derivative[1] of f, $D^k f$, and $k!$ (k factorial)[1] is the product of the first k positive counting numbers.

❑

In terms of the Taylor polynomials, we have shown that if $f^{(n+1)}$ is continuous at each point of the closed interval $[0, b]$, and

$$L_{n+1} \le f^{(n+1)} \le U_{n+1} \text{ on } [0; b],$$

then for all t in $[0; b]$

$$L_{n+1}\frac{t^{n+1}}{(n+1)!} \le f(t) - T_n(t) \le U_{n+1}\frac{t^{n+1}}{(n+1)!}. \tag{6.8}$$

Taylor's theorem in this form is a little inconvenient to use. With the help of the result that if a function is continuous at each point of a closed interval, $[a, b]$, then this function achieves its maximum and minimum on that interval (which we investigated in the advanced section starting page 63), and the intermediate value theorem for continuous functions, Theorem 3.5 on page 75, we can put Taylor's theorem in a more convenient form. We may assume that the values L_{n+1} and U_{n+1} and respectively the minimum and maximum of $f(t)$ for $0 \le t \le b$; choosing $t = b$ in inequalities 6.8, we rewrite them as

$$\min_{0 \le t \le b} f^{(n+1)}(t) \le \frac{f(b) - T_n(b)}{b^{n+1}/(n+1)!} \le \max_{0 \le t \le b} f^{(n+1)}(t).$$

Then by the intermediate value theorem, the quantity in the middle must be the value of $f^{(n+1)}$ somewhere inside $[0, b]$; call this value ϑb, $0 < \vartheta < 1$. Hence we may write

$$\frac{f(b) - T_n(b)}{b^{n+1}/(n+1)!} = f^{(n+1)}(\vartheta b).$$

---

1. Recall from Definition 2.5 that the k-th derivative is defined inductively, namely $D^0 f = f$, and for all counting numbers, k, $D^{k+1} f = D(D^k f)$. Similarly, $k!$ (*k factorial*) is defined by $0! = 1$, and for each counting number, k, $(k+1)! = (k+1)k!$.

Cross-multiplying and adding $T_n(b)$ to both sides yields almost the form we want. We will state a slightly more general form, whose derivation involves replacing the value 0 by any real number, denoted by x_o, and which also allows the interval on which the result holds to have x_o as its right endpoint. These minor details will be left for the reader to fill in.

Theorem 6.2: Taylor's theorem in one variable

Suppose f is a real valued function of a real variable (Definition 2.1 on page 21) possessing a continuous $n + 1$st derivative (Definition 2.6 on page 34 and Definition 2.5 on page 30) at each point in the closed interval, (Definition 3.11 on page 88), from x_o to $x_o + h$ (where h can be negative). Then

$$f(x_o + h) = \sum_{k=0}^{n} f^{(k)}(x_o)\frac{h^k}{k!} + f^{(n+1)}(x_o + \vartheta h)\frac{h^{n+1}}{(n+1)!},$$

where ϑ is some (usually unknown) value between 0 and 1 (determined by f, x_o, n, and h). The last term in the above equality is referred to as the *error* or *remainder term*.

❑

Note that for $n = 0$ Taylor's theorem is just the *mean value theorem*, Theorem 3.1 on page 62.

We can see immediately that Taylor's theorem allows arbitrarily accurate estimates of functions whose n + 1-th derivatives do not grow too fast with n. Specifically, if B_{n+1} is any upper bound[1] for $|f^{(n+1)}(x)|$, for x between x_o and $x_o + h$, then Taylor's theorem shows that

$$\left| f(x_o + h) - \sum_{k=0}^{n} f^{(k)}(x_o)\frac{h^k}{k!} \right| \le B_{n+1}\frac{|h^{n+1}|}{(n+1)!}. \qquad \textbf{6.9}$$

So, if as n grows large, the quantity $B_{n+1}\frac{|h^{n+1}|}{(n+1)!}$ approaches 0 for some fixed value of h, the n-th degree Taylor polynomial $\sum_{k=0}^{n} f^{(k)}(x_o)\frac{h^k}{k!}$ above yields an arbitrarily accurate estimate of $f(x_o + h)$.
This result should not be too surprising, since if B_{n+1} were 0, the function $f^{(n+1)}$ would be 0, and hence the function f would be an n-th degree polynomial[2]. So, if B_{n+1} is close to 0 you would expect f to be close to an n-th degree polynomial, as indicated by inequality 6.9.

---

1. Chapter 3 of *Modern University Calculus*, by Bell et al., [1], deals extensively with finding upper bounds for simple algebraic expressions.
2. Just antidifferentiate the equation $f^{(n+1)} = 0$ $n + 1$ times, using the fact that two functions with the same derivative differ by a constant.

As we will see in the next two illustrations, Taylor's theorem provides a simple way to obtain arbitrarily accurate estimates of the trigonometric and exponential functions.

Example 6.1: Taylor theorem approximation to *sin* at $x_0 = 0$

From Theorem 5.17 on page 164 and Problem 5.12.2 on page 166 we know that $sin' = cos$, $cos' = -sin$, $sin(0) = 0$, and $cos(0) = 1$. Hence, we may write

$$\begin{aligned} sin(0) &= 0 \\ sin'(0) &= cos(0) = 1 \\ sin''(0) &= -sin(0) = 0 \\ sin^{(3)}(0) &= -cos(0) = -1 \\ sin^{(4)}(0) &= sin(0) = 0 \\ &\text{etc.} \end{aligned}$$

Because these values are alternately 0, with the nonzero values changing sign as shown, the Taylor polynomial for sin about 0 is given by

$$T_{2n-1}(h) = T_{2n}(h) = h - \frac{h^3}{3!} + \frac{h^5}{5!} - \cdots \frac{(-1)^{n+1}}{(2n-1)!} h^{2n-1} \qquad n = 1, 2, \ldots .$$

For fixed h, the magnitude of the remainder,

$$\left| f^{(2n+1)}(\vartheta h) \frac{h^{2n+1}}{(2n+1)!} \right| \le \frac{|h|^{2n+1}}{(2n+1)!}$$

can be seen to approach 0 as n becomes large, as follows.

Since $|f^{(2n+1)}|$ is either $|sin|$ or $|cos|$, neither exceeding 1, as n increases by 1, the factor

$$\frac{|h|^{2n+1}}{(2n+1)!} \quad \text{is multiplied by} \quad \frac{h^2}{(2n+2)(2n+3)}$$

which clearly becomes small for large n. So

$$\lim_{n \to \infty} T_{2n}(h) = \lim_{n \to \infty} \sum_{j=1}^{n} (-1)^{j+1} \frac{h^{2j-1}}{(2j-1)!} = sin(h).$$

If, for instance, we want to determine $sin(1)$ to an accuracy of .000001, we need only choose n so that

$$\frac{1^{2n+1}}{(2n+1)!} < 0.000001 ,$$

a simple matter. The value $n = 5$ is seen to be adequate.

The value

$$T_5(1) = 1 - \frac{1}{3!} + \frac{1}{5!} - \frac{1}{7!} + \frac{1}{9!} - \frac{1}{11!} \cong 0.84171.$$ [1]

■

Example 6.2: Taylor theorem approximation to *exp* at $x_0 = 0$

From Theorem 5.6 on page 140 we know that the k-th derivative of the exponential function at 0 satisfies

$$D^k exp(0) = exp^{(k)}(0) = 1 \quad k = 0,1,\dots .$$

So the Taylor polynomial, $T_n(h)$ for exp about $x_0 = 0$ is given by

$$T_n(h) = \sum_{k=0}^{n} \frac{h^k}{k!} \qquad \textbf{\textit{6.10}}$$

Suppose we want to calculate $exp(3)$ to within .0001. This presumably can be accomplished with the approximation $T_n(3)$, by choosing n large enough so that the error bound in Taylor's theorem satisfies the condition

$$B_{n+1}\frac{|3^{n+1}|}{(n+1)!} < 0.0001.$$

For actual implementation of this inequality, we need an upper bound B_{n+1} for

$$|D^k exp(x)| = |exp^{(k)}(x)| = exp(x) = e^x,$$

$0 \le x \le 3$. Using the result of exercise 6.3.1 on page 182, that $e < 3$, we find that we may take

$$B_{n+1} = 3^3.$$

Hence we need only choose n in $T_n(3)$ to satisfy

$$\frac{|3^{n+4}|}{(n+1)!} < 0.0001.$$

It is not hard to verify that $n = 15$ does the job.

■

---

1. The *Mathematica* commands to do this evaluation are N[Limit[Normal[Series[Sin[x],{x,0,11}]],x->1], where Normal converts the desired symbolic sum to a usual *Mathematica* expression. Limit here, together with x->1, evaluates it at 1. You might want to examine the pieces of this command separately, to see what each generates.

Exercises 6.3

1. Show that $e < 3$, by using numerical integration (see Theorem 4.6 on page 117), to show that

$$ln(3) = \int_1^3 \frac{dx}{x} > 1 .$$

2. Simplify the determination of the error bound in Example 6.2 on page 181 by determining instead the value $exp(-3)$ to the appropriate accuracy.

3. Find $e = exp(1)$ to an accuracy of .000000001.

4. How large an n will ensure that the Taylor polynomial $T_n(\pi/2)$ about $x_o = 0$ for $sin(\pi/2)$ will be within .001 of $sin(\pi/2)$?

5. Find the Taylor polynomial $T_n(h)$ about $x_o = 0$ for $cos(h)$.

6. Find the Taylor polynomial $T_n(x)$ about $x_o = 1$ for $ln(x)$.

7. Establish the general version of Taylor's theorem, Theorem 6.2 on page 179, using the results preceding its statement.

Problem 6.4

This problem illustrates that even for simple looking functions, Taylor's theorem may be more of a theoretical tool than a practical one for accurate function calculation. The simple approach to determining the Taylor polynomials for this and many other functions makes use of the geometric series. We will be investigating this in the chapter to follow. Some important practical uses of Taylor's theorem will be introduced in the next section.

For

$$f(x) = \frac{1}{1 - x^2}$$

find the Taylor polynomials T_1, T_2, T_3, T_4 about $x_o = 0$.

3 Applications of Taylor's Theorem

If all that Taylor's theorem accomplished was allowing tabulation of important functions to any specified accuracy, it would be useful, somewhat limited (see Problem 6.4 on page 182), but not terribly exciting. Much more impressive is the dramatic improvement in the ease of carrying out numerical integration and numerical solution of differential equations that Taylor's theorem allows. The examples to be presented should provide some insight into these applications of Taylor's theorem. But the available space and the level of this monograph permit only an introduction to the uses of Taylor's theorem for extensive investigation as it occurs in most challenging scientific applications. A more detailed development of the type of numerical analysis problems which depend heavily on Taylor's theorem can be found in the monograph by Ralston, [14].

Example 6.5: Numerical evaluation of $\int_0^b f$

In order to investigate efficient numerical evaluation of $\int_0^b f$, we will consider the situation in which the real valued function, f, of a real variable has a continuous $n + 1$st derivative for some value of n. The key to our approach is treating this integral as the value $y(b)$ of the solution y of the differential equation

$$y' = f, \quad y(0) = 0.$$

Because $f^{(n+1)}$ is assumed continuous on the closed interval $[0, b]$, it achieves a maximum and minimum on this interval (see parts c and d in the several stages to establishing the mean value theorem, starting on page 63), Denote by B_{n+1} the larger magnitude of these two numbers. Now subdivide the interval from 0 to b into N equal-length subintervals, each of length $h = b/N$. To get a preliminary feeling for this area, we will first look at the case $n = 2$ before proceeding to the general situation. Justified by Taylor's theorem, we may write

$$y((j+1)h) = y(jh) + y'(jh)h + y''(jh)\frac{h^2}{2} + e_j(h), \qquad \textbf{6.11}$$

where (see inequality 6.9 on page 179) for any upper bound B_3 for $|y'''|$ on [0,b]

$$|e_j(h)| \le B_3\frac{h^3}{3!}, \quad j = 0, \ldots, N-1 \qquad \textbf{6.12}$$

(the upper limit for j is chosen so that at this value $(j+1)h = b$).

Choosing $j = N - 1$, and substituting into equation 6.11, we obtain

$$y(b) = y(b-h) + y'(b-h)h + y''(b-h)\frac{h^2}{2} + e_{N-1}(h). \qquad \textbf{6.13}$$

Now set $j = N - 2$ in equation 6.11, so that $(j + 1)h = b - h$, yielding

$$y(b-h) = y(b-2h) + y'(b-2h)h + y''(b-2h)\frac{h^2}{2} + e_{N-2}(h)$$

and substitute this for the first term to the right of the = sign in equation 6.13 to obtain

$$\begin{aligned} y(b) &= y(b-2h) + [y'(b-2h) + y'(b-h)]h + \\ &\quad [y''(b-2h) + y''(b-h)]\frac{h^2}{2} + e_{N-1}(h) + e_{N-2}(h). \end{aligned} \qquad \textbf{6.14}$$

Next we use equation 6.11 with j set equal to $N - 3$ and continue on in the same fashion, repeating this process, inductively, to obtain the equation

$$y(b) = y(0) + \sum_{j=1}^{N} y'(b-jh)h + \sum_{j=1}^{N} y''(b-jh)\frac{h^2}{2} + \sum_{j=1}^{N} e_{N-j}$$

or

$$y(b) - y(0) = \sum_{j=1}^{N} y'(b-jh)h + \sum_{j=1}^{N} y''(b-jh)\frac{h^2}{2} + \sum_{j=1}^{N} e_{N-j}. \qquad \textbf{6.15}$$

Note that the first two sums in this equation have known terms, since $y' = f$ and $y'' = f'$ are known. However, from the fundamental evaluation formula for integrals, equation 4.14 on page 103, we know that

$$y(b) - y(0) = \int_0^b y'.$$

Replacing $y(b) - y(0)$ in equation 6.15 by this integral, and then everywhere in the resulting equation substituting $y' = f$ and $y'' = f'$ yields

$$\int_0^b f = \sum_{j=1}^{N} f(b-jh)h + \sum_{j=1}^{N} f'(b-jh)\frac{h^2}{2} + \sum_{j=1}^{N} e_{N-j}.$$

But from inequality 6.12 it is seen that

$$\left|\sum_{j=1}^{N} e_{N-j}\right| \le NB_3\frac{h^3}{3!} = NB_3\left(\frac{b}{N}\right)^3\frac{1}{6} = B_3\frac{b^3}{6N^2},$$

where B_3 is an upper bound for the magnitude of $y'''(x) = f''(x)$ for $0 \le x \le b$.

Combining these last two results yields the inequality

$$\left| \int_0^b f - \left[\sum_{j=1}^{N} f(b - jh)h + \sum_{j=1}^{N} f'(b - jh)\frac{h^2}{2} \right] \right| \le B_3 \frac{b^3}{6N^2}. \qquad \textbf{6.16}$$

To illustrate the improvement this can yield over previous results, suppose we want to calculate

$$\int_0^1 exp(x^2)dx.$$

Here

$$\begin{aligned} f(x) &= exp(x^2) \\ f'(x) &= 2x\ exp(x^2) \\ f''(x) &= (2 + 4x^2)exp(x^2). \end{aligned}$$

Hence, using the result that $exp(1) = e < 3$ (see exercise 6.3.1 on page 182)

$$\text{for} \qquad 0 \le x \le 1 = b, \quad |f''(x)| \le 18 = B_3.$$

Then the error when using

$$\sum_{j=1}^{N} exp\left[\left(1 - j\frac{1}{N}\right)^2\right]\frac{1}{N} + \sum_{j=1}^{N} 2\left(1 - j\frac{1}{N}\right)exp\left[\left(1 - j\frac{1}{N}\right)^2\right]\frac{1}{2N^2}$$

to estimate

$$\int_0^1 exp(x^2)dx$$

does not exceed

$$B_3\frac{b^3}{6N^2} = 18 \times \frac{1^3}{6N^2} = \frac{3}{N^2}.$$

If we desire that the error not exceed .0003, we need only take $N = 100$. Compare this to the result of Theorem 4.6 on page 117, concerning the integral of a monotone function, where using a simple Riemann sum we can only guarantee an error not exceeding

$$\frac{1}{N}(3 - 1) = \frac{2}{N}$$

for the same partition. So using a simple Riemann sum would need a partition of the interval [0, 1] into

$$N = \frac{2}{0.0003} = 6{,}666$$

equal-length intervals.

■

We leave the proof of the general case, which we now state formally, to the reader.

Theorem 6.3: Numerical integration of function with continuous n+1-st derivative

Suppose that for some positive integer, n, f is a real valued function of a real variable (Definition 2.1 on page 21) with a continuous $n + 1$st derivative, $f^{(n+1)}$ (Definition 2.5 on page 30 and Definition 2.6 on page 34) with domain $[a;b]$ (Definition 3.11 on page 88). Then

$$\left| \int_a^b f - \sum_{i=0}^{n} \left[\sum_{j=1}^{N} f^{(i)}\left(b - j\frac{b-a}{N}\right) \frac{([b-a]/N)^i}{i!} \right] \right| \le B_{n+1} \frac{(b-a)^{n+1}}{N^n},$$

where B_{n+1} is an upper bound (Definition 3.10 on page 88) for $|f^{(n+1)}(x)|$, for $a \le x \le b$.

❑

The next applications of Taylor's theorem, to the numerical solution of more general differential equations, does not here yield a theoretical result of the type we just saw; but it does provide convincing evidence of the power of Taylor's theorem.

Example 6.6: Numerical solution of $x'' + x = 0, x(0) = 0, x'(0) = 1$

To apply Taylor's theorem most easily to this differential equation, it is common to convert it into a *system* of differential equations as follows:

Let $z = x'$ and write the above equations as

$$\begin{aligned} x' &= z & x(0) &= 0 \\ z' &= -x & z(0) &= 1. \end{aligned} \qquad \textbf{\textit{6.17}}$$

It will be evident why it's easier to apply Taylor's theorem to this system (whose solution, as you know from the results of Example 5.13 on page 166 is $x = sin$, $z = cos$). The general use of Taylor's theorem to first order equations (equations involving only first derivatives of the unknown functions) allows a great improvement over the simple approximation of derivatives by difference quotients (as in approximations 5.43 and 5.44 on page 155) for the numerical solution of the equations treated here. It is equally applicable for numerical solution of kinetic equations (see Example 4.5 on page 96).

For the numerical solution we seek, we first treat the simplest non-trivial case, namely, use Taylor's theorem to write

$$\begin{aligned} x(t+h) &\cong x(t) + x'(t)h + x''(t)\frac{h^2}{2} \\ z(t+h) &\cong z(t) + z'(t)h + z''(t)\frac{h^2}{2}, \end{aligned} \qquad \textbf{\textit{6.18}}$$

where the error in these approximations goes to 0 at a rate proportional to h^3 as h approaches 0. Notice that if equations 6.17 have a solution, the derivatives in approximations 6.18 exist. Now using equations 6.17, substitute for x' and z'

$$z' = -x \qquad \text{and hence for } y'', z'' \qquad z'' = -x' = -z$$
$$x' = z \qquad\qquad\qquad\qquad\qquad x'' = z' = -x$$

in approximations 6.18 above, to obtain

$$x(t+h) \cong x(t) + z(t)h - x(t)\frac{h^2}{2},$$

$$z(t+h) \cong z(t) - x(t)h - z(t)\frac{h^2}{2},$$

or

$$x(t+h) \cong (1 - h^2/2)x(t) + hz(t),$$
$$z(t+h) \cong (1 - h^2/2)z(t) - hx(t). \qquad \textbf{6.19}$$

Next, similarly to what we did in Example 6.5, starting on page 183, choose a fixed small value for h and successively substitute $t = 0, h, 2h,....$ in approximations 6.19, to obtain

$$x([j+1]h) \cong (1 - h^2/2)x(jh) + hz(jh),$$
$$z([j+1]h) \cong (1 - h^2/2)z(jh) - hx(jh). \qquad j = 0,1,... \qquad \textbf{6.20}$$

Let x_k denote our approximation to $x(kh)$ and z_k our approximation to $z(kh)$. We rewrite the above approximations as

$$x_{j+1} = (1 - h^2/2)x_j + hz_j$$
$$z_{j+1} = (1 - h^2/2)z_j - hx_j \qquad j = 0,1,2,... \qquad \textbf{6.21}$$

with initial conditions $x_0 = 0$, $z_0 = 1$ (corresponding to the original initial conditions $x(0) = 0$, $x'(0) = z(0) = 1$). With $h = .1$ we use *Mathematica* to solve equations 6.21,[1] finding

$$sin(2) = x(2) \cong x_{20} = 0.90813$$
$$cos(2) = z(2) \cong z_{20} = -0.41927.$$

To five-digit accuracy $sin(2) = .90930$, $cos(2) = -.41615$.

If we use the simplest first-order approach to finding $sin(2)$ numerically, namely, equations 5.47 on page 156, and using the same value, $h = .1$ as

1. The *Mathematica* commands we used to solve system 6.21 are
For[s=0;c=1;k=0,k<20,k++,stemp=.995s+.1c;c=.995c - .1s;s=stemp;
s>>>"sout2" ;c>>>"cout2"]
Output for $x_1,....x_{20}$ is in the file sout2, and for $z_1,....z_{20}$ is in cout2.

above, we obtain $sin(2) \cong 1.00745$. So we see that by using a second-order Taylor approximation, 6.20, rather than a simple linear approximation, the error magnitude at $t = 2$ decreases from

$$|sin(2) - 1.00745| = |.90930 - 1.00745| = .09815$$

to

$$|sin(2) - .90813| = |.90930 - .90813| = .00117,$$

a substantial error reduction at a trifling extra cost in computation.

■

Example 6.7: Glycerol kinetics

This situation arose in the research of our metabolism group. We used the two-compartment model indicated below in Figure 6-2, which is a specific case of the model discussed in Example 4.5 on page 96.

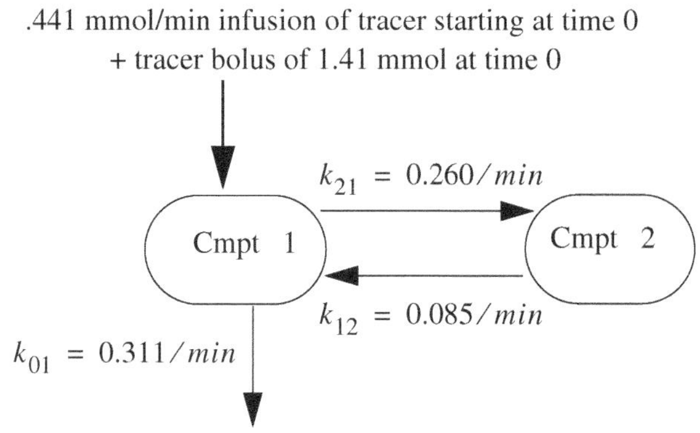

Figure 6-2

Specializing equations 4.5 on page 96, where $x_i(t)$ is the amount of tracer in compartment i at time t, and assuming no tracer present prior to time 0, this tracer system is represented by the specific equations

$$\begin{aligned} x_1'(t) &= -0.571x_1(t) + 0.085x_2(t) + 0.441, & x_1(0) &= 1.41 \\ x_2'(t) &= 0.260x_1(t) - 0.085x_2(t), & x_2(0) &= 0. \end{aligned} \qquad \textbf{6.22}$$

The simplest approximation for obtaining a numerical solution replaces

$$x_1'(t) \text{ by } \frac{x_1(t+h) - x_1(t)}{h} \quad \text{and} \quad x_2'(t) \text{ by } \frac{x_2(t+h) - x_2(t)}{h}$$

(which is just a *first-order* Taylor approximation). The resulting approximations are solved for $x_1(t+h)$ in terms of $x_1(t)$ and $x_2(t)$; and similarly for $x_2(t+h)$.

For carrying out this task we used the *MLAB* computer package, constructing a so-called *Do-File* of *MLAB* commands, GLYCEROL1.do, whose code is presented in Appendix 5 on page 781. The approximate values at
$t = 40$ for a step size of $h = 1$ were

$$x_1(40) \cong 1.3070, \; x_2(40) \cong 3.6479. \qquad \textbf{6.23}$$

To obtain a numerical solution using a second-order Taylor expansion (as in Example 6.6 on page 186) we created the *MLAB Do-File* GLYCEROL2.do which also appears in Appendix 5.

From this we obtained values

$$x_1(40) \cong 1.3023, \; x_2(40) \cong 3.6155 \qquad \textbf{6.24}$$

Running GLYCEROL2.do, but with a step size of $h = .01$ we obtained

$$x_1(40) \cong 1.3026, \; x_2(40) \cong 3.6208 \qquad \textbf{6.25}$$

Next we ran the MLAB differential equation solver, INTEGRATE, in the *MLAB Do-File* GLYCEROLMLI.do (the last item in Appendix 5) The values it returned were

$$x_1(40) \cong 1.3026, \; x_2(40) \cong 3.6210 \qquad \textbf{6.26}$$

Finally, using the formula for the theoretical solution that we will show how to obtain in the next section, in Example 6.11 on page 206,

$$\begin{aligned} x_1(t) &= 1.41801 + \frac{0.639695}{exp(0.612867t)} - \frac{0.647701}{exp(0.0431334t)} \\ x_2(t) &= 4.33743 - \frac{0.315081}{exp(0.612867t)} - \frac{4.02235}{exp(0.0431334t)}, \end{aligned} \qquad \textbf{6.27}$$

we find that approximations 6.26 are accurate to the 5 digits displayed.

The computations leading to approximations 6.24 and 6.25 illustrate the improvements that Taylor's theorem yields in numerical solution of differential equations.

■

Exercises 6.8

For those with access to *Mathematica* or similar programs (*Maple*, *Mathcad*, etc.)

1. Write programs to carry out solution of the equations of 6.22 on page 188, using $h = .1, .05, .005,$ for times $0,1,...,200$.

2. a. Graph the solutions of the previous question, using any graphics program you are comfortable with. It takes a bit of work to get *Mathematica* to do this. If you have the file of ordered pairs, 1 per line, in a file named pairs, *Mathematica* will plot this via the commands

   ```
   y = ReadList["pairs",Number,RecordLists->True]
   ListPlot[y,PlotJoined -> True]
   ```

   b. Once you have the graphs of x_1 and x_2, try to explain the behavior you observe by direct examination of the differential equations 6.22 themselves.

3. A commonly used single-compartment model for kinetics where saturation occurs — i.e., when there is a *clogging* effect, yielding diminishing returns from an increase in the amount, $y(t)$, in the compartment at time t — is the *Michaelis-Menten* model given by

$$y'(t) = -\frac{Vy(t)}{K + y(t)} + R_a(t), \qquad V > 0,\ K > 0.$$

   Take $V = K = 1$ and the infusion rate function

$$R_a(t) = exp(-t),$$

   a choice which arises in modeling an extravascular route of infusion, (See Godfrey, [7]) Solve the equation numerically for $0 \le t \le 10$ using the methods developed here, with a second order Taylor expansion, for $h = .01$, and $h = .001$. To check your answers, you can solve it using *Mathematica*'s NDSolve function.[1]

Some comments are in order on the method used in this section. There is a large body of literature concerning numerical integration, and numerical solution of differential equations; and no claim is made here that the methods based on Taylor's theorem are anywhere near the most efficient in terms of the amount of computation needed to obtain a specified accuracy. However, the method is simple, and with the aid of today's computers and their capabilities for symbolic manipulation, may very well be close to the most efficient in terms of the overall effort (programmer's and computer's resources).

---

1. It takes a bit or work extracting the answer in a convenient way. - NDSolve[{y'[t] == -y[t]/(1+y[t]) + Exp[-t], y[0] == 0},y,{t,0,10}] <Return> Output was {{y_> Interpolating Function....}}. Typing u=%;w = u[[1]]; v[t_] = y[t]/.%[[1]] <Return> yielded a function v which at any numerical value t evaluates to the desired solution. We can then save v in a file, say V, via Save["V",v], and retrieve it later via <<V <Return>

Knowledge of Taylor's theorem is essential in any case, since it forms the basis for so many theoretical investigations arising in applied mathematics.

No account was taken of the effects of roundoff on the error in numerical integration or numerical solutions of differential equations. Roundoff can produce a considerable portion of the total error in numerical techniques, and hence should be taken into account in any serious investigation of accuracy. The tools we have introduced are the ones for such investigations, which, however, do not constitute the basic aim of this monograph.

4 The Connection Between *exp, cos* and *sin*: Euler's Formula and Complex Numbers

In Section 2 we developed the Taylor polynomials for *exp, sin* and *cos,* shown below

Function	Taylor Polynomial, $T_n(t)$	
exp	$1 + t + \frac{t^2}{2!} + \frac{t^3}{3!} + \frac{t^4}{4!} + \frac{t^5}{5!} + \cdots$	
cos	$1 - \frac{t^2}{2!} + \frac{t^4}{4!} - \cdots$	***6.28***
sin	$t - \frac{t^3}{3!} + \frac{t^5}{5!} - \cdots .$	

As you should suspect, such similarities are not usually just coincidences, and they often yield extremely powerful tools to those perceptive enough to investigate them. The similarities above prompt us to enlarge our number system, with enormous benefit.

Just as the rational numbers didn't include $\sqrt{2}$, the real numbers don't include a square root of -1. Although it was the Pythagorean Theorem that initially prompted the enlargement of the number system from the rationals to the reals, the benefits of this enlargement extended much farther. In fact, without the real numbers, further development of mathematics seems difficult to imagine. (Just think how difficult a time we would be having if we were restricted to Roman numerals.) It is unfortunate that the *appended* quantity ***i***, whose square is -1, was labeled *imaginary,* because it obscures its really practical uses. True, ***i*** does not correspond to a measurement such as length or weight — but this doesn't make it *imaginary* - just different. However **the real proof that it's worth introducing is the simplification and the convenience that it provides,** which will become apparent very shortly.

Looking at the Taylor polynomials in expressions 6.28 above, notice that if in the Taylor polynomial for *exp* we substitute it for t, where $i^2 = -1$, then

$$T_n(it) = 1 + it + \frac{(it)^2}{2!} + \frac{(it)^3}{3!} + \frac{(it)^4}{4!} + \frac{(it)^5}{5!} + \cdots$$

$$= 1 - \frac{t^2}{2!} + \frac{t^4}{4!} - \cdots$$

$$+ i\left(t - \frac{t^3}{3!} + \frac{t^5}{5!} - \cdots\right).$$

That is, it looks like

$$exp(it) = cos(t) + i\, sin(t). \qquad \textbf{6.29}$$

This innocent looking equation is **Euler's formula.** It suggests a number of interesting results. For instance, if in this enlarged number system we are about to introduce, it's still true that

$$exp(a + b) = exp(a)\, exp(b),$$

then[1]

$$\begin{aligned} exp(i[x + y]) &= exp(ix)exp(iy) \\ &= [cos(x) + i\, sin(x)][cos(y) + i\, sin(y)] \\ &= cos(x)cos(y) - sin(x)sin(y) + \\ &\quad i[cos(x)sin((y) + cos(y)sin(x))]. \end{aligned}$$

But if equation 6.29 holds, then

$$exp(i[x + y]) = cos(x + y) + i\, sin(x + y).$$

So we might conjecture that

$$\begin{aligned} cos(x + y) &= cos(x)cos(y) - sin(x)sin(y) \\ sin(x + y) &= cos(x)sin(y) + cos(y)sin(x). \end{aligned} \qquad \textbf{6.30}$$

We should certainly be able to verify whether or not these equations hold. (They do, as we will see shortly.)

The extension of the real number system that we are introducing consists of **ordered pairs,** (x,y) **of real numbers,** where any pair of the form $(x,0)$

---

1. There are other extensions — to matrices, for which it is false that $exp(A + B) = exp(A)exp(B)$), because, while $A + B = B + A$, $exp(A)exp(B) \neq exp(B)exp(A)$, see Exercises 13.15.6 on page 567. So one should not naively believe that all formulas holding for some system will remain valid for all objects to which they may be applied when the system is enlarged.

represents the former real number x. The pair (x,y) is usually written $x + iy$ which makes it easy to remember the rule for multiplication of these *complex numbers* — namely,

$$(x, y)*(v, w) = (xv - yw, xw + yv), \qquad \mathbf{6.31}$$

because, assuming the distributive law, $a(b + c) = ab + ac$, holds for complex numbers, replacing i^2 by -1, and collecting together terms with a single factor of i yields

$$\begin{aligned}(x + iy)(u + iv) &= xv + ixw + iyv + i^2 yw\\ &= xv - yw + i(xw + yv).\end{aligned}$$

Notice, that as claimed earlier, the complex numbers of the form $(x,0)$ behave just like the real numbers, and will often simply be denoted by x.

The ordered pair notation for complex numbers is useful in suggesting that they can be considered as points in a plane,[1] as illustrated below.

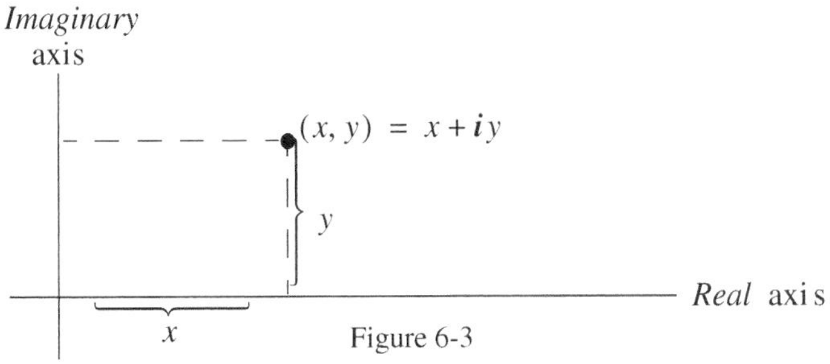

Figure 6-3

Two complex numbers $a + ib$ and $c + id$ are **equal** precisely when $a = c$ and $b = d$. The magnitude of $x + iy$, as is the case with the real numbers, is taken to be its distance from the origin (which is now the complex number $(0,0)$). Motivated by the Pythagorean theorem, we therefore define the **magnitude**, $|x + iy|$ of $x + iy$ by

$$|x + iy| = \sqrt{x^2 + y^2}. \qquad \mathbf{6.32}$$

Addition of complex numbers is defined by

$$(x + iy) + (u + iv) = x + u + i(y + v). \qquad \mathbf{6.33}$$

Multiplication of complex numbers was already defined by **equation 6.31**. We let **equation 6.29** be our definition of e^{it} Finally for the complex number $z = x + iy$, we define

$$e^z = e^{x + iy} = e^x e^{iy}. \qquad \mathbf{6.34}$$

1. That is, as vectors [see the discussion of Lines, starting on page 23] with the usual definition of addition, Definition 2.3 on page 24, but with a very special definition of multiplication, given by equation 6.31.

To see that this works out neatly we have a few tasks to carry out:

- We must extend *division* to complex numbers.
- We must show that equations 6.30 hold.
- We must verify that the *law of exponents*:

$$e^z e^w = e^{z+w} \qquad \textbf{6.35}$$

  holds for complex numbers z,w.
- We need to supply a geometric meaning to the arithmetic operations on complex numbers.

To handle the question of *division* of complex numbers, we need only show how to compute reciprocals. To determine the reciprocal of $z = (x,y) = x + \boldsymbol{i}y$, provided that z is not 0 (i.e., the complex number $(0,0)$ corresponding to the real number 0), we want to find the complex number (a,b) satisfying

$$(a + \boldsymbol{i}b)(x + \boldsymbol{i}y) = 1 = 1 + \boldsymbol{i}0 = (1,0).$$

Using the definition of equality of complex numbers, and the ordered pair notation for the product on the left, we see that we want to choose real numbers a,b such that

$$ax - by = 1 \quad \text{and} \quad bx + ay = 0. \qquad \textbf{6.36}$$

Divide the second of these equations by one of the nonzero values x or y, (for definiteness assume it's y), to obtain

$$a = -b\frac{x}{y}. \qquad \textbf{6.37}$$

Substitute this into the first of equations 6.36, getting

$$-\frac{x^2}{y}b - yb = 1$$

or

$$b = \frac{-1}{y + \dfrac{x^2}{y}} = \frac{-y}{x^2 + y^2}. \qquad \textbf{6.38}$$

Substituting then for b in equation 6.37 yields

$$a = \frac{x}{x^2 + y^2}. \qquad \textbf{6.39}$$

Notice that

$$(x + \boldsymbol{i}y)\left(\frac{x}{x^2 + y^2} + \boldsymbol{i}\frac{-y}{x^2 + y^2}\right) = \frac{x^2}{x^2 + y^2} + \frac{y^2}{x^2 + y^2} + \boldsymbol{i}0 = (1,0) = 1,$$

so that

$$\frac{x}{x^2+y^2} + i\frac{-y}{x^2+y^2} \text{ is seen to be the reciprocal of } x + iy. \qquad \textit{6.40}$$

This formula is most easily remembered via the computations

$$\frac{1}{x+iy} = \frac{x-iy}{x-iy}\,\frac{1}{x+iy} = \frac{x}{x^2+y^2} + i\frac{-y}{x^2+y^2}.$$

Next we show how to establish equations 6.30 on page 192. The first of these equations, written with independent variables (t, τ) replacing (x, y) is

$$cos(t + \tau) = cos(t)cos(\tau) - sin(t)sin(\tau).$$

To verify this we make use of the uniqueness of the solution of the equations $x'' + x = 0$, $x(0) = a$, $x'(0) = b$, see Example 5.13 on page 166. Namely, we let

$$S_t(\tau) = cos(t + \tau), \quad G_t(\tau) = cos(t)cos(\tau) - sin(t)sin(\tau).$$

We now let

$$x = S_t - G_t$$

and verify that $x(0) = 0$, $x'(0) = 0$, and $x'' + x = 0$. (This is easy to do and will be left to the reader.) Now we can verify that the function x given by $x(\tau) = 0$ for all τ is a solution of this system, and from Example 5.13 mentioned above is the unique solution. So

$$S_t(\tau) - G_t(\tau) = 0 \quad \text{for all } \tau$$

showing

$$S_t(\tau) = G_t(\tau) \quad \text{for all } \tau,$$

establishing the desired result.

Proof of the second of equations 6.30 can be carried out similarly, or can be accomplished by treating the variable t as a constant and differentiating the equation just derived.

Because the law of exponents has now been established for pure imaginary exponents, it is now really trivial to establish this law for complex exponents, and will be left to the reader.

As previously mentioned, we already have a geometric meaning associated with addition of complex numbers, namely, vector addition (the Lines discussion, starting page 23). To provide a geometric interpretation for complex number multiplication, for $|x + iy| \neq 0$ we write

$$x + iy = z = |z|\frac{z}{|z|} = |z|\left(\frac{x}{|z|} + i\frac{y}{|z|}\right) = |z|\left(\frac{x}{\sqrt{x^2 + y^2}} + i\frac{y}{\sqrt{x^2 + y^2}}\right).$$

But the complex number on the right inside the parenthesis is the vector $(cos(\vartheta), sin(\vartheta))$, illustrated in Figure 6-4; and from the discussion following equation 6.33 on page 193 (using equation 6.29 on page 192), we see that this vector is $e^{i\vartheta}$. So **in summary** we may say that the complex number z may be written as

$$z = |z|e^{i\vartheta}, \text{ where } e^{i\vartheta} = \frac{z}{|z|}. \qquad \textbf{6.41}$$

We refer to this representation of the complex number z as its **polar form.**

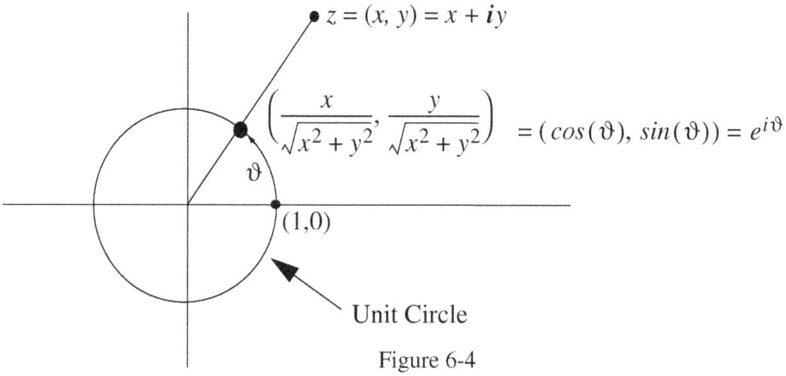

Figure 6-4

Because the law of exponents still holds, the polar form of a complex number allows both a simple geometric interpretation of multiplication of complex numbers, and allows a much simpler determination of their reciprocals; because if $z = |z|e^{i\vartheta}$ and $w = |w|e^{i\xi}$, then

$$z\,w = |z||w|e^{i(\vartheta + \xi)}. \qquad \textbf{6.42}$$

This shows that the distance of the product, $z\,w$, from the origin is just the product of their distances from the origin, and the angle made by the line from the origin to the product with the *real axis* is just the sum of their separate angles. Multiplication of a complex number by e^{it} just rotates this number about the origin through an angle of t radians. *Division* is allowed provided the denominator is not 0 (which occurs so long as the denominator has a non-zero magnitude).

It's necessary to point out that because $e^{i(\vartheta + 2k\pi)} = e^{i\vartheta}$ for each integer k, the polar representation is not unique. When writing a complex number in this form, it is common to use the angle satisfying the restriction $0 \le \vartheta < 2\pi$, which provides a unique polar representation for any nonzero complex number.

which provides a unique polar representation for any nonzero complex number.

It is not hard to see that each nonzero complex number has precisely n distinct n-th roots; because, using the representation just introduced, we may write

$$z = re^{i\vartheta},$$

where r is the magnitude of z. One n-th root of z is $\sqrt[n]{r}e^{i\vartheta/n}$. The remaining $n-1$ distinct other n-th roots are the points obtained by rotating this root about the origin through angles of $2\pi/n, 4\pi/n, ..., (n-1)2\pi/n$, respectively. That is, the roots are numbers of the form $\sqrt[n]{r}e^{i(\vartheta/n + k2\pi/n)}$.

The most significant result relevant to our solution of differential equations arises from the following.

Theorem 6.4: Fundamental theorem of algebra

Let n be any positive integer, and z represent a complex number (formally defined in Definition 6.8 on page 199). Suppose that $p(z)$ is an n-th *degree polynomial in z* — i.e., there are specified complex numbers $a_0, a_1, ..., a_n$, with $a_n \neq 0$, such that

$$p(z) = a_0 + a_1 z + \cdots + a_{n-1} z^{n-1} + a_n z^n \quad ,$$

Then there are complex numbers $\alpha_1, \alpha_2, ..., \alpha_n$ such that

$$p(z) = a_n (z - \alpha_1)(z - \alpha_2) \cdots (z - \alpha_n)$$

where the complex numbers α_i, determined by p are unique except for their order.

❑

The proof of this fairly easy to accept result, which we do not feel warrants presentation here, is widely available in books on mathematical analysis.[1] This theorem simply claims that n-th degree polynomials can be factored uniquely into linear factors of the form above. The result requires the complex numbers, since, for example, the polynomial $z^2 + 1$ can only be factored as $(z + i)(z - i)$. The **significance** of this result arises from the fact that the *shift rule,* Theorem 5.11 on page 146, holds for $D - \alpha_i$, for complex α_i. In order to see how this is implemented we must next take up the topic of differentiation of complex valued functions of a real variable.

1. In particular, in *Modern University Calculus* by Bell et al. [1]

Definition 6.5: g' for complex g

Suppose[1] $g = g_R + ig_I$, where both g_R and g_I are differentiable real valued functions of a real variable (Definition 2.1 on page 21 and Definition 2.5 on page 30) with the same domain (Definition 1.1 on page 2). Then we define the derivative g' by the equation

$$g' = g_R' + ig_I'$$

❑

It is still true that

$$g'(t) = \lim_{h \to 0} \frac{g(t+h) - g(t)}{h},$$

i.e., $g'(t)$ is the value that $\frac{g(t+h) - g(t)}{h}$ gets close to for small h, provided we use the following definition of distance, motivated by the Pythagorean theorem.

Definition 6.6: Distance between complex numbers, *z* and *w*

Let $z = z_R + iz_I$ and $w = w_R + iw_I$ be complex numbers. The *distance* between z and w is defined to be

$$|z - w| = \sqrt{(z_R - w_R)^2 + (z_I - w_I)^2} \qquad \textbf{6.43}$$

as illustrated below.

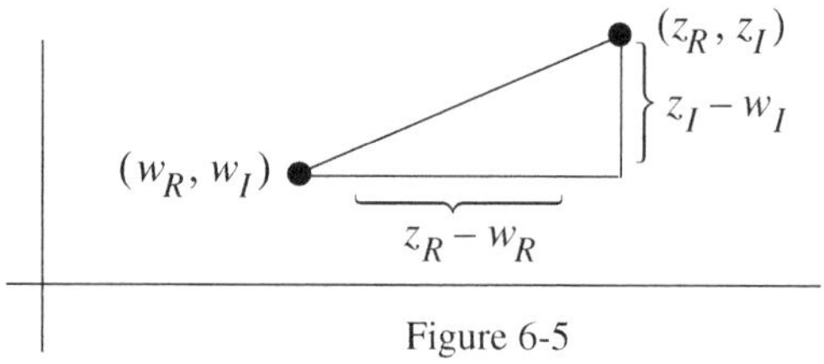

Figure 6-5

❑

1. When we write an equation like this, we are usually assuming that g_R and g_I are real valued functions of a real variable, with g_R called the *real part* of g, and g_I its *imaginary part.*

Theorem 6.7: $(e^{zt})'$

The complex valued function g, of a real variable, given by $g(t) = e^{zt}$ *where* t is real, and z is a fixed complex number, (see equations 6.29 on page 192 and 6.34 on page 193, or Definition 6.9 on page 200), is differentiable with

$$g'(t) = z e^{zt}.$$

❑

We leave the verification of Theorem 6.7 to the reader.

Exercises 6.9

1. Supply the details for the verification of equations 6.30 on page 192.
2. Suppose $f = f_R + i f_I$ is a given complex valued function of a real variable, and $y = y_R + i y_I$ is a complex valued function of a real variable satisfying the differential equation
$$y'' + ay' + by = f$$
for given real numbers a and b. Convince yourself that
$$y_R'' + a y_R' + by = f_R \quad \text{and} \quad y''_I + a y_I' + by_I = f_I.$$
Show that this result fails if we allow a or b to be complex.
3. Show that if z and w are complex numbers whose product $zw = 0$, then at least one of these numbers is 0, assuming this result holds for real numbers. Hint: Use the polar representation, equation 6.41 on page 196.
4. Show that the *shift rule*, Theorem 5.11 on page 146, still holds for complex k, and that the sum and product rules (equations 2.28 and 2.29 on page 53) hold for complex valued functions of a real variable.

5 Properties of Complex Numbers

For reference purposes we collect the various properties of complex numbers and complex valued functions of a real variable that have been developed so far, all together in this one section.

Definition 6.8: Complex numbers

A **complex number**, z, is an ordered pair, (x, y) of real numbers, usually written as $x + i y$; x is called the **real part** and y the **imaginary part** of z. The number $\bar{z} = x - i y$ is called the **complex conjugate of** z.

$(x, y) = (u, v)$ means $x = u$ and $y = v$.

Multiplication by a real: $a(x, y) = (ax, ay)$

Addition: $(x, y) + (u, v) = (x + u, y + v)$

Multiplication of complex numbers:

The distributive, associative and commutative laws of arithmetic are assumed, with i^2 replaced by -1, and resultant terms having the factor i constituting the imaginary part of a number formed by arithmetic operations. Hence

$$(x, y)(u, v) = (x + i\,y)(u + i\,v) = xu - yv + i(xv + yu)$$
$$= (xu - yv, xv + yu).$$

Real numbers may be thought of as complex numbers of the form $(x, 0)$.

❑

Motivated by Taylor's Theorem, for t real, we introduce

Definition 6.9: Imaginary and complex exponentials

$$exp(it) = e^{it} = cos(t) + i\,sin(t)$$

and

$$e^{x+iy} = e^x e^{iy}$$

for x,y and t real.

Hence $\quad sin(t) = (e^{it} - e^{-it})/(2i), \quad cos(t) = (e^{it} + e^{-it})/2.$

❑

Theorem 6.10: The law of exponentials

For complex z, w

$$exp(z + w) = e^{z+w} = e^z e^w = exp(z)\,exp(w)$$

still holds.

❑

Motivated by the Pythagorean theorem, we define magnitude and distance between complex numbers as follows.

Definition 6.11: Magnitude of *z*, distance between *z* and *w*

The **magnitude**, $|z|$, of $z = x + i\,y$; is defined by

$$|z| = \sqrt{x^2 + y^2}$$

The **distance,** $d(z, w)$**,** between z and w is defined by

$$d(z, w) = |z - w|.$$

❑

Definition 6.12: Polar form of nonzero z

The **polar form** of $z \neq (0, 0)$ is given by

$$z = re^{i\vartheta}$$

where

$$r = |z| \quad \text{and} \quad e^{i\vartheta} = \frac{z}{|z|}.$$

❑

Theorem 6.13: Polar form properties

If $z = re^{i\vartheta}$ and $w = Re^{i\varphi}$, then $zw = rRe^{i(\vartheta + \varphi)}$.

If $z \neq (0, 0)$, then $z = re^{i\vartheta}$ has reciprocal $\frac{1}{r}e^{-i\vartheta}$.

❑

Definition 6.14: Limit of a complex sequence

The sequence $z_1, z_2, z_3, \ldots$ of complex numbers, is said to have **limit the complex number** L, written, as before

$$\lim_{n \to \infty} z_n = L$$

provided that z_n gets and stays close to L when n gets large — i.e., provided $|z_n - L|$ is small for all sufficiently large n.

❑

Definition 6.15: Definition, limits and continuity of complex valued functions of a complex variable

A function g, (Definition 1.1 on page 2), is said to be **complex [real] valued** ***if its range consists of complex*** **[*real*]** ***numbers*** (Definition 6.8 on page 199), and be **a function of a complex [real] variable** ***if its domain consists of complex*** **[*real*]** ***numbers.***

If $g(z)$ is defined for all $z \neq z_0$ satisfying $|z - z_0| \leq h$, where h is some positive number, and $g(z)$ is *close* to the complex value L provided z is close enough to z_0 (i.e., provided $|z - z_0|$ is small enough), we say that

$g(z)$ has limit L as z approaches z_0, as before, written

$$\lim_{z \to z_0} g(z) = L$$

The function g is said to be **continuous at the point** z_0 in its domain, if for all z in the domain of g that are within a sufficiently small positive distance of z_0, $g(z)$ is close to $g(z_0)$ ❑

The full power of the following definition will not be used until later, but it is simpler to introduce it now, rather than just provide the restricted version we will be using for a while.

Definition 6.16: Derivative of complex valued function of a real or complex variable

A complex valued function, g, of a real or complex variable (Definition 6.15 on page 201) is said to be **differentiable at the point** z_0 in its domain, with derivative $g'(z_0)$ if

$$\lim_{h \to 0} \frac{g(z_0 + h) - g(z_0)}{h} = g'(z_0)$$

In particular, a complex valued function of a real variable, $g = g_R + i g_I$, where g_R, g_I are real valued functions of a real variable with the same domain, is **differentiable** with derivative $g' = g'_R + i g'_I$ if and only if g_R and g_I are differentiable.

❑

Theorem 6.17: Derivative of complex exponential

If $g(t) = e^{zt}$ (Definition 6.9 on page 200 t real, z complex, then

$$g'(t) = z e^{zt}$$

(see Definition 6.16 above).

❑

Theorem 6.18: Differentiation rules

The product and sum rules of differentiation (Theorem 2.13 on page 53) hold for complex valued functions of a real or complex variable (Definition 6.15 on page 201) and the derivative of g^n is what you expect (provided that it makes sense), for n a positive or negative integer. The shift rule (Theorem 5.11 on page 146) still holds for complex shifts.

❑

Definition 6.19: Integral of complex valued function of a real variable

For a complex valued function, g, of a real variable, with integrable real and imaginary parts (see Definition 6.5 on page 197 and Definition 4.2 on page 102), we define

$$\int_a^b g = \int_a^b g_R + i\int_a^b g_I .$$

❑

Theorem 6.20: Extension of fundamental evaluation theorem

The fundamental evaluation theorem (see Theorem 4.3 on page 103) still holds for complex valued functions of a real variable (Definition 6.15 on page 201) whose real and imaginary parts are continuous.

❑

Theorem 6.21: Complex valued functions with equal derivatives

Complex valued functions of a real variable (Definition 6.15 on page 201) with the same derivatives (Definition 6.16 on page 202) differ by some complex constant.

❑

Even though the following definition is a special case of the more general concept of a bounded set of vectors, Definition 10.43 on page 406, it's worth introducing separately, for its specialized uses.

Definition 6.22: Bounded set of complex numbers

The set A of complex numbers is said to be **bounded** if there is some real number $M > 0$ such that for all z in M, the length $|z|$ of z, Definition 6.11 on page 200, satisfies

$$|z| \leq M .$$

❑

6 Applications of Complex Exponentials

When we examined a spring-mass system with friction taken into account (Example 5.15 on page 168), and when we treated glycerol kinetics (Example 6.7 on page 188), we relied solely on numerical approximations, with no theoretical supporting results. So at the time we had no way to check on the accuracy of the numerical results. We are now in a position to supply this needed theory in a natural way. As usual, we will introduce this investigation by looking at a simple specific example, namely the spring-mass system with friction treated numerically in Example 5.15.

Example 6.10: Theoretical solution of $x'' + 0.1x' + x = 0$

We assume initial conditions of the form $x(0) = 0,\ x'(0) = 1$ and rewite this system as

$$(D^2 + 0.1D + 1)x = 0, \quad x(0) = 0,\ x'(0) = 1.$$

We will *complete the square* of the polynomial $D^2 + 0.1D + 1$, by writing

$$\begin{aligned} D^2 + 0.1D + 1 &= D^2 + 0.1D + (0.1/2)^2 - (0.1/2)^2 + 1 \\ &= (D + 1/20)^2 + 399/400 \end{aligned}$$

so that

$$(D^2 + 0.1D + 1)x = [(D + 1/20)^2 + 399/400]x.$$

Now this isn't simply formal manipulation with no content, because

$$[(D + 1/20)^2 + 399/400]x$$

really means

$$\begin{aligned} &(D + 1/20)[(D + 1/20)x] + \frac{399}{400}x \\ &= (D + 1/20)\left[x' + \frac{1}{20}x\right] + \frac{399}{400}x \\ &= x'' + \frac{1}{20}x' + \frac{1}{20}x' + \frac{1}{400}x + \frac{399}{400}x \\ &= x'' + 0.1x' + x. \end{aligned}$$

In a similar fashion it is meaningful to write

$$[(D + 1/20)^2 + 399/400]x = \left[D + \frac{1}{20} + i\sqrt{\frac{399}{400}}\right]\left[D + \frac{1}{20} - i\sqrt{\frac{399}{400}}\right]x.$$

(Expand this out to see that it is all legitimate.)

So we see that the differential equation

$$x'' + 0.1x' + x = 0,\ x(0) = 0,\ x'(0) = 1$$

can be written in the easier-to-manipulate form

$$(D - \alpha_1)[(D - \alpha_2)x] = 0,\ x(0) = 0,\ x'(0) = 1,$$

where $\alpha_1 = \alpha_{1R} + i\alpha_{1I}$ and $\alpha_2 = \alpha_{2R} + i\alpha_{2I}$ are complex constants.

Since x is not yet known, neither is $(D-\alpha_2)x$. So for the moment, let us denote it by y; i.e., the differential equation $(D-\alpha_1)[(D-\alpha_2)x] = 0$ may be written as

$$(D-\alpha_1)y = 0, \quad y = (D-\alpha_2)x. \qquad \textbf{6.44}$$

We will determine x by first determining y from the first of the differential equations in 6.44. (Notice that this first equation is rather simpler than the differential equation we started with in this example.) What we do is to apply the shift rule to the left equation above, $(D-\alpha_1)y = 0$, justified by the results of Exercise 6.9.4 on page 199, to see that this equation is equivalent to

$$e^{\alpha_1 t} D(e^{-\alpha_1 t} y) = 0,$$

or, since $e^{\alpha_1 t} = e^{\alpha_{1R} t}\, e^{i\alpha_{1I} t} = e^{\alpha_{1R} t}[\cos(\alpha_{1I} t) + i \sin(\alpha_{1I} t)] \neq 0$,

$$D(e^{-\alpha_1 t} y) = 0.$$

So $e^{-\alpha_1 t} y$ must be equal to some (complex) constant, c, and hence

$$y = ce^{\alpha_1 t}.$$

But we may substitute for y in this equation from the right-hand equation of equation 6.44 above, to obtain

$$(D-\alpha_2)x = ce^{\alpha_1 t}$$

to which we may again apply the shift rule, obtaining, as before,

$$D(e^{-\alpha_2 t} x) = e^{-\alpha_2 t} ce^{\alpha_1 t} = ce^{(\alpha_1 - \alpha_2)t}.$$

Provided $\alpha_1 - \alpha_2 \neq 0$, which is the case here, we find, using Theorem 6.7 on page 198, that

$$e^{-\alpha_2 t} x = \frac{ce^{(\alpha_1 - \alpha_2)t}}{\alpha_1 - \alpha_2} + d,$$

where d is another complex constant,

or

$$x = \frac{ce^{\alpha_1 t}}{\alpha_1 - \alpha_2} + de^{\alpha_2 t}.$$

The net result of this discussion is that if

$$(D-\alpha_1)[(D-\alpha_2)x] = 0 \text{ with } \alpha_1 \neq \alpha_2,$$

then

$$x = Ce^{\alpha_1 t} + de^{\alpha_2 t},$$

where $C = \dfrac{c}{\alpha_1 - \alpha_2}$ and d are complex constants.

The conditions $x(0) = 0,\ x'(0) = 1$ allow evaluation of C and d as follows.

$x(0) = 0$ yields $C + d = 0$, so that we may write

$$x = C(e^{\alpha_1 t} - e^{\alpha_2 t}),$$

and then applying the condition $x'(0) = 1$ yields $C(\alpha_1 - \alpha_2) = 1$. Thus, substituting for C yields

$$x = \frac{e^{\alpha_1 t} - e^{\alpha_2 t}}{\alpha_1 - \alpha_2}.$$

Recall that here

$$\alpha_1 = -\frac{1}{20} - i\sqrt{\frac{399}{400}},\quad \alpha_2 = -\frac{1}{20} + i\sqrt{\frac{399}{400}},\quad \alpha_1 - \alpha_2 = -2i\sqrt{\frac{399}{400}}.$$

Using Euler's formula, equation 6.29 on page 192, we conclude that

$$x(t) = e^{-t/20}\frac{sin(\sqrt{399/400}\ t)}{\sqrt{399/400}}. \qquad \textbf{\textit{6.45}}$$

Here it is now convenient to be able to check on the correctness of this theoretical answer. This can be done via *Mathematica*, as shown in the footnote[1] below. Comparisons of numerical solutions of this equation to the theoretical solution above is left to the problems which conclude this section.

■

The next example, which shows one way of obtaining the theoretical solution of the glycerol kinetics Example 6.7 on page 188, illustrates how the theory just developed can be used to yield theoretical solutions of systems of differential equations.

Example 6.11: Theoretical solution of glycerol kinetics

The equations of Example 6.7 were

$$\begin{aligned} x_1'(t) &= -0.571x_1(t) + 0.085x_2(t) + 0.441, & x_1(0) &= 1.41 \\ x_2'(t) &= 0.260x_1(t) - 0.085x_2(t), & x_2(0) &= 0. \end{aligned} \qquad \textbf{\textit{6.46}}$$

---

1. DSolve[{y''[t] + .1 y'[t] + y[t] == 0,y[0] == 0,y'[0] == 1},y[t],t]

We can reverse the type of process used in Example 6.6 which led us from a second-order differential equation to the system of first order equations, 6.17 on page 186, to obtain two second-order equations, one for x_1 and the other for x_2, as follows.

Solve the first of the differential equations in 6.46 for x_2, obtaining

$$x_2(t) = \frac{x_1'(t) + 0.571x_1(t) - 0.441}{0.085}.$$

Now differentiate this last equation, obtaining

$$x_2'(t) = \frac{x_1''(t) + 0.571x_1'(t)}{0.085}.$$

Substitute these expressions into the second of equations 6.46 to obtain

$$x_1''(t) + 0.656x_1'(t) + 0.026435x_1(t) = 0.037485. \qquad \textbf{\textit{6.47}}$$

Similarly, solving for $x_1(t)$ from the second of the differential equations in 6.46, differentiating it and substituting these expressions into the first of the differential equations of 6.46 yields an equation similar to the one just derived. We leave its derivation to the reader, Exercise 6.12.3 on page 208.

Next we need to supply initial conditions to x_1, x_1', x_2, x_2' for these second- order equations. Those for x_1 and x_2 come directly from 6.46. To obtain those for x_1' and x_2', just differentiate the differential equations in 6.46 and set $t = 0$, to obtain the values for $x_1'(0)$ and $x_2'(0)$.

We leave details of solving equation 6.47 and its counterpart for x_2 as exercises.

■

Remark on solving $P(D)x = f$

At this point the reader should have little difficulty in determining how to determine the theoretical solution of a differential equation of the form

$$P(D)x = f$$

where $P(D)$ is an n-th degree polynomial in the derivative operator, D, and f is a given continuous function — using the assumed factorization of $P(D)$ as indicated by the fundamental theorem of algebra, Theorem 6.4 on page 197.

Some of the exercises to follow will provide practice in this area. For actual numerical factorization of the polynomial $P(x)$ (i.e., finding numerical approximations to the α_i asserted by the fundamental theorem of algebra), the *Mathematica* program has very effective methods — using their function NSolve, together with their function N which allows arbitrary precision arithmetic in the calculations.

Exercises 6.12

1. Verify Theorem 6.7 on page 198.
2. Referring to Example 6.10 starting on page 204, find the solution of the differential equation $(D-\alpha)^2 x = 0$; i.e., handle the *repeated root* case. If a theoretical differential equation solver is available, check your answer (*Mathematica* uses DSolve, see Footnote 1 on page 206.)
3. Referring to the glycerol kinetics example, Example 6.11 on page 206,
   a. Using the methods developed in Example 6.10 on page 204, solve equation 6.47 on page 207, (it should be the same as the formula presented in equation 6.27 on page 189).
   b. Derive the second order differential equation for x_2, and solve it as in part a.
4. If you have *Mathematica* or a similar package, numerically solve equation on page 204
   a. Using a second-order Taylor approximation with $h = .01$, for $0 \le t \le 5$. Compare this to the result from equation 6.45 on page 206 at $t = 1.4$.
   b. Using a second-order Taylor approximation with $h = .001$ for $0 \le t \le 5$. Compare this to the result from equation 6.45 on page 206 at $t = 1.4$.
   c. Compare the above results to the corresponding ones obtained from Example 5.15 on page 168.
5. a. Solve the equations of Exercise 5.16.2 on page 169 numerically via a second-order Taylor polynomial approximation with $h = .05, .01$ on $[0; 50]$.
   b. Solve these equations theoretically using the methods of Example 6.10 on page 204, and compare to the numerical solutions just generated, as well as to those from numerical answers to Exercise 5.16.2 on page 169 at $t = 1,5,10,25,50$

6. Derive the results of Theorems 5.18 on page 171 and 5.19 on page 173 using the methods developed in Example 6.10 on page 204. Hints: It's probably simplest to look at the equation $x'' + x = e^{it}$ and use the results of Exercise 6.9.2 on page 199. Alternatively you could look at either

$$x'' + x = sin \quad \text{or} \quad x'' + x = \frac{(e^{it} - e^{-it})}{2i}.$$

7. a. Find the formula for the solution of the differential equation $(D^2 + 1)(D + 1)x = 0$.

   b. Using DSolve of *Mathematica* (see Footnote 1 on page 206) or the equivalent in a similar package, check your answer.

8. Same as Exercise 7 for $(D^2 + 1)(D + 1)x = cos$.

9. Find the formula solutions for the differential equations of Exercises 5.18 on page 174.

10. Find the formula solving $x'' + qx' + rx = cos(kt)$, $x(0) = a$, $x'(0) = b$.

11. If $\alpha_1, ..., \alpha_n$ are all distinct, show that the general solution of $(D - \alpha_1)\cdots(D - \alpha_n)x = 0$ is of the form $x(t) = \sum_{\text{all } j} a_j e^{\alpha_j t}$ where the a_j are arbitrary constants.

Problems 6.13

1. Looking at the differential equation $x'' + qx' + rx = 0$, $x(0) = a$ $x'(0) = b$, where q and r are positive (representing a spring-mass system with friction), what condition must q and r satisfy so that the solution has no oscillating (*sin* or *cos*) component? Hint: Look at $(D^2 + qD + r)x = 0$ and complete the square, as in Example 6.10 on page 204.

2. Show that a unique solution of the differential equation $(D - \alpha_1)(D - \alpha_2)\cdots\cdot(D - \alpha_n)x = f$, where f is a given continuous function, is specified by requiring satisfaction of n initial conditions (conditions on x and its derivatives at time 0. There are other similar specifications, in particular so-called *boundary conditions,* which are often used to determine unique solutions in applied situations).

3. Find the general formula for the solution of the differential equation $(D - \alpha)^n x = 0$.

4. Suppose $P(z)$ is an n-th degree polynomial with real coefficients in the complex variable z.

   a. If $P(z) = a_n(z - \alpha_1)(z - \alpha_2)\cdots(z - \alpha_n)$ with distinct α_i, show that any nonreal complex α occurs as one of a **complex**

conjugate pair — i.e., if $\alpha_i = \alpha_{iR} + i\alpha_{iI}$, $\alpha_{iI} \neq 0$, another α, $\alpha_j = \alpha_{iR} - i\alpha_{iI}$.

Hints: Every solution of the differential equation $P(D)x = 0$ is a weighted sum of terms $e^{\alpha_i t}$. But we know (Exercise 6.9.2 on page 199 and its obvious extension to n-th order constant coefficient differential equations) that the real and the imaginary parts (see Footnote 1 on page 198) of $e^{\alpha_i t}$ both satisfy this differential equation.

b. Using the results of Exercise 3 above, show that the conclusion of part a. holds even for repeated roots — i.e., even when the α_i are not distinct.

Chapter 7

Infinite Series

1 Introduction

Quite a few times in previous chapters we have solved problems by generating a sequence of numbers or functions which approached the solution we were seeking; for instance, bisection and Newton's method in Section 4 of Chapter 3, numerical approximation to integrals with successively smaller interval lengths in Section 6 of Chapter 4, increasingly accurate numerical solutions of differential equations in Chapter 5, and sequences of increasing order Taylor polynomial approximations to functions in Chapter 6.

In this chapter we will formalize the study of the processes already encountered, in the general study of the limiting behavior of sequences, $a_1, a_1 + a_2, a_1 + a_2 + a_3, ...$ of sums. That is, we will examine when a sequence of such sums approaches a limit — i.e., gets close to some specific value as a limit (recall Definition 6.14 on page 201). Here the terms a_i may be real or complex numbers, or real or complex-valued functions.

It might be asked why we look at sequences of ***sums*** — why not just look at sequences $A_1, A_2, A_3, ...$ of numbers, such as arise in Newton's method, to study such limiting behavior? The answer is, it doesn't really matter, because a sequence, $A_1, A_2, A_3, ...$ can always be written in the form $a_1, a_1 + a_2, a_1 + a_2 + a_3, ...$, where

$$A_1 = a_1, \quad A_2 = A_1 + (A_2 - A_1) = a_1 + a_2, ... \text{ etc.}$$

The advantage of representing the elements of a sequence as *partial sums* $a_1 + a_2 + \cdots + a_n$ is that frequently the behavior of such a sequence is most conveniently examined by looking at the terms $a_1, a_2, a_3, ...$. To name but a few algorithms for determining sequence behavior based on the terms of the partial sums, we have

- the *comparison test*, Theorem 7.13 on page 228,
- the *ratio test*, Theorem 7.14 on page 231,
- the *alternating series test*, Theorem 7.12 on page 228.

and the most important approximating sequences of all,

- the Taylor polynomials for a given function, f,

$$T_n(t) = \sum_{k=0}^{n} f^{(k)}(0)\frac{t^k}{k!}$$

which occur as sums right from the start.

We will find conditions under which the sequence of partial sums, $S_1, S_2, S_3, \ldots$ has a limit, where

$$S_n = \sum_{j=0}^{n} a_j .$$

Often we'll want to determine this limit to some specified accuracy. If a function S is constructed as a limit

$$S(t) = \lim_{n \to \infty} S_n(t) = \lim_{n \to \infty} \sum_{k=1}^{n} f_k(t),$$

where the f_k are differentiable, we will want to know not only

- the t values for which $\sum_{k=1}^{n} f_k(t)$ converges (has a limit),
- the values of $\lim_{n \to \infty} \sum_{k=1}^{n} f_k(t)$ to some specified accuracy,

but also

- whether S is differentiable

and if so

- whether S' can be calculated as you would expect if S were a finite sum, namely,

$$S' = \lim_{n \to \infty} \sum_{k=0}^{n} f_k'(t) .$$

The answers to these questions are important when seeking the solution of differential equations in the form of infinite series — a very powerful and general approach to this problem.

To make reading easier, in particular to avoid excessive page flipping, some definitions and results presented earlier may be repeated.

2 Preliminaries

The definitions of limit of a sequence given earlier (e.g., Definition 6.14 on page 201) was somewhat informal (a euphemism for *imprecise*). Because we will currently be concerned with computing limits to specified accuracies, we will now give a more precise meaning to the definition

$$\lim_{n \to \infty} x_n = L,$$

i.e., we will give a precise version of the phrase *if all values* $x_n, x_{n+1},...$ *can be made as close to L as desired by choosing* n *sufficiently large.*

Definition 7.1: Limit of a sequence (precise version)

Using Definition 6.11 of $|z|$ on page 201, we say the (possibly complex valued) sequence $x_1, x_2,...$ *converges* to the number L or *have limit* L,

$$\text{written } \lim_{n \to \infty} x_n = L$$

if for each positive number ε there is an integer N_ε such that

$$\text{whenever } n \ge N_\varepsilon \text{ we have } |x_n - L| \le \varepsilon .$$

❑

Here ε represents any *desired closeness to L*, and $n \ge N_\varepsilon$ is what is meant by the phrase *by choosing* n *sufficiently large* in the less precise definition of limit given in Definition 3.7 on page 85 and extended in Definition 6.14 on page 201.

Much of the effort of applied mathematics concerns determination of N_ε for a specific value of ε — the desired accuracy in determining $\lim_{n \to \infty} x_n$. Having found, say, $N_{0.001}$, then we know that for each integer $n \ge N_{0.001}$

$$|x_n - L| \le 0.001,$$

so that as an estimate of $L = \lim_{n \to \infty} x_n$, such x_n is accurate to within .001.

Since we will be needing it soon anyway, now is a reasonable time to introduce the corresponding precise definition of *continuity* and *limit of a function.*

Definition 7.2: $\lim_{x \to x_0} G(x)$, continuity of G at x (precise)

Suppose G is a (possibly complex valued) function (of a possibly complex variable) defined for all x satisfying the condition $0 < |x - x_0| \le h$ where h is some given positive number. G is said to have limit L as x approaches x_0, written

$$\lim_{x \to x_0} G(x) = L,$$

if, for each positive number ε there is an a positive real number δ_ε such that

$$\text{whenever } |x - x_0| \le \delta_\varepsilon, \text{ we have } |G(x) - L| \le \varepsilon.$$

G is said to be *continuous* at x_0 if it is defined at x_0, and for all x in the domain of G (Definition 1.1 on page 2), for each positive ε there is a positive δ_ε such that $|G(x_0) - G(x)| \le \varepsilon$ whenever $|x - x_0| \le \delta_\varepsilon$.

❑

Important properties of limits that we will be using are summarized in the following theorem, which strongly resembles the continuity results of Theorem 2.10 on page 38.

Theorem 7.3: Limit properties

Suppose $\boldsymbol{x}$ and $\boldsymbol{y}$ are sequences with limits L and M, respectively (Definition 7.1 on page 213), and G and H are (possibly complex valued) functions (possibly of a complex variable) with common domain, and

$$\lim_{x \to x_0} G(x) = P, \quad \lim_{x \to x_0} H(x) = Q$$

(see Definition 7.2 on page 213).

Then, if a and b are any constants, the following are all meaningful and true:

$$\lim_{n \to \infty} [ax_n + by_n] = aL + bM$$

$$\lim_{n \to \infty} [x_n y_n] = LM$$

$$\lim_{n \to \infty} \frac{1}{x_n} = \frac{1}{L} \quad \text{provided} \quad L \ne 0$$

and similarly

$$\lim_{x \to x_0} [aG(x) + bH(x)] = aP + bQ, \qquad \lim_{x \to x_0} [G(x)H(x)] = PQ$$

$$\lim_{x \to x_0} \frac{1}{G(x)} = \frac{1}{P} \quad \text{provided} \quad P \ne 0.$$

❑

Proof follows the lines in the optional discussion following the statement of the continuity theorems, Theorem 2.10 on page 38.

Since our investigations from here on will deal with sequences of sums, we introduce some convenient notations.

Definition 7.4: Infinite series, convergence and sum of a series

Let $a_0, a_1, a_2,...$ be a given sequence of real [complex, (Definition 6.8 on page 199)] numbers. The sequence $S_0, S_1, S_2,...$ defined by the formula

$$S_n = \sum_{j=0}^{n} a_j \quad n = 0, 1, 2, ...$$

is called a real [complex] *infinite series*, and will in this work be denoted by

$$\sum_{j=0 \to \infty} a_j \;.$$

If the sequence $S_0, S_1, S_2, \ldots$ has a limit, L, we say that $\sum_{j=0 \to \infty} a_j$ *converges* to L, and write

$$L = \sum_{j=0}^{\infty} a_j,$$

which we call the *sum* of the series $\sum_{j=0 \to \infty} a_j$.

A series which fails to converge is said to *diverge*.[1]

❑

Note that the infinite series $\sum_{j=0 \to \infty} a_j$ is well defined even when it fails to converge.

Most material on series fails to distinguish between the series and its sum, using the symbol

$$\sum_{j=0}^{\infty} a_j$$

for both objects.

Definition 7.5: Infinite series of functions

If $G_0, G_1, G_2, \ldots$ is a sequence of complex valued functions of a (possibly complex variable, see Definition 6.15 on page 201) with a common domain, then

$\sum_{j=0 \to \infty} G_j$ is the sequence of functions whose n-th element is $\sum_{j=0}^{n} G_j$.

❑

Remark:

It should be evident what is meant by $\sum_{j=k \to \infty} G_j$ and $\sum_{j=k}^{\infty} G_j$ so we won't belabor the issue.

Example 7.1: Obviously convergent series

Let $a_j = \frac{1}{j+1} - \frac{1}{j+2}$, $j = 0, 1, 2, \ldots$. Then

---

1. If the partial sums become and stay arbitrarily large, we say the series diverges to ∞; similarly for divergence to $-\infty$. There are many other types of divergent behavior.

$$S_n = \sum_{j=0}^{n} a_j = \sum_{j=0}^{n}\left(\frac{1}{j+1} - \frac{1}{j+2}\right)$$

$$= \left(\frac{1}{1} - \frac{1}{2}\right) + \left(\frac{1}{2} - \frac{1}{3}\right) + \cdots + \left(\frac{1}{n+1} - \frac{1}{n+2}\right)$$

$$= 1 - \frac{1}{n+2}$$

so

$$\lim_{n \to \infty} S_n = 1,$$

i.e., $\sum_{j=0 \to \infty}\left(\frac{1}{j+1} - \frac{1}{j+2}\right)$ converges to $\sum_{j=0}^{\infty}\left(\frac{1}{j+1} - \frac{1}{j+2}\right) = 1.$ ■

Example 7.2: Clearly divergent series

Let $a_j = 1$ for $j = 0,1,2,\ldots$. Then $S_n = \sum_{j=0}^{n} a_j = n+1$ and hence $\sum_{j=0 \to \infty} a_j$ diverges (to ∞). ■

Exercises 7.3

Which of the following converge?

1. $\sum_{j=0 \to \infty} (-1)^j$
2. $\sum_{j=0 \to \infty} (0.2)^j$ Hint: Recall the geometric series, Theorem 1.9 on page 16.
3. $\sum_{j=0 \to \infty} (-0.9)^j$ Hint: see previous hint.
4. $\sum_{j=0 \to \infty} 5^j$

A simple but vital result needed for the work to follow is the *triangle inequality,* which simplifies determination of upper bounds for complicated expressions.

Theorem 7.6: Triangle inequality

Recall Definition 6.11 on page 200, of the magnitude of and distance between complex numbers.

The inequality $|v + w| \le |v| + |w|$ holds for all complex numbers v, w.

❑

This result, illustrated in Figure 7-1, just asserts the well-known believable result that the shortest path between two points is the straight line connecting them.

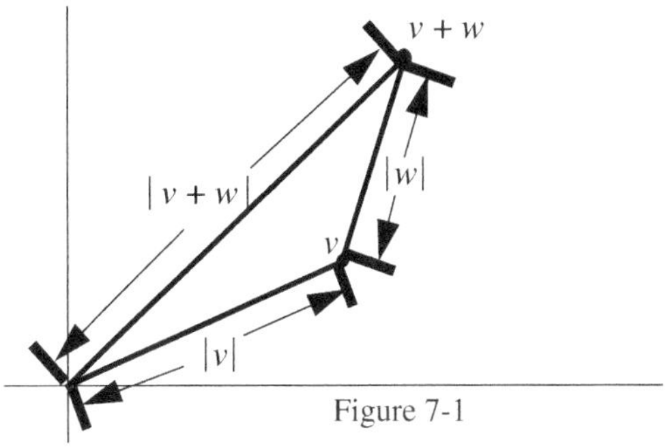

Figure 7-1

The proof here is relegated to a footnote.[1] Later in Theorem 10.20 on page 350, a more illuminating proof will be presented.

As it stands, the triangle inequality seems usable mainly for obtaining upper bounds — i.e., if we know upper bounds for $|v|$ and $|w|$,

$$|v| \le u_v \text{ and } |w| \le u_w,$$

then we find

$$|v + w| \le |v| + |w| \le u_v + u_w .$$

That is,

$$u_v + u_w \text{ is an upper bound for } |v + w| .$$

As will now be shown, the triangle inequality can be rewritten to make it a convenient tool for finding lower bounds. To do this, just let $v + w = z$. Then the triangle inequality becomes

---

1. Let $v = (v_x, v_y)$, $w = (w_x, w_y)$ and let the symbol $\Leftrightarrow$ be an abbreviation for the phrase *is equivalent to*. Then, using Definition 6.11 on page 200,

$$|v + w| \le |v| + |w| \Leftrightarrow \sqrt{(v_x + w_x)^2 + (v_y + w_y)^2} \le \sqrt{v_x^2 + v_y^2} + \sqrt{w_x^2 + w_y^2}$$

$$\Leftrightarrow (v_x + w_x)^2 + (v_y + w_y)^2 \le v_x^2 + v_y^2 + w_x^2 + w_y^2 + 2\sqrt{v_x^2 + v_y^2}\sqrt{w_x^2 + w_y^2}$$

$$\Leftrightarrow 2v_x w_x + 2v_y w_y \le 2\sqrt{(v_x^2 + v_y^2)(w_x^2 + w_y^2)} \qquad \text{(done if left side is } \le 0 \text{)}$$

$$\Leftrightarrow v_x^2 w_x^2 + v_y^2 w_y^2 + 2v_x w_x v_y w_y \le v_x^2 w_x^2 + v_y^2 w_y^2 + v_y^2 w_x^2 + v_x^2 w_y^2$$

$$\Leftrightarrow (v_y w_x - v_x w_y)^2 \ge 0, \text{ which is true. Hence, } |v + w| \le |v| + |w| \text{ holds.}$$

$$|z| \le |v| + |z - v|$$

or

$$|z - v| \ge |z| - |v| .$$

But since $|z - v| = |v - z|$, this yields the following result.

Theorem 7.7: Reverse triangle inequality

Recall Definition 6.11 on page 200. The inequality

$$|z - v| \ge \big|\, |z| - |v| \,\big|$$

holds for all complex numbers, z, v.

❑

3 Tests for Convergence, Error Estimates

In much of applied mathematics, attempts are made to construct solutions of differential equations and other problems, using a series of the form

$$\sum_{j=0 \to \infty} a_j f_j$$

where the f_j are given functions and the a_j are coefficients to be determined by requiring the sum of the above series to be the solution to the differential equation or of whatever other problem is being considered. Once the a_j have been determined, it is necessary to determine

- for which values of t the series $\sum_{j=0 \to \infty} a_j f_j(t)$ converges;
- whether the derivative of $\sum_{j=0}^{\infty} a_j f_j$ exists for those t, and, if so, its value — so that we can determine whether this sum satisfies the differential equation which led to its construction. Furthermore, if indeed this is the case, we want to be able to calculate this function to any desired accuracy and determine its interesting properties.

For these tasks we need easy-to-apply tests for convergence and error estimates, as well as usable techniques for manipulating and checking the properties of infinite series. Most of the remainder of this chapter is motivated by these goals.

One obvious problem with the definition of convergence of an infinite series is that it would seem that you need to know the sum of the series in order to establish its convergence. The Cauchy criterion that we are about to introduce shows that this is not really the case.

Theorem 7.8: Cauchy criterion for series convergence

If $a_0, a_1, a_2,...$ is a sequence of complex numbers (Definition 6.8 on page 199) then

$$\sum_{j=0 \to \infty} a_j$$

converges (Definition 7.4 on page 214), if and only if

for each given positive number ε, there is a real M_ε for which all finite sums $\left|\sum_{j=m}^{k} a_j\right| \le \varepsilon$, whenever $k \ge m$ are integers $\ge M_\varepsilon$.

This latter condition is *informally stated* as:

$$\left|\sum_{j=m}^{k} a_j\right| \quad \text{is } \textbf{\textit{small}} \text{ for all sufficiently large } m,k.$$

❑

Before establishing the Cauchy criterion, let's look at how it is used.

Example 7.4 $\displaystyle\sum_{j=0 \to \infty} \frac{0.7^j}{j!}$

Here, using the geometric series formula, Theorem 1.9 on page 16, we find

$$\begin{aligned}\left|\sum_{j=M}^{M+k} \frac{0.7^j}{j!}\right| &\le \sum_{j=M}^{M+k} 0.7^j \\ &\le 0.7^M \sum_{j=0}^{k} 0.7^j \\ &\le 0.7^M \frac{1-0.7^{k+1}}{1-0.7} \\ &\le 0.7^M/0.3.\end{aligned}$$

Evidently we can guarantee that $0.7^M \le \varepsilon$ by making $ln(0.7^M) \le ln(\varepsilon)$ (because the ln function is increasing, see Exercise 5.2.14 on page 142), or $M\,ln(0.7) \le ln(\varepsilon)$ (see Exercise 5.2.15 on page 142). That is $M \ge \dfrac{ln(\varepsilon)}{ln(0.7)}$ which follows because $ln(0.7) < 0$, $ln(1) = 0$ and ln is increasing.

So we may take $M_\varepsilon = ln(\varepsilon)/ln(0.7)$ in the Cauchy criterion, to prove convergence of this series.

■

It is evident from the Cauchy Criterion that if the terms of $\sum_{j=0 \to \infty} a_j$ don't approach 0, then this series must diverge. But, as we will soon see, there are many series $\sum_{j=0 \to \infty} a_j$ which diverge even though $\lim_{j \to \infty} a_j = 0$.

The Cauchy criterion is not usually used directly to establish convergence; rather it is one of the major theoretical tools used to develop convenient tests for convergence.

Although the optional part to follow should help develop a firmer understanding of the inner workings of the number system and some of its important properties, its omission should not prove a handicap to progress in later chapters.

Establishing the Cauchy criterion

The following lemma (helper theorem) is useful in establishing the Cauchy criterion.

Theorem 7.9: Convergence of real and imaginary parts of convergent series

The sequence $z_1, z_2, \ldots$ of complex numbers (Definition 6.8 on page 199) converges to L if and only if

$$\lim_{n \to \infty} z_{n,R} = L_R \quad \textit{and} \quad \lim_{n \to \infty} z_{n,I} = L_I, \qquad \textbf{7.1}$$

where $z_{n,R}[z_{n,I}]$ and $L_R[L_I]$ are the real [imaginary] parts of z_n and L respectively.

The Cauchy criterion is satisfied by the complex series $\sum_{j=0 \to \infty} a_j$ if and only if it is satisfied by the *real-part* and *imaginary-part* series

$$\sum_{j=0 \to \infty} a_{j,R} \quad \text{and} \quad \sum_{j=0 \to \infty} a_{j,I},$$

where, as above, $a_{j,R}[a_{j,I}]$ are the real [imaginary] parts of a_j.

❑

From the geometrical meaning of convergence, the assertions concerning equations 7.1 are easy to accept. Algebraically they follow from the inequalities

$$|z_{n,R} - L_R| \le |z_n - L|, \; |z_{n,I} - L_I| \le |z_n - L|$$

and

$$|z_n - L| \le |z_{n,R} - L_R| + |z_{n,I} - L_I|$$

(which follows from the triangle inequality, Theorem 7.6 on page 216), making use of the definition of convergence.

For the remaining assertions, it is sufficient to note that because

$$\left|\sum_{j=m}^{k} a_j\right| = \left|\sum_{j=m}^{k} a_{j,R} + i\sum_{j=m}^{k} a_{j,I}\right| \le \left|\sum_{j=m}^{k} a_{j,R}\right| + \left|\sum_{j=m}^{k} a_{j,I}\right|, \qquad \textit{7.2}$$

(again justified by the triangle inequality), if $\left|\sum_{j=m}^{k} a_{j,R}\right|$ and $\left|\sum_{j=m}^{k} a_{j,I}\right|$ are both small for sufficiently large k, then so is $\left|\sum_{j=m}^{k} a_j\right|$.

On the other hand, from the definition of the magnitude $|z|$ of a complex number z (see Section 5 of Chapter 6), we see that

$$\left|\sum_{j=m}^{k} a_j\right| \ge \left|\sum_{j=m}^{k} a_{j,R}\right|, \text{ and } \left|\sum_{j=m}^{k} a_j\right| \ge \left|\sum_{j=m}^{k} a_{j,I}\right|$$

so that if $\left|\sum_{j=m}^{k} a_j\right|$ is small for sufficiently large m,k, then so are both

$$\left|\sum_{j=m}^{k} a_{j,R}\right| \text{ and } \left|\sum_{j=m}^{k} a_{j,I}\right|.$$

From this last result we see that we need only establish the Cauchy criterion for series with real terms. At the very least this simplifies the illustrative diagrams.

We will first prove the easy part of the Cauchy criterion — that (for real a_j) if $\sum_{j=0\to\infty} a_j$ converges, then for all sufficiently large m,k, $\left|\sum_{j=m}^{k} a_j\right|$ is small — for if not, there would be some positive number ε, such that no matter how large we restrict them to be, there are numbers, m,k such that

$$\left|\sum_{j=m}^{k} a_j\right| > \varepsilon. \qquad \textit{7.3}$$

But since $\sum_{j=0\to\infty} a_j$ converges, for n sufficiently large n (all $n \ge N_{\varepsilon/2}$)

$$\left|\sum_{j=0}^{n} a_j - L\right| \le \varepsilon/2. \qquad \textit{7.4}$$

Now choose both $m-1, k-1 > N_{\varepsilon/2}$, so that inequality 7.3 also holds. It is evident that it is not possible for inequality 7.3 to hold and for inequality 7.4 to hold for $n = m - 1$ and $n = k$, because since

$$\left|\sum_{j=m}^{k} a_j\right| > \varepsilon \quad \text{and} \quad \sum_{j=0}^{k} a_j = \sum_{j=0}^{m-1} a_j + \sum_{j=m}^{k} a_j,$$

as we see from Figure 7-2, if, say, $\sum_{j=0}^{m-1} a_j$ was in the interval $[L-\varepsilon/2, L+\varepsilon/2]$,

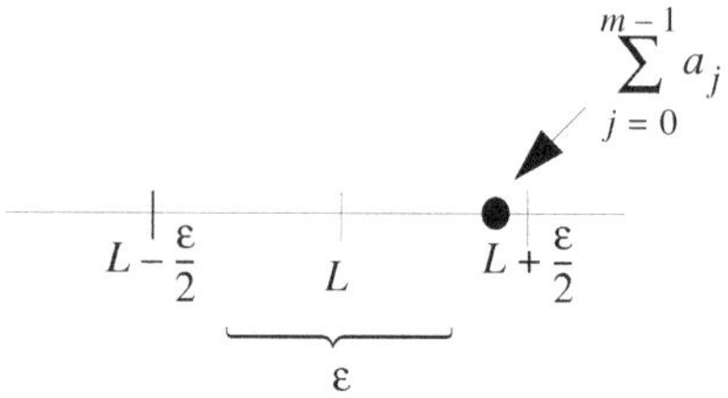

Figure 7-2

then $\sum_{j=0}^{k} a_j$ could not be in this interval (this is algebraically provable using the reverse triangle inequality (Theorem 7.7 on page 218) violating inequality 7.4 for $n = k$, which was, however, guaranteed by the assumed convergence of

$$\sum_{j=0 \to \infty} a_j .$$

So we see that if $\sum_{j=0 \to \infty} a_j$ converges, then it satisfies the Cauchy criterion.

In order to establish convergence of $\sum_{j=0 \to \infty} a_j$ when the Cauchy criterion is satisfied, we must really first construct its sum, and then show that this number is the desired limit. Here's how this is done.

When the Cauchy criterion is satisfied, there are numbers $u_r = N_{1/2^r}$, $n = 1,2,...$ such that

$$\left|\sum_{j=m}^{k} a_j\right| \le \frac{1}{2^r} \quad \text{as long as } m, k \ge N_{1/2^r}.$$

Define the sums $S_r = \sum_{j=0}^{u_r} a_j$, $r=1,2,...$ and the intervals $I_r = \left[S_r - \frac{1}{2^r},\ S_r + \frac{1}{2^r}\right]$.

The sequence of values, S_r is such that for all $k \ge 0$ all *later sums*, $\sum_{j=0}^{u_r + k} a_j$, lie in the interval I_r, as illustrated in Figure 7-3.

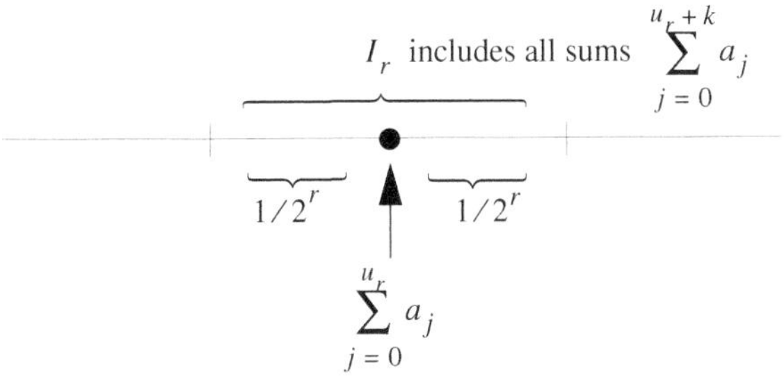

Figure 7-3

Let J_r be the points of I_r which are also in I_{r-1}, so that now J_r is a subset of J_{r-1}. We illustrate the sequence J_r of intervals in Figure 7-4.

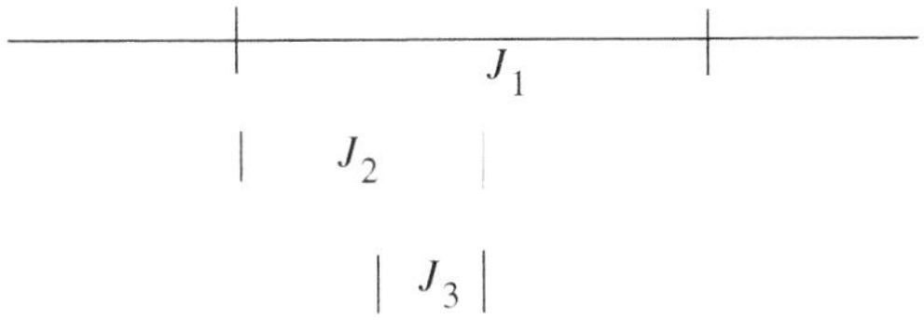

Figure 7-4

What we have constructed is a sequence of intervals, $J_1 = I_1, J_2, J_3,...$ such that

- J_r is a subset of J_{r-1} for all $r > 1$,
- The length of J_r doesn't exceed $2\frac{1}{2^r} = \frac{1}{2^{r-1}}$,
- Each interval J_r has positive length,
- One delicate issue — we chose the intervals J_r to include their endpoints, without which the result claimed to follow fails.

Under these conditions there is precisely one number, call it L, in all of the intervals,[1] J_r, $r = 1,2,...$.

We now claim that $\sum_{j=0 \to \infty} a_j$ converges to L. That is, given any $\varepsilon > 0$ we want to show that there is a value N_ε such that

$$\left| \sum_{j=0}^{n} a_j - L \right| \le \varepsilon \quad \text{so long as} \quad n \ge N_\varepsilon. \qquad \textit{7.5}$$

This follows because for all values $n \ge u_r$, $\sum_{j=0}^{n} a_j$ lies in J_r, L lies in J_r, and the length of J_r does not exceed $1/2^{r-1}$.

1. If this argument isn't convincing enough, what is needed is the precise development of the real number system given in a rigorous analysis course, such as that presented by Bell et al. [1].

Hence for such values of n

$$\left|\sum_{j=0}^{n} a_j - L\right| \le \frac{1}{2^{r-1}}.$$

So to prove our assertion, all we need do is choose r large enough so that $\frac{1}{2^{r-1}} \le \varepsilon$, which surely can be done.

To recap, assuming that the Cauchy criterion holds, if we choose $n \ge u_r = M_{1/2^r}$, ($M_{1/2^r}$ of the Cauchy Criterion), and r large enough so that $1/2^r \le \varepsilon$, then conditions 7.5 are satisfied, yielding convergence of $\sum_{j=0\to\infty} a_j$ to L, which establishes Theorem 7.8.

Problems 7.5

1. Show, by providing an example, that if the end points are not included in the intervals J_r in the proof of convergence when the Cauchy Criterion is satisfied, that there may be no points at all in all of these intervals. Hint: try $(0, 1/2^r)$.
2. Show that if $\sum_{j=0\to\infty} a_j$ converges, then $\sum_{j=0}^{\infty} a_j = \sum_{j=0}^{m-1} a_j \; \sum_{j=m}^{\infty} a_j$.

Here is an illuminating illustration of the use of the Cauchy criterion.

Example 7.6: Divergence of the harmonic series $\sum_{j=1\to\infty} 1/j$

The terms, $1/j$, of this series approach 0 as j grows large — so it might be conjectured that this series converges. It doesn't, as can be seen by examining

$$\sum_{j=n+1}^{2n} 1/j.$$

No matter how large n,

$$\sum_{j=n+1}^{2n} 1/j \ge \sum_{j=n+1}^{2n} \frac{1}{2n} = n \times \frac{1}{2n} = \frac{1}{2},$$

so that $\left|\sum_{j=m}^{k} a_j\right|$ is ***not*** *small for all sufficiently large m and k,* as it would have to be, by the Cauchy criterion, Theorem 7.8 on page 219, for convergence. So the harmonic series diverges.

■

The following theorem, informally stated earlier, ***sometimes*** provides a simple means of showing that a series fails to converge.

Theorem 7.10: n-th term test for divergence

If the sequence of *terms*, $a_1, a_2, \ldots$ of the series $\sum_{j=0\to\infty} a_j$ fails to have limit 0 (Definition 7.1 on page 213), then the series $\sum_{j=0\to\infty} a_j$ diverges.

❑

The proof is an immediate consequence of the Cauchy criterion, Theorem 7.8 on page 219. Note that even if the sequence $a_1, a_2, \ldots$ has limit 0, Example 7.6 on page 224 shows that the series $\sum_{j=1\to\infty} a_j$ can diverge.

Example 7.7: Convergence of $\sum_{j=1\to\infty} \frac{(-1)^{j+1}}{j}$

A small amount of experimentation leads to the following argument:

Write $\sum_{j=m}^{k} \frac{(-1)^{j+1}}{j} = (-1)^{m+1} \sum_{j=m}^{k} \frac{(-1)^{j-m}}{j}$

If $\sum_{j=m}^{k} \frac{(-1)^{j-m}}{j}$ has an even number of terms, it must be positive, since

$$\sum_{j=m}^{k} \frac{(-1)^{j-m}}{j} = \left(\frac{1}{m} - \frac{1}{m+1}\right) + \left(\frac{1}{m+2} - \frac{1}{m+1}\right) + \cdots + \left(\frac{1}{k-1} - \frac{1}{k}\right)$$

But also,

$$\sum_{j=m}^{k} \frac{(-1)^{j-m}}{j} = \frac{1}{m} - \frac{1}{k} + \sum_{j=m+1}^{k-1} \frac{(-1)^{j-m}}{j},$$

and the second summation here is negative (same reasoning as that showing the previously considered summation to be positive). So for an even number of terms, $k > m$,

$$0 < \sum_{j=m}^{k} \frac{(-1)^{j-m}}{j} \le \frac{1}{m} - \frac{1}{k}.$$

If $\sum_{j=m}^{k} \frac{(-1)^{j-m}}{j}$ has an odd number of terms, we may write

$$\sum_{j=m}^{k} \frac{(-1)^{j-m}}{j} = \sum_{j=m}^{k-1} \frac{(-1)^{j-m}}{j} + \frac{1}{k},$$

and then using the previously derived result we find

$$\left|\sum_{j=m}^{k} \frac{(-1)^{j-m}}{j}\right| \le \frac{1}{m} - \frac{1}{k-1} + \frac{1}{k}.$$

Thus in all cases

$$\left|\sum_{j=m}^{k} \frac{(-1)^{j+1}}{j}\right| \le \frac{1}{m} + \frac{1}{k}. \qquad 7.6$$

and therefore

$$\left|\sum_{j=m}^{k} \frac{(-1)^{j+1}}{j}\right|$$

surely gets small for sufficiently large m, k. Hence, by the Cauchy criterion, Theorem 7.8 on page 219,

$$\sum_{j=1\to\infty} \frac{(-1)^{j+1}}{j}$$

converges.

■

Here's an innocent-looking almost self-evident result which is very useful.

Theorem 7.11: Convergence of $\sum_{j=k\to\infty} a_j$ and $\sum_{j=r\to\infty} a_j$

If $a_0, a_1, a_2, \ldots$ is any given sequence of complex numbers, and k,r are nonnegative integers with $r > k$, then[1]

$$\sum_{j=r\to\infty} a_j \quad \text{converges if and only if} \quad \sum_{j=k\to\infty} a_j \quad \text{converges.}$$

Furthermore

$$\sum_{j=k}^{\infty} a_j = \sum_{j=k}^{r-1} a_j + \sum_{j=r}^{\infty} a_j.$$

❑

The first assertion follows instantly from the Cauchy Criterion.

The second assertion comes from the fact that for $m > r$

$$\sum_{j=k}^{m} a_j = \sum_{j=k}^{r-1} a_j + \sum_{j=r}^{m} a_j.$$

---

1. Recalling Definition 7.4 on page 214 of convergence.

From the already established first assertion we know that the limit as $m \to \infty$ exists for each of these three sums, and since the limit of the sum (of sequences) is the sum of their limits, we have

$$\lim_{m \to \infty} \sum_{j=k}^{m} a_j = \lim_{m \to \infty} \sum_{j=k}^{r-1} a_j + \lim_{m \to \infty} \sum_{j=r}^{m} a_j$$

or

$$\sum_{j=k}^{\infty} a_j = \sum_{j=k}^{r-1} a_j + \sum_{j=r}^{\infty} a_j,$$

(the second summation being a constant sequence with regard to the variable m).

Example 7.8: Estimating $\sum_{j=1}^{\infty} \frac{(-1)^{j+1}}{j}$

We can now get a good estimate of the error in approximating

$$\sum_{j=1}^{\infty} \frac{(-1)^{j+1}}{j} \quad \text{by} \quad \sum_{j=1}^{m-1} \frac{(-1)^{j+1}}{j},$$

because, from Theorem 7.11

$$\sum_{j=1}^{\infty} \frac{(-1)^{j+1}}{j} = \sum_{j=1}^{m-1} \frac{(-1)^{j+1}}{j} + \sum_{j=m}^{\infty} \frac{(-1)^{j+1}}{j}$$

and from Example 7.7, inequality 7.6 on page 226, we know that

$$\left| \sum_{j=m}^{k} \frac{(-1)^{j+1}}{j} \right| \le \frac{1}{m} + \frac{1}{k}.$$

Letting $k \to \infty$ in this inequality yields

$$\left| \sum_{j=m}^{\infty} \frac{(-1)^{j+1}}{j} \right| \le \frac{1}{m}.$$

So the magnitude of the error in approximating $\sum_{j=1}^{\infty} \frac{(-1)^{j+1}}{j}$ by $\sum_{j=1}^{m-1} \frac{(-1)^{j+1}}{j}$ does not exceed $1/m$.

■

A generalization of this result, established using the same reasoning is the following *alternating series theorem.*

Theorem 7.12: Alternating series

If $a_0, a_1, a_2, \ldots$ is a nonincreasing sequence of positive real numbers with limit 0, then the series (Definition 7.4 on page 214)

$$\sum_{j=0\to\infty} (-1)^{j+1} a_j$$

converges, and

$$\left| \sum_{j=0}^{\infty} (-1)^{j+1} a_j - \sum_{j=0}^{m-1} (-1)^{j+1} a_j \right| \le a_m .$$

❑

The most powerful practical techniques for determining convergence and estimating approximation error in series are furnished by the comparison test.

Theorem 7.13: Comparison test for series convergence

Suppose $\sum_{j=0\to\infty} z_j$ is a complex infinite series and $\sum_{j=0\to\infty} w_j$ is a real infinite series (see Definition 7.4 on page 214) with

$$|z_j| \le w_j \quad \text{for all integers } j \ge 0. \qquad \textit{7.7}$$

If $\sum_{j=0\to\infty} w_j$ converges, then so do $\sum_{j=0\to\infty} z_j$ and $\sum_{j=0\to\infty} |z_j|$.

Furthermore, the estimation error

$$\left| \sum_{j=0}^{\infty} z_j - \sum_{j=0}^{m-1} z_j \right| \le \sum_{j=0}^{\infty} w_j . \qquad \textit{7.8}$$

❑

The proof stems from the triangle inequality, Theorem 7.6 on page 216, which, using inequality 7.7 above, shows that for all nonnegative integers *k, m*

$$\left| \sum_{j=m}^{k} z_j \right| \le \sum_{j=m}^{k} |z_j| \le \sum_{j=m}^{k} w_j . \qquad \textit{7.9}$$

Since $\sum_{j=0\to\infty} w_j$ converges, the right-hand term above is small for all sufficiently large *k, m* (by the Cauchy criterion, Theorem 7.8 on page 219) making this true for the left-hand sums. So, again using the other part of the Cauchy criterion, it follows that $\sum_{j=0\to\infty} z_j$ converges.

Taking the now allowable limits as $k \to \infty$ in inequalities 7.9 establishes inequality 7.8.

The comparison test is only useful if we have some useful *comparison series,* $\sum_{j=0 \to \infty} w_j$. The most useful such series is the Geometric series

$$\sum_{j=0 \to \infty} z^j, \quad 0 < z < 1 .$$

Rather than giving specific examples, we will derive a general result which stems from comparison with the geometric series. The geometric series is so useful for comparison purposes because the ratio of successive terms

$$\frac{z^{j+1}}{z^j}$$

is just z. It's easy to see that the geometric series converges for $-1 < z < 1$, because Theorem 1.9 on page 16 established that

$$\sum_{j=0}^{n} z^j = \frac{1 - z^{n+1}}{1 - z}, \qquad \textit{7.10}$$

for all numbers[1] z different from 1. Since $\lim_{n \to \infty} z^{n+1} = 0$ for all z satisfying the condition $|z| < 1$, taking limits in inequality 7.10 shows **convergence of the geometric series,** $\sum_{j=0 \to \infty} z^j$ **to the sum** $\frac{1}{1-z}$ **for** $|z| < 1$.

So another way to phrase the convergence of the geometric series is that the *geometric series converges if the ratio of successive terms has magnitude less than* 1. It will be easy to show that if any series is such that the ratio of its successive terms has magnitude not exceeding some constant $c < 1$, then by comparison with the geometric series

$$\sum_{j=0}^{n} c^j$$

the given series, too, will converge.

To supply the needed details, suppose the (complex) series $\sum_{j=0 \to \infty} a_j$ satisfies the condition

$$\left|\frac{a_{j+1}}{a_j}\right| \le c < 1 \quad \text{for all } j \ge j_0 \qquad \textit{7.11}$$

Then for all $j \ge j_0$,

---

1. The original proof specified only real numbers — since the complex numbers had not yet been introduced. However, the proof extends without change to include the complex numbers.

$$|a_{j+1}| \le c|a_j|$$

$$|a_{j+2}| \le c|a_{j+1}| \le c^2|a_j|$$

$$\vdots$$

and generally (by induction) for all $j \ge j_0$

$$|a_{j+q}| \le c^q|a_j| \quad \text{for all nonnegative integers } q. \qquad \textbf{\textit{7.12}}$$

Hence for $m, k \ge j_0$

$$\left|\sum_{j=m}^{k} a_j\right| \le \sum_{j=m}^{k} |a_j| \qquad \text{(by the triangle inequality, Theorem 7.6 on page 216)}$$

$$\le \sum_{j=m}^{k} |a_m| c^{j-m} \qquad \text{(by inequality 7.12 above)}$$

$$= |a_m| \sum_{q=0}^{k-m} c^q \le |a_{j_0}| c^{m-j_0} \sum_{q=0}^{k-m} c^q \qquad \text{(again using 7.12)}$$

$$= |a_{j_0}| c^{m-j_0} \frac{1-c^{k-m+1}}{1-c} \qquad \text{(by Theorem 1.9 on page 16);}$$

i.e.,

$$\left|\sum_{j=m}^{k} a_j\right| \le |a_{j_0}| c^{m-j_0} \frac{1-c^{k-m+1}}{1-c}.$$

From this it follows that for all m, k sufficiently large, $\left|\sum_{j=m}^{k} a_j\right|$ is small. So by the Cauchy criterion, Theorem 7.8 on page 219, the series $\sum_{j=0 \to \infty} a_j$ converges. Moreover,

$$\left|\sum_{j=m}^{\infty} a_j\right| \le |a_{j_0}| \frac{c^{m-j_0}}{1-c} \quad \text{for} \quad m \ge j_0$$

so that, using Theorem 7.11 on page 226,

$$\left|\sum_{j=0}^{\infty} a_j - \sum_{j=0}^{m-1} a_j\right| \le |a_{j_0}| \frac{c^{m-j_0}}{1-c}. \qquad \textbf{\textit{7.13}}$$

So we have established the following theorem.

Theorem 7.14: Ratio test and remainder estimate

The complex infinite series $\sum_{j=0\to\infty} a_j$ converges (Definition 7.4 on page 214) if for all integers $j \geq j_0$

$$\left|\frac{a_{j+1}}{a_j}\right| \leq c < 1,$$

where j_0 is some given nonnegative integer and c is a constant.

With regard to error estimation, inequality 7.13 on page 230 is satisfied for all $m \geq j_0$.

❑

Example 7.9: Convergence of the exponential series, $\sum_{j=0\to\infty} x^j/j!$

Here x is some fixed constant and $a_j = x^j/j!$. For $x \neq 0$

$$\left|\frac{a_{j+1}}{a_j}\right| = \left|\frac{x^{j+1}/(j+1)!}{x^j/j!}\right| = \left|\frac{x}{j+1}\right|.$$

Let j_0 be the smallest integer $\geq |x|$. Then for $j \geq j_0$, $x \neq 0$

$$\left|\frac{a_{j+1}}{a_j}\right| \leq \frac{|x|}{|x+1|} < 1.$$

So the exponential series converges for all (complex) nonzero x. Clearly it converges for $x = 0$, and hence for all complex x.

■

Exercises 7.10

In the following determine which series converge, and for these, how large to take m so that $\sum_{j=0}^{m-1} a_j$ is within .0001 of $\sum_{j=0}^{\infty} a_j$.

1. $\displaystyle\sum_{j=1\to\infty} \frac{(-1)^{j+1}}{\sqrt{j}}$

2. $\displaystyle\sum_{j=1\to\infty} \frac{(-1)^{j+1}}{ln(j+1)}$

3. $\displaystyle\sum_{j=1\to\infty} (-1)^{j+1} sin(1/j)$

4. $\sum_{n=0 \to \infty} n^4/2^n$

5. $\sum_{n=0 \to \infty} (-1)^n \frac{ln(n)}{n}$

6. $\sum_{n=0 \to \infty} sin(1/2^n)$ Hint: Use Taylor's theorem to write $sin(x)$ as x + remainder.

7. $\sum_{n=0 \to \infty} n^n/n!$

8. $\sum_{n=1 \to \infty} sin(n!/n^n)$

9. $\sum_{n=1 \to \infty} 1/n^2$ Hint: $\frac{1}{n^2} < \frac{1}{(n-1)n} = \frac{1}{n-1} - \frac{1}{n}$

10. $\sum_{n=1 \to \infty} 1/n^{3/2}$ Hint: $\frac{1}{n^{3/2}} \leq \int_{n-1}^{n} \frac{dx}{x^{3/2}}$

11. The *p-series* $\sum_{n=1 \to \infty} 1/n^p$ where p is a positive real

12. $\sum_{n=2 \to \infty} \frac{1}{n\ ln(n)}$ Hint: $\frac{1}{n\ ln(n)} > \int_{n}^{n+1} \frac{dx}{x\ ln(x)}$

13. $\sum_{n=2 \to \infty} \frac{1}{n[ln(n)]^2}$ See hints for Exercises 10 and 12.

Exercises 10 through 13 above illustrate the very useful *integral comparison test* which we will formally introduce in Chapter 12 when *improper Riemann integrals* are developed.

4 Uniform Convergence and Its Applications

As mentioned earlier, in many problems we look for a solution as the sum of an infinite series, $\sum_{j=0 \to \infty} a_j f_j$, of functions. Power series, $\sum_{j=0 \to \infty} a_j t^j$, play a major role in solving differential equations, which should not be too surprising in light of Taylor's theorem. (In the Taylor series for a function g about 0, $\sum_{j=0 \to \infty} \frac{g^{(j)}(0)}{j!} t^j$, we may let $f_j(t) = t^j$ and $a_j = \frac{g^{(j)}(0)}{j!}$.)

In trying to use series to solve differential equations we introduce the notion of uniform convergence, which is central to determining whether a series can be integrated and differentiated, and calculating these quantities. To see how we get to this concept in a natural fashion, it's easiest to begin by dealing with integration of a series of functions. We start by introducing the idea of the *convergence set* of a series of functions.

Definition 7.15: Convergence set of a series of functions

Let G_j $j = 0,1,....$ be a sequence of (possibly complex valued) functions (of a possibly complex variable) with a common domain, (see Definition 6.15 on page 201). The set of values, z for which $\sum_{j=0\to\infty} G_j(z)$ converges is called the **convergence set** of

$$\sum_{j=0\to\infty} G_j$$

or (treating z as the appropriate identity function) of $\sum_{j=0\to\infty} G_j(z)$. ❑

Example 7.11: Convergence set and sum of geometric series

For $G_j(z) = z^j$, since $\sum_{j=0}^{m} z^j = \frac{1 - z^{m+1}}{1 - z}$ for $z \neq 1$ (by Theorem 1.9 on page 16, whose proof is still valid for complex numbers) and $\sum_{j=0}^{m} z^j = m+1$ for $z = 1$, it is evident that ***the convergence set of*** $\sum_{j=0\to\infty} z^j$ ***consists of*** precisely those ***complex z with*** $|z| < 1$, and ***the sum of this series is*** $\frac{1}{1-z}$.

■

Example 7.12: Convergence set of $\sum_{j=1\to\infty} x^j/j$, x real

By the ratio test, Theorem 7.14 on page 231, it follows that this series converges for $-1 < x < 1$. From the Alternating series test, Theorem 7.12 on page 228, this series is seen to converge for $x = -1$, while from the results on the harmonic series, Example 7.6 on page 224, it diverges for $x = 1$; we can see that it diverges for $|x| > 1$ as follows.

$$\left|\frac{x^{j+1}/(j+1)}{x^j/j}\right| = \frac{j}{j+1}|x|.$$

Thus for fixed $|x| > 1$ the terms x^j/j do not approach 0 as j grows large, which we know is necessary for convergence (as has already been seen to follow from the Cauchy criterion, page 219). Hence this series converges for $-1 \leq x < 1$ and diverges for all other reals.

■

Returning to the original discussion, since integration of series of functions is our first goal, let us assume that the functions G_j are real or complex valued functions of a real variable with common domain being the closed bounded interval $[a; b]$ with $a \le b$. Also assume that $[a; b]$ is the convergence set for $\sum_{j=0 \to \infty} G_j$. In order for $\sum_{j=0}^{\infty} G_j$ to have an integral from a to b, we know from Theorem 4.3 on page 103 and Definition 6.19, on page 202, that it is sufficient for $\sum_{j=0}^{\infty} G_j$ to be continuous on $[a; b]$. So we ask for a reasonable condition ensuring that $\sum_{j=0}^{\infty} G_j(t+h)$ will be close to $\sum_{j=0}^{\infty} G_j(t)$ when $t,\ t+h$ are in $[a; b]$, t fixed and h small enough.

Since the sum of a fixed finite number of continuous functions is continuous (Theorem 2.10 on page 38) it is evident that if $\left|\sum_{j=k}^{\infty} G_j(x)\right|$ can be made small (within as small a positive distance from 0 as desired), for all x in $[a; b]$,with a single restriction[1] on how large k must be, then $\sum_{j=0}^{\infty} G_j$ will be continuous at each x in $[a; b]$.

This discussion leads us to the notion of uniform convergence (which is a slightly stronger condition than we have seen is needed) that will ensure continuity of $\sum_{j=0}^{\infty} G_j$ when the G_j are continuous.

Definition 7.16: Uniform convergence of $\sum_{j=0 \to \infty} G_j$ on a set B

Suppose G_j, $j = 0,1,...$ are (possibly complex valued) functions (Definition 6.15 on page 201) with a common domain. The series $\sum_{j=0 \to \infty} G_j$ (Definition 7.4 on page 214) is said to **converge uniformly on a subset B** of this common domain if K can be chosen so that the sums $\left|\sum_{j=k}^{\infty} G_j(x)\right|$ are simultaneously small for all x in B and all $k \ge K$ — i.e., a single value of K can be chosen as a restriction on k $(k \ge K)$ to get all of these sums to be and stay within any specified positive distance from 0 (K, of course, depending on the set B, the G_j, and the specified distance).

❑

1. Quantitatively, if for the given G_j and reals a,b there is a non-negative integer, K_ε, determined solely by ε such that

$$\left|\sum_{j=K_\varepsilon}^{\infty} G_j(x)\right| \le \varepsilon \quad \text{for all } x \text{ in } [a;b].$$

Based on what led up to the definition of uniform convergence,[1] we have established the following result.

Theorem 7.17: Uniform convergence and continuity

If $\sum_{j=0\to\infty} G_j$ is a series of continuous, possibly complex valued, functions, uniformly convergent (Definition 7.16 on page 234) on the closed bounded real interval $[a; b]$, $a < b$ (Definition 3.11 on page 88) then

$$\sum_{j=0}^{\infty} G_j$$

is continuous at each x in $[a; b]$.

❑

As a result of this theorem, we can now prove that under the circumstances we are considering,

$$\int_a^b \sum_{j=0}^{\infty} G_j = \sum_{j=0}^{\infty} \int_a^b G_j. \qquad \textbf{7.14}$$

From the assumed convergence of $\sum_{j=0\to\infty} G_j$, we may write

$$\sum_{j=0}^{\infty} G_j(t) = \sum_{j=0}^{k-1} G_j(t) + \sum_{j=k}^{\infty} G_j(t)$$

for all t in $[a; b]$ (see Theorem 7.11 on page 226). From Theorem 7.17 we know the left-hand sum to be continuous, while the continuity of the G_j shows that the finite sum to the right of the integral is continuous (Theorem 2.10 on page 38). Hence, again using Theorem 2.10, we see that the rightmost sum is

---

1. In the previous discussion we saw that it was not necessary that $\left|\sum_{j=k}^{\infty} G_j(x)\right| \le \varepsilon$ for **all** $k \ge K_\varepsilon$, but for each ε just for $k = K_\varepsilon$ and all x. However, there is little harm in requiring it, since this condition will be satisfied in most cases of interest to us, and is usually easy to verify. The difference is a subtle one, and of little importance, but this result is a bit strange, since for any fixed x, we know from the assumed convergence of $\sum_{j=0\to\infty} G_j$ that $\left|\sum_{j=k}^{\infty} G_j(x)\right|$ gets and remains small as k grows large — i.e., for each positive ε there is $K_{\varepsilon,x}$ such that for all $k \ge K_{\varepsilon,x}$,

$$\left|\sum_{j=k}^{\infty} G_j(x)\right| \le \varepsilon.$$

also continuous. Thus all three sums are integrable (Theorem 4.3 on page 103), and (using Theorem 4.5 on page 107)

$$\int_a^b \sum_{j=0}^{\infty} G_j = \int_a^b \sum_{j=0}^{k-1} G_j + \int_a^b \sum_{j=k}^{\infty} G_j .$$

Again using Theorem 4.5 we see that

$$\int_a^b \sum_{j=0}^{\infty} G_j = \sum_{j=0}^{k-1} \int_a^b G_j + \int_a^b \sum_{j=k}^{\infty} G_j .$$

From the uniform convergence of $\sum_{j=0 \to \infty} G_j$ we know that the integrand of the rightmost term above becomes uniformly small (small for all points in $[a, b]$ simultaneously), for all k sufficiently large; so as k grows large, the right hand integral approaches 0 (again see Theorem 4.5). Since the left-side integral does not depend on k, and the limit of the sums is the sum of the limits (Theorem 7.3 on page 214), we see that equation 7.14 above holds. We now state it formally for reference purposes.

Theorem 7.18: Interchange of order of infinite series summation and integration

If $\sum_{j=0 \to \infty} G_j$ is an infinite series of continuous real or complex valued functions of a real variable which converges uniformly (Definition 7.16 on page 234), on the closed and bounded interval $[a; b]$ (Definition 3.11 on page 88), then

$$\int_a^b \sum_{j=0}^{\infty} G_j = \sum_{j=0}^{\infty} \int_a^b G_j .$$

❑

So under uniform convergence, the rule that the integral of the sum is the sum of the integrals extends to infinite series of continuous functions; which means that uniform convergence allows us to treat the sum of an infinite series much like a finite sum, as far as integration is concerned.

None of this discussion would get us very far if we didn't have some easy ways to verify uniform convergence. The first way that comes to mind is comparison with a series of constants — i.e.,

if for all x in $[a; b]$, $|G_j(x)| \le M_j$, where $\sum_{j=0 \to \infty} M_j$ converges, then evidently $\sum_{j=0\to\infty} G_j$ converges uniformly on $[a; b]$.

Formally, and more generally for G_j real or complex valued functions of a real or complex variable, we have the following comparison test result.

Theorem 7.19: Uniform convergence by *comparison*

Suppose that for all x in B $|G_j(x)| \le H_j(x)$ and $\sum_{j=0\to\infty} H_j$ converges uniformly on B (Definition 7.16 on page 234). Then

$$\sum_{j=0\to\infty} G_j$$

converges uniformly on B.

In particular, this result holds if M_j are constants with $|G_j(x)| \le M_j$ for all x in B and $\sum_{j=0\to\infty} M_j$ converges. (This latter test for uniform convergence is called the **Weierstrass *M*-test.**)

❑

Example 7.13: $\sum_{k=0\to\infty} k x^{k-1}$

By the ratio test, Theorem 7.14 on page 231, $\sum_{k=0\to\infty} k x^{k-1}$ converges for all x satisfying $-1 < x < 1$. If $|x| \le c < 1$, then $|k x^{k-1}| \le k c^{k-1}$, so $\sum_{=0\to\infty} k x^{k-1}$ converges uniformly for $|x| \le c < 1$. Hence by Theorem 7.18

$$\int_0^t \sum_{k=0}^{\infty} k x^{k-1} dx = \sum_{k=0}^{\infty} \int_0^t k x^{k-1} dx = \sum_{k=0}^{\infty} x^k \Big|_0^{x=t} = \sum_{k=0}^{\infty} t^k = \frac{1}{1-t},$$

for $0 \le t < 1$, the last step following from the geometric series theorem, Theorem 1.9 on page 16.

Now looking only at the left- and right-most terms, take the derivative (using the fundamental theorem of calculus, Theorem 4.10 on page 127 for the leftmost term) to obtain

$$\sum_{k=0}^{\infty} k t^{k-1} = \frac{1}{(1-t)^2}, \text{ or } \sum_{k=0}^{\infty} D(t^k) = D \sum_{k=0}^{\infty} t^k$$

(again using the geometric series theorem).

■

This is an example of a situation where

$$D\sum_{k=0}^{\infty} f_k = \sum_{k=0}^{\infty} D f_k,$$

as well as of our ability to obtain a simple formula for the sum of a series (here $\sum_{k=0}^{\infty} k t^{k-1}$), if the series for its antiderivative (here $\sum_{k=0}^{\infty} t^k$) has a simple formula. From the above derivation we see that this is no isolated accident. We will state the more general result immediately, leaving its derivation (which mirrors the computations in the example just presented) to the optional portion which follows.

Theorem 7.20: Interchange of order of infinite series summation and differentiation

Suppose G_j, $j = 0,1,2,...$ are real or complex valued functions of a real variable with common domain (Definition 6.15 on page 201) and

- $\sum_{j=0 \to \infty} G_j$ converges on the closed bounded interval $[a;b]$ (Definition 7.15 on page 233 and Definition 3.11 on page 88),

- The G_j have continuous derivatives (Definition 6.15 on page 201 and Definition 6.16 on page 202),

- $\sum_{j=0 \to \infty} G_j'$ converges uniformly on $[a;b]$, (Definition 7.16 on page 234).

Then $\sum_{j=0}^{\infty} G_j$ is differentiable with

$$D\sum_{j=0}^{\infty} G_j = \sum_{j=0}^{\infty} DG_j.$$

❑

⌈

The proof of Theorem 7.20 is as follows: Applying Theorem 7.18 on page 236 to $\sum_{j=0}^{\infty} G_j'$, replacing b by t, (the identity function on $[a, b]$) we have

$$\int_a^t \sum_{j=0}^{\infty} G_j' = \sum_{j=0}^{\infty} \int_a^t G_j'.$$

Applying the fundamental integral evaluation formula of Theorem 4.3 on page 103 to the right-hand integrals, we find

$$\int_a^t \sum_{j=0}^{\infty} G_j' = \sum_{j=0}^{\infty}[G_j(t) - G_j(a)].$$

Now differentiate the above, applying the fundamental theorem of calculus, Theorem 4.10 on page 127, to the left side, to obtain

$$\sum_{j=0}^{\infty} DG_j(t) = D\sum_{j=0}^{\infty}[G_j(t) - G_j(a)]. \qquad \textbf{7.15}$$

Given that $\sum_{j=0\to\infty} G_j$ converges on $[a; b]$, it is not hard to show[1] that

$$\sum_{j=0}^{\infty}[G_j(t) - G_j(a)] = \sum_{j=0}^{\infty} G_j(t) - \sum_{j=0}^{\infty} G_j(a). \qquad \textbf{7.16}$$

(Note that we had to assume the convergence of $\sum_{j=0\to\infty} G_j$ separately from that of $\sum_{j=0\to\infty} G_j'$, as can be seen by noting that if $G_j(t) = t^j + 5$, $-1 < t < 1$, $\sum_{j=0\to\infty} G_j'$ converges, while $\sum_{j=0\to\infty} G_j$ does not.) Substituting equation 7.16 into equation 7.15 concludes the proof.

$\lfloor$

Exercises 7.14

1. Show that for t the identity function on the open interval $(-1; 1)$, and all positive integers, k

$$D^k \sum_{k=0}^{\infty} t^j = \sum_{k=0}^{\infty} D^k t^j;$$

specifically that

---

1. For each value of t and large enough k, $\sum_{j=0}^{k} G_j(t)$ is close to $\sum_{j=0}^{\infty} G_j(t)$ by the assumed convergence of $\sum_{j=0\to\infty} G_j$, this holding in particular for $t = a$; hence $\sum_{j=0}^{k} [G_j(t) - G_j(a)]$ is close to $\sum_{j=0}^{\infty} G_j(t) - \sum_{j=0}^{\infty} G_j(a)$; since the left side is close to $\sum_{j=0}^{\infty}[G_j(t) - G_j(a)]$ the assertion follows.

$$\frac{1}{(1-t)^2} = D\frac{1}{1-t} = \sum_{j=0}^{\infty} j\,t^{\,j-1}$$

and
$$\frac{2}{(1-t)^3} = D\frac{1}{(1-t)^2} = \sum_{j=0}^{\infty} j(j-1)\,t^{\,j-2}.$$

2. Suppose that $\sum_{k=0\to\infty} a_k x^k$ is known to converge by the ratio test. Show that if $f(x) = \sum_{k=0}^{\infty} a_k x^k$, then f is differentiable with the sum of the term by term derivatives you expect.
Hint: If

$$\left|\frac{a_{k+1}x^{k+1}}{a_k x^k}\right| < c < 1 \text{ for large } k, \text{ what about}$$

$$\left|\frac{(k+1)a_{k+1}x^{k+1}}{k\,a_k x^k}\right|, \text{ the ratio for } \sum D a_k x^k\,?$$

3. Find a convergent infinite series for the standard normal distribution function (used in probability and statistics),

$$\Phi(t) = \frac{1}{2} + \frac{1}{\sqrt{2\pi}}\int_0^t e^{-x^2}dx,$$ and determine how many terms in this series will ensure determination of $\Phi(2)$ to within .00001.

4. Determine whether or not $\sum_{j=1\to\infty} (-1)^{j+1}\frac{1}{j}\sqrt[j]{x+1}$ is uniformly uniformly convergent for $-1 \le x \le c$, where c is any real number exceeding -1.

5. a. Show that $\sum_{n=1\to\infty} \frac{cos(nt)}{n^4}$ is uniformly convergent, where t is the identity function on the reals.

b. Decide whether $f = \sum_{n=1}^{\infty}\frac{cos(nt)}{n^4}$ and $g = \sum_{n=1}^{\infty}\frac{sin(nt)}{n^2}$ are differentiable.

Problems 7.15

1. Let the function f be given by $f(x) = \sqrt{1+x}$, x real $x > -1$.
   a. Determine formulas for $D^k f(x)$, $k = 1,2,\dots$.
   b. Show that the Taylor series about 0,

$$\sum_{j=0\to\infty} f^{(k)}(0)\frac{x^k}{k!}$$

converges for $-1 < x < 1$, and converges uniformly for $-c < x < c$ where c is any constant with $0 \le c < 1$.
Find a usable upper bound for the remainder,

$$\sum_{k=m}^{\infty} f^{(k)}(0)\frac{x^k}{k!} \quad \text{for } -1 < x < 1.$$

c. Now for the hard part. Show that the Taylor series sum

$$T(x) = \sum_{k=0}^{\infty} f^{(k)}(0)\frac{x^k}{k!} = \sqrt{1+x}.$$

Hints: First show that f itself satisfies the differential equation $2(1+x)f'(x) = f(x)$, $f(0) = 1$.
Next show that if g is any function satisfying this differential equation, i.e., for which $2(1+x)g'(x) = g(x)$, $g(0) = 1$ then $g(x) = f(x)$ — by dividing this last differential equation (for g) by the preceding one (for f) legitimate because neither f nor f' are 0 in the domain under consideration, and then recalling the formula for $(g/f)'$.
Finally, show that T satisfies this differential equation and initial condition (qualifying it as a g.)

2. Extend the result to m-th roots - i.e. to

$$f(x) = \sqrt[m]{1+x} = (1+x)^{1/m}, \quad -1 < x < 1.$$

3. Show that for all real x,

a. $$e^x = \sum_{j=0}^{\infty} \frac{x^j}{j!}$$

b. $$sin(x) = \sum_{j=0}^{\infty} (-1)^j \frac{x^{2j+1}}{(2j+1)!}$$

c. $$cos(x) = \sum_{j=0}^{\infty} (-1)^j \frac{x^{2j}}{(2j)!}$$

Hints: Make use of the remainder term in Taylor's theorem, Theorem 6.2 on page 179, the fact that $e < 3$ (see Problems 7.15 on page 240) and the facts that because $sin^2 + cos^2 = 1$ (see equation 5.58 on page 161 etc.) $|sin| \le 1, |cos| \le 1$.

Theorem 7.21: Power series convergence properties

If $\sum_{k=0\to\infty} a_k x^k$ converges at the complex value $x = x_0$, (Definition 6.8 on page 199 and Definition 7.4 on page 214), then it converges uniformly (Definition 7.16 on page 234) for all x with $|x| \le x_1$, where x_1 is any fixed positive value with $x_1 < |x_0|$.

❑

To establish this result, we note that from the convergence of the given series at x_0, by the n-th term test, Theorem 7.10 on page 225, the sequence whose n-th term is $a_n x_0^n$ must have limit 0. Also, for $|x| \le x_1 < |x_0|$,

$$|a_n x^n| = |a_n x_0^n| \left|\frac{x}{x_0}\right|^n \le |a_n x_0^n| \left|\frac{x_1}{x_0}\right|^n.$$

Because $\lim_{n\to\infty} a_n x_0^n = 0$, this sequence of terms is bounded, say $|a_n x_0^n| \le B$ for all n. Now, by the *Weierstrass M-test*, Theorem 7.19 on page 237, comparison with the series

$$\sum_{n=0\to\infty} B|x_1/x_0|^n,$$

known to converge by Example 7.11 on page 233, establishes the asserted uniform convergence, proving Theorem 7.21.

5 Power Series Solution of Differential Equations

In this work we can only sample lightly from the wide variety of phenomena which are described by differential equations. These include descriptions of the flow of heat in a system,[1] of the mechanical behavior of systems, such as automobiles and air and computer data traffic,[2] as well as the probability distributions describing the motion of atmospheric pollutants.

We have already touched on numerical solution of such differential equations, and developed some of the theory for constant coefficient differential equations.

Problems involving more than one independent variable — for instance, the effect of a change in both time and position [independent variables] on temperature [dependent variable] — lead to *partial differential equations*. These often lead back to differential equations with only one independent

1. Necessary, for example, to enable design which avoids meltdowns.
2. Needed for highway, airport, and computer designs, so as to avoid congestion and unnecessary waiting time.

variable (so-called *ordinary differential equations*) of the type we have been treating. These equations are not usually constant coefficients differential equations, but often these ordinary differential equations are *linear* in the dependent variable and its derivatives — i.e., of the form

$$x^{(n)} + b_{n-1}x^{(n-1)} + \cdots + b_1 x = f, \qquad \textbf{7.17}$$

where $b_1, b_2, \ldots, b_{n-1}$ and f are all known (possibly complex valued) functions of a real variable, all having power series representations (the b_j frequently being polynomials). Taylor's theorem, Theorem 6.2 on page 179, makes it reasonable to try for a power series solution of the form[1]

$$x(t) = \sum_{j=0}^{\infty} a_j t^j.$$

The approach that we will soon take is first to substitute the sum above into the differential equation to be solved, as if it is legitimate to write

$$Dx(t) = x'(t) = \sum_{j=0}^{\infty} D(a_j t^j) = \sum_{j=0}^{\infty} j a_j t^{j-1}$$

$$D^2 x(t) = \quad \ldots \quad = \sum_{j=0}^{\infty} j(j-1) a_j t^{j-2}$$

etc.,

then algebraically rearrange the terms in the differential equation as if there were only a finite number of them,[2] finally writing the resultant equation in the form

---

1. Given initial conditions $x^{(j)}(0) = x_j$, $j = 0,1,\ldots,n-1$, one reasonable approach is to use the differential equation directly to successively calculate $x_{(j)}(0)$ for all $j \geq n$. We will pursue an alternative approach, leaving the reader to experiment with this one, to see how it compares.
2. That is, treating the power series for $b_1, b_2, \ldots, b_{n-1}$ and f and the dependent variable $x(t)$ as if they were polynomials.

$$\sum_{k=0}^{\infty} B_k t^k = 0.$$

Each of the B_k will be known functions of the unknown a_j. In order to determine the a_j, we will need to know that if a power series

$$\sum_{k=0}^{\infty} B_k t^k$$

is 0 for all t in some nonzero length interval, $[0; b]$, then $B_k = 0$ for all values of $k = 1,2,...$. (This will be proved in the next section, Theorem 7.31 on page 260.)

Thus we will end up solving for the a_j by solving the equations

$$\begin{aligned} B_1(a_0, a_1, a_2,...) &= 0 \\ B_2(a_0, a_1, a_2,...) &= 0 \\ &\ \vdots \end{aligned} \qquad \textbf{\textit{7.18}}$$

Each of the equations in 7.18 above will usually involve only a finite number of the a_j, making them *difference equations*; in the cases that arise, initial conditions of the form

$$x^{(j)}(0) = x_j, \quad j = 0, 1, ... ,n-1 \qquad \textbf{\textit{7.19}}$$

together with equations 7.18 lead to solution of these difference equations — i.e., to unique determination of all of the a_j. It is then incumbent on us to verify that indeed $\sum_{j=0\to\infty} a_j t^j$ converges and that $x(t) = \sum_{j=0}^{\infty} a_j t^j$ is a solution of equation 7.17 satisfying the initial condition equations 7.19. This doesn't quite end the investigation unless we know that there is one and only one solution of equation 7.17 satisfying initial conditions 7.19.

Before proceeding with a few illustrative examples, some comments are in order about the relatively cavalier approach being taken. First, it would be satisfying to know in advance that we can treat power series as if they were polynomials, insofar as the manipulations we are performing on them in the process of getting to a candidate for the solution being sought. However, this is not absolutely necessary, since only after we have found an actual power series candidate for the solution do we have to verify that it converges and its sum is a solution. It is here that we must be able to justify the various operations that arise — rearranging the order of addition, multiplying series together, etc. Finally, as far as establishing that what has been obtained is the only solution of the differential equation satisfying the given initial conditions, since its proof is beyond the scope we have set for this work, we will state some relevant theorems and provide references for more detailed treatment of other aspects of the subject of power series and related solutions of differential equations.

We will first proceed with a few illustrative examples, postponing the most basic theoretical underpinnings concerning algebraic manipulation of series to the optional section which concludes this chapter.

Example 7.16: Power series solution of $x' + t^2 x = 0, \quad x(0) = 1$

We try $x(t) = \sum_{j=0}^{\infty} a_j t^j$, assuming $x'(t) = \sum_{j=1}^{\infty} j a_j t^{j-1}$.

Then

$$x'(t) + t^2 x(t) = \sum_{j=1}^{\infty} j a_j t^{j-1} + t^2 \sum_{j=0}^{\infty} a_j t^j$$

$$= \sum_{j=1}^{\infty} j a_j t^{j-1} + t^2 a_0 + \sum_{j=1}^{\infty} a_j t^{j+2}$$

$$= \sum_{j=0}^{\infty} (j+1) a_{j+1} t^j + t^2 a_0 + \sum_{j=3}^{\infty} a_{j-2} t^j$$

$$= a_1 + 2a_2 t + 3a_3 t^2$$

$$+ \sum_{j=3}^{\infty} (j+1) a_{j+1} t^j + t^2 a_0 + \sum_{j=3}^{\infty} a_{j-2} t^j .$$

But because $x(0) = 1$, we have $a_0 = 1$. Hence

$$x'(t) + t^2 x(t) = a_1 + 2a_2 t + (3a_3 + 1) t^2 + \sum_{j=3}^{\infty} [(j+1) a_{j+1} + a_{j-2}] t^j$$

If $x'(t) + t^2 x(t)$ is to be 0 on some interval including 0, by Theorem 7.31 on page 260, the coefficients of all powers of t above must be 0. This leads to a difference equation

$$a_{j+1} = \frac{-a_{j-2}}{j+1} \quad \text{for } j \geq 3$$

with initial conditions

$$a_1 = 0, a_2 = 0, a_3 = -\frac{1}{3}.$$

From these we see immediately that

$$a_{3k-1} = 0 \quad \text{and} \quad a_{3k-2} = 0$$

for all $k = 1,2,...$ as well as

$$a_{3k} = \frac{-a_{3(k-1)}}{3k} \text{ for } k = 2, 3, \ldots .$$

This leads to

$$a_3 = -\frac{1}{3}, a_6 = \frac{1}{3 \cdot 6}, a_9 = \frac{-1}{3 \cdot 6 \cdot 9}$$

and, by induction

$$a_{3k} = (-1)^k \frac{1}{3^k \cdot k!} = \frac{(-1/3)^k}{k!} .$$

So

$$x(t) = 1 + \sum_{k=1}^{\infty} \frac{(-1/3)^k}{k!} t^{3k} = \sum_{k=0}^{\infty} \frac{(-t^3/3)^k}{k!} t^{3k} = e^{-t^3/3} ;$$

i.e., the function we have found is

$$x(t) = e^{-t^3/3} ,$$

which indeed does satisfy

$$x'(t) + t^2 x(t) = 0, x(0) = 1.$$

(This could have been derived with the methods used to solve Problem 5.10.4 on page 153.) Theorem 7.22 on page 246, referred to in Golomb and Shanks [8], shows that this is the only solution of these equations, (something that also follows for this situation from the method used to solve Problem 5.10.4).

■

At this point we could continue presenting examples of the power series approach to solution of differential equations. Before doing this it's really essential to state some theorems (whose proofs are beyond the scope of this work) which provide conditions which ensure that the power series approach provides the solution we are seeking. A readable reference for this topic is Golomb and Shanks [8].

Theorem 7.22: Power series solutions

Suppose that $b_1, b_2, \ldots, b_{n-1}$ and f are known real valued functions of a real variable with power series expansions all converging for $|t| < R$; i.e.,

$$b_j(t) = \sum_{k=0}^{\infty} a_{jk} t^j , \qquad f(t) = \sum_{k=0}^{\infty} a_k t^k$$

are valid representations for $|t| < R$ (see Definition 7.4 on page 214). Then every solution of the differential equation

$$x^{(n)} + b_{n-1} x^{(n-1)} + \cdots + b_1 x = f \qquad \textit{7.20}$$

for $|t| < R$ can be expressed as a power series.[1]

For given real values, $x_0, x_1, ..., x_{n-1}$, there is precisely one such solution satisfying

$$x^{(j)}(0) = x_j, \quad j = 0, 1, ..., n-1..$$

Every solution of the *homogeneous reduced* equation

$$x^{(n)} + b_{n-1}x^{(n-1)} + \cdots + b_1 x = 0 \qquad \textbf{7.21}$$

is a weighted sum

$$\sum_{j=1}^{n-1} c_j X_j$$

where the c_j are constants and X_j is the unique solution of equation 7.21 satisfying the initial conditions

$$X_j^{(j)}(0) = 1, \qquad X_k^{(j)}(0) = 0 \quad \text{if} \quad j \neq k, \quad j,k = 0,...,n-1.$$

If x_p is any fixed solution of equation 7.20, then **every** solution of equation 7.20 is of the form $x_p + x_h$ where x_h is one of the solutions of equation 7.21.

❑

These results show that when the given coefficient functions b_j and forcing function f are known to have power series expansions about 0, we expect the power series approach to work, provided the manipulations we need (such as grouping terms, and multiplying power series together to form the product power series) are legal — which we will verify in the next section.

Often the series we obtain for the desired solution will not be a function that we recognize, and in fact, many of the important functions of applied mathematics arise as power series solutions of differential equations.

In some cases of great interest, we cannot write the reduced (homogeneous) differential equation in the form of equation 7.21 where each of the known functions has a power series expansion about $t_0 = 0$, but rather in the form

$$t^n x^{(n)}(t) + t^{n-1} b_{n-1} x^{(n-1)}(t) + \cdots + b_0 x(t) = f(t),$$

where the b_j all have power series representations about 0. The theory gets trickier here, but with appropriate modifications can often be extended in

1. We have chosen to phrase everything in terms of power series about $t_0 = 0$. This is not restrictive, since a change of independent variable to T, where $T = t - t_0$ reduces other cases to $T = T_0 = 0$.

important cases — beyond the scope of this book — e.g., Bessel's equations, arising in the theory of mechanical vibrations,

$$t^2 x''(t) + tx'(t) + (t^2 - n^2)x(t) = 0,$$

where we seek solutions of the form

$$x(t) = t^r \sum_{k=0}^{\infty} a_k t^k, \quad a_0 \neq 0.$$

As already mentioned the power series approach generally leads to difference equations for determination of the coefficients a_k, and it is necessary to determine something about their solution to find out how fast $a_k t^k$ approaches 0 as k grows large — to calculate the solution we seek with sufficient accuracy. We illustrate this process in the example to be presented next. In contrast to the situation in Example 7.16 on page 245 which did not really require the power series method, this next example doesn't seem amenable to the other theoretical approaches we have presented.

Example 7.17: Solution of $x'' + tx = 0 \quad x(0) = 1 \quad x'(0) = 0$

Trying $x(t) = \sum_{j=0}^{\infty} a_j t^j$, we are led to the equation

$$\sum_{j=2}^{\infty} j(j-1)\, a_j t^{j-2} + \sum_{j=0}^{\infty} a_j t^{j+1} = 0,$$

or, replacing $j - 2$ by J in the first sum and $j + 1$ by J in the second sum

$$\sum_{J=0}^{\infty} (J+2)(J+1)a_{J+2}\, t^J + \sum_{J=1}^{\infty} a_{J-1}\, t^J = 0.$$

The coefficient of t^0 is $2a_2$. The coefficient of t^J for any $J > 0$ is $(J+2)(J+1)a_{J+2} + a_{J-1}$. So we have

$$2a_2 + \sum_{J=1}^{\infty} [(J+2)(J+1)a_{J+2} + a_{J-1}]t^J = 0$$

Now $a_0 = x(0) = 1$, and $a_1 = x'(0) = 0$.

By Theorem 7.31 on page 260 in the next section, in order for this equality to hold on any nonzero length interval about 0, the coefficient of each power of t must be 0. So we require

$$a_2 = 0 \text{ and for each } J = 1,2,\ldots \quad (J+2)(J+1)a_{J+2} + a_{J(-1)} = 0,$$

or

$$a_{J+2} = \frac{-1}{(J+2)(J+1)} a_{J-1}.$$

Because a_1 and a_2 are both 0, it's not hard to see from this last equation that

$$0 = a_1 = a_4 = \cdots = \quad \text{and} \quad 0 = a_2 = a_5 = \cdots = ,$$

i.e.,

$$a_{3k+1} = 0 \text{ and } a_{3k+2} = 0 \text{ for all integers } k \geq 0.$$

Similarly, identifying $J + 2$ with $3k$ (and hence $J - 1$ with $3(k - 1)$) we find

$$a_{3k} = \frac{-1}{3k(3k-1)} a_{3(k-1)}.$$

Hence, because $a_0 = 1$, we have

$$a_3 = \frac{-1}{3 \cdot 2}, \quad a_6 = \frac{-1}{6 \cdot 5 \cdot 3 \cdot 3}, \quad a_9 = \frac{1}{9 \cdot 8 \cdot 6 \cdot 5 \cdot 3 \cdot 2}, \ \cdots \ .$$

i.e., for all positive integers, k

$$a_{3k} = \frac{(-1)^k}{(3k)(3k-1)(3[k-1])(3[k-1]-1)\cdots\cdots(3 \cdot 2)}. \qquad \textbf{7.22}$$

The sum

$$\sum_{j=0}^{3k} a_j t^j = \sum_{r=0}^{k} a_{3r} t^{3r} = \sum_{r=0}^{k} a_{3r} (t^3)^r \qquad \textbf{7.23}$$

has terms whose magnitudes are decreasing as r increases for those r satisfying the condition

$$\left| \frac{a_{3(r+1)}(t^3)^{r+1}}{a_{3r}(t^3)^r} \right| = \frac{|t^3|}{3(r+1)[3(r+1)-1]} < 1 \qquad \textbf{7.24}$$

which, for fixed t, occurs for large enough r. From equation 7.22 we see that the alternating series test, Theorem 7.12 on page 228, then lets us determine how many terms are needed in the sum(s) in 7.23 to obtain specified accuracy for any given t. For example, to obtain accuracy .001 at $t = 2$, we note that at $t = 2$, the ratio in inequality 7.24 is less than 1 for all positive r, and by trial and error, we find easily that $|a_{3k}| \leq 0.001$ for $k = 3$.

■

Exercises 7.18

1. Repeat Example 7.17 on page 248 for $x(0) = 0, x'(0) = 1$.
2. Repeat Example 7.17 on page 248 for $x(0) = 1, x'(0) = 1$.

Problem 7.19

Use the power series approach to solve

$$x'' + tx = e^t,\ x(0) = 1, x'(0) = 0$$

and see if you can determine how many terms to take so that you are assured an accuracy of .001 in the computation of $x(2)$. Hints: You can obtain equations very similar to many of the equations in Example 7.17. While you're not likely to find a simple formula for the a_j, with effort you can get useful upper bounds for $|a_j|$, by noting that

$$|a_{j+2}| \le \frac{|a_{j-1}| + 1/j!}{(j+1)(j+2)},\ |a_{j+3}| \le \frac{|a_{j-1}| + 1/(j+1)!}{(j+2)(j+3)}.$$

Now let us define the sequence c_j by the equations

$$c_k = |a_k|$$

for some fixed k and thereafter

$$c_{j+3} = \frac{c_j + 1/(j+1)!}{(j+2)(j+3)}$$

We claim that if $c_j \ge \frac{1}{(j+1)!}$ then $c_{j+3} \ge \frac{1}{(j+4)!}$.

This follows from the rather crude reasoning that $c_j \ge 0$, and

$$c_{j+3} \ge \frac{1}{(j+3)(j+2)(j+1)!} = \frac{1}{(j+3)!} > \frac{1}{(j+4)!}.$$

Hence, as defined, if ever $c_J \ge \frac{1}{(J+1)!}$, then for all $j > J$

$$c_{j+3} \le \frac{2c_j}{(j+2)(j+3)}.$$

Now using the easily established fact that for $j \ge k$, $|a_j| \le c_j$ we see how to get more easily computable bounds for the values of $|a_j|$. The details get a bit messy — but note, that if we never have $c_j \ge 1/(j+1)!$, then we already have good bounds for $|a_j|$, and otherwise, we can get decent bounds (though not quite as good), by the reasoning just used.

Remarks on the power series method

This method appears to be extremely general and powerful. However, while it yields a formula for the solution, the properties of the solution that are of interest to us are not likely to be immediately obvious from this formula. We may be able to determine some of these properties from the differential equation, and guess at some of the others by doing numerical calculations directly from the series, and plotting the solution. However, in many cases, a great deal more work needs to be done to extract the information we may want.

6 Operations on Infinite Series

In the previous section we performed various operations on the sums of infinite series as if they were sums of a finite number of terms. The purpose of this section is to investigate the legitimacy of these operations. The operations of interest will be those of

- rearranging the order of addition and/or grouping of the terms of series sums as if they were finite sums;
- adding or multiplying series sums as if they were finite sums.

The first point to realize is that rearranging the order of the terms of an infinite series may very well affect its convergence.

Example 7.20: Rearranging $\sum_{j=1\to\infty} \frac{(-1)^{j+1}}{j}$

By the alternating series test, Theorem 7.12 on page 228, this series converges. Now look at only the negative terms in this series,

$$-\frac{1}{2}, -\frac{1}{4}, -\frac{1}{6}, \ldots;$$

for any value of k, the sum

$$\sum_{j=k}^{2k} -\frac{1}{2j} \leq -\frac{1}{2}\sum_{j=k}^{2k-1} \frac{1}{2k} = -\frac{1}{4k}k = -\frac{1}{4}$$

So a rearrangement which *piles up* sufficiently many negative terms immediately after each positive term can be formed which subtracts at least 1/4 from each positive term. Since the positive terms go to 0 for large j, the partial sums will become arbitrarily far negative — causing the rearranged series to *diverge to* $-\infty$.

■

This example shows that conditions are needed which will reliably inform us if it is the case that rearranging the order of the terms in a series does not alter its sum. We must first define precisely what we mean by a rearrangement of a series.

Definition 7.23: Rearrangement of series $\sum_{j=0\to\infty} a_j$

The series $\sum_{j=0\to\infty} b_j$ (Definition 7.4 on page 214), is said to be a **rearrangement of** $\sum_{j=0\to\infty} a_j$ if for each nonnegative integer j,

$$b_j = a_{r(j)},$$

where r is some sequence, (Definition 1.2 on page 3)

- whose domain consists of the nonnegative integers;
- whose range consists of the nonnegative integers;
- which is *one-to-one* — i.e., for any given integer $k \geq 0$ there is exactly one j such that $r(j) = k$. (We already know from the previous condition that there is at least one j such that $r(j) = k$, see Definition 1.1 on page 2, of *range* and Definition 1.2 on page 3 of *sequence*.)

The reason for the last two conditions is to ensure that each position in the b sequence corresponds to precisely one position in the a sequence, and each position in the a sequence corresponds to precisely one position in the b sequence.

❑

In line with the goal we are seeking, we introduce the notion of unconditional convergence.

Definition 7.24: Unconditional convergence of $\sum_{j=0\to\infty} a_j$

The series $\sum_{j=0\to\infty} a_j$ of (possibly complex) numbers (Definition 7.4 on page 214) is said to be **unconditionally convergent**, if all rearrangements of $\sum_{j=0\to\infty} a_j$, (Definition 7.23 above) converge.

A series which converges, but fails to converge unconditionally, is said to be **conditionally convergent**.

❑

To determine some reasonable conditions under which we obtain unconditional convergence, notice that in Example 7.20 on page 251, it was evident that the series consisting solely of the negative terms diverged to $-\infty$. Since the original series converges, it must be the case that the series consisting of just the positive terms must diverge to $+\infty$, in order to overcome the effect of the negative terms in the initial arrangement. This gives us the clue we need — namely, that we can ensure unconditional convergence if the two series — that of just the positive terms and that of just the negative terms, both converge. This leads us to the following definition.

Definition 7.25: Absolute convergence

An infinite series, $\sum_{j=0 \to \infty} a_j$ (Definition 7.4 on page 214), is said to be **absolutely convergent** if the series $\sum_{j=0 \to \infty} |a_j|$ is convergent (see Definition 6.11 on page 200).

❑

We now investigate the implications of absolute convergence. The arguments are somewhat delicate, and not critical for understanding what remains. So, the next material is made optional.

┌

Suppose that $\sum_{j=0 \to \infty} a_j$ is absolutely convergent, and $\sum_{j=0 \to \infty} b_j$ is any specified rearrangement of $\sum_{j=0 \to \infty} a_j$. By the definition of rearrangement, we may write that there is a sequence, r, such that for each nonnegative integer, j,

$$b_j = a_{r(j)},$$

where r satisfies the conditions given in Definition 7.23. For this particular sequence, we claim that as j grows large, so does $r(j)$ — i.e., for any given value of n we can find J_n such that

$$\text{if } j > J_n, \text{ then } r(j) > n \qquad \textit{7.25}$$

because each position in the b sequence corresponds to (has its element transferred from) some one position in the a sequence; the collection of all a positions mapped into b positions 1 through n is a finite collection; its largest element is the position, J_n, that we seek; examine Figure 7-5 to convince yourself of the validity of this argument.

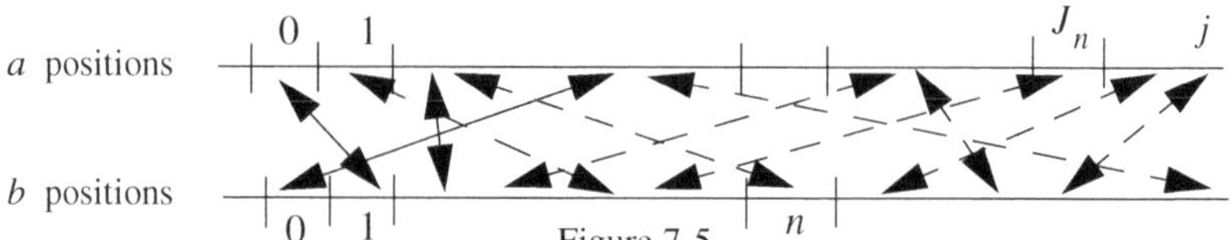

Figure 7-5

Now to show that $\sum_{j=0 \to \infty} b_j$ converges. The Cauchy criterion (Theorem 7.8 on page 219) suggests examining $\left|\sum_{j=m}^{k} b_j\right|$ to see if it becomes small for all sufficiently large m,k. We find

$$\left|\sum_{j=m}^{k} b_j\right| \le \sum_{j=m}^{k} |b_j| = \sum_{j=m}^{k} |a_{r(j)}|$$

Because $\sum_{j=0 \to \infty} a_j$ converges absolutely, we know from the Cauchy criterion that

$$\sum_{J=M}^{K} |a_J|$$

gets small as M, K_k become large. Now simply choose $m \ge J_{M_0}$ (so that all $r(j)$ in the sum $\sum_{j=m} |a_{r(j)}|$ satisfy the condition $r(j) \ge M_0$ where M_0 is sufficiently large so that for all $M, K \ge M_0$

$$\sum_{J=M}^{K} |a_J| \text{ is small.} \qquad \textit{7.26}$$

Then $\sum_{j=m}^{k} |a_{r(j)}|$ is a sum of a_i s with all $i \ge M_0$ and no indices repeated, so that

$$\sum_{j=m}^{k} |a_{r(j)}| \le \sum_{J=min(r(j) \text{ with } m \le j \le k)}^{max(r(j) \text{ with } m \le j \le k)} |a_J| .$$

The right-hand side of this inequality is small, being a sum of the form in expression 7.26 and hence, for all sufficiently large, m,k, the sum

$$\sum_{j=m}^{k} |b_j| = \sum_{j=m}^{k} |a_{r(j)}|$$

is small.

So we have shown that for all m,k sufficiently large, the sums

$$\sum_{j=m}^{k} |b_j| \text{ and } \left|\sum_{j=m}^{k} b_j\right|$$

are small; hence by the Cauchy criterion, Theorem 7.8 on page 219, and the definition of absolute convergence, Definition 7.25 on page 253, the infinite series

$$\sum_{j=0 \to \infty} b_j$$

converges and converges absolutely.

Next we show that all rearrangements of $\sum_{j=0 \to \infty} a_j$ converge to the same value (i.e., have the same sum).

First, choose m_o sufficiently large so that for all $m \geq m_o$, $n \geq m_o$,

$$\sum_{j=0}^{m-1} a_j \text{ and } \sum_{j=0}^{m-1} b_j \text{ are close to } A = \sum_{j=0}^{\infty} a_j \text{ and } B = \sum_{j=0}^{\infty} b_j \text{, respectively.}$$

(This can be done because both associated series have been shown to converge.) Hence it follows that for all such m,M

$$\sum_{j=0}^{m-1} a_j - \sum_{j=0}^{M-1} b_j \quad \text{is close to } A - B.$$

The details of proof of this last assertion, using the triangle inequality, are left to the reader.

Now further restrict m_o to be large enough so that for all $m \geq m_o$, $k \geq m$, the quantity

$$\sum_{j=m}^{k} |b_j|$$

is small; this is implied by the Cauchy Criterion, because, as we showed earlier in this proof, $\sum_{j=0 \to \infty} b_j$ is absolutely convergent.

Finally, choose m large enough so that for any fixed M satisfying the condition $M \geq m_o$ it is also the case that every one of the terms in the sum $\sum_{j=1}^{M-1} b_j$ is included in the sum $\sum_{j=0}^{m-1} a_j$; this is achievable because $\sum_{j=0 \to \infty} b_j$ is a rearrangement of $\sum_{j=0 \to \infty} a_j$. But then, the difference

$$\sum_{j=0}^{m-1} a_j - \sum_{j=0}^{M-1} b_j$$

is a finite sum of b_j s with $j > M - 1$, i.e.,

$$\left| \sum_{\substack{\text{some finite \#} \\ \text{of } j \text{ with } j > M-1}} b_j \right| \leq \sum_{j=M}^{K} |b_j| \text{ for sufficiently large } K.$$

The last displayed sum is known to be small, and hence

$$\left| \sum_{j=0}^{m-1} a_j - \sum_{j=0}^{M-1} b_j \right|$$

is small; the difference of these sums is close to A -- B, and thus $|A - B|$ is small. But $|A - B|$ is a fixed number, and if it is *small* (i.e., can be made as close to 0 as desired by suitable choice of m_o, m), it follows that $|A - B| = 0$. i.e.,

$$\sum_{j=0}^{\infty} a_j = \sum_{j=0}^{\infty} b_j.$$

So we have proved the following result.

Theorem 7.26: Absolute and unconditional convergence

If an infinite series is absolutely convergent (Definition 7.25 on page 253), then it is unconditionally convergent (Definition 7.24 on page 252), and all rearrangements are convergent and absolutely convergent, and all rearrangements converge to the same sum.

❑

We next consider multiplication of infinite series.

Definition 7.27: Products of series, Cauchy product

Suppose $\sum_{j=0 \to \infty} a_j$ and $\sum_{j=0 \to \infty} b_j$ are infinite series of complex numbers (Definition 7.4 on page 214).

A product of these series is any rearrangement of a series whose terms consist of ***all*** pairwise products, $a_i b_j$, of elements of the individual series.

One such product, the **Cauchy product**, denoted by

$$\left(\sum_{i=0 \to \infty} a_i \right)\left(\sum_{j=0 \to \infty} b_j \right) \qquad \textbf{\textit{7.27}}$$

is the series whose terms are, respectively, $a_0 b_0, a_0 b_1, a_1 b_0, a_0 b_2, a_1 b_1, a_2 b_0, \ldots$ — i.e., the series obtained by expansion of the terms of

$$\sum_{i=0 \to \infty} a_i \left(\sum_{j=0}^{\infty} b_{j-i} \right).$$

❑

The basic result on multiplication of series is as follows.

Theorem 7.28: Convergence of product series

If $\sum_{i=0 \to \infty} a_i$ and $\sum_{j=0 \to \infty} b_j$ are absolutely convergent series of complex numbers (Definition 7.25 on page 253) then their Cauchy product (Definition 7.27 above) and hence every product, is absolutely convergent, with sum

$$\left(\sum_{i=0}^{\infty} a_i \right)\left(\sum_{j=0}^{\infty} b_j \right).$$

❑

We establish this result by examining the subsequence of partial sums of the *absolute* Cauchy product of the form

$$S_n = \sum_{i=0}^{n}\left(\sum_{j=0}^{i}|a_j||b_{i-j}|\right), \qquad n = 0, 1, \ldots .$$

If T_n is the partial sum consisting of the first n terms of the Cauchy product

$$\left(\sum_{i=0\to\infty}|a_i|\right)\left(\sum_{j=0\to\infty}|b_j|\right).$$

then T_n represents a nondecreasing sequence and satisfies

$$T_n \le S_n \le \sum_{i=0}^{n-1}\left(\sum_{j=0}^{n-1}|a_i||b_j|\right) = \left(\sum_{i=0}^{n-1}|a_i|\right)\left(\sum_{j=0}^{n-1}|b_j|\right) \le \left(\sum_{i=0}^{\infty}|a_i|\right)\left(\sum_{j=0}^{\infty}|b_j|\right).$$

That is, T_n represents a bounded and monotone nondecreasing sequence. Hence, we see from Theorem 3.8 on page 86 that this sequence converges, showing the Cauchy product to be absolutely convergent, and hence, by Theorem 7.26 just proved, to be unconditionally convergent, as are all of its rearrangements, to the same sum. This sum is, in fact,

$$\left(\sum_{i=0}^{\infty}a_i\right)\left(\sum_{j=0}^{\infty}b_j\right)$$

which we see as follows.

The quantities

$$W_n = \left(\sum_{i=0}^{n-1}a_i\right)\left(\sum_{j=0}^{n-1}b_j\right), \qquad n = 1,2,3,\ldots$$

form a *subsequence* of the sequence of partial sums for many products of the series

$\sum_{i=0\to\infty} a_i$ and $\sum_{j=0\to\infty} b_j$; that is, there is a product series whose partial sums are $V_1, V_2, \ldots$, and for each positive integer n, W_n is one of the V_j; specifically

$$W_n = V_{J(n)}$$

where $J(n)$ increases with n. Since the V_j are approaching a limit as j gets large, $W_n = V_{J(n)}$ must be approaching that same limit. But

$$\lim_{n\to\infty} W_n = \left(\sum_{i=0}^{\infty}a_i\right)\left(\sum_{j=0}^{\infty}b_j\right).$$

Hence

$$\lim_{j\to\infty} V_j = \left(\sum_{i=0}^{\infty}a_i\right)\left(\sum_{j=0}^{\infty}b_j\right)$$

which establishes the result we were after.

Two more minor results, whose proofs are left to the reader, are needed.

Theorem 7.29: Linear operator property of $\sum_{i=0}^{\infty}$

If $\sum_{i=0\to\infty} a_i$ and $\sum_{j=0\to\infty} b_j$ are convergent series of complex numbers, (Definition 7.4 on page 214), and A and B are complex constants, then the series

$$\sum_{i=0\to\infty} (Aa_i + Bb_i)$$

converges and its sum is $A\sum_{i=0}^{\infty} a_i + B\sum_{i=0}^{\infty} b_i$ ❑

Theorem 7.30: Grouping and Convergence

Suppose that $0 = n_0, n_1, n_2, \ldots$ represents any increasing sequence of positive integers. Let $\sum_{i=0\to\infty} a_i$ be a convergent series of complex numbers, (Definition 7.4 on page 214), and let

$$b_j = \sum_{i=n_j}^{n_{j+1}-1} a_i \qquad j = 0, 1, 2, \ldots .$$

Then $\sum_{j=0\to\infty} b_j$ converges and $\sum_{j=0}^{\infty} b_j = \sum_{i=0}^{\infty} a_i$ ❑

This theorem states that a convergent series remains convergent, and to the same sum, if adjacent terms are grouped and added together in some manner to form the new series, for example,

$$\sum_{k=0}^{\infty} x^k = \sum_{j=0}^{\infty} (x^{2j} + x^{2j+1}).$$

Exercises 7.21

1. Find a product series for $\left(\sum_{j=0}^{\infty} x^j\right)^2$ for $-1 < x < 1$, showing directly that

$$\frac{1}{(1-x)^2} = \sum_{k=1}^{\infty} kx^{k-1} \quad \text{for} \quad -1 < x < 1.$$

We obtained this result earlier, using term-by-term differentiation — see Exercises 7.14 on page 239.)

Hint: Look at $\sum_{j=0\to\infty}\left(\sum_{i=0}^{j} a_i\, b_{j-i}\right)$ Note that this series is not itself a product series, but is a series generated from a product series by grouping terms.

2. Find a simple product series for $\left(\sum_{j=0\to\infty}\frac{x^j}{j!}\right)\left(\sum_{i=0\to\infty}\frac{x^i}{i!}\right)$

Hint: You may find the binomial theorem useful — namely for

$$\binom{n}{j} = \frac{n!}{j!(n-j)!},\text{where } 0! = 1,\ \sum_{j=0}^{n}\binom{n}{j}a^j b^{n-j} = (a+b)^n$$

There are lots of ways to establish the binomial theorem — a direct (not especially enlightening) inductive proof, or as an application of Taylor's theorem to $f(x) = (1+x)^n$, where $x = b/a$.

3. For x such that $\sum_{k=0\to\infty} x^k$ is convergent, if $S = \sum_{k=0}^{\infty} x^k$, show that $1 + Sx = S$, and from this find the simple formula for S.

Problems 7.22

1. Show that if $\sum_{j=0\to\infty} a_j$ is conditionally convergent, then $\sum_{j=0\to\infty} a_j^+$ and $\sum_{j=0\to\infty} a_j^-$ are divergent to ∞, where

$$a_j^+ = \begin{cases} a_j & \text{if } a_j > 0 \\ 0 & \text{otherwise} \end{cases} \quad \text{and } a_j = a_j^+ - a_j^-.$$

2. Using the results of the previous problem, show that by suitable rearrangement, a conditionally convergent series can be *rearranged* to converge to any real number, diverge to ∞ or $-\infty$, or oscillate aimlessly. This property of conditionally convergent series makes them very useful in *search problems*, where we want to find an object which can be quite distant, but want to *zero in* on it.

3. Show that if the (possibly complex) power series $\sum_{j=0\to\infty} a_j x^j$ converges uniformly for $|x| < r$, where $r > 0$, then it then it converges absolutely for $|x| < r$.
Hints: First establish this result when the series converges for $|x| \le r$. . To do this, note that for such x

$$\sum_{j=m}^{k} |a_j x^j| = \sum_{j=m}^{k} |a_j r^j| c^j \quad \text{where } c = \left|\frac{x}{r}\right| < 1,$$ noting that the factor $|a_j r^j|$ goes to 0 by the assumed convergence. Then use the comparison test, Theorem 7.13 on page 228. Now this result can be extended even if the original series is not known to converge for $|x| = r$, because it converges for $|x| \le r_o$ for every nonnegative $r_o < r$.

The last result to be established in this chapter concerns uniqueness of the power series expansion about some point. This result was used in all of the work we did on power series solutions of differential equations.

Theorem 7.31: Uniqueness of power series coefficients

If $\sum_{j=0}^{\infty} a_j x^j = 0$, (Definition 7.4 on page 214), for all x with $0 \le x \le b$, $b > 0$, then $a_j = 0$ for all nonnegative integers *j*.

Hence if $\sum_{j=0}^{\infty} a_j x^j = \sum_{j=0}^{\infty} A_j x^j$ for all x with $0 \le x \le b$, $b > 0$, then $a_j = A_j$ for all nonnegative integers *j*.

❑

The proof is fairly straightforward — namely, since $\sum_{j=0 \to \infty} a_j x^j$ converges for $0 \le x \le b$, for $0 \le c < b$ the series $\sum_{j=0 \to \infty} a_j c^j$ converges, and hence $|a_j c^j|$ becomes small as j grows large. Thus for all j sufficiently large, simultaneously for all x satisfying $-c/2 \le x \le c/2$, we have

$$|a_j x^j| = \left|a_j c^j \left(\frac{x}{c}\right)^j\right| \le \frac{1}{2^j}.$$

Hence by the comparison test, Theorem 7.13 on page 228, $\sum_{j=0 \to \infty} a_j x^j$ converges uniformly in x in $[-c/2, c/2]$. Let

$$W(x) = \sum_{j=0}^{\infty} a_j x^j \quad \text{for } -c/2 \le x \le c/2.$$

Since $W(0) = a_0$ and $W(0) = 0$, it follows that $a_0 = 0$.

The derived series, $\sum_{j=1 \to \infty} j a_j x^{j-1}$, is seen to converge uniformly for $-c/2 \le x \le c/2$ by comparison with $\sum_{j=1 \to \infty} j \frac{1}{2^j}$ (which itself is easily seen to converge by the ratio test).

Hence, from Theorem 7.20 on page 238, we see that $W(x) = \sum_{j=1}^{\infty} j a_j x^{j-1}$ is the derivative of $\sum_{j=0}^{\infty} a_j x^j$ for $-c/2 \le x \le c/2$.

But $W'(0) = a_1$ and $W'(0) = 0$, because $\frac{W(h) - W(0)}{h} = 0$ for all positive $h < c/2$. So $a_1 = 0$. This argument can be repeated indefinitely (more precisely, using induction) showing that all $a_j = 0$, proving this theorem.

Chapter 8

Multivariable Differential Calculus

1 Introduction

Up to this point we have restricted our attention to problems involving functions of only one variable — that is, functions whose inputs and outputs are real numbers. As we'll see, this background is essential for the proper study of functions whose inputs consist of more than just numbers. Such functions come up often enough to warrant the extensive treatment being introduced in this chapter; because just about all systems we'll ever meet, from electronic to biological and economic ones, are influenced by many (input) factors. To list a few examples,

- The exact location of a space satellite is most strongly influenced by the sequence of thrusts provided by its engines, as well as the winds it encounters, the air densities it meets, etc.
- The weight of a person is influenced by the food consumed, exercise levels, genetic factors, etc.
- The cost of living is affected by the Federal Reserve discount rates, consumer demand, and so forth.
- To adequately predict the weather in any area requires a great deal of input, from knowledge of the trade winds and jet stream behaviors to an enormous number of other measurements on temperature, pressure, humidity, over many different regions.

There are so many factors that affect every phenomenon we want to predict and control, that we must be very selective and consider only those which seem very important, if we expect any success in our efforts. To proceed systematically we start by introducing the following definition.

Definition 8.1: Function of n real variables, vector

A **function of n real variables** is a function whose domain (Definition 1.1 on page 2) consists of sequences $(x_1, x_2, ...,x_n)$ of real numbers. Such sequences are also referred to as (*real*) ***n*-dimensional vectors**, or (real) ***n*-tuples**. If the range (outputs) consists of real numbers, the function is called **real-valued**.

❑

Figure 8-1 illustrates Definition 8.1.

Function of n variables

Figure 8-1

Example 8.1: Energy as a function of current and resistance

If I represents electrical current and R represents electrical resistance, then the heat energy generated per unit time, W (in *watts*) is described by the equation

$$W = f(I, R) = I^2 R$$

■

We can picture a two or three dimensional vector either as a point in 2 or 3 dimensional space, or as an arrow in two or three dimensions, with a specified length and direction, as illustrated in Figure 8-2.

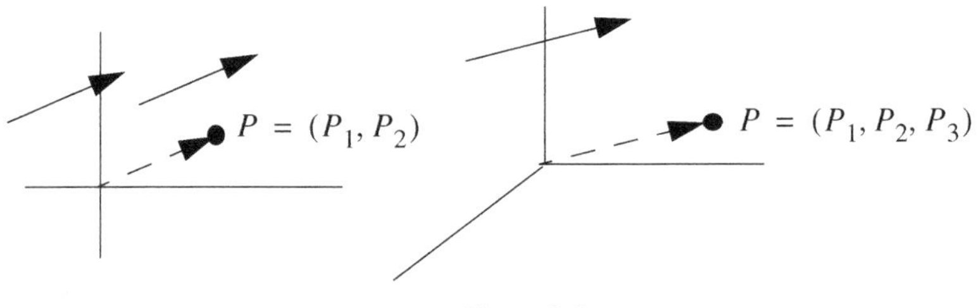

Figure 8-2

Which representation of a vector we choose depends on the particular situation we're in. Our ***pictures***, even of three-dimensional vectors, on paper are necessarily drawn in two dimensions - but ***will be used to illustrate*** the general case ***and suggest results*** of a general nature. The ***proofs of any such results must be carried out algebraically***; the reasoning, however, is most often motivated by the picture, but ***not relying on the picture for its validity***.

The **fundamental idea** of our study of functions of n variables is to **convert** our questions and descriptions **to** questions and descriptions of **functions of one variable,** i.e., to functions we are familiar with. To see how to do this, we examine the ***graph*** of a function f — that is, we examine the set of points of the form

$$(x_1, \ldots, x_n, f(x_1, \ldots, x_n)).$$

Our picture will be for the case $n = 2$ (see Figure 8-3). Here the graph of the function f is pictured as a ***surface*** whose height above the point $(x_1, x_2, 0)$ in the base plane is $f(x_1, x_2)$.

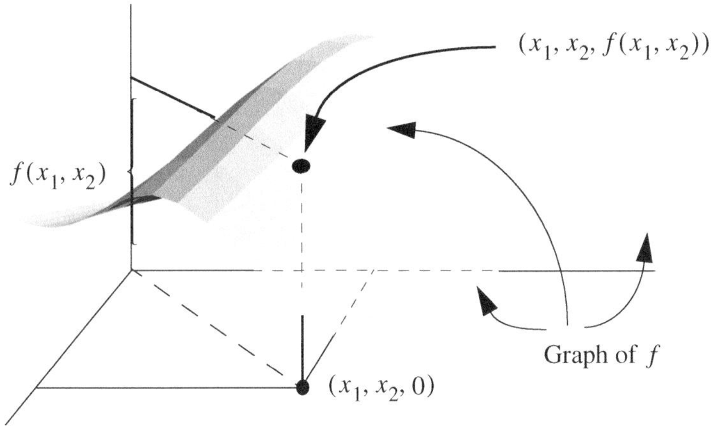

Figure 8-3

Before proceeding any further we'll introduce some notation to keep our expressions from getting too complicated looking.

Definition 8.2: Vector notation

In our work with real valued functions of n real variables, (Definition 8.1 on page 261), ***n*-dimensional vectors** which are finite sequences of numbers, (Definition 8.1 on page 261) are denoted as follows:

$$\boldsymbol{x} = (x_1, x_2, \dots, x_n), \ \boldsymbol{x}_0 = (x_{01}, x_{02}, \dots, x_{0n}), \ \boldsymbol{h} = (h_1, h_2, \dots, h_n), \text{ etc.,}$$

so that a point $(x_1, x_2, \dots, x_n, f(x_1, x_2, \dots, x_n))$ is denoted[1] by $(\boldsymbol{x}, f(\boldsymbol{x}))$

and denote the **standard unit length coordinate axis vectors**, by $\boldsymbol{e}_k$, $k = 1, \dots, n$, where the j-th coordinate of $\boldsymbol{e}_k$ is given by

$$e_{kj} = \begin{cases} 1 & \text{if } j = k \\ 0 & \text{if } j \neq k \end{cases}$$

Also if t is a real number, $\boldsymbol{x} + t\boldsymbol{h} = (x_1 + th_1, \dots, x_n + th_n)$. (Actually, this was already the case, because a sequence is a function, and we already have a meaning assigned to $\boldsymbol{x} + \boldsymbol{y}$ and $t\boldsymbol{x}$ where t is a number, namely (see Definition 1.2 on page 3 and Definition 2.7 on page 35),

1. There is a slight notational inconsistency, which shouldn't bother anyone — namely, $(\boldsymbol{x}, f(\boldsymbol{x}))$ is really $((x_1, x_2, \dots, x_n), f(x_1, x_2, \dots, x_n))$.

$$x + y = (x_1 + y_1, \dots, x_n + y_n) \qquad cx = (cx_1, \dots, cx_n)$$

The operations above have a geometrical interpretation that was provided in Figures 2-5 through 2-7 starting on page 24.

❑

2 Local Behavior of Function of *n* Variables

Suppose now that we want to investigate the local behavior of the function f of n real variables, *near* the point $\boldsymbol{x}$, *that is, at points displaced just a short distance from* $\boldsymbol{x}$. As we can see from Figure 8-4, in order to follow the plan presented in the introduction, of converting such an investigation to one involving functions of one variable, **we restrict the displacement from $\boldsymbol{x}$ to be in a fixed direction specified by some nonzero length vector**. For convenience we choose this vector to be of unit length, and denote it by $\boldsymbol{u}$.

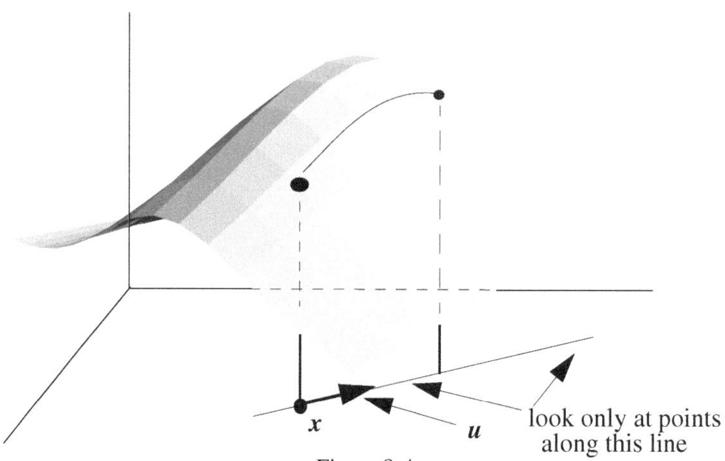

Figure 8-4

We are thus restricting consideration to the function of one variable obtained by restricting inputs to be on the line through $\boldsymbol{x}$ in the direction specified by the unit length vector, $\boldsymbol{u}$.

Extending the reasoning that generated equation 2.2 on page 25, we see that this line consists of all points of the form $\boldsymbol{x} + t\boldsymbol{u}$, where t may be any real number, as illustrated in Figure 8-5.

For completeness and reference purposes, we introduce Definition 8.3.

Definition 8.3: Line in n dimensions

The **line through the point** $\boldsymbol{x}$ (see Definition 8.2 on page 263) in the direction specified by the unit length vector $\boldsymbol{u}$ is the set of all points $\boldsymbol{x} + t\boldsymbol{u}$ for t any real number.

❑

The function of one variable that is obtained by restricting the inputs of the real valued function, f, of n real variables to the line whose formula is $\boldsymbol{x} + t\boldsymbol{u}$ will be denoted by G, that is,

$$G(t) = f(\boldsymbol{x} + t\,\boldsymbol{u}). \qquad \textit{8.1}$$

The function G is illustrated in Figure 8-5, the portion of the graph of f

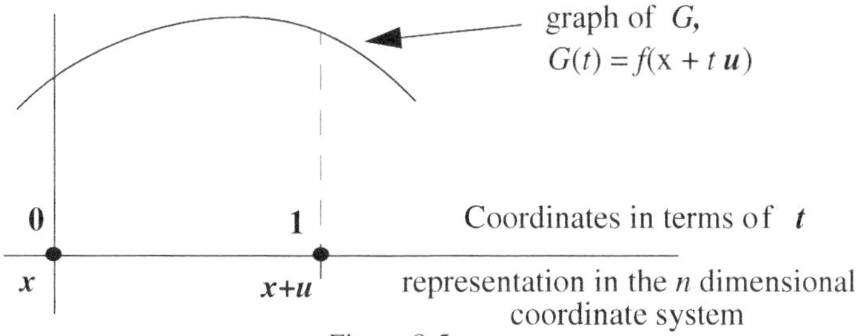

Figure 8-5

from which it came appearing above the line $\boldsymbol{x} + t\boldsymbol{u}$, of Figure 8-4.

We can now make use of our *one-variable* results, by introducing the concept of *directional derivative.*

Definition 8.4: Directional derivative

If the derivative (Definition 2.5 on page 30) $G'(0)$ of the function G specified by equation 8.1 exists, it is called the **directional derivative of f at $\boldsymbol{x}$ in the direction specified by the unit vector $\boldsymbol{u}$.** This directional derivative will be denoted by

$$D_{\boldsymbol{u}} f(\boldsymbol{x}).$$

❑

Notice that if we want to approximate $f(x + h, y + k)$ in a manner similar to the fundamental approximation, equation 2.7 on page 31, we simply use

$$f(x + h, y + k) \cong f(x, y) + D_{\boldsymbol{u}} f(x, y)\|(h, k)\|,$$

where $\|(h, k)\| = \sqrt{h^2 + k^2}$ is just the length of (h,k) (for the general formal definition, see Definition 8.6 on page 270) and

$$\boldsymbol{u} = \frac{(h, k)}{\|(h, k)\|}.$$

Before developing the necessary formulas for simple mechanical computation of the directional derivative, we illustrate its computation directly from the definition.

Example 8.2 Finding $D_{\boldsymbol{u}} f(x)$ for $f(x) = f(x_1, x_2) = x_1 exp(3x_2)$

Here

$$G(t) = f(\boldsymbol{x} + t\boldsymbol{u}) = f(x_1 + tu_1, x_2 + tu_2)$$
$$= (x_1 + tu_1)\, exp(3(x_2 + tu_2)).$$

Noting that x_1, x_2, u_1, u_2 all stand for fixed constants in the current context, we see from the rules for differentiation of functions of one variable (Theorem 2.13 on page 53) that

$$G'(t) = u_1 exp(3(x_2 + tu_2)) + 3u_2(x_1 + tu_1) exp(3(x_2 + tu_2))$$
$$= (u_1 + 3u_2(x_1 + tu_1)) exp(3(x_2 + tu_2)).$$

So finally we see that

$$D_{\boldsymbol{u}} f(\boldsymbol{x}) = G'(0) = (u_1 + 3u_2 x_1)\, exp(3x_2).$$

■

Exercises 8.3

1. Find $D_{\boldsymbol{u}} f(\boldsymbol{x})$ for $f(x_1, x_2) = cos(x_1^2) + x_1^2 x_2$. Evaluate this for $(x_1, x_2) = \left(\frac{\pi}{2}, 2\right)$ $(u_1, u_2) = \left(\frac{1}{\sqrt{5}}, \frac{2}{\sqrt{5}}\right)$.
2. Use the previous result to find an approximation to
$$f\left(\frac{\pi}{2} + 0.02, 2.04\right)$$
3. Make up your own functions, and compute various directional derivatives of your own choosing, checking the results using *Mathematica* or *Maple*, or the equivalent.

Directional derivatives are not very much used for practical computations; rather, as we will shortly see, they will be the starting point for development of the important practical tools of multivariable calculus. So we will not place a great deal of emphasis here on routine computations with them, but will immediately continue with developing their important properties. First we will concern ourselves with finding a simpler way to compute directional derivatives. It turns out that the formula we will develop will greatly increase our understanding of what's really going on.

We start by observing that there are certain special directional derivatives which can usually be computed by inspection — namely, those for which $\boldsymbol{u}$ is a **coordinate axis vector**; that is (already introduced in Definition 8.2 on page 263), $\boldsymbol{u}$ is one of the vectors $\boldsymbol{e}_i$ whose j-th coordinate, e_{ij}, is given by

$$e_{ij} = \begin{cases} 1 & \text{if} \quad j = i \\ 0 & \text{if} \quad j \neq i \end{cases} \qquad \textit{8.2}$$

In three dimensions

$$\boldsymbol{e}_1 = (1, 0, 0), \quad \boldsymbol{e}_2 = (0, 1, 0), \quad \boldsymbol{e}_3 = (0, 0, 1).$$

The significance of being able to compute these directional derivatives almost by inspection arises because any vector can be written as a sum of vectors in the coordinate axis directions — so that we will be able to write all directional derivatives very easily in terms of these simple ones. The ease of computation of these *coordinate axis directional derivatives* follows because we can compute $D_{e_i} f(\boldsymbol{x})$ by treating all x_j with $j \neq i$ as fixed constants with x_i being the variable of interest in the function of one variable being differentiated.

Example 8.4: $D_{e_1} x_1^2 e^{x_1 x_2 x_3}$

Here, to obtain the desired answer, we act as if x_2 and x_3 are fixed constants, and take the derivative of the resulting function of the one variable, x_1, to obtain that if

$$f(x_1, x_2, x_3) = x_1^2 e^{x_1 x_2 x_3},$$

then

$$D_{e_1} f(x_1, x_2, x_3) = 2x_1 e^{x_1 x_2 x_3} + x_1^2 x_2 x_3 e^{x_1 x_2 x_3} = (2x_1 + x_1^2 x_2 x_3)\, e^{x_1 x_2 x_3}.$$

■

Because they come up so frequently, the special directional derivatives, $D_{e_i} f(\boldsymbol{x})$, have special names and notations.

Definition 8.5: Partial derivatives

The directional derivative $D_{e_i} f(\boldsymbol{x})$, where the vector $\boldsymbol{e}_i$ is given in Definition 8.2 on page 263, is called the **partial derivative of f at $\boldsymbol{x}$ with respect to its i-th coordinate**, and is denoted[1] by $D_i f(\boldsymbol{x})$.

---

1. Sometimes $D_{e_i} f(\boldsymbol{x})$ is denoted by $f_i(\boldsymbol{x})$. The most common notation for $D_{e_i} f(\boldsymbol{x})$, and by far the one ***worth avoiding*** if at all possible, is $\frac{\partial}{\partial x_i} f(\boldsymbol{x})$, since ***it has no redeeming virtues of any kind.***

D_i is called the **partial derivative operator with respect to the i-th coordinate**. It is a function whose domain (inputs) and range (outputs) consist of functions.

❑

Exercises 8.5

Compute $D_i f(\boldsymbol{x})$ for the functions of Example 8.2 on page 266 and of Exercises 8.3 on page 266.

As mentioned earlier, one can always proceed from the point $\boldsymbol{x}$ to the point $\boldsymbol{x} + t\boldsymbol{u}$ along segments parallel to the coordinate axes, as illustrated in Figure 8-6.

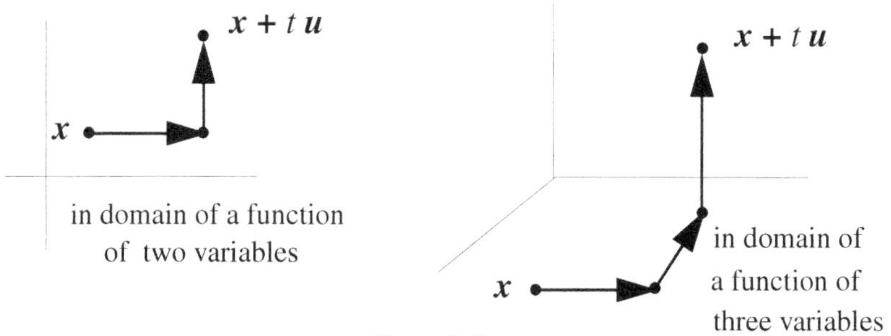

Figure 8-6

Now to get to the details of computing any directional derivative from the partial derivatives. We will illustrate the derivation of the general result via the case $n = 3$. We first rewrite the difference quotient $[f(\boldsymbol{x} + t\boldsymbol{u}) - f(\boldsymbol{x})]/t$ as a sum of function differences, in each of which the changes in the argument is restricted to precisely one coordinate. Specifically we write

8.3

$$\frac{f(\boldsymbol{x} + t\boldsymbol{u}) - f(\boldsymbol{x})}{t} = \frac{f(x_1 + tu_1, x_2 + tu_2, x_3 + tu_3) - f(x_1, x_2, x_3)}{t}$$

$$= \frac{f(x_1 + tu_1, x_2 + tu_2, x_3 + tu_3) - f(x_1 + tu_1, x_2 + tu_2, x_3}{t}$$

$$+ \frac{f(x_1 + tu_1, x_2 + tu_2, x_3) - f(x_1 + tu_1, x_2, x_3)}{t}$$

$$+ \frac{f(x_1 + tu_1, x_2, x_3) - f(x_1, x_2, x_3)}{t}.$$

Most readers should not have much difficulty applying the mean value theorem (Theorem 3.1 on page 62) to each of the last three ratios above, to obtain the result we present shortly in equation 8.7. The precise arguments are presented in the optional segment to follow.

Look at the numerator of the second line of equations 8.3. It is

$$f(\boldsymbol{x}+t\boldsymbol{u})-f(\boldsymbol{x}+t\boldsymbol{u}-tu_3\boldsymbol{e}_3) = f(z+tu_3\boldsymbol{e}_3)-f(z), \qquad 8.4$$

where

$$z = \boldsymbol{x}+t\boldsymbol{u}-tu_3\boldsymbol{e}_3 .$$

To be able to legitimately apply the mean value theorem we let

$$G(\tau) = f(z+\tau\boldsymbol{e}_3). \qquad 8.5$$

We see that the expression in equation 8.4 can be expressed in the simpler form

$$G(tu_3)-G(0),$$

which, from the mean value theorem, Theorem 3.1 on page 62, is equal to $tu_3G'(\vartheta_3 tu_3)$, where ϑ_3 lies between 0 and 1 (assuming the hypotheses of the mean value theorem are satisfied). But using equation 8.5 above, we see that

$$\begin{aligned} G'(\tau) &= \lim_{k\to 0}\frac{f([z+\tau\boldsymbol{e}_3]+k\boldsymbol{e}_3)-f(z+\tau\boldsymbol{e}_3)}{k} \\ &= D_3 f(z+\tau\boldsymbol{e}_3). \end{aligned}$$

So

$$tu_3G'(\vartheta_3 tu_3) = tu_3D_3f(z+\vartheta_3 tu_3\boldsymbol{e}_3). \qquad 8.6$$

That is, the first numerator in the final part of equation 8.3 on page 268 is given by the right side of equation 8.6. This gives the first ratio of the final part of equation 8.3 as

$$u_3D_3f(x_1+tu_1, x_2+tu_2, x_3+\vartheta_3 tu_3\boldsymbol{e}_3).$$

In a similar fashion the remaining two ratios in the final part of equation 8.3 may be written as

$$u_2D_2f(x_1+tu_1, x_2+\vartheta_2 tu_2, x_3)$$

and

$$u_1D_1f(x_1+\vartheta_1 tu_1, x_2, x_3),$$

respectively.

So we find that under the appropriate circumstances allowing application of the mean value theorem, Theorem 3.1 on page 62, we have

$$\begin{aligned}\frac{f(\boldsymbol{x}+t\boldsymbol{u})-f(\boldsymbol{x})}{t} &= u_1D_1f(x_1+\vartheta_1tu_1, x_2, x_3)\\ &+ u_2D_2f(x_1+tu_1, x_2+\vartheta_2tu_2, x_3)\\ &+ u_3D_3f(x_1+tu_1, x_2+tu_2, x_3+\vartheta_3tu_3\boldsymbol{e}_3)\end{aligned} \qquad 8.7$$

where the ϑ_i are numbers between 0 and 1. To obtain the desired directional derivative, $D_{\boldsymbol{u}}f(\boldsymbol{x})$, we must take the limit as t approaches 0. Provided that the partial derivatives D_jf at the indicated values above for t near 0 get close to their values at $t = 0$ (i.e., at $\boldsymbol{x}$), we would expect this limit to be

$$\sum_{j=1}^{3} u_jD_jf(\boldsymbol{x})$$

A precise statement of the result just derived requires at least a minimal discussion of *continuity* for functions of several variables. All of the material to be presented below is a natural extension of previously introduced concepts related to continuity and *closeness*. We return from this detour at Theorem 8.9.

To begin we extend the definition of length and distance between vectors from the two-dimensional case, Definition 6.11 on page 200, to n-dimensional vectors in the following.

Definition 8.6: Length of vector, distance between vectors

The **length** (or **norm**) $\|\boldsymbol{x}\|$, **of the n-dimensional vector** $\boldsymbol{x} = (x_1, x_2,...,x_n)$ is defined[1] by

$$\|\boldsymbol{x}\| = \sqrt{\sum_{i=1}^{n} x_i^2}; \qquad 8.8$$

the (**Euclidean**) **distance between** $\boldsymbol{x} = (x_1, x_2,...,x_n)$ and $\boldsymbol{y} = (y_1, y,...,y_n)$ is defined to be

$$\|\boldsymbol{x} - \boldsymbol{y}\| \qquad 8.9$$

❑

Definition 8.7: Continuity of function of n variables, (precise)

We say that the real valued function, g, of n real variables (Definition 8.1 on page 261) is **continuous at** $\boldsymbol{x}_0$ in the domain of g, if, for $\boldsymbol{x}$ in the domain of g, $g(\boldsymbol{x}) - g(\boldsymbol{x}_0)$ gets close to 0 as $\|\boldsymbol{x}_0 - \boldsymbol{x}\|$ gets small — i.e.,

1. This is motivated by the Pythagorean theorem.

small changes to the input, $\boldsymbol{x}_0$, yield small changes to the output $g(\boldsymbol{x}_0)$. More precisely, keeping in mind Definition 8.6 on page 270,

> Given any positive number, ε, there is a positive number, δ, (determined by g, ε and $\boldsymbol{x}_0$) such that
>
> whenever $\boldsymbol{x}$ is in the domain of g and $\|\boldsymbol{x}_0 - \boldsymbol{x}\| \le \delta$,
>
> we have $|g(\boldsymbol{x}_0) - g(\boldsymbol{x})| \le \varepsilon$.

❑

The elementary properties of real valued continuous functions of n real variables are essentially the same as those of *one-variable* functions, and are derived in the same way. We extend the continuity theorem, Theorem 2.10 on page 38, and Theorem 2.11 on page 42 (continuity of constant and identity functions) to functions of n variables.

Theorem 8.8: Continuity of functions of *n* variables

Assuming the real valued functions f and g of n real variables (Definition 8.1 on page 261) have the same domain and are continuous at $\boldsymbol{x}_0$ (Definition 8.7 on page 270) the sum, $f + g$, product, fg, and *ratio*, f/g (assuming it is defined) are continuous at $\boldsymbol{x}_0$; if the real valued function, R, of a single real variable is continuous and its domain includes the range of f, then the composition[1] $R(f)$ is continuous at $\boldsymbol{x}_0$.

The *coordinate functions*, c_j, defined via

$$c_j(\boldsymbol{x}) = c_j(x_1, x_2, \ldots, x_n) = x_j \qquad \textbf{\textit{8.10}}$$

are continuous at all $\boldsymbol{x}$.

❑

So, as seen earlier, for functions of one variable, ***we can often recognize continuity of a function of several variables from its structure***.

Now that we have available the concept of continuity of a function of several variables, the above discussion yields the following result.

Theorem 8.9: Determination of directional derivatives

If f is a real valued function of n real variables (Definition 8.1 on page 261) all of whose partial derivatives (Definition 8.5 on page 267) exist *at and near* $\boldsymbol{x}_0$ (i.e., for all $\boldsymbol{x}$ within some positive distance from $\boldsymbol{x}_0$ (see Definition 8.6 on page 270) and are continuous at $\boldsymbol{x}_0$ (Definition 8.7 on page 270) then the directional derivative (Definition 8.4 on page 265) $D_u f(\boldsymbol{x}_0)$ exists and is given by the formula

---

1. Whose value at $\boldsymbol{x}$ is $R(f(\boldsymbol{x}))$.

$$D_{\boldsymbol{u}} f(\boldsymbol{x}_0) = \sum_{j=1}^{n} u_j D_j f(\boldsymbol{x}_0). \qquad 8.11$$

❑

Exercises 8.6

Repeat Exercises 8.3 on page 266 using equation 8.11.

In the formula just developed, a very important type of expression appears, namely a number of the form

$$\sum_{j=1}^{n} x_j y_j$$

This expression is called the *dot product* of the real vector $\boldsymbol{x} = (x_1, x_2, ..., x_n)$ with the real vector $\boldsymbol{y} = (y_1, y_2, ..., y_n)$. We see, rather generally, that the directional derivative, $D_{\boldsymbol{u}} f(\boldsymbol{x}_0)$, is the dot product of the unit vector $\boldsymbol{u}$ with the vector whose components are the partial derivatives of f at $\boldsymbol{x}_0$. This latter vector and dot products are important enough to warrant the formal definitions we now present.

Definition 8.10: Gradient vector, dot product of two vectors

Let f be a real valued function of n real variables (Definition 8.1 on page 261) all of whose partial derivatives (Definition 8.5 on page 267) exist at $\boldsymbol{x}$. The vector

$$(D_1 f(\boldsymbol{x}), D_2 f(\boldsymbol{x}), ..., D_n f(\boldsymbol{x}))$$

is called the **gradient of f at $\boldsymbol{x}$**. It will be denoted[1] by

$$\textbf{\textit{grad}}\ f(\boldsymbol{x}) \quad \text{or} \quad \boldsymbol{D}f(\boldsymbol{x}).$$

The vector-valued function whose value at $\boldsymbol{x}$ is $\boldsymbol{D}f(\boldsymbol{x})$ is denoted by

$$\boldsymbol{D}f \quad \text{or} \quad \textbf{\textit{grad}}\, f$$

and is referred to as the **gradient function**, or, when there is no danger of ambiguity, simply as the **gradient**, or **gradient vector**.

The **dot product**, $\boldsymbol{x} \cdot \boldsymbol{y}$, **of the real n-dimensional vectors**

$$\boldsymbol{x} = (x_1, x_2, ..., x_n) \quad \text{and} \quad \boldsymbol{y} = (y_1, y_2, ..., y_n)$$

1. It is also written $\nabla f(\boldsymbol{x})$.

is defined by the equation[1]

$$\boldsymbol{x} \cdot \boldsymbol{y} = \sum_{j=1}^{n} x_j y_j . \qquad 8.12$$

❑

The result concerning directional derivative computation can now be neatly and formally stated in the following theorem.

Theorem 8.11: Computation of directional derivatives

Let f be a real valued function of n real variables (Definition 8.1 on page 261), whose gradient vector components (Definition 8.10 on page 272) exist and are continuous (Definition 8.7 on page 270) at $\boldsymbol{x}$ and at all points within some positive distance (Definition 8.6 on page 270) of $\boldsymbol{x}$. Then

$$D_{\boldsymbol{u}} f(\boldsymbol{x}) = \boldsymbol{u} \cdot \boldsymbol{D} f(\boldsymbol{x}), \qquad 8.13$$

also written $D_{\boldsymbol{u}} f = (\boldsymbol{u} \cdot \boldsymbol{D}) f$ or $D_{\boldsymbol{u}} = \boldsymbol{u} \cdot \boldsymbol{D}$.

That is, when the partial derivatives of f are continuous at and near $\boldsymbol{x}$, the directional derivative, $D_{\boldsymbol{u}} f(\boldsymbol{x})$, is the dot product of $\boldsymbol{u}$ with ***grad*** $f(\boldsymbol{x})$.

This result is sometimes written

$$D_{\boldsymbol{u}} f(\boldsymbol{x}) = (\boldsymbol{u} \cdot \boldsymbol{D}) f(\boldsymbol{x}),\ D_{\boldsymbol{u}} f = (\boldsymbol{u} \cdot \boldsymbol{D}) f, \text{ or } D_{\boldsymbol{u}} = \boldsymbol{u} \cdot \boldsymbol{D}, \qquad 8.14$$

where

$$\boldsymbol{D} = (D_1, D_2, \ldots, D_n)$$

is the vector of partial derivative operators[2] (the **gradient operator**).

❑

The fundamental approximation for functions of one variable, equation 2.7 on page 31,

$$f(x_0 + h) \cong f(x_0) + f'(x_0) h, \qquad 8.15$$

---

1. When $\boldsymbol{x}$ and $\boldsymbol{y}$ are vectors of complex numbers, this definition needs to be modified, to

$$\boldsymbol{x} \cdot \boldsymbol{y} = \sum_{j=1}^{n} x_j \overline{y_j}$$

where $\overline{y_j}$ is the complex conjugate of y_j, (see Definition 6.8 on page 199) in order that $\boldsymbol{x} \cdot \boldsymbol{x} = \|\boldsymbol{x}\|^2$ be real and nonnegative.

2. This notation will prove extremely useful, as we will see later.

which was essential to studying functions of one variable, can now be generalized to functions of several variables in a natural way. From the meaning of directional derivative we see from approximation 8.15 that

$$f(\boldsymbol{x}_0 + \boldsymbol{h}) \cong f(\boldsymbol{x}_0) + D_{\boldsymbol{u}} f(\boldsymbol{x}_0) \|\boldsymbol{h}\|, \qquad \textit{8.16}$$

where $\boldsymbol{u}$ is the unit vector in the direction of $\boldsymbol{h}$, that is

$$\boldsymbol{u} = \frac{\boldsymbol{h}}{\|\boldsymbol{h}\|}.$$

Substituting $\boldsymbol{h}/\|\boldsymbol{h}\|$ for $\boldsymbol{u}$ in equation 8.13 and substituting the result into the right side of approximation 8.16 yields the desired extension of the fundamental approximation for functions of one variable — namely, the following.

Fundamental multivariate approximation

If f is a real valued function of n real variables (Definition 8.1 on page 261) whose gradient vector (Definition 8.10 on page 272) exists at $\boldsymbol{x}_0$, then

$$f(\boldsymbol{x}_0 + \boldsymbol{h}) \cong f(\boldsymbol{x}_0) + \boldsymbol{h} \cdot \boldsymbol{D}f(\boldsymbol{x}_0) = f(\boldsymbol{x}_0) + \boldsymbol{h} \cdot \boldsymbol{grad}\boldsymbol{f}(\boldsymbol{x}_0) \qquad \textit{8.17}$$

(where the *dot product* symbol is also defined in Definition 8.10).

❑

In this generalization, ***the derivative***, Df, ***is replaced by the gradient vector***, $\boldsymbol{D}f$, and ***ordinary multiplication here turns into dot product multiplication***.

Evidently the gradient vector and dot product occupy a central role in the study of functions of several variables, so now seems a reasonable time to seek a clear understanding of the dot product formula, equation 8.12 on page 273 and formula 8.13 on page 273 for $D_{\boldsymbol{u}} f(\boldsymbol{x})$.

Since the formula for $D_{\boldsymbol{u}} f(\boldsymbol{x})$ involved the dot product of a unit length vector, $\boldsymbol{u}$, with $\boldsymbol{D}f$, we'll first examine dot products of the form

$$\boldsymbol{u} \cdot \boldsymbol{X}$$

in three dimensions, where $\boldsymbol{u} = (u_1, u_2, u_3)$ and $\boldsymbol{X} = (X_1, X_2, X_3)$.

First look at the simplest unit vector, namely, $\boldsymbol{u} = (1,0,0)$. Then

$$\boldsymbol{u} \cdot \boldsymbol{X} = X_1.$$

So in this case, the dot product of $\boldsymbol{X}$ with the unit vector along the positive first coordinate axis is just the first coordinate of $\boldsymbol{X}$.

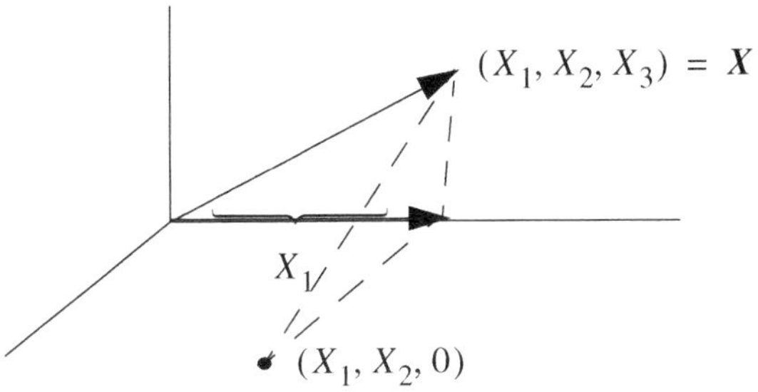

Figure 8-7

Looking at Figure 8-7, the vector of length X_1 is the vector along the first axis (whose endpoint is) closest to (the endpoint of) the vector $\boldsymbol{X}$. This follows

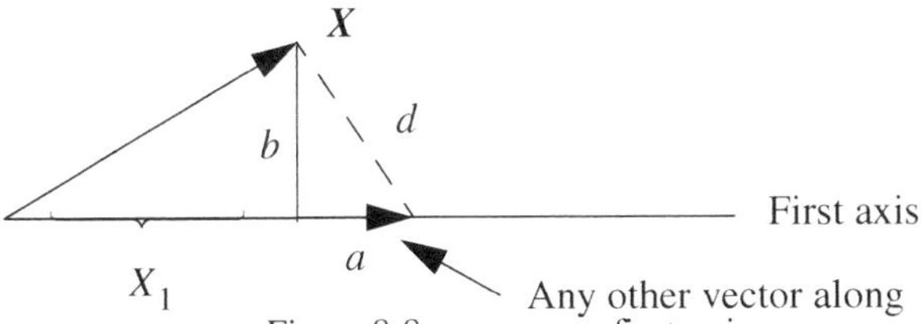

Figure 8-8

because, as seen in Figure 8-8, from the definition of distance (Definition 8.6 on page 270) the distance, d, of any other vector along the first axis from the endpoint of $\boldsymbol{X}$ (here $d = \sqrt{a^2 + X_2^2 + X_3^2}$) exceeds $b = \sqrt{X_2^2 + X_3^2}$.

This vector, $(X_1, 0, 0)$, which is the *first-axis vector closest to* $\boldsymbol{X}$, is called the *projection* of $\boldsymbol{X}$ on $\boldsymbol{u} = (1,0,0)$ (or on the first axis). What we have just seen suggests the possibility that generally $|\boldsymbol{u} \cdot \boldsymbol{X}|$ is the magnitude of the projection of the vector $\boldsymbol{X}$ along the axis specified by any unit length vector, $\boldsymbol{u}$.

We'll investigate this conjecture, and then for reference purposes, sum up the results in a formal definition and theorem.

To determine the projection of the n-dimensional vector $\boldsymbol{X}$ along the axis specified by the unit length vector $\boldsymbol{u}$, we must find the vector $\alpha \boldsymbol{u}$ closest to $\boldsymbol{X}$. That is, we must find the real number, α, which minimizes the quantity

$\|X - \alpha u\|$. To make this easier, we'll do the equivalent task of choosing α to minimize

$$G(\alpha) = \|X - \alpha u\|^2 = \sum_{j=1}^{n} (X_j - \alpha u_j)^2. \qquad 8.18$$

To find the desired α we could solve the equation $G'(\alpha) = 0$. We will leave this to the reader, and instead complete the square, by first writing

$$G(\alpha) = \sum_{j=1}^{n} (X_j - \alpha u_j)^2 = \sum_{j=1}^{n} X_j^2 - 2\left(\sum_{j=1}^{n} X_j u_j\right)\alpha + \left(\sum_{j=1}^{n} u_j^2\right)\alpha^2$$

$$= (u \cdot u)\alpha^2 - 2(X \cdot u)\alpha + X \cdot X$$

$$= \alpha^2 - 2(X \cdot u)\alpha + X \cdot X \quad \text{because } u \text{ is of unit length.}$$

Now we *complete the square* of the last expression, by adding $(X \cdot u)^2$ immediately preceding the last term in the final expression above, to obtain a perfect square, and then compensating, to preserve the equality, obtaining

$$G(\alpha) = \alpha^2 - 2(X \cdot u)\alpha + (X \cdot u)^2 - (X \cdot u)^2 + X \cdot X$$

$$= (\alpha - X \cdot u)^2 - (X \cdot u)^2 + X \cdot X.$$

It is now evident that $G(\alpha)$ is minimized by choosing

$$\alpha = X \cdot u.$$

We can now supply the promised definition and theorem.

Definition 8.12: Projection of X on Y

Let X and Y be any real n-dimensional vectors (see Definition 8.1 on page 261). The **projection,** $P_Y(X)$, **of X on Y** is the vector βY (where β is a real number) closest to X, i.e., minimizing $\|X - \beta Y\|$ (see Definition 8.6 on page 270). $P_Y(X)$ is sometimes referred to as the **component of X in the Y direction**.

❑

Theorem 8.13: Projection of X on Y

Let $X = (X_1, X_2, ..., X_n)$ and $Y = (X_1, X_2, ..., X_n)$ be any real n-dimensional vectors. If $Y = \mathbf{0}$ (i.e., $Y_i = 0$ for all i), then $P_Y(X) = \mathbf{0}$. Otherwise (see Definition 8.10 on page 272)

$$P_Y(X) = \frac{X \cdot Y}{Y \cdot Y} Y. \qquad 8.19$$

❑

The proof is trivial when $Y = \mathbf{0}$, because in this case ***all*** vectors $\beta Y = \mathbf{0}$, and so the closest of these vectors to X is clearly $\mathbf{0}$, as asserted.

If $Y \neq \mathbf{0}$, we may write

$$\beta Y = \beta\|Y\|\frac{Y}{\|Y\|} = \beta\|Y\|u.$$

Choosing $\alpha = \beta\|Y\|$, we know from the previous *completion of the square* argument that the vector $\alpha u = \alpha Y/\|Y\|$ closest to X is

$$(X \cdot u)u = \left(X \cdot \frac{Y}{\|Y\|}\right)\frac{Y}{\|Y\|} = \frac{X \cdot Y}{\|Y\|^2}Y = \frac{X \cdot Y}{Y \cdot Y}Y$$

establishing Theorem 8.13.

It is evident that we can write any vector X as the sum of two vectors, one of which is the component of X in the Y direction, illustrated in Figure 8-9.

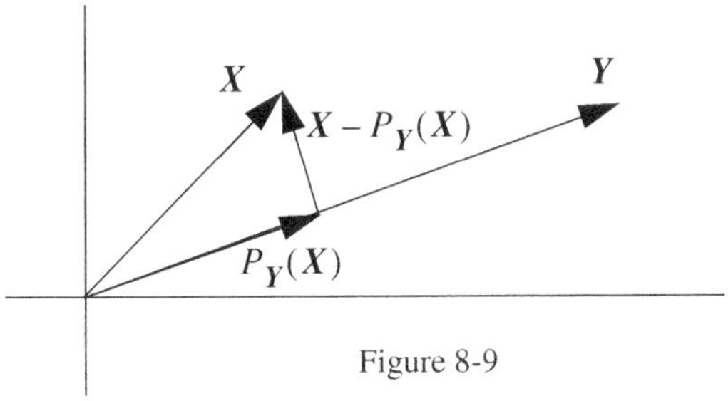

Figure 8-9

That is, we can write

$$X = P_Y(X) + [X - P_Y(X)]. \qquad \textit{8.20}$$

It would appear that the second vector in this *decomposition* has a *zero component* in the Y direction — which we easily verify (using equation 8.19), via

$$\begin{aligned} P_Y(X - P_Y(X)) &= P_Y(X) - P_Y(P_Y(X)) \\ &= P_Y(X) - P_Y(X) \qquad \textit{8.21} \\ &= 0. \end{aligned}$$

A decomposition such as given by equation 8.20 is called an **orthogonal decomposition**, or an **orthogonal breakup**. Orthogonal breakups are extremely important, because they produce *zeros*. We'll see later on just how useful this is in solving systems of equations. We're just starting to investigate this concept, and we begin with a formal definition followed by a simple operational characterization.

Definition 8.14: Orthogonality

The real n-dimensional vector X is said to be **orthogonal** (or **perpendicular**) to the real n-dimensional vector Y *if*

$$P_Y(X) = \mathbf{0} \quad \text{also written} \quad X \perp Y$$

(see Definition 8.12 on page 276).

❑

Equation 8.21 makes it reasonable to refer to $X - P_Y(X)$ as **the component of X perpendicular to Y.**

An immediate consequence of Definition 8.14 and Theorem 8.13 is the following.

Theorem 8.15: Orthogonality of X and Y

$X \perp Y$ if and only if $X \cdot Y = 0$. Hence $X \perp Y$ if and only if $Y \perp X$

❑

We can see immediately from this result and formula 8.13 on page 273 for directional derivatives, that $D_u f(x) = 0$ precisely when $u \perp \mathbf{grad} f(x)$. This suggests that this gradient vector specifies the direction of greatest rate of change of the function f at x, a result we shall soon establish, and one which will evidently be critical in studying the behavior of functions of several variables.

From equations 8.20 and 8.21 we know that $X = P_Y(X) + [X - P_Y(X)]$ is a breakup into perpendicular components. It turns out to be important that the converse also holds. We state these results in Theorem 8.16.

Theorem 8.16: Orthogonal Breakup

If X and Y are real n-dimensional vectors (see Definition 8.1 on page 261) then

$$X = P_Y(X) + [X - P_Y(X)] \qquad 8.22$$

(where $P_Y(X)$ is defined in Definition 8.12 on page 276) is an orthogonal breakup — i.e.,

$$P_Y(X) \perp X - P_Y(X) \qquad 8.23$$

(see Theorem 8.15 on page 278).

Moreover, if

$$X = V_Y + V_{Y\perp} \qquad 8.24$$

where V_Y is a scalar multiple of Y, and $V_{Y\perp}$ is orthogonal to Y, then this orthogonal breakup is unique — i.e., must be the one given by equation 8.22; so that we must have

$$V_Y = P_Y(X) \quad \text{and} \quad V_{Y\perp} = X - P_Y(X). \qquad 8.25$$

❑

We have already shown that equation 8.23 holds. So we need only prove that latter part of this theorem. We first handle the trivial case where $Y = \mathbf{0}$. In this case we must have

$$V_Y = \mathbf{0} = P_Y(X),$$

and then from equation 8.24 it follows that

$$V_{Y\perp} = X = X - P_Y(X).$$

This establishes equation 8.25 for the case $Y = \mathbf{0}$.

If $Y \neq \mathbf{0}$, then from equation 8.24 and the definition of V_Y we may write

$$X = V_Y + V_{Y\perp} = \alpha Y + V_{Y\perp}. \qquad 8.26$$

Now take the dot product of the left and right sides of this equation with Y, and use the assumed perpendicularity of $V_{Y\perp}$ and Y, to obtain

$$X \cdot Y = \alpha(Y \cdot Y),$$

from which it follows that $\alpha = X \cdot Y / Y \cdot Y$, and hence

$$\alpha Y = P_Y(X).$$

Now substituting this into equation 8.26, we obtain the desired equation 8.25, proving Theorem 8.16. To state this theorem in English, ***any vector, X, can be written* in one and only one way *as the sum of two vectors — one of which is in the direction of a specified vector Y and the other perpendicular to Y. The component in the direction of Y is the projection of X on Y.*** All of this is suggested by Figure 8-9 on page 277, but established algebraically using the definitions of dot product and projection.

We never used the Pythagorean theorem directly in our proofs, but rather, used it to suggest the proper definition of the length of a vector. With the machinery we have built up, it is a natural corollary.

Theorem 8.17: Corollary — Pythagorean theorem

For any real n-dimensional vectors, X and Y

$$\|X\|^2 = \|P_Y(X)\|^2 + \|X - P_Y(X)\|^2 \qquad 8.27$$

(see Definition 8.12 on page 276 and Definition 8.6 on page 270).

Furthermore

$$\|P_Y(X)\| \le \|X\|, \qquad 8.28$$

equality always holding in the trivial case $X = \mathbf{0}$, and in the nontrivial case, $X \neq \mathbf{0}$, if and only if Y is a nonzero scalar multiple of X.

❑

To prove these results we write

$$\begin{aligned}\|X\|^2 &= X \cdot X \\ &= (P_Y(X) - [X - P_Y(X)]) \cdot (P_Y(X) - [X - P_Y(X)]).\end{aligned}$$

Now expand out[1] and use the orthogonality of $P_Y(X)$ and $X - P_Y(X)$ (result 8.23 on page 278 and Theorem 8.15 on page 278) to obtain equation 8.27. Inequality 8.28 follows immediately from equation 8.27.

Finally, if Y is a nonzero scalar multiple of X, we know from Theorem 8.13 on page 276 that $X = P_Y(X)$. On the other hand, if $\|X\| = \|P_Y(X)\|$, we see from the Pythagorean theorem equality, 8.27, that

$$\|X - P_Y(X)\|^2 = 0,$$

which implies that

$$X = P_Y(X),$$

which, in the nontrivial case we are considering, $X \neq \mathbf{0}$, using the projection formulas from Theorem 8.13 on page 276 we see that X must be a nonzero scalar multiple of Y, and Y must be a nonzero scalar multiple of X. These arguments establish Theorem 8.17.

Theorem 8.17 allows us to establish our earlier conjecture, that a nonzero gradient vector specifies the direction of greatest change of the associated function. So suppose f is a real valued function of n real variables which allows application of formula 8.13 on page 273 for directional derivatives. Let $\boldsymbol{u}$ be any unit length n dimensional vector which ***is not a scalar multiple of grad*** $f(\boldsymbol{x})$. Then

$$\begin{aligned}|D_{\boldsymbol{u}} f(\boldsymbol{x})| &= |\boldsymbol{u} \cdot \boldsymbol{grad}\ f(\boldsymbol{x}).| \\ &= \left|\boldsymbol{u}_g \cdot \boldsymbol{grad}\ f(\boldsymbol{x}) + \boldsymbol{u}_{g\perp} \cdot \boldsymbol{grad}\ f(\boldsymbol{x})\right|\end{aligned}$$

where $\boldsymbol{u}_g = P_{\boldsymbol{grad} f(\boldsymbol{x})}(\boldsymbol{u})$, $\boldsymbol{u}_{g\perp} = \boldsymbol{u} - \boldsymbol{u}_g \perp \boldsymbol{grad}\ f(\boldsymbol{x})$ and $\|\boldsymbol{u}_g\| < 1$, justified by Theorem 8.16 on page 278 and Theorem 8.17 on page 279.

---

1. Using the distributivity of dot products — i.e., $(u+v) \cdot w = u \cdot w + v \cdot w$, etc.

Hence

$$|D_u f(x)| = u_g \cdot \boldsymbol{grad}\ f(x).$$

But because u_g is in the $\boldsymbol{grad}\ f(x)$ direction, but has length < 1, we can write $u_g = c u_G$ where u_G is the unit length vector in the direction of $\boldsymbol{grad}\ f(x)$ and $|c| < 1$.

Hence

$$|D_u f(x)| = |c|\, u_G \cdot \boldsymbol{grad}\ f(x).$$
$$= |c| D_{u_G} f(x) < D_{u_G} f(x).$$

That is, if u is any unit vector not parallel to the nonzero gradient vector, $\boldsymbol{grad}\ f(x)$, and u_G is the unit vector in the direction of $\boldsymbol{grad}\ f(x)$, then

$$|D_u f(x)| < D_{u_G} f(x),$$

establishing that under the given conditions, the gradient specifies the direction of greatest rate of change of *f*. Note that $-\boldsymbol{grad}\ f(x)$ is the direction of the most negative rate of change of *f*. We summarize the various properties that we have established about the gradient vector as follows.

Theorem 8.18: Gradient vector properties

Suppose the real valued function, *f*, of *n* real variables (Definition 8.1 on page 261) has partial derivatives (Definition 8.5 on page 267) defined for all values within some positive distance (Definition 8.6 on page 270) of the vector x, and these partial derivatives are continuous (Definition 8.7 on page 270) at x. If the unit vector $u \perp \boldsymbol{grad}\ f(x)$ (Definition 8.14 on page 278 and Definition 8.10 on page 272) then $D_u f(x) = 0$ (see Definition 8.4 on page 265). If $\boldsymbol{grad}\ f(x) \neq \mathbf{0}$, $D_u f(x)$ is a maximum for

$$u = u_G = \frac{\boldsymbol{grad}\ f(x)}{\|\boldsymbol{grad}\ f(x)\|}$$

and a minimum for

$$u = -u_G = -\frac{\boldsymbol{grad}\ f(x)}{\|\boldsymbol{grad}\ f(x)\|}.$$

❑

From this result it should be evident that the principal *nice* places to hunt for maxima and minima of functions of several variables are those values of x at which $\boldsymbol{grad}\ f(x) = \mathbf{0}$. Finding such x involves solving a system of (usually nonlinear) equations, and we have not yet developed the tools for doing this. Furthermore, once we have solved these equations, to handle this subject properly, sufficient conditions to determine whether the solution corresponds to a maximum, minimum or neither must be developed. Since all of this depends on the linear algebra to be introduced in the next chapters, we postpone any further discussion of this subject until the necessary tools are available.

A useful corollary to the Pythagorean theorem, Theorem 8.17 on page 279, is the Schwartz inequality or vectors.

Theorem 8.19: The Schwartz inequality for vectors

Recall Definition 8.10 on page 272 of dot product of two vectors. If $\boldsymbol{X}$ and $\boldsymbol{Y}$ are *real* n-dimensional vectors, *then*

$$(\boldsymbol{X} \cdot \boldsymbol{Y})^2 \le (\boldsymbol{X} \cdot \boldsymbol{X})(\boldsymbol{Y} \cdot \boldsymbol{Y}) \qquad 8.29$$

with equality only if one of these vectors is a scalar multiple of the other.

❑

To establish this version of the Schwartz inequality, we note that if $\boldsymbol{Y} = \boldsymbol{0}$ then inequality 8.29 is certainly true. Thus we need only consider the case $\boldsymbol{Y} \neq \boldsymbol{0}$. For this we note that inequality 8.28 on page 280 is equivalent to

$$\|P_{\boldsymbol{Y}}(\boldsymbol{X})\|^2 \le \|\boldsymbol{X}\|^2. \qquad 8.30$$

But when $\boldsymbol{Y} \neq \boldsymbol{0}$

$$\begin{aligned}\|P_{\boldsymbol{Y}}(\boldsymbol{X})\|^2 &= \left\|\frac{\boldsymbol{X} \cdot \boldsymbol{Y}}{\boldsymbol{Y} \cdot \boldsymbol{Y}}\boldsymbol{Y}\right\|^2 = \left(\frac{\boldsymbol{X} \cdot \boldsymbol{Y}}{\boldsymbol{Y} \cdot \boldsymbol{Y}}\boldsymbol{Y}\right) \cdot \left(\frac{\boldsymbol{X} \cdot \boldsymbol{Y}}{\boldsymbol{Y} \cdot \boldsymbol{Y}}\boldsymbol{Y}\right) \\ &= \frac{(\boldsymbol{X} \cdot \boldsymbol{Y})(\boldsymbol{X} \cdot \boldsymbol{Y})}{\boldsymbol{Y} \cdot \boldsymbol{Y}}.\end{aligned} \qquad 8.31$$

Substituting $\|\boldsymbol{X}\|^2 = \boldsymbol{X} \cdot \boldsymbol{X}$ and the last expression from equation 8.31 into equation 8.30 yields the asserted equation 8.29, proving the Schwartz inequality for vectors.

We may think of the Schwartz inequality as the assertion that the magnitude

$$|cos(\boldsymbol{X}, \boldsymbol{Y})| = \frac{\|P_{\boldsymbol{Y}}(\boldsymbol{X})\|}{\|\boldsymbol{X}\|} \qquad 8.32$$

of the cosine[1] determined by the vectors $\boldsymbol{X}$ and $\boldsymbol{Y}$ does not exceed 1; here writing $P_{\boldsymbol{Y}}(\boldsymbol{X}) = \beta\boldsymbol{Y}$ (see equation 8.19 on page 276 yielding the formula $\beta = (\boldsymbol{X} \cdot \boldsymbol{Y})/(\boldsymbol{Y} \cdot \boldsymbol{Y})$). we provide a sign for $cos(\boldsymbol{X}, \boldsymbol{Y})$ by requiring

$$cos(\boldsymbol{X}, \boldsymbol{Y}) < 0 \text{ if } \beta < 0 \text{ and } cos(\boldsymbol{X}, \boldsymbol{Y}) > 0 \text{ if } \beta > 0,$$

---

1. Note that here we are not referring to the cosine of an angle, but the ratio of the signed leg (projection) to the hypotenuse length, a geometric configuration not requiring consideration of angles.

as illustrated in Figure 8-10; this yields the property that the *cosine* is negative when the vector Y and the projection of X on Y point in the opposite directions (i.e., are negative multiples of each other).

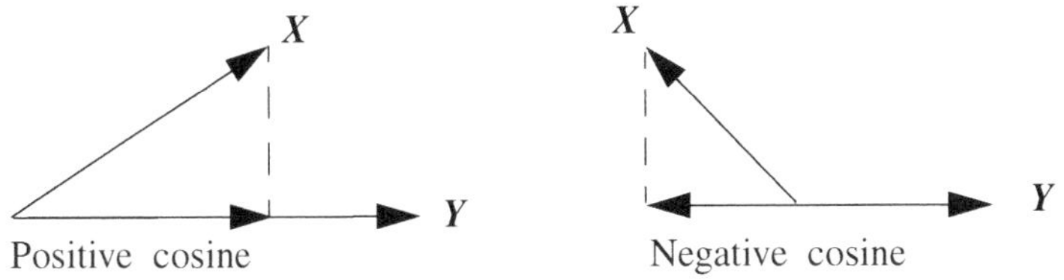

Figure 8-10

It's useful to introduce the *signum* function at this point to write down a convenient representation of the dot product.

Definition 8.20: $sgn(a)$

We define the **signum function**, *sgn*, of a real variable, by

$$sgn(a) = \begin{cases} 1 & \text{if } a > 0 \\ 0 & \text{if } a = 0 \\ -1 & \text{if } a < 0. \end{cases}$$

❑

From the arguments above, in particular equation 8.31, we see that

$$\begin{aligned} cos(\mathbf{X},\mathbf{Y}) &= sgn(\mathbf{X}\cdot\mathbf{Y})\frac{\|P_{\mathbf{Y}}(\mathbf{X})\|}{\|\mathbf{X}\|} \\ &= sgn(\mathbf{X}\cdot\mathbf{Y})\frac{|\mathbf{X}\cdot\mathbf{Y}|}{\|\mathbf{X}\|\|\mathbf{Y}\|} = \frac{\mathbf{X}\cdot\mathbf{Y}}{\|\mathbf{X}\|\|\mathbf{Y}\|}. \end{aligned}$$

That is, for neither $\mathbf{X}$ nor $\mathbf{Y}$ the $\mathbf{0}$ vector

$$cos(\mathbf{X},\mathbf{Y}) = \frac{\mathbf{X}\cdot\mathbf{Y}}{\|\mathbf{X}\|\|\mathbf{Y}\|} \quad \text{or equivalently} \quad \mathbf{X}\cdot\mathbf{Y} = \|\mathbf{X}\|\|\mathbf{Y}\|\,cos(\mathbf{X},\mathbf{Y}). \qquad \textit{8.33}$$

We should think of equation 8.33 as the **definition** of the cosine defined by two vectors, representing the ratio of the signed leg (projection) to the hypotenuse length; i.e., the algebraic realization of a geometric concept.

The Schwartz inequality may be extended to integrals, which is the setting where it finds most of its application. The precise statement is as follows:

Theorem 8.21: Schwartz inequality for integrals

If f and g are real valued functions of a real variable (Definition 8.1 on page 261) continuous at each x (see Definition 7.2 on page 213), in the closed bounded interval $[a; b]$ (Definition 3.11 on page 88) then (Definition 4.2 on page 102)

$$\left(\int_a^b f g\right)^2 \le \left(\int_a^b f^2\right)\left(\int_a^b g^2\right). \qquad 8.34$$

❑

This follows by applying the Schwartz inequality for vectors (Theorem 8.19 on page 282) to

$$\boldsymbol{X} = (f_0, f_1, \ldots, f_{n-1}) \quad \text{and} \quad \boldsymbol{Y} = (g_0, g_1, \ldots, g_{n-1})$$

where

$$f_k = f(a + k\,h), \qquad g_k = g(a + k\,h), \quad \text{and} \quad h = \frac{b-a}{n},$$

which yields

$$\left(h \sum_{k=0}^{n-1} f(a + k\,h) g(a + k\,h)\right)^2 \le \left(h \sum_{k=0}^{n-1} f^2(a + k\,h)\right)\left(h \sum_{k=0}^{n-1} g^2(a + k\,h)\right).$$

Since these sums are Riemann approximating sums (Definition 4.1 on page 101) for the integrals of inequality 8.34, letting n grow arbitrarily large in the inequality above leads to the sought-after establishment of inequality 8.34. As before, because of the assumed continuity, equality holds only if one of the functions involved is a scalar multiple of the other.

This version of the Schwartz inequality can sometimes be applied to obtain simple useful bounds for integrals which are hard to evaluate.

Example 8.7: Upper bound for $\int_1^2 \frac{1}{\sqrt{2\pi}} e^{-x^2/2} dx$

Write the integrand as

$$\frac{1}{\sqrt{2\pi x}} \sqrt{x}\, e^{-x^2/2}.$$

So

$$\left(\int_1^2 \frac{1}{\sqrt{2\pi}} e^{-x^2/2} dx\right)^2 \le \left(\frac{1}{2\pi}\int_1^2 \frac{dx}{x}\right)\left(\int_1^2 x e^{-x^2} dx\right)$$

$$= \frac{ln(2)}{2\pi}\left(\frac{1}{2}[e^{-1} - e^{-4}]\right) = (0.1389)^2.$$

The actual value of the integral of interest is 0.1359.

■

Note that the *more nearly parallel* the two factors you choose, the better the bound you get.

Exercises 8.8

Use the Schwartz inequality for integrals, Theorem 8.21, to find upper bounds for the following. If possible, use some numerical integration scheme to determine how good these bounds are.

1. $\int_1^2 e^{\sqrt{x}}\, dx$

2. $\int_1^{1.5} \sqrt{sin(x^3)}\, dx$

3. $\int_1^2 \frac{dx}{\sqrt{x^{0.5}+1}}$

Chapter 9

Coordinate Systems — Linear Algebra

1 Introduction

To introduce the study of choosing useful coordinate systems for dealing with functions of several variables, let's return to consideration of the local description of such functions. To emphasize that we want to concentrate on the behavior of a function f near a fixed input $\boldsymbol{x}_0$, we write the graph of f as the set of all points of the form

$$(\boldsymbol{x}_0 + \boldsymbol{h},\ f(\boldsymbol{x}_0 + \boldsymbol{h})) \qquad \textit{9.1}$$

as $\boldsymbol{h}$ varies over its possible values.

Now recall fundamental approximation 8.17 on page 274, which we will write as

$$f(\boldsymbol{x}_0 + \boldsymbol{h}) \cong f(\boldsymbol{x}_0) + \boldsymbol{h} \cdot \boldsymbol{grad}\ f(\boldsymbol{x}_0).$$

From this we see that for a well-behaved function, f, our approximate (local) description of the graph of f near $\boldsymbol{x}_0$ is the set of points of the form

$$(\boldsymbol{x}_0 + \boldsymbol{h},\ f(\boldsymbol{x}_0) + \boldsymbol{h} \cdot \boldsymbol{grad}\ f(\boldsymbol{x}_0)). \qquad \textit{9.2}$$

For $n = 2$, this set of points is what we call the *tangent plane* to f at $\boldsymbol{x}_0$; in higher dimensions it is called the *tangent hyperplane* to f at $\boldsymbol{x}_0$.

To obtain the insight we need to make effective use of our local description, we must refine expression 9.2 substantially. As we'll see, there are two approaches which suggest themselves. They are complementary, and we'll be working with both of them. Each involves constructing a more suitable coordinate system than the original one involving the coordinate axis vectors, $\boldsymbol{e}_k$ whose j-th coordinate is given by, (see Definition 8.2 on page 263),

$$e_{kj} = \begin{cases} 1 & \text{if} \quad j = k \\ 0 & \text{if} \quad j \neq k \end{cases}$$

in which expression 9.2 was developed. In this original coordinate system, any vector $\boldsymbol{x} = (x_1, x_2, \ldots, x_n)$ may be written as the sum

$$\boldsymbol{x} = \sum_{k=1}^{n} x_k \boldsymbol{e}_k. \qquad \textit{9.3}$$

We refer to x_k as the **k-th coordinate of x with respect to the original coordinate axis vector, e_k**. ***This original coordinate system is often the one most suitable for problem formulation — but frequently is inappropriate for problem solution.***

To specify a new coordinate system, we choose new coordinate axis vectors, say $\boldsymbol{v}_1, \boldsymbol{v}_2,...,\boldsymbol{v}_k$, and analogous to equation 9.3, express any vector, $\boldsymbol{x}$, in the form

$$\boldsymbol{x} = \sum_{k=1}^{n} a_k(\boldsymbol{x})\,\boldsymbol{v}_k. \qquad \textbf{9.4}$$

Assuming we can do this, we refer to the real number, $a_k(\boldsymbol{x})$, as the coordinate of $\boldsymbol{x}$ with respect to the vector $\boldsymbol{v}_k$; $a_k(\boldsymbol{x})$ is determined by $\boldsymbol{x}$.

The proper choice of coordinate system often makes apparently difficult problems easy to solve. From a geometric viewpoint, it would appear obvious that a coordinate system whose origin is at the point of tangency, $\boldsymbol{x}_0$, and whose axes (vectors) lie either in the tangent hyperplane, or perpendicular to it, would be more suitable than the original coordinate system, as illustrated in Figure 9-1.

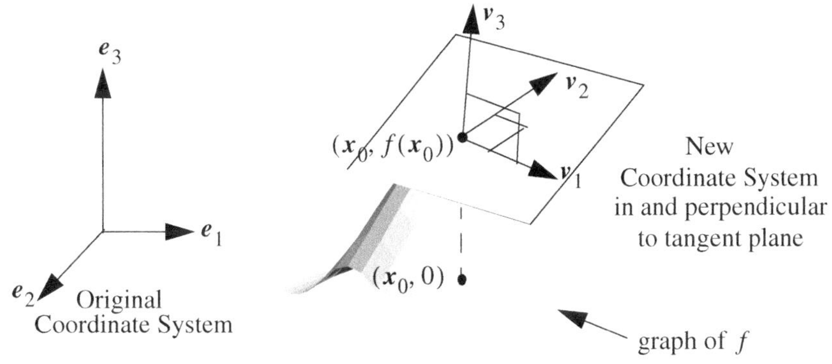

Figure 9-1

This chapter is devoted to the systematic development of appropriate coordinate systems. In both of the approaches we will be considering, we first rewrite expression 9.2 as the set of points

$$(\boldsymbol{x}_0, f(\boldsymbol{x}_0)) + (\boldsymbol{h}, \boldsymbol{h} \cdot \boldsymbol{grad}\ f(\boldsymbol{x}_0)) \qquad \textbf{9.5}$$

as $\boldsymbol{h}$ varies over its possible values.

In this form, the preferred origin, $(\boldsymbol{x}_0, f(\boldsymbol{x}_0))$, is separated out from the vectors to be added to it to yield the tangent hyperplane. In our ***first approach*** we construct our new coordinate system directly from vectors of the form $(\boldsymbol{h}, \boldsymbol{h} \cdot \boldsymbol{grad}\ f(\boldsymbol{x}_0))$ where we need only consider vectors $\boldsymbol{h} = \boldsymbol{e}_k$. That is, we construct our new coordinate system from vectors of the form $(\boldsymbol{e}_k, \boldsymbol{e}_k \cdot \boldsymbol{grad}\ f(\boldsymbol{x}_0))$, $k = 1,...,n$, because every vector $(\boldsymbol{h}, \boldsymbol{h} \cdot \boldsymbol{grad}\ f(\boldsymbol{x}_0))$ may be written as

$$\sum_{k=1}^{n} h_k(\boldsymbol{e}_k, \boldsymbol{e}_k \cdot \boldsymbol{grad}\ f(\boldsymbol{x}_0)). \qquad \textbf{9.6}$$

The vectors $(\boldsymbol{e}_k, \boldsymbol{e}_k \cdot \boldsymbol{grad}\ f(\boldsymbol{x}_0))$ $k = 1,...,n$ may themselves not be especially suitable as coordinate axis vectors. This is because, as we'll see, there are a great many advantages to having the coordinate axis vectors *orthogonal* (Definition 8.14 on page 278) to each other,[1] and frequently, the vectors $(\boldsymbol{e}_k, \boldsymbol{e}_k \cdot \boldsymbol{grad}\ f(\boldsymbol{x}_0))$ don't have this property. So we need to determine how to create an equivalent orthogonal sequence of vectors (a sequence with the properties we want), from the original sequence of vectors (in this case the vectors $(\boldsymbol{e}_k, \boldsymbol{e}_k \cdot \boldsymbol{grad}\ f(\boldsymbol{x}_0))$. We will accomplish this shortly by means of the *Gram-Schmidt orthogonalization procedure*, which is just an extension of the orthogonal breakup of Theorem 8.16 on page 278.

In the other approach that we will use to choose a better coordinate system for the tangent hyperplane, we note that in contrast to thinking of a plane as being determined by the vectors in this plane, we can characterize a plane by specifying one point in the plane, and, as you can see from the geometry, the direction of a single vector perpendicular to the plane — just examine Figure 9-1, where the single vector being referred to is $\boldsymbol{v}_3$. So let's see if we can determine a vector perpendicular to all of the vectors $(\boldsymbol{h}, \boldsymbol{h} \cdot \boldsymbol{grad}\ f(\boldsymbol{x}_0))$. Denote the vector we're looking for by

$$\boldsymbol{v} = (q_1, q_2,...,q_n, b)$$

and try to find its coordinates, $q_1, q_2,...,q_n, b$, such that

$$\boldsymbol{v} \perp (\boldsymbol{h}, \boldsymbol{h} \cdot \boldsymbol{grad}\ f(\boldsymbol{x}_0)), \qquad \textbf{9.7}$$

for all $\boldsymbol{h} = \boldsymbol{e}_k$, $k = 1,...,n$. Using the dot product condition for orthogonality, Theorem 8.15 on page 278, we find that the conditions 9.7 lead to the following equations:

$$\begin{aligned} q_1 + bD_1 f(\boldsymbol{x}_0) &= 0 \\ q_2 + bD_2 f(\boldsymbol{x}_0) &= 0 \\ &\cdot \\ &\cdot \\ &\cdot \\ q_n + bD_n f(\boldsymbol{x}_0) &= 0. \end{aligned} \qquad \textbf{9.8}$$

1. Among other things, it greatly simplifies determining the new coordinates of specified vectors with respect to the new axis vectors if these new coordinate axis vectors are orthogonal to each other.

The value of b doesn't really matter (as long as it is nonzero), since multiplying b by any nonzero value would simply result in multiplication of the other coordinates of $\boldsymbol{v}$ by this same value, not changing its direction in any essential way. So choose $b = -1$, to find that

$$\boldsymbol{v} = (\boldsymbol{grad}\ f(\boldsymbol{x}_0), -1) \tag{9.9}$$

is essentially the unique solution to equations 9.8, and hence specifies the unique direction (determined by a vector unique except for magnitude and algebraic sign) orthogonal to the tangent plane.

To show the converse, namely, that every vector perpendicular to $(\boldsymbol{grad}\ f(\boldsymbol{x}_0), -1)$ has the form $(\boldsymbol{h}, \boldsymbol{h} \cdot \boldsymbol{grad}\ f(\boldsymbol{x}_0))$, let $\boldsymbol{h} = (x_1, x_2, ..., x_n)$, and let $(x_1, x_2, ..., x_n, x_{n+1}) = (\boldsymbol{h}, x_{n+1})$ be any vector perpendicular to $(\boldsymbol{grad}\ f(\boldsymbol{x}_0), -1)$; then from this assumed orthogonality we have

$$\boldsymbol{h} \cdot \boldsymbol{grad}\ f(\boldsymbol{x}_0) - x_{n+1} = 0.$$

So any vector $(\boldsymbol{h}, x_{n+1})$ orthogonal to $(\boldsymbol{grad}\ f(\boldsymbol{x}_0), -1)$ must have $x_{n+1} = \boldsymbol{h} \cdot \boldsymbol{grad}\ f(\boldsymbol{x}_0)$; i.e., must have the form $(\boldsymbol{h}, \boldsymbol{h} \cdot \boldsymbol{grad}\ f(\boldsymbol{x}_0))$ as asserted.

We have thus shown the following.

Theorem 9.1: Characterization of tangent hyperplane

For well-behaved real valued functions f of n real variables — i.e., functions with continuous partial derivatives everywhere, (Definition 8.5 on page 267 and Definition 8.7 on page 270), the set of vectors which get added to the point of tangency, (see Definition 8.2 on page 263), to form the tangent hyperplane, consists of those vectors perpendicular to, (Definition 8.14 on page 278) the vector $(\boldsymbol{grad}\ f(\boldsymbol{x}_0), -1)$.

❑

This result may seem a bit strange — but a small amount of investigation can clarify its meaning. Let

$$\boldsymbol{x} = (x_1, x_2, ..., x_n), \quad \boldsymbol{X} = (\boldsymbol{x}, x_{n+1}) = (x_1, x_2, ..., x_n, x_{n+1}), \tag{9.10}$$

and

$$g(\boldsymbol{X}) = f(\boldsymbol{x}) - x_{n+1}. \tag{9.11}$$

Then we may write the graph of f as the set of points, $\boldsymbol{X}$, such that

$$g(\boldsymbol{X}) = 0. \tag{9.12}$$

Definition 9.2: Implicit representation

When we specify the graph of some function f by means of one or more equations, [1] we refer to this set of equations as an **implicit representation of the function** f.

❑

Letting $X_0 = (x_0, f(x_0))$, we see that $g(X_0) = 0$, and the vector

$$(\boldsymbol{grad}\ f(x_0), -1) = \boldsymbol{grad}\ g(X_0).$$

That is, ***the tangent hyperplane to*** f ***at*** $X_0 = (x_0, f(x_0))$ ***is the set of vectors gotten by adding to*** X_0 ***all vectors perpendicular to grad*** $g(X_0)$, ***where*** $g(X) = 0$ ***is an implicit representation of*** f.

But we should be able to convince you more generally of the following.

Assertion 9.3: Tangent hyperplane from implicit representation

If $g(X) = 0$ is any *reasonable* implicit representation (see Definition 9.2 above) of the well-behaved[2] real valued function f, of n real variables then we expect to obtain the tangent hyperplane to f at $X_0 = (x_0, f(x_0))$ by adding to X_0 all vectors perpendicular to $\boldsymbol{grad}\ g(X_0)$ (Definition 8.10 on page 272 , provided that $\boldsymbol{grad}\ g(X_0) \neq 0$, and g has partial derivatives at and near X_0 which are continuous (Definition 8.7 on page 270) at X_0.

❑

Here's the reasoning. Assuming applicability of the formula for computing directional derivatives as dot products, Theorem 8.11 on page 273, and using Theorem 8.15 on page 278 (expressing orthogonality in terms of dot products), if U is any $n+1$-dimensional unit length vector, then $D_U(g(X_0)) = 0$ if and only if

$$U \perp \boldsymbol{grad}\ g(X_0). \qquad \textbf{\textit{9.13}}$$

That is, small displacements from X_0 in directions U perpendicular to $\boldsymbol{grad}\ g(X_0)$ are the only ones which essentially do not alter the value of g. So, assuming that $g(X) = 0$ is a reasonably behaved implicit representation of the function f and $g(X_0) = 0$, the set of all points of the form $X_0 + tU$

---

1. Like equation 9.12, but not assuming that f necessarily appears in the formula for g exactly as given in equation 9.11.
2. For example, f has continuous partial derivatives everywhere.

where $U \perp \boldsymbol{grad}\ g(X_0)$ approximates the set of X with $g(X) = 0$, at least *locally*, i.e., for X *near* X_0 (that is, for small t).

Provided that the set of all points of the form $X_0 + tU$, for all real t and all U satisfying condition 9.13 represents a hyperplane,[1] we thus see that these points form the tangent hyperplane to f at $X_0 = (\boldsymbol{x}_0, f(\boldsymbol{x}_0))$. At least in three dimensions, ($n = 2$), it's easy to convince yourself that the set of all points $X_0 + tU$, where $U \perp \boldsymbol{grad}\ g(\boldsymbol{x}_0)$, is a plane — as illustrated in Figure 9-2.

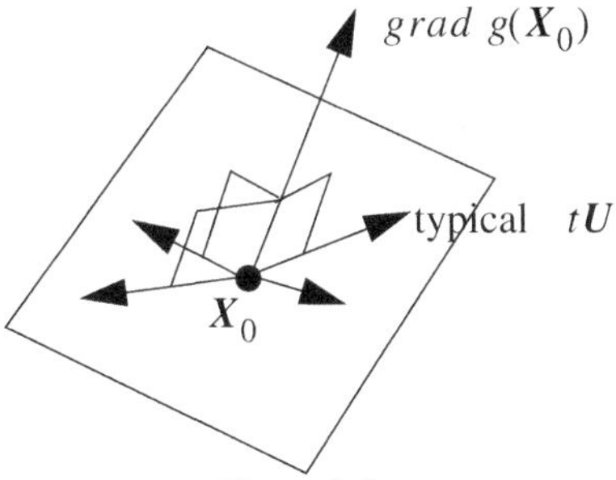

Figure 9-2

All of the material on tangent hyperplanes constructed from implicit representations of a function probably seems very abstract. So we conclude this presentation with a few concrete examples, which should help make this subject clearer. Note how natural the implicit representation below is.

Example 9.1: Tangent to sphere $x^2 + y^2 + z^2 = 2^2$ at north pole

A sphere of radius 2 about the origin is the set of points (x,y,z) for which

$$x^2 + y^2 + z^2 = 2^2,$$

i.e. the set of points (x,y,z) for which

$$g(x, y, z) = x^2 + y^2 + z^2 - 4 = 0.$$

One point on this sphere is $X_0 = (0,0,2)$ — the *north pole*. Intuitively the tangent plane should be the set of (x,y,z) for which $z = 2$. To derive this more algebraically, note that

$$\boldsymbol{grad}\ g(x, y, z) = (2x, 2y, 2z).$$

1. Shortly we will give the precise definition of hyperplane.

Hence

$$\textit{\textbf{grad}}\ g(0,0,2) = (0,0,4).$$

The unit length vectors perpendicular to $(0,0,4)$ are all unit length vectors of the form $\boldsymbol{U} = (U_1, U_2, 0)$, (see Theorem 8.15 on page 278); hence the set of all vectors $t\boldsymbol{U}$ simply consists of all vectors of the form $(X,Y,0)$. From Assertion 9.3, the desired tangent plane consists of all points

$$\boldsymbol{X}_0 + t\boldsymbol{U} = (X,Y,2)$$

which is the same as the set of all (x,y,z) for which $z = 2$, as claimed.

■

Example 9.2: Tangent to sphere $x^2 + y^2 + z^2 = 2^2$ at $(1,1,\sqrt{2})$

Here we just search for the vectors $\boldsymbol{H}$, perpendicular to the gradient; the $\boldsymbol{H}$ needn't be unit vectors, because any vector, $t\boldsymbol{U}$, with $\boldsymbol{U}$ a unit vector perpendicular to the gradient, is just an arbitrary vector $\boldsymbol{H}$, perpendicular to the gradient. By Assertion 9.3 on page 291, the tangent plane to the sphere given by

$$g(x,y,z) = x^2 + y^2 + z^2 - 4 = 0$$

at $(1,1,\sqrt{2})$ consists of all points of the form $(1,1,\sqrt{2}) + \boldsymbol{H}$, where

$$\boldsymbol{H} \perp \textit{\textbf{grad}}\ g(1,1,\sqrt{2}). \qquad \textbf{\textit{9.14}}$$

Since, as we already saw, $\textit{\textbf{grad}}\ g(x,y,z) = (2x, 2y, 2z)$, we have

$$\textit{\textbf{grad}}\ g(1,1,\sqrt{2}) = (2,2,2\sqrt{2}).$$

From Theorem 8.15 on page 278 we see that orthogonality condition 9.14 gets translated to the equation

$$\boldsymbol{H} \cdot \textit{\textbf{grad}}\ g(1,1,\sqrt{2}) = 0,$$

or, using the definition of dot product, Definition 8.10 on page 272,

$$2H_1 + 2H_2 + 2\sqrt{2}\,H_3 = 0, \qquad \textbf{\textit{9.15}}$$

where

$$\boldsymbol{H} = (H_1, H_2, H_3). \qquad \textbf{\textit{9.16}}$$

Equation 9.15 yields

$$H_3 = -\frac{H_1 + H_2}{\sqrt{2}}. \qquad \textbf{\textit{9.17}}$$

Letting $t = \|\boldsymbol{H}\|$, and $\boldsymbol{U} = \dfrac{\boldsymbol{H}}{\|\boldsymbol{H}\|}$, we see that the desired tangent plane to the sphere $g(x, y, z) = x^2 + y^2 + z^2 - 4 = 0$ at $(1,1,\sqrt{2})$ consists of all points of the form

$$(1,1,\sqrt{2}) + t\,\boldsymbol{U} = (1,1,\sqrt{2}) + \boldsymbol{H}, \qquad \textbf{\textit{9.18}}$$

where $\boldsymbol{H}$ ***satisfies*** equations 9.16 and 9.17.

It is somewhat illuminating to write this as the set of all points of the form

$$(1,1,\sqrt{2})+(H_1,H_2,H_3)=(1,1,\sqrt{2})+H_1\left(1,0,\frac{-1}{\sqrt{2}}\right)+H_2\left(0,1,\frac{-1}{\sqrt{2}}\right),$$

in which we may think of the vectors $\left(1, 0, \frac{-1}{\sqrt{2}}\right)$ and $\left(0, 1, \frac{-1}{\sqrt{2}}\right)$ as *non-orthogonal* coordinate axis vectors.

■

Exercises 9.3

Using Example 9.2 as a model, construct the following:

1. The tangent line to the ellipse $x^2 + 4y^2 = 9$ at the points
   a. $(0,3/2)$
   b. $(0,-3/2)$
   c. $(3,0)$
   d. $(1,2)$
   e. $(-1,-2)$
2. The tangent plane to the ellipsoid $x^2 + 2y^2 + 3z^2 = 6$
   a. at $(1,1,1)$
   b. at $(1,-1,-1)$
   c. Approximately what is the value of $-\sqrt{(6-x^2-2y^2)}$ for $x = 1.1,\ y = -1.2$?

2 Tangent Hyperplane Coordinate Systems

Our investigations have led us to the problem of constructing coordinate systems for certain sets of vectors. So far these sets have come in two forms — the first being one consisting of all vectors of the form

$$(\boldsymbol{x}_0, f(\boldsymbol{x}_0)) + (\boldsymbol{h}, \boldsymbol{h} \cdot \boldsymbol{grad}\ f(\boldsymbol{x}_0))$$

and the second form consisting of all vectors of the form

$$(\boldsymbol{x}_0, f(\boldsymbol{x}_0)) + \boldsymbol{H}$$

where $\boldsymbol{H}$ is orthogonal to $\boldsymbol{grad}\ f(\boldsymbol{x}_0)$ (or more generally to every vector in some set S).

The second type of coordinate system construction will be treated when we deal with the topic of orthogonal complements. But before we can do this properly, we have to build the framework for handling construction of useful coordinate systems for sets of vectors such as those of the form

$$(\boldsymbol{h}, \boldsymbol{h} \cdot \boldsymbol{grad}\ f(\boldsymbol{x}_0)) = \left(\sum_{k=1}^{n} h_k \boldsymbol{e}_k, \sum_{k=1}^{n} h_k \boldsymbol{e}_k \cdot \boldsymbol{grad}\ f(\boldsymbol{x}_0) \right).$$

The first point to realize is that such a collection consists of an infinite number of vectors — one for each possible choice of $\boldsymbol{h}$. However, as we already observed, the structure of this set is not infinitely complicated, because each such vector can be written in the form

$$\begin{aligned}(\boldsymbol{h}, \boldsymbol{h} \cdot \boldsymbol{grad}\ f(\boldsymbol{x}_0)) &= \left(\sum_{k=1}^{n} h_k \boldsymbol{e}_k, \sum_{k=1}^{n} h_k \boldsymbol{e}_k \cdot \boldsymbol{grad}\ f(\boldsymbol{x}_0) \right) \\ &= \sum_{k=1}^{n} h_k (\boldsymbol{e}_k, \boldsymbol{e}_k \cdot \boldsymbol{grad}\ f(\boldsymbol{x}_0)),\end{aligned}$$

where the vectors $\boldsymbol{e}_k$ are the standard unit vectors, $\boldsymbol{e}_1 = (1, 0, 0, ..., 0)$, etc.

That is, the collection of vectors we're interested in, is the set of all *linear combinations,* (weighted sums)

$$(\boldsymbol{h}, \boldsymbol{h} \cdot \boldsymbol{grad}\ f(\boldsymbol{x}_0)) = \sum_{k=1}^{n} h_k \boldsymbol{v}_k \qquad \textbf{\textit{9.19}}$$

of the fixed vectors

$$\boldsymbol{v}_k = (\boldsymbol{e}_k, \boldsymbol{e}_k \cdot \boldsymbol{grad}\ f(\boldsymbol{x}_0)), \quad k = 1, 2, ..., n.$$

We will see that often the vectors $\boldsymbol{v}_1, \boldsymbol{v}_2, ..., \boldsymbol{v}_n$ form what is called a basis for a coordinate system for the set of vectors of the form $(\boldsymbol{h}, \boldsymbol{h} \cdot \boldsymbol{grad}\ f(\boldsymbol{x}_0))$. The values $h_1, h_2, ..., h_n$ are the *coordinates* of the vector $(\boldsymbol{h}, \boldsymbol{h} \cdot \boldsymbol{grad}\ f(\boldsymbol{x}_0))$ with respect to the basis vectors $\boldsymbol{v}_1, \boldsymbol{v}_2, ..., \boldsymbol{v}_n$, meaning simply that equation 9.19 is satisfied.

Rather shortly, we will become more precise about the general meaning of a *basis* and *coordinates.*

It will frequently be the case that the interesting vector collections will be the set of all linear combinations of some given sequence, $\boldsymbol{v}_1, \boldsymbol{v}_2, ..., \boldsymbol{v}_j$, of vectors. For this reason we formally introduce the following notation.

Definition 9.4: Linear combination, subspace

Let $\boldsymbol{v}_1, \boldsymbol{v}_2,...,\boldsymbol{v}_j$ be some given sequence of n-dimensional vectors (see Definition 8.1 on page 261).

Any vector of the form

$$\sum_{i=1}^{j} a_i \boldsymbol{v}_i ,$$

(see Definition 8.2 on page 263),where the a_i are real numbers is called a (real) **linear combination** of the vectors $\boldsymbol{v}_1, \boldsymbol{v}_2,...,\boldsymbol{v}_j$.

The set of all linear combinations

$$\sum_{i=1}^{j} a_i \boldsymbol{v}_i$$

where the a_i are arbitrary real numbers, is called the **subspace generated by** (or **spanned by**) $\boldsymbol{v}_1, \boldsymbol{v}_2,...,\boldsymbol{v}_j$, and is denoted here by the symbol

$$[[\boldsymbol{v}_1, \boldsymbol{v}_2,...,\boldsymbol{v}_j]]. \qquad \textit{9.20}$$

❑

Notation 9.5

When we write out a vector $\boldsymbol{v}_k$ as an n-tuple, for readability, and convenience, we may write it in *column* form — i.e., write it as

$$\begin{pmatrix} v_{k1} \\ v_{k2} \\ . \\ . \\ v_{kn} \end{pmatrix}, \text{ or } \begin{pmatrix} a \\ b \\ c \end{pmatrix} \text{ rather than } (v_{k1}, v_{k2},....,v_{kn}) \text{ or } (a, b, c).$$

The less sloppy way of doing this will be introduced when we get to matrices and proper matrix notation.

❑

Example 9.4: Subspace spanned by $\boldsymbol{e}_1, \boldsymbol{e}_2$

Let $S = [[\begin{pmatrix} 1 \\ 0 \\ 0 \end{pmatrix}, \begin{pmatrix} 0 \\ 1 \\ 0 \end{pmatrix}]]$. Geometrically this is the set of all three-dimensional vectors of the form

$$x\begin{pmatrix}1\\0\\0\end{pmatrix} + y\begin{pmatrix}0\\1\\0\end{pmatrix} = \begin{pmatrix}x\\y\\0\end{pmatrix},$$

that is, the *x-y base plane.* This is the plane through the origin $\begin{pmatrix}0\\0\\0\end{pmatrix}$ and the points $\begin{pmatrix}1\\0\\0\end{pmatrix}$ and $\begin{pmatrix}0\\1\\0\end{pmatrix}$.

■

Generally we tend to picture a subspace as a line or a plane through the origin in three-dimensional space, although this is only suggestive in higher dimensions (i.e., when dealing with vectors having more than three coordinates). If we are given a subspace, $[[\boldsymbol{v}_1, \boldsymbol{v}_2,...,\boldsymbol{v}_j]]$ of n-dimensional vectors, as already mentioned, we may find this original representation of S inconvenient for some problems. We will now make a start on changing our coordinate systems for subspaces, to more convenient ones. First, we observe that if one of the $\boldsymbol{v}_i$ is itself a linear combination of the others, then it can be discarded from the sequence defining S. We state this formally as follows.

Theorem 9.6: Redundant vectors in subspace determination

Given the subspace $S = [[\boldsymbol{v}_1, \boldsymbol{v}_2,...,\boldsymbol{v}_j]]$ of real n-dimensional vectors (Definition 9.4 on page 296), if for some subscript i, and real numbers b_k, $k \neq i,\ 1 \leq k \leq j,$ we have

$$\boldsymbol{v}_i = \sum_{k \neq i} b_k \boldsymbol{v}_k, \qquad \textbf{9.21}$$

then S is also generated by the sequence obtained by deleting the i-th element $\boldsymbol{v}_i$ from the sequence $\boldsymbol{v}_1, \boldsymbol{v}_2,...,\boldsymbol{v}_j$.

□

You should be able to convince yourself of the validity of this result, by noting that for any vector, $\boldsymbol{v}$ in S, we may write

$$\boldsymbol{v} = \sum_{k=1}^{j} a_k \boldsymbol{v}_k = \sum_{k \neq i} a_k \boldsymbol{v}_k + a_i \sum_{k \neq i} b_k \boldsymbol{v}_k = \sum_{k \neq i} (a_k + a_i b_k) \boldsymbol{v}_k.$$

This discussion leads to the following definition.

Definition 9.7: Linear dependence, linearly independent sequence

If $\boldsymbol{v}_1, \boldsymbol{v}_2,...,\boldsymbol{v}_j$ is a given sequence of real n-dimensional vectors (Definition 8.1 on page 261), and for some subscript i, and real numbers b_k, $k \neq i,\ 1 \leq k \leq j,$ we have

$$v_i = \sum_{k \neq i} b_k v_k,$$

we say that the vector v_i is **linearly dependent** on the vectors v_k, $k \neq i$. If none of the v_k is linearly dependent on the remaining ones from the sequence $v_1, v_2, ..., v_j$, then the (entire) sequence $v_1, v_2, ..., v_j$ is said to be **linearly independent** — (abbreviated LI). Otherwise, the sequence $v_1, v_2, ..., v_j$ is called **linearly dependent** (LD).

❑

It seems reasonable to have as few vectors as possible in generating a subspace *S*. That implies that we should attempt to work solely with linearly independent generating sequences. In line with the feeling that the preferred generating sequences are those with the fewest vectors, we define the dimension of a subspace *S* as the smallest number of vectors a generating sequence can have. More precisely, dimension is defined as follows.

Definition 9.8: Dimension of a subspace

Suppose *S* is the subspace (Definition 9.4 on page 296) generated by a sequence $v_1, v_2, ..., v_j$ of real *n*-dimensional vectors. If *S* does not consist solely of the **0** vector, the **dimension of** *S*, written

$$dim\ S$$

is the smallest positive integer, q, for which

$$S = [[w_1, ..., w_q]].$$

If *S* consists solely of the **0** vector, then *dim S* is defined to be 0.

❑

So ***dim S* is the smallest number of vectors which can be used to form a coordinate system for *S*.**

Definition 9.9: Coordinates

If v is in the subspace *S* generated by the sequence $v_1, v_2, ..., v_j$ (Definition 9.4 on page 296), of real *n*-dimensional vectors and v can be written

$$v = \sum_{k=1}^{j} a_k v_k,$$

the a_k are called **coordinates of v with respect to the generating sequence $v_1, v_2, ..., v_j$**.

❑

We note that if $\boldsymbol{v}_1, \boldsymbol{v}_2,...,\boldsymbol{v}_j$ is a linearly independent sequence, then the coordinates of each $\boldsymbol{v}$ in S with respect to $\boldsymbol{v}_1, \boldsymbol{v}_2,...,\boldsymbol{v}_j$ are unique, because suppose we had two coordinates sequences for $\boldsymbol{v}$:

$$\boldsymbol{v} = \sum_{k=1}^{j} a_k \boldsymbol{v}_k \quad \text{and} \quad \boldsymbol{v} = \sum_{k=1}^{j} b_k \boldsymbol{v}_k .$$

Then

$$\mathbf{0} = \boldsymbol{v} - \boldsymbol{v} = \sum_{k=1}^{j} (a_k - b_k)\boldsymbol{v}_k . \qquad \textit{9.22}$$

But if the two coordinate sequences were not identical, then for at least one value of k, say k_o, we would have

$$a_{k_o} \neq b_{k_o} .$$

Using equation 9.22 we could then write

$$\boldsymbol{v}_{k_o} = \sum_{k \neq k_o} \frac{a_k - b_k}{a_{k_o} - b_{k_o}} \boldsymbol{v}_k ,$$

which implies that $\boldsymbol{v}_1, \boldsymbol{v}_2,...,\boldsymbol{v}_j$ could not be a linearly independent sequence, as originally assumed. To avoid this contradiction, we must admit that all coordinate sequences for $\boldsymbol{v}$ with respect to linearly independent $\boldsymbol{v}_1, \boldsymbol{v}_2,...,\boldsymbol{v}_j$ are the same; i.e., coordinates with respect to $\boldsymbol{v}_1, \boldsymbol{v}_2,...,\boldsymbol{v}_j$ are unique. The property of uniqueness of coordinates is usually desirable — in the sense that a lack of uniqueness tends to muddy the waters. We now state the above result formally.

Theorem 9.10: Coordinate uniqueness for LI sequences

Let $\boldsymbol{v}_1, \boldsymbol{v}_2,...,\boldsymbol{v}_j$ be a linearly independent sequence of real n-dimensional vectors (Definition 9.7 on page 297). The coordinates of any vector $\boldsymbol{v}$ in the subspace $[[\boldsymbol{v}_1, \boldsymbol{v}_2,...,\boldsymbol{v}_j]]$ with respect to the sequence $\boldsymbol{v}_1, \boldsymbol{v}_2,...,\boldsymbol{v}_j$ (see Definition 9.9 on page 298) are unique.

❑

Definition 9.11: Basis

The sequence $\boldsymbol{v}_1, \boldsymbol{v}_2,...,\boldsymbol{v}_j$ of real n-dimensional vectors is called a **basis for the subspace** S (Definition 9.4 on page 296), if $S = [[\boldsymbol{v}_1, \boldsymbol{v}_2,...,\boldsymbol{v}_j]]$ and $\boldsymbol{v}_1, \boldsymbol{v}_2,...,\boldsymbol{v}_j$ is a linearly independent sequence (Definition 9.7 on page 297).

❑

We now come to a result which you may have already taken for granted, but which, in fact, isn't immediately obvious, namely, that all bases for a subspace S consist of the same number of vectors — this number being *dim S*.

Note that if we can establish that all bases for any subspace S have the same number of vectors, then this number must be $dim\ S$ — because from Definition 9.8 on page 298, of $dim\ S$, there is some sequence $\boldsymbol{w}_1, \ldots, \boldsymbol{w}_q$ such that

$$S = [\![\boldsymbol{w}_1, \ldots, \boldsymbol{w}_q]\!] \text{ and } q = dim\ S. \qquad \textbf{9.23}$$

But $w_1, \ldots, w_q$ must be a linearly independent sequence (Definition 9.7 on page 297) or we would be able to write for some r that

$$\boldsymbol{w}_r = \sum_{i=1}^{q} c_i \boldsymbol{w}_i.$$

Were this to hold, we know from Theorem 9.6 on page 297 that $\boldsymbol{w}_r$ could be eliminated from the sequence $\boldsymbol{w}_1, \ldots, \boldsymbol{w}_q$, with the vectors remaining still generating S, thus showing that $dim\ S < q$. This would contradict the valid equations 9.23. The only way to eliminate this contradiction is to admit that **if** all bases for S have the same number of vectors, this number must be $dim\ S$.

It remains to show that all bases for any given subspace have the same number of vectors. The proof we are presenting depends on the following ***innocent looking lemma***.

Lemma 9.12

If $\boldsymbol{v}_1, \boldsymbol{v}_2, \ldots, \boldsymbol{v}_j$ is a linearly dependent sequence, (Definition 9.7 on page 297) and $\boldsymbol{v}_1 \neq \boldsymbol{0}$, then at least one of the $\boldsymbol{v}_k$ is linearly dependent on the $\boldsymbol{v}_i$ **preceding** it in the sequence $\boldsymbol{v}_1, \boldsymbol{v}_2, \ldots, \boldsymbol{v}_j$ — i.e., for some $k > 1$,

$$\boldsymbol{v}_k = \sum_{i<k} a_i \boldsymbol{v}_i. \qquad \textbf{9.24}$$

❑

Before establishing Lemma 9.12, we give the following illustrative example, which is just a miniature version of the general proof to be presented immediately following.

Example 9.5: Special case of Lemma 9.12

Suppose we know that $\boldsymbol{v}_1, \boldsymbol{v}_2, \boldsymbol{v}_3, \boldsymbol{v}_4, \boldsymbol{v}_5$ is a linearly dependent sequence because $\boldsymbol{v}_2$ is linearly dependent on the remaining $\boldsymbol{v}_k$, say,

$$\boldsymbol{v}_2 = 3\boldsymbol{v}_1 - 2\boldsymbol{v}_3 - 5\boldsymbol{v}_4.$$

In this case, solve for $\boldsymbol{v}_4$, obtaining

$$v_4 = \frac{3}{5}v_1 - \frac{1}{5}v_2 - \frac{2}{5}v_3,$$

in the form desired.

■

The general proof of Lemma 9.12 is as follows:

Since $\boldsymbol{v}_1, \boldsymbol{v}_2,...,\boldsymbol{v}_j$ is a linearly dependent sequence, from the definition of linear dependence, Definition 9.7 on page 297, for some i

$$\boldsymbol{v}_i = \sum_{k \neq i} b_k \boldsymbol{v}_k \qquad \textbf{9.25}$$

If all of the b_k are 0, then we know that $i \neq 1$, because the hypotheses of Lemma 9.12 exclude the case $\boldsymbol{v}_1 = \boldsymbol{0}$; So if all b_k are 0, we may conclude that for some $k > 1$, $\boldsymbol{v}_k = \boldsymbol{0}$, in which case it is certainly true that equation 9.24 holds, just by taking all $a_i = 0$ for $i < k$. If the b_k are not all 0, then examining equation 9.25 we let k_o denote the largest subscript k for which $b_k \neq 0$. If $k_o < i$ then equation 9.25 is already the desired representation. If $k_o > i$, then solve equation 9.25 for $\boldsymbol{v}_{k_o}$ to obtain

$$\boldsymbol{v}_{k_o} = \frac{1}{b_{k_o}}\boldsymbol{v}_i - \sum_{k < k_o} \frac{b_k}{b_{k_o}}\boldsymbol{v}_k,$$

thus establishing Lemma 9.12.

We can now go on to establish that all bases for a subspace S consist of *dim S* vectors. All we need show is that if both

$$S = [[\boldsymbol{v}_1, \boldsymbol{v}_2...,\boldsymbol{v}_j]] \quad \text{and} \quad S = [[\boldsymbol{w}_1, \boldsymbol{w}_2,...,\boldsymbol{w}_m]]$$

with $\boldsymbol{v}_1, \boldsymbol{v}_2,...,\boldsymbol{v}_j$ and $\boldsymbol{w}_1, \boldsymbol{w}_2,...,\boldsymbol{w}_m$ both being linearly independent sequences, then $j = m$.

To do this we will show that neither $j > m$ nor $j < m$ can occur. If it were the case the $j > m$, then since $\boldsymbol{v}_m$ is a vector in $S = [[\boldsymbol{w}_1, \boldsymbol{w}_2,...,\boldsymbol{w}_m]]$ (see Definition 9.4 on page 296), it follows that

$$S = [[\boldsymbol{v}_m, \boldsymbol{w}_1, \boldsymbol{w}_2,...,\boldsymbol{w}_m]].$$

But since $S = [[\boldsymbol{w}_1, \boldsymbol{w}_2,...,\boldsymbol{w}_m]]$ and $\boldsymbol{v}_m$ is in S, the sequence $\boldsymbol{v}_m, \boldsymbol{w}_1, \boldsymbol{w}_2,...,\boldsymbol{w}_m$ is linearly dependent (Definition 9.7 on page 297). Furthermore $\boldsymbol{v}_m \neq \boldsymbol{0}$ (or $\boldsymbol{v}_1, \boldsymbol{v}_2,...,\boldsymbol{v}_j$ would be a linearly dependent sequence. Thus by Lemma 9.12 at least one of the $\boldsymbol{w}_i$, say $\boldsymbol{w}_r$, is linearly dependent on the vectors preceding it in the sequence $\boldsymbol{v}_m, \boldsymbol{w}_1, \boldsymbol{w}_2,...,\boldsymbol{w}_m$, and thus may be discarded, without any loss — i.e., still

$$S = [[\boldsymbol{v}_m, \boldsymbol{w}_1, \boldsymbol{w}_2,...,\boldsymbol{w}_{r-1}, \boldsymbol{w}_{r+1},...,\boldsymbol{w}_m]].$$

We next *prepend* $\boldsymbol{v}_{m-1}$ to the sequence

$$\boldsymbol{v}_m, \boldsymbol{w}_1, \boldsymbol{w}_2,...,\boldsymbol{w}_{r-1}, \boldsymbol{w}_{r+1},...,\boldsymbol{w}_m,$$

obtaining, as before

$$S = [[\boldsymbol{v}_{m-1}, \boldsymbol{v}_m, \boldsymbol{w}_1, \boldsymbol{w}_2,...,\boldsymbol{w}_{r-1}, \boldsymbol{w}_{r+1},...,\boldsymbol{w}_m]].$$

Since the first two vectors in the above sequence are linearly independent (because $v_1, v_2,...,v_j$ is assumed linearly independent) just as before, we can eliminate another of the w_i. We continue this process, until all of the w_i are eliminated, leaving us with the result that

$$S = [\![v_1, v_2,...,v_m]\!].$$

Thus, if $j > m$, our initial assumptions that $v_1, v_2,...,v_j$ is a basis for S would imply that

$$S = [\![v_1, v_2,...,v_m]\!] \quad \text{as well as} \quad S = [\![v_1, v_2,...,v_j]\!].$$

But this would imply that v_j is linearly dependent on $v_1, v_2,...,v_m$, which for $j > m$ is not consistent with the linear independence of $v_1, v_2,...,v_j$. So we have to discard the assumption $j > m$. The case $j < m$ is identical. Thus we conclude that $j = m$.

It can now be shown that if *dim* $S = k$, and $V_1,...,V_k$ is a linearly independent sequence of vectors from S, then $V_1,...,V_k$ is a basis for S — by prepending one at a time the vectors $V_1,...,V_k$ to the basis $v_1,...,v_k$, and eliminating all the v_i, just as we did previously, obtaining $S = [\![V_1,...,V_k]\!]$.

Theorem 9.13: Properties of bases

Referring to Definition 9.4 on page 296 (subspace) Definition 9.8 on page 298 (dimension, *dim S*, of a subspace), and Definition 9.11 on page 299 (basis of a subspace),

> All bases for any given subspace made up of n-dimensional real vectors consist of sequences made up of precisely *dim S* vectors. If *dim* $S = k$, and $V_1,...,V_k$ is a linearly independent sequence of vectors from S, then $V_1,...,V_k$ is a basis for S.

❑

About the only *immediate* result that we can derive from this theorem is that the dimension of n-dimensional space is n. (We'll be a little more formal about that shortly.) So the material we have gone through so far in this chapter has very little practical application *at this instant.* As we will soon see, it serves as the absolutely necessary foundation required to investigate the main question of interest — *how to construct a good coordinate system.*

Definition 9.14: Real and Euclidean n-dimensional space

The set of all real n-dimensional vectors,[1] $x = (x_1, x_2, ... ,x_n)$ is called

---

1. Where such a vector is just a real valued sequence — which is a real valued function with integer domain — so that standard arithmetical operations involving functions are assumed — in particular multiplication by a real scalar and addition of vectors.

Real n-dimensional space, and is frequently denoted by

$$\mathcal{R}_n .$$

Real n-dimensional space with the definitions of distance and length (norm), given by Definition 8.6 on page 270, is called **(real) Euclidean n-dimensional space**, and is frequently denoted by.

$$\mathcal{E}_n .$$

❑

Recalling Definition 9.8 on page 298, and using Definition 9.14 above, it's now easy to use Theorem 9.13 to show the following.

Theorem 9.15: Dimension of $\mathcal{R}_n$

$$dim\,\mathcal{R}_n = n.$$

❑

To show that the dimension of real n-dimensional space is n we note simply that from equation 9.3 on page 287 it follows that

$$\mathcal{R}_n = [\![\boldsymbol{e}_1, \boldsymbol{e}_2, ..., \boldsymbol{e}_n]\!]$$

and it really is *obvious* that this sequence, $\boldsymbol{e}_1, \boldsymbol{e}_2, ..., \boldsymbol{e}_n$, of standard unit vectors (Definition 8.2 on page 263) is a linearly independent one — just try constructing any of the $\boldsymbol{e}_i$ as a linear combination of the remaining ones. Now apply Theorem 9.13.

One desirable property we would like for any sequence of basis vectors is that the coordinates of all vectors with respect to this basis be easy to determine. It will turn out that ***orthogonality of the basis vectors will ensure that this is the case***. First we show the following.

Theorem 9.16: Independence of nonzero orthogonal vectors

If $\boldsymbol{w}_1, \boldsymbol{w}_2, ..., \boldsymbol{w}_m$ is a sequence of nonzero pairwise orthogonal vectors (Definition 8.14 on page 278), then it is a linearly independent sequence (Definition 9.7 on page 297).

❑

This follows from the fact that if one of these vectors happened to be linearly dependent on the remaining ones, say,

$$\boldsymbol{w}_r = \sum_{k \neq r} a_k \boldsymbol{w}_k, \qquad \textbf{\textit{9.26}}$$

then by taking dot products of both sides of this equation with an arbitrary one of the $\boldsymbol{w}_k$, say, $\boldsymbol{w}_q$, using the dot product condition for orthogonality, Theorem 8.15 on page 278, we have

$$\boldsymbol{w}_r \cdot \boldsymbol{w}_q = a_q \boldsymbol{w}_q \cdot \boldsymbol{w}_q.$$

From the assumed pairwise orthogonality, the left side is 0, while the assumption that the $\boldsymbol{w}_i$ are nonzero implies that $\boldsymbol{w}_q \cdot \boldsymbol{w}_q \neq 0$. We therefore conclude that for each q different from r, $a_q = 0$, from which it would follow from equation 9.26 that $\boldsymbol{w}_r = \mathbf{0}$. The assumption that led to this untenable result was that the sequence $\boldsymbol{w}_1, \boldsymbol{w}_2,...,\boldsymbol{w}_m$ could be linearly dependent. Thus Theorem 9.16 must be true.

Now we can show how easy it is to determine the coordinates of any vector with respect to an orthogonal basis. Suppose $\boldsymbol{w}_1, \boldsymbol{w}_2,...,\boldsymbol{w}_m$ is an orthogonal basis[1] for the subspace S. If $\boldsymbol{v}$ is an element of S, then we may write

$$\boldsymbol{v} = \sum_{i=1}^{m} a_i \boldsymbol{w}_i .$$

Take dot products of both sides of this equation, with an arbitrary one of the $\boldsymbol{w}_i$, say, $\boldsymbol{w}_q$ (which we can do, since it is assumed that we know the coordinates of all of these vectors in the original coordinate system) making use of the dot product condition for orthogonality, Theorem 8.15 on page 278, to conclude that

$$a_i = \frac{\boldsymbol{v} \cdot \boldsymbol{w}_i}{\boldsymbol{w}_i \cdot \boldsymbol{w}_i}.$$

Note that the denominator above cannot be 0, because $\boldsymbol{w}_1, \boldsymbol{w}_2,...,\boldsymbol{w}_m$ is assumed to be a basis for S, see Definition 9.11 on page 299.

We summarize the results of this discussion in the theorem to follow.

Theorem 9.17: Coordinates relative to orthogonal basis

If $S = [\![\boldsymbol{w}_1, \boldsymbol{w}_2,...,\boldsymbol{w}_m]\!]$ (Definition 9.4 on page 296) where the vectors $\boldsymbol{w}_i$ are mutually orthogonal (see Theorem 8.15 on page 278), and nonzero, and $\boldsymbol{v}$ is any vector in S, then the i-th coordinate of $\boldsymbol{v}$ with respect to the basis $\boldsymbol{w}_1, \boldsymbol{w}_2,...,\boldsymbol{w}_m$ is $\boldsymbol{v} \cdot \boldsymbol{w}_i / \boldsymbol{w}_i \cdot \boldsymbol{w}_i$; that is, if

$$\boldsymbol{v} = \sum_{i=1}^{m} a_i \boldsymbol{w}_i, \text{ then } a_i = \frac{\boldsymbol{v} \cdot \boldsymbol{w}_i}{\boldsymbol{w}_i \cdot \boldsymbol{w}_i}. \qquad \textbf{9.27}$$

❑

Now that we know what types of bases we're looking for, we're in a position to see how to construct them. The idea is to construct the desired basis, which we'll call $\boldsymbol{w}_1, \boldsymbol{w}_2,...,\boldsymbol{w}_p$, from a given generating sequence $\boldsymbol{v}_1, \boldsymbol{v}_2,...,\boldsymbol{v}_p$, one vector at a time, with $\boldsymbol{w}_1 = \boldsymbol{v}_1$ and for $j = 2 ,..., p,$ $\boldsymbol{w}_j$

1. That is, a sequence of pairwise orthogonal vectors which is a basis (Definition 9.11 on page 299) for the subspace S.

being the *component* of $\boldsymbol{v}_j$ perpendicular to $[\![\boldsymbol{v}_1, \boldsymbol{v}_2,...,\boldsymbol{v}_{j-1}]\!]$, as illustrated in Figure 9-3.

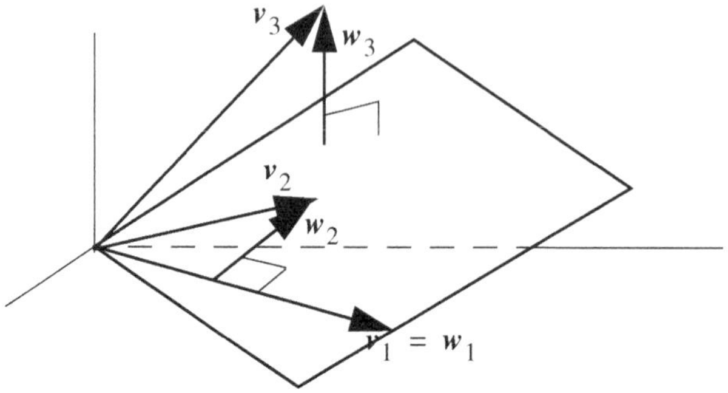

Figure 9-3

What we're doing here is getting the next vector by knocking off any component of the current old vector parallel to the previous ones. Because of the orthogonality of the previously constructed vectors, the component of $\boldsymbol{v}_3$ in the subspace (plane) generated by $\boldsymbol{w}_1$ and $\boldsymbol{w}_2$ turns out to be just the sum

$$P_{\boldsymbol{w}_1}(\boldsymbol{v}_3) + P_{\boldsymbol{w}_2}(\boldsymbol{v}_3).$$

of the projections of v_3 on w_1 and w_2.

The method just illustrated is called *Gram-Schmidt orthogonalization* and replaces each vector by the component of that vector which is perpendicular to all of the previous vectors in the original sequence. This is done by subtracting off from this vector the component in the hyperplane generated by the previous vectors. Algebraically the Gram-Schmidt procedure is defined as follows.

Definition 9.18: Gram-Schmidt orthogonalization of $\boldsymbol{v}_1, \boldsymbol{v}_2...,\boldsymbol{v}_p$

Given the sequence $\boldsymbol{v}_1, \boldsymbol{v}_2,...,\boldsymbol{v}_p$, of real n-dimensional vectors (see Definition 8.1 on page 261), recalling Definition 8.12 on page 276 of the projection $P_{\boldsymbol{Y}}(\boldsymbol{X})$, of $\boldsymbol{X}$ on $\boldsymbol{Y}$, let

$$\boldsymbol{w}_1 = \boldsymbol{v}_1 \quad \text{and} \quad \boldsymbol{w}_{j+1} = \boldsymbol{v}_{j+1} - \sum_{i \le j} P_{\boldsymbol{w}_i}(\boldsymbol{v}_{j+1}) \qquad \textit{9.28}$$

$$\text{for } j = 1,2,...,p-1.$$

This process which generates the sequence $\boldsymbol{w}_1, \boldsymbol{w}_2,...,\boldsymbol{w}_p$ is called **Gram-Schmidt orthogonalization of** $\boldsymbol{v}_1, \boldsymbol{v}_2,...,\boldsymbol{v}_p$. Computation of the projection $P_{\boldsymbol{Y}}(\boldsymbol{X})$ is given by Theorem 8.13 on page 278.

❑

Notice that this procedure is the natural extension of the procedure of the *orthogonal breakup* of Theorem 8.16 on page 278.

Example 9.6: Orthogonalization of $[\![\begin{pmatrix}1\\1\\1\end{pmatrix}, \begin{pmatrix}1\\0\\1\end{pmatrix}, \begin{pmatrix}0\\1\\0\end{pmatrix}]\!]$

Here

$$v_1 = \begin{pmatrix}1\\1\\1\end{pmatrix},\ v_2 = \begin{pmatrix}1\\0\\1\end{pmatrix} \text{ and } v_3 = \begin{pmatrix}0\\1\\0\end{pmatrix}.$$

$$w_1 = v_1 = \begin{pmatrix}1\\1\\1\end{pmatrix}, \qquad w_2 = v_2 - P_{v_2}(v_2) = \begin{pmatrix}1\\0\\1\end{pmatrix} - \frac{v_2 \cdot w_1}{w_1 \cdot w_1} w_1$$

$$= \begin{pmatrix}1\\0\\1\end{pmatrix} - \frac{2}{3}\begin{pmatrix}1\\1\\1\end{pmatrix} = \begin{pmatrix}1/3\\-2/3\\1/3\end{pmatrix} = \frac{1}{3}\begin{pmatrix}1\\-2\\1\end{pmatrix}$$

and

$$w_3 = v_3 - P_{w_1}(v_3) - P_{w_2}(v_3)$$

$$= \begin{pmatrix}0\\1\\0\end{pmatrix} - \frac{v_3 \cdot w_1}{w_1 \cdot w_1} w_1 - \frac{v_3 \cdot w_2}{w_2 \cdot w_2} w_2$$

$$= \begin{pmatrix}0\\1\\0\end{pmatrix} - \frac{1}{3}\begin{pmatrix}1\\1\\1\end{pmatrix} - \frac{1}{3}\begin{pmatrix}1\\2\\-1\end{pmatrix} = \begin{pmatrix}0\\0\\0\end{pmatrix}$$

Note that $w_1 \perp w_2$ and w_3 is orthogonal to all vectors.

■

For many purposes it is desirable to discard any w_j that are $\mathbf{0}$ and *normalize* the remaining w_j — i.e., replace them by unit length vectors in the same direction. This prompts Definition 9.19.

Definition 9.19: Gram-Schmidt orthornomalization

If the Gram-Schmidt orthogonalization procedure, Definition 9.18 on page 305, is modified by discarding all of the newly constructed w_j which are $\mathbf{0}$ (renumbering what vectors remain) and replacing each nonzero w_j with

$$u_j = \frac{w_j}{\|w_j\|} = \frac{w_j}{\sqrt{w_j \cdot w_j}},$$

(recall Definition 8.10 on page 272) the resulting procedure is called **Gram-Schmidt orthonormalization**, or simply **orthonormalization**.

❑

The advantages of an orthonormal basis are most easily seen from the simplification that the normalization provides in determining coordinates with respect to this basis — see equation 9.27 on page 304, where the denominator $\boldsymbol{w}_i \cdot \boldsymbol{w}_i$ can be eliminated because it equals 1.

Exercises 9.7

1. Orthogonalize and orthonormalize the following sequences

a. $\begin{pmatrix}1\\1\\1\end{pmatrix}, \begin{pmatrix}0\\4\\2\end{pmatrix}, \begin{pmatrix}-1\\-5\\-3\end{pmatrix}$

b. $\begin{pmatrix}3\\-4\\3\end{pmatrix}, \begin{pmatrix}2\\3\\2\end{pmatrix}, \begin{pmatrix}1\\0\\-1\end{pmatrix}$

c. $\begin{pmatrix}1\\1\\1\end{pmatrix}, \begin{pmatrix}0\\2\\1\end{pmatrix}, \begin{pmatrix}3\\1\\-1\end{pmatrix}$

2. What is the dimension of the subspaces generated by the vector sequences of the previous exercises?

We have already indicated that the Gram-Schmidt orthogonalization procedure (at least in theory) produces orthogonal vectors $\boldsymbol{w}_1, \boldsymbol{w}_2,...,\boldsymbol{w}_p$ from the sequence $\boldsymbol{v}_1, \boldsymbol{v}_2...,\boldsymbol{v}_p$. Furthermore, we shall see that, excluding the trivial case $\boldsymbol{v}_1 = \boldsymbol{0}$, $\boldsymbol{w}_j = \boldsymbol{0}$ if and only if $\boldsymbol{v}_j$ is linearly dependent on $\boldsymbol{v}_1, \boldsymbol{v}_2,...,\boldsymbol{v}_{j-1}$. Finally, we shall show that for each j, the sequences $\boldsymbol{v}_1, \boldsymbol{v}_2...,\boldsymbol{v}_j$ and $\boldsymbol{w}_1, \boldsymbol{w}_2,...,\boldsymbol{w}_j$ generate the same subspace, as we would expect from the geometric picture of the Gram-Schmidt construction.

So let's now establish the asserted theoretical properties of the Gram-Schmidt procedure.

Theorem 9.20: Properties of the Gram-Schmidt procedure

The vectors, $\boldsymbol{w}_1, \boldsymbol{w}_2,...,\boldsymbol{w}_p$ generated by the Gram-Schmidt procedure, Definition 9.18 on page 305, are mutually orthogonal (Definition 8.14 on page 278 and Theorem 8.15 on page 278). For each $j = 1,...,p$

$$[\![\boldsymbol{v}_1, \boldsymbol{v}_2...,\boldsymbol{v}_j]\!] = [\![\boldsymbol{w}_1, \boldsymbol{w}_2,...,\boldsymbol{w}_j]\!] \qquad \textit{9.29}$$

and for $1 \le j < p$,

$w_{j+1} = 0$ if and only if v_{j+1} belongs to $[\![v_1, v_2 ..., v_j]\!]$ (Definition 9.4 on page 296).

❑

We will first establish equation 9.29.

Certainly $[\![v_1]\!] = [\![w_1]\!]$, since $v_1 = w_1$.

Next we show that for any $j = 1,2,...,p-1$ for which

$$[\![v_1, v_2 ..., v_j]\!] = [\![w_1, w_2, ..., w_j]\!] \qquad \textit{9.30}$$

we have

$$w_{j+1} \text{ belongs to } [\![v_1, v_2, ..., v_j, v_{j+1}]\!] \qquad \textit{9.31}$$

and

$$v_{j+1} \text{ belong to } [\![w_1, w_2, ..., w_j, w_{j+1}]\!], \qquad \textit{9.32}$$

from which, equation 9.29 will follow by induction.[1]

To accomplish this, we repeat equation 9.28

$$w_{j+1} = v_{j+1} - \sum_{i \le j} P_{w_i}(v_{j+1})$$

and apply the formula for projection, Theorem 8.13 on page 276, to the w_i above

$$P_{w_i}(v_{j+1}) = \begin{cases} \dfrac{v_{j+1} \cdot w_i}{w_i \cdot w_i} & \text{if } w_i \neq \mathbf{0} \\ \mathbf{0} & \text{otherwise.} \end{cases} \qquad \textit{9.33}$$

which indeed shows that condition 9.31 follows from equation 9.30.

On the other hand, rewriting the second of equations 9.28 on page 305 in the form

$$v_{j+1} = w_{j+1} + \sum_{i \le j} P_{w_i}(v_{j+1})$$

and substituting from the right side of equation 9.33 leads immediately to the conclusion that result 9.32 follows from equation 9.30, thus fully establishing equation 9.29.

Next we show that $w_i \perp w_j$ whenever $i \neq j$, as follows:

---

1. Equations 9.31 and 9.32 are usually abbreviated to $w_{j+1} \in [\![v_1, v_2, ..., v_j, v_{j+1}]\!]$ and $w_{j+1} \in [\![v_1, v_2, ..., v_j, v_{j+1}]\!]$, a practice we shall follow when it is desirable.

First recall Theorem 8.15 on page 278, relating orthogonality to the dot product being 0, and note that since all vectors are perpendicular to the $\mathbf{0}$ vector, we need only consider nonzero $\boldsymbol{w}$s. Since the proof will be inductive, we first compute

$$\begin{aligned}
\boldsymbol{w}_1 \cdot \boldsymbol{w}_2 &= \boldsymbol{v}_1 \cdot (\boldsymbol{v}_2 - P_{\boldsymbol{w}_1}(\boldsymbol{v}_2)) \\
&= \boldsymbol{v}_1 \cdot \left(\boldsymbol{v}_2 - \frac{\boldsymbol{v}_2 \cdot \boldsymbol{v}_1}{\boldsymbol{v}_1 \cdot \boldsymbol{v}_1}\right)\boldsymbol{v}_1 \\
&= \boldsymbol{v}_1 \cdot \boldsymbol{v}_2 - \frac{\boldsymbol{v}_2 \cdot \boldsymbol{v}_1}{\boldsymbol{v}_1 \cdot \boldsymbol{v}_1}\boldsymbol{v}_1 \cdot \boldsymbol{v}_1 \\
&= 0.
\end{aligned}$$

Now examining any $m = 2,3,...,$ for which $\boldsymbol{w}_1, \boldsymbol{w}_2,...,\boldsymbol{w}_m$ are known to be mutually perpendicular, we see that for $j \le m$ and $\boldsymbol{w}_j \neq \mathbf{0}$

$$\boldsymbol{w}_{m+1} \cdot \boldsymbol{w}_j = \boldsymbol{v}_{m+1} \cdot \boldsymbol{w}_j - \sum_{\substack{i \le m \\ \boldsymbol{w}_i \neq \mathbf{0}}} \frac{\boldsymbol{v}_{m+1} \cdot \boldsymbol{w}_i}{\boldsymbol{w}_i \cdot \boldsymbol{w}_i} \boldsymbol{w}_i \cdot \boldsymbol{w}_j .$$

The only nonzero term in the right hand sum is the one with $i = j,$ because of the (inductively assumed) orthogonality of $\boldsymbol{w}_1, \boldsymbol{w}_2,...,\boldsymbol{w}_m.$ So

$$\boldsymbol{w}_{m+1} \cdot \boldsymbol{w}_j = \boldsymbol{v}_{m+1} \cdot \boldsymbol{w}_j - \frac{\boldsymbol{v}_{m+1} \cdot \boldsymbol{w}_j}{\boldsymbol{w}_j \cdot \boldsymbol{w}_j} \boldsymbol{w}_j \cdot \boldsymbol{w}_j = 0,$$

showing that whenever $\boldsymbol{w}_1, \boldsymbol{w}_2,...,\boldsymbol{w}_m$ are mutually orthogonal, so are $\boldsymbol{w}_1, \boldsymbol{w}_2,...,\boldsymbol{w}_m, \boldsymbol{w}_{m+1}.$ Hence, it follows by induction that $\boldsymbol{w}_1, \boldsymbol{w}_2,...,\boldsymbol{w}_p$ are mutually orthogonal.

To complete the proof we show that for $1 \le j < p$

$$\boldsymbol{w}_{j+1} = 0 \text{ if and only if } \boldsymbol{v}_{j+1} \text{ belongs to } [\![\boldsymbol{v}_1, \boldsymbol{v}_2...,\boldsymbol{v}_j]\!]$$

(see Definition 9.4 on page 296) as follows.

From equations 9.28 on page 305, if $\boldsymbol{w}_{j+1} = 0,$

$$\boldsymbol{v}_{j+1} = \sum_{i \le j} P_{\boldsymbol{w}_i}(\boldsymbol{v}_{j+1}) = \sum_{\substack{i \le j \\ \boldsymbol{w}_i \neq \mathbf{0}}} \frac{\boldsymbol{v}_{j+1} \cdot \boldsymbol{w}_i}{\boldsymbol{w}_i \cdot \boldsymbol{w}_i} \boldsymbol{w}_i .$$

The right side is an element of $[\![\boldsymbol{w}_1, \boldsymbol{w}_2,...,\boldsymbol{w}_j]\!],$ which we have already shown equal to $[\![\boldsymbol{v}_1, \boldsymbol{v}_2...,\boldsymbol{v}_j]\!].$ Thus if $\boldsymbol{w}_{j+1} = 0,$ then

$$\boldsymbol{v}_{j+1} \in [\![\boldsymbol{v}_1, \boldsymbol{v}_2,...,\boldsymbol{v}_j]\!],$$

meaning that v_{j+1} is linearly dependent on $v_1, v_2, \ldots, v_j$.

Going the other way (the *if* part) if v_{j+1} is linearly dependent on $[\![v_1, v_2 \ldots, v_j]\!]$, then by what was already proved, v_{j+1} also belongs to $[\![w_1, w_2, \ldots, w_j]\!]$, so that it is correct to write

$$w_{j+1} = \sum_{i \le j} b_i w_i \quad \text{for some real } b_i .$$

Since $w_1, w_2, \ldots, w_p$ are mutually orthogonal, the only way this can occur is for $w_{j+1} = 0$ (just take dot products, to show all b_i are 0).

This establishes the validity of Theorem 9.20.

As an immediate corollary to Theorem 9.20, we have the following.

Theorem 9.21: Orthogonality to previous subspaces

In the Gram-Schmidt orthogonalization procedure (see Definition 9.18 on page 305) for each $j = 1,2,\ldots,p$, the vector w_j is orthogonal to every element of $[\![v_1, v_2, \ldots, v_{j-1}]\!]$ (see Definition 9.4 on page 296 and Theorem 8.15 on page 278).

❑

This follows because w_j is orthogonal to every element of the sequence $w_1, w_2, \ldots, w_{j-1}$ and hence to every element of $[\![w_1, w_2, \ldots, w_{j-1}]\!] = [\![v_1, v_2 \ldots, v_{j-1}]\!]$.

So far the development of the Gram-Schmidt procedure has been carried out without taking account of practical considerations. Computationally this procedure as described in Definition 9.18 on page 305 tends to deteriorate badly as the number of vectors to be orthogonalized grows large. This is due to the buildup of roundoff errors inherent in digital computers. Later, in Theorem 10.31 on page 366 ff., we shall present a few schemes each of which is easy to implement, and which often effectively control this error buildup.

The Gram-Schmidt procedure is singularly unsuited for hand computation[1] — in any cases where it is likely to be useful, carrying it out by hand would make routine assembly line work seem exciting. However, the availability of fast inexpensive computers has made this procedure an extremely practical and effective one for solution of a variety of problems in mathematics and statistics. The *Mathematica*, *MLab* and *Splus* computer packages have procedures to carry out the Gram-Schmidt procedure, buried under the headings "QR Matrix Decomposition," "QRDecomposition" and

1. Which explains the scarcity of exercises of such a nature in this work.

"QRFAC" respectively. However, of the three only Splus seems to make it relatively easy to determine the orthogonal vectors $\boldsymbol{w}_1, \boldsymbol{w}_2, ..., \boldsymbol{w}_p$ by means of its qr.Q function.[1]

Exercises 9.8

1. Verify Theorem 9.21 on the vectors from Exercises 9.7 on page 307.
2. If you have access to a computer package which does orthogonalization and /or orthonormalization, see if you can do the previous exercise, as well as Exercises 9.7 on page 307 using such a package.

With the machinery that we have now developed, we are in a position to provide a theoretical solution to a very important problem in scientific data analysis — namely, that of *linear modeling*.[2] Linear modeling is concerned with furnishing a useful description of the relationships between measurable quantities; usually there are some (relatively) easily accessible quantities, $x_1, x_2, ..., x_n$ (*independent* variables) such as blood pressure, temperature, velocity, acceleration, etc., and another quantity, y (the *dependent* variable) which is harder to come by when it is needed. The dependent variable may not even be obtainable at the time of measurement of the independent variables — for instance, the time of the next earthquake, based on available seismograph measurements, or it may be very expensive — such as a definitive diagnosis of a condition suggested by inexpensive lab tests (the independent variables).

The difficulty in coming up with a useful description of the relationship between the independent variables $x_1, x_2, ..., x_n$ and the dependent variable y is that there may be other factors besides $x_1, x_2, ..., x_n$, factors not taken explicitly into account, which also influence y. For example, we may want to determine the effect of daily caloric intake on weight change over a specific time period. But other factors, such as exercise level and basal metabolism, which we may not be able to control or measure, may have a significant effect on weight. So, because it tends to ignore some factors influencing the dependent variables, whatever description we generate will likely be imperfect. Even such an imperfect description may be quite useful — e.g., to help determine the caloric intake needed to achieve a specific weight. To arrive at a suitable description, we usually gather data.

---

1. Those who are interested in using the Internet to obtain, at no cost, a Gram-Schmidt program for use on a PC may contact the author via e-mail at judah.rosenblatt@utmb.edu.
2. Also known as *regression analysis*, or mundanely as *curve fitting*.

Example 9.9: Weight versus calories

In order to try to determine the relation between weight and caloric intake, ignoring other factors influencing weight, we might gather data, $(c_i, \Delta w_i)$, $i = 1,...,n$, from n people representative of some specific population; here c_i could be the daily caloric intake from person i, and Δw_i the total weight change of person i over the given time period. A graph of the data might appear as the xs in Figure 9-4. If there were no factors but calories

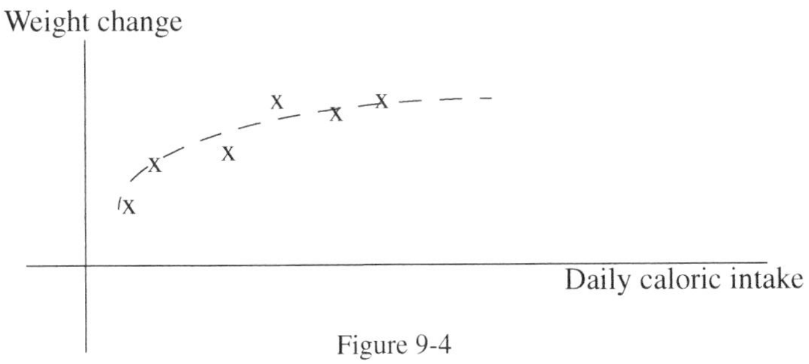

Figure 9-4

contributing to weight change, we might expect the data to lie on some smooth curve, such as is given by the dashed line in Figure 9-4, e.g., a straight line, or the graph of some polynomial. If we are trying to fit a second-degree polynomial to such data, we would be trying to find values a_0, a_1, a_2 such that for all caloric intakes, c, of interest, the corresponding weight change

$\Delta w \cong a_0 + a_1 c + a_2 c^2$, based on data $(c_1, \Delta w_1), (c_2, \Delta w_2),...,(c_n, \Delta w_n)$.

Letting

$$\boldsymbol{\Delta w} = \begin{pmatrix} \Delta w_1 \\ \cdot \\ \cdot \\ \cdot \\ \Delta w_n \end{pmatrix}, \quad \boldsymbol{c} = \begin{pmatrix} c_1 \\ \cdot \\ \cdot \\ \cdot \\ c_n \end{pmatrix}, \quad \boldsymbol{c}^2 = \begin{pmatrix} c_1^2 \\ \cdot \\ \cdot \\ \cdot \\ c_n^2 \end{pmatrix},$$

and $\mathbf{1}$ stand for the n-dimensional vector consisting of all 1s, we are trying to approximate the vector $\boldsymbol{\Delta w}$ by a vector $a_0\mathbf{1} + a_1\boldsymbol{c} + a_2\boldsymbol{c}^2$. Keeping in mind that the vectors $\mathbf{1}, \boldsymbol{c}, \boldsymbol{c}^2$ are known, we see that we are trying to approximate the weight change vector $\boldsymbol{\Delta w}$ by some element from the subspace $S = [\![\mathbf{1}, \boldsymbol{c}, \boldsymbol{c}^2]\!]$. A not unreasonable choice for the approximation to $\Delta \boldsymbol{w}$ is the element of S that is closest to $\Delta \boldsymbol{w}$. This is called the least squares approximation to $\Delta \boldsymbol{w}$ because we want to choose constants a_0, a_1, a_2 to minimize the squared distance

$$\left\|\Delta \boldsymbol{w} - (a_0 + a_1 \boldsymbol{c} + a_2 \boldsymbol{c}^2)\right\|^2 = \sum_{i=1}^{n} [\Delta w_i - (a_0 + a_1 c + a_2 c^2)]^2 .$$

This leads us naturally to extend the definition of projection, Definition 8.12 on page 276, to its extension, which follows.

Definition 9.22: Projection of a vector on a subspace

Given an n-dimensional real vector, $\boldsymbol{w}$, and a subspace, S (Definition 9.4 on page 296) generated by the real n-dimensional vectors $\boldsymbol{v}_1,...,\boldsymbol{v}_k$, the vector $\boldsymbol{v}$ in S is said to be a **projection of $\boldsymbol{w}$ on S** if no vector, V, in S is closer[1] to $\boldsymbol{w}$ than $\boldsymbol{v}$ is.

❑

We will see shortly that the projection of $\boldsymbol{w}$ on S exists and is unique. We will also present a practical computer method of finding it. We illustrate the concept of the projection of $\boldsymbol{w}$ on S in Figure 9-5.

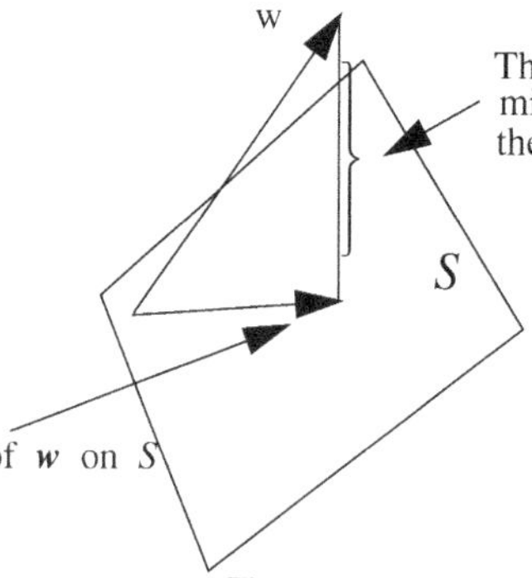

Figure 9-5

Consider the Gram-Schmidt procedure, Definition 9.18 on page 305; you may recall that $\boldsymbol{w}_2 = \boldsymbol{v}_2 - P_{\boldsymbol{w}_1}(\boldsymbol{v}_2)$. Letting $S_j = [\![\boldsymbol{v}_1, \boldsymbol{v}_2 ..., \boldsymbol{v}_j]\!]$, $j = 1,2,...,$ from the equation

$$\boldsymbol{w}_{j+1} = \boldsymbol{v}_{j+1} - \sum_{i \le j} P_{\boldsymbol{w}_i}(\boldsymbol{v}_{j+1}),$$

it is reasonable to guess that $\sum_{i \le j} P_{\boldsymbol{w}_i}(\boldsymbol{v}_{j+1})$ is just the projection of $\boldsymbol{v}_{j+1}$ on S_j. A look at Figure 9-3 on page 305 might help in this regard.

1. In the sense that $\|V - w\| < \|v - w\|$, see Definition 8.6 on page 272.

We will verify this conjecture algebraically in a little while. Before providing the simple solution to finding the projection of $\boldsymbol{w}$ on a subspace S, let's first see how we might approach the problem *head-on*. We want to choose $a_1,\dots,a_k$ to minimize

$$G(a_1,\dots,a_k) = \left\| \boldsymbol{w} - \sum_{i=1}^{k} a_i \boldsymbol{v}_i \right\|^2 = \sum_{j=1}^{n} \left[w_j - \sum_{i=1}^{k} a_i v_{ji} \right]^2, \qquad \textbf{\textit{9.34}}$$

where v_{ji} stands for the i-th coordinate of the vector $\boldsymbol{v}_j$. It's pretty clear from our previous experience with minimizing functions of a single real variable, that in order to achieve the desired result, we would compute the partial derivatives, $D_i G(a_1,\dots,a_k)$, $i = 1,2,..,k$ (Definition 8.5 on page 267) and then solve for $a_1,\dots,a_k$ satisfying

$$\begin{aligned} D_1 G(a_1,\dots,a_k) &= 0 \\ &\;\;. \\ &\;\;. \\ &\;\;. \\ D_k G(a_1,\dots,a_k) &= 0. \end{aligned} \qquad \textbf{\textit{9.35}}$$

This course of action would lead you to the so-called *normal equations*, which we leave to Exercises 9.10 below. If you try these exercises, there are two features that will probably stand out. First, the calculations themselves are a trifle treacherous, and any sloppiness in the bookkeeping is likely to result in failure. Secondly, even having derived the normal equations, their solution to find the projection you want is by no means trivial. Later on (see Theorem 10.30 on page 363 ff.), we will present a much simpler derivation and solution of the normal equations.

Exercises 9.10

1. Derive the normal equations for the case $k = 3$, by applying equations 9.35 to the function G given by equation 9.34.
2. For those with a great deal of energy and patience, derive the normal equations for the general case by applying equations 9.35 to equation 9.34. Compare the obtained result specialized to the case $k = 3$, to that obtained from the previous exercise.

We now return to what we feel is the simplest approach to finding the projection of a vector $\boldsymbol{w}$ on a subspace S. To do this we will work in a convenient coordinate system — namely, the one based on the nonzero vectors, $\boldsymbol{w}_1, \boldsymbol{w}_2,\dots,\boldsymbol{w}_m$, which we obtain from the vectors $\boldsymbol{v}_1, \boldsymbol{v}_2\dots,\boldsymbol{v}_k$ via the Gram-

Schmidt orthonormalization procedure (Definition 9.19 on page 306). In this coordinate system, the projection of $\mathbf{w}$ on S is the vector

$$\sum_{i=1}^{m} b_i \mathbf{w}_i \quad \text{minimizing} \quad H(b_1,...,b_m) = \left\| \mathbf{w} - \sum_{i=1}^{m} b_i \mathbf{w}_i \right\|^2$$
$$= \left(\mathbf{w} - \sum_{i=1}^{m} b_i \mathbf{w}_i\right) \cdot \left(\mathbf{w} - \sum_{i=1}^{m} b_i \mathbf{w}_i\right)$$

Using the orthonormality of the sequence $\mathbf{w}_1, \mathbf{w}_2,...,\mathbf{w}_m$ (see Theorem 8.15 on page 278 and Definition 9.19 on page 306) and evident properties of dot products, such as distributivity of multiplication over addition[1] and commutativity, we see that

$$\left(\mathbf{w} - \sum_{i=1}^{m} b_i \mathbf{w}_i\right)\left(\mathbf{w} - \sum_{i=1}^{m} b_i \mathbf{w}_i\right) = \mathbf{w} \cdot \mathbf{w} - 2\sum_{i=1}^{m} b_i(\mathbf{w}_i \cdot \mathbf{w}) + \sum_{i=1}^{m} b_i^2. \qquad \mathbf{9.36}$$

One approach that might be taken is to solve for the values of b_i which make all partial derivatives of H equal to 0 (a natural extension of the methods of functions of one variable, see page 72) — i.e., solve the equations

$$D_i H(b_1,...,b_m) = 0, \quad i = 1,...,m.$$

While this would lead easily to the correct answer, we would have to establish that the values we got really achieved a minimum — requiring substantial theoretical development, which is better postponed. Instead, we take the following purely elementary algebraic approach.

By completion of the square (see discussion on page 276), we find

$$\mathbf{w} \cdot \mathbf{w} - 2\sum_{i=1}^{m} b_i(\mathbf{w}_i \cdot \mathbf{w}) + \sum_{i=1}^{m} b_i^2 = \sum_{i=1}^{m} (b_i - \mathbf{w}_i \cdot \mathbf{w})^2 + \mathbf{w} \cdot \mathbf{w} - \sum_{i=1}^{m} (\mathbf{w}_i \cdot \mathbf{w})^2. \qquad \mathbf{9.37}$$

It is evident from the right side of this equation that the expression in equation 9.37 is minimized by choosing $b_i = \mathbf{w}_i \cdot \mathbf{w}$. This shows that

---

1. $\mathbf{x} \cdot (\mathbf{y} + \mathbf{z}) = \mathbf{x} \cdot \mathbf{y} + \mathbf{x} \cdot \mathbf{z}$ and $\mathbf{x} \cdot \mathbf{y} = \mathbf{y} \cdot \mathbf{x}$, respectively. The commutativity must be modified when dot products are extended to complex valued vectors.

$$\sum_{i=1}^{m} (w_i \cdot w) w_i$$

is the projection of w on S.

At this point it pays to introduce some notation to simplify summarizing the results we have just established.

Definition 9.23: $P_S(w)$

Let S be a subspace of real n-dimensional space (Definition 9.4 on page 296 and Definition 9.14 on page 302), and let w be a real n-dimensional vector. The projection of w on S (Definition 9.22 on page 313) is denoted by $P_S(w)$.

❑

We now summarize the results that have just been developed.

Theorem 9.24: Computation of $P_S(w)$

Let S be the subspace of real n-dimensional space (see Definition 9.4 on page 296 and Definition 9.14 on page 302) generated by the sequence $v_1,...,v_k$ of real n-dimensional vectors (see Definition 8.1 on page 261) and let w represent an arbitrary n-dimensional real vector. To compute the projection, $P_S(w)$ of w on S (Definition 9.22 on page 313) orthonormalize the sequence $v_1,...,v_k$ (see Definition 9.19 on page 306) to generate the sequence $w_1, w_2,...,w_m$ of orthonormal vectors. Then

$$P_S(w) = \sum_{i=1}^{m} P_{w_i}(w) = \sum_{i=1}^{m} (w_i \cdot w) w_i.$$

❑

3 Solution of Systems of Linear Equations

The method presented at the end of the previous section may be considered as an alternative to solving the *normal equations* referred to on page 314, and in Exercises 9.10 which follow that discussion. This suggests that orthogonalization might itself be useful in solving general systems of linear equations. As we'll see, this is the case. Orthogonalization is probably the most frequently used practical method of solution of linear systems on a computer and furthermore, is a powerful tool for gaining insight into what's really going on.

Before introducing the application of orthogonalization methods to solve systems of linear equations, we look at some of the problems which lead to such systems.

We already know that if

$$\boldsymbol{y} = \begin{pmatrix} y_1 \\ \cdot \\ \cdot \\ y_n \end{pmatrix}$$

then the coordinates, $a_1,...,a_n$, of $\boldsymbol{y}$ relative to the given linearly independent vectors $\boldsymbol{v}_1,...,\boldsymbol{v}_n$ are given as the solution of the system

$$\sum_{i=1}^{n} a_i \boldsymbol{v}_i = \boldsymbol{y}. \qquad \textbf{\textit{9.38}}$$

Note here that the *original* coordinates of the vector $\boldsymbol{y}$ are furnished by the sequence $y_1,....,y_n$. So systems of linear equations arise naturally when we want to change coordinates.

Systems of linear equations also arise very generally when we want to solve systems of nonlinear equations — for example, in the study of maxima and minima of functions of several variables using an extension of Newton's method.

To get an idea of how this happens, let f be a real valued function of n real variables, with continuous partial derivatives (see Definition 8.5 on page 267 and Definition 8.7 on page 270). To find maxima and minima of f, just as for functions of a single variable, it is natural to look for points in the domain where the (partial) derivatives are all 0. (Recall the discussion on page 72.) Hence, we let g_i denote the partial derivative of g with respect to its *i-th* coordinate (see Definitions 8.5 and 8.4 on page 267 and page 265, respectively) and want to examine those sequences $x_1, x_2, ... ,x_n$ for which

$$D_i f(x_1, x_2, ... ,x_n) = g_i(x_1, x_2, ... ,x_n) = 0 \quad \text{all } i = 1,...,n \qquad \textbf{\textit{9.39}}$$

Usually this system of equations is *nonlinear* — in that it is only rarely the case that the functions g_i are all of the form

$$g_i(x_1, x_2, ... ,x_n) = a_{i,1}x_1 + a_{i,2}x_2 + \cdots + a_{i,n}x_n,$$

where the $a_{i,j}$ are constants.

To extend Newton's method (recall the discussion following equation 3.18 on page 79), we denote our first estimate for the point we are seeking by $\boldsymbol{x}_0$ (in this case, a vector) and denote the successive points to be generated by $\boldsymbol{x}_1, \boldsymbol{x}_2,...$. To see how to generate $\boldsymbol{x}_{k+1}$, which we will call $\boldsymbol{x}_k + \boldsymbol{h}_k$, from

$\boldsymbol{x}_k$, we'll make use of the fundamental multivariate approximation on page 274, which applied to this situation yields

$$g_i(\boldsymbol{x}_k + \boldsymbol{h}_K) \cong g_i(\boldsymbol{x}_k) + \boldsymbol{h} \cdot \boldsymbol{grad}\, g_i(\boldsymbol{x}_k). \qquad \textit{9.40}$$

We would like to choose $\boldsymbol{h}_k$ so that

$$g_i(\boldsymbol{x}_{k+1}) = g_i(\boldsymbol{x}_k + \boldsymbol{h}_K), \quad i = 1, \dots, n$$

are all simultaneously closer to 0 than were the $g_i(\boldsymbol{x}_k)$.

To try to accomplish this, justified by approximation 9.40 above, we instead go for the easier task of trying to determine $\boldsymbol{h}_k$ satisfying

$$g_i(\boldsymbol{x}_k) + \boldsymbol{h}_k \cdot \boldsymbol{grad}\, g_i(\boldsymbol{x}_k) = 0, \quad i = 1, \dots, n. \qquad \textit{9.41}$$

For given $\boldsymbol{x}_k$ and g_i $i = 1,\dots,n$ equations 9.41 specifying the n-dimensional version of Newton's method constitute a system of n linear equations in the n unknown coordinates of the vector $\boldsymbol{h}_k$. Unfortunately at this point we have to suspend further discussion of Newton's method until the necessary matrix notation and results concerning solution of systems of linear equations have been developed.

Even though Newton's method will not be further developed until later, the effort put into developing the related equations at this time is not wasted — because it's worth examining equations 9.41 just derived for comparison with the first example used to illustrate systems of equations, namely, equations 9.38. To do this we rewrite equations 9.41 in the form

$$\boldsymbol{grad}\, g_i(\boldsymbol{x}_k) \cdot \boldsymbol{h}_k = g_i(\boldsymbol{x}_k) \quad i = 1,\dots,n. \qquad \textit{9.42}$$

Now in equations 9.38 we are concentrating on representing the right side as a linear combination of column vectors, $\boldsymbol{v}_i$, while in equations 9.42 the right hand side is represented in terms of dot products. When matrix notation is introduced we'll see that this difference is carried through to viewing a matrix as either a sequence of column vectors, or (with the dot product representation in mind), as a sequence of row vectors. These two viewpoints are complementary, and both are needed in order to develop an adequate understanding of systems of linear equations.

Exercises 9.11

1. Write equations 9.38 in terms of dot products.
2. Write equations 9.42 in terms of linear combinations of vectors.

For the first approach to solving a system of linear equations we'll represent the system of equations in the form suggested by equations 9.38; namely, given the n-dimensional vectors $\boldsymbol{v}_1, \boldsymbol{v}_2 \dots, \boldsymbol{v}_k$ and the real n-dimensional vector, $\boldsymbol{y}$, we seek real numbers $a_1,\dots,a_k$ satisfying

$$\sum_{i=1}^{k} a_i \boldsymbol{v}_i = \boldsymbol{y}. \qquad \textbf{9.43}$$

We know that if any of the $\boldsymbol{v}_j$ is a linear combination of those $\boldsymbol{v}_i$ with $i < j$, then we may choose $a_j = 0$; that is, if there is a solution to equations 9.43 ($a_1,...,a_k$ for which equations 9.43 hold true), then there is one with $a_j = 0$. In orthogonalizing $\boldsymbol{v}_1, \boldsymbol{v}_2..., \boldsymbol{v}_k$ to $\boldsymbol{w}_1',....,\boldsymbol{w}_k'$, we can determine the $\boldsymbol{v}_j$ referred to (there are the ones with $\boldsymbol{w}_j' = \boldsymbol{0}$, see Theorem 9.20 on page 307). So for the purpose of keeping notation simple, it will be assumed that these $\boldsymbol{v}_j$ have been identified, that the corresponding a_j have been set equal to 0, and that the remaining $\boldsymbol{v}_i$ have been relabeled sequentially. So we assume that we are trying to solve equation system 9.43, where the sequence of vectors $\boldsymbol{v}_1, \boldsymbol{v}_2..., \boldsymbol{v}_k$ is linearly independent. Now **orthonormalize** $\boldsymbol{v}_1, \boldsymbol{v}_2..., \boldsymbol{v}_k$ to $\boldsymbol{w}_1,....,\boldsymbol{w}_k$ and recall (Theorem 9.21 on page 310), that for each j,

$$\boldsymbol{w}_j \perp \boldsymbol{v}_1, \boldsymbol{v}_2..., \boldsymbol{v}_{j-1}$$

(i.e., $\boldsymbol{w}_j$ is orthogonal to every vector of the sequence $\boldsymbol{v}_1, \boldsymbol{v}_2..., \boldsymbol{v}_{j-1}$). So if we take the dot product of w_k with both sides of equation 9.43 we obtain

$$a_k \boldsymbol{v}_k \cdot \boldsymbol{w}_k = \boldsymbol{y} \cdot \boldsymbol{w}_k. \qquad \textbf{9.44}$$

This equation can be solved for a_k, because we can show $\boldsymbol{v}_k \cdot \boldsymbol{w}_k \neq 0$ as follows:

If $\boldsymbol{v}_k \cdot \boldsymbol{w}_k$ were equal to 0, because

$$\boldsymbol{v}_k \in [\![\boldsymbol{v}_1, \boldsymbol{v}_2..., \boldsymbol{v}_k]\!] = [\![\boldsymbol{w}_1, \boldsymbol{w}_2,..., \boldsymbol{w}_k,]\!]$$

we would have

$$\boldsymbol{v}_k \in [\![\boldsymbol{v}_1, \boldsymbol{v}_2..., \boldsymbol{v}_{k-1}]\!] = [\![\boldsymbol{w}_1, \boldsymbol{w}_2,..., \boldsymbol{w}_{k-1}]\!]$$

which has been excluded by our procedure.

So we find

$$a_k = \frac{\boldsymbol{y} \cdot \boldsymbol{w}_k}{\boldsymbol{v}_k \cdot \boldsymbol{w}_k}. \qquad \textbf{9.45}$$

Now rewrite equations 9.43 as

$$\sum_{i=1}^{k-1} a_i \boldsymbol{v}_i = \boldsymbol{y} - \frac{\boldsymbol{y} \cdot \boldsymbol{w}_k}{\boldsymbol{v}_k \cdot \boldsymbol{w}_k} \boldsymbol{v}_k \qquad \textbf{9.46}$$

and notice that the right side of this equation is known. Thus we can repeat the process we just went through, but taking the dot product of both sides of equation 9.46 with $\boldsymbol{w}_{k-1}$, this time determining a_{k-1}. Continue on in this fashion, inductively, until all of the a_i are determined.

It's best to take stock of what's actually been done, before continuing this development.

We've constructed an algorithm for finding a solution of equations 9.43 ***provided that a solution exists.*** But the algorithm will provide the sequence $a_1,...,a_k$ whether or not there is an actual solution of equations 9.43. So if our interest is in finding a solution of equations 9.43, following determination of $a_1,...,a_k$ it is necessary to check whether we got what we wanted.

The simplest way conceptually of determining whether or not we have really generated a solution of equations 9.43 with the a_i just calculated is to compute

$$\sum_{i=1}^{k} a_i \boldsymbol{v}_i$$

and see if it yields the vector $\boldsymbol{y}$. If so, we have a solution. If not, there is no solution. There may, of course, be more than one solution. When we treat *orthogonal complements* (Chapter 11), we'll find out how to determine all solutions of linear systems like those of equations 9.43 (the so-called *general solution*).

We summarize the algorithm just developed in the following definition.

Definition 9.25: Attempted orthogonalization solution

to $$\sum_{i=1}^{k} a_i \boldsymbol{v}_i = \boldsymbol{y}$$

Let $\boldsymbol{v}_1, \boldsymbol{v}_2 ..., \boldsymbol{v}_k$ be a given sequence of real n-dimensional vectors (Definition 8.1 on page 261). Orthogonalize $\boldsymbol{v}_1, \boldsymbol{v}_2 ..., \boldsymbol{v}_k$ to $\boldsymbol{w}_1', ..., \boldsymbol{w}_k'$ (see Definition 9.18 on page 305); discard all $\boldsymbol{v}_j$ with indices j for which $\boldsymbol{w}_j' = \boldsymbol{0}$ (see Theorem 9.20 on page 307) and relabel the remaining $\boldsymbol{v}_j$ as $\boldsymbol{v}_1, \boldsymbol{v}_2 ..., \boldsymbol{v}_k$. Now orthonormalize the (relabeled) $\boldsymbol{v}_1, \boldsymbol{v}_2 ..., \boldsymbol{v}_k$ to $\boldsymbol{w}_1, ..., \boldsymbol{w}_k$ (see Definition 9.19 on page 306).

Let

$$a_k = \frac{\boldsymbol{y} \cdot \boldsymbol{v}_k}{\boldsymbol{v}_k \cdot \boldsymbol{w}_k} \qquad \textbf{\textit{9.47}}$$

and, inductively having determined $a_k, a_{k-1}, ..., a_{j+1}$, define

$$a_j = \frac{\left(\boldsymbol{y} - \sum_{i=j+1}^{k} a_i \boldsymbol{v}_i\right) \cdot \boldsymbol{w}_j}{\boldsymbol{v}_j \cdot \boldsymbol{w}_j} \qquad j = k-1, ..., 1. \qquad \textbf{\textit{9.48}}$$

❑

Here is an illustrative example of the process we have just described.

Example 9.12: Solution by orthogonalization

We examine the system

$$\begin{aligned} a_1 + a_2 \qquad &= 2 \\ a_1 \quad + \quad a_3 &= 1 \\ 2a_1 + a_2 + a_3 &= 3. \end{aligned} \qquad \textit{9.49}$$

We rewrite this as

$$a_1\begin{pmatrix}1\\1\\2\end{pmatrix} + a_2\begin{pmatrix}1\\0\\1\end{pmatrix} + a_3\begin{pmatrix}0\\1\\1\end{pmatrix} = \begin{pmatrix}2\\1\\3\end{pmatrix}.$$

Using the Gram-Schmidt procedure, Definition 9.18 on page 305, we find

$$\boldsymbol{w}_1' = \begin{pmatrix}1\\1\\2\end{pmatrix}, \quad \boldsymbol{w}_2' = \begin{pmatrix}1/2\\-1/2\\0\end{pmatrix}, \quad \boldsymbol{w}_3' = \begin{pmatrix}0\\0\\0\end{pmatrix}$$

Since $w_3' = \mathbf{0}$, we know that $\boldsymbol{v}_3$ is a linear combination of $\boldsymbol{v}_1$ and $\boldsymbol{v}_2$. (In fact we find easily that $\boldsymbol{v}_3 = \boldsymbol{v}_1 - \boldsymbol{v}_2$.)

We need only look for a solution of

$$a_1\begin{pmatrix}1\\1\\2\end{pmatrix} + a_2\begin{pmatrix}1\\0\\1\end{pmatrix} = \begin{pmatrix}2\\1\\3\end{pmatrix}. \qquad \textit{9.50}$$

Taking dot products of both sides with $\boldsymbol{w}_2'$, or, to make life simpler, with $2\,\boldsymbol{w}_2'$, we find

$$a_2 \times 1 = 1 \quad \text{or} \quad a_2 = 1\,.$$

Substituting this into equation 9.50, we find

$$a_1\begin{pmatrix}1\\1\\2\end{pmatrix} + 1\begin{pmatrix}1\\0\\1\end{pmatrix} = \begin{pmatrix}2\\1\\3\end{pmatrix} \quad \text{or} \quad a_1\begin{pmatrix}1\\1\\2\end{pmatrix} = \begin{pmatrix}1\\1\\2\end{pmatrix}.$$

Of course, we recognize immediately that $a_1 = 1$ — if we didn't, we could take the dot product of both sides with $\boldsymbol{w}_1'$ to find a_1. It's easy to verify that $a_1 = 1, a_2 = 1, a_3 = 0$ is a solution of equations 9.49, and if the right-hand side of equations 9.49 were replaced by

$$\begin{pmatrix} 2 \\ 2 \\ 3 \end{pmatrix},$$

there would be no solution.

■

As already mentioned, the method we've just gone through is generally practical only for computer implementation, and then only with certain essential modifications. Besides modifying the Gram-Schmidt computations to counteract the effects of roundoff error, account must be taken of errors which arise when checking whether any of the $\boldsymbol{w}'_j$ are $\mathbf{0}$; we rarely expect to obtain an exact 0 on the computer. Rather we should check on whether the length, $\|\boldsymbol{w}'_j\|$, is *sufficiently small.*

When we get to the material on orthogonal complements in Section 4 of Chapter 11, we'll present a simpler way than the one already given, to check by computer on the existence of at least one solution of a linear system, (sequence $a_1,...,a_k$), satisfying equations such as 9.43 on page 319.

Problem 9.13

In attempting to use the algorithm given by Definition 9.25 on page 320 to solve a system such as 9.43 on page 319, what is the meaning of the obtained value

$$\sum_{i=1}^{k} a_i \boldsymbol{v}_i$$

when there is no solution?

Hint: Let $S = [\![\boldsymbol{v}_1, \boldsymbol{v}_2 ..., \boldsymbol{v}_k]\!]$ and write $\boldsymbol{y} = [\boldsymbol{y} - P_S(\boldsymbol{y})] + P_S(\boldsymbol{y})$, noting from Theorem 9.24 on page 316 that

$$\boldsymbol{w}_j \cdot [\boldsymbol{y} - P_S(\boldsymbol{y})] = \boldsymbol{w}_j \cdot \left[\boldsymbol{y} - \sum_i P_{\boldsymbol{w}_i}(\boldsymbol{y})\right] = \boldsymbol{w}_j \cdot \left[\boldsymbol{y} - \sum_i (\boldsymbol{w}_i \cdot \boldsymbol{y})\boldsymbol{w}_i\right]$$
$$= 0.$$

$j = k, k\text{-}1, ..., 1.$

which shows that the algorithm from Definition 9.25 yields the same a_j as when $\boldsymbol{y}$ is replaced by $P_S(\boldsymbol{y})$ in equations 9.43. But the equations

$$\sum_{i=1}^{k} a_i \boldsymbol{v}_i = P_S(\boldsymbol{y})$$

have a unique solution and yield $P_S(\boldsymbol{y})$.

Exercises 9.14

In the following, find a solution, or show that none exists.

1. $$\begin{aligned} a_1 \qquad - \quad a_3 &= -1 \\ a_1 + 4a_2 - 5a_3 &= 3 \\ a_1 + 2a_2 - 3a_3 &= 1 \end{aligned}$$

2. $$\begin{aligned} a_1 \qquad - \quad a_3 &= -1 \\ a_1 + 4a_2 - 5a_3 &= 3 \\ a_1 + 2a_2 - 3a_3 &= 1 \end{aligned}$$

3. $$\begin{aligned} 3a_1 + 2a_2 + a_3 &= 6 \\ -4a_1 + 3a_2 \qquad &= -1 \\ 3a_1 + 2a_2 - a_3 &= 7 \end{aligned}$$

While we won't devote very much of our effort to the standard way of solving systems of linear equations, we should at least outline this method, because of its suitability for hand implementation. Here is the essence of *Gaussian elimination*. We write equations 9.43 on page 319 as

$$v_{11}a_1 + v_{12}a_2 + \cdots + v_{1k}a_k = y_1$$

$$\begin{matrix} \cdot \\ \cdot \\ \cdot \end{matrix}$$

$$v_{n1}a_1 + v_{n2}a_2 + \cdots + v_{nk}a_k = y_n.$$

We solve the first equation (if possible) for a_1 — i.e., express a_1 in terms of the other variables in the first equation. The obtained expression is then substituted for a_1 in the remaining equations. (This can be accomplished by multiplying the first equation successively by $-v_{ji}/v_{11}$ and adding the resulting equation to the j-th equation, $j = 1,2,...,n$.)

Next solve the new second equation for a_2, substituting into the remaining equations. (This can be accomplished as above.)

Continue this process until it is no longer possible to continue it further; it is not required that a_i be solved for at the i-th stage; in fact, that may not even be possible. (If a_i isn't available in the i-th equation, solve for some other a_j using the i-th equation.)

When this process terminates, there are three possibilities:

First, there is the possibility of a system which is recognizably unsolvable, such as

$$a_1 + a_2 = 3$$
$$0 = 2$$

Second is the possibility that some of the variables a_j can be assigned arbitrarily, determining the remaining ones uniquely.

Third is the case where there is precisely one solution.

The first case corresponds to the situation in equations 9.43 where $\boldsymbol{y}$ is not an element of the subspace $[\![\boldsymbol{v}_1, \boldsymbol{v}_2,...,\boldsymbol{v}_k]\!]$.

The second corresponds to $\boldsymbol{y} \in [\![\boldsymbol{v}_1, \boldsymbol{v}_2,...,\boldsymbol{v}_k]\!]$ but $\boldsymbol{v}_1, \boldsymbol{v}_2,...,\boldsymbol{v}_k$ are linearly dependent.

The third corresponds to $\boldsymbol{y} \in [\![\boldsymbol{v}_1, \boldsymbol{v}_2,...,\boldsymbol{v}_k]\!]$ but $\boldsymbol{v}_1, \boldsymbol{v}_2,...,\boldsymbol{v}_k$ are linearly independent

The next examples illustrate the various possibilities.

Example 9.15: Gaussian elimination for inconsistent system

$$2a_1 + 3a_2 = 3$$
$$4a_1 + 6a_2 = 2$$

Multiply the first equation by -2 and add to the second equation, to obtain

$$2a_1 + 3a_2 = 3$$
$$0 = 4.$$

demonstrably inconsistent.

■

Example 9.16: Another inconsistent system

$$2a_1 + 3a_2 + 4a_3 = 3$$
$$6a_1 + 9a_2 + 12a_3 = 5$$

Multiply the first equation by -3 and add to the second one, to obtain

$$2a_1 + 3a_2 + 4a_3 = 3$$
$$0 = -4.$$

also inconsistent.

■

Example 9.17: Gaussian elimination leading to many solutions

$$2a_1 + 3a_2 + 4a_3 = 3$$
$$6a_1 + 8a_2 + a_3 = 5$$

Proceed as in Example 9.16, to obtain

$$2a_1 + 3a_2 + 4a_3 = 3$$
$$-a_2 - 11a_3 = -4.$$

Now multiply the second of these equations by 3 and add to the first, to obtain

$$2a_1 \quad + \quad 37a_3 = -9$$
$$-a_2 - 11a_3 = -4$$

It is easily seen that a_3 may be assigned arbitrarily, and this will determine both a_1 and a_2 uniquely.

■

Example 9.18: Gaussian elimination unique solution

$$a_1 + a_2 = 3$$
$$3a_1 - a_2 = 5$$
$$5a_1 + a_2 = 11$$

Multiply the first equation by -3 and add to the second equation; then multiply the first equation by -5 and add it to the third one, to obtain

$$a_1 + a_2 = 3$$
$$-4a_2 = -4$$
$$-4a_2 = -4$$

The third equation is redundant and thus may be dropped. It is immediately seen that $a_2 = 1$, and hence, substituting this into the first equation, yields $a_1 = 2$.

■

The final example in this chapter illustrates why Gaussian elimination is more difficult to implement on a computer than solution by orthogonalization.

Example 9.19

$$\begin{aligned} a_2 + a_3 &= 1 \\ a_1 \quad + \quad a_3 &= 0 \\ a_1 - a_2 - 2a_3 &= 2 \end{aligned}$$

Because we cannot operate on *autopilot* by mindlessly using the first equation to eliminate a_1 from the system, nor the second equation to eliminate a_2 from the system, the programming complexity for automatic computer implementation of Gaussian elimination is considerably greater than for orthogonalization. This system is easy to solve by hand. Adding the first equation to the third yields

$$a_1 - a_3 = 3.$$

Using the second equation to substitute a_1 for $-a_3$ above, yields $a_1 = 3/2$, then using the second equation gets $a_3 = -3/2$, and finally, substituting these two values into the third equation yields $a_2 = 5/2$.

■

Exercises 9.20

Repeat Exercises 9.14 but using Gaussian elimination.

Chapter 10

Matrices

1 Introduction

Up to this point we have avoided the use of matrix ideas and notation in treating systems of linear equations, in the belief that it's better to stay in a simple setting as long as it continues to yield substantial results without strain. The derivations we are about to present now depend crucially on matrix notation and the idea of a matrix as representing a function.

Initially we introduce matrix notation simply to avoid the effort of writing out a system of linear equations as

$$\sum_{i=1}^{k} a_i \boldsymbol{v}_i = \boldsymbol{y}\,. \qquad \textbf{\textit{10.1}}$$

where $\boldsymbol{y}$ and $\boldsymbol{v}_1, \boldsymbol{v}_2,...,\boldsymbol{v}_k$ are n-dimensional real vectors (see Definition 8.1 on page 261) and the a_i are real numbers. Toward this end, we write

$$\boldsymbol{v}_i = \begin{pmatrix} v_{1i} \\ v_{2i} \\ \cdot \\ \cdot \\ v_{ni} \end{pmatrix}, \qquad i = 1,...,k$$

and line up the vectors $\boldsymbol{v}_1, \boldsymbol{v}_2,...,\boldsymbol{v}_k$ in a rectangular array,[1] denoted by

$$\mathbf{V} = [\boldsymbol{v}_1, \boldsymbol{v}_2,...,\boldsymbol{v}_k] = \begin{bmatrix} v_{11} & v_{12} & \cdots & v_{1k} \\ v_{21} & v_{22} & \cdots & v_{2k} \\ \cdot & \cdot & \cdots & \cdot \\ \cdot & \cdot & \cdots & \cdot \\ v_{n1} & v_{n2} & \cdots & v_{nk} \end{bmatrix}. \qquad \textbf{\textit{10.2}}$$

1. As occurred earlier, our notation isn't entirely consistent, with parentheses, commas, and brackets omitted or used capriciously.

Definition 10.1: Matrix, matrix notation for equations

The array $\mathbf{V}$ given in equations 10.2 is called an n by k (or $n \times k$) matrix. To indicate the number of rows and columns, $\mathbf{V}$ is sometimes written $\mathbf{V}^{n \times k}$. Equations 10.1 are conveniently written as

$$\mathbf{V}\boldsymbol{a} = \mathbf{y}. \qquad 10.3$$

$\mathbf{V}\boldsymbol{a}$ is called the **product** of the matrix $\mathbf{V}$ with the k-dimensional vector

$$\boldsymbol{a} = \begin{pmatrix} a_1 \\ \cdot \\ \cdot \\ a_k \end{pmatrix},$$ so that

$$\mathbf{V}\boldsymbol{a} = \sum_{i=1}^{k} a_i \mathbf{v}_i. \qquad 10.4$$

We can also write

$$\mathbf{V}\boldsymbol{a} = \begin{pmatrix} V_1 \cdot \boldsymbol{a} \\ \cdot \\ \cdot \\ V_n \cdot \boldsymbol{a} \end{pmatrix}, \qquad 10.5$$

where the *dot product* is defined in Definition 8.10 on page 272, and V_r stands for the r-th row of the matrix $\mathbf{V}$, i.e.,

$$V_r = (v_{r1}, v_{r2}, \ldots, v_{rk}).$$

❑

As we shall see when we define multiplication of two matrices, when we write a vector in matrix notation, we distinguish a *column vector* (a $1 \times n$ matrix), from the corresponding row vector (an $n \times 1$ matrix).

Exercises 10.1

1. Verify that the two representations of $\mathbf{V}\boldsymbol{a}$, those given by equations 10.4 and 10.5, yield the same result.
2. Familiarize yourself with multiplication of a matrix times a vector, by making up as many drill exercises as you need, and checking your results with *Mathematica*, *Maple*, or *Splus*. In *Mathematica* you must use the *dot* notation, because *Mathematica* does not have a special class of matrices, while *Splus* does.

2 Matrices as Functions, Matrix Operations

When we write a system of linear equations in the form given by equations 10.3, the matrix $\mathbf{V}$ determines a function, $\mathcal{V}$, as shown in Figure 10-1.

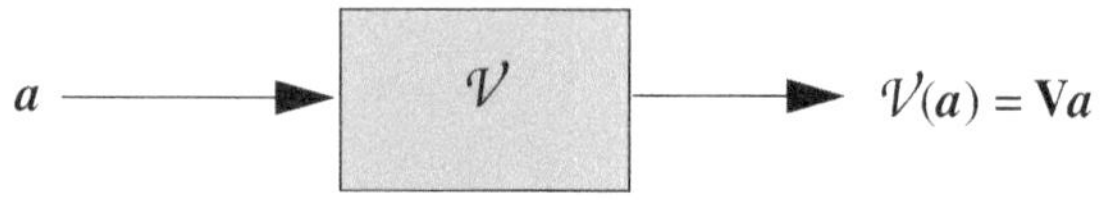

Figure 10-1

Thinking of a matrix $\mathbf{V}$ as representing a function leads us immediately to the definition of matrix multiplication, as the matrix operation corresponding to composition of functions — instead of as an unmotivated definition, to be justified by later applications. To see how this comes about, suppose that $\boldsymbol{a}$ is a real k-dimensional vector, and $\mathbf{V}$ is a real $n \times k$ matrix. Then $\mathbf{y} = \mathbf{V}\boldsymbol{a}$ is a real n-dimensional vector. If $\mathbf{W}$ is a real $r \times n$ matrix, then $\mathbf{W}\boldsymbol{y}$ is a real r-dimensional vector. The vector $\mathbf{W}\boldsymbol{y}$ corresponds to the composition $\mathcal{W}(\mathcal{V}(y))$. Successive matrix applications corresponding to composition of the associated functions come up frequently in applications.

Example 10.2: Examples of successive matrix multiplications

When we solved a system of linear equations by orthogonalization, starting with equation 9.49 on page 321, which we can write in the matrix-vector form

$$\mathbf{V}\boldsymbol{a} = \mathbf{y}$$

for each k, we took the dot product of both sides of this equation with the orthonormal vector $\boldsymbol{w}_k$, getting

$$(\mathbf{V}\boldsymbol{a}) \cdot \boldsymbol{w}_k = \mathbf{y} \cdot \boldsymbol{w}_k .$$

But using the dot product form for multiplying a vector by a matrix (equation 10.3, but with $\mathbf{V}\boldsymbol{a}$ *here* playing the role of $\mathbf{V}$ *there*, and $\boldsymbol{w}_k$ *here* playing the role of $\boldsymbol{a}$ *there*), this sequence of equations (one for each k) is seen to be the equivalent of multiplying the vectors on both sides of equation 9.49 by the matrix whose k-th row is $\boldsymbol{w}_k$. Calling this matrix $\mathbf{W}^{\mathrm{T}}$ (notation for the transpose of the matrix $\mathbf{W}$, for reasons soon to be presented) this process may be abbreviated as

$$\mathbf{W}^{\mathrm{T}}(\mathbf{V}\boldsymbol{a}) = \mathbf{W}^{\mathrm{T}} y.$$

The left side is the result of successively *multiplying* the vector $\boldsymbol{a}$ by the matrix $\mathbf{V}$ and then multiplying this resultant vector by the matrix $\mathbf{W}^{\mathrm{T}}$.

We will see in Theorem 10.12 on page 339 that the Gaussian elimination procedure of Examples 9.15, 9.16 and 9.17, starting on page 324, can be considered as a sequence of successive matrix-vector multiplications (which you might want to verify now for yourself).

When we get to the actual implementation of the multidimensional Newton's method, this will also be seen as a sequence of matrix-vector multiplications.

■

Since successive multiplications of a vector by a sequence of matrices arises quite often, it seems reasonable to investigate whether the composition of functions corresponding to matrices corresponds itself to a matrix. That is, *is there a matrix, which we might want to call* **WV**, *corresponding to the composition* $\mathcal{W}(\mathcal{V})$? To see rather easily that there is, we first note the immediately verifiable result.

Theorem 10.2: Linear operator property of matrix × vector

Let **V** be a real $n \times k$ matrix (see Definition 10.1 on page 328) $\boldsymbol{v}$ and $\boldsymbol{w}$ be real k-dimensional vectors, and a and b be real numbers. Then

$$\mathbf{V}(a\boldsymbol{v} + b\boldsymbol{w}) = a\mathbf{V}(\boldsymbol{v}) + b\mathbf{V}(\boldsymbol{w}),$$

where, to keep matters clear,[1] the left side represents the matrix **V** times the vector $a\boldsymbol{v} + b\boldsymbol{w}$, and the right side is the sum of a times the vector $\mathbf{V}\boldsymbol{v}$ and b times the vector $\mathbf{V}\boldsymbol{w}$.

❑

To find out how to define **WV** to correspond to the composition $\mathcal{W}(\mathcal{V})$ (see Figure 10-1 on page 329) note that, using the linear operator property above,

$$\mathbf{W}(\mathbf{V}\boldsymbol{a}) = \mathbf{W}\left(\sum_{i=1}^{k} a_i \boldsymbol{v}_i\right)$$

$$= \sum_{i=1}^{k} a_i \mathbf{W}(\boldsymbol{v}_i).$$

But then using the row-oriented version of a matrix times a vector, as indicated by equations 10.5 on page 328, we see that

1. All of this would be unnecessary if parentheses were used only for grouping, and brackets were used to enclose function arguments, as in *Mathematica*. Referring to Definition 8.2 on page 263, note that the expression $a\boldsymbol{v} + b\boldsymbol{w}$ has a clearly defined meaning.

$$\sum_{i=1}^{k} a_i \mathbf{W}(\mathbf{v}_i) = \sum_{i=1}^{k} a_i \begin{pmatrix} \mathbf{W}_1 \cdot \mathbf{v}_i \\ \cdot \\ \cdot \\ \cdot \\ \mathbf{W}_r \cdot \mathbf{v}_i \end{pmatrix} \qquad \textit{10.6}$$

where, recall, $\mathbf{W}_j$ is the j-th row of **W**, and $\mathbf{v}_i$ is the i-th column of **V**. But this indicates that $\mathcal{W}(\mathcal{V})$ itself corresponds to the matrix whose i-th column is given on the right of equation 10.6, and whose j-i element (the one in the j-th row, i-th column) is therefore the dot product $\mathbf{W}_j \cdot \mathbf{v}_i$.

This discussion leads to the definition of matrix product.

Definition 10.3: Product, WV, of matrices

Let **W** be a real $r \times n$ matrix and **V** be a real $n \times k$ matrix (Definition 10.1 on page 328) with associated functions $\mathcal{W}$ and $\mathcal{V}$ respectively (see Definition 10.1 on page 328 and Figure 10-1 on page 329). The matrix product **WV** corresponding to the composition $\mathcal{W}(\mathcal{V})$ is defined to be the $r \times n$ matrix whose j-i element (element in j-th row, i-th column) is given by

$$(\mathbf{WV})_{ji} = \mathbf{W}_j \cdot \mathbf{v}_i \qquad \textit{10.7}$$

(see Definition 8.10 on page 272) where $\mathbf{W}_j$ is the j-th row of **W** and $\mathbf{v}_i$ is the i-th column of **V**.

It's worth noting that the i-th column of **WV** is

$$\begin{pmatrix} \mathbf{W}_1 \cdot \mathbf{v}_i \\ \cdot \\ \cdot \\ \cdot \\ \mathbf{W}_r \cdot \mathbf{v}_i \end{pmatrix} \qquad \textit{10.8}$$

and the j-th row of **VW** is

$$(\mathbf{W}_j \cdot \mathbf{v}_1, \ldots, \mathbf{W}_j \cdot \mathbf{v}_k), \qquad \textit{10.9}$$

both of these assertions following easily from equation 10.7.

❑

We note that if $k \neq r$, the product **VW** is not even defined, and even if $k = r = n$, it is not usually the case that **VW = WV**. That is, **matrices do not satisfy the commutative law of multiplication**.

Definition 10.4: Scalar multiple of a matrix, sum of matrices

Let **V** and **W** be matrices (Definition 10.1 on page 328) whose *j-i* , (*j*-th row, *i*-th column) elements are $\mathbf{V}_{ji}$ and $\mathbf{W}_{ji}$ respectively.

If a is a real number and **V** is a real $n \times k$ matrix, then $a\mathbf{V}$ is the matrix whose *j-i* element is $a\mathbf{V}_{ji}$.

If **V** and **W** are real $n \times k$ matrices, then **V+W** is the matrix whose *j-i* element is $\mathbf{V}_{ji} + \mathbf{W}_{ji}$.

These definitions just correspond to the functions $a\mathcal{V}$ and $\mathcal{V} + \mathcal{W}$.

If $\mathbf{X}_1, \mathbf{X}_2, \ldots$ is a sequence of real $n \times k$ matrices, the sum $\sum_{i=1}^{k} \mathbf{X}_i$ is defined in the usual way, by induction (see Definition 1.7 on page 12).

❑

Exercises 10.3

1. As in Exercise 10.1.2, familiarize yourself with the operations of matrix multiplication and addition just introduced.
2. Show that matrix multiplication is not commutative by computing **VW** and **WV** for

$$\mathbf{V} = \begin{bmatrix} 1 & 2 \\ 3 & 4 \end{bmatrix} \text{ and } \mathbf{W} = \begin{bmatrix} 5 & 6 \\ 7 & 8 \end{bmatrix}.$$

   It's not necessary to compute the entire matrices **VW** and **WV**.
3. Show that **matrix multiplication is associative**:

$$\mathbf{A(BC) = (AB)C}$$

   and distributive

$$\mathbf{A(B+C) = AB + AC}.$$

4. Find a matrix **A** such that $\mathbf{AV} = a\mathbf{V}$, where **V** is a given matrix and a is a real number.

3 Rudimentary Matrix Inversion

Many reasons for wanting to solve systems of linear equations have been provided earlier, along with some methods of implementing a solution. For theoretical work and to develop a deeper understanding of such solutions, and in the practical situation where it is necessary to obtain numerical solutions of a system of n equations in n unknowns, $\mathbf{V}\boldsymbol{a} = \boldsymbol{y}$, for given **V** and many different values of $\boldsymbol{y}$, it is important to introduce solution of this system by *matrix inversion.* The inverse of an $n \times n$ matrix **V** turns out to be the unique matrix corresponding to the function $\mathcal{V}^{-1}$ provided this inverse function exists. Recall Definition 2.8 on page 37 of the inverse of a function. We repeat this definition, with current symbols, to have it close by.

Definition 10.5: Inverse function of function $\mathcal{V}$

Suppose that $\mathcal{V}$ is a function (Definition 1.1 on page 2) with some specified domain, *dmn* $\mathcal{V}$ and range, *rng* $\mathcal{V}$. If there is a function $\mathcal{F}$ whose ***domain*** is *rng* $\mathcal{V}$ and whose ***range*** is *dmn* $\mathcal{V}$, satisfying the equation

$$\mathcal{F}(\boldsymbol{x}) = y \quad \text{for all } \boldsymbol{y} \text{ such that } \boldsymbol{y} = \mathcal{V}(\boldsymbol{x}) \text{ for some } \boldsymbol{x} \text{ in dmn } \mathcal{V},$$

then the function $\mathcal{F}$ is called the *inverse function* of the function $\mathcal{V}$, and is commonly denoted by the symbol $\mathcal{V}^{-1}$. Note that the equation

$$\mathcal{V}^{-1}(\mathcal{V}(\boldsymbol{x})) = \boldsymbol{x} \text{ for all } \boldsymbol{x} \textit{ in dmn } \mathcal{V} \qquad \textbf{10.10}$$

can therefore be considered the defining equation for $\mathcal{V}^{-1}$.

❑

It is ***not*** immediately evident that there exists a matrix corresponding to the inverse $\mathcal{V}^{-1}$, assuming this inverse function exists. However, in the situation under consideration, it is not hard to believe that there is an inverse matrix, because it is easy to see (and will be shown shortly) that the operations used in Gaussian elimination (see the paragraph following Exercises 9.14, on page 323) to solve a system of equations can be accomplished by matrix multiplication.

It is evident there is no hope of finding an inverse $\mathcal{V}^{-1}$ of the function $\mathcal{V}$ unless the columns of the corresponding matrix $\mathbf{V}$ are linearly independent (Definition 9.7 on page 297) Because in the case of ***linearly dependent columns*** of $\mathbf{V}$, suppose the system $\mathbf{V}\boldsymbol{a} = \boldsymbol{y}$ has some solution, $\boldsymbol{a} = (a_1,...,a_n)$ and column j of $\mathbf{V}$ is linearly dependent on the remaining columns. Then it's easy to see that if the coefficient a_j is replaced by any other real value whatsoever, suitable changes in the other a_i can be made so that, calling the altered vector $\boldsymbol{a}^*$, we still have $\mathbf{V}\boldsymbol{a}^* = \boldsymbol{y}$. But then there would be two distinct values, $\boldsymbol{a}$ and $\boldsymbol{a}^*$ with $\mathcal{V}(\boldsymbol{a}) = \mathcal{V}(\boldsymbol{a}^*)$. It would follow, using Definition 10.5, that $\mathcal{V}$ cannot have an inverse function (for then we would have to have both

$$\mathcal{V}^{-1}(\boldsymbol{y}) = \boldsymbol{a} \quad \text{and} \quad \mathcal{V}^{-1}(\boldsymbol{y}) = \boldsymbol{a}^*,$$

which is not possible when $\boldsymbol{a}$ is different from $\boldsymbol{a}^*$).

So we know that ***the*** $n \times n$ ***matrix*** $\mathbf{V}$ ***cannot have an inverse matrix unless its columns are linearly independent. But this doesn't yet show that if its columns are linearly independent that it does possess an inverse matrix.***

If there is an inverse matrix, $\mathbf{V}^{-1}$, then corresponding to the defining equation 10.10, we must have

$$\mathbf{V}^{-1}\mathbf{V}\boldsymbol{x} = \boldsymbol{x}$$

for all n-dimensional real vectors $\boldsymbol{x}$. Choosing $\boldsymbol{x}$ to be successively the standard unit vectors $\boldsymbol{e}_1, \boldsymbol{e}_2,...,\boldsymbol{e}_n$ (see Definition 8.2 on page 263) and using the fact that for each $n \times n$ matrix $\mathbf{M}$, the product $\mathbf{M}\boldsymbol{e}_k$ is the k-th column of $\mathbf{M}$ (just apply equation 10.4 on page 328 with $\mathbf{M}$ replacing $\mathbf{V}$ and $\boldsymbol{e}_k$ replacing $\boldsymbol{a}$) we see that the k-th column of $\mathbf{M} = \mathbf{V}^{-1}\mathbf{V}$ must be $\boldsymbol{e}_k$. That is, the $n \times n$ real matrix $\mathbf{V}$ has an inverse matrix, $\mathbf{F}$, if and only if

$$\mathbf{F}\,\mathbf{V} = [\boldsymbol{e}_1, \boldsymbol{e}_2,...,\boldsymbol{e}_n].$$

Definition 10.6: The $n \times n$ identity matrix I

The matrix $[\boldsymbol{e}_1, \boldsymbol{e}_2,...,\boldsymbol{e}_n]$, whose j-th column is the standard unit vector, $\boldsymbol{e}_j$ (see Definition 8.2 on page 263) is called the $n \times n$ *identity matrix*, and is denoted by the symbol $\mathbf{I}$, or sometimes by $\mathbf{I}^{n \times n}$.

The $n \times n$ identity matrix is one whose diagonal elements are all 1, and whose *off-diagonal* elements are all 0 — e.g., the 3×3 identity matrix is

$$\begin{bmatrix} 1 & 0 & 0 \\ 0 & 1 & 0 \\ 0 & 0 & 1 \end{bmatrix}$$

❑

Definition 10.7 Inverse,[1] $\mathbf{V}^{-1}$, of a matrix, V

Let $\mathbf{V}$ be a real $n \times n$ matrix (Definition 10.1 on page 328). Recalling the definition of matrix multiplication (Definition 10.3 on page 331) if it exists, the **inverse** $\mathbf{V}^{-1}$, of $\mathbf{V}$ is the unique matrix satisfying the equation

$$\mathbf{V}^{-1}\mathbf{V} = \mathbf{I}^{n \times n}. \qquad \textit{10.11}$$

❑

What we'll do next is construct a plausible candidate for the inverse of an $n \times n$ matrix $\mathbf{V}$ with linearly independent columns, and then show that this candidate does the job. From Theorem 9.13 on page 302, together with Theorem 9.15 on page 303, assuming them to be linearly independent, the

1. While we could define the inverse of a matrix $\mathbf{V}$ as the matrix corresponding to the inverse function, $\mathcal{V}^{-1}$ of the function $\mathcal{V}$ which corresponds to $\mathbf{V}$ (if it exists) this complicates matters too much.

columns of $\mathbf{V}$ form a basis for the space of n-dimensional real vectors, and hence the system

$$\mathbf{V}\boldsymbol{a} = \boldsymbol{y}$$

has a unique solution for each choice of the n-dimensional vector $\boldsymbol{y}$. So choosing $\boldsymbol{y}$ successively to be the standard unit vectors $\boldsymbol{e}_1, \boldsymbol{e}_2, ..., \boldsymbol{e}_n$ (Definition 8.2 on page 263) we let $\boldsymbol{a}_i$ be the unique solution of

$$\mathbf{V}\boldsymbol{a}_i = \boldsymbol{e}_i \qquad \textit{10.12}$$

Now let the $n \times n$ matrix $\mathbf{A}$ be the matrix whose i-th column is $\boldsymbol{a}_i$ — i.e.,

$$\mathbf{A} = [\boldsymbol{a}_1, ..., \boldsymbol{a}_n]. \qquad \textit{10.13}$$

It follows that

$$\mathbf{VA} = \mathbf{I}^{n \times n}. \qquad \textit{10.14}$$

So it *looks like* $\mathbf{A} = \mathbf{V}^{-1}$; and in fact, that is the case. But equation 10.14 provides a **right inverse**, not a **left inverse** that is required by equation 10.11. So we still have some work[1] to do, to show that the matrix $\mathbf{A}$ is also a left inverse. First we show the columns of $\mathbf{A}$ to be linearly independent; for if not, from Definition 9.7 on page 297 of linear independence, for some sequence of real numbers, $b_1, ..., b_n$, not all of which are 0, the equation

$$\sum_{i=1}^{n} b_i \boldsymbol{a}_i = \mathbf{0}$$

would hold. But then, using the linear operator property, Theorem 10.2 on page 330, it would follow that

$$\mathbf{V}\sum_{i=1}^{n} b_i \boldsymbol{a}_i = \sum_{i=1}^{n} b_i \mathbf{V}(\boldsymbol{a}_i) = \sum_{i=1}^{n} b_i \boldsymbol{e}_i = \mathbf{0},$$

which isn't possible unless all of the b_i are 0. We can only conclude that the columns of $\mathbf{A}$ are linearly independent. Under these circumstances a **right inverse**, $\mathbf{C}$, of $\mathbf{A}$, satisfying the equation

$$\mathbf{AC} = \mathbf{I}^{n \times n} \qquad \textit{10.15}$$

can be constructed, via the same method used to construct $\mathbf{A}$. Then using the associativity of matrix multiplication (Exercise 10.3.3 on page 332),

$$\mathbf{C} = \mathbf{IC} = (\mathbf{VA})\mathbf{C} = \mathbf{V}(\mathbf{AC}) = \mathbf{VI} = \mathbf{V},$$

1. This is not totally trivial, first because we know that matrix multiplication is not generally commutative (see Exercise 10.3.2 on page 332). Second, as we will see later on, Problems 11.5 on page 422, a real $n \times k$ matrix $\mathbf{V}$ with linearly independent columns will have many left inverses and no right inverse if $n > k$ (more rows than columns).

i.e., $$\mathbf{C} = \mathbf{V}.$$

Substituting $\mathbf{V}$ for $\mathbf{C}$ in equation 10.15 yields the result

$$\mathbf{AV} = \mathbf{I},$$

showing that the originally constructed right inverse, $\mathbf{A}$, is also a left inverse of $\mathbf{V}$. (See Problems 10.4 on page 337 for an alternate more straightforward argument establishing this result.)

Finally, we show that the just constructed matrix $\mathbf{A}$ is the unique left inverse of $\mathbf{V}$. For if $\mathbf{BV} = \mathbf{I}$, then

$$\mathbf{B} = \mathbf{BI} = \mathbf{B(VA)} = \mathbf{(BV)A} = \mathbf{IA} = \mathbf{A}.$$

We summarize the results derived in the preceding discussion.

Theorem 10.8: Inverse of $n \times n$ matrix

The real $n \times n$ matrix $\mathbf{V}$ (Definition 10.1 on page 328) has an inverse matrix, $\mathbf{V}^{-1}$ (Definition 10.7 on page 334) if and only if its columns are linearly independent (Definition 9.7 on page 297). The i-th column, $\boldsymbol{a}_i$, of $\mathbf{V}^{-1}$ is the solution of

$$\mathbf{V}\boldsymbol{a}_i = \boldsymbol{e}_i,$$

where $\boldsymbol{e}_i$ are the standard unit vectors, defined in Definition 8.2 on page 263.
The matrices $\mathbf{V}$ and $\mathbf{V}^{-1}$ are left and right inverses of each other, i.e.,

$$\mathbf{V}^{-1}\mathbf{V} = \mathbf{I} = \mathbf{V}\mathbf{V}^{-1}.$$

❑

Definition 10.9: Invertibility, singularity

If an $n \times n$ matrix $\mathbf{V}$ has an inverse (see Definition 10.7 on page 334 and Theorem 10.8 above) it is said to be **invertible** or **nonsingular**. If it fails to have an inverse, it is called **singular.**

❑

The following evident result is very useful for theoretical development.

Theorem 10.10: System solution and matrix inversion

The system of n equations in n unknowns

$$\mathbf{V}^{n \times n} \boldsymbol{a} = \boldsymbol{y}^{n \times 1} \qquad \textit{10.16}$$

has as its solution a unique vector $\boldsymbol{a}$ if and only if the $n \times n$ matrix (Definition 10.1 on page 328) $\mathbf{V}^{n \times n}$ is invertible (Definition 10.9 above) i.e., if and only if the columns of $\mathbf{V}$ are linearly independent (Definition 9.7 on page 297). In this situation the system is solvable for $\boldsymbol{a}$ via multiplication of both sides of equation 10.16 by $\mathbf{V}^{-1}$ to obtain

$$\boldsymbol{a} = \mathbf{V}^{-1} \boldsymbol{y} \cdot \qquad \textit{10.17}$$

❑

While equation 10.17 is usable to solve system 10.16, since determination of $\mathbf{V}^{-1}$ is equivalent to solving n such systems (one for each choice of $\boldsymbol{y} = \boldsymbol{e}_i$), ***it doesn't pay to compute*** $\mathbf{V}^{-1}$ ***to solve only*** **one** ***such system***.[1]

One final, easily established result is needed for reference later on.

Theorem 10.11: Characterization of the identity, I

If $\mathbf{A}$ is an $n \times n$ matrix (Definition 10.1 on page 328) such that $\mathbf{A}\boldsymbol{x} = \boldsymbol{x}$ for all n-dimensional vectors $\boldsymbol{x}$, then $\mathbf{A} = \mathbf{I}^{n \times n}$ (Definition 10.6 on page 334).

❑

Problems 10.4

1. Establish Theorem 10.8 by the somewhat more straightforward argument that since $\mathbf{V}\boldsymbol{a}_i = \boldsymbol{e}_i$ (equation 10.12 on page 335), from equation 10.13 on page 335 (the definition of $\mathbf{A}$) $(\mathbf{AV})\boldsymbol{a}_i = \boldsymbol{a}_i$ for all $i = 1,...,n$. Using the already established linear independence of the $\boldsymbol{a}_i$, then show that the equation $(\mathbf{AV})\boldsymbol{e}_i = \boldsymbol{e}_i$ must hold for all i. Conclude therefore that $\mathbf{AV}$ must be the identity matrix, and that therefore $\mathbf{A}$ is the left inverse of $\mathbf{V}$.

2. Establish directly from the definition of the matrix $\mathbf{A}$, equation 10.13 on page 335, that $\boldsymbol{x} = \mathbf{A}\boldsymbol{y}$ is a solution of the equation $\mathbf{V}\boldsymbol{x} = \boldsymbol{y}$ by calculating $\mathbf{V}(\mathbf{A}\boldsymbol{y})$ using the linear operator property, Theorem 10.2 on page 330. This result is related to the *impulse response* approach to solving differential equations, to be treated in Section 8 of Chapter 15 (see also Theorem 15.46 on page 741).

---

1. Reminding us of the joke whose punch line is *"The chef says that for one crummy sandwich it don't pay to kill no elephant."*

3. Establish Theorem 10.11.
   Hint: By successively choosing $\boldsymbol{x}$ to be $\boldsymbol{e}_1, \boldsymbol{e}_2, ..., \boldsymbol{e}_n$, we see that $\mathbf{AI} = \mathbf{I}$. But evidently, $\mathbf{AI} = \mathbf{A}$.

Exercises 10.5

Invert the following matrices if possible. Check your results using some reliable computer package, such as *Maple, Mathematica, Mathcad, Splus.*

1. $$\begin{bmatrix} 1 & 2 & 3 \\ 1 & 1 & 1 \\ 1 & 1 & -1 \end{bmatrix}$$

2. $$\begin{bmatrix} 1/2 & 1/3 & 1/4 \\ 1/3 & 1/4 & 1/5 \\ 1/4 & 1/5 & 1/6 \end{bmatrix}$$

3. $$\begin{bmatrix} 2 & 1 & 4 \\ 3 & 3.5 & .5 \\ 6 & 8 & 5 \end{bmatrix}$$

4. $$\begin{bmatrix} 6 & 4 & 3 \\ 20 & 15 & 12 \\ 15 & 12 & 10 \end{bmatrix}$$

Problems 10.6

1. Given an invertible matrix $\mathbf{V}$, what is the effect of replacing the j-th row of $\mathbf{V}$ by a nonzero multiple of this row on $\mathbf{V}^{-1}$?
2. Given an invertible matrix $\mathbf{V}$, what is the effect of replacing the i-th column of $\mathbf{V}$ by a nonzero multiple of this column on $\mathbf{V}^{-1}$?

The most widely used methods of matrix inversion by hand computation are those based on Gaussian elimination. These methods make repeated use of the following easy-to-establish result, which can be verified using expressions 10.8 and 10.9 from Definition 10.3 on page 331.

Theorem 10.12: Operations on product of matrices

Suppose that **V**, **W**, and **X** are matrices (Definition 10.1 on page 328) for which the equation **V** = **WX** holds (see Definition 10.3 on page 331). Then the effect on **V** of adding a multiple of a given row of **W** to some other specified row of **W** is the same as adding this multiple of the same row of **V** to the same specified row of **V**. Interchanging rows of **W** has the effect of interchanging the corresponding rows of **V**. Multiplying a row of **W** by a real number multiplies the same row of **V** by this number.

The effect on **V** of adding a multiple of a given column of **X** to some other specified column of **X** is the same as adding this multiple of the same column of **V** to the same specified column of **V**. Interchanging columns of **X** has the effect of interchanging the corresponding columns of **V**. Multiplying a column of **X** by a given number has the effect of multiplying the same column of **V** by this number.

❑

To use Theorem 10.12, we start with the equation

$$\mathbf{V} = \mathbf{IV}. \qquad \textit{10.18}$$

On **V** we now successively perform the row operations indicated in Theorem 10.12 so as to reduce the left side of equation 10.18 to the identity matrix — preserving the equality by performing these operations on the left factor of the right-hand side. When we are done, we have an equation of the form

$$\mathbf{I} = \mathbf{FV},$$

and it follows from Theorem 10.8 on page 336 that **F** is the desired inverse. Notice that we need not keep the equal sign or the matrix **V** on the right-hand side in implementing these calculations. Here is one example, omitting unnecessary symbols.

Example 10.7: Inverting $\mathbf{V} = \begin{bmatrix} 1 & 1 & 1 \\ 1 & 1 & 2 \\ 1 & 0 & 3 \end{bmatrix}$

Start by writing **V** and **I** (side by side):

$$\begin{bmatrix} 1 & 1 & 1 \\ 1 & 1 & 2 \\ 1 & 0 & 3 \end{bmatrix} \qquad \begin{bmatrix} 1 & 0 & 0 \\ 0 & 1 & 0 \\ 0 & 0 & 1 \end{bmatrix}$$

Now add $-1*$row 1 to row 2 to obtain

$$\begin{bmatrix} 1 & 1 & 1 \\ 0 & 0 & 1 \\ 1 & 0 & 3 \end{bmatrix} \qquad \begin{bmatrix} 1 & 0 & 0 \\ -1 & 1 & 0 \\ 0 & 0 & 1 \end{bmatrix}$$

Next add $-1*$row 1 to row 3 to obtain

$$\begin{bmatrix} 1 & 1 & 1 \\ 0 & 0 & 1 \\ 0 & -1 & 2 \end{bmatrix} \qquad \begin{bmatrix} 1 & 0 & 0 \\ -1 & 1 & 0 \\ -1 & 0 & 1 \end{bmatrix}$$

Now add $-1*$row 2 to row 1 to get

$$\begin{bmatrix} 1 & 1 & 0 \\ 0 & 0 & 1 \\ 0 & -1 & 2 \end{bmatrix} \qquad \begin{bmatrix} 2 & -1 & 0 \\ -1 & 1 & 0 \\ -1 & 0 & 1 \end{bmatrix}$$

Then add $-2*$row 2 to row 2, getting

$$\begin{bmatrix} 1 & 1 & 0 \\ 0 & 0 & 1 \\ 0 & -1 & 0 \end{bmatrix} \qquad \begin{bmatrix} 2 & -1 & 0 \\ -1 & 1 & 0 \\ 1 & -2 & 1 \end{bmatrix}$$

Next add row 3 to row 1

$$\begin{bmatrix} 1 & 0 & 0 \\ 0 & 0 & 1 \\ 0 & -1 & 0 \end{bmatrix} \qquad \begin{bmatrix} 3 & -3 & 1 \\ -1 & 1 & 0 \\ 1 & -2 & 1 \end{bmatrix}$$

Multiply row 3 by -1, obtaining

$$\begin{bmatrix} 1 & 0 & 0 \\ 0 & 0 & 1 \\ 0 & 1 & 0 \end{bmatrix} \qquad \begin{bmatrix} 3 & -3 & 1 \\ -1 & 1 & 0 \\ -1 & 2 & -1 \end{bmatrix}$$

Finally, interchange rows 2 and 3, to yield

$$\begin{bmatrix} 1 & 0 & 0 \\ 0 & 1 & 0 \\ 0 & 0 & 1 \end{bmatrix} \qquad \begin{bmatrix} 3 & -3 & 1 \\ -1 & 2 & -1 \\ -1 & 1 & 0 \end{bmatrix}$$

This establishes that

$$\mathbf{V}^{-1} = \begin{bmatrix} 1 & 1 & 1 \\ 1 & 1 & 2 \\ 1 & 0 & 3 \end{bmatrix}^{-1} = \begin{bmatrix} 3 & -3 & 1 \\ -1 & 2 & -1 \\ -1 & 1 & 0 \end{bmatrix}$$

Carrying out the multiplication

$$\begin{bmatrix} 3 & -3 & 1 \\ -1 & 2 & -1 \\ -1 & 1 & 0 \end{bmatrix} \begin{bmatrix} 1 & 1 & 1 \\ 1 & 1 & 2 \\ 1 & 0 & 3 \end{bmatrix}$$

confirms the computation.

■

Exercises 10.8

Invert the matrices of Exercises 10.5 on page 338.

We now show how to use orthogonalization methods in the inversion of matrices. To try to invert the matrix

$$\mathbf{V} = [v_1,...,v_n], \qquad \textit{10.19}$$

we first attempt to orthonormalize $v_1,...,v_n$ (Definition 9.19 on page 306). If we fail, because for some $j = 2,...,n,$ either $\|v_j\| = 0$ or

$$\frac{\left\| v_j - P_{[\![w_1, w_2,...,w_{j-1}]\!]} \right\|}{\|v_j\|}$$

is too close to 0, we agree to say that $\mathbf{V}$ is not invertible, because column j is either linearly dependent on the previous columns or is too close to being linearly dependent on them (see Theorem 9.20 on page 307) and then the arguments on page 333 concerning the inability to invert matrices with linearly dependent columns apply.

The next step for inversion, suggested by the technique of solution of systems of equations by orthogonalization (Definition 9.25 on page 320) requires formation of the dot product of one of the orthonormalized vectors with one of the original ones. In order to carry this out neatly using matrix notation we need the concept of the transpose of a matrix. The transpose $\mathbf{W}^T$ of the matrix $\mathbf{W}$ is the matrix whose j-th column has the same elements as the j-th row of the matrix $\mathbf{W}$.

Definition 10.13: Transpose of a matrix

If $\mathbf{W}$ is a rectangular matrix (Definition 10.1 on page 328) whose i-j element is W_{ij}, its transpose, $\mathbf{W}^{\mathrm{T}}$, is the rectangular matrix whose i-j element is the j-i element, W_{ji} of $\mathbf{W}$. If $\mathbf{W}$ is an $m \times n$ matrix, then $\mathbf{W}^{\mathrm{T}}$ is an $n \times m$ matrix.

❑

If $\mathbf{W}$ is the matrix consisting of the orthonormalized columns of the matrix $\mathbf{V}$ (see Definition 9.19 on page 306) note that the i-j element of the matrix product $\mathbf{W}^{\mathrm{T}}\mathbf{V}$ is $w_i \cdot v_j$ where w_i is the i-th column of $\mathbf{W}$ and v_j the j-th column of $\mathbf{V}$; from the orthonormality $w_i \cdot v_j = 0$ for $i > j$, and hence, in theory, the matrix $\mathbf{U} = \mathbf{W}^{\mathrm{T}}\mathbf{V}$ is *upper triangular* — as illustrated in Figure 10-2, meaning that below the diagonal there are only

$$\mathbf{U} = \begin{bmatrix} a & b & c & d \\ 0 & e & f & g \\ 0 & 0 & h & i \\ 0 & 0 & 0 & j \end{bmatrix}$$

Upper triangular 4×4 matrix

Figure 10-2

zeros.

Since the columns of $\mathbf{W}$ are orthonormal, it follows that

$$\mathbf{W}^{\mathrm{T}}\mathbf{W} = \mathbf{I}^{n \times n}$$

or

$$\mathbf{W}^{\mathrm{T}} = \mathbf{W}^{-1}. \qquad \textit{10.20}$$

Thus we may write, using the fact that $\mathbf{W}^{\mathrm{T}}$ is also a right inverse of $\mathbf{W}$ (see Theorem 10.8 on page 336)

$$\mathbf{V} = \mathbf{W}(\mathbf{W}^{\mathrm{T}}\mathbf{V}) = \mathbf{WU}, \qquad \textit{10.21}$$

where $\mathbf{U}$ is upper triangular, a property which will prove very useful.

At this point we need an easily established result, namely, the following.

Theorem 10.14: Inverse of matrix product

The product, $\mathbf{AB}$ (Definition 10.3 on page 331) of two $n \times n$ invertible matrices (Definition 10.9 on page 336) $\mathbf{A}$ and $\mathbf{B}$ is itself invertible, with

$$(\mathbf{AB})^{-1} = \mathbf{B}^{-1}\mathbf{A}^{-1}.$$

❑

This result follows from the associativity of matrix multiplication (see Exercise 10.3.3 on page 332) simply multiplying **AB** by $\mathbf{B}^{-1}\mathbf{A}^{-1}$. It's easy to remember by recalling that you usually take off your socks *after* taking off your shoes, because you put on the socks *before* putting on your shoes.

Because $\mathbf{U} = \mathbf{W}^{\mathrm{T}}\mathbf{V}$ as given above in equation 10.21, from Theorem 10.14 it follows that $\mathbf{U}$ is invertible; applying Theorem 10.14 to the left and right of equation 10.21, we find, using equation 10.20, that

$$\mathbf{V}^{-1} = \mathbf{U}^{-1}\mathbf{W}^{-1} = \mathbf{U}^{-1}\mathbf{W}^{\mathrm{T}} . \qquad \mathit{10.22}$$

To complete the task of determining $\mathbf{V}^{-1}$ in the method depending on orthogonalization, we need only see how to compute $\mathbf{U}^{-1}$. Notice, that once the Gram-Schmidt procedure has successfully produced n nonzero orthonormal vectors, we know that $\mathbf{V}$ and $\mathbf{U}$ have inverses. There are then no problems related to some coefficients being zero — a problem that does arise in simple Gaussian elimination, which complicates the computer code. Before proceeding with the general algebraic approach to inverting the upper triangular matrix $\mathbf{U}$, we look at the particular case

$$\mathbf{U} = \begin{bmatrix} a & b & c & d \\ 0 & e & f & g \\ 0 & 0 & h & i \\ 0 & 0 & 0 & j \end{bmatrix}.$$

Let the inverse, $\mathbf{U}^{-1}$, be represented as

$$\mathbf{U}^{-1} = \begin{bmatrix} A & B & C & D \\ E & F & G & H \\ J & K & L & M \\ N & P & Q & R \end{bmatrix}$$

Then, because

$$\mathbf{U}\mathbf{U}^{-1} = \mathbf{I}^{4\times 4},$$

we must have

$$\begin{bmatrix} a & b & c & d \\ 0 & e & f & g \\ 0 & 0 & h & i \\ 0 & 0 & 0 & j \end{bmatrix}\begin{bmatrix} A \\ E \\ J \\ N \end{bmatrix} = \begin{bmatrix} 1 \\ 0 \\ 0 \\ 0 \end{bmatrix}.$$

Looking at the last row, we see that $jN = 0$, and since $j \neq 0$, it follows that $N = 0$. Looking at the next to last row, we see that $iJ + jN = 0$, which, using the fact that $N = 0$ and $i \neq 0$, shows that $J = 0$, and in the same manner, we find $H = 0$. The final equation for the first column is $aA + bE + cJ + dN = 1$, which, utilizing the previous results, yields

$$A = 1/d.$$

In the same way, going on to the second column shows that $P = 0$, $K = 0$, and

$$F = 1/e, \quad B = -\frac{b}{a}F = -\frac{b}{ae}.$$

It should be evident how to continue this process, which shows that $\mathbf{U}^{-1}$ is upper triangular, and allows us to determine all of its elements, one at a time. This procedure generalizes algebraically as follows:

To invert the invertible lower triangular matrix $\mathbf{U}$, denote the *i-j* element of $\mathbf{U}$ by u_{ij}, and let m_{ij} denote the *i-j* element of $\mathbf{M} = \mathbf{U}^{-1}$. We want to determine the rows $\boldsymbol{M}_1,...,\boldsymbol{M}_n$ of $\mathbf{M}$ to satisfy

$$\boldsymbol{M}_i \cdot \boldsymbol{u}_j = \begin{cases} 1 & \text{if} \quad i = j \\ 0 & \text{if} \quad i \neq j, \end{cases} \qquad \textit{10.23}$$

where $\boldsymbol{u}_j$ is the *j*-th column of **U.** Written out in terms of coordinates, this is

$$\sum_{k=1}^{n} m_{ik}u_{kj} = \begin{cases} 1 & \text{if} \quad i = j \\ 0 & \text{if} \quad i \neq j \end{cases} \qquad \textit{10.24}$$

Since $\mathbf{U}$ is upper triangular, $u_{kj} = 0$ if $k > j$. Hence we may rewrite equation as

$$\sum_{k=1}^{j} m_{ik}u_{kj} = \begin{cases} 1 & \text{if} \quad i = j \\ 0 & \text{if} \quad i \neq j, \end{cases} \qquad \textit{10.25}$$

where we note that the upper limit of summation is what has changed. Next we show that $\mathbf{M}$ must be *upper triangular,*[1] to obtain even further simplification. (This result would have been apparent had we done some simple computations

---

1. 0s below the diagonal.

on 3×3 matrices.) The upper triangularity of **M** follows from equation 10.23 because

$$\begin{aligned} \boldsymbol{M}_i \cdot \boldsymbol{u}_1 &= 0 \\ &\cdot \\ &\cdot \\ \boldsymbol{M}_i \cdot \boldsymbol{u}_{i-1} &= 0 \end{aligned} \qquad \textbf{\textit{10.26}}$$

for each $i > 1$.

But

$$\boldsymbol{u}_1 = \begin{pmatrix} u_{11} \\ 0 \\ \cdot \\ \cdot \\ 0 \end{pmatrix}$$

so that the first of the equations 10.26 yields $m_{i1}u_{11} = 0$. But $u_{11} \neq 0$, because if this were not so, **U** would not be invertible. Hence $m_{i1} = 0$ for $i > 1$.

Since

$$\boldsymbol{u}_2 = \begin{pmatrix} u_{12} \\ u_{22} \\ 0 \\ \cdot \\ \cdot \\ 0 \end{pmatrix}.$$

Now $u_{22} \neq 0$, because if it were, $\boldsymbol{u}_1$ and $\boldsymbol{u}_2$ would be linearly dependent, again leading to the *singularity* (Definition 10.9 on page 336) of **U**, which is not possible. But now the second of equations 10.26, provided $i > 2$,

$$0 = \boldsymbol{M}_i \cdot \boldsymbol{u}_2 = m_{i1}u_{21} + m_{i2}u_{22}$$

together with the previously proved $m_{i1} = 0$ for $i > 1$, shows that $m_{i2} = 0$ for $i > 2$, Continuing on inductively, we find

$$m_{ik} = 0 \quad \text{for} \quad k < i, \qquad \textbf{\textit{10.27}}$$

showing that $\mathbf{M} = \mathbf{U}^{-1}$ is also upper triangular.

So we may rewrite equations 10.25, the system to be solved for the elements m_{ik} of $\mathbf{M} = \mathbf{U}^{-1}$ as

$$\sum_{k=1}^{j} m_{ik}u_{kj} = \begin{cases} 1 & \text{if} \quad i = j \\ 0 & \text{if} \quad i \neq j \end{cases}, \text{ with } m_{ik} = 0 \quad \text{for} \quad k < i, \qquad \textbf{\textit{10.28}}$$

or, using the conditions associated with equations 10.28,

$$\sum_{k=i}^{j} m_{ik}u_{kj} = \begin{cases} 1 & \text{if} \quad i = j \\ 0 & \text{if} \quad i \neq j. \end{cases} \qquad \textbf{\textit{10.29}}$$

These are the equations we will actually use to find the elements of **M**.

Taking $j = i$ in these equations yields

$$m_{ii} = \frac{1}{u_{ii}} \quad \text{for} \quad i = 1, 2,...,n. \qquad \textbf{\textit{10.30}}$$

To obtain the remaining elements of **M**, we choose $k = j$ and solve equations 10.29 for m_{ij}, to obtain

$$m_{ij} = -\frac{\sum_{k=i}^{j-1} m_{ik}u_{kj}}{u_{jj}} \quad \text{for } j > i. \qquad \textbf{\textit{10.31}}$$

Once again, we note that

$$m_{ij} = 0 \quad \text{for} \quad j < i. \qquad \textbf{\textit{10.32}}$$

Notice that having orthonormalized the columns of **V** to be the columns of **W**, letting **U** be defined as $\mathbf{W}^T\mathbf{V}$, with *i-j* elements u_{ij}, and the inverse **M** of **U** having *i-j* elements m_{ij}, which can be calculated from equations 10.30, 10.31 and 10.32, we have all the information needed to use equation 10.22 on page 343 to calculate $\mathbf{V}^{-1}$. We summarize this as follows.

Theorem 10.15: Matrix inversion using orthogonalization

Let **V** be a real $n \times n$ matrix (Definition 10.1 on page 328). To invert **V** (Definition 10.9 on page 336), using orthogonalization methods

- Orthonormalize the columns of **V** putting them in a matrix **W** (see Definition 9.19 on page 306). Admit defeat if for some $j = 2,...,n$, either $\|\boldsymbol{v}_j\| = 0$, or

$$\frac{\left\|\boldsymbol{v}_j - P_{[\![\boldsymbol{w}_1, \boldsymbol{w}_2,...,\boldsymbol{w}_{j-1}]\!]}\right\|}{\|\boldsymbol{v}_j\|} = 0.$$

- If the first step has succeeded, form the upper triangular matrix $\mathbf{U} = \mathbf{W}^{\mathrm{T}}\mathbf{V}$, where the transpose $\mathbf{W}^{\mathrm{T}}$ of $\mathbf{W}$ is defined in Definition 10.13 on page 342.
- Compute the inverse, $\mathbf{M}$, of $\mathbf{U}$ (the i-j elements of $\mathbf{M}$ and $\mathbf{U}$ are denoted by m_{ij} and u_{ij}, respectively) from the following equations:

$$m_{ii} = \frac{1}{u_{ii}} \quad \text{for } i = 1, 2, \ldots, n$$

$$m_{ij} = -\frac{\sum_{k=i}^{j-1} m_{ik}u_{kj}}{u_{jj}} \quad \text{for } j > i$$

$$m_{ij} = 0 \quad \text{for } j < i$$

- The inverse of $\mathbf{V}$ is given by

$$\mathbf{V}^{-1} = \mathbf{M}\mathbf{W}^{\mathrm{T}},$$

(see Definition 10.3 on page 331).

❑

Exercises 10.9

Invert the matrices of Exercises 10.5 on page 338 and the matrix in Example 10.7 on page 339 using the methods just developed. If necessary, try a few by hand, but if possible use a computer to carry out the various computations.

4 Change of Coordinates and Rotations by Matrices

If $\mathbf{y}$ stands for an arbitrary real n-tuple (sequence of n reals, see Definition 1.2 on page 3) the matrix equation

$$\mathbf{Va} = \mathbf{y}$$

(see equation 10.3 on page 328) can be interpreted as equations to express $\mathbf{y}$ in terms of the columns of $\mathbf{V}$ (see equation 10.4 on page 328). If $\mathbf{V}$ is invertible (see Theorem 10.10 on page 337) the unique solution $\mathbf{a} = \mathbf{V}^{-1}\mathbf{y}$ of $\mathbf{Va} = \mathbf{y}$ yields the coordinates of $\mathbf{y}$ in terms of the columns of $\mathbf{V}$ (see Definition 9.9 on page 298 and generally, Section 2 of Chapter 9).

Thus, $\mathbf{V}$ can be thought of as specifying a change of coordinates for a vector, originally expressed in terms of the vectors $\boldsymbol{e}_j$ (see Definition 8.2 on page 263). If the columns of $\mathbf{V}$ are orthogonal (Theorem 8.15 on page 278) and of unit length (Definition 8.6 on page 270) then $\mathbf{V}$ is called *orthonormal* and specifies a change of coordinates to an orthogonal coordinate system (see Theorem 9.17 on page 304). In particular, if the original coordinate axes can be *rotated* to obtain new axes, this change of coordinates can be specified by an orthonormal matrix.

Alternatively, we can look at the matrix **V** as specifying a function whose value at **a** is $(\mathbf{y} = \mathbf{Va})$. An orthonormal **V** will correspond to a transformation which is a rotation, together with possible reflections — e.g., in two dimensions the transformation

$$\mathbf{V}\begin{pmatrix} x_1 \\ x_2 \end{pmatrix} = \begin{bmatrix} 1 & 0 \\ 0 & -1 \end{bmatrix}\begin{pmatrix} x_1 \\ x_2 \end{pmatrix} = \begin{pmatrix} x \\ -x_2 \end{pmatrix}$$

reflects the second axis about the origin and leaves the first axis unchanged. (As part of a three-dimensional picture, it is a rotation about the third (z) axis, but in two dimensions it isn't a rotation.) ***Rotations are important for implementing many computer graphics***.

For reference we now summarize some of the previous discussion.

Definition 10.16: Orthogonal and orthonormal matrices

An $m \times n$ matrix (Definition 10.1 on page 328) is called **orthogonal** if its ***columns*** are mutually orthogonal (see Theorem 8.15 on page 278). A matrix is called **orthonormal** if it is orthogonal and its ***columns*** are of unit length (Definition 8.6 on page 270).

❑

Orthonormal matrix properties

If **V** is an $n \times n$ orthonormal matrix (see definition above) the system $\mathbf{Va} = \mathbf{y}$ (see Definition 10.1 on page 328) represents a change of coordinates consisting of a rotation and some reflections (reversal of axes); if the original coordinates are the elements of **y**, the corresponding new coordinates are the elements of **a**. Every rotation can be represented by an $n \times n$ orthonormal matrix.

Problems 10.10

1. Show that if **V** is an $n \times n$ orthonormal matrix (Definition 10.16 above) then $\mathbf{V}^T\mathbf{V} = \mathbf{I}^{n \times n}$ (see Definition 10.13 on page 342, Definition 10.6 on page 334, and Definition 10.3 on page 331).

2. Show that if **V** is an $n \times n$ orthonormal matrix (Definition 10.16 above), then its **rows** are mutually orthogonal and of unit length. Hint: If $\mathbf{V}^T\mathbf{V} = \mathbf{I}$, then from Theorem 10.8 on page 336 $\mathbf{V}\mathbf{V}^T = \mathbf{I}$.

3. Show that an orthonormal matrix, considered as a coordinate change, preserves distance —i.e., $\|\mathbf{Ax} - \mathbf{Ay}\| = \|\mathbf{x} - \mathbf{y}\|$ (see Definition 8.6 on page 270)

From the results of Problem 1 above, we see that in matrix terms, Theorem 9.17 on page 304 becomes the following.

> If $\mathbf{V}$ is an $n \times n$ orthonormal matrix (see definition above) and $\mathbf{Va} = \mathbf{y}$, then $\mathbf{a} = \mathbf{V}^{\mathrm{T}}\mathbf{y}$ (where $\mathbf{V}^{\mathrm{T}}$ is the transpose of $\mathbf{V}$, see Definition 10.13 on page 342).

Exercises 10.11

1. Write down the matrix specifying a 45-degree counterclockwise rotation in two dimensions.
2. Write down the matrix specifying a change of coordinates to axes in two dimensions which consist of axes 45 degrees counterclockwise of the original axes.
3. Explain the relationship of the matrices in the two previous exercises.

5 Matrix Infinite Series — Theory

The matrix geometric series is useful for error estimation involving matrices in much the same way the numerical geometric series is for analyzing relative error, (recall Exercises 1.10.1 on page 17 and Exercises 1.12.4 on page 20). In order to present the arguments in a simple manner that paves the way to handling the matrix geometric series similarly to our handling of the scalar geometric series, it's convenient to build up some useful machinery. To start we introduce a matrix norm and its properties. For reference purposes we precede this with the following.

Notation 10.17: $\sum\limits_{i,\,j}$

If $g_{i,\,j}$ represents a quantity defined for some specific set, P, of pairs, $(i,\,j)$, when there is no danger of ambiguity the symbol

$$\sum_{i,\,j} g_{i,\,j}$$

stands for the sum of the $g_{i,\,j}$ over all of the pairs (i,j) in P.

❑

Definition 10.18: Norm $\|\mathbf{A}\|$, of the matrix A

If $\mathbf{A}$ is a real matrix (Definition 10.1 on page 328) with elements A_{ij}, the *Frobenius* **norm**, $\|\mathbf{A}\|$ of $\mathbf{A}$, is given by

$$\|\mathbf{A}\| = \sqrt{\sum_{i,\,j} |A_{ij}|^2},$$

where the sum is over all i and j corresponding to elements of $\mathbf{A}$ (see Notation 10.17 above).

❑

Assertion 10.19

We see that if $\|\mathbf{A}^{n \times k}\| < c$, then every element of $\mathbf{A}$ has magnitude less than c, while if every element of $\mathbf{A}$ has magnitude less than d, $\|\mathbf{A}^{n \times k}\| < nkd$; so that knowing something about the norm of a matrix gives information about each of its elements, and conversely.

❑

Note that $\|\mathbf{A}\|$ is just the length of the vector whose elements are the elements of $\mathbf{A}$. In light of this fact, we need to prove the general triangle inequality for vectors, an extension of the triangle inequality for complex numbers (which are vectors themselves); see Theorem 7.6 on page 216. The proof of this more general case is easier and more understandable than that of Theorem 7.6, because of the availability here of the Schwartz inequality.

Theorem 10.20: Triangle inequality for *n*-dimensional vectors

For $\boldsymbol{v}$ and $\boldsymbol{w}$ any n-dimensional vectors (Definition 8.1 on page 261)

$$\|\boldsymbol{v} + \boldsymbol{w}\| \le \|\boldsymbol{v}\| + \|\boldsymbol{w}\|$$

(see Definition 10.18 above).

❑

Using the associativity of dot products (see Footnote 1 on page 217) and the obvious commutativity of real dot products (see Definition 8.10 on page 272) and then the Schwartz inequality (Theorem 8.19 on page 282) it follows that

$$\begin{aligned}\|\boldsymbol{v} + \boldsymbol{w}\|^2 = (\boldsymbol{v} + \boldsymbol{w}) \cdot (\boldsymbol{v} + \boldsymbol{w}) &= \boldsymbol{v} \cdot \boldsymbol{v} + \boldsymbol{w} \cdot \boldsymbol{w} + 2\boldsymbol{v} \cdot \boldsymbol{w} \\ &\le \boldsymbol{v} \cdot \boldsymbol{v} + \boldsymbol{w} \cdot \boldsymbol{w} + 2\|\boldsymbol{v}\|\|\boldsymbol{w}\| \\ &= (\|\boldsymbol{v}\| + \|\boldsymbol{w}\|)^2.\end{aligned}$$

proving Theorem 10.20.

Note that **this result applies to matrix norms**, because of their vector norm interpretation. Also, the **reverse triangle inequality**, Theorem 7.7 on page 218, still holds for n-dimensional vectors (proof being the same as that of Theorem 7.7).

The next result, together with the triangle inequality, is what lets us treat infinite series of matrices almost identically to infinite series of numbers. When the two matrices in this theorem consist of one row and one column, respectively, it is just the Schwartz inequality.

Theorem 10.21: Product norm inequality

If $\mathbf{A}$ and $\mathbf{B}$ are two matrices whose product $\mathbf{AB}$ (Definition 10.3 on page 331) is meaningful, then, recalling Definition 10.18 on page 350,

$$\|\mathbf{AB}\| \leq \|\mathbf{A}\| \, \|\mathbf{B}\| .$$

❑

We establish Theorem 10.21 by letting $(\mathbf{AB})_{i,j}$ denote the i,j element of $\mathbf{AB}$, recalling that A_i denotes the i-th row of $\mathbf{A}$ and b_j the j-th column of $\mathbf{B}$; then, using the Schwartz inequality, Theorem 8.19 on page 282, and Notation 10.17 on page 349, we have

$$\begin{aligned}
\|\mathbf{AB}\|^2 = \sum_{i,j} (\mathbf{AB})_{i,j}^2 &= \sum_{i,j} (A_i \cdot b_j)_{i,j}^2 \\
&\leq \sum_{i,j} (A_i \cdot A_i)(b_j \cdot b_j) \text{ by the Schwartz inequality,} \\
&= \left[\sum_i (A_i \cdot A_i)\right]\left[\sum_j (b_j \cdot b_j)\right] \\
&= \|\mathbf{A}\|^2 \|\mathbf{B}\|^2.
\end{aligned}$$

Note that in this computation we used the fact that $\sum_i (A_i \cdot A_i)$ is just the sum of squares of all of the elements of $\mathbf{A}$, done row by row, while $\sum_j (b_j \cdot b_j)$ is the sum of squares of all elements of $\mathbf{B}$, but done by columns.

Next we introduce matrix infinite series.

Definition 10.22: $\sum_{j=r \to \infty} \mathbf{M}_j, \ \sum_{j=r}^{\infty} \mathbf{M}_j$, r an integer

For each integer, $j \geq r$, suppose that $\mathbf{M}_j$ is a real $n \times k$ matrix (Definition 10.1 on page 328). The sequence of partial sums

$$\sum_{j=r}^{n} \mathbf{M}_j, \quad n = r, r+1,...$$

denoted[1] by

$$\sum_{j = r \to \infty} \mathbf{M}_j$$

is said to **converge[2] to the matrix $\mathbf{M}$**, called the **sum** of this series, and denoted by

$$\sum_{j=r}^{\infty} \mathbf{M}_j,$$

if

$$\lim_{n \to \infty} \left\| \left(\sum_{j=r}^{n} \mathbf{M}_j \right) - \mathbf{M} \right\| = 0$$

(see Definition 10.18 on page 350).

❑

From Assertion 10.19 on page 350, we see that the series $\sum_{j = r \to \infty} \mathbf{M}_j$ converges if and only if, for each i,k, the sequence of i,k elements of the matrices in this matrix series converges to the i,k element of its sum, i.e., if and only if for each i,k $\sum_{j = r \to \infty} (\mathbf{M}_j)_{i,k}$ converges to $\mathbf{M}_{i,k}$.

Hence by the Cauchy criterion, Theorem 7.8 on page 219, again using Assertion 10.19, it follows that $\sum_{j = r \to \infty} \mathbf{M}_j$ converges if and only if

$$\left\| \sum_{j=m}^{n} \mathbf{M}_j \right\|$$

becomes small for all sufficiently large m,n with $m < n$ (this is the *matrix Cauchy criterion*).

Now suppose the scalar series $\sum_{j = r \to \infty} \|\mathbf{M}_j\|$ converges. It then follows

1. As with scalars (see Definition 7.4 on page 214).
2. In fact, any sequence of $k \times m$ matrices, $\mathbf{B}_1, \mathbf{B}_2, \ldots$ is said to *converge to the limit,* $\mathbf{B}$, (a $k \times m$ matrix) if $\lim_{n \to \infty} \|\mathbf{B}_n - \mathbf{B}\| = 0$. This also defines convergence of a sequence of n-dimensional vectors (which can be considered as $n \times 1$ or $1 \times n$ matrices) and the limit of a convergent sequence of vectors.

from the triangle inequality, Theorem 10.20 on page 350, that

$$\left\| \sum_{j=m}^{n} \mathbf{M}_j \right\| \le \sum_{j=m}^{n} \|\mathbf{M}_j\| \quad \text{for } r \le m \le n. \qquad \textit{10.33}$$

From the assumed convergence of $\sum_{j=r\to\infty} \|\mathbf{M}_j\|$, and (again) the (scalar) Cauchy criterion, the right side of inequality 10.33, and hence the left side, gets small for all sufficiently large m,n. Hence, by the matrix Cauchy criterion above, the series $\sum_{j=r\to\infty} \|\mathbf{M}_j\|$ converges. (This is the *matrix comparison test.*)

We leave it to the reader to verify that if $\sum_{j=r\to\infty} \mathbf{M}_j$ converges for some specific integer $r = r_o$, then it converges for all integers r for which this series is defined, and

$$\sum_{j=r}^{\infty} \mathbf{M}_j = \sum_{j=r}^{r+s} \mathbf{M}_j + \sum_{j=r+s+1}^{\infty} \mathbf{M}_j. \qquad \textit{10.34}$$

Furthermore, if $\sum_{j=r\to\infty} \|\mathbf{M}_j\|$ converges, then the *error*

$$\left\| \sum_{j=r}^{\infty} \mathbf{M}_j - \sum_{j=r}^{n} \mathbf{M}_j \right\| \le \sum_{j=n+1}^{\infty} \|\mathbf{M}_j\|. \qquad \textit{10.35}$$

(Verification of equation 10.34 and inequality 10.35 is similar to the proofs of Theorem 7.11 on page 226 and Theorem 7.13 on page 228, respectively.)

For reference purposes we summarize these results as follows.

Theorem 10.23: Criteria for matrix series convergence.

The matrix series $\sum_{j=r\to\infty} \mathbf{M}_j$ (Definition 10.22 on page 351) converges if and only if, for each i,k, the sequence of i,k elements of the matrices in this matrix series converges to the i,k element of its sum, i.e., if and only if for each i,k

$$\sum_{j=r\to\infty} (\mathbf{M}_j)_{i,k} \quad \text{converges to} \quad \mathbf{M}_{i,k}. \qquad \textit{10.36}$$

The matrix series $\sum_{j=r\to\infty} \mathbf{M}_j$ converges if and only if

$$\left\| \sum_{j=m}^{n} \mathbf{M}_j \right\| \text{ becomes small} \qquad \textbf{\textit{10.37}}$$

(see Definition 10.18 on page 350) for all sufficiently large m,n with $m < n$ (*matrix Cauchy criterion*).

If $\sum_{j = r_o \to \infty} \mathbf{M}_j$ converges, then so does $\sum_{j = r \to \infty} \mathbf{M}_j$ for every integer for which this series is defined; and for $s \geq r$

$$\sum_{j=r}^{\infty} \mathbf{M}_j = \sum_{j=r}^{s} \mathbf{M}_j + \sum_{j=s+1}^{\infty} \mathbf{M}_j . \qquad \textbf{\textit{10.38}}$$

If $\sum_{j = r \to \infty} \|\mathbf{M}_j\|$ converges, then so does $\sum_{j = r \to \infty} \mathbf{M}_j$ and the *error*

$$\left\| \sum_{j=r}^{\infty} \mathbf{M}_j - \sum_{j=r}^{n} \mathbf{M}_j \right\| \leq \sum_{j=n+1}^{\infty} \|\mathbf{M}_j\|. \qquad \textbf{\textit{10.39}}$$

(This last result is the *matrix comparison test.*)

❑

These results pave the way for application of the results developed for scalar infinite series in Chapter 7. In particular, we can now handle the matrix geometric series very neatly.

Exercises 10.12

1. Show that the strict inequality is possible in $\|\mathbf{AB}\| \leq \|\mathbf{A}\| \|\mathbf{B}\|$.
2. Verify equation 10.34 and inequality 10.35.

6 The Matrix Geometric Series

The results we're aiming at arise from the same type of algebraic equality which held for numbers — namely,

$$1 + x + x^2 + \cdots + x^m = \frac{1 - x^{m+1}}{1 - x} .$$

However, a few obvious changes have to be made. First, we multiply through by $1 - x$ and we replace x by the $n \times n$ matrix $\mathbf{X}$ and the number 1 by the $n \times n$ identity matrix, **I.** We define nonnegative integer powers of an $n \times n$ matrix in the same way as for numbers, as follows.

Definition 10.24: $\mathbf{X}^k$, $\mathbf{X}$ a square matrix, integer $k \geq 0$

If $\mathbf{X}$ is a real $n \times n$ matrix (Definition 10.1 on page 328) then we define

$$\mathbf{X}^0 = \mathbf{I}^{n \times n}$$

(Definition 10.6 on page 334) and, inductively, for all integers $k \geq 0$

$$\mathbf{X}^{k+1} = \mathbf{X}\mathbf{X}^k.$$

❑

Associativity of matrix multiplication, see Exercise 10.3.3 on page 332, shows via mathematical induction that for all nonnegative integers j,k

$$\mathbf{X}^{j+k} = \mathbf{X}^j\mathbf{X}^k \qquad \mathit{10.40}$$

We can now state the extension of Theorem 1.9 on page 16.

Theorem 10.25: The finite matrix geometric series $\sum_{j=0}^{m} \mathbf{X}^j$

Let $\mathbf{X}$ be a real $n \times n$ matrix (Definition 10.1 on page 328). Then, recalling Definition 10.6 on page 334, Definition 10.3 on page 331, and Definition 10.24 above, if m is any nonnegative integer,

$$(\mathbf{I}^{n \times n} - \mathbf{X}) \sum_{j=0}^{m} \mathbf{X}^j = \mathbf{I}^{n \times n} - \mathbf{X}^{m+1}.$$

❑

It should be convincing enough to note that the distributive law allows the left side of the above equation to be written as

$$\sum_{j=0}^{m} \mathbf{X}^j - \sum_{j=0}^{m} \mathbf{X}^{j+1},$$

which is a *telescoping series* (easily seen by writing out the two series one below the other) leading to the result on the right. A rigorous proof is by induction.

As we will see, if $\mathbf{X}$ is small enough (as specified by the norm) then just as in the scalar geometric series, it will turn out that $\sum_{j=0 \to \infty} \mathbf{X}^j$ will converge, its sum being the inverse, $(\mathbf{I}^{n \times n} - \mathbf{X})^{-1}$.

So, let $\mathbf{A}$ be a real $n \times n$ matrix whose norm is less than 1. Then, using Theorem 1.9 on page 16, concerning the scalar geometric series and the results of Example 7.11 on page 233, we find from Theorem 10.23 on page 353, using

Definition 10.24 of integral powers of a square matrix, that the matrix geometric series, $\sum_{k=0\to\infty} \mathbf{A}^k$ converges, and the error

$$\left\| \sum_{k=0}^{\infty} \mathbf{A}^k - \sum_{k=0}^{n} \mathbf{A}^k \right\| \le \frac{\|\mathbf{A}\|^{n+1}}{1-\|\mathbf{A}\|}.$$

It's now easy to see that the sum of this series is $(\mathbf{I}-\mathbf{A})^{-1}$, where $\mathbf{I}$ is the $n\times n$ identity matrix, because, using the triangle inequality, Theorem 10.20 on page 350, the product norm inequality, Theorem 10.21 on page 351, the result specified by equation 10.38 on page 354 and the distributive law, we find

$$\begin{aligned}
\left\| \left(\sum_{k=0}^{\infty} \mathbf{A}^k\right)(\mathbf{I}-\mathbf{A})-\mathbf{I} \right\| &= \left\| \left(\sum_{k=0}^{m} \mathbf{A}^k\right)(\mathbf{I}-\mathbf{A}) + \left(\sum_{k=m+1}^{\infty} \mathbf{A}^k\right)(\mathbf{I}-\mathbf{A}) - \mathbf{I} \right\| \\
&\le \left\| \left(\sum_{k=0}^{m} \mathbf{A}^k\right)(\mathbf{I}-\mathbf{A}) - \mathbf{I} \right\| + \left\| \left(\sum_{k=m+1}^{\infty} \mathbf{A}^k\right)(\mathbf{I}-\mathbf{A}) \right\| \\
&= \left\| -\mathbf{A}^{m+1} \right\| + \left\| \left(\sum_{k=m+1}^{\infty} \mathbf{A}^k\right)(\mathbf{I}-\mathbf{A}) \right\| \\
&\le \|\mathbf{A}\|^{m+1} + \left(\sum_{k=m+1}^{\infty} \|\mathbf{A}\|^k\right)\|\mathbf{I}-\mathbf{A}\|.
\end{aligned}$$

Since the last line can be made arbitrarily small by choosing m sufficiently large, it follows that

$$\left\| \left(\sum_{k=0}^{\infty} \mathbf{A}^k\right)(\mathbf{I}-\mathbf{A})-\mathbf{I} \right\| = 0.$$

From this it can be seen that

$$\left(\sum_{k=0}^{\infty} \mathbf{A}^k\right)(\mathbf{I}-\mathbf{A}) = \mathbf{I},$$

establishing that

$$\sum_{k=0}^{\infty} \mathbf{A}^k = (\mathbf{I}-\mathbf{A})^{-1}.$$

Summarizing the matrix geometric series results, we have the following.

Theorem 10.26: Matrix geometric series

Let $\mathbf{A}$ be a real $n \times n$ matrix with norm $\|\mathbf{A}\| < 1$, and $\mathbf{I}$ the $n \times n$ identity matrix (see Definition 10.1 on page 328, Definition 10.18 on page 350, and Definition 10.6 on page 334). The matrix geometric series

$$\sum_{k = 0 \to \infty} \mathbf{A}^k$$

(see Definition 10.22 on page 351) converges.

The error $\left\| \sum_{k=0}^{\infty} \mathbf{A}^k - \sum_{k=0}^{n} \mathbf{A}^k \right\|$ in approximating its sum by the first $n + 1$ terms, $\sum_{k=0}^{n} \mathbf{A}^k$ does not exceed $\dfrac{\|\mathbf{A}\|^{n+1}}{1 - \|\mathbf{A}\|}$.

The sum of the geometric series

$$\sum_{k=0}^{\infty} \mathbf{A}^k = (\mathbf{I} - \mathbf{A})^{-1}.$$

❑

Exercises 10.13

1. Suppose that the matrix $\tilde{\mathbf{A}}$ is an approximate inverse of the matrix $\mathbf{A}$, in the sense that the product $\tilde{\mathbf{A}}\mathbf{A}$ (Definition 10.3 on page 331) is close to the identity matrix (Definition 10.6 on page 334). Precisely, assume that

$$\tilde{\mathbf{A}}\mathbf{A} = \mathbf{I} - \boldsymbol{\varepsilon},$$

where $\|\boldsymbol{\varepsilon}\| < 1$ (Definition 10.18 on page 350). Show how to use this information to obtain a better inverse from $\tilde{\mathbf{A}}$ by multiplying both sides of the above equation on the left by just a few terms from the matrix geometric series (Theorem 10.26 above).

2. Try applying the results of the previous exercise to the following:

$$\mathbf{A} = \begin{bmatrix} 463 & 375 & 463 \\ 288 & 451 & 0.998 \\ 0.000341 & 0.000248 & 0.000439 \end{bmatrix}$$

$$\tilde{\mathbf{A}} = \begin{bmatrix} 0.0309 & -0.0078 & -32540 \\ -0.0197 & 0.00719 & 20740 \\ -0.0128 & 0.00197 & 15740 \end{bmatrix}$$

An important area of application of the matrix geometric series is that of determining the effect of roundoff or statistical estimation errors in solving systems of equations. In these circumstances we would like to be solving a system of equations

$$\mathbf{A}\boldsymbol{x} = \boldsymbol{y}, \qquad \textbf{\textit{10.41}}$$

where $\boldsymbol{y}$ are measured values, and the matrix $\mathbf{A}$ describes the behavior of the system of interest. What often prevents the achievement of our aim is that the matrix $\mathbf{A}$ may be only *imperfectly available* — for example, we may be forced to estimate $\mathbf{A}$ from measured data, using instead, an estimate $\mathbf{A}_0$. Under these circumstances we actually end up solving

$$\mathbf{A}_0\boldsymbol{x}_0 = \boldsymbol{y} \qquad \textbf{\textit{10.42}}$$

for the solution, $\boldsymbol{x}_0$, rather than solving equation 10.41 for its solution, $\boldsymbol{x}$.

If we define the *error matrix* $\boldsymbol{\varepsilon}$ by the equation

$$\mathbf{A} = \mathbf{A}_0 - \varepsilon, \qquad \textbf{\textit{10.43}}$$

we may be able to make some reasonable assumptions about $\boldsymbol{\varepsilon}$ from our knowledge of the situation under which the measurements on $\mathbf{A}$ were made (possibly a bound on $\|\varepsilon\|$, or assumptions concerning the statistical nature of the elements ε_{ij}).

With such assumptions it may sometimes be possible to legitimately write equation 10.41 as

$$\begin{aligned}\boldsymbol{x} &= \mathbf{A}^{-1}\boldsymbol{y} = (\mathbf{A}_0-\varepsilon)^{-1}\boldsymbol{y}\\ &= [\mathbf{A}_0(\mathbf{I} - \mathbf{A}_0^{-1}\varepsilon)]^{-1}\boldsymbol{y}\\ &= (\mathbf{I} - \mathbf{A}_0^{-1}\varepsilon)^{-1}\mathbf{A}_0^{-1}\boldsymbol{y},\end{aligned}$$

or, assuming $\left\|A_0^{-1}\varepsilon\right\|$ small enough, using Theorem 10.26,

$$\boldsymbol{x} = \sum_{k=0}^{\infty}(\mathbf{A}_0^{-1}\varepsilon)^k\mathbf{A}_0^{-1}\boldsymbol{y}. \qquad \textbf{\textit{10.44}}$$

Combining equation 10.44 with $\boldsymbol{x}_0 = \mathbf{A}_0^{-1}\boldsymbol{y}$ (assuming $\mathbf{A}_0$ invertible) which is obtained from equation 10.42, we find

$$\boldsymbol{x} - \boldsymbol{x}_0 = \left[\sum_{k=1}^{\infty}(\mathbf{A}_0^{-1}\varepsilon)^k\right]\mathbf{A}_0^{-1}\boldsymbol{y}, \qquad \textbf{\textit{10.45}}$$

which would be a reasonable starting point for determining the effect of the error $\boldsymbol{\varepsilon}$ on the solution — in particular, determining how far the desired solution $\boldsymbol{x}$ is from the calculated solution $\boldsymbol{x}_0$.

Problem 10.14

Suppose the elements of the estimate

$$\mathbf{A}_0 = \begin{bmatrix} 1.1 & 0.75 & 0.2 \\ 0.63 & 0.84 & -1.91 \\ 1.2 & -0.96 & 6.4 \end{bmatrix}$$

are all within .08 of the corresponding values in $\mathbf{A}$. Provide an upper bound for the error that this can inflict on the solution when $\boldsymbol{y} = (5, 1.2, 4)$.

Hint: Find an upper bound for the error $\|\boldsymbol{x} - \boldsymbol{x}_0\|$.

It's worth noting that those who have reached this mathematical level should be developing their own exercises, making use of whatever computer packages are available, in order to get a *feel* for the material. For matrix manipulation on a personal computer, the *Mathcad* computer package is fairly simple to use, as is *Splus*. *Mathematica* will also serve, although a matrix in *Mathematica* is represented as a *list* (*column*) of *lists* (*rows*), and is a trifle cumbersome.

An important theoretical result, whose establishment does not seem obvious by other methods, is already evident from this last discussion — namely that if the matrix $\mathbf{A}$ is invertible and the matrix ε is sufficiently small, then $\mathbf{A} - \varepsilon$ is invertible. In fact, from the invertibility of $\mathbf{A}$, we may write

$$\mathbf{A} - \boldsymbol{\varepsilon} = \mathbf{A}(\mathbf{I} - \mathbf{A}^{-1}\varepsilon). \qquad \textbf{\textit{10.46}}$$

But for sufficiently small ε it follows from the product norm inequality, Theorem 10.21 on page 351, that the norm of $\mathbf{A}^{-1}\varepsilon$ is less than 1 if the norm of ε is less than $1/\|\mathbf{A}\|$. But then, from the matrix geometric series theorem, Theorem 10.26 on page 357, the rightmost factor in equation 10.46, $\mathbf{I} - \mathbf{A}^{-1}\varepsilon$, is invertible. Then from Theorem 10.14 on page 342, concerning the inverse of a matrix product, it follows that the matrix $\mathbf{A} - \boldsymbol{\varepsilon}$ is invertible with inverse $(\mathbf{I} - \mathbf{A}^{-1}\varepsilon)^{-1}\mathbf{A}^{-1}$. We summarize these results as follows.

Theorem 10.27: Continuity of matrix inversion

If $\mathbf{A}$ is an invertible real $n \times n$ matrix (Definition 10.1 on page 328 and Definition 10.9 on page 336) and ε is a real $n \times n$ matrix with $\|\varepsilon\| < 1/\|\mathbf{A}\|$ (Definition 10.18 on page 350) the matrix $\mathbf{A} - \varepsilon$ is invertible with

$$(\mathbf{A}-\varepsilon)^{-1} = (\mathbf{I}-\mathbf{A}^{-1}\varepsilon)^{-1}\mathbf{A}^{-1} = \sum_{k=0}^{\infty}(\mathbf{A}^{-1}\varepsilon)^k\mathbf{A}^{-1} \qquad \textit{10.47}$$

(see Definition 10.24 on page 355 and Definition 10.22 on page 351).

Using the triangle inequality and the product norm inequality, Theorem 10.20 on page 350 and Definition 10.22 on page 351, and the geometric series results (see Example 7.11 on page 233) it follows that

$$\left\|(\mathbf{A}-\boldsymbol{\varepsilon})^{-1}-\mathbf{A}^{-1}\right\| \le \sum_{k=1}^{\infty}\left\|\mathbf{A}^{-1}\varepsilon\right\|^k\left\|\mathbf{A}\right\|^{-1} = \frac{\left\|\mathbf{A}^{-1}\varepsilon\right\|\left\|\mathbf{A}\right\|^{-1}}{1-\left\|\mathbf{A}^{-1}\varepsilon\right\|}. \qquad \textit{10.48}$$

❑

Exercises 10.15

1. What can be said about $\left\|(\mathbf{A}-\boldsymbol{\varepsilon})^{-1}-\mathbf{A}^{-1}\right\|$ if $\|\mathbf{A}\|^{-1} = 2$ and $\|\varepsilon\| = 0.15$?
2. Generate some examples satisfying the conditions of the previous exercise, and compare your general answer with them.
3. Determine whether any conclusions can be drawn about $\left\|(\mathbf{A}-\boldsymbol{\varepsilon})^{-1}-\mathbf{A}^{-1}\right\|$ knowing only that $\|\mathbf{A}\| = 100$ and $(\|\varepsilon\| = 0.15)$.
   Hints: Even assuming **A** is invertible, it's not hard to show that we can find **A** and ε satisfying the above conditions, but $\mathbf{A}-\boldsymbol{\varepsilon}$ is not invertible.

We now take up the topic of the *normal equations.* Recall from equations 9.35 on page 314 that the normal equations arise when we try to solve the problem of finding the projection of a vector on a subspace by setting the appropriate partial derivatives equal to 0. Our purpose for investigating the normal equations here is to improve the accuracy of the Gram-Schmidt orthonormalization process. However, rather than approaching these equations from a *partial derivative* viewpoint (which is needlessly complicated) they will arise from a simple, intuitively appealing geometric characterization. Namely, we will first show that the projection, $P_S(\boldsymbol{y})$, of the vector $\boldsymbol{y}$ on the subspace S is that vector $\boldsymbol{w}$ in S such that

$$\boldsymbol{y}-\boldsymbol{w} \perp \boldsymbol{v}$$

for all vectors $\boldsymbol{v}$ in S — or, more briefly, $P_S(\boldsymbol{y})$ is the vector in S such that

$$y - P_S(y) \perp S$$

as illustrated in Figure 10-3.

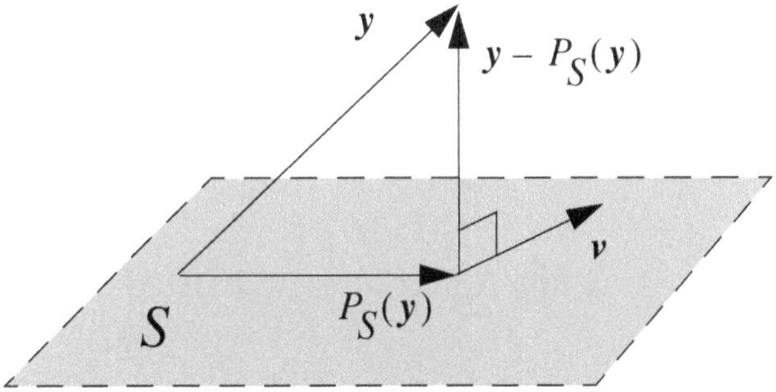

Figure 10-3

Theorem 10.28: Characterization of $P_S(y)$

Let S be a subspace of Euclidean n-dimensional space (Definition 9.4 on page 296) and $\boldsymbol{y}$ a given n-dimensional vector (Definition 8.1 on page 261). Recall the definition of projection on a subsbpace (Definition 9.22 on page 313) and orthogonality ($\perp$, see Theorem 8.15 on page 278 and also Theorem 9.24 on page 316). Then for each vector $\boldsymbol{v}$ in S,

$$\boldsymbol{y} - P_S(\boldsymbol{y}) \perp \boldsymbol{v}. \qquad \textbf{\textit{10.49}}$$

If $\boldsymbol{w}$ is a vector in S such that

$$\boldsymbol{y} - \boldsymbol{w} \perp \boldsymbol{v} \quad \text{for every} \quad \boldsymbol{v} \quad \text{in} \quad S, \qquad \textbf{\textit{10.50}}$$

then

$$\boldsymbol{w} = P_S(\boldsymbol{y}).$$

❑

Just from Figure 10-3, the assertion of this theorem is believable. A proof is straightforward, since from Theorem 9.20 on page 307, for any given subspace, S, we can choose an orthonormal basis, $\boldsymbol{w}_1, \boldsymbol{w}_2, ..., \boldsymbol{w}_j$, for S by means of the Gram-Schmidt procedure applied to the vectors originally defining S. We may then write, using Theorem 9.17 on page 304 (the formula for coordinates relative to an orthogonal basis) and the normalization which yields $\boldsymbol{w}_i \cdot \boldsymbol{w}_i = 1$,

$$\left(P_S(\boldsymbol{y}) = \sum_{i=1}^{j} (\boldsymbol{w}_i \cdot \boldsymbol{y})\boldsymbol{w}_i \right).$$

Since every vector $\boldsymbol{w}$ in S may be written in the form

$$w = \sum_{i=1}^{j} a_i w_i$$

to establish equation 10.49, all we need show is that for each $i = 1,2,...,j$

$$w_i \perp y - P_S(y).$$

This is easy enough to verify, and is left to the reader.

To establish that when equation 10.50 holds, then $w = P_S(y)$, let w represent any vector in S such that

$$y - w \perp v \quad \text{for every} \quad v \quad \text{in} \quad S.$$

To show that $w = P_S(y)$, write

$$y - w = [y - P_S(y)] + [P_S(y) - w]$$

and note that the second bracketed term is in S. Now take the dot product of both sides of this equation with $P_S(y) - w$. The left side of the resulting equation is 0 from equation 10.50, while the first term on the right side also is 0 (from equation 10.49). Thus we are left with the equation

$$[P_S(y) - w] \cdot [P_S(y) - w] - 0$$

or

$$\| P_S(y) - w \|^2 = 0,$$

proving that $w = P_S(y)$ and establishing Theorem 10.28.

To return to our development of the normal equations, note that if

$$S = [\![v_1, v_2 ..., v_k]\!]$$

and y is an arbitrary n-dimensional vector, we know that for suitable choice of k-dimensional column vector β (unique only if $v_1, v_2 ..., v_k$ is a linearly independent sequence)

$$P_S(y) = \mathbf{V}\beta, \qquad \textbf{\textit{10.51}}$$

where $\mathbf{V}$ is the matrix $[v_1, v_2 ..., v_k]$.

In fact, each vector v in S may be written in the form

$$v = \mathbf{V}a, \qquad \textbf{\textit{10.52}}$$

where a is a k-dimensional column vector determined by v (again, uniquely only if $v_1, v_2 ..., v_k$ are linearly independent). Writing the condition

$$y - P_S(y) \perp v$$

of Theorem 10.28 in terms of equations 10.51 and 10.52 shows that any k-dimensional vector β for which $P_S(y) = \mathbf{V}\beta$ must satisfy the equation

$$y - \mathbf{V}\beta \perp \mathbf{V}a \qquad \textit{10.53}$$

for *all* k-dimensional vectors, a.

It's most convenient to phrase equation 10.53 completely in matrix terminology, treating a and β as $k \times 1$ matrices. Then using the dot product condition for orthogonality, Theorem 8.15 on page 278, we see that equation 10.53 may be written as

$$(\mathbf{V}a)^{\mathrm{T}}(y - \mathbf{V}\beta) = 0, \qquad \textit{10.54}$$

where $(\mathbf{V}a)^{\mathrm{T}}$ denotes the transpose of $\mathbf{V}a$, see Definition 10.13 on page 342.

To make effective use of this last equation, we need the following theorem, whose proof is left to Exercise 10.16.1 on page 366.

Theorem 10.29: Transpose of a product

If $\mathbf{A}$ and $\mathbf{B}$ are two matrices whose product $\mathbf{AB}$ (Definition 10.3 on page 331) is meaningful, then

$$(\mathbf{AB})^{\mathrm{T}} = \mathbf{B}^{\mathrm{T}}\mathbf{A}^{\mathrm{T}},$$

where T is the transpose operator (Definition 10.13 on page 342).

❑

Applying Theorem 10.29 to equation 10.54 yields

$$a^{\mathrm{T}}\mathbf{V}^{\mathrm{T}}(y - \mathbf{V}\beta) = 0 \qquad \textit{10.55}$$

for all choices of a. Since one choice of a is $a = \mathbf{V}^{\mathrm{T}}(y - \mathbf{V}\beta)$, we see that the only way for equation 10.55 to hold for all choices of a is if

$$\mathbf{V}^{\mathrm{T}}(y - \mathbf{V}\beta) = 0.$$

That is, in order for equation 10.55 to hold for all choices of a, β must be chosen so that

$$\mathbf{V}^{\mathrm{T}}\mathbf{V}\beta = \mathbf{V}^{\mathrm{T}}y.$$

These are the *normal equations*, and we summarize our results as follows.

Theorem 10.30: Normal equations for $P_S(y)$

Let the subspace S (Definition 9.4 on page 296) of Euclidean n-dimensional space be defined by

$$S = [\![v_1, v_2 \ldots, v_k]\!],$$

let $P_S(y)$ be the projection of y on S (Definition 9.22 on page 313) and let $\mathbf{V}$ be the matrix (Definition 10.1 on page 328) whose columns are specified by

$$\mathbf{V} = [v_1, v_2 \ldots, v_k].$$

Writing $P_S(y)$ as a linear combination of the columns of $\mathbf{V}$,

$$P_S(y) = \mathbf{V}\beta,$$

the coefficient vector, β, must satisfy the *normal equations,*

$$\mathbf{V}^T\mathbf{V}\beta = \mathbf{V}^T y, \qquad \textbf{\textit{10.56}}$$

where $\mathbf{V}^T$ is the transpose of $\mathbf{V}$ (see Definition 10.13 on page 342).

❑

As previously noted, the normal equations are just orthogonality relations. With the availability of the normal equations we can modify the Gram-Schmidt procedure to control better the buildup of roundoff error effects, that would otherwise grow too fast as the number of vectors to be orthonormalized becomes large.

Recall how the Gram-schmidt *orthonormalization* procedure generates an orthonormal basis, $w_1, w_2 \ldots, w_j$, for $S = [\![v_1, v_2 \ldots, v_k]\!]$ (see Definition 9.19 on page 306).

First, we let $w_1' = v_1$; then, if $w_1' = \mathbf{0}$, it is discarded, while if $w_1' \neq \mathbf{0}$ we let

$$w_1 = \frac{w_1'}{\|w_1'\|}.$$

Assume inductively that $w_1, w_2, \ldots, w_i$, orthogonal and of unit length have been generated and the vectors suitably relabeled so that we want to generate w_{i+1} from v_{i+1} and $w_1, w_2, \ldots, w_i$.

We let

$$w_{i+1}' = v_{i+1} - \sum_{k=1}^{i} P_{w_k}(v_{i+1}), \qquad \textbf{\textit{10.57}}$$

discarding if this is $\mathbf{0}$, and relabeling, or normalizing otherwise. We continue in this fashion until the full sequence of orthonormalized vectors is generated.

But by Theorem 9.24 on page 316, equation 10.57 is equivalent to

$$w_{i+1}' = v_{i+1} - P_{[\![w_1, w_2, \ldots, w_i]\!]}(v_{i+1}) \qquad \textbf{\textit{10.58}}$$

when $w_1, w_2,...,w_i$ are mutually orthogonal. If they are close to being orthogonal, but, due to roundoff error, they are not exactly orthogonal, then it is preferable to compute the next orthonormal vector in the sequence using the normal equations to generate the last term in equation 10.58. That may seem like a pretty difficult way to solve the roundoff error problem, but as we will see, the matrix geometric series can be used for this solution, making it quite simple on a computer. The improved orthogonality that this approach generates tends to make the dot products $w_{i+1} \cdot w_k$, $k < j + 1$, closer to 0 — helping to better eliminate terms which are *supposed* to vanish when solving systems of equations.

To see how to implement equation 10.58, we define the $n \times i$ matrix $\mathbf{W}_i$ by

$$\mathbf{W}_i = [w_1,...,w_i] \qquad \textbf{\textit{10.59}}$$

and write

$$P_{[\![w_1, w_2,...,w_i]\!]}(v_{i+1}) = \mathbf{W}_i\beta_i, \qquad \textbf{\textit{10.60}}$$

where from the normal equations, Theorem 10.30 on page 363, with $\mathbf{V}$ replaced by $\mathbf{W}_i$, β_i in place of β, and v_{i+1} in place of y, β_i is determined by the equation

$$\mathbf{W}_i^{\mathrm{T}}\mathbf{W}_i\beta_i = \mathbf{W}_i^{\mathrm{T}}v_{i+1}. \qquad \textbf{\textit{10.61}}$$

The simplification of solution of equation 10.61 comes about if the $w_1,...,w_i$ are close enough to being orthonormal, because then the matrix

$$\mathbf{W}_i^{\mathrm{T}}\mathbf{W}_i$$

will be *close* to the $i \times i$ identity matrix, and hence will be easy to invert using just a few terms of the matrix geometric series. Then we may solve equation 10.61 for β_i, obtaining

$$\beta_i = (\mathbf{W}_i^{\mathrm{T}}\mathbf{W}_i)^{-1}\mathbf{W}_i^{\mathrm{T}}v_{i+1} \qquad \textbf{\textit{10.62}}$$

and substitute this in equation 10.60, yielding

$$P_{[\![w_1, w_2,...,w_i]\!]}(v_{i+1}) = \mathbf{W}_i(\mathbf{W}_i^{\mathrm{T}}\mathbf{W}_i)^{-1}\mathbf{W}_i^{\mathrm{T}}v_{i+1}. \qquad \textbf{\textit{10.63}}$$

From this equation we may substitute in equation 10.58, to obtain the final form of the improved Gram-Schmidt procedure. We summarize these results in the following theorem.

Theorem 10.31: Improved Gram-Schmidt procedure

To orthonormalize the sequence $v_1, v_2 \dots, v_k$ of nonzero real n-dimensional vectors (Definition 8.1 on page 261) proceed as follows:

Let $w_1' = v_1$ and $w_1 = \dfrac{w_1'}{\|w_1'\|}$ (see Definition 8.6 on page 270)

and having reached the stage where the orthonormal vectors $w_1, \dots, w_i$ have been determined (discarding vectors found to be effectively $\mathbf{0}$, and suitably relabeling), let

$$\mathbf{W}_i = [w_1, \dots, w_i],$$

defining

$$w_{i+1}' = v_{i+1} - \mathbf{W}_i(\mathbf{W}_i^{\mathrm{T}}\mathbf{W}_i)^{-1}\mathbf{W}_i^{\mathrm{T}} v_{i+1},$$

discarding w_{i+1}' if it is very close to $\mathbf{0}$ (and relabeling as appropriate).; otherwise normalize

$$w_{i+1} = \frac{w_{i+1}'}{\|w_{i+1}'\|}.$$

This process is relatively easy to implement, since the near orthonormality of $w_1, \dots, w_i$ ensures that

$$\mathbf{W}_i^{\mathrm{T}}\mathbf{W}_i = \mathbf{I}^{i \times i} - \varepsilon_i,$$

where $\|\varepsilon_i\|$ is small, allowing a good approximation to $(\mathbf{W}_i^{\mathrm{T}}\mathbf{W}_i)^{-1}$ using just a few, say r, terms of the geometric series (see Theorem 10.26 on page 357)

$$\sum_{m=0}^{r} \varepsilon_i^m = \sum_{m=0}^{r} (\mathbf{I}^{i \times i} - \mathbf{W}_i^{\mathrm{T}}\mathbf{W}_i)^m.$$

❑

Exercises 10.16

1. Establish that the formula for the transpose (Definition 10.13 on page 342) of the product of two matrices (Definition 10.3 on page 331) is given by

$$(\mathbf{AB})^{\mathrm{T}} = \mathbf{B}^{\mathrm{T}}\mathbf{A}^{\mathrm{T}}.$$

Hints: First show that if $\mathbf{V}$ is a matrix with k columns and $\boldsymbol{a}$ is a column vector with k coordinates, then

$$(\mathbf{V}\boldsymbol{a})^{\mathrm{T}} = \boldsymbol{a}^{\mathrm{T}}\mathbf{V}^{\mathrm{T}}$$

by writing

$$\mathbf{V}\boldsymbol{a} = \sum_{j=1}^{k} a_j v_j$$

where v_j is the j-th column of **V.**

Then you may write

$$(\mathbf{AB}v)^{\mathrm{T}} = [\mathbf{A}(\mathbf{B}v)]^{\mathrm{T}} = (\mathbf{B}v)^{T}\mathbf{A}^{\mathrm{T}} = v^{\mathrm{T}}\mathbf{B}^{\mathrm{T}}\mathbf{A}^{\mathrm{T}}$$

and

$$(\mathbf{AB}v)^{\mathrm{T}} = v^{\mathrm{T}}(\mathbf{AB})^{\mathrm{T}}.$$

The result now follows because these must hold for all v.

2. If you have the matrix manipulation facilities, try comparing the simple Gram-Schmidt orthonormalization procedure, Definition 9.19 on page 306, with the improved Gram-Schmidt procedure above, for some value of n (say $n = 9$) on the columns of the $n \times n$ Hilbert matrix, whose i-j element is given by $1/(i+j)$.

There are other modifications to improve the Gram-Schmidt procedure's accuracy. They are available in Golub and Van Loan [9].

One topic which we do not cover here is that of pseudo-inverses. The Moore-Penrose pseudo-inverse is a matrix representation which yields the projection operator, P_S, as follows. Let $\mathbf{X}$ be an $n \times p$ matrix whose columns span the subspace S. The pseudo-inverse, $\tilde{\mathbf{X}}$, is a $p \times n$ matrix such that the product $\mathbf{X}\tilde{\mathbf{X}}\mathbf{Y}^{n \times 1}$ is the projection of $\mathbf{Y}$ on S. With the background developed in this book, those who want to should not have a great deal of difficulty learning more about this topic from the available literature, or even developing it from what has been presented here.

7 Taylor's Theorem in n Dimensions

A good deal of theoretical work on functions of several variables makes use of the n-dimensional version of Taylor's theorem. With the tools already available we can establish this result with a minimum of effort.[1] To express $f(\boldsymbol{x}+\boldsymbol{h})$ in terms of f and its various partial derivatives at $\boldsymbol{x}$, we simply let

$$g(t) = f(\boldsymbol{x}+t\boldsymbol{u}), \qquad \textbf{\textit{10.64}}$$

where $\boldsymbol{u} = \boldsymbol{h}/\|\boldsymbol{h}\|$ and expand g in a one-dimensional Taylor series about $t = 0$; we then rewrite this expression (including the error bound) in terms of the directional derivatives of f at $\boldsymbol{x}$ in the direction of $\boldsymbol{u} = \boldsymbol{h}/\|\boldsymbol{h}\|$, finally expressing these directional derivatives in terms of the partial derivatives of f.

---

1. The connection between the n-dimensional Taylor's theorem and matrices will not be evident until the next section, where we make use of Taylor's theorem to establish conditions which help determine whether we have achieved a maximum, a minimum, or neither, at points where the partial derivatives of f are all 0.

Throughout the remainder of the discussion leading to the formal statement of the multivariate Taylor theorem, we will be assuming that all of the partial derivatives being dealt with are continuous at the points being looked at. This condition will ensure that results which make use of the mean value theorem will be true (recall the discussion leading up to Theorem 8.9 on page 271 concerning the determination of directional derivatives).

Provided its derivatives exist, we may apply Taylor's theorem (Theorem 6.2 on page 179) identifying g here with f of Theorem 6.2, 0 here with x_o of Theorem 6.2, and t here with h of Theorem 6.2, to obtain

$$g(t) = \sum_{k=0}^{n} g^{(k)}(0)\frac{t^k}{k!} + g^{(n+1)}(\vartheta t)\frac{t^{n+1}}{(n+1)!},$$

where ϑ is some (usually unknown) value between 0 and 1. For our purposes it is best to write this result using the D notation — where $Df = f'$, $D^2 f = f''$, etc. This yields

$$g(t) = \sum_{k=0}^{n} (D^k g)(0)\frac{t^k}{k!} + (D^{n+1} g)(\vartheta t)\frac{t^{n+1}}{(n+1)!}. \qquad \mathit{10.65}$$

But we know from the definition of directional derivative (Definition 8.4 on page 265) that

$$(D^k g)(0) = (D^k_{\boldsymbol{u}} f)(\boldsymbol{x}), \quad \text{where} \quad \boldsymbol{u} = \boldsymbol{h}/\|\boldsymbol{h}\|. \qquad \mathbf{\mathit{10.66}}$$

Furthermore, from Theorem 8.11 on page 273, regarding computation of directional derivatives, we may replace $D^k_{\boldsymbol{u}}$ by $(\boldsymbol{u} \cdot \boldsymbol{D})^k$, where

$$\boldsymbol{D} = (D_1, D_2, \ldots, D_n)$$

is the vector of partial derivative operators. Substituting all of the above, including equation 10.64, into equation 10.65 yields

$$f(\boldsymbol{x} + t\boldsymbol{h}/\|\boldsymbol{h}\|) = \sum_{k=0}^{n} \left(\frac{\boldsymbol{h} \cdot \boldsymbol{D}}{\|\boldsymbol{h}\|}\right)^k f(x)\frac{t^k}{k!} + \left(\frac{\boldsymbol{h} \cdot \boldsymbol{D}}{\|\boldsymbol{h}\|}\right)^{n+1} f\left(\boldsymbol{x} + \vartheta t\frac{\boldsymbol{h}}{\|\boldsymbol{h}\|}\right)\frac{t^{n+1}}{(n+1)!},$$

where we should have parentheses around the factors $(\boldsymbol{u} \cdot \boldsymbol{D})^k f$ and $(\boldsymbol{u} \cdot \boldsymbol{D})^{n+1} f$ but have omitted them for readability.[1] By choosing $t = \|\boldsymbol{h}\|$, we can rewrite this in the more compact and readable form

$$f(\boldsymbol{x} + \boldsymbol{h}) = \sum_{k=0}^{n} \frac{(\boldsymbol{h} \cdot \boldsymbol{D})^k f}{k!}(\boldsymbol{x}) + \frac{(\boldsymbol{h} \cdot \boldsymbol{D})^{n+1} f}{(n+1)!}(\boldsymbol{x} + \vartheta \boldsymbol{h}), \qquad \mathbf{\mathit{10.67}}$$

---

1. Realize that $(\boldsymbol{u} \cdot \boldsymbol{D})^k f(\boldsymbol{x})$ is the k-th directional derivative of f in the $\boldsymbol{u}$ direction, evaluated at the point $\boldsymbol{x}$.

where the $(\boldsymbol{x})$ [or $(\boldsymbol{x} + \vartheta\boldsymbol{h})$] following the function symbolized by

$$\frac{(\boldsymbol{h} \cdot \boldsymbol{D})^k f}{k!}$$

indicates that this function is to be evaluated at the point $\boldsymbol{x}$ [or $(\boldsymbol{x} + \vartheta\boldsymbol{h})$].

Note that by $(\boldsymbol{h} \cdot \boldsymbol{D})g$ we mean $\sum_{k=1}^{b} h_i D_i g$, a weighted sum of partial derivatives. By

$$(\boldsymbol{h} \cdot \boldsymbol{D})^k f,$$

we mean the quantity defined inductively by the equations

$$(\boldsymbol{h} \cdot \boldsymbol{D})^1 f = (\boldsymbol{h} \cdot \boldsymbol{D}) f$$

and for each $j > 1$

$$(\boldsymbol{h} \cdot \boldsymbol{D})^{j+1} f = (\boldsymbol{h} \cdot \boldsymbol{D}) Q_j, \quad \text{where} \quad Q_j = (\boldsymbol{h} \cdot \boldsymbol{D})^j f.$$

As it stands, the definition of $(\boldsymbol{h} \cdot \boldsymbol{D})^k f$ seems a bit inflexible. It would be convenient if we could treat the quantity $(\boldsymbol{h} \cdot \boldsymbol{D})^k$ as if it were an ordinary polynomial — for example, in two dimensions, writing

$$(\boldsymbol{h} \cdot \boldsymbol{D})^2 = (h_1 D_1 + h_2 D_2)^2 = h_1^2 D_1^2 + 2h_1 h_2 D_1 D_2 + h_2^2 D_2^2$$

instead of

$$h_1^2 D_1^2 + h_1 h_2 D_1 D_2 + h_2 h_1 D_2 D_1 + h_2^2 D_2^2$$

as required by the above inductive definition. This, of course, requires that $D_1 D_2 f$ be the same as $D_2 D_1 f$. Try computing these second partial derivatives for some functions of your own choosing. You'll see that they usually come out the same. However, it is not always the case that $D_1 D_2 = D_2 D_1$, as can be seen from Problem 10.17 on page 372. But when all of the partial derivatives involved exist and are continuous, then we can show the following.

Theorem 10.32: Commutativity of D_i and D_j

If f is a real valued function of n real variables (Definition 8.1 on page 261) such that $D_i D_j f$ and $D_j D_i f$ (Definition 8.5 on page 267) both exist at and near the point $\boldsymbol{x}_o$ (i.e., for all $\boldsymbol{x}$ satisfying $\|\boldsymbol{x} - \boldsymbol{x}_o\| \le a$ (Definition

8.6 on page 270) where a is some positive number) and are continuous at $\boldsymbol{x}_o$ (Definition 8.7 on page 270) then

$$(D_i D_j f)(\boldsymbol{x}_o) = (D_j D_i f)(\boldsymbol{x}_o).$$

❑

We indicate how this is proved for the two-dimensional case, since no real generality is lost. To avoid subscripts we will replace $\boldsymbol{x}_o$ by (x, y), omit the parentheses around $D_1 D_2 f$, and then show that under the given conditions

$$D_1 D_2 f(x, y) = D_2 D_1 f(x, y).$$

The quantity

$$\begin{aligned} &D_1 D_2 f(x, y) \\ &= \lim_{h \to 0} \frac{D_2 f(x+h, y) - D_2 f(x, y)}{h} \\ &= \lim_{h \to 0} \frac{\lim_{k \to 0} \dfrac{f(x+h, y+k) - f(x+h, y)}{k} - \lim_{k \to 0} \dfrac{f(x, y+k) - f(x, y)}{k}}{h} \\ &= \lim_{h \to 0} \left[\lim_{k \to 0} \frac{[f(x+h, y+k) - f(x+h, y)] - [f(x, y+k) - f(x, y)]}{kh} \right]. \end{aligned}$$

If we let

$$g(t) = f(x+t, y+k) - f(x+t, y), \qquad \textbf{10.68}$$

then we find

$$D_1 D_2 f(x, y) = \lim_{h \to 0} \left[\lim_{k \to 0} \frac{g(h) - g(0)}{kh} \right].$$

This suggests using the mean value theorem (Theorem 3.1 on page 62) which yields

$$D_1 D_2 f(x, y) = \lim_{h \to 0} \left[\lim_{k \to 0} \frac{g'(c)}{k} \right], \qquad \textbf{10.69}$$

where c is some number between 0 and h, which is, however, dependent on the value of k.

From equation 10.68 we see that

$$g'(t) = D_1 f(x+t, y+k) - D_1 f(x+t, y).$$

Substituting this into equation 10.69, and applying the mean value twice, letting d denote some value between 0 and k, yields

$$
\begin{aligned}
D_1D_2f(x, y) &= \lim_{h \to 0}\left[\lim_{k \to 0} \frac{D_1f(x+c, y+k) - D_1f(x+c, y)}{k}\right] \\
&= \lim_{h \to 0}\left[\lim_{k \to 0} D_2D_1f(x+c, y+d)\right] \\
&= D_2D_1f(x, y),
\end{aligned}
$$

this last step following from the continuity of D_2D_1f at (x, y).[1] This establishes Theorem 10.32.

We can now provide a formal statement of the n-dimensional version of Taylor's theorem for later reference.

Theorem 10.33: Multivariable Taylor's theorem

Let f be a real valued function of n real variables (Definition 8.1 on page 261) and suppose that all of the partial derivatives (Definition 8.5 on page 267) referred to below exist for all $\boldsymbol{x}$ satisfying $\|\boldsymbol{x} - \boldsymbol{x}_o\| \le a$ (Definition 8.6 on page 270), where a is some positive number, and are continuous (see Theorem 8.8 on page 271) at $\boldsymbol{x}_o$. Then for $\|\boldsymbol{h}\| \le a$

$$
f(\boldsymbol{x}_0 + \boldsymbol{h}) = \sum_{k=0}^{n} \left(\frac{(\boldsymbol{h} \cdot \boldsymbol{D})^k f}{k!}\right)(\boldsymbol{x}_0) + \left(\frac{(\boldsymbol{h} \cdot \boldsymbol{D})^{n+1} f}{(n+1)!}\right)(\boldsymbol{x}_0 + \vartheta \boldsymbol{h}),
$$

where

ϑ is some number between 0 and 1

$\boldsymbol{D} = (D_1, D_2, ..., D_n)$ is the vector of partial derivative operators (see Definition 8.5 on page 267)

$$
\boldsymbol{h} \cdot \boldsymbol{D} = \sum_{i=1}^{n} h_i D_i
$$

$(\boldsymbol{h} \cdot \boldsymbol{D})^k$ is the operator obtained by treating

$$
(\boldsymbol{h} \cdot \boldsymbol{D})^k = \left(\sum_{i=1}^{n} h_i D_i\right)^k
$$

as if it were an ordinary sum raised to the k-th power, and

---

1. It might appear that the middle line above should simply be $\lim_{h \to 0} D_2D_1f(x+c, y)$. This is not allowed because the quantity c in the first line depends on k. Think about it.

the $(\boldsymbol{x}_0)$ [or $(\boldsymbol{x}_0 + \vartheta \boldsymbol{h})$] following the function symbolized by

$$\frac{(\boldsymbol{h} \cdot \boldsymbol{D})^k f}{k!}$$

indicates that this function is to be evaluated at the point $\boldsymbol{x}_0$ [or $(\boldsymbol{x}_0 + \vartheta \boldsymbol{h})$].

❑

By contrast with the case $n = 1$, Taylor's theorem for $n > 1$ is used almost exclusively for establishing important general results, and not for direct computation. For this reason no exercises are supplied at this point.

Problem 10.17

Show that the conclusion of Theorem 10.32 on page 369 need not follow if the continuity of the second partial derivatives at $\boldsymbol{x}_o$ fails to hold, by examining the function f defined by

$$f(x, y) = \begin{cases} 0 \quad \text{for} \quad x = y = 0, \\ xy\dfrac{x^2 - y^2}{x^2 + y^2} \quad \text{otherwise.} \end{cases}$$

8 Maxima and Minima in Several Variables

We can see that if the function f of n real variables has all first partial derivatives, $D_i f$ at and near $\boldsymbol{x}_o$, these derivatives being continuous at $\boldsymbol{x}_o$, then f cannot possibly achieve either a maximum or a minimum at x_o unless

$$D_i f(\boldsymbol{x}_o) = 0 \quad \text{for all } i = 1,...,n,$$

i.e., unless

$$\boldsymbol{grad}\, f(\boldsymbol{x}_o) = \boldsymbol{0},$$

because it is evident that if f achieves a maximum or minimum at x_o, then under the given conditions, all directional derivatives, $D_{\boldsymbol{u}} f(\boldsymbol{x}_o)$, must be 0 (a picture should be convincing enough). But if $\boldsymbol{grad}\, f(\boldsymbol{x}_o) \neq \boldsymbol{0}$, say $D_j f(\boldsymbol{x}_o) \neq 0$, just choose $u_j = 1$ and $u_i = 0$ for all other $i \neq j$. For this choice of $\boldsymbol{u}$, $D_{\boldsymbol{u}} f(\boldsymbol{x}_o) = D_j f(\boldsymbol{x}_o) \neq 0$.

What we want to develop are easily verifiable conditions indicating what kind of behavior we have when $\boldsymbol{grad}\, f(\boldsymbol{x}_o) = \boldsymbol{0}$ is already known to hold. Here again it's simplest to return to directional derivatives to obtain a simple answer. We know from Theorem 3.4 on page 68 that if a real valued function of a real variable has a positive second derivative, the graph of this function lies above the tangent line except at the point of tangency; if the derivative is 0 at the point of tangency, then the point of tangency yields a minimum, as illustrated in Figure 10-4. Similarly, a negative second derivative yields a maximum.

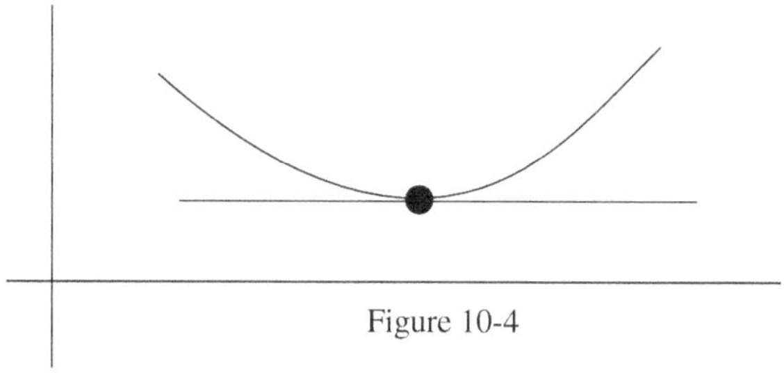

Figure 10-4

This suggests that for a local minimum at $\boldsymbol{x}_o$ of a function of several variables whose gradient is $\mathbf{0}$ at $\boldsymbol{x}_o$, all *second directional derivatives* $D_{\mathbf{u}}^2 f(\boldsymbol{x}_o) > 0$ should suffice, and similarly, for a local maximum at $\boldsymbol{x}_o$, $D_{\mathbf{u}}^2 f(\boldsymbol{x}_o) < 0$ for all $\boldsymbol{u}$ should be sufficient. This intuition is correct, but there are two aspects which require some development. First, it is not a totally trivial task to establish this result. Second is the practical difficulty of determining when it is the case that these directional derivative conditions are satisfied. We now tackle these two issues.

To concentrate on the important ideas in establishing the second directional derivative condition presented above, we will make the simplifying assumption that all second partial derivatives, $D_i D_j f(\boldsymbol{x})$, exist and are continuous at and near $\boldsymbol{x}_o$ — i.e., for all $\boldsymbol{x}$ satisfying $\|\boldsymbol{x} - \boldsymbol{x}_o\| < a$, where a is some positive number. From Theorem 8.8 on page 271, on continuity of functions of several variables,Theorem 8.11 on page 273, on computation of directional derivatives, and the application of Theorem 10.32 on page 369 to computation of higher directional derivatives, this assures that all first and second directional derivatives exist and are continuous at and near $\boldsymbol{x}_o$.

The first result we want to prove is that if

$$D_{\boldsymbol{u}}^2 f(\boldsymbol{x}_o) > 0 \quad \text{for all unit length vectors, } \boldsymbol{u}, \qquad \textbf{\textit{10.70}}$$

then under the above conditions, there is some positive number, p, such that

$$D_{\boldsymbol{u}}^2 f(\boldsymbol{x}_o) > p. \qquad \textbf{\textit{10.71}}$$

(A similar result holds when $D_{\boldsymbol{u}}^2 f(\boldsymbol{x}_o) < q$ where $q < 0$.)

If this were not the case, then there would be a sequence, $\boldsymbol{u}_1, \boldsymbol{u}_2, ...$ of unit length vectors such that

$$\lim_{k \to \infty} D_{\boldsymbol{u}_k}^2 f(\boldsymbol{x}_o) = 0.$$

From the sequence $\boldsymbol{u}_1, \boldsymbol{u}_2,...$ we can extract a subsequence with the property that the sequence of first components of these vectors converges — just use bisection to choose the subinterval in which an infinite number of values lie, from the previously chosen subinterval; that is, since these are unit length vectors, all of the original first coordinates lie in the interval $[-1;1]$. An infinite number of these coordinates must lie in at least one of the subintervals $[-1;0]$, $[0;1]$. Choose one of these intervals (the choice being arbitrary if both contain an infinite number of the first coordinates). This process yields a subsequence whose first coordinates converge. Starting with this subsequence, extract a sub-subsequence whose second coordinates also converge. Continue in this fashion until a subsequence (of the original sequence) which converges, is generated. Call this convergent unit length vector subsequence $\boldsymbol{v}_1, \boldsymbol{v}_2,...,$ where each of the $\boldsymbol{v}_i$ comes from the sequence $\boldsymbol{u}_1, \boldsymbol{u}_2,...$, the value $\boldsymbol{v}_{i+1}$ being from a *later* $\boldsymbol{u}_j$ than that of $\boldsymbol{v}_i$. Let

$$\boldsymbol{V} = \lim_{i \to \infty} \boldsymbol{v}_i.$$

We can see that $\boldsymbol{V}$ is a unit length vector as follows: From the convergence of the $\boldsymbol{v}_i$ to $\boldsymbol{V}$, $\lim_{i \to \infty} \|\boldsymbol{v}_i - \boldsymbol{V}\| = 0$. Hence

$$\sum_{k=1}^{n} (v_{i,k} - V_k)^2 = \|\boldsymbol{v}_i - \boldsymbol{V}\|^2 \text{ approaches } 0 \text{ as } i \text{ becomes large.}$$

By choosing i sufficiently large but fixed, we can make $v_{i,k}$ close to V_k for all $k = 1,2,...,n$. But since $\boldsymbol{v}_i$ is of unit length, this shows that $\boldsymbol{V}$ has length which must be arbitrarily close to 1 — hence $\|\boldsymbol{V}\| = 1$. From the continuity of $D_{\boldsymbol{u}}^2 f(\boldsymbol{x}_o)$ with respect to $\boldsymbol{u}$ (which follows from the formula for the second directional derivative, valid under the partial derivative continuity assumptions we have made) $D_{\boldsymbol{v}_i}^2 f(\boldsymbol{x}_o)$ must be close to $D_{\boldsymbol{V}}^2 f(\boldsymbol{x}_o)$ for large i. Since these values are approaching 0, this would yield

$$D_{\boldsymbol{V}}^2 f(\boldsymbol{x}_o) = 0,$$

which contradicts the assumed inequality 10.70. So we have, with some effort, shown the following result.

Theorem 10.34: Positive $D_{\boldsymbol{u}}^2 f(\boldsymbol{x}_o)$ bounded away from 0

Let f be a real valued function of n real variables (Definition 8.1 on page 261) all of whose second partial derivatives, $D_i D_j f(\boldsymbol{x})$, exist and are continuous (see Definition 8.5 on page 267 and Theorem 8.8 on page 271) for all $\boldsymbol{x}$ satisfying $\|\boldsymbol{x} - \boldsymbol{x}_o\| < a$ (Definition 8.6 on page 270) where a is some positive number.

Then if the second directional derivative, $D_{\boldsymbol{u}}^2 f(\boldsymbol{x}_o) > 0$ (see Theorem 8.11 on page 273) for all unit length vectors, $\boldsymbol{u}$, there is a positive number p such that

$$D_{\boldsymbol{u}}^2 f(\boldsymbol{x}_o) > p \text{ for all unit length vectors, } \boldsymbol{u}.$$

If $D_{\boldsymbol{u}}^2 f(\boldsymbol{x}_o) < 0$ for all unit length vectors, $\boldsymbol{u}$, there is a negative number, q, such that

$$D_{\boldsymbol{u}}^2 f(\boldsymbol{x}_o) < q \text{ for all unit length vectors, } \boldsymbol{u}.$$

❑

Under the conditions of Theorem 10.34, again using the formula for directional derivative, Theorem 8.11 on page 273, we see that $D_{\boldsymbol{u}}^2 f(\boldsymbol{x})$ is a continuous function of the variable $(\boldsymbol{u},\boldsymbol{x})$; So it follows that

$$D_{\boldsymbol{u}}^2 f(\boldsymbol{x}) - D_{\boldsymbol{u}}^2 f(\boldsymbol{x}_o) = \sum_{i,j} u_i u_j [D_i D_j f(x) - D_i D_j f(x_o)]$$

is simultaneously small for all unit length vectors, $\boldsymbol{u}$, and all $\boldsymbol{x}$ sufficiently close to $\boldsymbol{x}_o$, because the factors in the square brackets are uniformly small for such $\boldsymbol{x}$ due to the continuity of the partial derivatives at $\boldsymbol{x}_o$, and the boundedness of the $u_i u_j$. This fact, together with the reverse triangle inequality, Theorem 7.7 on page 218, shows that under the conditions of Theorem 10.34, there is a positive number, b, such that if $D_{\boldsymbol{u}}^2 f(\boldsymbol{x}_o) > 0$ for all unit length vectors, $\boldsymbol{u}$, then for all $\boldsymbol{x}$ satisfying $\|\boldsymbol{x} - \boldsymbol{x}_o\| < b$, and all unit length vectors, $\boldsymbol{u}$, there is a positive number P (any number smaller than the value p of Theorem 10.34 will do), such that

$$D_{\boldsymbol{u}}^2 f(\boldsymbol{x}) > P.$$

We need to have this condition satisfied — i.e., that ***everywhere nearby to*** $\boldsymbol{x}_o$ the second directional derivative is positive, so that for ***all*** such $\boldsymbol{x}$, the graph of f lies above the tangent, as illustrated in Figure 10-5.

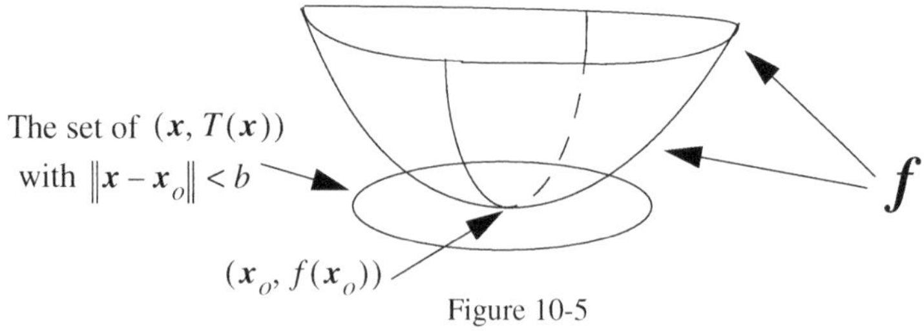

Figure 10-5

This result then guarantees that if $D_{\boldsymbol{u}}^2 f(\boldsymbol{x}_o) > 0$ and we also have $\boldsymbol{grad}\, f(\boldsymbol{x}_o) = \boldsymbol{0}$, then for all $\boldsymbol{x} \neq \boldsymbol{x}_o$ satisfying $\|\boldsymbol{x} - \boldsymbol{x}_o\| < b$ we have

$$f(\boldsymbol{x}) > f(\boldsymbol{x}_o).$$

That is, f has a strong relative (i.e., *local*) minimum at $\boldsymbol{x}_o$. We state this result formally as our next theorem.

Theorem 10.35: Conditions for relative extrema

Let f be a real valued function of n real variables (Definition 8.1 on page 261) all of whose second partial derivatives, $D_i D_j f(\boldsymbol{x})$ exist and are continuous (see Definition 8.5 on page 267 and Theorem 8.8 on page 271) for all $\boldsymbol{x}$ satisfying $\|\boldsymbol{x} - \boldsymbol{x}_o\| < a$ (Definition 8.6 on page 270) where a is some positive number, and for which

$$\boldsymbol{grad}\, f(\boldsymbol{x}_o) = \boldsymbol{0}$$

(Definition 8.10 on page 272).

Then if $D_{\boldsymbol{u}}^2 f(\boldsymbol{x}_o) > 0$, $[D_{\boldsymbol{u}}^2 f(\boldsymbol{x}_o) < 0]$ for all unit length vectors, $\boldsymbol{u}$, f has a strong relative minimum [maximum] at $\boldsymbol{x}_o$ — that is, there is a positive number, b, such that for all $\boldsymbol{x}$ satisfying $\|\boldsymbol{x} - \boldsymbol{x}_o\| < b$,

$$f(\boldsymbol{x}) > f(\boldsymbol{x}_o) \quad [f(x) < f(\boldsymbol{x}_o)].$$

❑

This may seem like a practical theorem, and it is, provided we have some way of determining conditions under which the inequality

$$D_{\boldsymbol{u}}^2 f(\boldsymbol{x}_o) > 0 \qquad \textbf{\textit{10.72}}$$

is satisfied for all unit length vectors, $\boldsymbol{u}$.

There is a very slick approach to solving this problem based on eigenvalues, in Chapter 13; this approach has to be postponed until we develop that area. However, there is a straightforward, albeit more tedious, approach which does provide a satisfying solution — namely, using completion of the square, and we proceed to this approach now.

Under the conditions of Theorem 10.35, from Theorem 8.11 on page 273, we may write

$$D_{\boldsymbol{u}}^2 f(\boldsymbol{x}_o) = \sum_{i,\,j} u_i u_j D_i D_j f(\boldsymbol{x}_o).$$

Let us introduce the following matrix abbreviation.

Definition 10.36: Hessian matrix

The matrix **H** (Definition 10.1 on page 328) whose i-j element is given by

$$H_{i,j} = D_i D_j f(\boldsymbol{x}_o)$$

(Definition 8.5 on page 267) is called the **Hessian** of the function f (at $\boldsymbol{x}_o$).

❑

Our object is to determine conditions under which, for all unit length vectors, $\boldsymbol{u}$,

$$\sum_{i,j} u_i u_j H_{i,j} > 0. \qquad \textbf{\textit{10.73}}$$

We see at once, that if for some i, $H_{i,i} \le 0$, then equation 10.73 must fail to hold for at least one unit vector, $\boldsymbol{u}$, namely, for $u_i = 1$, $u_j = 0$ for $j \ne i$, because clearly here the length $\|\boldsymbol{u}\| = 1$ and

$$\sum_{i,j} u_i u_j H_{i,j} = u_1^2 H_{1,1} = H_{1,1} \le 0.$$

So in our search for conditions under which equation 10.73 holds for all unit length vectors, $\boldsymbol{u}$, we may limit ourselves to the situation $H_{i,i} > 0$ for all i. Also, we can relax the restriction to vectors of unit length, so long as their lengths are not 0 — since if equation 10.73 holds for all unit length vectors, $\boldsymbol{u}$, then for any nonzero length vector, $\boldsymbol{h}$, we let $\boldsymbol{u} = \boldsymbol{h}/\|\boldsymbol{h}\|$, and if equation 10.73 holds for all unit length vectors, then

$$\sum_{i,j} \frac{h_i h_j}{\|\boldsymbol{h}\|^2} H_{i,j} = \sum_{i,j} u_i u_j H_{i,j} > 0$$

so that for all nonzero length vectors $\boldsymbol{h}$

$$\sum_{i,j} h_i h_j H_{i,j} > 0.$$

Now rewrite

$$\sum_{i,j} h_i h_j H_{i,j} = H_{1,1} h_1^2 + \sum_{\substack{i,j \\ i=1,\, j\ne 1}} h_i h_j H_{i,j}$$

$$+ \sum_{\substack{i,j \\ i\ne 1,\, j=1}} h_i h_j H_{i,j} + \sum_{\substack{i,j \\ i>1,\, j>1}} h_i h_j H_{i,j} ,$$

which we next rewrite as

$$H_{1,1}\left(h_1^2 + \sum_{j>1} h_1 h_j \frac{H_{1,j}}{H_{1,1}} + \sum_{i>1} h_i h_1 \frac{H_{i,1}}{H_{1,1}}\right) + \sum_{\substack{i,j \\ i>1,\, j>1}} h_i h_j H_{i,j}$$

or

$$H_{1,1}\left(h_1^2 + \sum_{j>1} h_1 h_j \frac{H_{1,j}+H_{j,1}}{H_{1,1}}\right) + \sum_{\substack{i,j \\ i>1,\, j>1}} h_i h_j H_{i,j}$$

$$= H_{1,1}\left(h_1^2 + \left(\sum_{j>1} h_j \frac{H_{1,j}+H_{j,1}}{H_{1,1}}\right) h_1\right) + \sum_{\substack{i,j \\ i>1,\, j>1}} h_i h_j H_{i,j}.$$

We now complete the square (recall the discussion on page 276) to see that

$$\sum_{i,j} h_i h_j H_{i,j}$$

$$= H_{1,1}\left(h_1^2 + \left(\sum_{j>1} h_j \frac{H_{1,j}+H_{j,1}}{H_{1,1}}\right) h_1 + \frac{1}{4}\left(\sum_{j>1} h_j \frac{H_{1,j}+H_{j,1}}{H_{1,1}}\right)^2\right)$$

$$-H_{1,1}\frac{1}{4}\left(\sum_{j>1} h_j \frac{H_{1,j}+H_{j,1}}{H_{1,1}}\right)^2 + \sum_{\substack{i,j \\ i>1,\, j>1}} h_i h_j H_{i,j}$$

$$= H_{1,1}\left(h_1 + \frac{1}{2}\sum_{j>1} h_j \frac{H_{1,j}+H_{j,1}}{H_{1,1}}\right)^2$$

$$-H_{1,1}\frac{1}{4}\left(\sum_{j>1} h_j \frac{H_{1,j}+H_{j,1}}{H_{1,1}}\right)^2 + \sum_{\substack{i,j \\ i>1,\, j>1}} h_i h_j H_{i,j}.$$

By choosing

$$h_1 = -\frac{1}{2}\left(\sum_{j>1} h_j \frac{H_{1,j}+H_{j,1}}{H_{1,1}}\right),$$

it is evident that the first term in the last expression preceding can be made equal to 0, no matter what the values of h_j, $j > 1$, Thus the *quadratic form*

$$\sum_{i,j} h_i h_j H_{i,j} > 0$$

for all nonzero vectors $\boldsymbol{h}$ if and only if the *smaller-sized* quadratic form

$$-H_{1,1}\frac{1}{4}\left(\sum_{j>1} h_j \frac{H_{1,j}+H_{j,1}}{H_{1,1}}\right)^2 + \sum_{\substack{i,j \\ i>1,\, j>1}} h_i h_j H_{i,j}$$

is positive for all nonzero choices of $\boldsymbol{h}_{-1} = (h_2, h_3, \ldots, h_n)$. We can continually repeat this process, starting with this reduced size quadratic form, until we get to the point where the answer becomes clear (it's an inductive process). The easiest way to understand this process is by means of examples, of which we provide two. Following these examples we will provide a more formal statement of the process we have just developed. To make the computations more readable, we will use (x, y, z) in place of (x_1, x_2, x_3).

Example 10.18: $x^2 + 2y^2 + z^2 + 3xy - 4xz - yz$

We rewrite the above quadratic form as

$$\begin{aligned}
&(x^2 + (3y-4z)x) + 2y^2 + z^2 - yz \\
&= \left(x^2 + (3y-4z)x + \frac{(3y-4z)^2}{4}\right) - \frac{(3y-4z)^2}{4} + 2y^2 + z^2 - yz \\
&= \left(x + \frac{3y-4z}{2}\right)^2 - \frac{y^2}{4} - 3z^2 + 5yz.
\end{aligned}$$

Now setting

$$x = -\frac{3y-4z}{2},$$

we need only examine

$$-\frac{y^2}{4} - 3z^2 + 5yz$$

for positivity for all nonzero (y, z). Actually we need go no further, because the coefficients of the squared terms are negative, and we already know that for positivity in all cases of nonzero (y, z), these coefficients must be positive. Also this quadratic form is easily seen to become negative for $y = 1, z = 0$.

Thus we see that the quadratic form

$$x^2 + 2y^2 + z^2 + 3xy - 4xz - yz$$

fails to be positive for all nonzero vectors (x,y,z).

■

Our next example requires going through to the bitter end.

Example 10.19: $x^2 + 2y^2 + z^2 + \frac{xy}{10} - \frac{xz}{10} - \frac{yz}{10}$

As before, we complete the square to rewrite the above expression as

$$\left(x^2 + \left[\frac{y}{10} - \frac{z}{10}\right]x\right) + 2y^2 + z^2 - \frac{yz}{10}$$

$$= \left(x^2 + \frac{1}{20}(y-z)\right)^2 - \frac{(y-z)^2}{400} + 2y^2 + z^2 - \frac{yz}{10}$$

$$= \left(x^2 + \frac{1}{20}(y-z)\right)^2 + 1.9975y^2 + 0.9975z^2 - 0.095yz.$$

As before, we need only examine the quadratic form

$$1.9975y^2 + 0.9975z^2 - 0.095yz$$

for positivity for all nonzero vectors (y,z). Completing the square again, we find

$$1.9975y^2 + 0.9975z^2 - 0.095yz$$

$$= 1.9975\left(y^2 - \left[\frac{0.095}{1.9975}z\right]y\right) + 0.9975z^2$$

$$= 1.9975\left(y - \left[\frac{0.095}{2 \times 1.9975}z\right]\right)^2 + \left(0.9975 - \frac{0.095 \times 0.095}{4 \times 1.9975}\right)z^2.$$

The coefficient of z^2 is positive, so that if z is nonzero, this last expression must be positive. If z is 0, then for y nonzero, we see that this assertion remains true. Finally, for both y and z zero, but x nonzero, from the previous expression

$$\left(x^2 + \frac{1}{20}(y-z)\right)^2 + 1.9975y^2 + 0.9975z^2 - 0.095yz,$$

which equals the original quadratic form, we see that whenever (x,y,z) is not the zero vector, the quadratic form

$$x^2 + 2y^2 + z^2 + \frac{xy}{10} - \frac{xz}{10} - \frac{yz}{10}$$

is positive.

■

Given these examples, if either you feel you need some practice to gain a better understanding of this area or if you have masochistic tendencies, you can make up some quadratic forms of your own. You will be able to check your results using *Mathematica*'s Eigenvalue function or *Splus*'s eigenfunction, once we have treated the subject of eigenvalues (see Theorem 13.26 on page 588).

We see that to determine places to look for relative maxima and minima, of a function f, we first look for *interior* extrema — i.e., f values for which

$$\textbf{\textit{grad}} f(\boldsymbol{x}_o) = \boldsymbol{0}, \qquad \textbf{\textit{10.74}}$$

and then examine the Hessian matrix, $\mathbf{H}$ (see Definition 10.36 on page 377) to determine whether for all nonzero vectors, $\boldsymbol{h}$, the quadratic form

$$\sum_{i,j} h_i h_j H_{i,j}$$

is always of one algebraic sign. We have just seen one method of carrying out this latter task, but as yet have not introduced any technique for determining the vectors, $\boldsymbol{x}_o$, satisfying equation 10.74. The most common means of doing this, multidimensional Newton's method, will be presented in the next section. In this regard, several remarks are worth making. First, none of the methods developed by mathematicians thus far work in all cases. Second, no mention has been made of extrema which occur at boundary points. Finally, after this process, we only have generated what we know to be relative maxima or minima. So, we may have found smaller peaks or larger valleys than the ones being sought. (So, instead of Everest, we may have obtained Mount Rainier or Beacon Hill.) In general there is no completely satisfactory resolution of these difficulties, although with some effort, we can often determine whether we have found what we want in important specific cases.

Despite the fact that our results are far from complete, because of the complexity that is already evident, we summarize what has been derived in Theorem 10.37.

Theorem 10.37: Finding and examining relative extrema

Suppose f is a real valued function of n real variables (Definition 8.1 on page 261) and we are seeking relative maxima or minima, at points $\boldsymbol{x}_o$ which are *interior* to the domain of f, i.e., points $\boldsymbol{x}_o$ such that $f(\boldsymbol{x})$ is defined for all $\boldsymbol{x}$ within some positive distance[1] of $\boldsymbol{x}_o$. Assuming that the function f has second derivatives which are continuous at all points interior to its domain (see Definition 8.5 on page 267 and Theorem 8.8 on page 271) in order that f

1. This distance (Definition 8.6 on page 270) determined by $\boldsymbol{x}_o$.

have a strong relative minimum [maximum] at x_o, i.e., $f(x) > f(x_o)$ [$f(x) < f(x_o)$] for all $x \neq x_o$ within some positive distance of x_o, it is necessary that the gradient vector (Definition 8.10 on page 272) satisfy

$$\boldsymbol{grad} f(\boldsymbol{x}_o) = \boldsymbol{0}.$$

At such a point, if the Hessian matrix, **H** (see Definition 10.36 on page 377) satisfies the condition that its associated quadratic form is **positive definite** [**negative definite**], i.e.,

$$\sum_{i,j} h_i h_j H_{i,j} > 0 \ [< 0] \qquad \textit{10.75}$$

for all nonzero real n-dimensional vectors $\boldsymbol{h} = (h_1, ..., h_n)$, then

$f(x_o)$ is a strong relative minimum [maximum].

In order for inequality 10.75 to be satisfied, it is necessary that

$$H_{i,i} 0 > \ [< 0] \text{ for all } i.$$

Rewriting $\sum_{i,j} h_i h_j H_{i,j}$ via completion of the square, we find that inequality 10.75 holds if and only if the reduced size quadratic form

$$-H_{1,1} \frac{1}{4} \left(\sum_{j>1} h_j \frac{H_{1,j} + H_{j,1}}{H_{1,1}} \right)^2 + \sum_{\substack{i,j \\ i>1,\, j>1}} h_i h_j H_{i,j}.$$

is positive [negative] for all nonzero vectors $(h_2, ..., h_n)$.

This latter result permits a recursive reduction of the problem until, at worst, we get to the point where the quadratic form being examined is based on one-dimensional vectors (where the result is sure to be evident).

❑

Problems 10.20

1. A real-valued function of two real variables is said to have a **saddle point** at x_o if $f(x_o)$ is a relative maximum in one direction, and a relative minimum in the perpendicular direction. Construct such a function. Hint: Try some simple polynomials first.
2. Apply Theorem 10.37 to the function you constructed.
3. Apply Theorem 10.37 to the function given by

$$f(x, y) = xe^{-x^2} + ye^{-y^2}.$$

4. Apply Theorem 10.37 to the function given by

$$f(x, y) = xe^{-x^2} - ye^{-y^2}.$$

9 Newton's Method in n Dimensions

The widespread availability of digital computers is what is largely responsible for the feasibility of numerical solution of equations

$$\textbf{\textit{grad}}\, f(\boldsymbol{x}) = \mathbf{0} \qquad \textit{10.76}$$

used in maxima and minima problems. Up to this point no methods for solving this system of equations have been introduced. We will see how essential Taylor's theorem is for establishing the important properties of the methods to be introduced.

Newton's method for real valued functions of a real variable seeks a real number, r, satisfying

$$g(r) = 0.$$

by starting with an initial guess, and hopefully generating better and better estimates, by *linearizing* at each stage — i.e., finding where the tangent line intersects the x axis to generate the next estimate, see Figure 3-16 on page 78. We know that the problem of determining maxima and minima of a real valued function, f, of n real variables commonly leads us to look for solutions of equation 10.76 above (see Theorem 10.37 on page 381) i.e., solution of the system of equations

$$g_1(\boldsymbol{x}) = D_1 f(\boldsymbol{x}) = 0$$

$$\cdot$$

$$\cdot$$

$$g_n(\boldsymbol{x}) = D_n f(\boldsymbol{x}) = 0.$$

So we start this section by extending Newton's method to solve a system of equations for a value, $\boldsymbol{r}$, of the n-dimensional real vector $\boldsymbol{x}$ satisfying

$$\begin{pmatrix} g_1(\boldsymbol{r}) \\ \cdot \\ \cdot \\ g_n(\boldsymbol{r}) \end{pmatrix} = \boldsymbol{g}(\boldsymbol{r}) = \mathbf{0},$$

where the g_k are real valued functions of n real variables (Definition 8.1 on page 261) assumed to have continuous second partial derivatives (see Definition 8.5 on page 267 and Theorem 8.8 on page 271). The only tools available are the ability to compute the function and its partial derivatives at any finite number of x values. The linearization that will be used here just arises from use of the first two terms of the multidimensional Taylor expansion for the g_k. Specifically, from Taylor's theorem for such functions (Theorem 10.33 on page 371) we may write

$$g_k(\boldsymbol{x}+\boldsymbol{h}) = g_k(\boldsymbol{x}) + \sum_{i=1}^{n} h_i D_i g_k(\boldsymbol{x}) + \sum_{i,j=1}^{n} h_i h_j D_i D_j g_k(\boldsymbol{x}+\vartheta_k h) \qquad \textbf{\textit{10.77}}$$

where $0 \le \vartheta_k \le 1$, ϑ_k depending on $\boldsymbol{x}$ and $\boldsymbol{h}$.

Under the continuity assumption, when $\boldsymbol{h}$ is small, it is reasonable to use the *fundamental multivariate approximations*

$$g_k(\boldsymbol{x}+\boldsymbol{h}) \cong g_k(\boldsymbol{x}) + \sum_{i=1}^{n} h_i D_i g_k(\boldsymbol{x}), \quad k = 1,...,n \qquad \textbf{\textit{10.78}}$$

(see Approximation 8.17 on page 274).

We begin the multivariate process by choosing an initial guess,[1] $\boldsymbol{x}_o$ for $\boldsymbol{r}$. Of course, we would like to choose the increment vector, $\boldsymbol{h}_o = (h_{o,1},...,h_{o,n})$ so that the $g_k(\boldsymbol{x}_o + \boldsymbol{h}_o)$ are closer to 0 than are the $g_k(\boldsymbol{x}_o)$. In hopes of accomplishing this, we instead go for the relatively simple task of trying to choose $\boldsymbol{h}_o$ to satisfy the system

$$g_k(\boldsymbol{x}_o) + \sum_{i=1}^{n} h_i D_i g_k(\boldsymbol{x}_o) = 0, \quad k = 1,...,n. \qquad \textbf{\textit{10.79}}$$

Notation 10.38: (..) [..] {..} concatenated matrices

In what follows **in this section only** we will adhere strictly to notation made necessary for proper interpretation all of the symbols being strung together in what remains of this section, namely,

- **Function arguments** (inputs) will be enclosed in **curly braces** { },
- **Parentheses** ( ) will be used for **grouping**, and an expression of the form $(...)\{y\}$ denotes that the function indicated in the parentheses is to be evaluated at its argument, $\boldsymbol{y}$.

1. As in the one-dimensional case, if this initial guess is far from $\boldsymbol{r}$, this process may be, and in fact usually is, unsuccessful in real-life problems.

- **Square brackets surrounding an array** such as

$$\begin{bmatrix} a & b & c \\ d & e & f \end{bmatrix}$$

  indicate a **matrix** — but a symbol for a matrix (**D** below) is not enclosed in square brackets, as when we let

$$\mathbf{D} = \begin{bmatrix} D_1 \\ D_2 \end{bmatrix}$$

- Concatenation of symbols for matrices indicates matrix multiplication (Definition 10.3 on page 331) — so that if $(\mathbf{HG}^{\mathbf{T}})\{x\}$ is the value of the matrix function $\mathbf{HG}^{\mathbf{T}}$ at the vector x — i.e., $(\mathbf{HG}^{\mathbf{T}})\{x\}$ is a matrix, and **B** as well as **h** are appropriate size matrices, the symbol

$$(\mathbf{HG}^{\mathbf{T}})\{x\}\,\mathbf{Bh}$$

  represents the product of these three matrices. Note that the transpose operator superscript, **T**, applies only to the matrix **G**, and not to **HG**, it being necessary to write $(\mathbf{HG})^{\mathbf{T}}$ for the transpose of **HG** (Definition 10.13 on page 342).

❑

Let

$$\mathbf{D} = \begin{bmatrix} D_1 \\ \cdot \\ \cdot \\ D_n \end{bmatrix}$$

be the gradient $(n \times 1)$ matrix operator (see Definition 8.10 on page 272 and Footnote 2 on page 273) and let **g** be the column matrix function defined by

$$\mathbf{g} = \begin{bmatrix} g_1 \\ \cdot \\ \cdot \\ g_n \end{bmatrix}.$$

Then, with $\mathbf{g}^{\mathrm{T}}$ being the transpose of $\mathbf{g}$ (see Definition 10.13 on page 342) we have

$$(\mathbf{Dg}^{\mathrm{T}})^{\mathrm{T}} = \left(\begin{bmatrix} D_1 \\ \cdot \\ \cdot \\ D_n \end{bmatrix} \begin{bmatrix} g_1 & \cdot\cdot & g_n \end{bmatrix}\right)^{\mathrm{T}} = \begin{bmatrix} D_1 g_1 & \cdot\ \cdot & D_1 g_n \\ \cdot & \cdot\ \cdot & \cdot \\ \cdot & \cdot\ \cdot & \cdot \\ D_n g_1 & \cdot\ \cdot & D_n g_n \end{bmatrix}^{\mathrm{T}}$$

$$= \begin{bmatrix} D_1 g_1 & \cdot\ \cdot & D_n g_1 \\ \cdot & \cdot\ \cdot & \cdot \\ \cdot & \cdot\ \cdot & \cdot \\ D_1 g_n & \cdot\ \cdot & D_n g_n \end{bmatrix}$$

so that[1]

$$(\mathbf{Dg}^{\mathrm{T}})^{\mathrm{T}}\{\boldsymbol{x}_o\} = \begin{bmatrix} (D_1 g_1)\{\boldsymbol{x}_o\} & \cdot\ \cdot & (D_n g_1)\{\boldsymbol{x}_o\} \\ \cdot & \cdot\ \cdot & \cdot \\ \cdot & \cdot\ \cdot & \cdot \\ (D_1 g_n)\{\boldsymbol{x}_o\} & \cdot\ \cdot & (D_n g_n)\{\boldsymbol{x}_o\} \end{bmatrix}.$$

Again note that $(D_i g_k)\{\boldsymbol{x}_o\}$ stands for the function $D_i g_k$ evaluated at $\boldsymbol{x}_o$.

The system given in equations 10.79 on page 384 may be written concisely in matrix notation as

$$(\mathbf{Dg}^{\mathrm{T}})^{\mathrm{T}}\{\boldsymbol{x}_o\}\mathbf{h}_o = -\mathbf{g}\{\boldsymbol{x}_o\} \qquad \textbf{\textit{10.80}}$$

where $\mathbf{h}_o$ is a column $(n \times 1)$ matrix.

If the Jacobian matrix $(\mathbf{Dg}^{\mathrm{T}})^{\mathrm{T}}\{\boldsymbol{x}_o\}$ is invertible (see Theorem 10.8 on page 336) then the solution, $\mathbf{h}_o$, of system 10.80 is given by

$$\mathbf{h}_o = -\left((\mathbf{Dg}^{\mathrm{T}})^{\mathrm{T}}\{\boldsymbol{x}_o\}\right)^{-1}\mathbf{g}\{\boldsymbol{x}_o\}.$$

---

1. The matrix $(\mathbf{Dg}^{\mathrm{T}})^{\mathrm{T}}$ is sometimes called the *Jacobian* matrix, denoted by J.

For defining the n-dimensional Newton's method, this process is then continually repeated, first with $\boldsymbol{x}_o$ replaced by $\boldsymbol{x}_1 = \boldsymbol{x}_o + \mathbf{h}_o$ as the next guess, leading to

$$\mathbf{h}_1 = -\left((\mathbf{Dg}^{\mathrm{T}})^{\mathrm{T}}\{\boldsymbol{x}_1\}\right)^{-1}\mathbf{g}\{\boldsymbol{x}_1\}$$

and so forth.[1] This is the multidimensional Newton's method, which we summarize formally as follows.

Definition 10.39: Multidimensional Newton's method

Let $\mathbf{g}$ be an $n \times 1$ matrix of real valued functions of n real variables (Definition 8.1 on page 261) whose partial derivatives exist (Definition 8.5 on page 267) at all $\boldsymbol{x}$ considered below.

Let the gradient operator matrix,[2] $\mathbf{D}$, be an $n \times 1$ matrix given by

$$\mathbf{D} = \begin{bmatrix} D_1 \\ \cdot \\ \cdot \\ D_n \end{bmatrix}, \qquad \textit{10.81}$$

where D_i is the partial derivative operator with respect to the i-th coordinate (see Definition 8.5 on page 267). With T being the matrix transpose operator (which interchanges rows and columns on the matrix of which it is a superscript, see Definition 10.13 on page 342) we see that

$$(\mathbf{Dg}^{\mathrm{T}})^{\mathrm{T}}\{\boldsymbol{x}\} = \begin{bmatrix} (D_1 g_1)\{\boldsymbol{x}\} & .\,. & (D_n g_1)\{\boldsymbol{x}\} \\ \cdot & .\,. & \cdot \\ \cdot & .\,. & \cdot \\ (D_1 g_n)\{\boldsymbol{x}\} & .\,. & (D_n g_n)\{\boldsymbol{x}\} \end{bmatrix}.$$

---

1. There is a slight inconsistency in our notation here, since the vectors $\boldsymbol{x}_i$ are being treated as a column matrices where the context makes it fairly obvious.
2. Often written ∇.

The sequence of Newton iterates, $\mathbf{x}_o, \mathbf{x}_1,...,$ used in trying to find a sequence which converges to a solution of $\mathbf{g}\{\mathbf{x}\} = \mathbf{0}$, i.e.. a vector $\mathbf{r}$ such that

$$\mathbf{g}\{\mathbf{r}\} = \mathbf{0}$$

is defined as follows.

The first element is the arbitrarily chosen $n \times 1$ (column) matrix, $\mathbf{x}_o$. Given any $\mathbf{x}_m$, the next element $\mathbf{x}_{m+1}$ is given by

$$\mathbf{x}_m + \mathbf{h}_m \qquad \textit{10.82}$$

where

$$\mathbf{h}_m = -\left((\mathbf{D}\mathbf{g}^{\mathrm{T}})^{\mathrm{T}}\{\mathbf{x}_m\}\right)^{-1}\mathbf{g}\{\mathbf{x}_m\} \qquad \textit{10.83}$$

provided this inverse exists and $\mathbf{x}_m + \mathbf{h}_m$ is in the domain of (an allowable input to) $\mathbf{g}$; i.e.,

$$\mathbf{x}_{m+1} = \mathbf{x}_m - \left((\mathbf{D}\mathbf{g}^{\mathrm{T}})^{\mathrm{T}}\{\mathbf{x}_m\}\right)^{-1}\mathbf{g}\{\mathbf{x}_m\} \qquad \textit{10.84}$$

❑

Notice the striking similarity of this equation to equation 3.18 on page 79, which specifies the one-dimensional Newton's method. (Of course, when $n = 1$, it specifies the one-dimensional Newton's method. But what we're stressing here is that the notation needn't change greatly when we generalize to n dimensions.) The condition that $(\mathbf{D}\mathbf{g}^{\mathrm{T}})^{\mathrm{T}}\{\mathbf{x}_m\}$ be invertible is just the extension of the one-dimensional requirement that $f'(x_m) \neq 0$. It will be shown soon that when the initial guess $\mathbf{x}_o$ is close enough to a sought-for solution, $\mathbf{r}$, under very reasonable conditions, the sequence of Newton iterates converges to $\mathbf{r}$. In practice, however, it is rare that theoretical considerations are used to determine an m for which $\mathbf{x}_m$ is sufficiently close to $\mathbf{r}$ — the iteration continues until one or all of the following conditions are satisfied:

- The computed value of the distance (see Definition 8.6 on page 270)

  $$\|\mathbf{x}_m - \mathbf{x}_{m-1}\|$$

  is sufficiently small,
- $\mathbf{g}(\mathbf{x}_m)$ is sufficiently close to $\mathbf{0}$ (i.e., $\|\mathbf{g}(\mathbf{x}_m)\|$ is small enough),
- The sequence of iterates is behaving badly,
- Exhaustion has set in.

In just about all computer packages which implement Newton's method, it is not unusual to encounter situations in which the method fails because of a poor choice of starting values. So operating on *autopilot* when using Newton's method is not a wise practice.

We present one example of successful use of Newton's method.

Example 10.21: Primed constant infusion estimation

Examination of the single-compartment model introduced in Example 5.8 on page 149 revealed that under this model, if some drug is being infused at a constant rate into the bloodstream, then the amount of drug, $F(t)$, in the bloodstream at time t must have a formula of the following form:

$$F(t) = \frac{I}{k} + de^{-kt},$$

where I is the infusion rate, which may be the sum of an internal production rate and an external infusion. The value k is a constant reflecting the properties of the compartmental system, and d is a constant which can be chosen to match the amount of drug in the bloodstream at some specific time point. In many situations where this model applies, the external infusion rate is under the experimenters control, and the only other available information consists of data of the form

$$(t_i, y_i) = (t_i, F(t_i)), \quad i = 1,...,n$$

or, more commonly, *noisy data*

$$(t_i, y_i) = (t_i, F(t_i) + error_i), \quad i = 1,...,n,$$

where $error_i$ is often assumed to have completely known statistical characteristics. Usually we want to obtain point or interval estimates of the unknown parameters I, k, and d from this data. The most widely available procedure for such estimation attacks the problem looking for the triple (a, d, k) which minimizes the expression

$$g(a, d, k) = \sum_{i=1}^{n} \left(y_i - \left[a + de^{-kt_i}\right]\right)^2. \qquad \textbf{10.85}$$

This is referred to as *nonlinear least squares*, because the unknowns being sought occur nonlinearly — i.e., not as simple multiplicative weights. In this case it is the parameter k which occurs nonlinearly, being in the exponent.

In this example we will apply Newton's method to obtain the desired result. There are, however, several other important approaches to this minimization problem, which we will treat briefly later on.

Surprisingly enough, the Newton's method approach even for this simple example, is quite tedious. This phenomenon of complexity even in simple examples is pervasive in much of advanced applied mathematics — a fact which makes the construction of useful illustrative examples quite difficult at this level. Most such problems are handled by sophisticated computer packages, such as *Splus*, *Mathematica*, *Maple*, *MLab*, etc., with the user well shielded from the gory details. The disadvantage of being so shielded is that many users don't have a good gut feel for what's going on. In order to overcome this difficulty for this example, a good deal of detail is provided. Since it takes up so much space, the Newton method of minimization of the expression in equation 10.85 is relegated to Appendix 6 starting on page 783. Should you decide to look at this material, you will have to judge for yourself whether the detail is beneficial. If you feel that you understand the difficulties and pitfalls, skim over the details and just hit the highlights. Since the implementation is in the *Splus* language, unless you are familiar with this language, it may be rough going.

■

Comments on good starting values and other issues

There are no universal rules for choosing good starting values for Newton's or most other approximation methods. Some starting values do not yield convergence of the successive parameter estimates at all, and others yield convergence, but to a wrong value. There are a few important situations in which good starting values are available. In particular, in trying to estimate the parameters a_1, b_1, a_2, b_2 to minimize a sum such as

$$\sum_{j=1}^{n}\left(y_j - \left[a_1 e^{-b_1 t_j} + a_2 e^{-b_2 t_j}\right]\right)^2,$$

if b_1 and b_2 are far enough apart, the contribution of the larger exponent is negligible for large values of t, so that the smaller exponential parameter can be estimated by the large t data values. The corresponding weight can easily be estimated. Once this is done, the larger exponential parameter and its corresponding weight can be estimated from the smaller t data values (by subtracting off the estimated contribution just obtained, from the smaller data values). Another, more sophisticated method makes use of the fact that the above regression function satisfies a linear constant coefficient differential equation. If the data allow a reasonably accurate reconstruction of the true regression function and its antiderivatives by some simple curve-fitting routine, then linear regression methods often allow estimation of good starting values, see Rosenblatt [15].

Often an alternate method can be used to get close enough to starting values suitable for minimization by Newton's method. One such method is *steepest descent,* which will be briefly discussed in the next section.

Alternatively, methods such as *steepest descent,* which find extrema directly, can be used as the sole approach to solving the equation

$$\mathbf{g}(\mathbf{x}) = \mathbf{0}, \text{ by direct minimization of } \|\mathbf{g}(\mathbf{x})\|^2.$$

In our treatment of Newton's method, we determined formulas for the derivatives needed in this algorithm. There are some symbolic manipulation programs which can do this (*Splus*, *Mathematica*, and *Maple* allow this). But many computer packages compute the needed derivatives numerically (unless this symbolic capability is specifically required in the program). This can lead to inaccuracies and decreased efficiency.

Here are a few exercises which you might try, assuming you have access to computer packages which include matrix operations. Try out any available Newton's method or direct minimization methods as well.

Exercises 10.22

1. Solve the system of equations

$$x - 0.68\,sin(x) - 0.19\,cos(y) = 0$$
$$y - 0.68\,cos(x) + 0.19\,sin(y) = 0$$

   starting off with $(x,y) = (0,0)$.

2. In 10.21 on page 389 we treated the general case of an infusion of a substrate also produced by the body. In such a case, we do not know the substrate level in the blood at the time the infusion is started. This forces us to estimate three parameters a, d, k (or equivalently I, k, and d). If instead we infuse a tracer whose blood level in the body is 0 when the infusion starts, it's not hard to see that the solution we seek is of the form $a(1 - e^{-kt})$. Now choose values a and $k > 0$, and values $t_1, t_2,...,t_n$ and set up Newton's method to estimate a and k for this situation, both when the generated data is noise-free, and in the presence of noise (generated from some convenient random number generator).

3. Choose values a, b and values $t_1, t_2,...,t_n$. Assuming data of the form ae^{-bt_i}, use Newton's method to carry out least squares estimation of a and b.

4. Same as previous question, but with noisy data, generated by some available random number generator.

5. Same as the two previous questions, but with $a_1e^{-b_1t_j} + a_2e^{-b_2t_j}$.

6. Check your results with computer packages, such as *Solver* (in *Microsoft Excel*, which can be set up for minimization, maximization, or equation solution), *Mlab*, *Mathematica*, etc.

We now continue with the main (and only) theoretical result that we will present concerning the multidimensional Newton's method. It's pretty tough sledding, and those who don't at the moment feel up to going through the details may skip to the summary, Theorem 10.41 on page 398.

As we shall now show, we can only guarantee convergence of Newton's method to the desired result if our initial guess is close enough to this result. The better computer packages for minimization go to quite a bit of effort to overcome this deficiency, but none make any claim to being perfect — which means that ***after obtaining an answer, this answer must be examined carefully to see whether it makes sense.***

The behavior of the multidimensional Newton iterate sequence $\mathbf{x}_m$ given in Definition 10.39 on page 387 is essentially the same as that of one-dimensional Newton iterates (see Theorem 3.6 on page 82). However, without some preliminary spade work rewriting various multidimensional versions of Taylor's theorem in matrix notation, the details can be overwhelming. The first step in this simplification process is to write equations 10.77 on page 384 in matrix terms. To see how this is done, first note that from the assumed continuity of the second partial derivatives, we may write

$$D_i D_j g_k\{\boldsymbol{x} + \vartheta_k \boldsymbol{h}\} = (D_i D_j g_k)\{\boldsymbol{x}\} + \varepsilon_{k,\,i,\,j}\{\boldsymbol{x}, \boldsymbol{h}\}$$

where for each fixed $\boldsymbol{x}$,

$$\lim_{\|\boldsymbol{h}\| \to 0} \varepsilon_{k,\,i,\,j}\{\boldsymbol{x}, \boldsymbol{h}\} = 0 .$$

Hence we may rewrite the first-order Taylor theorem, equation 10.77 on page 384 as

$$g_k\{\boldsymbol{x} + \boldsymbol{h}\} = g_k\{\boldsymbol{x}\} + \sum_{i=1}^{n} h_i (D_i g_k)\{\boldsymbol{x}\} + \sum_{i,\,j=1}^{n} h_i h_j (D_i D_j g_k)\{\boldsymbol{x}\} \qquad \textbf{10.86}$$

$$+ \sum_{i,\,j=1}^{n} h_i h_j \varepsilon_{k,\,i,\,j}\{\boldsymbol{x}, \boldsymbol{h}\}.$$

Now with $\mathbf{D}$ an $n \times 1$ matrix, we have

$$\mathbf{DD}^{\mathrm{T}} = \begin{bmatrix} D_1^2 & D_1 D_2 & \dots & D_1 D_n \\ \cdot & \cdot & \cdot & \cdot \\ \cdot & \cdot & \cdot & \cdot \\ D_n D_1 & D_n D_2 & \dots & D_n^2 \end{bmatrix}$$

so that $\sum_{i,j=1}^{n} h_i h_j (D_i D_j g_k)\{\boldsymbol{x}\}$ may be written as

$$\mathbf{h}^{\mathrm{T}}(\mathbf{DD}^{\mathrm{T}} g_k)\{x\}\mathbf{h}.$$

(You may need to write this out longhand to convince yourself of the validity of this last assertion.) Similarly, the last term $\sum h_i h_j \varepsilon_{k,i,j}\{\boldsymbol{x}, \boldsymbol{h}\}$ of equation 10.86 may be written as

$$\mathbf{h}^{\mathrm{T}} \varepsilon_k\{\boldsymbol{x}, \boldsymbol{h}\}\mathbf{h}$$

where $\varepsilon_k\{\boldsymbol{x}, \boldsymbol{h}\}$ is an $n \times n$ matrix whose *i-j* element is $\varepsilon_{k,i,j}\{\boldsymbol{x}, \boldsymbol{h}\}$.

So we may abbreviate the entire set of first-order Taylor theorem equations in 10.77 on page 384 for $k = 1,\dots,n$ in the matrix form

$$\mathbf{g}\{\mathbf{x}+\mathbf{h}\} = \mathbf{g}\{\mathbf{x}\} + (\mathbf{Dg}^{\mathrm{T}})^{\mathrm{T}}\{\mathbf{x}\}\mathbf{h} + \mathbf{Q_1}\{\mathbf{x}, \mathbf{h}\} + \mathbf{R_1}\{\mathbf{x}, \mathbf{h}\},$$

where both $\mathbf{x}$ and $\mathbf{h}$ are $n \times 1$ matrices, $\mathbf{Q_1}\{\mathbf{x}, \mathbf{h}\}$ is an $n \times 1$ matrix whose k-th element is $\mathbf{h}^{\mathrm{T}}(\mathbf{DD}^{\mathrm{T}} g_k)\{x\}\mathbf{h}$ and $\mathbf{R_1}\{\mathbf{x}, \mathbf{h}\}$ is an $n \times 1$ matrix whose k-th element is $\mathbf{h}^{\mathrm{T}} \varepsilon_k\{\boldsymbol{x}, \boldsymbol{h}\}\mathbf{h}$.

Similarly, if $\mathbf{H}$ is a $q \times 1$ matrix whose elements H_j are real valued functions of n real variables with continuous first partial derivatives, from Taylor's theorem, Theorem 10.33 on page 371, we may write

$$H_j\{\mathbf{x}+\mathbf{h}\} = H_j\{\mathbf{x}\} + \sum_{i=1}^{n} h_i (D_i H_j)\{\mathbf{x} + \eta_j \mathbf{h}\},$$

where $0 \le \eta_j \le 1$, $j = 1,\dots,q$, (η_j depending on $\mathbf{x}, \mathbf{h}$).

From the assumed continuity of the $D_i H_j$ s, we may write

$$(D_i H_j)\{\mathbf{x} + \eta_j \mathbf{h}\} = (D_i H_j)\{\mathbf{x}\} + e_{j,i}\{\mathbf{x}, \mathbf{h}\},$$

where for each fixed $\mathbf{x}$,

$$\lim_{\|\boldsymbol{h}\| \to 0} e_{j,i}\{\mathbf{x}, \mathbf{h}\} = 0.$$

Just as with the first order Taylor expansion, we may write the zeroth order Taylor expansion as

$$\mathbf{H}\{\mathbf{x}+\mathbf{h}\} = \mathbf{H}\{\mathbf{x}\} + (\mathbf{DH}^{\mathrm{T}})^{\mathrm{T}}\{\mathbf{x}\}\mathbf{h} + \mathbf{R}_0\{\mathbf{x},\mathbf{h}\}\mathbf{h}, \qquad \textit{10.87}$$

where $\mathbf{R}_0\{\mathbf{x},\mathbf{h}\}$ is a $q \times n$ matrix whose *k-i* element is $e_{k,i}\{\mathbf{x},\mathbf{h}\}$.

Remark

If $\mathbf{H}$ is an $n \times s$ matrix, this last result can still be applied by treating $\mathbf{H}$ as a vector with $q = ns$ elements, and after computing the last two terms in equation 10.85 above, rearranging their ns elements to form the appropriate matrix.

For reference purposes we next summarize the matrix versions of the multidimensional Taylor theorems needed in the material to follow in this chapter. This summary makes use of the strict notation specified in Notation 10.38 on page 384, with all vectors interpreted as column $(n \times 1)$ matrices.

Theorem 10.40: Taylor theorem; first and zeroth order matrix versions

Let g_k, H_k, $k = 1,...,n$ *be real valued functions of* n real variables with continuous second and first partial derivatives, respectively, for all values within some specified positive distance, *d,* of the point $\mathbf{x}$.

Let $\mathbf{D}$ be the $n \times 1$ gradient operator

$$\begin{bmatrix} D_1 \\ \cdot \\ \cdot \\ D_n \end{bmatrix}$$

and $\mathbf{g}, \mathbf{H}$ be the $n \times 1$ matrices with k-th elements g_k, H_k, respectively.

Then for all $\|h\| < d$

$$\mathbf{g}\{\mathbf{x}+\mathbf{h}\} = \mathbf{g}\{\mathbf{x}\} + (\mathbf{Dg}^{\mathrm{T}})^{\mathrm{T}}\{\mathbf{x}\}\mathbf{h} + \mathbf{Q}_1\{\mathbf{x},\mathbf{h}\} + \mathbf{R}_1\{\mathbf{x},\mathbf{h}\} \qquad \textit{10.88}$$

where both $\mathbf{x}$ and $\mathbf{h}$ are $n \times 1$ matrices, $\mathbf{Q}_1\{\mathbf{x},\mathbf{h}\}$ is an $n \times 1$ matrix whose k-th element is $\mathbf{h}^{\mathrm{T}}(\mathbf{DD}^{\mathrm{T}}g_k)\{x\}\mathbf{h}$, and $\mathbf{R}_1\{\mathbf{x},\mathbf{h}\}$ is an $n \times 1$ matrix whose k-th element is $\mathbf{h}^{\mathrm{T}}\varepsilon_k\{\boldsymbol{x},\boldsymbol{h}\}\mathbf{h}$, with $\varepsilon_k\{\boldsymbol{x},\boldsymbol{h}\}$ being an $n \times n$ matrix whose *i-j* elements, $\varepsilon_{k,i,j}\{\boldsymbol{x},\boldsymbol{h}\}$, all satisfy the conditions that for each fixed $\boldsymbol{x}$,

$$\lim_{\|\boldsymbol{h}\|\to 0} \varepsilon_{k,i,j}\{\boldsymbol{x},\boldsymbol{h}\} = 0 .$$

Also,

$$\mathbf{H}\{\mathbf{x}+\mathbf{h}\} = \mathbf{H}\{\mathbf{x}\} + (\mathbf{D}\mathbf{H}^{\mathrm{T}})^{\mathrm{T}}\{\mathbf{x}\}\mathbf{h} + \mathbf{R}_0\{\mathbf{x},\mathbf{h}\}\mathbf{h}, \qquad \textit{10.89}$$

where $\mathbf{R}_0\{\mathbf{x},\mathbf{h}\}$ is a $q \times n$ matrix whose *k-i* elements $e_{k,i}\{\mathbf{x},\mathbf{h}\}$ all satisfy the conditions that for each fixed $\mathbf{x}$

$$\lim_{\|\boldsymbol{h}\|\to 0} e_{k,i}\{\mathbf{x},\mathbf{h}\} = 0.$$

By treating a matrix as a vector for this computation, this result applies to matrices.

Note that $(\mathbf{D}\mathbf{H}^{\mathrm{T}})^{\mathrm{T}}$ and $\mathbf{D}\mathbf{D}^{\mathrm{T}}g_k$ are, respectively, $n \times n$ and $n \times 1$ matrix valued functions, whose values at $\mathbf{x}$ are $(\mathbf{D}\mathbf{H}^{\mathrm{T}})^{\mathrm{T}}\{\mathbf{x}\}$ and $(\mathbf{D}\mathbf{D}^{\mathrm{T}}g_k)\{\mathbf{x}\}$, respectively. T is the transpose operator, transposing the matrix immediately preceding it (see Definition 10.13 on page 342) — so that $\mathbf{D}\mathbf{H}^{\mathrm{T}}$ is the gradient matrix *times* (operating on) the matrix $\mathbf{H}^{\mathrm{T}}$.

❑

We now have the tools needed to examine the behavior of the multidimensional Newton's method. It would be tempting to try for an obvious, simple extension of the method used to establish the one-dimensional Newton's method behavior,[1] but this does not seem to me to be feasible. Instead we look back to the one-dimensional case, and extend the following one-dimensional Taylor theorem approach (which wasn't even available in the derivation of the one-dimensional error estimate for Newton's method). The version of the one-dimensional Taylor's theorem that we will use is a special case of Theorem 6.2 on page 179. Assuming that the real valued function g of a real variable has continuous second derivatives:

$$g(x+h) = g(x) + g'(x)h + g''(x)\frac{h^2}{2} + \varepsilon_2(x,h)h^2, \qquad \textbf{10.90}$$

where for each fixed x, $\lim_{h\to 0}\varepsilon_2(x,h) = 0\cdot$

If we now let $x_n = x+h$ and $x = r$ and rewrite equation 10.90 with these substitutions, we obtain

$$g(x_n) = g(r) + g'(r)(x_n - r) + g''(r)\frac{(x_n - r)^2}{2} + \varepsilon_2(r, x_n - r)(x_n - r)^2$$

---

1. You might look over the proof for one dimension that was given for Theorem 3.6 on page 82, to see the obstacles to a simple extension to n dimensions.

or

$$g(x_n) - g(r) = g'(r)(x_n - r) + g''(r)\frac{(x_n - r)^2}{2} + \varepsilon_2(r, x_n - r)(x_n - r)^2. \qquad \textbf{\textit{10.91}}$$

Recall once again from equation 3.18 on page 79 that Newton's method iteration for functions, g, of one variable is given by

$$x_{n+1} = x_n - \frac{g(x_n)}{g'(x_n)}.$$

Thus, assuming $g(r) = 0$,

$$x_{n+1} - r = x_n - r - \frac{g(x_n) - g(r)}{g'(x_n)}.$$

Substituting from equation 10.91, we find

$$x_{n+1} - r = x_n - r - \frac{g'(r)(x_n - r) + g''(r)\frac{(x_n - r)^2}{2} + \varepsilon_2(r, x_n - r)(x_n - r)^2}{g'(x_n)}$$

$$= \frac{(g'(x_n) - g'(r))(x_n - r) - g''(r)\frac{(x_n - r)^2}{2} - \varepsilon_2(r, x_n - r)(x_n - r)^2}{g'(x_n)}$$

If we similarly apply Taylor's theorem for g',

$$g'(x_n) - g'(r) = g''(r)(x_n - r) + \varepsilon_1(r, x_n - r)(x_n - r)$$

and substitute this in the equation just derived, we obtain, after some standard algebraic simplification that

$$x_{n+1} - r = \frac{g''(r)\frac{(x_n - r)^2}{2} + [\varepsilon_1(r, x_n - r) + \varepsilon_2(r, x_n - r)](x_n - r)^2}{g'(x_n)}.$$

Now use the continuity of $1/g'$ at r, together with the triangle inequality, Theorem 7.6 on page 216 or Theorem 10.20 on page 350, to conclude that there is a constant, C, such that for all x_n sufficiently close to r,

$$|x_{n+1} - r| \le C|x_n - r|^2.$$

This establishes that if the starting value, x_0, is close enough to r, then Newton's method works quite well in one dimension. This proof has the advantage that it extends well to proving its extension to higher dimensions. However, without a great deal more work, this extension is only of theoretical importance in showing the efficiency of Newton's method. It is of little practical computational use, and may be omitted without disrupting later development.

The multidimensional Newton's method, Definition 10.39 on page 387, for solving the system of equations

$$\mathbf{g}(\mathbf{r}) = \mathbf{0} \tag{10.92}$$

is specified by a starting value, $\mathbf{x}_0$, and iterations (see equation 10.84 on page 388)

$$\mathbf{x}_{m+1} = \mathbf{x}_m - \left((\mathbf{Dg}^T)^T\{\mathbf{x}_m\}\right)^{-1}\mathbf{g}\{\mathbf{x}_m\}.$$

Assuming that equation 10.92 really holds, we may write

$$\begin{aligned}\mathbf{x}_{m+1} - \mathbf{r} &= \mathbf{x}_m - \mathbf{r} - \left((\mathbf{Dg}^T)^T\{\mathbf{x}_m\}\right)^{-1}(\mathbf{g}\{\mathbf{x}_m\} - \mathbf{g}\{\mathbf{r}\}) \\ &= \left((\mathbf{Dg}^T)^T\{\mathbf{x}_m\}\right)^{-1}\left((\mathbf{Dg}^T)^T\{\mathbf{x}_m\}(\mathbf{x}_m - \mathbf{r}) - (\mathbf{g}\{\mathbf{x}_m\} - \mathbf{g}\{\mathbf{r}\})\right).\end{aligned} \tag{10.93}$$

Now identifying the pair $(\mathbf{x}, \mathbf{x}+\mathbf{h})$ in the Taylor equation 10.88 on page 394 with $(\mathbf{r}, \mathbf{x}_m)$, so that $\mathbf{x} = \mathbf{r}$ and $\mathbf{h} = \mathbf{x}_m - \mathbf{r}$, equation 10.88 can be written for use here as

$$\mathbf{g}\{\mathbf{x}_m\} - \mathbf{g}(\mathbf{r}) = (\mathbf{Dg}^T)^T\{\mathbf{r}\}(\mathbf{x}_m - \mathbf{r}) + \mathbf{Q}_1(\mathbf{r}, \mathbf{x}_m - \mathbf{r}) + \mathbf{R}_1(\mathbf{r}, \mathbf{x}_m - \mathbf{r}).$$

Substituting this into equation 10.93 above yields

$$\begin{aligned}\mathbf{x}_{m+1} - \mathbf{r} = \Big(&\left((\mathbf{Dg}^T)^T\{\mathbf{x}_m\}\right)^{-1}\left((\mathbf{Dg}^T)^T\{\mathbf{x}_m\} - (\mathbf{Dg}^T)^T\{\mathbf{r}\}\right)(\mathbf{x}_m - \mathbf{r}) \\ &- \mathbf{Q}_1(\mathbf{r}, \mathbf{x}_m - \mathbf{r}) - \mathbf{R}_1(\mathbf{r}, \mathbf{x}_m - \mathbf{r})\Big).\end{aligned} \tag{10.94}$$

From this, using the triangle inequality, Theorem 10.20 on page 350, and the product norm inequality, Theorem 10.21 on page 351, we see that

$$\begin{aligned}\|\mathbf{x}_{m+1} - \mathbf{r}\| \le \left\|\left((\mathbf{Dg}^T)^T\{\mathbf{x}_m\}\right)^{-1}\right\| \Big(&\left\|(\mathbf{Dg}^T)^T\{\mathbf{x}_m\} - (\mathbf{Dg}^T)^T\{\mathbf{r}\}\right\| \|\mathbf{x}_m - \mathbf{r}\| \\ &+ \|\mathbf{Q}_1(\mathbf{r}, \mathbf{x}_m - \mathbf{r})\| + \|\mathbf{R}_1(\mathbf{r}, \mathbf{x}_m - \mathbf{r})\|\Big).\end{aligned}$$

But using the first-order Taylor theorem result, equation 10.87 on page 394, identifying the matrix $(\mathbf{Dg}^T)^T$ with $\mathbf{H}$ (interpreting $(\mathbf{Dg}^T)^T$ as a vector, as mentioned in the remark following equation 10.87) and then again using Theorems 10.20 and 10.21, it follows that

$$\left\|(\mathbf{Dg}^T)^T\{\mathbf{x}_m\} - (\mathbf{Dg}^T)^T\{\mathbf{r}\}\right\| = \left\|(\mathbf{DH}^T)^T\{\mathbf{r}\}(\mathbf{x}_m - \mathbf{r}) + \mathbf{R}_0(\mathbf{r}, \mathbf{x}_m - \mathbf{r})(\mathbf{x}_m - \mathbf{r})\right\|$$

$$\leq \left\|(\mathbf{DH}^T)^T\{\mathbf{r}\}\right\| \|\mathbf{x}_m - \mathbf{r}\| + \|\mathbf{R}_0(\mathbf{r}, \mathbf{x}_m - \mathbf{r})\| \|\mathbf{x}_m - \mathbf{r}\|.$$

Thus,

$$\|\mathbf{x}_{m+1} - \mathbf{r}\| \leq \left\|\left((\mathbf{Dg}^T)^T\{\mathbf{x}_m\}\right)^{-1}\right\| \Big(\left\|(\mathbf{DH}^T)^T\{\mathbf{r}\}\right\| \|\mathbf{x}_m - \mathbf{r}\|^2$$

$$+ \|\mathbf{R}_0(\mathbf{r}, \mathbf{x}_m - \mathbf{r})\| \|\mathbf{x}_m - \mathbf{r}\|^2 + \|\mathbf{Q}_1(\mathbf{r}, \mathbf{x}_m - \mathbf{r})\| + \|\mathbf{R}_1(\mathbf{r}, \mathbf{x}_m - \mathbf{r})\| \Big).$$

As long as $\|\mathbf{x}_m - \mathbf{r}\|$ is close enough to 0, using the definition of $\| \;\|$ and the product norm inequality (Theorem 10.21 on page 351) as needed, we see that

- $\|\mathbf{R}_0(\mathbf{r}, \mathbf{x}_m - \mathbf{r})\|$ is near 0 (see equation 10.89 on page 395 ff.)
- $\|\mathbf{Q}_1(\mathbf{r}, \mathbf{x}_m - \mathbf{r})\| \leq n \max_k \|\mathbf{DD}^T g_k\{\mathbf{r}\}\| \|\mathbf{x}_m - \mathbf{r}\|^2$
- $\|\mathbf{R}_1(\mathbf{r}, \mathbf{x}_m - \mathbf{r})\| \leq n \max_k \|\varepsilon_k(\mathbf{r}, \mathbf{x}_m - \mathbf{r})\| \|\mathbf{x}_m - \mathbf{r}\|^2$, with the middle factor here being small because of the assumed smallness of $\|\mathbf{x}_m - \mathbf{r}\|$.

Furthermore, by the continuity of matrix inversion (Theorem 10.27 on page 359) for $\|\mathbf{x}_m - \mathbf{r}\|$ close enough to 0,

$$\left\|\left((\mathbf{Dg}^T)^T\{\mathbf{x}_m\}\right)^{-1}\right\| \text{ is close to } \left\|\left((\mathbf{Dg}^T)^T\{\mathbf{r}\}\right)^{-1}\right\|.$$

Together, these results show that for $\|\mathbf{x}_m - \mathbf{r}\|$ close enough to 0 (i.e., $\mathbf{x}_m$ close enough to $\mathbf{r}$), there is a constant, c, such that

$$\|\mathbf{x}_{m+1} - \mathbf{r}\| \leq c\|\mathbf{x}_m - \mathbf{r}\|^2.$$

$\llcorner$

The results whose proof we have outlined (albeit with some of the finer details probably missing) is worth summarizing as

Theorem 10.41: Convergence of multidimensional Newton method

Let $\mathbf{g}$ be an $n \times 1$ matrix of real valued functions of n real variables (Definition 8.1 on page 261) all of whose elements have continuous partial derivatives (see Definition 8.5 on page 267 and Theorem 8.8 on page 271) for all $\mathbf{x}$ within some positive distance d of the vector $\mathbf{r}$ (Definition 8.6 on page 270) for which $\mathbf{g}(\mathbf{r}) = \mathbf{0}$. Referring to the multidimensional Newton method (Definition 10.39 on page 387) if $(\mathbf{Dg}^T)^T\{\mathbf{r}\}$ is invertible (Definition 10.9 on page 336) there is a positive value δ and a constant C such that whenever $\|\mathbf{x}_m - \mathbf{r}\| \le \delta$, then next Newton iterate, $\mathbf{x}_{m+1}$, given by

$$\mathbf{x}_{m+1} = \mathbf{x}_m - \left((\mathbf{Dg}^T)^T\{\mathbf{x}_m\}\right)^{-1} \mathbf{g}(\mathbf{x}_m)$$

is meaningful and satisfies the inequality

$$\|\mathbf{x}_{m+1} - \mathbf{r}\| \le C\|\mathbf{x}_m - \mathbf{r}\|^2.$$

This implies very rapid convergence of Newton's method (see Definition 10.22 and Footnote 2 on page 352) if the initial guess, $\mathbf{x}_0$, is close enough to $\mathbf{r}$.

❑

10 Direct Minimization by Steepest Descent

The approach to minimizing a real valued function, f, of several variables used so far has been by way of attempted solution of the equation

$$\textbf{\textit{grad}}\, f(\boldsymbol{x}) = \mathbf{0}. \qquad \textbf{\textit{10.95}}$$

And for this task the only method as of now is Newton's method. But for Newton's method to have much assurance of success, it's usually necessary that the initial guess be chosen close enough to the solution.

For this reason alone we need a method which brings us close to the desired solution from a distant starting point. The method we will look at in this section, the very natural direct method of *steepest descent*, makes use of the fact that for fixed $\boldsymbol{x}$, the directional derivative $D_u f(\boldsymbol{x})$ is smallest (most negative) for[1]

---

1. As an additional bonus, reasoning in the reverse direction, if we can find some direct successive approximation method for minimizing $f(\boldsymbol{x})$, then we can solve the system $\boldsymbol{H}(\boldsymbol{x}) = \mathbf{0}$ (or at least get reasonably close to solving this system) via a procedure for minimizing $f(x) = \|H(\boldsymbol{x})\|^2$. Thus the problem of minimizing a function of n variables is essentially equivalent to that of solving a system of n (possibly nonlinear) equations in n unknowns. This allows us the flexibility of using either method at any stage of either problem.

$$u = \frac{-\textbf{\textit{grad}}\ f(x)}{\|\textbf{\textit{grad}}\ f(x)\|}$$

(see Theorem 8.18 on page 281). It is assumed that our computing is restricted to calculation of f and its gradient at any finite set of points — since all we are likely to have available is a formula for $f(x)$. With these restrictions we present.

Definition 10.42: Method of steepest descent

The method of steepest descent starts out with an initial guess, $\boldsymbol{x}_0$, for where the minimum of $f(\boldsymbol{x})$ is achieved. The ***next iterate, $\boldsymbol{x}_1$, is generated by a (one-dimensional) search for a minimum along the domain line in the direction of the negative of the gradient of f*** (Definition 8.10 on page 272) ***at $\boldsymbol{x}_0$***, using any one-dimensional search method chosen. ***This procedure is then repeated starting at $\boldsymbol{x}_1$, and continued inductively.***

❑

If for each $n = 0,1,2,...$, $\boldsymbol{x}_n$ exists with $\textbf{\textit{grad}}\ f(\boldsymbol{x}_n) \neq \boldsymbol{0}$, it's not hard to see that $f(\boldsymbol{x}_0), f(\boldsymbol{x}_1),...$ generated by *steepest descent* is a decreasing sequence.

This procedure cannot proceed past the n-th step if $\textbf{\textit{grad}}\ f(\boldsymbol{x}_n) = \boldsymbol{0}$, and in this case there is certainly no guarantee that f has a local minimum at $\boldsymbol{x}_n$. To see this just examine

$$f(x, y) = x^2 - y^2, \text{ with } \boldsymbol{x}_0 = (2, 0).$$

Here $\textbf{\textit{grad}}\ f(x, y) = (2x, 2y)$ and $-\textbf{\textit{grad}}\ f(2, 0) = (-(4, 0))$, which leads to

$$\boldsymbol{x}_1 = (0, 0),$$

which is neither a local minimum or a local maximum, although

$$\textbf{\textit{grad}}\ f(0,0) = (0,0).$$

If you think about it, you can see that even if the sequence $\boldsymbol{x}_0, \boldsymbol{x}_1,...$ does not terminate, it need not converge, and if it converges (see Definition 10.22 and Footnote 2 on page 352) the limit might not yield even a local minimum, and if it yields a local minimum, this needn't be an existing absolute minimum. We leave the verification of these assertions to the reader.

If the sequence $\boldsymbol{x}_0, \boldsymbol{x}_1,...$ fails to converge, but is bounded (i.e., for all n, there is a constant, k, such that $\|\boldsymbol{x}_n\| \leq k$), then it is still possible to extract a convergent subsequence, using bisection on each coordinate. This, however, is a nontrivial computational task. At the limit of this sequence, provided it converges to a point in the domain of f, the results of Theorem 10.35 on page

376 will often enable us to verify whether or not a local minimum has been achieved at $\lim_{n \to \infty} \boldsymbol{x}_n$. Not a very satisfactory theory, but often adequate.

The next example shows that even when steepest descent leads to a minimum in the limit, it may be somewhat inefficient.

Example 10.23: Steepest descent *hemstitching*

We examine the very simple function given by

$$f(x, y) = x^2 + (10y)^2, \qquad \textbf{\textit{10.96}}$$

which evidently has an absolute minimum at (0,0), and let our initial guess be of the form

$$\boldsymbol{x}_0 = a_0(k_0, \sqrt{1 - k_0^2}) \ .$$

This is a polar coordinate representation, useful because the distance of $\boldsymbol{x}_0$ from the desired minimum is just $|a_0|$. The graph of f has an elongated oval, bowl-shaped appearance, as illustrated in Figure 10-6.

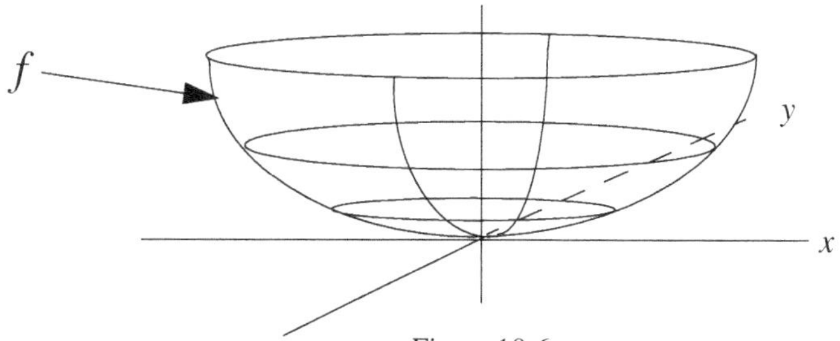

Figure 10-6

From equation 10.96 it follows that $\textbf{\textit{grad}}\ f(x, y) = (2x, 200y)$, and hence

$$\textbf{\textit{grad}}\ f(\boldsymbol{x}_0) = 2a_0(k_0, 100\sqrt{1 - k_0^2})$$

so for $k_0 > 0$, the slope of the line in the direction of $-\textbf{\textit{grad}}\ f(\boldsymbol{x}_0)$ is

$$s_0 = \frac{100\sqrt{1 - k_0^2}}{k_0}. \qquad \textbf{\textit{10.97}}$$

With the aid of some elementary algebra, we find

$$k_0 = \frac{100}{\sqrt{100^2 + s_0^2}} \quad \text{and} \quad \sqrt{1 - k_0^2} = \frac{s_0}{\sqrt{100^2 + s_0^2}}. \qquad \textbf{\textit{10.98}}$$

The slope of the line from $\boldsymbol{x}_0$ to (0,0) (the direction we *should* be taking) is

$$S_0 = \frac{\sqrt{1-k_0^2}}{k_0} = 0.01 s_0,$$

showing that, at least at the first step, steepest descent is way off in its direction. We illustrate what's happening in Figure 10-7 below.

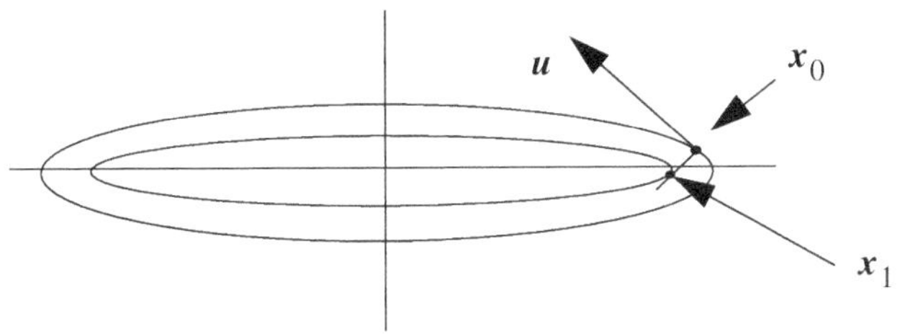

Figure 10-7

The ellipses above are domain curves of constant height (*level curves* of f) and the gradient of f at $\boldsymbol{x}_0$ is perpendicular to the outer-drawn ellipse, because the directional derivative, in the direction of this ellipse is 0 (the height of f not changing along this ellipse) but this directional derivative has the formula (see equation 8.13 on page 273)

$$D_u f(\boldsymbol{x}) = \boldsymbol{u} \cdot \boldsymbol{D}f(\boldsymbol{x}),$$

where $\boldsymbol{u}$ lies in the direction of this ellipse. So $\boldsymbol{u} \cdot \boldsymbol{D}f(\boldsymbol{x}) = \boldsymbol{u} \cdot \boldsymbol{grad}\ f(\boldsymbol{x}) = 0$, which (see Theorem 8.15 on page 278) shows that $\boldsymbol{u} \perp \boldsymbol{grad}\ f(\boldsymbol{x}_0)$

You might think that the minimum of f along the direction of $-\boldsymbol{grad}\ f(\boldsymbol{x}_0)$ would be at the x (first coordinate) axis, but it doesn't unless $k_0 = 0, 1$ or -1, because, as seen in Figure 10-7, the value of f along the gradient line is even smaller at the point, $\boldsymbol{x}_1$, at which it is tangent to an ellipse corresponding to a smaller value of f than where the gradient line hits the x axis.

To obtain quantitative results for this example, we note that the first steepest descent line has the equation

$$y = (x - k_0 a_0)s_0 + \sqrt{1-k_0^2}\ a_0 \qquad \textit{10.99}$$

(because it must go through the point $\boldsymbol{x}_0 = (k_0 a_0, \sqrt{1-k_0^2}\, a_0)$, and have slope s_0). To find the point $\boldsymbol{x}_1$ along this line at which $f(\boldsymbol{x})$ is minimized, substitute for y from equation 10.99, and find the value of x minimizing the resulting expression

$$x^2 + 100y^2 = x^2 + 100[(x - k_0 a_0)s_0 + \sqrt{1-k_0^2}\, a_0]^2 .$$

We leave it to the reader to verify that the vector $\boldsymbol{x}_1$ is given by

$$\boldsymbol{x}_1 = \frac{99 s_0 a_0}{\sqrt{100^2 + s_0^2}\,(1 + 100\, s_0^2)}(100\, s_0, -1). \qquad \textbf{\textit{10.100}}$$

We now rewrite $\boldsymbol{x}_1$ so that it resembles $\boldsymbol{x}_0$ — i.e., write it in the form

$$\boldsymbol{x}_1 = a_1(k_1, \sqrt{1-k_1^2}), \qquad \textbf{\textit{10.101}}$$

where $(a_1 = -\|\boldsymbol{x}_1\|)$ so that $|a_1| = \|\boldsymbol{x}_1\|$. From equation 10.100, we find

$$a_1 = \frac{-99 s_0 a_0 \sqrt{1 + (100 s_0)^2}}{\sqrt{100^2 + s_0^2}\,(1 + 100\, s_0^2)}. \qquad \textbf{\textit{10.102}}$$

Once again, using equation 10.100, together with the first coordinate given in equation 10.101, we find that

$$a_1 k_1 = \frac{99 s_0 a_0}{\sqrt{100^2 + s_0^2}\,(1 + 100\, s_0^2)} 100\, s_0 ,$$

and substituting into this equation from equation 10.102 yields

$$k_1 = \frac{-100\, s_0}{\sqrt{1 + (100 s_0)^2}} \qquad \textbf{\textit{10.103}}$$

Recalling the formula for k_0 (see equations 10.98)

$$k_0 = \frac{100}{\sqrt{100^2 + s_0^2}}, \qquad \textbf{\textit{10.104}}$$

if we set $k_1 = -k_0$ in order to set up a repetitive pattern, we find that we need

$$\frac{(100s_0)^2}{1+(100s_0)^2} = \frac{100^2}{100^2+s_0^2} \quad \text{or} \quad \frac{1}{1+(100s_0)^2} = \frac{1}{100^2+s_0^2}$$

which allows us to choose $s_0 = 1$. From the previous two equations we have

$$k_0 = \frac{100}{\sqrt{10{,}001}} = -k_1 .$$

Substituting $s_0 = 1$ into equation 10.102, we find

$$a_1 = \frac{-99}{101} a_0 .$$

By the same reasoning that led to equation 10.97 on page 401 for s_0, we find

$$s_1 = \frac{100\sqrt{1-k_1^2}}{k_1} = \frac{100\sqrt{1-k_0^2}}{-k_0} = -1 .$$

This leads to the repetitive situation in which the gradient line slopes alternate in sign and have constant magnitude 1, and

$$|a_{n+1}| = \|\boldsymbol{x}_{n+1}\| = \frac{99}{101}\|\boldsymbol{x}_n\| .$$

Thus the progress of steepest descent at each stage is excruciatingly slow in this example and we see that steepest descent can zigzag in a very inefficient way. This behavior, illustrated somewhat imperfectly in Figure 10-8, is sometimes called **hemstitching**.

The choice of the function given by $f(x, y) = x^2 + (10y)^2$ is not especially artificial — since the multidimensional Taylor's theorem, Theorem 10.33 on page 371, shows that the departure from linearity, at least locally, is quadratic, and the above formula is just a particular positive definite quadratic form, see Theorem 10.37, equation 10.75 on page 382, yielding a local minimum.

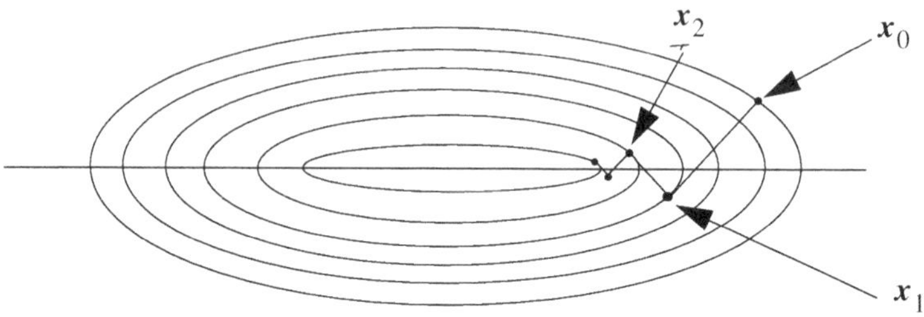

Figure 10-8

■

As has been mentioned already, the steepest descent procedure need ***not*** lead to convergence to a local minimum. The example just given can be modified to lead to this situation by instead letting

$$f(x, y) = \begin{cases} x^2 + (10y)^2 & \text{for } x \geq 0, \\ -x^2 + (10y)^2 & \text{for } x < 0. \end{cases}$$

Then ***grad*** $f(x,y)$ exists and is continuous at all (x,y), and starting at $\boldsymbol{x}_0 = a_0(k_0, \sqrt{1 - k_0^2})$ with $a_0 > 0$, $0 < k_0 < 1$ still yields convergence to (0,0), which in this case is not a local minimum. The Hessian matrix, see Definition 10.36 on page 377, is not defined at (0,0) for this modification, but this illustration can be further modified so that the behavior just observed still occurs, and the Hessian matrix exists.

Luckily, the sufficient conditions for a minimum can often be used to establish the success of such a search; see Theorem 10.37 on page 381.

The behavior exhibited in Example 10.23 on page 401 arose because of the strong elliptical nature of the **level curves** of the function f (**domain curves of** f **on which the value of** f **is constant**). It seems reasonable that the relative change in the magnitudes of the steepest descent steps would give a clue as to how oval these ellipses are. Using such information in some of the more sophisticated approaches to minimization, the axes of these ellipses (or ellipsoids in more than two dimensions) are rescaled, to make them more circular and cut down on the hemstitching. An understanding of the literature on how this might be attempted requires the knowledge of eigenvalues and eigenvectors, to be presented in Chapter 13.

For a thorough treatment of steepest descent (as well as many other aspects of optimization) see Luenberger [11], p. 148 ff.

Many computer programs for minimization start out with some sort of direct approach, e.g., a modified version of steepest descent, and switch to Newton's method for greater efficiency when it seems that a minimum is close by. But as stressed earlier, no procedures that can be depended on to work in all cases that may arise in practice have yet been developed.

Comments on reading the optimization literature

When and if you try to pursue the subject of optimization further, in monographs, (such as Luenberger [11]), or articles you may encounter, you will likely be confronted with some apparently unfamiliar concepts and results, which can make progress very difficult. Most of these items are generalizations of ideas that are already fairly familiar.

There is the concept of a closed bounded set — which is just a generalization of a closed bounded interval. Its importance lies in the fact that a continuous function with a closed bounded set as its domain achieves a maximum and a minimum (just as we showed to be the case for a continuous function whose domain is a closed bounded interval, in establishing the proof of the mean value theorem in the optional section on page 63; in fact, the proofs are almost identical). For future reference we provide the following formal definition.

Definition 10.43: Closed set, bounded set, in n dimensions

A set A of real n-dimensional vectors (Definition 8.1 on page 261) is said to be **closed** if for each convergent sequence, $\boldsymbol{x}_1, \boldsymbol{x}_2,...$ of elements of A (see Definition 10.22 on page 351 and Footnote 2 on page 352)

$$\lim_{n \to \infty} \boldsymbol{x}_n \text{ belongs to } A.$$

The set A of real n-dimensional vectors is said to be **bounded** if there is a fixed number $M > 0$, such that for all $\boldsymbol{x}$ in A, the norm $\|\boldsymbol{x}\|$ (Definition 8.6 on page 270) satisfies

$$\|\boldsymbol{x}\| \le M .$$

❑

Note that being closed is a property possessed by closed intervals, but not by nonempty bounded open ones (intervals which do not include their end points).

In some investigations a bounded sequence of n-dimensional vectors is generated, and it is necessary to extract a convergent subsequence from it by discarding some of its terms (maybe even an infinite number of these terms, but still letting an infinite number of elements remain). This can always be done — essentially using bisection for each coordinate, leading to the often used result that *every bounded sequence has a convergent subsequence.*

Proving the existence of an absolute minimum of some continuous real valued function of n real variables is often accomplished by establishing that outside of some n-dimensional sphere (the set of $\boldsymbol{x}$ for which $\|\boldsymbol{x}\| > r$) all values $f(x)$ exceed $f(\boldsymbol{x}_o)$, where $\|\boldsymbol{x}_o\| \le r$. Since the sphere consisting of all $\boldsymbol{x}$ with $\|\boldsymbol{x}\| \le r$ is closed and bounded, $f(x)$ achieves a minimum over this sphere — i.e.,

there is a value

$$\boldsymbol{x}^* \text{ with } \|\boldsymbol{x}^*\| \le r$$

such that

$$f(\boldsymbol{x}^*) \le f(\boldsymbol{x}) \text{ for all } \boldsymbol{x} \text{ with } \|\boldsymbol{x}\| \le r.$$

But then $f(\boldsymbol{x}^*)$ is the overall absolute minimum of $f(\boldsymbol{x})$ — i.e.,

$$f(\boldsymbol{x}^*) \le f(\boldsymbol{x}) \text{ for all } n\text{-dimensional vectors } \boldsymbol{x}.$$

Proofs involving the existence of maxima and minima of functions usually make use of one or more of the following definitions.

Definition 10.44: inf, sup

Let S be some nonempty set of real numbers bounded below (i.e., all exceeding some fixed value) then

inf S is the unique real number such that

$inf\ S \le s$ for every element s in S,
and no larger number has this property.

inf S is called the **infimum**, or **greatest lower bound** (**glb**) of S.

If S is some nonempty set of real numbers, bounded above (i.e., all below some fixed value) then

sup S is the unique real number such that

$sup\ S \ge s$ for every element s in S
and no smaller number has this property.

sup S is called the **supremum**, or **least upper bound** (**lub**) of S.

❑

The existence of least upper bounds and greatest lower bounds of bounded sets of real numbers is not difficult to establish, if real numbers are defined as nested sequences (see Appendix 1 on the real numbers starting on page 757).

In terms of these concepts in the method of establishing the existence of maxima and minima just outlined above, the set S may be taken as the set of all values $f(x)$ with $\|\boldsymbol{x}\| > r$, and the assertion that for some $\boldsymbol{x}_o$ with $\|\boldsymbol{x}_o\| \le r$

$$f(\boldsymbol{x}_o) \le f(\boldsymbol{x}) \quad \text{for all } \|\boldsymbol{x}\| > r$$

may be conveniently written as

$$f(\boldsymbol{x}_o) \le \mathit{inf}\ S.$$

For later reference we need to extend the concept of a bounded function (Definition 3.10 on page 88) to complex valued functions, and introduce the *sup* and *inf* of a bounded real-valued function.

Definition 10.45: Bounded function, sup *f*, [inf *f*]

A real (or complex) valued function, f, is said to be **bounded on a set** A in its domain (Definition 1.1 on page 2) is there is some real number B such that

$$|f(x)| \le B \quad \text{for all } x \text{ in } A;$$

The magnitude, $|\boldsymbol{y}|$, of a complex number, $\boldsymbol{y}$ (which includes the case where this number is real) is given in Definition 6.11 on page 200.

If f is a real-valued function whose range is bounded from above [below], then by the **supremum [infimum] of** f is meant *sup rng f* [*inf rng f*], where *rng f* stands for the range of the function f (again recall Definition 1.1 on page 2).

❑

It's not feasible here to adequately cover the obstacles that may arise when trying to understand the more advanced aspects of this subject. But most of them arise as fairly immediate extensions of already familiar concepts — so that the obstacles to be overcome are those of finding the definitions of the unfamiliar terms and determining what has been generalized. (For instance, one generalization of *interval* is that of a *convex set of* n-dimensional vectors. Such a set is one which includes the line segment connecting any pair of its points. When the domain of a real valued function of n real variables is a convex set, this allows one-dimensional searches of the type used in steepest descent. It explains the restriction to convex domains in certain modifications of the Newton's method, where if a step fails to yield a decrease, it can be cut in half, still staying in the domain.) When all else fails, try for assistance from a friendly knowledgeable mathematician.

Chapter 11

Orthogonal Complements

1 Introduction

In this chapter we will see how the concept of the orthogonal complement of a subspace allows determination of the general form of a system of linear equations, and aids in the solution of optimization problems under constraints. In addition it provides a much better understanding of the nature of systems of linear equations.

2 General Solution Structure

The starting point is a system of n linear equations in m unknowns, specified by the matrix equation[1] (see Definition 10.1 on page 328)

$$\mathbf{Ax} = \mathbf{y}, \qquad 11.1$$

where $\mathbf{A}$ is a given $n \times m$ real matrix, $\mathbf{y}$ is a given $n \times 1$ real matrix (a *column* matrix - or *column* vector); by the *solution* of this system, we mean the collection of all $m \times 1$ matrices $\mathbf{x}$ for which equation 11.1 holds. For given $\mathbf{A}$ and $\mathbf{y}$ we want to determine the nature of this solution. Information of this type will provide insight into the physical systems described by such equations, and enable us to extract quantitative information about these systems.

Suppose that by some means or other, you are able to find what is called a *particular* solution of equation 11.1 — i.e., a specific vector $\boldsymbol{x}_p$ for which

$$\mathbf{Ax}_p = \mathbf{y}. \qquad 11.2$$

From the *linear operator property* of matrix multiplication (see Theorem 10.2 on page 330) if $\mathbf{x}$ is any arbitrary specific solution of equation 11.1, we find that

$$\mathbf{A}(\mathbf{x} - \mathbf{x}_p) = (\mathbf{Ax}) - (\mathbf{Ax}_p) = \mathbf{y} - \mathbf{y} = \mathbf{0}.$$

1. Such systems, in which the number of equations need not be the same as the number of unknowns, have important applications in the study of objects described by differential equations, ranging from the numerical solution of such differential eqations (recall Example 6.6 on page 186) to the development of *impulse response* in Section 6 of this chapter.

The vector

$$\mathbf{x}_H = \mathbf{x} - \mathbf{x}_p$$

is a solution of the *homogeneous* system

$$\mathbf{A}\mathbf{x}_H = \mathbf{0}. \qquad 11.3$$

So, from the linear operator property of matrix multiplication we have shown the following.

Theorem 11.1: General solution of $\mathbf{Ax} = \mathbf{y}$

Let $\mathbf{A}$ be a given $n \times m$ real matrix and $\mathbf{y}$ a given $n \times 1$ real matrix (Definition 10.1 on page 328). Given any specific vector $\boldsymbol{x}_p$ for which

$$\mathbf{A}\mathbf{x}_p = \mathbf{y} \qquad 11.4$$

(i.e., any **particular solution** of $\mathbf{Ax} = \mathbf{y}$) then every vector $\mathbf{x}$ such that $\mathbf{Ax} = \mathbf{y}$ may be written as

$$\mathbf{x} = \mathbf{x}_p + \mathbf{x}_H, \qquad 11.5$$

where

$$\mathbf{A}\mathbf{x}_H = \mathbf{0}, \qquad 11.6$$

i.e., $\mathbf{x}_H$ is some solution of the **homogeneous system**, equation 11.6, determined by $\mathbf{x}$ and $\boldsymbol{x}_p$.

❑

Thus the problem of finding all vectors $\mathbf{x}$ ***satisfying equation* 11.1 *breaks down into two problems:***

- ***finding any one particular solution,*** $\boldsymbol{x}_p$

and

- ***finding all solutions of the homogeneous equation* 11.6**.

These two problems will be handled separately.

Remark

This same structure applies to linear differential operators such as $D^2 + tD - 2$, where, if y is a function with second derivatives,

$$((D^2 + tD - 2)y)(t) = y''(t) + ty'(t) - 2y(t).$$

Here if y is some given reasonably well-behaved function and y_p is some specific function such that

$$y_p'' + ty_p' - 2y = f$$

then every function y such that for all t

$$y''(t) + ty'(t) - 2y(t) = f(t)$$

may be written as

$$y = y_p + y_H,$$

where (for all t)

$$y_H''(t) + ty_H'(t) - 2y_H(t) = 0.$$

From an abstract viewpoint, this follows because the linear differential operator, $D^2 + tD - 2$, satisfies the linear operator property. From a more concrete viewpoint it isn't surprising either, since we know that many differential equations can be approximated by systems of linear algebraic equations (again recall Example 6.6 on page 186). In fact, as we will see in the advanced section starting on page 433, we may think of the derivative and expressions like $D^2 + tD - 2$ as *essentially* big matrices. ***So it is because the relationship between systems of algebraic equations like* 11.1 *and a great many differential equations* is more than just a superficial abstract resemblance, *that so much effort is spent studying these algebraic systems.***

3 Homogeneous Solution

Making use of Theorem 11.1 to find all solutions of equation 11.1, we first turn our attention to ***finding all solutions*** of the homogeneous equation 11.6.

Notice that using the *row version* of a matrix times a vector (equation 10.5 on page 328) we may write equation 11.6 in the form[1]

$$\begin{aligned} \boldsymbol{A}_1 \cdot \boldsymbol{x}_H &= 0 \\ &\vdots \\ \boldsymbol{A}_n \cdot \boldsymbol{x}_H &= 0. \end{aligned} \qquad \textit{11.7}$$

Here $\boldsymbol{A}_i$ represents the i-th row of $\mathbf{A}$, treated as a vector with m coordinates.

1. To stress the geometric aspect of these equations, we use the vector form, rather than the matrix notation.

From Theorem 8.15 on page 278 it follows that the set of all solutions of equations 11.7 (i.e., all vectors $\mathbf{x}_H$ such that $A_i \cdot x_H = 0$, all $i = 1,2,...,n$) is the set of all vectors orthogonal to each row of $\mathbf{A}$. It is not hard to see that this is the same as the set of all vectors orthogonal to every element of the **row space of A**, i.e. to the subspace

$$\mathbf{A}_{row} = [\![A_1,...,A_n]\!] \qquad \textit{11.8}$$

generated by the rows of $\mathbf{A}$ (see Definition 9.4 on page 296). This leads us to the following.

Definition 11.2: Orthogonal complement

Let S be a subspace (Definition 9.4 on page 296) of real Euclidean n-dimensional space (Definition 9.14 on page 302). The **orthogonal complement of** S, denoted by $S^{\perp}$, is the set of all real n-dimensional vectors orthogonal to all vectors in S (see Theorem 8.15 on page 278).

❑

Note that if $S = [\![w_1,...,w_k]\!]$, then to show that v belongs to $S^{\perp}$ it is sufficient to show that $v \perp w_i$, i.e., $v \cdot w_i = 0$, for each $i = 1,2,...,k$.

We introduce the following definition for brevity.

Definition 11.3: General solution

For any given function f and object y, the **general solution of**

$$f(x) = y \qquad \textit{11.9}$$

is the set of *all* elements, x, in the domain of f (Definition 1.1 on page 2) such that equation 11.9 holds.

❑

Note that this definition applies to the system $\mathbf{Ax} = \mathbf{y}$, by letting

$$x = \mathbf{x},\ y = \mathbf{y},\ f(x) = \mathbf{Ax}.$$

With this terminology, we see that the general solution of the system 11.7,

$$\mathbf{Ax}_H = \mathbf{0}$$

is the orthogonal complement (Definition 11.2 above),

$$\mathbf{A}_{row}^{\perp} = [\![A_1,...,A_n]\!]^{\perp},$$

of the subspace spanned by the A_j, where A_j is the j-th row of $\mathbf{A}$.

Our next task is that of explicitly representing $\mathbf{A}_{row}^{\perp}$. In three dimensions it is easy enough to determine the orthogonal complement of a set consisting of a single vector — for example, the orthogonal complement of $(0,0,1)$ is just the x-y plane. Conversely, the orthogonal complement of the x-y plane is the z axis.

To determine the orthogonal complement of a more general set of vectors, first choose a coordinate system for the subspace generated by this set (using, say, the Gram-Schmidt procedure, see Definition 9.19 on page 306 or Theorem 10.31 on page 366, in the obvious way). Now extend this coordinate system so that it includes the entire Euclidean space in which these vectors lie (again use the Gram-Schmidt procedure in a way you might want to figure out by yourself before we present it). The coordinate vectors arising from this extension will turn out to generate the desired orthogonal complement.

To be specific, if we want to find $[[\boldsymbol{A}_1,...,\boldsymbol{A}_n]]^{\perp}$, where the $\boldsymbol{A}_j$ are the rows of the $n \times m$ matrix $\mathbf{A}$, we apply the improved Gram-Schmidt procedure, Theorem 10.31 on page 366, to the sequence $\boldsymbol{A}_1,...,\boldsymbol{A}_n, \boldsymbol{e}_1,...,\boldsymbol{e}_m$, where the $\boldsymbol{e}_k$, $k = 1,...,m$ are the m-dimensional standard unit vectors, with the j-th coordinate of $\boldsymbol{e}_k$ given by

$$e_{kj} = \begin{cases} 1 & \text{if} \quad j = k \\ 0 & \text{if} \quad j \neq k. \end{cases} \qquad \textbf{\textit{11.10}}$$

Recalling Definition 9.14 on page 302, of Euclidean space, since $\boldsymbol{e}_1,...,\boldsymbol{e}_m$ generate Euclidean m-dimensional space, E_m, we must have

$$[[\boldsymbol{A}_1,...,\boldsymbol{A}_n, \boldsymbol{e}_1,...,\boldsymbol{e}_m]] = E_m.$$

Now apply the Gram-Schmidt orthonormalization procedure, Theorem 10.31 on page 366, and suppose we obtain the orthonormal sequence $\boldsymbol{w}_1,...,\boldsymbol{w}_q, \boldsymbol{w}_{q+1},...,\boldsymbol{w}_m$, where $\boldsymbol{w}_1,...,\boldsymbol{w}_q$ arise from $\boldsymbol{A}_1,...,\boldsymbol{A}_n$ and $\boldsymbol{w}_{q+1},...,\boldsymbol{w}_m$ arise from continuing the process on the remaining vectors $\boldsymbol{e}_1,...,\boldsymbol{e}_m$. A little reflection should convince you that

$$[[\boldsymbol{A}_1,...,\boldsymbol{A}_n]] = [[\boldsymbol{w}_1,...,\boldsymbol{w}_q]] \qquad \textbf{\textit{11.11}}$$

and

$$[[\boldsymbol{A}_1,...,\boldsymbol{A}_n]]^{\perp} = [[\boldsymbol{w}_{q+1},...,\boldsymbol{w}_m]]. \qquad \textbf{\textit{11.12}}$$

If this isn't adequate, here's some more detail. From Theorem 9.20 on page 307, equation 11.11 follows immediately. Since equation 11.11 implies that

$$[[\boldsymbol{A}_1,...,\boldsymbol{A}_n]]^{\perp} = [[\boldsymbol{w}_1,...,\boldsymbol{w}_q]]^{\perp},$$

all we need show is that

$$[\![\boldsymbol{w}_1,...,\boldsymbol{w}_q]\!]^\perp = [\![\boldsymbol{w}_{q+1},...,\boldsymbol{w}_m]\!]. \qquad \textbf{\textit{11.13}}$$

But because $\boldsymbol{w}_1,...,\boldsymbol{w}_m$ is an orthonormal sequence, each of the vectors $\boldsymbol{w}_{q+1},...,\boldsymbol{w}_m$ is orthogonal to all of the vectors $\boldsymbol{w}_1,...,\boldsymbol{w}_q$. So every element of $[\![\boldsymbol{w}_{q+1},...,\boldsymbol{w}_m]\!]$ is in the set $[\![\boldsymbol{w}_1,...,\boldsymbol{w}_q]\!]^\perp$, i.e.,[1]

$$[\![\boldsymbol{w}_{q+1},...,\boldsymbol{w}_m]\!] \subseteq [\![\boldsymbol{w}_1,...,\boldsymbol{w}_q]\!]^\perp.$$

So to verify equation 11.13 all we need show is that

$$[\![\boldsymbol{w}_1,...,\boldsymbol{w}_q]\!]^\perp \subseteq [\![\boldsymbol{w}_{q+1},...,\boldsymbol{w}_m]\!],$$

i.e., that if $\boldsymbol{v}$ is orthogonal to each of the $\boldsymbol{w}_1,...,\boldsymbol{w}_q$, then $\boldsymbol{v}$ is in $[\![\boldsymbol{w}_{q+1},...,\boldsymbol{w}_m]\!]$.

This follows from the fact that because

$$E_m = [\![\boldsymbol{w}_1,...,\boldsymbol{w}_m]\!]$$

we may write

$$\boldsymbol{v} = \sum_{i=1}^{m} a_i \boldsymbol{w}_i.$$

Now since

$$\boldsymbol{v} \perp \boldsymbol{w}_j, \; j = 1,...,q$$

it follows from Theorem 8.15 on page 278 that

$$\boldsymbol{v} \cdot \boldsymbol{w}_j = 0 \quad \text{for } j = 1,...,q.$$

Hence for $j = 1,...,q$

$$0 = \boldsymbol{v} \cdot \boldsymbol{w}_j = \left(\sum_{i=1}^{m} a_i \boldsymbol{w}_i\right) \cdot \boldsymbol{w}_j = \sum_{i=1}^{m} a_i \boldsymbol{w}_i \cdot \boldsymbol{w}_j = a_j \boldsymbol{w}_j \cdot \boldsymbol{w}_j = a_j.$$

Thus $\boldsymbol{v} = \sum_{i=q+1}^{m} a_i \boldsymbol{w}_i$, which shows that $\boldsymbol{v}$ is in $[\![\boldsymbol{w}_{q+1},...,\boldsymbol{w}_m]\!]$.

---

1. The symbol $\subseteq$ is read *is a subset of.* If we write $A \subseteq B$, we mean that every element of the set A is also an element of the set B. A and B represent the same set of objects if and only if $A \subseteq B$ and $B \subseteq A$.

This establishes the following.

Theorem 11.4: Basis for $[[A_1,...,A_n]]^{\perp}$, solution of $\mathbf{Ax} = \mathbf{0}$

Let $A_1,...,A_n$ be a given sequence of real m-dimensional vectors (Definition 8.1 on page 261) and $e_1,...,e_m$ the standard m-dimensional unit vectors (see equation 11.10 on page 413). Use the Gram-Schmidt orthonormalization procedure, Theorem 10.31 on page 366, to orthonormalize the sequence $A_1,...,A_n, e_1,...,e_m$, obtaining $w_1,...,w_q, w_{q+1},...,w_m$, where

$$[[w_1,...,w_q]] = A_1,...,A_n,$$

(see Definition 9.4 on page 296 for the definition of $[[w_1,...,w_q]]$).

Then if $A_1,...,A_n$ are the rows of the matrix $\mathbf{A}$, the general solution (Definition 11.3 on page 412) of

$$\mathbf{Ax} = \mathbf{0}$$

is

$$[[w_{q+1},...,w_m]] = [[A_1,...,A_n]]^{\perp}.$$

❑

If, as sometimes occurs in statistical problems, the matrix $\mathbf{A}$ has many more rows than columns, it is very inefficient to use Theorem 11.4 directly to determine $[[A_1,...,A_n]]^{\perp}$ or $[[A_1,...,A_n]]$. The following result shows how to cut down the labor somewhat, and most likely, to improve the accuracy

Theorem 11.5: Row space of A

Let $\mathbf{A}$ be an $n \times p$ matrix (Definition 10.1 on page 328) with rows $A_1,...,A_n$. The subspace (Definition 9.4 on page 296)

$$\mathbf{A}_{row} = [[A_1,...,A_n]]$$

generated by $A_1,...,A_n$ is the same as the subspace generated by the rows of the product

$$\mathbf{A}^{\mathrm{T}}\mathbf{A}$$

i.e.,

$$\mathbf{A}_{row} = (\mathbf{A}^{\mathrm{T}}\mathbf{A})_{row}.$$

❑

The significance of this result is that computation of $\mathbf{A}^T\mathbf{A}$ is quite simple, and if $\mathbf{A}$ has many more rows than columns, the number of rows of $\mathbf{A}^T\mathbf{A}$ will evidently be much smaller than the number of rows of $\mathbf{A}$ (the number of columns being the same). So the amount of effort applying Theorem 11.4 to $\mathbf{A}^T\mathbf{A}$ will be considerably less than applying it to $\mathbf{A}$ directly, but with the same final result. We leave the proof to Problem 11.2.

Exercises 11.1

Find the general solutions of $\mathbf{Ax} = \mathbf{0}$ for the following choices of $\mathbf{A}$:

1. $\begin{bmatrix} 1 & 1 & 1 \\ 1 & 1 & 0 \\ 1 & 0 & 0 \end{bmatrix}$
2. $\begin{bmatrix} 1 & 1 & 0 \\ 1 & 1 & 0 \\ 1 & 0 & 1 \end{bmatrix}$
3. $\begin{bmatrix} 1 \\ 1 \\ 1 \end{bmatrix}$
4. $\begin{bmatrix} 1 & 1 & 1 \\ 0 & 0 & 0 \\ 1 & -1 & 0 \end{bmatrix}$
5. $\begin{bmatrix} 1 & 3 & 5 \\ 2 & 4 & 8 \\ 3 & 5 & 11 \end{bmatrix}$

Problem 11.2

Establish Theorem 11.5.

Hints: Show instead that $A_{row}^{\perp} = (\mathbf{A}^T\mathbf{A})_{row}^{\perp}$. It's very evident that every element of the left side belongs to the right side, because if $\mathbf{Ax} = \mathbf{0}$, then $\mathbf{A}^T\mathbf{Ax} = \mathbf{0}$. But if $\mathbf{A}^T\mathbf{Ax} = \mathbf{0}$, then $\mathbf{y} = \mathbf{Ax}$ is in the space spanned by the columns of $\mathbf{A}$ and is orthogonal to every column of $\mathbf{A}$. We know that $\mathbf{y}$ must be $\mathbf{0}$, because $\mathbf{y} = \sum_j x_j A_j$ and $\mathbf{y} \cdot \mathbf{y} = \mathbf{y} \cdot \sum_j x_j A_j = \sum_j x_j (\mathbf{y} \cdot A_j) = 0$.

4 Particular and General Solution of $\mathbf{Ax} = \mathbf{y}$

Determination of whether there exists a solution of the system

$$\mathbf{Ax} = \mathbf{y} \qquad \textit{11.14}$$

for given matrix $\mathbf{A}$ (see Definition 10.1 on page 328) and given vector $\mathbf{y}$ is easy using orthogonalization, because if equation 11.14 holds, then $\mathbf{y}$ must be a linear combination (Definition 9.4 on page 296) of the columns of $\mathbf{A}$. Denote these columns by $\boldsymbol{a}_1,...,\boldsymbol{a}_m$. To see if $\mathbf{y}$ is in $[[\boldsymbol{a}_1,...,\boldsymbol{a}_m]]$, just orthogonalize the sequence $\boldsymbol{a}_1,...,\boldsymbol{a}_m, \mathbf{y}$, using, say, a modification of Theorem 10.31 on page 366, in which you let any $\mathbf{0}$ vectors remain, obtaining $\boldsymbol{w}_1,...,\boldsymbol{w}_{m+1}$. Only if $\boldsymbol{w}_{m+1} = \mathbf{0}$ is $\mathbf{y}$ a linear combination of the $\boldsymbol{a}_1,...,\boldsymbol{a}_m$ (see Theorem 9.20 on page 307). So we have

Theorem 11.6: Existence of solution of $\mathbf{Ax} = \mathbf{y}$

Let $\mathbf{A}$ be a real $n \times m$ matrix (Definition 10.1 on page 328) and $\mathbf{y}$ be a real n-dimensional vector.[1] Let the columns of $\mathbf{A}$ be denoted by $\boldsymbol{a}_1,...,\boldsymbol{a}_m$. There is at least one m-dimensional vector, $\mathbf{x}$, such that $\mathbf{Ax} = \mathbf{y}$ if and only if, in orthogonalizing $\boldsymbol{a}_1,...,\boldsymbol{a}_m, \mathbf{y}$ to $\boldsymbol{w}_1,...,\boldsymbol{w}_{m+1}$ (Definition 9.18 on page 305, or Theorem 10.31 on page 366) the final vector

$$\boldsymbol{w}_{m+1} = \mathbf{0}.$$

❑

We can actually obtain a particular solution of $\mathbf{Ax} = \mathbf{y}$ (having shown that it existed) by orthogonalization, as was shown in Definition 9.25 on page 320.

We summarize the entire set of results on solving the system $\mathbf{Ax} = \mathbf{y}$ in

Theorem 11.7: General solution of $\mathbf{Ax} = \mathbf{y}$

Let $\mathbf{A}$ be a given $n \times m$ real matrix (Definition 10.1 on page 328) with rows (m-dimensional vectors) denoted by $\boldsymbol{A}_1,...,\boldsymbol{A}_n$ and columns (n-dimensional vectors) denoted by $\boldsymbol{a}_1,...,\boldsymbol{a}_m$.

- There exists at least one m-dimensional vector $\mathbf{x}_p$ for which $\mathbf{Ax}_p = \mathbf{y}$ if and only if when orthogonalizing $\boldsymbol{a}_1,...,\boldsymbol{a}_m, \mathbf{y}$ to $\boldsymbol{w}_1,...,\boldsymbol{w}_{m+1}$, the vector $\boldsymbol{w}_{m+1} = \mathbf{0}$; if so, such an $\mathbf{x}_p$ can be found via the method[2] of Definition 9.25 on page 320 or, more accurately, using Theorem 10.31 on page 366.

---

1. Represented here as an $n \times 1$ matrix (column matrix), not that it makes any difference.

- Every $\mathbf{x}$ for which $\mathbf{Ax} = \mathbf{y}$ can be written as

  $$\mathbf{z} = \mathbf{x}_p + \mathbf{x}_H$$

  where $\mathbf{x}_p$ is the particular solution just referred to and $\mathbf{x}_H$ is one of the elements of the subspace (Definition 9.4 on page 296)

  $$\mathbf{A}_{row}^{\perp} = [\![\boldsymbol{A}_1,...,\boldsymbol{A}_n]\!]^{\perp},$$

  a basis for which is explicitly supplied by the method given in Theorem 11.4 on page 415.

❑

An alternative to using the method just given above for checking on whether there is a solution to the system $\mathbf{Ax} = \mathbf{y}$, and which can sometimes be useful in computation, employs the following approach related to orthogonal complements — namely, that a vector lies in a given subspace, S, if and only if it is orthogonal to every element of the orthogonal complement of S. Implementation of this relies on orthogonalizing $\boldsymbol{a}_1,...,\boldsymbol{a}_m$, $\boldsymbol{e}_1,...,\boldsymbol{e}_n$ (where $\boldsymbol{a}_1,...,\boldsymbol{a}_m$ are the n-dimensional column vectors of $\mathbf{A}$ and $\boldsymbol{e}_1,...,\boldsymbol{e}_n$ are the standard n-dimensional unit vectors, Definition 8.2 on page 263 or equation 11.10 on page 413). Suppose this is done, yielding $\boldsymbol{w}_1,...,\boldsymbol{w}_n$, where

$$[\![\boldsymbol{w}_1,...,\boldsymbol{w}_q]\!] = [\![\boldsymbol{a}_1,...,\boldsymbol{a}_m]\!]$$

and hence

$$[\![\boldsymbol{w}_{q+1},...,\boldsymbol{w}_n]\!] = [\![\boldsymbol{w}_1,...,\boldsymbol{w}_q]\!]^{\perp} = [\![\boldsymbol{a}_1,...,\boldsymbol{a}_m]\!]^{\perp}.$$

Then there is a solution to $\mathbf{Ax} = \mathbf{y}$ if and only if $\mathbf{y} \perp \boldsymbol{w}_j$ for each $j = q + 1,...,n$.

We summarize this in a slightly more general form, useful for checking certain so-called *estimability* conditions in statistics — i.e., for checking on whether or not it is possible to obtain reasonable estimates for certain unknown parameters involved in curve fitting.

---

2. We will present a slightly different approach to finding a particular solution $\mathbf{x}_p$ in Section 6 of this chapter on impulse response.

Theorem 11.8: Characterizing the inclusion $S \subseteq T$

Let $S = [\![x_1,...,x_r]\!]$ and $T = [\![v_1,...,v_k]\!]$ be two subspaces (Definition 9.4 on page 296) of n-dimensional Euclidean space (Definition 9.14 on page 302) and let $e_1,...,e_n$ be the standard n-dimensional unit vectors (see Definition 8.2 on page 263). Using the improved Gram-Schmidt orthonormalization procedure (Theorem 10.31 on page 366) orthonormalize $v_1,...,v_k, e_1,...,e_n$ to $w_1,...,w_n$ with $T = [\![w_1,...,w_q]\!]$.

Then

$$S \subseteq T \quad (S \text{ is a subset of } T, \text{ see Footnote 1 on page 414})$$

if and only if

$$x_i \cdot w_k = 0$$

(see Definition 8.10 on page 272), for each $i = 1,...,r$ and each $k = q+1,....,n$.

❑

The proof is left as Problem 11.3.1.

Problems 11.3

1. Supply the details of the proof of Theorem 11.8.
2. Show how to determine whether $S \subseteq T$ for $S = [\![x_1,...,x_r]\!]$ and $T = [\![v_1,...,v_k]\!]$ by orthogonalizing $x_1,...,x_r, v_1,...,v_k$.
3. Suppose that S and T are subspaces of n-dimensional Euclidean space, E_n, with $S \subseteq T$.
   a. Show that it is possible to choose an orthonormal coordinate system $w_1,...,w_n$ for E_n such that

   $$S = [\![w_1,...,w_r]\!] \text{ and } T = [\![w_1,...,w_r,w_{r+1},...,w_q]\!].$$

   b. Use the result of part a to show that if $S \subseteq T$, then $P_S(y) = P_S(P_T(y))$. Draw an illustrative picture.
   Both of these last results are important in statistical regression (curve-fitting) theory.

Visualizing a subspace as a plane or a line through the origin, you should not be surprised to learn that the intersection[1] of two subspaces is also a subspace. However, with the definition of subspace that we have adopted

1. The **intersection**, $S \cap T$, of two sets, S and T, is the collection of objects belonging to both of these sets.

(Definition 9.4 on page 296) this result would have been difficult for us to establish prior to the introduction of orthogonal complements. We supply the details of its establishment solely to provide more experience dealing with manipulations involving orthogonal complements.

Suppose $S = [\![\boldsymbol{x}_1,...,\boldsymbol{x}_r]\!]$ and $T = [\![\boldsymbol{v}_1,...,\boldsymbol{v}_k]\!]$, where we may assume both $\boldsymbol{x}_1,...,\boldsymbol{x}_r$ and $\boldsymbol{v}_1,...,\boldsymbol{v}_k$ to be orthonormal sequences. By use of the Gram-Schmidt procedure on $\boldsymbol{x}_1,...,\boldsymbol{x}_r,\boldsymbol{e}_1,...,\boldsymbol{e}_n$ and $\boldsymbol{v}_1,...,\boldsymbol{v}_k,\boldsymbol{e}_1,...,\boldsymbol{e}_n$, we can construct the orthonormal bases $\boldsymbol{x}_{r+1},...,\boldsymbol{x}_n$, and $\boldsymbol{v}_{k+1},...,\boldsymbol{v}_n$ for $S^{\perp}$ and $T^{\perp}$, respectively.

Now if $\boldsymbol{w}$ is an element of the intersection of S and T, then, because $\boldsymbol{w}$ is in both S and T,

$$\boldsymbol{w} \perp \boldsymbol{v}_i \quad \text{for} \quad i = k+1,...,n$$

and

$$\boldsymbol{w} \perp \boldsymbol{x}_i \quad \text{for} \quad i = r+1,...,n.$$

So $\boldsymbol{w}$ is orthogonal to every element of $\boldsymbol{x}_{r+1},...,\boldsymbol{x}_n,\boldsymbol{v}_{k+1},...,\boldsymbol{v}_n$.

Thus if $\boldsymbol{w}$ is in $S \cap T$ then $\boldsymbol{w}$ belongs to $[\![\boldsymbol{x}_{r+1},...,\boldsymbol{x}_n,\boldsymbol{v}_{k+1},...,\boldsymbol{v}_n]\!]^{\perp}$.

Conversely, if $\boldsymbol{w}$ is a vector in $[\![\boldsymbol{x}_{r+1},...,\boldsymbol{x}_n,\boldsymbol{v}_{k+1},...,\boldsymbol{v}_n]\!]^{\perp}$, then $\boldsymbol{w}$ is orthogonal to every element of the subspace $[\![\boldsymbol{x}_{r+1},...,\boldsymbol{x}_n,\boldsymbol{v}_{k+1},...,\boldsymbol{v}_n]\!]$. This implies that $\boldsymbol{w}$ belongs to S and $\boldsymbol{w}$ belongs to T— i.e., $\boldsymbol{w}$ belongs to $S \cap T$.

Hence we see that if $\boldsymbol{w}$ belongs to $[\![\boldsymbol{x}_{r+1},...,\boldsymbol{x}_n,\boldsymbol{v}_{k+1},...,\boldsymbol{v}_n]\!]^{\perp}$, then $\boldsymbol{w}$ is an element of $S \cap T$ — which establishes that

$$S \cap T = [\![\boldsymbol{x}_{r+1},...,\boldsymbol{x}_n,\boldsymbol{v}_{k+1},...,\boldsymbol{v}_n]\!]^{\perp}.$$

Thus we have shown the following representation of subspace intersection.

Theorem 11.9: Characterization of $S \cap T$

If S and T are subspaces (Definition 9.4 on page 296) of Euclidean n-dimensional space (Definition 9.14 on page 302) with orthogonal complements (Definition 11.2 on page 412)

$$S^{\perp} = [\![\boldsymbol{x}_{r+1},...,\boldsymbol{x}_n]\!] \quad \text{and} \quad T^{\perp} = [\![\boldsymbol{v}_{k+1},...,\boldsymbol{v}_n]\!]$$

then their intersection, $S \cap T$ (see Footnote 1 on page 419) is a subspace, characterized by the equation

$$S \cap T = [\![\boldsymbol{x}_{r+1},...,\boldsymbol{x}_n,\boldsymbol{v}_{k+1},...,\boldsymbol{v}_n]\!]^{\perp}.$$

❑

Problems 11.4

1. Prove that if S is a subspace of E_n, then $(S^{\perp})^{\perp} = S$.
2. Show that if the subspace S is a subset of the subspace T, then the subspace $T^{\perp}$ is a subset of the subspace $S^{\perp}$.
3. Draw a picture of two intersecting planes through the origin to convince yourself of the validity of Theorem 11.9.

The inverse of a square matrix can be given a very neat characterization in terms of orthogonal complements, as follows.

Since $\mathbf{A}^{-1}$ is a matrix, $\mathbf{B}$ (Definition 10.1 on page 328) such that

$$\mathbf{B}\,\mathbf{A} = \mathbf{I}$$

(see Definition 10.3 on page 331 and Definition 10.6 on page 334) recalling equation 10.7 on page 331, we see that if $\boldsymbol{B}_i$ is the i-th row of $\mathbf{B}$ and $\boldsymbol{a}_j$ is the j-th column of $\mathbf{A}$,

$$\boldsymbol{B}_i \cdot \boldsymbol{a}_j = \begin{cases} 0 & \text{if } i \neq j \\ 1 & \text{if } i = j. \end{cases} \qquad \textit{11.15}$$

Hence, by Theorem 8.15 on page 278, the i-th ***row***, $\boldsymbol{B}_i$, of $\mathbf{B}$ is orthogonal to every column $\boldsymbol{a}_j$ except for column $\boldsymbol{a}_i$. Therefore the inverse, $\mathbf{B} = \mathbf{A}^{-1}$ can be constructed row by row using one of the variants of the Gram Schmidt procedure (say, Theorem 10.31 on page 366) by first applying it to

$$\boldsymbol{a}_1,\ldots,\boldsymbol{a}_{i-1}, \boldsymbol{a}_{i+1},\ldots,\boldsymbol{a}_n,\boldsymbol{e}_1,\ldots,\boldsymbol{e}_n$$

(where $\boldsymbol{e}_1,\ldots,\boldsymbol{e}_n$ are the standard n-dimensional unit vectors, see Definition 8.2 on page 263); the single (nonzero) vector arising from $\boldsymbol{e}_1,\ldots,\boldsymbol{e}_n$ suitably scaled (i.e., multiplied by a real number) so as to satisfy the last equation in 11.15, is $\boldsymbol{B}_i$. We summarize this in the following theorem.

Theorem 11.10: Geometric interpretation of $\mathbf{B} = \mathbf{A}^{-1}$

If $\mathbf{A}$ is an invertible $n \times n$ matrix (Definition 10.1 on page 328 and Definition 10.9 on page 336) its inverse matrix $\mathbf{B}$ (Definition 10.7 on page 334) is characterized by

- each row $\boldsymbol{B}_i$ having positive length (Definition 8.6 on page 270) orthogonal (see Theorem 8.15 on page 278) to every column of $\mathbf{A}$ except $\boldsymbol{a}_i$;
- $\boldsymbol{B}_i$ scaled (multiplied by a real factor) so that $\boldsymbol{B}_i \cdot \boldsymbol{a}_i = 1$.

❑

Problems 11.5

1. Where does the procedure leading to Theorem 11.10 fail if the columns $\boldsymbol{a}_1,...,\boldsymbol{a}_n$ are not linearly independent?
2. How can you construct $\mathbf{A}^{-1}$ a ***column*** at a time by an orthogonalization method similar to the one described for the proof of Theorem 11.10?
3. Suppose $\mathbf{A}$ is an $n \times m$ matrix with linearly independent columns.

   a. Show that $m \leq n$.

   b. Show how to construct an $m \times n$ matrix, $\mathbf{B}$, such that $\mathbf{BA} = \mathbf{I}^{m \times m}$.

   c. If $m \neq n$, determine all matrices $\mathbf{B}$ such that $\mathbf{BA} = \mathbf{I}^{m \times m}$. Show that for given $\boldsymbol{y}$, the solution, $\boldsymbol{x}$, to the system $\mathbf{A}\boldsymbol{x} = \boldsymbol{y}$ is unique (either empty, if y is not in A_{col}, the subspace generated by the columns of $\mathbf{A}$, or a unique *m*-dimensional vector otherwise),

   d. and, if the solution is nonempty, it is given by $\boldsymbol{x} = \mathbf{B}\boldsymbol{y}$, no matter which $\mathbf{B}$ is used.

   e. If $\mathbf{B}$ such that $\mathbf{BA} = \mathbf{I}^{m \times m}$ has rows all of which are in the subspace $\mathbf{A}_{col} = [\![\boldsymbol{a}_1,...,\boldsymbol{a}_n]\!]$ generated by the columns of $\mathbf{A}$, what does the product $\mathbf{B}\boldsymbol{y}$ represent when $\mathbf{A}\boldsymbol{x} = \boldsymbol{y}$ is inconsistent, i.e., when there is no vector $\boldsymbol{x}$ for which $\mathbf{A}\boldsymbol{x} = \boldsymbol{y}$? Hint: Write $y = \boldsymbol{y}_{\mathbf{A}_{col}} + \boldsymbol{y}_{\mathbf{A}_{col^\perp}}$.

Add some more problems regarding left and right inverses and for matrices which give solution to a system with many solutions. For problem 3 above show there is no right inverse if $m > n$.

5 Selected Applications

This section is devoted to some results encountered in applied areas.

The first results are related to the theory of maxima and minima under constraints — and provide the insight for understanding of the topic *Lagrange multipliers*. Our development, although incomplete, is essentially correct. More detailed proof, as well as further extensions are given in Luenberger [11].

The constrained maxima-minima problem is that of

- minimizing (or maximizing) $f(\boldsymbol{x})$

subject to the constraints

$$g_i(\boldsymbol{x}) = 0, \quad i = 1,...,k, \qquad \textbf{11.16}$$

where f and the g_i are real valued functions of n real variables, all having the same domain.

Example 11.6: Shortest distance between sun and earth

One of the simplest problems of this type is that of finding the closest that the earth gets to the sun, assuming that we know the orbit of the earth relative to the sun. In this case if we let (a,b,c) represent fixed coordinates for the sun in some sensibly chosen coordinate system, and $\boldsymbol{x} = (x_1, x_2, x_3)$ the possible earth coordinates in this system, the distance, $f(\boldsymbol{x})$, between the earth and the sun is given by

$$f(\boldsymbol{x}) = \sqrt{(x_1 - a)^2 + (x_2 - b)^2 + (x_3 - c)^2}.$$

The constraint is that the earth's coordinates must lie on its orbit — i.e., a constraint of the form

$$g(\boldsymbol{x}) = \alpha x_1^2 + \beta x_2^2 + \gamma x_3^2 + \delta x_1 x_2 + \phi x_1 x_3 + \kappa x_2 x_3 + \lambda x_1 x_2 x_3 - \rho^2 = 0$$

where $\alpha, \beta, \gamma, \delta, \phi, \kappa, \lambda, \rho$ are constants chosen to place the earth on its elliptical orbit. If the coordinate system is chosen sensibly, the equation above will turn out fairly simple.

■

Here are a few more examples of constrained optimization.

Example 11.7: Best linear unbiased estimate

If we make several measurements on some quantity of interest,[1] it's reasonable to try to combine our measurements in such a way as to obtain the *best* estimate of this quantity. For instance, x_1 might be an estimate of the unemployment rate based on company surveys, x_2 an unemployment rate estimate based on unemployment claims, and x_3 an estimate based on household surveys, and so forth. Under certain reasonable statistical assumptions this leads us to choose values $a_1, a_2,...,a_n$ to minimize the quantity

$$\sum_{j=1}^{n} a_j^2 \sigma_j^2$$

1. Such as the cost of living, the mileage of some make of automobile, or a person's red blood cell count.

subject to the constraint

$$\sum_{j=1}^{n} a_j = 1 \cdot$$

Here $\sigma_j = \sqrt{\sigma_j^2}$ is the so-called *standard deviation* of the measurement x_j, a measure of its variability, which may be assumed known in many situations. The constraint above is put on so that the estimate

$$\sum_{j=1}^{n} a_j x_j$$

is a *reasonable* (*unbiased*) one of the quantity of interest (to be continued, page 427).

■

Example 11.8: Meaning of the gradient vector

Earlier, see Theorem 8.18 on page 281, we showed that the gradient of the real valued function F of n real variables specified the direction of greatest change of F. The problem of determining the direction of greatest change of F could have been set up as follows:

For given n-dimensional vector x, find $\boldsymbol{u} = (u_1,...,u_n)$

- maximizing $\boldsymbol{u} \cdot \mathbf{D}F(\boldsymbol{x})$, $[f(\boldsymbol{u})]$

(see Theorem 8.11 on page 273),

- subject to $\|\boldsymbol{u}\|^2 = 1$ $\quad [g(\boldsymbol{u}) = \|\boldsymbol{u}\|^2 - 1 = 0]$.

■

Example 11.9: Extreme orbital distance

Find the minimum and maximum distances of the point $(1,2,3)$ from the ellipse

$$\frac{(x-6)^2}{2} + \frac{(y-7)^2}{3} + (z-5)^2 = 1. \qquad \textit{11.17}$$

Here, letting $\boldsymbol{X} = (x, y, z)$, we want to maximize and minimize

$$f(\boldsymbol{X}) = (x-1)^2 + (y-2)^2 + (z-3)^2 = \text{dist}^2$$

subject to constraint 11.17.

■

To handle the problem of optimization under constraints requires the development of some theory — theory which will require the use of orthogonal complements and the basic approximation equation 8.17 on page 274. Writing this approximation, but with the symbol f replaced by g_i, we obtain

$$g_i(\boldsymbol{x}+\boldsymbol{h}) \cong g_i(\boldsymbol{x}) + \boldsymbol{h} \cdot \mathbf{D}g_i(\boldsymbol{x}). \qquad \textbf{\textit{11.18}}$$

To work toward the solution, suppose $\boldsymbol{x}_o$ to be a point at which f has a (local) minimum,[1] subject to constraints 11.16 on page 423. That is, suppose

$$f(\boldsymbol{x}_o+\boldsymbol{h}) \leq f(\boldsymbol{x}_o) \qquad \textbf{\textit{11.19}}$$

for all sufficiently small $\boldsymbol{h}$ such that

$$0 = g_i(\boldsymbol{x}_o) = g_i(\boldsymbol{x}_o+\boldsymbol{h}), \qquad \text{all} \quad i = 1,...,k. \qquad \textbf{\textit{11.20}}$$

Now acting *as if* approximation 11.18 was an exact equality,[2] we replace constraints 11.20 by constraints which restrict consideration to those $\boldsymbol{h}$ for which

$$0 = g_i(\boldsymbol{x}_o) = g_i(\boldsymbol{x}_o+\boldsymbol{h}) = g_i(\boldsymbol{x}_o) + \boldsymbol{h} \cdot \mathbf{D}g_i(\boldsymbol{x}_o),$$

i.e., to those $\boldsymbol{h}$ for which

$$\boldsymbol{h} \cdot \mathbf{D}g_i(\boldsymbol{x}_o) = 0, \quad \text{all } i = 1,...,k.$$

Thus the constraints $g_i(\boldsymbol{x}) = 0$, $i = 1,...,k$, lead us to act as if we need only consider those $\boldsymbol{h}$ satisfying the membership requirement[3]

$$\boldsymbol{h} \in [\![\mathbf{D}g_1(\boldsymbol{x}_o),...,\mathbf{D}g_1(\boldsymbol{x}_o)]\!]^{\perp}. \qquad \textbf{\textit{11.21}}$$

Now once again acting as if the fundamental approximation, 8.17 on page 274 was exact, we see, examining inequality 11.19, that we are hunting for $\boldsymbol{x}_o$ such that among those $\boldsymbol{h}$ satisfying condition 11.21 we have satisfied the inequality

$$f(\boldsymbol{x}_o) \geq f(\boldsymbol{x}_o) + \boldsymbol{h} \cdot \mathbf{D}f(\boldsymbol{x}_o). \qquad \textbf{\textit{11.22}}$$

---

1. Of course, once you have a candidate for such a point, it remains to determine what kind of point it is. We will provide some theory to help do this in problem 11.11.1 on page 428.
2. This is the part of the derivation which needs patching with an appropriate *limit* argument.
3. Recall that the symbol $\in$ is read as *belongs to*, or *is a member of.*

But this implies that ***we must have***

$$\boldsymbol{h} \cdot \mathbf{D}f(\boldsymbol{x}_o) = 0 \qquad \textit{11.23}$$

for all $\boldsymbol{h}$ satisfying constraint 11.21 — by the following reasoning:

> Suppose there were an $\boldsymbol{h}^*$ satisfying constraint 11.21, such that equation 11.23 wasn't satisfied — i.e., such that
>
> $$\boldsymbol{h}^* \cdot \mathbf{D}f(\boldsymbol{x}_o) \neq 0;$$
>
> Observe that in order for inequality 11.22 to be valid, it must be the case that
>
> $$\boldsymbol{h}^* \cdot \mathbf{D}f(\boldsymbol{x}_o) < 0.$$
>
> But then $\boldsymbol{h} = -\boldsymbol{h}^*$ would still satisfy the *linear* (approximate) constraint 11.21, and hence still *essentially* satisfy the original constraint 11.20, but violate the *linear* (approximate) inequality, and hence *essentially violate* the original inequality 11.19.
>
> ***That is, if linear equation 11.23 weren't satisfied for all h satisfying linear constraint 11.21, then it* appears *that you'd violate the requirement 11.19 for some h satisfying the actual constraints 11.20.***

So, it seems that at all points where the formula $D_{\boldsymbol{u}} = \boldsymbol{u} \cdot \boldsymbol{D}$ for the directional derivative operator holds for f and all the g_i (see Theorem 8.11 on page 273) in order for f to achieve a local maximum or minimum, it is necessary that equation 11.23 hold for all $\boldsymbol{h}$ satisfying linear constraint 11.21. (Note that no conclusions are drawn about points for which the equation $D_{\boldsymbol{u}} = \boldsymbol{u} \cdot \mathbf{D}$ fails to hold. Such points, e.g., boundary points of the domain, must be examined separately.) Examination of conditions 11.21 and 11.23 shows that together they assert that

$\mathbf{D}f(\boldsymbol{x}_o)$ must be orthogonal to all $\boldsymbol{h}$ which are orthogonal to all $\mathbf{D}g_i(\boldsymbol{x}_o)$.

That is (see Problem 11.4.1 on page 421)

$\mathbf{D}f(\boldsymbol{x}_o)$ must belong to the subspace $[\![\mathbf{D}g_i(\boldsymbol{x}_o) \quad i = ,...,k]\!]$.

The arguments above lead to the following ***Lagrange multiplier theorem***.

Theorem 11.11: Extrema under constraints

Of the points at which the formula $D_{\boldsymbol{u}} = \boldsymbol{u} \cdot \boldsymbol{D}$ holds unrestrictedly for f and all the g_i (see Theorem 8.11 on page 273) the only ones that can possibly qualify as points where a minimum or maximum of f can be achieved, subject to the constraints $g_i(\boldsymbol{x}) = 0, \quad i = 1,...,k,$ are those $\boldsymbol{x}_o$ satisfying

$$g_i(\boldsymbol{x}_o) = 0, \qquad i = 1,...,k \tag{11.24}$$

and

$$\mathbf{D}f(\boldsymbol{x}_o) \in [[\mathbf{D}g_1(\boldsymbol{x}_o),...,\mathbf{D}g_1(\boldsymbol{x}_o)]], \tag{11.25}$$

where $\in$ denotes membership.[1]

Condition 11.25 is sometimes written as

$$\mathbf{D}f(\boldsymbol{x}_o) = \lambda_1 \mathbf{D}g_1(\boldsymbol{x}_o) + \cdots + \lambda_k \mathbf{D}g_k(\boldsymbol{x}_o). \tag{11.26}$$

The λ_i are called **Lagrange multipliers**, and turn out to have a useful interpretation in economic problems, see Luenberger [11], p. 231.

❑

Note that this theorem reduces to the first part of Theorem 10.37 on page 381, when there are no constraints. This might not appear to be the case, if you fail to recognize that by definition the subspace generated by no vectors at all, consists solely of the $\mathbf{0}$ vector.

We illustrate the Lagrange multiplier theorem to complete Example 11.7 which was started on page 423.

Example 11.10: Completion of Example 11.7

In this example we desired to minimize $\sum_{j=1}^{n} a_j^2 \sigma_j^2$ subject to the constraint $\sum_{j=1}^{n} a_j = 1$. So we may take

$$f(a_1,...,a_n) = \sum_{j=1}^{n} a_j^2 \sigma_j^2 \quad \text{and} \quad g(a_1,...,a_n) = \sum_{j=1}^{n} a_j - 1.$$

---

1. Again, recall that $\in$ is read *is a member of.* See Definition 9.4 on page 296 for the meaning of [[ ]]

(Note the minor notational changes, where we replace $\boldsymbol{x}_o$ by $\boldsymbol{a}$ and discard subscripts on the symbol λ because there is only one constraint.)

Since $\mathbf{D}f(\boldsymbol{a}) = (2a_1\sigma_1^2,...,2a_n\sigma_n^2)$ and $\mathbf{D}g(\boldsymbol{a}) = (1,...,1)$, it is seen that we are looking for a solution of the equations

$$\sum_{j=1}^{n} a_j - 1 = 0, \quad (2a_1\sigma_1^2,...,2a_n\sigma_n^2) = \lambda(1,...,1)$$

or

$$\sum_{j=1}^{n} a_j = 1, \qquad 2a_j\sigma_j^2 = \lambda \qquad \text{for} \quad j = 1,...,n.$$

To solve, rewrite these equations as

$$\sum_{j=1}^{n} a_j = 1, \qquad \textbf{\textit{11.27}}$$

$$a_j = \frac{\lambda}{2}\frac{1}{\sigma_j^2}. \qquad \textbf{\textit{11.28}}$$

Sum equations 11.28 over j, and apply equation 11.27 to obtain

$$\lambda = \frac{2}{\sum_{j=1}^{n} \frac{1}{\sigma_j^2}} \quad \text{and hence} \quad a_k = \frac{1}{\sigma_k^2} \frac{1}{\sum_{j=1}^{n} \frac{1}{\sigma_j^2}}.$$

Notice that a_k, the weight given to the k-th measurement, is inversely proportional to this measurement's variance, σ_k^2, which, on reflection, should not surprise you, since measurements with greater variability should have less influence. Notice also that the actual values of the σ_k^2 needn't be known; only the relative values σ_i^2/σ_j^2 are needed.

■

Problems 11.11

1. Derive sufficient conditions, similar to those in Theorem 10.37 on page 381, for minima and maxima under constraints.
   Hints: Suppose that we have found a point, $\boldsymbol{x}_o$, at which the

necessary Lagrange multiplier theorem conditions are satisfied. If $\mathbf{H}$ denotes the Hessian matrix of f at $\boldsymbol{x}_o$ (see Definition 10.36 on page 377) then if the quadratic form $\sum_{i,j} h_i h_j H_{i,j} > 0$ for all nonzero vectors $\boldsymbol{h}$ satisfying the constraints $g_i(\boldsymbol{x}_o + \boldsymbol{h}) = 0 \quad i = 1,...,k$, then, from Theorem 10.37 on page 381, we expect a minimum for f at $\boldsymbol{x}_o$. Now the constraint condition is replaced by its linearized version, requiring satisfaction of condition 11.21 on page 425, namely,

$$\boldsymbol{h} \in [[\mathbf{D}g_1(\boldsymbol{x}_o),...,\mathbf{D}g_1(\boldsymbol{x}_o)]]^{\perp}.$$

Using Theorem 11.4 on page 415, we can find a basis, $\boldsymbol{v}_1,...,\boldsymbol{v}_{n-k}$ for $[[\mathbf{D}g_1(\boldsymbol{x}_o),...,\mathbf{D}g_1(\boldsymbol{x}_o)]]^{\perp}$. Let $\mathbf{V}$ be the matrix whose columns are $\boldsymbol{v}_1,...,\boldsymbol{v}_{n-k}$ — i.e.,

$$\mathbf{V} = [\boldsymbol{v}_1,...,\boldsymbol{v}_{n-k}].$$

Then the linearized constraints can be accomplished by writing

$$\mathbf{h} = \mathbf{V}\mathbf{b}$$

noting that a quadratic form $F = \sum_{i,j} h_i h_j H_{i,j}$ may be written in the matrix symbolism (with T denoting transposition) as

$$\mathbf{h}^{\mathrm{T}}\mathbf{H}\mathbf{h}.$$

Thus what we need to determine is whether the quadratic form

$$(\mathbf{V}\mathbf{b})^{\mathrm{T}}\mathbf{H}\mathbf{V}\mathbf{b} > 0$$

for all nonzero vectors $\mathbf{b}$, i.e., using Exercise 10.16.1 on page 366, establishing whether the matrix $\mathbf{V}^{\mathrm{T}}\mathbf{H}\mathbf{V}$ is positive definite, say by the method of Theorem 10.37 on page 381.

2. Using the methods established in this section, find the maximum of the expression $2(xy + yz + xz)$ subject to the constraint $x + y + z = 3$.
3. Use Lagrange multipliers and the sufficient conditions developed in the solution to problem 1 above, to finish up Examples 11.8 and 11.9.

The next topic that we will treat in this section is that of the *rank* of a matrix. First we will define *row rank* and *column rank*, and then show, in an optional section that they are equal. This result arises just often enough in applications to want to have it available for reference. The proof is useful too, although not essential for later understanding, so it is presented in an optional portion.

Definition 11.12: Row rank, column rank, rank

Let $\mathbf{A}$ be a real $n \times m$ matrix (Definition 10.1 on page 328) whose columns are denoted by $\boldsymbol{a}_1,...,\boldsymbol{a}_m$ and whose rows are denoted by $\boldsymbol{A}_1,...,\boldsymbol{A}_n$. Recall the definition and notation for subspace (Definition 9.4 on page 296) and the meaning of subspace dimension (Definition 9.8 on page 298).

The **row rank** of $\mathbf{A}$ is the dimension of the subspace $[[\boldsymbol{A}_1,...,\boldsymbol{A}_n]]$.

The **column rank** of $\mathbf{A}$ is the dimension of the subspace $[[\boldsymbol{a}_1,...,\boldsymbol{a}_m]]$.

Because of Theorem 11.13 below, the common value of the row and column rank of the matrix $\mathbf{A}$ is simply called its *rank,* and is denoted by

$$rnk\mathbf{A}.$$

❑

In the statistical theory of linear models, we sometimes need to know the row rank of a model matrix, where the number of rows far exceeds the number of columns. It's often easy to see that the columns are linearly independent, which immediately tells us the row rank that is needed.

┌

Theorem 11.13: Equality of matrix row and column rank

If $\mathbf{A}$ is a real $n \times m$ matrix (Definition 10.1 on page 328) with rows $\boldsymbol{A}_1,...,\boldsymbol{A}_n$ and columns $\boldsymbol{a}_1,...,\boldsymbol{a}_m$ then (recalling Definition 9.4 on page 296 and Definition 9.8 on page 298)

$$dim[[\boldsymbol{A}_1,...,\boldsymbol{A}_n]] = dim[[\boldsymbol{a}_1,...,\boldsymbol{a}_m]].$$

❑

To establish this result, we need to recall various results from Chapter 9. The key to our arguments consists of the results derived about solving equations using columns, and the methods using rows.

Let r denote the column rank of $\mathbf{A}$, and R its row rank (Definition 11.12 on page 430). You should be able to convince yourself that interchanging any two columns of $\mathbf{A}$ will affect neither its column nor its row rank. This is left for the reader to verify. Accepting this, we can assume that the first r columns of $\mathbf{A}$ are linearly independent — all other columns being linearly dependent on $\boldsymbol{a}_1,...,\boldsymbol{a}_r$.

Now from Theorem 11.4 on page 415 the general solution of $\mathbf{Ax} = \mathbf{0}$ is

$$\mathbf{A}_{row}^{\perp} = [\![A_1,...,A_n]\!]^{\perp}.$$

and because the rows are m dimensional vectors,

$$dim\ \mathbf{A}_{row}^{\perp} = dim\ [\![A_1,...,A_n]\!]^{\perp} = m - dim[\![A_1,...,A_n]\!] = m - R.$$

We will now show by direct arguments on the columns of $\mathbf{A}$ that the dimension of the general solution of $\mathbf{Ax} = \mathbf{0}$ is $m - r$, from which the assertion $R = r$ will follow.

Looking at the system $\mathbf{Ax} = \mathbf{0}$ in terms of columns, we may write

$$\sum_{i=1}^{r} x_i \boldsymbol{a}_i + \sum_{i=r=1}^{n} x_i \boldsymbol{a}_i = \mathbf{0}.$$

We rewrite this system in matrix form as

$$\mathbf{A}_1^{(n \times r)} \boldsymbol{x}_1^{(r \times 1)} + \mathbf{A}_2^{(n \times [m-r])} \boldsymbol{x}_2^{([m-r] \times 1)} = \mathbf{0}^{(n \times 1)}, \qquad \textit{11.29}$$

where

$$\mathbf{A}_1 = [\boldsymbol{a}_1,...,\boldsymbol{a}_r] \quad \text{and} \quad \mathbf{A}_2 = [\boldsymbol{a}_{r+1},...,\boldsymbol{a}_m].$$

Since each of the columns of $\mathbf{A}$ beyond the r-th is (by our earlier maneuvering) linearly dependent on $\boldsymbol{a}_1,...,\boldsymbol{a}_r$, we may write that for each $j = 1,...,m - r$

$$\boldsymbol{a}_{r+j} = [\boldsymbol{a}_1,...,\boldsymbol{a}_r]\boldsymbol{b}_{r+j} = \mathbf{A}_1 \boldsymbol{b}_{r+j}$$

and hence

$$\mathbf{A}_2 = [\boldsymbol{a}_{r+1},...,\boldsymbol{a}_m] = \mathbf{A}_1[\boldsymbol{b}_{r+1},...,\boldsymbol{b}_m] = \mathbf{A}_1^{(n \times r)} \boldsymbol{B}^{(r \times [m-r])}. \qquad \textit{11.30}$$

So, using equations 11.29 and 11.30, we may rewrite the system $\mathbf{Ax} = \mathbf{0}$ as

$$\mathbf{A}_1^{(n \times r)} \boldsymbol{x}_1^{(r \times 1)} + \mathbf{A}_1^{(n \times r)} \mathbf{B}^{(r \times [m-r])} \boldsymbol{x}_2^{([m-r] \times 1)} = \mathbf{0}$$

or

$$\mathbf{A}_1^{(n \times r)}(\boldsymbol{x}_1 + \mathbf{B}\boldsymbol{x}_2) = \mathbf{0}. \qquad \textit{11.31}$$

Examining equation 11.31, we see that since the columns of $\mathbf{A}_1$ are linearly independent, each arbitrary choice of $\boldsymbol{x}_2$ determines a unique $\boldsymbol{x}_1$ (namely, $x_1 = -\mathbf{B}\boldsymbol{x}_2$ such that

$$\boldsymbol{x} = \begin{pmatrix} \boldsymbol{x}_1 \\ \boldsymbol{x}_2 \end{pmatrix} \text{ solves } \mathbf{Ax} = \mathbf{0}.$$

Conversely, each solution $\mathbf{x}$ of $\mathbf{Ax} = \mathbf{0}$ determines $\boldsymbol{x} = \begin{pmatrix} \boldsymbol{x}_1 \\ \boldsymbol{x}_2 \end{pmatrix}$ with $x_1 = -\mathbf{B}\boldsymbol{x}_2$. So the general solution of $\mathbf{Ax} = \mathbf{0}$ consists of all $\boldsymbol{x}$ of the form

$$\boldsymbol{x} = \begin{pmatrix} \mathbf{B}^{(r \times [m-r])}\boldsymbol{x}_2^{([m-r] \times 1)} \\ \boldsymbol{x}_2^{([m-r] \times 1)} \end{pmatrix} = \begin{bmatrix} \mathbf{B}^{(r \times [m-r])} \\ \mathbf{I}^{([m-r] \times [m-r])} \end{bmatrix} \boldsymbol{x}_2^{([m-r] \times 1)}$$

where $\boldsymbol{x}_2$ is arbitrary.

Thus the columns of the matrix $\mathbf{M} = \begin{bmatrix} \mathbf{B}^{(r \times [m-r])} \\ \mathbf{I}^{([m-r] \times [m-r])} \end{bmatrix}$ generate the subspace of solutions of $\mathbf{Ax} = \mathbf{0}$. But it's easy to see just from the presence of the $[m-r] \times [m-r]$ identity matrix part of $\mathbf{M}$ that there are $m - r$ linearly independent columns in $\mathbf{M}$, and thus the dimension of the solution space of $\mathbf{Ax} = \mathbf{0}$ is $m - r$. Since this dimension was also $m - R$, we see that $r = R$, as claimed, proving Theorem 11.13.

$\lfloor$

Exercises 11.12 Directly verify Theorem 11.13 for the following:

1. $\begin{bmatrix} 1 \\ 1 \\ 0 \end{bmatrix}$

2. $\begin{bmatrix} 1 & 1 \\ 1 & 0 \\ 0 & 1 \end{bmatrix}$

3. $\begin{bmatrix} 1 & 2 & 3 \\ 4 & 5 & 6 \\ 7 & 8 & 9 \end{bmatrix}$

6 Impulse Response

A problem of great interest in the physical sciences and engineering concerns the behavior of a system which is inactive until it is switched on at some specific time t_0. As we'll see in Chapter 15, this leads to the development of some very interesting mathematics needed to understand how to model such situations. In order to permit a smooth development in Chapter 15 , we now introduce the very simple finite dimensional sequence version of this problem, where it fits conveniently with the material already presented.

The problem of interest in Chapter 15 may be modeled as that of solving (in some sense) the differential equation

$$\left(D^k + \sum_{i=0}^{k-1} a_i D^i\right) x = g \qquad \textit{11.32}$$

where g and $a_0,\dots,a_{k-1}$ are known real valued functions of a real variable with $g(t) = 0$ for $t < 0$, subject to initial conditions of the form

$$x(0) = x'(0) = \cdots = x^{(k-1)}(0) = 0. \qquad \textit{11.33}$$

The corresponding finite dimensional problem that generalizes well is chosen to start with the solution of the system of algebraic equations

$$\mathbf{Ax} = \mathbf{y}. \qquad \textit{11.34}$$

The matrix A corresponds to the differential operator $D^k + \sum_{i=0}^{k-1} a_i D^i$.

The vector $\boldsymbol{y}$ corresponds to the *forcing function* g and the solution vector $\boldsymbol{x}$ to the solution function x.

It shouldn't be hard to convince yourself that a matrix can approximate a differential operator as follows.

$Dx = x'$ can be approximated by a vector

$$\boldsymbol{x}' = \begin{pmatrix} x'(0) \\ x'(h) \\ . \\ . \\ x'(nh) \end{pmatrix} \cong \begin{pmatrix} (x(h)-x(0))/h \\ (x(2h)-x(h))/h \\ . \\ . \\ (x((n+1)h)-x(nh))/h \end{pmatrix} = \frac{1}{h}\begin{bmatrix} -1 & 1 & 0 & . & 0 \\ . & . & . & . & . \\ . & . & . & . & . \\ . & . & . & . & . \\ 0 & . & 0 & -1 & 1 \end{bmatrix} \begin{pmatrix} x(0) \\ x(h) \\ . \\ . \\ x((n+1)h) \end{pmatrix}$$

$= \mathbf{D}\boldsymbol{x}$, where

$$\mathbf{D} = \frac{1}{h}\begin{bmatrix} -1 & 1 & 0 & . & 0 \\ . & . & . & . & . \\ . & . & . & . & . \\ . & . & . & . & . \\ 0 & . & 0 & -1 & 1 \end{bmatrix} \quad \text{and} \quad \boldsymbol{x} = \begin{pmatrix} x(0) \\ x(h) \\ . \\ . \\ x((n+1)h) \end{pmatrix}$$

i.e., the derivative operator, D, can be approximated by the matrix **D,** the function x by the vector $\boldsymbol{x}$, and the function $x' = Dx$ by the matrix vector product $\mathbf{D}\boldsymbol{x}$. With some effort all of this can be extended to higher derivatives and differential operators.

We need to be careful in defining system 11.34 to try somehow to build in conditions analogous to initial conditions 11.33 of the differential equations being approximated. The condition $x(0) = 0$ is simulated by choosing the first coordinate $x_1 = 0$. (Note the change of subscript from $x(0)$ to x_1. This is unfortunate, arising because most sequences conventionally start with x_1.) The condition $x'(0) = 0$ is approximated by the equation $(x(h) - x(0))/h = 0$, leading to $x(h) \cong x_2 = 0$, etc. So we want the first k coordinates of $\boldsymbol{x}$ to be 0; i.e., we want the condition

$$x_1 = x_2 = \cdots = x_k = 0. \qquad \textbf{\textit{11.35}}$$

to hold. To accomplish this we impose two conditions on system 11.34. We set

$$y_1 = y_2 = \cdots = y_k = 0. \qquad \textbf{\textit{11.36}}$$

We choose

the first k rows of **A** to have
1s on the main diagonal and 0s elsewhere; ***11.37***

which guarantees that any solution of system 11.33 satisfies conditions 11.35.

We will require

the matrix **A** to be $(k+m) \times (k+m)$
with the m columns linearly independent ***11.38***

to guarantee unique solution for any $(k+m) \times 1$ vector $\boldsymbol{y}$ whose first k coordinates are 0 (see Theorem 10.10 on page 337 to match what you expect from the differential equation 11.32). Nothing seems to be lost by setting all other coordinates of the first k columns of **A** to 0. For $i = k+1,...,k+m$, the unit vector

$$e_{ij}^{(k+m)\times 1} = \begin{cases} 1 & \text{if } j = i \\ 0 & \text{otherwise} \end{cases}$$

can be considered analogous to an impulse electrical or mechanical forcing function, which puts all of its emphasis at time i. Hence there is a unique solution

$$\mathbf{x} = \mathbf{G}_i^{(k+m)\times 1} \quad \text{to the system} \quad \mathbf{Ax} = \mathbf{e}_i. \qquad \textbf{\textit{11.39}}$$

Definition 11.14: Impulse response, Green's function

The sequence of vectors $\mathbf{G}_i$ given by equations 11.39 above is called the **Impulse Response** of the system described by equations 11.34, 11.36 and conditions 11.37, 11.38.

The significance of the impulse response is that every solution of $\mathbf{Ax} = \mathbf{y}$ subject to equations 11.36 and conditions 11.37 and 11.38 can be written as a weighted sum (superposition) of impulse responses; i.e. that the impulse response is a **Green's function** — one which generates a solution by ***superposition***. You might try establishing this yourself before reading further.

❑

To prove this claim, note that since $\boldsymbol{AG} = \boldsymbol{e}_i$ for $i = k+1,...,k+m$ from the linear operator property of matrices (Theorem 10.2 on page 330) it follows that

$$\mathbf{A}\sum_{i=k+1}^{k+m} y_i\mathbf{G}_i = \sum_{i=k+1}^{k+m} y_i\mathbf{AG}_i = \sum_{i=k+1}^{k+m} y_i\mathbf{e}_i = \begin{pmatrix} 0 \\ . \\ 0_k \\ y_{k+1} \\ . \\ . \\ y_{k+m} \end{pmatrix}.$$

We have thus shown the following matrix version of the impulse response solution.

Theorem 11.15: Impulse response form of solution to Ax = y

Let $\mathbf{A}$ be a $(k+m)\times(k+m)$ real matrix satisfying conditions 11.37and 11.38, and let $\mathbf{y}$ be any $(k+m)\times 1$ matrix (column vector) satisfying condition 11.36. The unique solution of $\mathbf{Ax} = \mathbf{y}$ satisfying conditions 11.35 is given by

$$\mathbf{x} = \sum_{\mathbf{i}=k+1}^{\mathbf{k}+m} \mathbf{y}_\mathbf{i}\mathbf{G}_i, \qquad \textit{11.40}$$

where the sequence of vectors $\mathbf{G}_i$ is the impulse response given in Definition 11.14 above. The right side of equation 11.40 is called the **impulse response form of the solution of** $\mathbf{Ax} = \mathbf{y}$, and the vector $(G_{k+1},...,G_{k+n})$ is, therefore, a Green's function.

Note that we can write the equations $\mathbf{Ax} = \mathbf{y}$ as

$$\sum_{i=1}^{k+m} a_{ij} x_j = y_i \qquad i = 1,\dots,k+m \qquad \textbf{\textit{11.41}}$$

and equations 11.40 as

$$\sum_{j=1}^{k+m} G_{ji} y_i = x_j. \qquad \textbf{\textit{11.42}}$$

❑

We'll see in Chapter 15 how this result extends to constant coefficient linear differential equations.

Chapter 12

Multivariable Integrals

1 Introduction

This chapter introduces some of the important several-variable extensions of the one-dimensional Riemann integral, together with their corresponding evaluation theorems, all of which may be viewed as generalizations of the fundamental evaluation theorems (see Theorem 1.8 on page 12 and Theorem 4.3 on page 103). Along with the development of the multiple integral, we will develop the basic result on change of variable in multiple integrals, which requires knowledge of the determinant[1] of an $n \times n$ matrix.

A precise, complete development of the material of this chapter is not compatible with the space limitations under which we are operating, and would not contribute that much in the way of model understanding. So our approach will concentrate on the ideas behind the important results, and a few essential tools.

2 Multiple Integrals

The most obvious extension of the one-dimensional Riemann integral would seem to be the one suggested by the area interpretation of $\int_a^b f(x)dx$; that is, for a function $f(x, y)$ of two variables, an integral represents the volume above or below some set A in the x-y plane as illustrated in Figure 12-1.

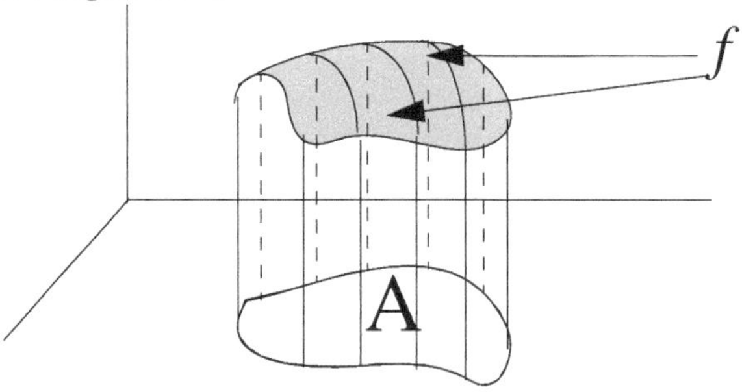

Figure 12-1

1. Contrary to much educational opinion, determinants are of almost no immediate use for practical solution of systems of equations; rather they are *essential for developing important results of a practical nature,* such as those on change of variable for integration here, and the characteristic equation for eigenvalues (see Theorem 13.3 on page 538).

For functions of more than two variables we will operate by analogy on the algebraic version of the two-variable integral.

Multiple integrals are needed for computations of such important quantities as

- total energy from an energy density defined at all points in some volume,
- the probability of an event related to n measurements $X_1,...,X_n$ described by the probability density function $f_{X_1,...,X_n}$,
- moments of inertia of three-dimensional physical objects, such as space vehicles, needed in the study of the positional behavior of these objects subject to various forces.

Our immediate aim is to explain what is meant by the multiple integral[1] $\int_A f$, $\int_A f dV_n$ or $\int_A f(\boldsymbol{x})d\boldsymbol{x}$, of a real valued function f of n real variables (Definition 8.1 on page 261) whose domain includes the set A; for the present the set A is assumed to be bounded (Definition 10.43 on page 406).

The most fundamental type of set in n dimensions is an interval, defined as follows.

Definition 12.1: n-Dimensional interval and its volume

An n-dimensional set, denoted by $[\boldsymbol{a};\boldsymbol{b}]$ where $\boldsymbol{a} = (a_1,...,a_n)$ and $\boldsymbol{b} = (b_1,...,b_n)$ are real n-dimensional vectors with $a_i \le b_i$ for all $i = 1,...,n$ is called an **n-dimensional interval** if it consists of all real n-dimensional vectors $\boldsymbol{x} = (x_1,...,x_n)$ whose coordinates satisfy

$$a_i \le x_i \le b_i \quad \text{for all} \quad i = 1,...,n.$$

The **n-dimensional volume**, $V(R)$ of $R = [\boldsymbol{a};\boldsymbol{b}]$ is the product

$$V(R) = (b_1 - a_1)(b_2 - a_2)\cdots(b_n - a_n).$$

❑

In two dimensions an interval is a rectangle.

1. More often written in the cumbersome form $\int_A \cdots \int f(x_1,...,x_n)dx_1 \cdots dx_n$.

We lay the foundation for the general definition of $\int_A f$ by first defining the multiple integral[1]

$$\int_{[a;b]} f$$

over the interval $[\boldsymbol{a};\boldsymbol{b}]$. We do this by first setting up an *n-dimensional partition* of $[\boldsymbol{a};\boldsymbol{b}]$, generalizing the previously introduced concept of *one-dimensional partition* (see Definition 4.1 on page 101) as follows.

Definition 12.2: Partition, P, of $[a;b]$, $\|P\|$, Riemann sum

Given an n-dimensional interval $[\boldsymbol{a};\boldsymbol{b}]$ (Definition 12.1 above) for each $i = 1,\ldots,n$ when $a_i < b_i$, define an increasing sequence

$$a_i = x_{i0} < x_{i1} < \cdots < x_{i,k_i} = b_i.$$

In two dimensions the **partition**, P, generated by these two sequences ***consists of*** all two-dimensional intervals of the form $(x_{1,j}, x_{2,q}), (x_{1,j+1}, x_{2,q+1})$ as illustrated in Figure 12-2.

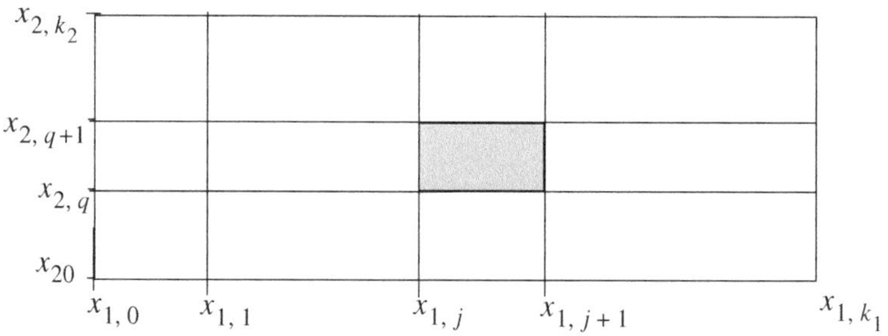

Figure 12-2

In n dimensions, the **partition** ***P consists of*** **all n-dimensional intervals of the form**

$$I_{j_1,\ldots,j_n} = [(x_{1,j_1}, x_{2,j_2},\ldots,x_{n,j_n});(x_{1,j_1+1}, x_{2,j_2+1},\ldots,x_{n,j_n+1})].$$

As in one-variable integrals, we let $X_{j_1,\ldots,j_n}$ be an arbitrary point in the n-dimensional interval $I_{j_1,\ldots,j_n}$ and form the **Riemann sum**

---

1. Also written $\int_a^b f$, $\int_{[a;b]} f(\boldsymbol{x})d\boldsymbol{x}$, $\int_{[a;b]} f dV_n$ or the most cumbersome $\int_{[a;b]} f(x_1,\ldots,x_n)\, dx_1 \cdots dx_n$.

$$R(X, P) = \sum_{\substack{\text{all intervals of} \\ \text{the partition } P}} f(X_{j_1,\dots,j_n}) I_{j_1,\dots,j_n}$$

where X is the sequence made up of the $X_{j_1,\dots,j_n}$ in the above sum (in the order corresponding to the partition elements).

We define the **norm** $\|P\|$ **of the partition** P as

$$\max_{i,\, j}(x_{i,\, j+1} - x_{i,\, j})$$

(the largest of all numbers $x_{i,\, j+1} - x_{i,\, j}$).

❑

We can now define the Riemann integral of f over an n-dimensional interval (a multidimensional, or multiple integral).

Definition 12.3: Riemann integral $\int_{[a;b]} f$

If there exists a single real number, v, such that all Riemann sums (Definition 12.2 on page 439) $R(X, P)$, independent of the choice of the $X_{j_1,\dots,j_n}$, become arbitrarily close to v, as the norm $\|P\|$ of the partition P gets close to 0 (i.e., if $\lim_{\|P\| \to 0} R(X, P) = v$) the value v is called the (multiple) **Riemann integral of f over the interval** $[\boldsymbol{a};\boldsymbol{b}]$, and is denoted by

$$\int_{\boldsymbol{a}}^{\boldsymbol{b}} f \;, \quad \int_{[\boldsymbol{a};\boldsymbol{b}]} f(\boldsymbol{x})d\boldsymbol{x} \;, \quad \int_{[\boldsymbol{a};\boldsymbol{b}]} f dV_n \quad \text{or} \quad \int_{[\boldsymbol{a};\boldsymbol{b}]} f(x_1,\dots,x_n)\, dx_1 \cdots dx_n .$$

❑

Note that existence of this integral implies the boundedness of f on $[\boldsymbol{a};\boldsymbol{b}]$, a result which we leave to the following problem.

Problem 12.1

Prove that if the function f is integrable on $[\boldsymbol{a};\boldsymbol{b}]$, then f is bounded (Definition 10.45 on page 408) there.

Our object here is to deviate as little as possible from the original concepts and notation for the one dimensional Riemann integral. So the integral over an n-dimensional interval is defined if all Riemann sums get close to one value as the size of ***all*** intervals defining the partition shrink uniformly and simultaneously in all directions to 0.

For our next result we need to extend the concept of monotonicity (see Definition 2.9 on page 37) to functions of n variables.

Definition 12.4: Monotone function of n variables

The real valued function, f, of n real variables (Definition 8.1 on page 261) is said to be **monotone** if all of the functions of one variable obtained from $f(\boldsymbol{x})$ by keeping all but one coordinate of $\boldsymbol{x}$ fixed are monotone. *That is, we will say f is monotone, if it is monotone in every coordinate axis direction.*

❑

We state without proof the extension of Theorems 4.6 on page 117 and 4.7 on page 119.

Theorem 12.5: Integral of monotone function of n variables

If the real valued function, f, of n real variables (Definition 8.1 on page 261) is defined at all points of the n-dimensional interval $[\boldsymbol{a};\boldsymbol{b}]$ (Definition 12.1 on page 438) and f is monotone (Definition 12.4 above) then $\int_{[\boldsymbol{a};\boldsymbol{b}]} f dV_n$ exists (Definition 12.3 on page 440). Furthermore, if the partition P_m is specified by partitioning each of the one dimensional intervals $[a_i;b_i]$ into m equal length subintervals, then

$$\left|\int_{[\boldsymbol{a};\boldsymbol{b}]} f dV_n - R(\boldsymbol{X}, P_m)\right| \leq \frac{n}{m}(b_1 - a_1)\cdots(b_n - a_n)[max\, f(\boldsymbol{x}) - min f(\boldsymbol{x})]$$

for all Riemann sums for f associated with the partition P_m.

❑

Proof can be found in *Modern University Calculus* [1].

The result is of very little computational value for large n, because while doubling the value of m (the number of intervals in the partitioning of each coordinate axis direction) cuts the above error bound in half, it increases the number of terms in the Riemann sum $R(\boldsymbol{X}, P_m)$ by a factor of 2^n, not a very good return on the extra effort invested.

Theorem 12.6: Integrals of continuous and of differentiable functions

Let f be a real valued continuous function of n real variables (see Definition 8.7 on page 270) whose domain includes the n-dimensional interval $[\boldsymbol{a};\boldsymbol{b}]$ (Definition 12.1 on page 438). Then $\int_{[\boldsymbol{a};\boldsymbol{b}]} f dV_n$ exists

(Definition 12.3 on page 440). Furthermore, if f has a continuous gradient vector (see Definition 8.10 on page 272 and Definition 8.7 on page 270) and $S(\boldsymbol{X}, P)$ is any Riemann sum associated with the partition P of $[\boldsymbol{a};\boldsymbol{b}]$,

$$\left| \int_{[\boldsymbol{a};\boldsymbol{b}]} f\,dV_n - S(\boldsymbol{X}, P_m) \right| \le \|P\| B \sqrt{n}\, V([\boldsymbol{a};\boldsymbol{b}])$$

where B is any upper bound for all values of $|D_i f(\boldsymbol{x})|$ for all $i = 1,\dots,n$ and all $\boldsymbol{x}$ in $[\boldsymbol{a};\boldsymbol{b}]$.

❑

As with the previous theorem, this result is not very useful computationally, and its proof is available in *Modern University Calculus* [1].

Numerical computation of $\int_{[\boldsymbol{a};\boldsymbol{b}]} f\,dV_n$ for large n is usually a formidable task, accounting for a large proportion of the efforts and obstacles in fields such as probability.

For many applications it is absolutely necessary to define integrals over sets much more general than n-dimensional intervals. For example, if we want to determine the total energy in some sphere, we must integrate the energy density over this sphere. Also, to calculate the probability that a sum of n random values governed by a probability density falls below some value, x, we must integrate the joint density, f, of these random variables over the set of points $(x_1,\dots,x_n)$ in n dimensions satisfying the inequality

$$\sum_{i=1}^{n} x_i \le x .$$

In order to define $\int_A f\,dV_n$, where A is some bounded n-dimensional set of points, we partition an n-dimensional interval in which A must lie, and form two Riemann sums — one based solely on those partition (n-dimensional) intervals lying entirely within A and the other obtained by appending all other partition subintervals which include at least one point from A. To keep things sensible, it's simplest to restrict consideration to sets A with the property that as the norm, $\|P\|$ of the partition approaches 0, the total n-dimensional volume of the appended intervals approaches 0. This restriction leads to the following

Definition 12.7: Jordan content of a set

Let A be a subset of the n-dimensional interval $[\boldsymbol{a};\boldsymbol{b}]$ (Definition 12.1 on page 438). Let P be a partition of $[\boldsymbol{a};\boldsymbol{b}]$ (see Definition 12.2 on page 439) let V_P be the total n-dimensional volume of those elements of P which are subsets of A, and let B_P be the total n-dimensional volume of all elements of P which include both points in A and points not in A.

If $\lim_{\|P\| \to 0} B_P$ exists and equals 0, then it can be shown (and it isn't hard to convince yourself, see Figure 12-3) that $\lim_{\|P\| \to 0} V_P$ exists. This value is called the **Jordan content of** A; it represents the n-dimensional volume of A.

❑

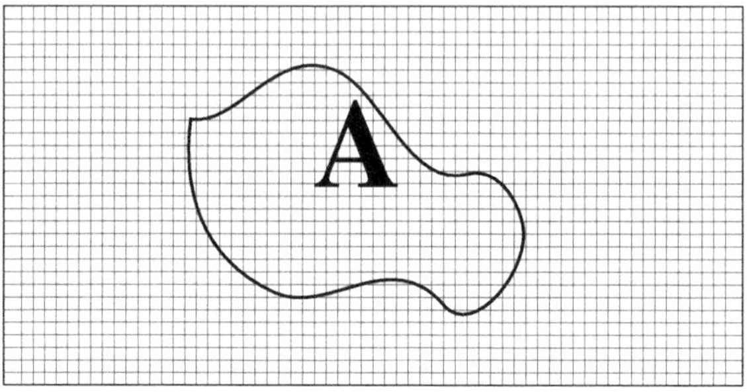

Figure 12-3

As you might anticipate, $\int_A f dV_n$ is well defined for monotone and for continuous functions if A is a closed bounded set with a well-defined Jordan content.

Definition 12.8: $\int_A f dV_n$ for sets A with Jordan content

Let A be a closed bounded subset (see Definition 10.43 on page 406) of n-dimensional vectors with a well defined Jordan content (Definition 12.7 above). Let P represent any partition (Definition 12.2 on page 439) of the n-dimensional interval $[\boldsymbol{a};\boldsymbol{b}]$ where A is a subset of $[\boldsymbol{a};\boldsymbol{b}]$. Let $S_f(X, P)$ represent any Riemann sum for f, which

- includes all elements of P completely in A,
- includes no elements of P whose intersection with A is empty (i.e., having no points in common with A).

We define **the integral of f over the set A, denoted as** $\int_A f dV_n$, **by**

$$\int_A f dV_n = \lim_{\|P\| \to 0} S_f(X, P),$$

provided this limit exists — i.e., provided that there is a value, v such that $S_f(X, P)$ is close to v whenever $\|P\|$ is small enough, independent of the choice of the values $X_{j_1,...,j_n}$ in the interval $I_{j_1,...,j_n}$ of the partition P.

❑

Theorem 12.9: $\int_A f dV_n$ for f monotone, and for f continuous

If f is a real valued function of n real variables (Definition 8.1 on page 261) and A is a closed bounded subset (Definition 10.43 on page 406) of *dmn* f with a well-defined Jordan content (Definition 12.7 on page 442) then

$\int_A f dV_n$ exists if f is monotone (see Definition 12.4 on page 441)

$\int_A f dV_n$ exists if f is continuous (see Definition 8.7 on page 270).

❑

Interestingly enough, while probability often requires computation of n-dimensional integrals, the use of Monte-Carlo simulation, which can be considered a part of the subject of probability, is one of the principal means of obtaining numerical estimates of n-dimensional integrals when n is large. The basic idea is as follows.

Suppose we want to determine $\int_A f dV_n$ where A is a subset of $[\boldsymbol{a};\boldsymbol{b}]$,

$f \geq 0$ and $\int_{[\boldsymbol{a};\boldsymbol{b}]} f dV_n > 0$ is known. The function

$$g = \frac{f}{\int_{[\boldsymbol{a};\boldsymbol{b}]} f dV_n}$$

is a probability density, and it is relatively easy to generate n-dimensional vectors, $\boldsymbol{X}_1,...,\boldsymbol{X}_k$ which for practical purposes behave as if they were statistically independent random vectors described by the density g — i.e., as if the probability that $\boldsymbol{X}_i$ belongs to A is

$$\int_A g dV_n = \frac{\int_A f dV_n}{\int_{[\boldsymbol{a};\boldsymbol{b}]} f dV_n}.$$

Statistically independent vectors $\boldsymbol{X}_i$ have the property that when a large number of them are generated, the proportion, p_n, of these generated vectors which lie in A is usually a good estimate of

$$\int_A g\,dV_n = \frac{\int_A f\,dV_n}{\int_{[\boldsymbol{a};\boldsymbol{b}]} f\,dV_n}.$$

So the proportion p_n multiplied by the (assumed) known value $\int_{[\boldsymbol{a};\boldsymbol{b}]} f\,dV_n$ estimate the desired integral $\int_A f\,dV_n$.

The big advantage of this approach is that the accuracy is largely independent of dimension, the error being roughly on the order of magnitude $1/\sqrt{k}$, where k is the number of random vectors that were generated. While this doesn't usually yield great accuracy (to double the accuracy one needs to increase the number of generated vectors by a factor of 4, to increase the accuracy one thousand-fold requires increasing the number of generated vectors by a factor of one million) it's much better than most other methods for large n. The assumption of knowledge of the value of $\int_{[\boldsymbol{a};\boldsymbol{b}]} f\,dV_n$ is satisfied in many important cases beyond the scope of our treatment.

3 Iterated Integrals

One intuitively reasonable way of evaluating $\int_{[\boldsymbol{a};\boldsymbol{b}]} f\,dV_n$ is suggested by Figure 12-4.

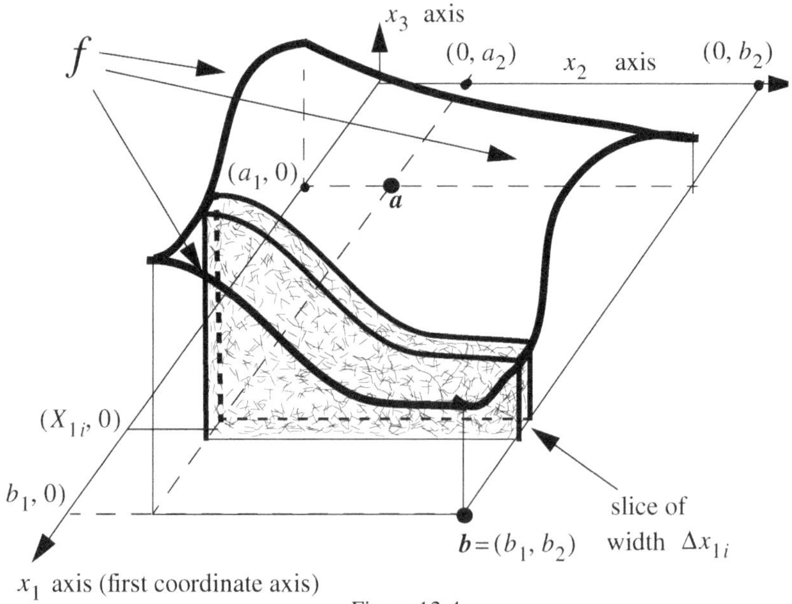

Figure 12-4

The volume of the indicated slice is approximately its face area,

$$\int_{a_2}^{b_2} f(x_{1i}, x_2)\,dx_2$$

times the slice width, Δx_{1i}.

So the total volume we want is approximately

$$\sum_i \left\{ \int_{a_2}^{b_2} f(x_{1i}, x_2)\,dx_2 \right\} \Delta x_{1i},$$

which is an approximating Riemann sum for the iterated integral

$$\int_{a_1}^{b_1} \left\{ \int_{a_2}^{b_2} f(x_{1i}, x_2)\,dx_2 \right\} dx_1.$$

Note that the symbol above indicates integration ***first*** (inner integral) of the function G_{x_1} given by

$$G_{x_1}(x_2) = f(x_1, x_2)$$

from a_2 to b_2, yielding a function H, given by

$$H(x_1) = \int_{a_2}^{b_2} G_{x_1}(x_2)dx_2.$$

Then H is integrated from a_1 to b_1.

In general we have the following result, which may not be the strongest we can get, but is adequate for most practical purposes.

Theorem 12.10: Evaluation of $\int_{[a;b]} f dV_n$ as an iterated integral

Suppose f is a real valued function of n real variables (Definition 8.1 on page 261) for which the multiple integral $\int_{[a;b]} f dV_n$ exists (Definition 12.3 on page 440).

If the *n-fold* ***iterated integral***

$$\int_{a_n}^{b_n} \left[\int_{a_{n-1}}^{b_{n-1}} \cdots\cdots \left[\int_{a_1}^{b_1} f(x_1,\ldots,x_n)dx_1 \right] dx_2 \cdots dx_{n-1} \right] dx_n$$

exists,[1] then

$$\int_{[a;b]} f dV_n = \int_{a_n}^{b_n}\left[\int_{a_{n-1}}^{b_{n-1}} \cdots \left[\int_{a_1}^{b_1} f(x_1,\ldots,x_n)dx_1\right]dx_2 \cdots dx_{n-1}\right]dx_n \,.$$

The order of integration may be chosen to try for the simplest evaluation. For instance, sometimes the order given in

$$\int_{a_1}^{b_1}\left[\int_{a_2}^{b_2} f(x_1, x_2)dx_1\right]dx_2$$

provides much simpler evaluation than that given in

$$\int_{a_2}^{b_2}\left[\int_{a_1}^{b_1} f(x_1, x_2)dx_2\right]dx_1 \,.$$

❑

Example 12.2: $\int_{[(0,0);(1,2)]} xy\, dV_2$

Here the expression xy represents the function f with $f(x, y) = xy$. Using Theorem 12.10, we can evaluate the integral above as an iterated integral, which, making use of the fundamental evaluation formula (see Theorem 4.3 on page 103) leads to

$$\int_{[(0,0);(1,2)]} xy\, dV_2 = \int_0^1\left[\int_0^2 xy dy\right]dx = \int_0^1\left[xy^2/2\Big|_0^{y=2}\right]dx$$

$$= \int_0^1 (2x - 0 \cdot x)dx = x^2\Big|_0^{x=1} = 1.$$

■

Exercises 12.3

Let $[a;b] = [(1, 2, 3), (3, 5, 7)]$. Evaluate $\int_{[a,b]} f dV_3$, where

1. $f(x, y, z) = yz^2 e^{xyz}$,

2. $f(x, y, z) = \dfrac{xz}{1 + xy}$,

3. $f(x, y, z) = xy^2 exp(xyz)$.

Overall hint: The order of integration can make quite a difference in the ease of evaluation.

---

1. Strictly speaking this iterated integral must be defined inductively.

Problem 12.4

Show that even if the iterated integral, $\int_{a_1}^{b_1}\left[\int_{a_2}^{b_2} f(x, y)dy\right]dx$ exists, the double integral $\int_{[a;b]} f dV_2$ may fail to exist.

Hint: Let $f(x, y) = \begin{cases} sin(x) & \text{for } y \text{ rational,} \\ -sin(x) & \text{for } y \text{ irrational.} \end{cases}$

We now want to extend the results on evaluation of multiple integrals by iterated integrals to more general sets. Because of the needs of certain applications (in particular, probability theory and transform theory) it's essential to precede this development by defining several extensions of Riemann integrals. These extensions are so-called *improper integrals*, and they are just limiting cases of ordinary Riemann integrals.

First we consider certain unbounded functions, such as given by

$$f(\tau) = \frac{1}{\sqrt{1-t^2}}, \quad -1 < \tau < 1 .$$

This particular function is already known to us as the derivative of the Arcsin function (see Definition 5.13 on page 160). The integral

$$\int_0^t \frac{1}{\sqrt{1-\tau^2}} d\tau = Arc sin(t)$$

and $Arc sin(t)$ is the (signed) length on the unit circle from the point (1,0) to the point $(t, \sqrt{1-t^2})$, as illustrated in Figure 12-5.

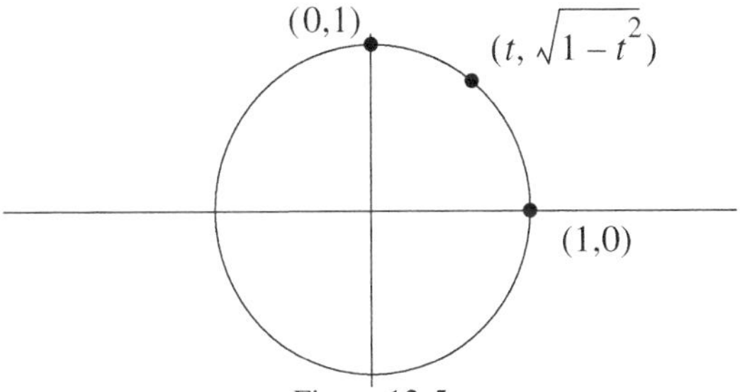

Figure 12-5

From this interpretation we can be fairly certain that

$$\lim_{\substack{t \to 1 \\ t < 1}} \int_0^t \frac{1}{\sqrt{1-\tau^2}} d\tau$$

exists (and equals $\pi/2$) even though $Arcsin'(t) = 1/(\sqrt{1-t^2})$ grows arbitrarily large as $t < 1$ approaches 1. So even though the Riemann integral of $Arcsin'$ from 0 to 1 fails to exist, it's reasonable to extend the integral concept for this function.

Similarly, because $\int_1^t \frac{d\tau}{\tau^2}$ gets close to 1 as t becomes large, it's convenient to define the improper integral $\int_1^\infty \frac{d\tau}{\tau^2}$ as $\lim_{t \to \infty} \int_1^t \frac{d\tau}{\tau^2}$.

The function given by

$$f(\tau) = \begin{cases} 1/\tau^2 & \text{for } \tau > 1, \\ 0 & \text{otherwise} \end{cases}$$

is usable as a probability density, with the probability of an outcome $\leq x$ being $\int_1^x f(\tau)\, d\tau$. Then the probability of obtaining some finite outcome, $P\{X < \infty\}$, equals 1, as it should.

We formalize this discussion as follows.

Definition 12.11: Improper Riemann integrals

Provided $\lim_{t \to \infty} \int_a^t f(\tau) d\tau$ exists (i.e., $\int_a^t f(\tau) d\tau$ gets arbitrarily close to some fixed value v as t becomes large) we let the **improper integral**

$$\int_a^\infty f(\tau) d\tau = \lim_{t \to \infty} \int_a^t f(\tau) d\tau.$$

Similarly for

$$\int_{-\infty}^b f(\tau) d\tau.$$

If both improper integrals

$$\int_0^\infty f(\tau) d\tau \quad \text{and} \quad \int_{-\infty}^0 f(\tau) d\tau$$

exist, we define the **improper integral**

$$\int_{-\infty}^\infty f(\tau) d\tau = \int_0^\infty f(\tau) d\tau + \int_{-\infty}^0 f(\tau) d\tau.$$

We similarly define improper integrals

$$\int_a^b f(\tau)d\tau,$$

when

$$\lim_{\substack{t \to b \\ t < b}} \int_a^t f(\tau)d\tau \quad \text{or} \quad \lim_{\substack{t \to a \\ t > a}} \int_t^b f(\tau)d\tau \ \text{exist.}$$

❑

One very handy application of improper integrals is their use in extending the comparion test for convergence of infinite series, (Theorem 7.13 on page 228). The motivation for this theorem is provided by Figure 12-6.

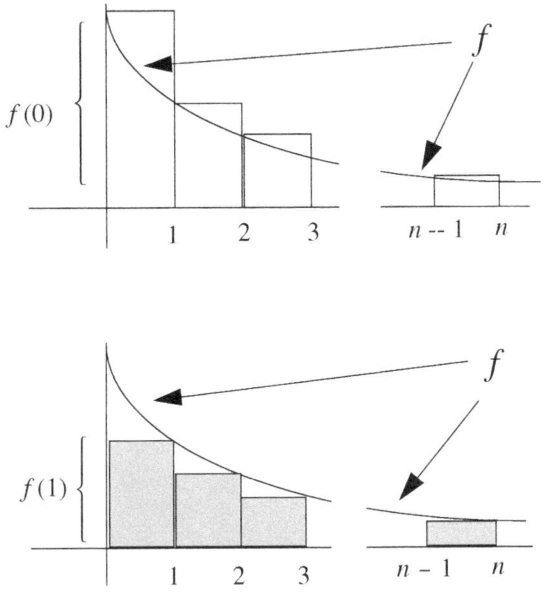

Figure 12-6

The precise statment of the result suggested by Figure 12-6 is as follows.

Theorem 12.12: Integral comparison test

Let f be a nonnegative nonincreasing real valued function of a real variable (see Definition 2.9 on page 37). The infinite series

$$\sum_{j = 0 \to \infty} f(j)$$

converges (Definition 7.4 on page 214) if the improper integral

$$\int_1^{\infty} f(x)dx$$

exists (Definition 12.11 above) while it diverges if this improper integral fails to exist. If the above integral exists, we have the inequality

$$\int_1^{\infty} f(x)dx \leq \sum_{k=n+1}^{\infty} f(k) \leq \int_n^{\infty} f(x)dx.$$

❑

Definition 12.13: Improper multiple integral over $\mathcal{E}_n$

Let f be a real valued function of n real variables (Definition 8.1 on page 261) such that $\int_{[\boldsymbol{a};\boldsymbol{b}]} f dV_n$ (Definition 12.3 on page 440) exists for all n-dimensional intervals $[\boldsymbol{a};\boldsymbol{b}]$. If $\int_{[\boldsymbol{a};\boldsymbol{b}]} |f| dV_n$ approaches a single value as all coordinates of $\boldsymbol{a}$ become algebraically very negative (***very small a***) and those of $\boldsymbol{b}$ become algebraically very positive (***very large b***) which we write as

$$\lim_{\substack{\boldsymbol{a} \to -\infty \\ \boldsymbol{b} \to \infty}} \int_{[\boldsymbol{a};\boldsymbol{b}]} |f| dV_n = v,$$

then $\lim_{\substack{\boldsymbol{a} \to -\infty \\ \boldsymbol{b} \to \infty}} \int_{[\boldsymbol{a};\boldsymbol{b}]} f\, dV_n$ will also exist (i.e., $\int_{[\boldsymbol{a};\boldsymbol{b}]} f dV_n$ will have a limit for small $\boldsymbol{a}$ and large $\boldsymbol{b}$), and we define the **improper integral of f over Euclidean n-dimensional space**, $\mathcal{E}_n$, as

$$\int_{\mathcal{E}_n} f dV_n = \lim_{\substack{\boldsymbol{a} \to -\infty \\ \boldsymbol{b} \to \infty}} \int_{[\boldsymbol{a};\boldsymbol{b}]} f\, dV_n.$$

❑

Remarks

- The proof that $\lim_{\substack{\boldsymbol{a} \to -\infty \\ \boldsymbol{b} \to \infty}} \int_{[\boldsymbol{a};\boldsymbol{b}]} f\, dV_n$ will exist under the the conditions given above is just the natural extension of Theorem 7.26 on page 256.

- Note that the definitions of $\int_{-\infty}^{\infty} f(\tau)d\tau$ and $\int_{\mathcal{E}_1} f dV_n$ aren't the same. This can be seen from the fact that $\int_{-\infty}^{\infty} \frac{\sin(\pi x)}{1+|x|} dx$ exists, but $\int_{\mathcal{E}_1} \frac{\sin(\pi x)}{1+|x|} dx$ is not defined because $\int_0^{\infty} \frac{|sin(\pi x)|}{1+|x|} dx$ is not defined (a fact you might want to verify).

We are now in a position to define the Riemann improper integral $\int_A f dV_n$ for certain sets in $\mathcal{E}_n$ which aren't bounded. These sets arise commonly in probability problems, where the above integral is the probability that a random vector, $\boldsymbol{X}$, whose probability density is f, assumes a value in the set A — e.g., that $X_1 + X_2$ satisfies the condition $X_1 + X_2 < 5$ (i.e., assumes a value in the set of points (x_1, x_2) for which $x_1 + x_2 < 5$, the shaded set indicated in Figure 12-7).

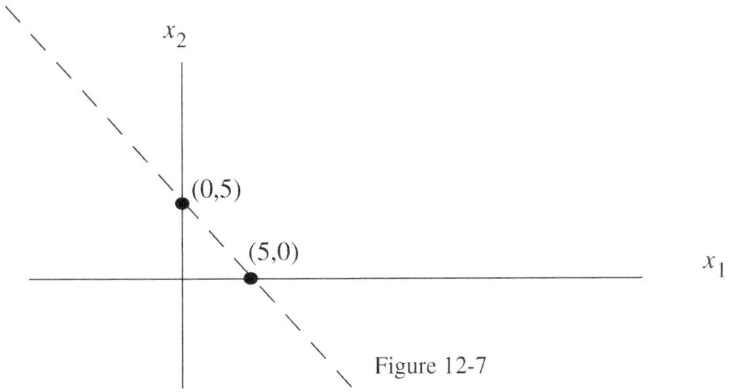

Figure 12-7

For this next improper integral, recall the definition of the intersection, $S \cap T$ of two sets, S and T (Footnote 1 on page 419).

Definition 12.14: Improper integral $\int_A f dV_n$

Suppose f is a real valued function of n real variables (Definition 8.1 on page 261) and A is a subset of n-dimensional space for which

- $[\boldsymbol{a};\boldsymbol{b}] \cap A$ has a well-defined Jordan content for each n-dimensional interval $[\boldsymbol{a};\boldsymbol{b}]$ (Definition 12.1 on page 438, and Definition 12.7 on page 442);
- $\int_{[\boldsymbol{a};\boldsymbol{b}] \cap A} f dV_n$ exists for each $[\boldsymbol{a};\boldsymbol{b}]$

  (Definition 12.8 on page 443)

and

- $\lim_{\substack{\boldsymbol{a} \to -\infty \\ \boldsymbol{b} \to \infty}} \int_{[\boldsymbol{a};\boldsymbol{b}] \cap A} f\, dV_n$, i.e., $\int_{[\boldsymbol{a};\boldsymbol{b}] \cap A} f\, dV_n$ approaches a fixed value for algebraically large b and algebraically small a.

Then the **improper multiple integral of f over the set A,**

$$\int_A f dV_n \text{ is defined as } \lim_{\substack{a \to -\infty \\ b \to \infty}} \int_{[a;b] \cap A} f\, dV_n .$$

Once again, proof of the existence of the last limit is an extension of Theorem 7.26 on page 256.

❑

A few of the expected properties of $\int_A f dV_n$ are now summarized.

Theorem 12.15: Properties of n-dimensional integrals

Let A, B be nonoverlapping sets in $\mathcal{E}_n$ for which $\int_A f dV_n$, $\int_A g dV_n$, $\int_B f dV_n$, and $\int_B g dV_n$ exist (proper or improper — see Definition 12.11 on page 449, Definition 12.13 on page 451, Definition 12.14 on page 452 as well as Definition 12.8 on page 443). Then

$$\int_A k f dV_n = k \int_A f dV_n \qquad \textit{12.1}$$

$$\int_A (f+g) dV_n = \int_A f dV_n + \int_A g dV_n . \qquad \textit{12.2}$$

If $f(\boldsymbol{x}) \le g(\boldsymbol{x})$ for all $\boldsymbol{x}$ in A, then

$$\int_A f dV_n \le \int_A g dV_n . \qquad \textit{12.3}$$

In particular

$$\left| \int_A f dV_n \right| \le \int_A |f| dV_n . \qquad \textit{12.4}$$

If $\mathcal{U}$ denotes the union[1] of the (nonoverlapping) sets A and B — i.e., the set of points in at least one of the sets A, B, then

---

1. The **union** of sets A,B is denoted as $A \cup B$. The union of an indexed collection of sets, A_t, for t belonging to a specified set T, is the set of points in at least one of the A_t, and is denoted by $\bigcup_{t \text{ in } T} A_t$.

$$\int_{\mathcal{U}} f\,dV_n = \int_{A \cup B} f\,dV_n = \int_A f\,dV_n + \int_B f\,dV_n. \qquad \textbf{12.5}$$

If $m \leq f(\boldsymbol{x}) \leq M$ for all $\boldsymbol{x}$ in $[\boldsymbol{a};\boldsymbol{b}]$, then

$$mV[\boldsymbol{a};\boldsymbol{b}] \leq \int_{[\boldsymbol{a};\boldsymbol{b}]} f\,dV_n \leq MV([\boldsymbol{a};\boldsymbol{b}]), \qquad \textbf{12.6}$$

where $V([\boldsymbol{a};\boldsymbol{b}])$ is the n-dimension volume of $[\boldsymbol{a}(;\boldsymbol{b})]$ (see Definition 12.1 on page 438).

4 General Multiple Integral Evaluation

As you might expect looking at Figure 12-1 on page 437 and Figure 12-4 on page 445, general multiple integrals, $\int_A f\,dV_n$, can often be evaluated as iterated integrals, as we did for rectangles in Theorem 12.10 on page 446, using the fundamental evaluation formula for functions of one variable (see Theorem 4.3 on page 103) as we did in Example 12.2 on page 447. To see how this is done, we need to introduce the following definition.

Definition 12.16: Sequential and general set notation

If A *is a subset of* n-dimensional Euclidean space, $\mathcal{E}_n$ (Definition 9.14 on page 302) we say that **A is given in sequential form** if

$$A = \{(x_1,...,x_n) \text{ in } \mathcal{E}_n{:}L_1 < x_1 < U_1;L_2(x_1) < x_2 < U_2(x_1);...$$
$$...;L_n(x_1,...,x_{n-1}) < x_n < U_n(x_1,...,x_{n-1})\}.$$

(Note the semicolon separators, which distinguish this notation.)

By this is meant the inductive definition that A consists of all points $(x_1,...,x_n)$ where

- the allowable first coordinates of these points are those x_1 which satisfy $L_1 < x_1 < U_1$
- For each allowable x_1, the allowable second coordinates, x_2, are those which satisfy $L_2(x_1) < x_2 < U_2(x_1)$

Here L_2 and U_2 are real valued functions of one real variable, and inductively, for each allowable sequence of values $x_1,...,x_{k-1}$ for the first $k-1$ coordinates, the allowable k-th coordinates, x_k, are those satisfying

$$L_k(x_1,...,x_{k-1}) < x_k < U_k(x_1,...,x_{k-1}),$$

where L_k and U_k are real valued function of $k-1 > 1$ real variables.

For the purpose of allowing various types of improper integrals, if L_1, U_1 or any of the L_k, U_k values can be arbitrarily large, they are associated with $< \infty$, and similarly for $> -\infty$.

❑

Now in most of the cases that we will meet, the n-dimensional set of interest is not originally given in sequential form, although, as we shall see, this form is necessary for setting the limits of integration of iterated integrals. In general a set A, of points in $\mathcal{E}_n$ is represented in the form

$$A = \{x \text{ in } \mathcal{E}_n : S(x)\} \qquad \textbf{\textit{12.7}}$$

where $S(x)$ is some set of properties (such as $\|x\| \le 4$, in which case the set A above represents the set of n-dimensional vectors whose length is less than or equal to 4). If $S(x)$ is of the form $S_1(x), S_2(x)$, the comma just stands for the word *and.*

The main result connecting multiple and iterated integrals is given below in Theorem 12.17. It is more restricted than it need be, but is adequate for most standard situations.

Theorem 12.17: Iterated evaluation of $\int_A f dV_n$

Suppose that f is a real valued function of n real variables with domain (Definition 1.1 on page 2) the closed bounded interval $[\boldsymbol{a};\boldsymbol{b}]$ (Definition 12.1 on page 438) and suppose A is a subset of $[\boldsymbol{a};\boldsymbol{b}]$ for which $\int_A f dV_n$ exists (see Definition 12.8 on page 443).

If A can be written in the sequential form (Definition 12.16 on page 454)

$$A = \{(x_1,...,x_n) \text{ in } \mathcal{E}_n : L_1 < x_1 < U_1; L_2(x_1) < x_2 < U_2(x_1);...$$
$$...; L_n(x_1,...,x_{n-1}) < x_n < U_n(x_1,...,x_{n-1})\}.$$

where the functions L_k and U_k $k = 2,...,n$ are all continuous (see Definition 8.7 on page 270) and the (inductively defined) iterated integral

$$\int_{L_1}^{U_1}\left[\int_{L_2(x_1)}^{U_2(x_1)} \cdots\cdots \left[\int_{L_n(x_1,...,x_{n-1})}^{U_n(x_1,...,x_{n-1})} f(x_1,...,x_n)dx_n\right]dx_{n-1}\cdots dx_1\right]dx_n$$

exists, then it equals $\int_A f dV_n$

This result extends to improper integrals over unbounded sets, A, when

$\int_{\mathcal{E}_n} f dV_n$, and $\int_{-\infty}^{\infty}\left[\int_{-\infty}^{\infty} \cdots \left[\int_{-\infty}^{\infty} f(x_1,...,x_n)dx_n\right]dx_{n-1}\cdots dx_1\right]dx_n$ exist with the latter equaling $\left|\int_{\mathcal{E}_n} |f|\,dV_n\right|$.

Note that any order of integration different from the one specified above is allowable. It often pays to search for the order leading to the simplest result.

❑

The real art of evaluating $\int_A f\,dV_n$ ***as an iterated integral consists mainly of rewriting the set A in sequential form*** (or as a nonoverlapping union of sets in sequential form) which is the aspect we will stress. The method used here illustrates one of the rare situations where an algebraic approach to the problem is preferable to a geometric one — mainly because a graphical approach doesn't seem to add much understanding, and does not extend to integrals over more than two-dimensional sets.

In general we try to

- Determine the values of x_1 for which there are points $(x_1,...,x_n)$ in A, provided this is an interval, $(L_1;U_1)$, (representing those x_1 satisfying $L_1 < x_1 < U_1$, for each x_1 in $(L_1;U_1)$;
- determine those values of x_2 for which there are points $(x_1, x_2,...,x_n)$ in A; provided this always results in an interval, omitting its endpoints, denote it by $(L_2(x_1);U_2(x_1))$.

Continue this process, if possible,

- at the k-th stage for any specific values of $x_1,...,x_{k-1}$ previously determined, find the set of x_k for which there are points $(x_1,...,x_k, ...,x_n)$ in A. Hopefully, this results in an interval, which we denote by $(L_k(x_1,...,x_{k-1});U_k(x_1,...,x_{k-1}))$.

The process is complete after $k = n$.

Note the use of the $<$ in the above inequalities. While many sets A would require the $\leq$ in writing them in sequential form, exclusion of the $=$ in the $\leq$ would not affect the value of the integral — e.g., the set of (x,y) satisfying $xy < 1$ is not the same as the set of (x,y) satisfying $xy \leq 1$. But this doesn't matter for integration purposes, since

$$\int_{\{(x,y) \text{ in } \mathcal{E}_2 : xy = 1\}} f\,dV_2 = 0.$$

We will therefore pay no attention to these minor obstacles. To take this out of the overly abstract stage we go through a few examples.

Example 12.5: $A = \{(x, y) \text{ in } \mathcal{E}_2 : x^2 + 3y^2 < 1\}$

Here A represents the inside of an ellipse. If (x, y) is in A, you can easily see that we must have $x^2 < 1$, i.e., $-1 < x < 1$. Furthermore, for any such value of x. the point $(x, 0)$ is in A. Thus we have[1] $(L_x;U_x) = (-1;1)$. Now for each x in $(-1;1)$, in order for (x,y) to be in A, we must have

$$3y^2 < 1 - x^2,$$

or

$$-\sqrt{\frac{1-x^2}{3}} < y < \sqrt{\frac{1-x^2}{3}}.$$

Hence

$$(L_y(x);U_y(x)) = \left(-\sqrt{\frac{1-x^2}{3}};\sqrt{\frac{1-x^2}{3}}\right).$$

■

This example could have been approached from a graphical point of view quite successfully. The following example clearly shows the superiority of the algebraic approach adopted here.

Example 12.6: $A = \{((w, x, y, z) \text{ in } \mathcal{E}_4 : w^2 - x + y^2 - 2z^2 < 4)\}$

Here we note that the second coordinate, x, can compensate for any values of the other coordinates — i.e., $-\infty < w < \infty$ is allowed, because for any w there are always values of x, y, z satisfying $w^2 - x + y^2 - 2z^2 < 4$ (just choose $y = z = 0$ and $x = w^2$). We have to be careful, because it's not the case that for any w, x there are always values of y,z for which $w^2 - x + y^2 - 2z^2 < 4$; just look at $w, x = w^2$, $y = 10$, $z = 0$, which cannot satisfy $w^2 - x + y^2 - 2z^2 < 4$.

But for each w in $(-\infty;\infty)$ and each y in $(-\infty;\infty)$ there are points (w,x,y,z) in A — choose $z = 0$ and $x = w^2 + y^2$. Thus we may write

$$A = \{(w, x, y, z) \text{ in } \mathcal{E}_4 : -\infty < w < \infty, -\infty < y < \infty, -\infty < z < \infty,\\ w^2 + y^2 + 2z^2 - 4 < x\}.$$

■

Sometimes we cannot write A in sequential form, but can write it as a union of nonoverlapping sets, each of which can be written in sequential form. Then from Theorem 12.15, equation 12.5 on page 454 we can apply the results of

1. Note the minor change of notation, because we are not using subscripts for the coordinates.

of Theorem 12.17 to determine the multiple integral of f over A as a sum of iterated integrals. To illustrate this we give the next example.

Example 12.7: $\int_A f dV_2$ **where** $A = \{(x, y) \text{ in } \mathcal{E}_2 : xy < 1\}$

We note that

- for $x > 0$ the inequality $xy < 1$ is equivalent to $y < 1/x$
- if $x = 0$ $xy < 1$ is equivalent to $-\infty < y < \infty$
- for $x < 0$ the inequality $xy < 1$ is equivalent to $y > 1/x$

Hence, as illustrated in Figure 12-8, A is the union of the nonoverlapping sets A_-, A_0, A_+, where

$$A_+ = \{(x, y) \text{ in } \mathcal{E}_2 : 0 < x < \infty ; -\infty < y < 1/x\}$$

$$A_0 = \{(x, y) \text{ in } \mathcal{E}_2 : x = 0 ; -\infty < y < \infty\}$$

$$A_- = \{(x, y) \text{ in } \mathcal{E}_2 : -\infty < x < 0 ; \; 1/x < y < \infty\}.$$

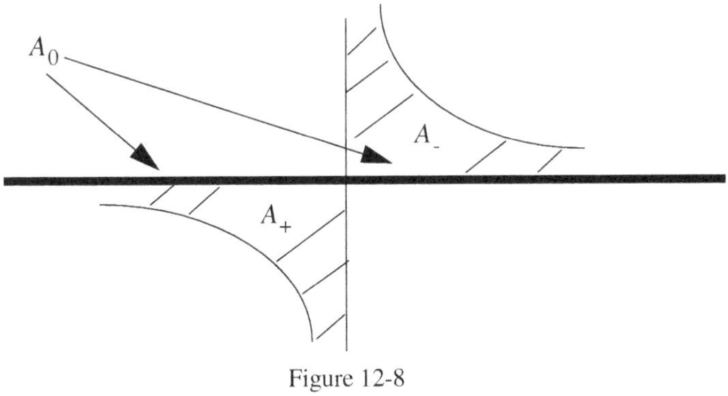

Figure 12-8

Thus, provided the relevant integrals exist, we have

$$\int_A f dV_2 = \int_{A_+} f dV_2 + \int_{A_0} f dV_2 + \int_{A_-} f dV_2 = \int_{A_+} f dV_2 + \int_{A_-} f dV_2$$

$$= \int_0^{\infty}\left[\int_{-\infty}^{1/x} f(x, y) dy\right] dx + \int_{-\infty}^{0}\left[\int_{1/x}^{\infty} f(x, y) dy\right] dx.$$

Examining Figure 12-8, you can see how these limits of integration could be determined geometrically. However, as you'll see in working some of the exercises to follow, the geometric approach won't work well in higher dimensions.

■

Exercises 12.8

Write the following sets, A, in sequential form, or as a union of a finite number of such sets, and indicate how to evaluate $\int_A f dV_n$ (for the appropriate n) as a single iterated integral, or as a sum of iterated integrals.

1. $\{(x, y, z) \text{ in } \mathcal{E}_3 : xyz < 2\}$

2. $\{(x, y) \text{ in } \mathcal{E}_2 : x^2 - y^2 < 5\}$

   a. with limits for y set first (the easier order)

   b. with limits for x set first.

3. $\{(x, y, z) \text{ in } \mathcal{E}_3 : x^2 - y^2 + z^2 < 5\}$
   Hint: See previous question.

4. $\{(x, y) \text{ in } \mathcal{E}_2 : x^2 < 1, y^2 < 1, x + y < 1\}$
   Note that this is not in sequential notation.
   Hint: $-1 < x < 1$ is allowable, but when $-1 < x < 0$ the restriction $x + y < 1$ is always satisfied for $y^2 < 1$. However, for $x > 0$ we need $y < 1 - x$.

5. $\{(x, y, z, u) \text{ in } \mathcal{E}_4 : xyu + z^3 < 13\}$

6. $\left\{(x, y, z) \text{ in } \mathcal{E}_3 : \frac{x^2}{4} + \frac{y^2}{3} + \frac{z^2}{7} < 13\right\}$

   Use this result to determine an iterated integral for the volume of this ellipsoid.

5 Multiple Integral Change of Variables

Sometimes for ease of evaluation of a multiple integral, and often to simplify theoretical development, it pays to change coordinates — i.e., use an equivalent but more suitable description of the integral under scrutiny. The coordinate changes we will be considering will always be provided by some function g whose domain and range both consist of n-dimensional vectors. If a point P in n dimensions originally has coordinates $x_1,...,x_n$, the coordinates under the coordinate change g are given by

$$(w_1,...,w_n) = g(x_1,...,x_n).$$

So g is an n-dimensional vector valued function of n real variables. If two points are different in the original coordinate system, generally, you wouldn't want them to have identical coordinates in the new coordinate system.[1] That is, for

$$w = g(x) \text{ and } z = g(y)$$

$$\text{if } w = z \quad \text{you generally want} \quad x = y.$$

This leads us to the following.

Definition 12.18: One-one function

A function g with domain A and range B (see Definition 1.1 on page 2) is called **one-one** ("one to one") if

$$\text{whenever } g(x) = g(y) \text{ it follows that } x = y.$$

We also refer to such a function as a **one-to-one mapping**.

❑

Next we introduce the rather restricted definition of coordinate change that will be adequate for our purposes in this chapter.

Definition 12.19: Change of coordinates in n-dimensions

Let A be a subset of n-dimensional space (Definition 9.14 on page 302). A function g (Definition 1.1 on page 2) whose domain is A and whose range is a subset of n-dimensional space will be called a **change of coordinates for A** if g is one-one (see Definition 12.18 above).

---

1. There are some important exceptions, e.g., polar coordinates $(0, \vartheta)$ for all possible ϑ correspond to only the x, y coordinates $(0,0)$. Cases like this can usually be handled with little trouble.

If the point P in A has original coordinates given by

$$\boldsymbol{x} = (x_1,...,x_n),$$

then

$$\boldsymbol{w} = (w_1,...,w_n) = \boldsymbol{g}(\boldsymbol{x}) = (g_{(1)}(\boldsymbol{x}),...,g_{(n)}(\boldsymbol{x}))$$

is the **vector of new coordinates of P under the coordinate change (specified by)** $\boldsymbol{g}$.

❑

We have already dealt with orthogonal matrices which represent coordinate changes — describing points using a new set of basis vectors (see Definition 9.9 on page 298, Definition 9.11 on page 299, Theorem 9.17 on page 304 and Chapter 10 Section 4 on page 347). In many cases the change of coordinates can be specified by an invertible matrix $\mathbf{G}$; here

$$\boldsymbol{w} = \mathbf{G}\boldsymbol{x} = \boldsymbol{g}(\boldsymbol{x}).$$

However, many such coordinate changes, such as those to polar and cylindrical coordinates, cannot be so described.

Now suppose $\boldsymbol{g}$ specifies a change of coordinates for the n-dimensional set A, and we wish to express $\int_A f dV_n$ in terms of the new coordinates for A. This integral is well approximated by Riemann sums, each of whose terms is of the form $f(\mathbf{X})V(R)$, so long as the partition norm, $\|P\|$, is sufficiently small; here R represents a typical n-dimensional interval generated by P, and $V(R)$ is the n-dimensional volume of R (see Definition 12.1 on page 438). We illustrate this situation in Figure 12-9.

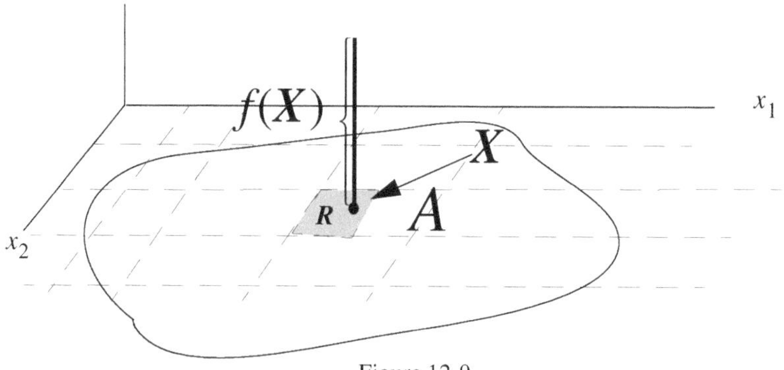

Figure 12-9

We need to examine the typical Riemann sum term, $f(\mathbf{X})V(R)$, in the new coordinate system given by the coordinate change function $\boldsymbol{g}$. Since $\boldsymbol{g}$ is one-one (being a coordinate change) it must have an inverse function, $\boldsymbol{g}^{-1}$ (see Appendix 2 on page 761). To keep the notation simple we let

$$\boldsymbol{H} = \boldsymbol{g}^{-1} \qquad \textbf{\textit{12.8}}$$

and note that

$$\text{if } \boldsymbol{w} = \boldsymbol{g}(\boldsymbol{x}) \quad \text{then} \quad \boldsymbol{x} = \boldsymbol{H}(\boldsymbol{w}). \qquad \textbf{\textit{12.9}}$$

Then

$$f(\mathbf{X})V(R) = f(\boldsymbol{H}(\mathbf{W}))V(R),$$

where

$$\boldsymbol{W} = \boldsymbol{g}(\boldsymbol{X}).$$

Since the desired integral, $\int_A f\,dV_n$ is to be expressed in the new coordinate system, the set over which we integrate must consist of those point $\boldsymbol{w} = \boldsymbol{g}(\boldsymbol{x})$ as $\boldsymbol{x}$ traverses A. We need a notation for this $\boldsymbol{w}$-set, and introduce it in the more general definition which follows.

Definition 12.20: Image of a set under function / matrix

If $\boldsymbol{G}$ is any function (Definition 1.1 on page 2) and A is any set of points from the domain of $\boldsymbol{G}$, the **image of A under $\boldsymbol{G}$** consists of all of the points $\boldsymbol{G}(\boldsymbol{x})$ arising from points $\boldsymbol{x}$ in A. The **image of A under $\boldsymbol{G}$ *is denoted by*** $\boldsymbol{G}(A)$.

Using the type of general set notation introduced in Definition 12.16 on page 454, we may write

$$\boldsymbol{G}(A) = \{\boldsymbol{y} : \boldsymbol{y} = \boldsymbol{G}(\boldsymbol{x}) \quad \text{for some } \boldsymbol{x} \text{ in } A\}.$$

If $\mathbf{M}$ is an $m \times n$ matrix (Definition 10.1 on page 328) and A is a set of n-dimensional vectors, the image $\mathbf{M}A$ of A under $\mathbf{M}$ is defined as

$$\mathbf{M}A = \{\boldsymbol{y} = \mathbf{M}\boldsymbol{x} \quad \text{for some } \boldsymbol{x} \text{ in } A\}.$$

(Here, we're just using the fact that an $m \times n$ matrix **$\mathbf{M}$ defines a function whose domain consists of** n-dimensional vectors, and whose range consists of m-dimensional vectors.)

❑

So, to express $\int_A f dV_n$ in the new w coordinate system, the w set over which we integrate is $\boldsymbol{g}(A)$. Since $f(\boldsymbol{X})V(R) = f(\boldsymbol{H}(\boldsymbol{W}))V(R)$, you might be tempted to believe that $\int_A f dV_n$ is equal to $\int_{\boldsymbol{g}(A)} f(\boldsymbol{H}) dV_n$, but this isn't usually the case, because the latter integral is approximated by $\sum f(\boldsymbol{H}(\boldsymbol{W}_i))V(R_i)$; the ***heights*** $f(\boldsymbol{H}(\boldsymbol{W}_i)) = f(\boldsymbol{X}_i)$ are the ***right ones***, but the coordinate change, even in very simple cases (such as shown in Figure 12-10) has changed the size of the *base* n-dimensional volumes. If

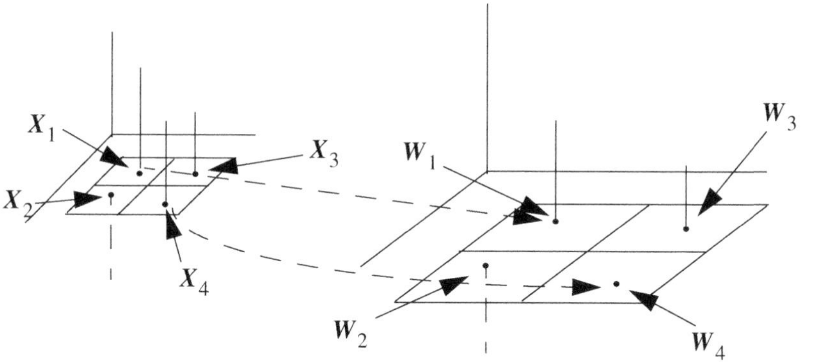

Figure 12-10

you want $\int_A f dV_n$ to equal some integral over $g(A)$ involving the composition $f(\boldsymbol{H})$, you must correct for this distortion. This correction factor will be denoted by $|J|$ (the magnitude of the so-called *Jacobian* determinant, J) so that we may write

$$\int_A f dV_n = \int_{g(A)} f(\boldsymbol{H})|J| dV_n \qquad \textbf{\textit{12.10}}$$

or, in more conventional notation suggesting how the integral is computed

$$\int_A f(\boldsymbol{x}) d\boldsymbol{x} = \int_{g(A)} f(\boldsymbol{H}(\boldsymbol{w}))|J(\boldsymbol{w})| d\boldsymbol{w}, \qquad \textbf{\textit{12.11}}$$

where for small n-dimensional intervals (Definition 12.1 on page 438) r, including the point $\boldsymbol{w}$ from $g(A)$, $J(\boldsymbol{w})$ is approximately the ratio $V(\boldsymbol{H}(r))/V(r)$ of n-dimensional volumes ($\boldsymbol{H}(r)$ being the image of r under $\boldsymbol{H} = \boldsymbol{g}^{-1}$).

Of course, in order for all of this to hold together, the coordinate change $\boldsymbol{g}$ and its inverse $\boldsymbol{H}$ must be reasonably well behaved; specifically, we shall assume that the elements $g_{(i)}$ of $\boldsymbol{g}$ and $H_{(j)}$ of $\boldsymbol{H}$ have continuous gradients (see Definition 8.10 on page 272 and Definition 8.7 on page 270). This will ensure that locally $\boldsymbol{g}$ and $\boldsymbol{H}$ behave like constant matrix transformations (just as differentiable functions of a single variable behave like straight line functions).

If r is an n-dimensional interval anchored at $\boldsymbol{w}$, we may write it as

$$r = \{\boldsymbol{w}^* \text{ in } \mathcal{E}_n : \boldsymbol{w}^* = \boldsymbol{w} + d\boldsymbol{w}^*, 0 \le d\boldsymbol{w}^* \le d\boldsymbol{w}\},$$

where $0 \le d\boldsymbol{w}^* \le d\boldsymbol{w}$ means that for all i, $0 \le dw_i^* \le dw_i$, see Figure 12-11.

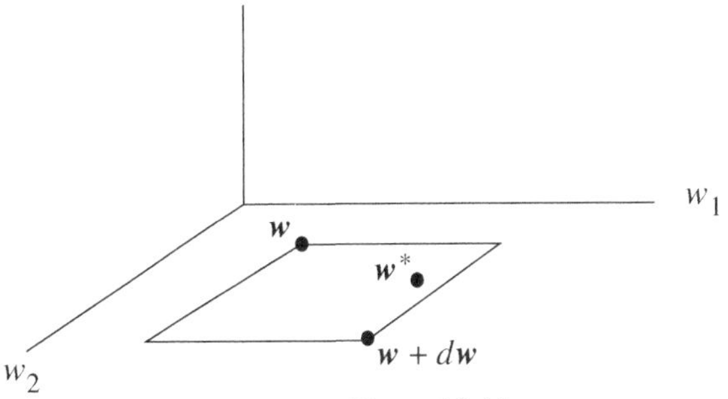

Figure 12-11

The image $R = \boldsymbol{H}(r)$ is given by

$$R = \{\boldsymbol{x}^* \text{ in } \mathcal{E}_n : \boldsymbol{x}^* = \boldsymbol{H}(\boldsymbol{w} + d\boldsymbol{w}^*),\ 0 \le d\boldsymbol{w}^* \le d\boldsymbol{w}\}.$$

From the fundamental approximation (equation 8.17 on page 274) for **small** $\|d\boldsymbol{w}\|$, R can be well approximated by the set of n-dimensional vectors, $\boldsymbol{x}^*$, whose coordinates satisfy

$$\begin{aligned} x_1^* &= H_{(1)}(\boldsymbol{w}) + d\boldsymbol{w}^* \cdot \textbf{\textit{grad}}\ H_{(1)}(\boldsymbol{w}) \\ &\ \ \vdots \\ x_n^* &= H_{(n)}(\boldsymbol{w}) + d\boldsymbol{w}^* \cdot \textbf{\textit{grad}}\ H_{(n)}(\boldsymbol{w}) \end{aligned} \qquad \textbf{\textit{12.12}}$$

for $\qquad 0 \le d\boldsymbol{w}^* \le d\boldsymbol{w}$,

i.e., using matrix notation

$$\boldsymbol{H}(r) = R \cong \left\{\boldsymbol{x}^* \text{ in } \mathcal{E}_n : \boldsymbol{x}^* = \boldsymbol{H}(\boldsymbol{w}) + \begin{bmatrix} \textbf{\textit{grad}}\ H_{(1)}(\boldsymbol{w}) \\ \cdot \\ \cdot \\ \textbf{\textit{grad}}\ H_{(n)}(\boldsymbol{w}) \end{bmatrix} d\boldsymbol{w}^*, 0 \le d\boldsymbol{w}^* \le d\boldsymbol{w}\right\},$$

where we choose to write $\boldsymbol{x}^*$ as a column matrix, forcing $\boldsymbol{H}(\boldsymbol{w})$ and $d\boldsymbol{w}^*$ to be column matrices, and $\textbf{\textit{grad}}\ H_{(i)}(\boldsymbol{w})$ to be row vectors (or row matrices). Letting $\mathbf{D}$ be the column matrix representing the gradient operator in its most abbreviated matrix form, equations 12.12 become

$$\mathbf{H}(r) = R \cong \{\boldsymbol{x}^* \text{ in } \mathcal{E}_n : \boldsymbol{x}^*=\mathbf{H}(\boldsymbol{w})+((\mathbf{DH}^{\mathrm{T}})(\boldsymbol{w}))^{\mathrm{T}}d\boldsymbol{w}^*,\ 0\le d\boldsymbol{w}^*\le d\boldsymbol{w}\}, \quad \textbf{\textit{12.13}}$$

see Section 9 of Chapter 10 for a similar derivation with more details.

Using the column form of $\mathbf{DH}(\boldsymbol{w})d\boldsymbol{w}^*$, rather than the original dot product form), we rewrite $\mathbf{H}(r)$ as

$$\mathbf{H}(r) \cong \{\boldsymbol{x}^* \text{ in } \mathcal{E}_n : \boldsymbol{x}^* = \mathbf{H}(\boldsymbol{w}) + \sum_{i=1}^{n} dw_i^* \begin{pmatrix} D_i H_{(1)}(\boldsymbol{w}) \\ \cdot \\ \cdot \\ D_i H_{(n)}(\boldsymbol{w}) \end{pmatrix}, 0 \le dw_i^* \le dw_i\},$$

in order to see that approximately $\mathbf{H}(r)$ is approximately a parallelepiped anchored at $\boldsymbol{x} = \mathbf{H}(\boldsymbol{w})$, whose edges are multiples (dw_i^*) of the vectors $D_i\mathbf{H}(\boldsymbol{w})$, as illustrated in Figure 12-12.

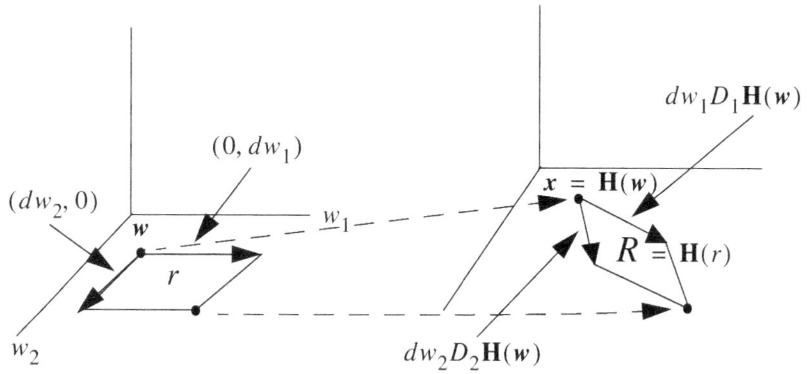

Figure 12-12

It shouldn't be difficult to convince yourself that for fixed w, $V(R)/V(r) = V(\mathbf{H}(r))/V(r)$ is closely determined by the matrix $\mathbf{DH}(\boldsymbol{w})$, unaffected by the values $\mathbf{H}(\boldsymbol{w})$, whatever the *positive* values of the coordinates of $d\boldsymbol{w}$ are. For this reasons we may as well let all of the coordinates of $d\boldsymbol{w}$ be specified equal to 1, and treat $\mathbf{H}(\boldsymbol{w})$ as if it were the **0** ***vector*** (or **0** column matrix). Then, because $V(r) = dw_1 dw_2 \cdots dw_n = 1$, and r is just the unit n-dimensional cube, r_u, given by

$$r_u = \{\boldsymbol{w}^* \text{ in } \mathcal{E}_n : 0 \le w_i \le 1 \text{ all } i = 1,\ldots,n\},$$

determining a good approximation to $V(\mathbf{H}(r))/V(r)$ for small n-dimensional intervals, r, reduces to ***determining the n-dimensional volume of the image***

of the unit cube, r_u, ***under the mapping specified by the matrix*** $\mathbf{DH}(w)$. You may think of this image as the parallelepiped *formed by* the columns of $\mathbf{DH}(w)$, as illustrated in Figure 12-13 for the two-dimensional case.

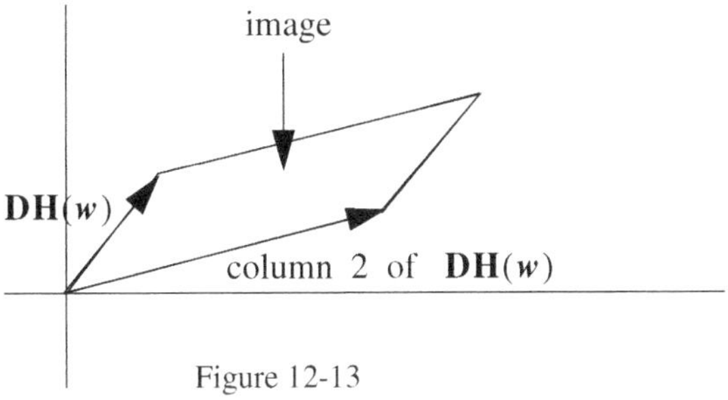

Figure 12-13

That is, we want the n-dimensional volume of the set

$$\left\{ x^* \text{ in } \mathcal{E}_n \; : \; x^* = \sum_{i=1}^{n} dw_i^* \mathbf{C}_i, \quad 0 \le dw_i^* \le 1 \right\},$$

where $\mathbf{C}_i = D_i\mathbf{H}(w)$ are, for this problem, fixed vectors (the columns of $\mathbf{DH}(w)$. That is, if we let $\mathbf{C}$ be the matrix whose i-th column is $\mathbf{C}_i$, $i = 1,...,n$, we want the volume of the columns of the set

$$\{x^* \text{ in } \mathcal{E}_n \; : \; x^* = \mathbf{C}dw^*, \; dw^* \text{ any vector in the unit cube}\}.$$

This is just the volume of the image of the unit cube of the function specified by the $n \times n$ matrix $\mathbf{C} = \mathbf{DH}(w)$ (or, as we said earlier, the volume of the parallelepiped formed by the columns of $\mathbf{C}$).

You might believe this to be simply a matter of setting limits of integration. Except for the word *simply*, this is true. However, attacked frontally, this is an intractable problem. Some preliminary simplification is necessary, and we leave these details to Appendix 7 on page 799. We summarize the results derived in this appendix by first formally introducing the definition of the determinant of an $n \times n$ matrix.

Definition 12.21: Determinant of an $n \times n$ real matrix

Let $\mathbf{B}$ be an $n \times n$ real matrix (Definition 10.1 on page 328) with columns $\mathbf{B}_i$. The **magnitude** of the **determinant of B** is the n-fold integral of the constant function whose value is 1, over the set A given by

$$A = \left\{ x^* \text{ in } \mathcal{E}_n \; : \; x^* = \sum_{i=1}^{n} dw_i^* \mathbf{B}_i, \quad 0 \le dw_i^* \le 1 \right\}.$$

(see Definition 10.4 on page 332). The determinant itself represents the **signed volume** of the parallelepiped formed by the columns of **B.**

The determinant of the matrix **B** is written *det* **B**, where the actual matrix is often written out following the word *det*. Sometimes we write *det*(**B**).

❑

The main result in Appendix 7, given in Theorem A7.2 on page 804, is restated in the following theorem.

Theorem 12.22: Evaluation of determinant

Let **B** be a real $n \times n$ matrix (Definition 10.1 on page 328). Via a sequence of elementary row operations (interchange of rows, or adding a multiple of one row to another) transform the matrix **B** to a diagonal matrix,[1] $\mathcal{D}$ (one whose elements $d_{ij} = 0$ for $i \neq j$). If the number of row interchanges is an *even* number, then the determinant of **B** is given by

$$det\,\mathbf{B} = \prod_{i=1}^{n} d_{ii} = d_{11} d_{22} \cdots d_{nn}.$$

If the number of row interchanges is an *odd* number, then

$$det\,\mathbf{B} = -\prod_{i=1}^{n} d_{ii}.$$

❑

We now have what is needed to state the main result on change of variable in multiple integration.

Theorem 12.23: Change of variable in multiple integration

Let f be a real valued function of n real variables, for which the proper or improper Riemann integral $\int_A f dV_n$ exists (see Definition 12.8 on page 443 and Definition 12.14 on page 452). Let $\boldsymbol{g}$ be a coordinate change (Definition 12.19 on page 460) whose elements have continuous gradients (see Definition 8.10 on page 272 and Theorem 8.8 on page 271) and let $\boldsymbol{H}$ be the inverse coordinate change (see the discussion leading to equations 12.8 and 12.9 beginning page 462). Then

$$\int_A f dV_n = \int_{\boldsymbol{g}(\boldsymbol{H})} f(\boldsymbol{H})|J| dV_n,$$

1. Alternatively, one can transform to either an upper [or a lower] triangular matrix (one whose elements $d_{ij} = 0$ above [below] the diagonal — i.e.,

$$d_{ij} = 0 \text{ for } i > j \ [i < j]$$

without affecting the results which follow.

where $\mathbf{g}(A)$ is the image of A under $\mathbf{g}$ (Definition 12.20 on page 462) and the **Jacobian determinant**, J, is given by

$$J = det(\mathbf{DH}^{\mathrm{T}})$$

(see Definition 12.21 on page 466 and Theorem 12.22 on page 467) and $\mathbf{D}$ is the column matrix representing the gradient operator, see Definition 8.10 on page 272 and Example 12.9 below).

❑

Here is an elementary illustration.

Example 12.9: $\int_A exp\left(-\frac{x^2+y^2}{2}\right)dV_2$, $A = \{(x,y)$ **in** $E_2 : 0 < x^2 + y^2 < 1\}$

Here what we want to evaluate is $\int_A f dV_2$ for

$$f(x, y) = exp\left(-\frac{x^2 + y^2}{2}\right).$$

This is a natural for change to **polar coordinates** where

$$\begin{pmatrix} x \\ y \end{pmatrix} = \mathbf{H}(r, \vartheta) = \begin{pmatrix} r\,cos(\vartheta) \\ r\,sin(\vartheta) \end{pmatrix}. \qquad \textit{12.14}$$

We see that $x^2 + y^2 = r^2 cos^2(\vartheta) + r^2 sin^2(\vartheta) = r^2$. Hence we may take

$$\mathbf{g}(A) = \{(r, \vartheta) : 0 < r^2 < 1, 0 < \vartheta \le 2\pi\},$$

the limits for the angle ϑ in $\mathbf{g}(A)$ to yield a set of (x,y) which cover A. We chose $0 < \vartheta \le 2\pi$ but $\pi < \vartheta \le 3\pi$ would do equally well.

Next we compute $\mathbf{DH}^{\mathrm{T}}$.

$$\mathbf{DH}^{\mathrm{T}} = \begin{bmatrix} D_1 \\ D_2 \end{bmatrix} \begin{bmatrix} H_{(1)} & H_{(2)} \end{bmatrix} = \begin{bmatrix} D_1 H_{(1)} & D_1 H_{(2)} \\ D_2 H_{(1)} & D_2 H_{(2)} \end{bmatrix}.$$

So from equation 12.14, $H_{(1)}(r, \vartheta) = r\,cos(\vartheta), H_{(2)}(r, \vartheta) = r\,sin(\vartheta)$, we have

$$(\mathbf{DH}^{\mathrm{T}})(r, \vartheta) = \begin{bmatrix} cos(\vartheta) & sin(\vartheta) \\ -r\,sin(\vartheta) & r\,cos(\vartheta) \end{bmatrix}.$$

Hence

$$\begin{bmatrix} \boldsymbol{grad} H_{(1)}(r, \vartheta) \\ \boldsymbol{grad} H_{(2)}(r, \vartheta) \end{bmatrix} = [(\mathbf{DH}^{\mathrm{T}})(r, \vartheta)]^{\mathrm{T}} = \begin{bmatrix} cos(\vartheta) & -r\,sin(\vartheta) \\ sin(\vartheta) & r\,cos(\vartheta) \end{bmatrix}.$$

Subtracting $\frac{sin(\vartheta)}{cos(\vartheta)}$*row 1 from row 2 yields

$$\begin{bmatrix} cos(\vartheta) & -r\,sin(\vartheta) \\ 0 & \frac{r\sin^2(\vartheta)}{cos(\vartheta)} + r\,cos(\vartheta) \end{bmatrix}$$

The product of the diagonal elements is $r(\sin^2(\vartheta) + \cos^2(\vartheta)) = r$. Hence the magnitude $|J(r, \vartheta)| = r$ and thus

$$\int_{\{(x,y) \text{ in } \mathcal{E}_2 : 0 < x^2 + y^2 < 1\}} exp\left(-\frac{x^2 + y^2}{2}\right) dV_2 = \int_0^{2\pi}\left[\int_0^1 exp\left(-\frac{r^2}{2}\right) r\,dr\right] d\vartheta$$

$$= \int_0^{2\pi} exp\left(-\frac{r^2}{2}\right)\Bigg|_0^{r=1} d\vartheta$$

$$= \int_0^{2\pi} (1 - exp[-1/2])\,d\vartheta$$

$$= 2\pi(1 - exp[-1/2])$$

■

Exercises 12.10

Most applications of Theorem 12.23 arise in theoretical investigations. So we provide only a small number of exercises for self-testing.

1 Evaluate $V = \int_{\mathcal{E}_1} exp(-x^2/2)dx$.

Hint: $V^2 = \int_{\mathcal{E}_2} exp\left(-\frac{x^2 + y^2}{2}\right) dy\,dx$

2. Evaluate $\int_0^1 \left[\int_{-\sqrt{1-y^2}}^{\sqrt{1-y^2}} sin(x^2 + y^2)dx\right] dy$

Hint: Consider this as the value of a double integral over the unit circle. Convert to polar coordinates (see equation 12.14).

3. Determine the volume of a sphere,
$S_R = \{(x, y, z) \text{ in } \mathcal{E}_3 : x^2 + y^2 + z^2 < R^2\}$, of radius R.

Hint: Use **three-dimensional polar coordinates**, given by

$$x = r\,cos(\vartheta)\,sin(\varphi),\ y = r\,cos(\vartheta)\,cos(\varphi),\ z = r\,sin(\vartheta)$$
$$0 \le \varphi < 2\pi,\ -\pi/2 \le \vartheta \le \pi/2,\ r \ge 0$$

as illustrated in Figure 12-14.

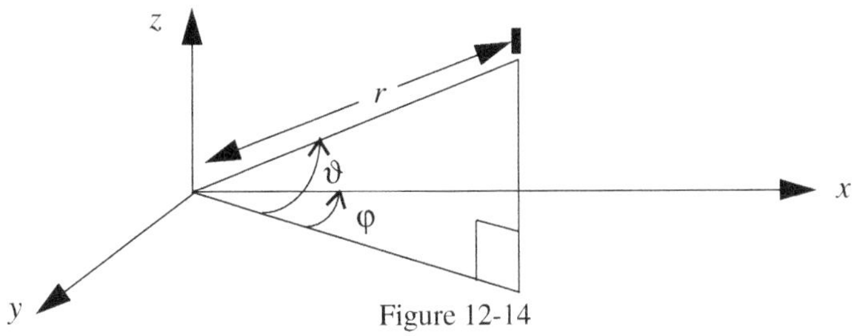

Figure 12-14

The final topic of this section is that of the *row expansion/column expansion* for determinants, needed for the development of the topic of eigenvalues. We introduce this idea by looking first at determinants of 2×2 and 3×3 matrices. For the 2×2 case below, in line with Theorem 12.22 on page 467, subtract $\frac{b}{a}$ *row 1 from row 2, to obtain

$$\begin{aligned} det\begin{bmatrix} a & c \\ b & d \end{bmatrix} &= det\begin{bmatrix} a & c \\ 0 & d - \frac{b}{a}c \end{bmatrix} \\ &= ad - bc \\ &= \begin{pmatrix} a \\ b \end{pmatrix} \cdot \begin{pmatrix} d \\ -c \end{pmatrix}. \end{aligned} \qquad \textbf{12.15}$$

For the 3×3 case, similarly (and with a good deal of algebraic manipulation, to use row operations to convert the 3×3 matrix to upper triangular form) we find

$$det\begin{bmatrix} a & d & g \\ b & e & h \\ c & f & i \end{bmatrix} = \begin{pmatrix} a \\ b \\ c \end{pmatrix} \cdot \begin{pmatrix} ei - fh \\ -di + gf \\ dh - eg \end{pmatrix}. \qquad \textbf{12.16}$$

Now note that

$$\left\| \begin{pmatrix} d \\ -c \end{pmatrix} \right\| = \left\| \begin{pmatrix} c \\ d \end{pmatrix} \right\| \quad \text{but} \quad \begin{pmatrix} d \\ -c \end{pmatrix} \perp \begin{pmatrix} c \\ d \end{pmatrix}.$$

Thus, also using equation 8.33 on page 283, we find that

$$\begin{aligned} det\begin{bmatrix} a & c \\ b & d \end{bmatrix} &= \begin{pmatrix} a \\ b \end{pmatrix} \cdot \begin{pmatrix} d \\ -c \end{pmatrix}. = \left\| \begin{pmatrix} a \\ b \end{pmatrix} \right\| \left\| \begin{pmatrix} d \\ -c \end{pmatrix} \right\| cos\left(\begin{pmatrix} a \\ b \end{pmatrix}, \begin{pmatrix} d \\ -c \end{pmatrix} \right) \\ &= \left\| \begin{pmatrix} a \\ b \end{pmatrix} \right\| \left\| \begin{pmatrix} c \\ d \end{pmatrix} \right\| sin\left(\begin{pmatrix} a \\ b \end{pmatrix}, \begin{pmatrix} c \\ d \end{pmatrix} \right). \end{aligned}$$

But the factors

$$\left\|\begin{pmatrix}a\\b\end{pmatrix}\right\| sin\left(\begin{pmatrix}a\\b\end{pmatrix},\begin{pmatrix}c\\d\end{pmatrix}\right) \quad \text{and} \quad \left\|\begin{pmatrix}c\\d\end{pmatrix}\right\|$$

are, respectively, the *height* and the *base length* of the parallelogram formed by the vectors $\begin{pmatrix}a\\b\end{pmatrix}$ and $\begin{pmatrix}c\\d\end{pmatrix}$, as seen in Figure 12-15.

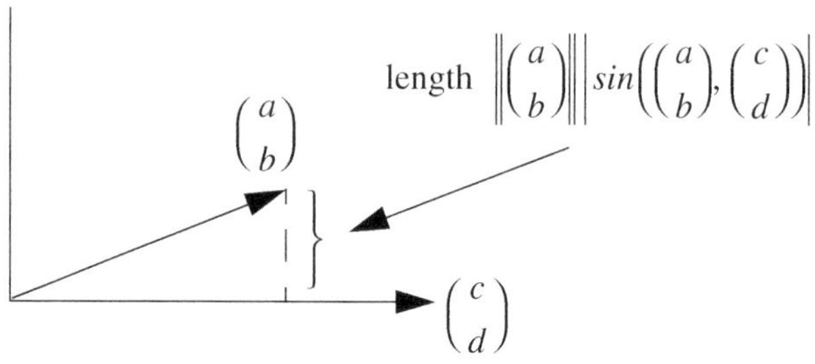

Figure 12-15

Thus, we see geometrically what we already knew, that this determinant is the signed area of the parallelogram formed by its columns.

Similarly, if we let

$$\begin{pmatrix}ei-fh\\-di+gf\\dh-eg\end{pmatrix} = \begin{pmatrix}A\\-B\\C\end{pmatrix},$$

then

$$det\begin{bmatrix}a & d & g\\b & e & h\\c & f & i\end{bmatrix} = \begin{pmatrix}a\\b\\c\end{pmatrix}\cdot\begin{pmatrix}A\\-B\\C\end{pmatrix},$$

and since we know that this determinant represents the signed volume of the figure formed by its columns, the vector

$$\begin{pmatrix}A\\-B\\C\end{pmatrix}$$

must be a vector whose magnitude is the *base area* of the parallelogram formed by the second and third columns, and, as a vector, it must be perpendicular to this parallelogram. This perpendicularity is made even more believable, by the fact that being the volume it is, the determinant is 0 if the first column is a linear combination of the remaining ones; this means that in this case, the indicated dot product must be 0, which will certainly be the case if the second vector in the dot product is perpendicular to all the vectors from which it was formed and the first column is a linear combination of these vectors.[1] But we also notice that A is the determinant of the matrix obtained from the original 3×3 by deleting from it the row and column of element a. Similarly, for B and C. This suggests that in general, if $\mathbf{A}$ is an $n \times n$ matrix with i-j element a_{ij}, we may write

$$det(\mathbf{A}) = \sum_{i=1}^{n} (-1)^{i+1} a_{i1} A_{i1},$$

where A_{i1} is the determinant of the matrix obtained from $\mathbf{A}$ by deleting its i-th row and first column. This is the case. We will state this result formally. for needed reference in the development of *eigenvalues*, in Chapter 13 on preferred coordinate systems. It is established in Appendix 8 on page 807.

Theorem 12.24: Column and row expansions for determinant

The determinant, *det*(A) (Definition 12.21 on page 466) of a real $n \times n$ matrix $\mathbf{A}$ can be represented by the *column expansion*

$$det(\mathbf{A}) = \sum_{i=1}^{n} (-1)^{i+1} a_{i1} A_{i1}, \qquad \mathit{12.17}$$

where A_{i1} is the determinant of the $(n-1) \times (n-1)$ matrix obtained from $\mathbf{A}$ by deleting its i-th row and first column. We need to initiate this definition by explicitly stating that if $\mathbf{A}$ is a 1×1 matrix, i.e., $\mathbf{A} = [a]$, then

$$det(\mathbf{A}) = a.$$

A similar *row expansion* is

$$det(\mathbf{A}) = \sum_{i=1}^{n} (-1)^{i+1} a_{1j} A_{1j},$$

1. As a by-product of this development, what we have just seen will lead directly to the definition of the *cross product* of two three-dimensional vectors, the vector $(A, -B, C)$ being the cross product of columns 2 and 3. The cross product will be needed later in this chapter for the development of Stoke's theorem.

where A_{1j}, **the 1,*j* minor of A**, is the determinant of the matrix obtained from **A** by deleting its first row and its *j*-th column.

Recall that $|det\,\mathbf{A}|$ represents the *n*-dimensional volume of the image under **A** (see Definition 12.20 on page 462) of the unit *n*-dimensional interval,

$$I_n = \{(x_1,...,x_n) \text{ in } \mathcal{E}_n : 0 \le x_i \le 1\}$$

❑

This theorem is of little practical computational use, but is critical for the theoretical development of eigenvalue theory.

Remark: Theorem 12.24 has an interesting interpretation, concerning the significance of the vector $\boldsymbol{A}_1$ whose *i*-th component is $(-1)^{i+1}A_{i1}$ — namely, if $\boldsymbol{a}_1$ denotes the first column of the $n \times n$ matrix **A**, Theorem 12.24 states that the signed volume of the figure formed by the columns of **A** is $\boldsymbol{a}_1 \cdot \boldsymbol{A}_1$. It therefore must be the case that $\boldsymbol{A}_1$ is a vector whose magnitude is the $n - 1$ dimensional volume of the figure formed by columns 2 to *n* of **A**, and which is orthogonal to the subspace formed by these columns; note it is orthogonal to this subspace, because if $\boldsymbol{a}_1$ is in the subspace formed by $\boldsymbol{a}_2,...,\boldsymbol{a}_n$, the parallelepiped formed by the columns of **A** must have 0 volume; i.e., $0 = det\,\mathbf{A} = \boldsymbol{a}_1 \cdot \boldsymbol{A}_1$.

Exercises 12.11

1. Establish equation 12.16.
2. Show that if the non-normalized Gram-Schmidt procedure, Definition 9.18 on page 305, is applied to the columns of the real $n \times n$ matrix **A**, to yield the matrix **G** of orthogonal columns, then
$$|det(\mathbf{A})| = \prod_{j=1}^{n} \|\boldsymbol{g}_j\|$$
where[1] $\|\boldsymbol{g}_j\|$ is the length of the *j*-th column of **G.**
Hint: The column operations of the Gram-Schmidt procedure don't alter the determinant. But the volume of the figure formed from orthogonal vectors is just the product of their lengths.
3. The **cross product** $\boldsymbol{v} \times \boldsymbol{w}$ of three-dimensional vectors with
$\boldsymbol{v} = \begin{pmatrix} d \\ e \\ f \end{pmatrix}, \boldsymbol{w} = \begin{pmatrix} r \\ s \\ t \end{pmatrix}$ is symbolically obtained as $det\begin{bmatrix} \boldsymbol{i} & d & r \\ \boldsymbol{j} & e & s \\ \boldsymbol{k} & f & t \end{bmatrix}$

---

1. The symbol $\prod$ denotes the product of the factors following it.

where $i = e_1 = \begin{pmatrix} 1 \\ 0 \\ 0 \end{pmatrix}$, $j = e_2 = \begin{pmatrix} 0 \\ 1 \\ 0 \end{pmatrix}$, $k = \begin{pmatrix} 0 \\ 0 \\ 1 \end{pmatrix}$

Show that $v \times w$ is orthogonal to v and w and that $\|v \times w\| = \|v\| \|w\| |sin(v, w)|$ — i.e., $\|v \times w\|$ is the unsigned area of the parallelogram formed by v and w. Hint: This can be done using the column expansion, or alternatively, using the Gram-Schmidt procedure (Definition 9.18 on page 305) to orthogonalize v and w, noting that the product of the lengths of the orthogonalized vectors must be the parallelogram area, and that the length of $v \times w$ is this product. To finish up using this approach, all you need verify is that $v \times w$ is orthogonal to v and w.

6 Some Differentiation Rules in n Dimensions

In this section we introduce the multivariate extension of the chain rule (see Theorem 2.13, equation 2.30 on page 53) and the integral extension of the derivative of an infinite series (see Theorem 7.20 on page 238). Since these results are established by methods we have already used, we will do little more than state the results, and provide a few useful illustrations of their application.

Theorem 12.25: Multivariate *chain rule*

Let f be a real valued function of n real variables (Definition 8.1 on page 261) and let $\mathbf{x}$ be an n-dimensional vector valued function of m real variables, treated as a *row* matrix (a $1 \times n$ matrix) such that the composition $f(\boldsymbol{x})$ (a real valued function of m real variables) is defined for all points close enough to the m dimensional real vector $\boldsymbol{t}$.

Assume that $\boldsymbol{grad}\, f = [D_1 f, ..., D_n f]$ is continuous at all points close enough to $\boldsymbol{x}(\boldsymbol{t})$ (Theorem 8.8 on page 271) and

$$\mathbf{D}\boldsymbol{x} = \begin{bmatrix} D_1 \\ \cdot \\ \cdot \\ D_m \end{bmatrix} \begin{bmatrix} x_{(1)}, & . & . & ., & x_{(n)} \end{bmatrix} = \begin{bmatrix} D_1 x_{(1)} & . & . & D_1 x_{(n)} \\ . & & . & . \\ . & & . & . \\ D_m x_{(1)} & . & . & D_m x_{(n)} \end{bmatrix}$$

exists, with all elements continuous at all points close enough to $\boldsymbol{t}$. Then the composition $f(\boldsymbol{x})$ has a gradient,

$$\mathbf{D}[f(\boldsymbol{x})] = [D_1 f(\boldsymbol{x}), ..., D_n f(\boldsymbol{x})],$$

which exists and is continuous at all points close enough to t, and satisfies

$$[D_1(f(\boldsymbol{x})),...,D_n(f(\boldsymbol{x}))] = [(D_1 f)(\boldsymbol{x}),...,(D_n f)(\boldsymbol{x})]\begin{bmatrix} D_1 x_{(1)} & \cdot\ \cdot & D_m x_{(1)} \\ \cdot & \cdot\ \cdot & \cdot \\ \cdot & \cdot\ \cdot & \cdot \\ D_1 x_{(n)} & \cdot\ \cdot & D_m x_{(n)} \end{bmatrix}$$

or, in more abbreviated form

$$\mathbf{D}^{\mathrm{T}}(f(\boldsymbol{x})) = (\mathbf{D}^{\mathrm{T}} f)(\boldsymbol{x})(\mathbf{D}\boldsymbol{x})^{\mathrm{T}}. \qquad \textit{12.18}$$

Except for the needed transpose indicator, T (Definition 10.13 on page 342) this abbreviated version looks identical to the one-dimensional chain rule. It can be implemented on a computer exactly as written in equation 12.18. For hand work it expands[1] out to

$$D_i(f(\boldsymbol{x})) = \sum_{j=1}^{n} (D_j f)(x_{(1)},...,x_{(n)}) D_i(x_{(j)}). \qquad \textit{12.19}$$

Note that $D_i(f(\boldsymbol{x}))$ is the partial derivative of the composition, $f(\boldsymbol{x})$, with respect to its i-th coordinate, $i = 1,...,m$, while $(D_j f)(\boldsymbol{x})$ is the composition, of the partial derivative of f with respect to its j-th coordinate $(j = 1,...,n)$ with the n-dimensional vector valued function $\boldsymbol{x}$.

❑

In establishing this result, we may as well assume that $m = 1$ (i.e., that the vector-valued function $\boldsymbol{x}$ is a function of a single real variable).

The proof makes extensive use of the mean value theorem, Theorem 3.1 on page 62, together with the definition of the derivative of $f(\boldsymbol{x})$ at t.

One simple illustration follows.

Example 12.12: Illustration of chain rule

Let $f(X_1, X_2) = X_1^2 + X_2^2$ for all real numbers, X_1, X_2, and let the functions x_1, x_2 be defined by the formulas $x_1(t, \tau) = t + \tau^2, x_2(t, \tau) = t\tau$.

Then

---

1. Equation 12.19 also can be written in the poorest notation as

$$\frac{\partial(f(\boldsymbol{x}))}{\partial t_i} = \sum_{j=1}^{n} \frac{\partial f}{\partial x_j}(\boldsymbol{x})\frac{\partial x_j}{\partial t_i}.$$

$$(f(\boldsymbol{x}))(t,\tau) = (f(x_1,x_2))(t,\tau) = (x_1(t,\tau))^2 + (x_2(t,\tau))^2$$
$$= (t+\tau^2)^2 + (t\tau)^2.$$

From this we can see directly that

$$(D_1 f(x_1,x_2))(t,\tau) = 2(t+\tau^2) + 2t\tau^2.$$

(In this case, one would not use the chain rule except for illustration. In general its main use is theoretical.)

Using the chain rule, we find

$$(\mathbf{D}^{\mathrm{T}} f)(x_1,x_2) = ([D_1, D_2]f)(x_1,x_2)$$
$$= (D_1 f, D_2 f)(x_1,x_2)$$
$$= (2x_1, 2x_2)$$

while

$$(\mathbf{D}\boldsymbol{x})^{\mathrm{T}} = \left(\begin{bmatrix} D_1 \\ D_2 \end{bmatrix}\begin{bmatrix} x_1 & x_2 \end{bmatrix}\right)^{\mathrm{T}} = \begin{bmatrix} D_1x_1 & D_2x_1 \\ D_1x_2 & D_2x_2 \end{bmatrix}.$$

Hence

$$(\mathbf{D}\boldsymbol{x})^{\mathrm{T}}(t,\tau) = \begin{bmatrix} 1 & 2\tau \\ \tau & t \end{bmatrix}.$$

Therefore

$$((\mathbf{D}^{\mathrm{T}} f)(\boldsymbol{x}))(t,\tau)(\mathbf{D}\boldsymbol{x})^{\mathrm{T}}(t,\tau) = [2(t+\tau^2), 2t\tau]\begin{bmatrix} 1 & 2\tau \\ \tau & t \end{bmatrix}$$
$$= \begin{bmatrix} 2(t+\tau^2) + 2t\tau & \ldots \\ \ldots & \ldots \end{bmatrix},$$

agreeing with the result we got directly.

■

Exercises 12.13

In these exercises, assume all the variables real, and all formulas defined wherever possible.

1. Let
$$f(X_1, X_2) = X_1 X_2^2$$
$$x_1(u,v,w) = uvw^2$$

$x_2(u, v, w) = cos(u).$

Find $D_i(f(\boldsymbol{x})), \quad i = 1, 2, 3$, where $\boldsymbol{x} = (x_1, x_2)$.

2. Same as previous question where $f(X_1, X_2) = ln(X_1) + exp(X_1, X_2)$.
3. Make up some similar exercises, and check them using *Maple* or *Mathematica* if they are available.
   Hints: This may be a bit tricky. At least in *Mathematica* you can accomplish the type of composition you want via lists.
   Say, define *f,x,y* in *Mathematica* by

   $f[\{x\_, y\_\}]{:=}x + y^2, x[t\_, u\_] = tu^2, y[t\_, u\_] = Exp[tu]$,

   corresponding to
   $f(X, Y) = X + Y^2, x(t, u) = tu^2, y(t, u) = exp(tu)$,
   then the formula for the derivative of the composition $f(x, y)$ with respect to its first variable, t, is obtainable from *Mathematica* via D[f[{x[t,u],y[t,u]}],t]. This gives you a way to check your previous answers.
4. Prove the multivariate chain rule (Theorem 12.25 on page 474).

The final topic of this section is that of finding the derivative G', where

$$G(t) = \int_a^b f(t, x)dx.$$

Since an integral is essentially a sum and the derivative of a sum is the sum of the derivatives, we expect

$$G'(t) = \int_a^b D_1 f(t, x)dx. \qquad \textbf{12.20}$$

The only problem is that since *an integral is not exactly a sum of a finite number of terms, we need to append conditions to ensure that it behaves enough like one* so that equation 12.20 holds. Since we have already encountered this problem when differentiating an infinite series of functions (see Theorem 7.20 on page 238) the theorem we state now should come as no surprise.

Theorem 12.26: $D\int_a^b f(t, x)dx$

Suppose f is a real valued function of two real variables (Definition 8.1 on page 261) with $f(\tau, x)$ defined for all x in the closed bounded interval, $[a\,; b]$, $a < b$ (Definition 3.11 on page 88) and all τ sufficiently close to t, such that

- $\int_a^b f(\tau, x)dx$ exists for all such τ (see Definition 4.2 on page 102).

- $D_1 f(t, x)$ exists for all x in $[a ; b]$ (see Definition 8.5 on page 267).

- $\dfrac{f(t+h, x) - f(t, x)}{h} \underset{h \to 0}{\to} D_1 f(t, x)$ uniformly for x

  in $[a ; b]$. (This condition is analogous to the third condition of Theorem 7.20 on page 238. It ***means that the left side gets close*** **simultaneously** ***for all x in [a ; b] to the right side***. In specific cases it is often established by the mean value theorem (Theorem 3.1 on page 62)).

Then the function G defined by $G(t) = \int_a^b f(t, x)dx$ is differentiable with

$$G'(t) = \int_a^b D_1 f(t, x)dx.$$

This result extends to improper integrals, such as

$$\int_{-\infty}^{\infty} f(t, x)dx,$$

if there is a positive function g, for which $\int_{-\infty}^{\infty} g(t, x)dx$ exists and for all nonzero h close enough to 0

$$\left|\frac{f(t+h, x) - f(t, x)}{h}\right| \le g(t, x).$$

❑

Proof is similar to that of Theorem 7.20 on page 238.

This theorem can be very useful for evaluation of certain integrals, as we see in the following example.

Example 12.14: Use of Theorem 12.26 to evaluate $\int_0^{10} x^2 e^x dx$

To evaluate this integral, define $F(t) = \int_0^{10} x^2 e^{tx} dx$. From Theorem 12.26 it's easy to see that

$$F(t) = G''(t),$$

where $G(t) = \int_0^{10} e^{tx} dx$. But using the fundamental evaluation formula

(Theorem 4.3, equation 4.14 on page 103) together with the known properties of the exponential function (Theorem 5.6 on page 140 and Theorem 5.10 on page 144)

$$G(t) = \frac{1}{t}e^{tx}\Big|_0^{x=10} = \frac{e^{10t}-1}{t}.$$

Hence using familiar differentiation rules, we have

$$G''(t) = \frac{100e^{10t} - 10e^{10t}}{t^2} - \frac{10e^{10t}t^2 - (e^{10t}-1)2t}{t^4}.$$

But the desired integral is $F(1) = G''(1) = 82e^{10} - 2$, a computation which is a lot less prone to error than a pair of integrations by parts (Theorem 4.9 on page 121).

■

Exercises 12.15

1. Let $F(t) = \int_0^3 e^{tx^2}\,dx$. Find an integral for $F'(t), F''(t)$.

2. Find an integral for $F'(t)$, where $F(t) = \int_{-\infty}^{\infty} e^{-tx^2}\,dx, \quad t > 0$.

3. Compute $\int_0^{\pi} x^2 cos(x)dx$

4. Determine $\int_0^{2\pi} x^3 sin(x^2)dx$.

5. Determine $\int_0^{2\pi} x^7 sin(x^2)dx$.

7 Line and Surface Integrals

The main applications of the results of this section occur in advanced parts of mathematical physics — electromagnetic theory, heat flow, etc. They are likely to be hard reading, and may be omitted or read casually by those with little interest in such applications.

Limits of sums arise in many forms from various applications. The line and surface integrals to be examined in this section are critical for the study of the partial differential equations of mathematical physics, which describe the behavior of objects subject to electrical and magnetic fields and the flow of heat — topics essential to modern technology. Line and surface integrals and their relationships represent a vast area in mathematics, parts of which can be quite complicated. A careful detailed study of this material is far beyond the scope of what can be presented here. About all that we can hope for in this section is to provide an idea of the nature of some of the results in this field, some intuition on why they hold, and a few of the applications, with some idea of the obstacles which must be overcome for a proper development.

The first type of integral, the *line integral,*[1] arises when we want to compute the work done (or recovered) by an object being moved along some curve, subjected to a force dependent on its position. So, suppose that an object, such as a magnet, is moved along a two-dimensional curve $\boldsymbol{x} = (x_1, x_2)$ (a pair of functions of one variable) starting at $\boldsymbol{x}(a)$ and ending at $\boldsymbol{x}(b)$, $a < b$ (see Appendix 4 Definition A4.1 on page 773 for more detail). Also suppose that the external force on this object at position $\boldsymbol{x}(t)$ is given by the force vector $\boldsymbol{F}(t) = (F_1(\boldsymbol{x}(t)), F_2(\boldsymbol{x}(t)))$, where F_1 and F_2 are real valued functions of two real variables (Definition 8.1 on page 261). It is assumed that only the component of the external force vector, $\boldsymbol{F}(t)$, in the direction of motion (i.e., tangent to the curve $\boldsymbol{x}$) contributes to the work.

We'll first consider the case where $\boldsymbol{x}$ is differentiable with $\boldsymbol{x}'(t) = (x_1'(t), x_2'(t)) \neq (0, 0)$. Looking at Figure 12-16, we see that the component of the force $\boldsymbol{F}(t)$ tangent to $\boldsymbol{x}$ at t is $\boldsymbol{F}(t) \cdot \dfrac{\boldsymbol{x}'(t)}{\|\boldsymbol{x}'(t)\|}$ (see Definition 8.12 on page 276 and the definition of the tangent to a curve, Definition A4.2 on page 774). The magnitude of the work done by the object in going from $\boldsymbol{x}(t)$ to $\boldsymbol{x}(t + h)$ is approximately the magnitude of this component multiplied by the arc length of $\boldsymbol{x}$ from t to $t + h$ (see Figure 12-16).

1. Which should really be called a *curve integral.*

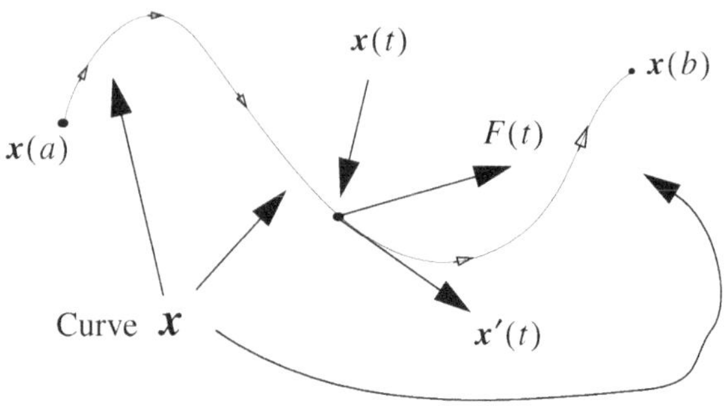

Figure 12-16

This latter is approximately

$$\sqrt{(x_1(t+h)-x_1(t))^2+(x_2(t+h)-x_2(t))^2} \cong \sqrt{((x_1'(t))^2+(x_2'(t))^2)}|h|$$
$$= \|\boldsymbol{x}'(t)h\|.$$

So the amount of energy that must be supplied by the internal mechanism moving the object is

$$-\boldsymbol{F}(\boldsymbol{x}(t))\cdot\boldsymbol{x}'(t)h$$

(so that, if the external force opposes the motion, the work (energy) supplied is positive). From this it follows that the total work done by the object in going along the curve $\boldsymbol{x}$ from $\boldsymbol{x}(a)$ to $\boldsymbol{x}(b)$ is

$$-\int_a^b \boldsymbol{F}(\boldsymbol{x}(t))\cdot\boldsymbol{x}'(t)dt. \qquad \textbf{12.21}$$

Note that written out longhand, the above integral is

$$-\int_a^b [F_1(\boldsymbol{x}(t))x_1'(t)+F_2(\boldsymbol{x}(t))x_2'(t)]dt\,.$$

The work done in moving along a three-dimensional curve is formally the same.

Definition 12.27: Line integrals

Let $\boldsymbol{F} = (F_1,...,F_n)$ be a real n-dimensional continuous vector valued function of n real variables (where this means that $\boldsymbol{F}$ has continuous components, $F_1,...,F_n$ (see Definition 8.7 on page 270)) and let $\boldsymbol{x} = (x_1,...,x_n)$ be an n-dimensional curve with domain $[a;b]$ and continuous derivative $\boldsymbol{x}' = (x_1',...,x_n')$ (i.e., the coordinates are continuous functions, see Theorem 2.10 on page 38). Recalling the definition of integral (Definition 4.2 on page 102) we define the line integrals along the curve $\boldsymbol{x}$, denoted, respectively, by

$$\int_{\boldsymbol{x}} \boldsymbol{F}(\boldsymbol{x})\cdot d\boldsymbol{x} \quad \text{and} \quad \int_{\boldsymbol{x}} F_k(\boldsymbol{x})dx_k \quad k = 1,...,n$$

by

$$\int_{x} F(x) \cdot dx = \int_a^b \sum_{k=1}^{n} F_k(x(t)) x_k'(t) dt \qquad \textbf{12.22}$$

and

$$\int_{x} F_k(x) dx_k = \int_a^b F_k(x(t)) x_k'(t)(dt). \qquad \textbf{12.23}$$

In two dimensions, the following somewhat less efficient notation is often used: $\boldsymbol{F}$ is often denoted by (f, g), and the curve (x_1, x_2) is often denoted by (x, y) or by C. Then

$$\int_C f(x, y) dx \text{ denotes } \int_a^b f(x(t), y(t)) x'(t) dt \qquad \textbf{12.24}$$

and

$$\int_C g(x, y) dy \text{ denotes } \int_a^b g(x(t), y(t)) y'(t) dt. \qquad \textbf{12.25}$$

Similarly in three dimensions often we see $F = (f, g, h)$, $C = (x, y, z)$ and $\int_C f(x, y, z) dx$, $\int_C g(x, y, z) dy$ and $\int_C h(x, y, z) dz$ are defined in the same manner.

For applications to most problems it is necessary to extend Definition 12.27 equations 12.22 and 12.23 to piecewise smooth curves (Definition 3.13 on page 89) i.e., curves with piecewise continuous derivatives, so that $\boldsymbol{x}$ can have sharp corners; this allows it to be formed by patching together several smooth curves.

❑

This definition of line integral is far from being as general as we might like, but is adequate for a reasonable introduction to this topic.

The concept of equivalent curves will be needed shortly, and is most naturally introduced here. Intuitively, we think of two curves as equivalent, if they trace out the same points, in the same order, but not necessarily at the same rate. So we'll think of $\boldsymbol{x}$ and $\boldsymbol{X}$ as equivalent piecewise smooth curves, if there is a monotone increasing differentiable real valued function, z, of a real variable with a differentiable inverse, such that $\boldsymbol{x} = \boldsymbol{X}(z)$. We see immediately from the chain rule (equation 2.30 on page 53) that

$$\int_a^b F_k(x(t)) x_k'(t) dt = \int_a^b F_k(X(z(t))) X_k'(z(t)) dt$$

and from the substitution rule (equations 4.28 and 4.29 on page 123) with substitution $u = z(t)$, that this latter integral is

$$\int_{z(a)}^{z(b)} F_k(X(u)) X_k'(u) du;$$

i.e.,

$$\int_a^b F_k(\boldsymbol{x}(t))x_k'(t)dt = \int_{z(a)}^{z(b)} F_k(X(u))X_k'(u)du. \qquad \textbf{12.26}$$

This discussion leads to the following definition.

Definition 12.28: Equivalent curves

If $\boldsymbol{x}$ and $\boldsymbol{X}$ are two piecewise smooth curves (see Definition 3.13 on page 89) related by the equation

$$\boldsymbol{x} = \boldsymbol{X}(z),$$

where z is a monotone increasing real valued function of a real variable (recall Definition 2.8 on page 37 and Definition 2.9 on page 37) and both z and its inverse function have continuous derivatives (see Definition 2.5 on page 30, Definition 2.6 on page 34, and Theorem 2.10 on page 38) then $\boldsymbol{x}$ with domain the closed bounded interval (see Definition 1.1 on page 2 and Definition 3.11 on page 88) $[a; b]$ and $\boldsymbol{X}$ with domain $[z(a); z(b)]$ are called *equivalent.*[1]

❑

We will need the result derived above fairly soon, so we state it formally as a theorem.

Theorem 12.29: Line integrals over equivalent curves.

If $\boldsymbol{x}$ and $\boldsymbol{X}$ are equivalent curves (Definition 12.28 above) then provided at least one of the following integrals exists, so does the other with

$$\int_a^b F_k(\boldsymbol{x}(t))x_k'(t)dt = \int_{z(a)}^{z(b)} F_k(\boldsymbol{X}(u))X_k'(u)du.$$

In two dimensions, if C represents $\boldsymbol{x}$ with domain $[a; b]$ and C^* represents the curve $\boldsymbol{X}$ ***with domain*** $[z(a); z(b)]$, in the less efficient notation represented by equations 12.24 and 12.25, the above conclusion reads

$$\int_C f(x, y)dx = \int_{C^*} f(x, y)dx$$

❑

The first of our extensions of the fundamental evaluation theorems (Theorem 1.8 on page 12 and Theorem 4.3, equation 4.14 on page 103) is motivated by considerations of flow of an incompressible fluid, such as blood flow in veins and arteries, water flow around submarines, or airflow around airplane wings, or in a tornado.2 Most generally we have a **vector field in space and time**, which is just

1. This definition too can be extended to require less of z. However, for this more general definition to be useful, line integrals would need the more general definition alluded to earlier, all of which would have taken a great deal more space, without contributing much to the understanding.
2. In the airflow cases we are assuming no effective compression of the air.

triple of functions $\boldsymbol{v} = (v_1, v_2, v_3)$, of four variables, x_1, x_2, x_3, t, where $\boldsymbol{v}(x_1, x_2, x_3, t)$ represents the velocity vector of the portion of the fluid at position (x_1, x_2, x_3) at time t.

We try to simplify examination of the fluid flow almost immediately. One way to do this is to examine steady state situations, where the velocity field does not vary with time (and even in a tornado, this may not be too poor a description over a short time, if we are traveling with the tornado). To further simplify, we might restrict to two-dimensional flow — the situation examined being one in which the velocity also does not vary with the third coordinate, x_3.

We can then picture the vector field as a collection of arrows, as shown in Figure 12-17, where any given arrow starting at $(x_1, x_2, 0)$ represents the velocity $\boldsymbol{v}(x_1, x_2) = (v_1(x_1, x_2), v_2(x_1, x_2), 0)$ of the fluid located at (x_1, x_2, x_3) for all x_3. We also assume continuity of (the components of) $\boldsymbol{v}$ (see Definition 8.7 on page 270).

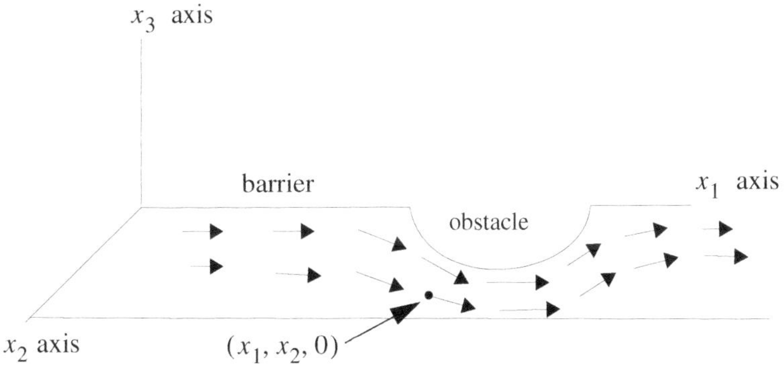

Figure 12-17

It's of some interest to determine from this velocity field, the net rate of flow of fluid across any *reasonable-looking*[1] curve, $\boldsymbol{x}$ — where we have to choose what we take to be a positive rate. In Figure 12-18 we could decide that the rate is

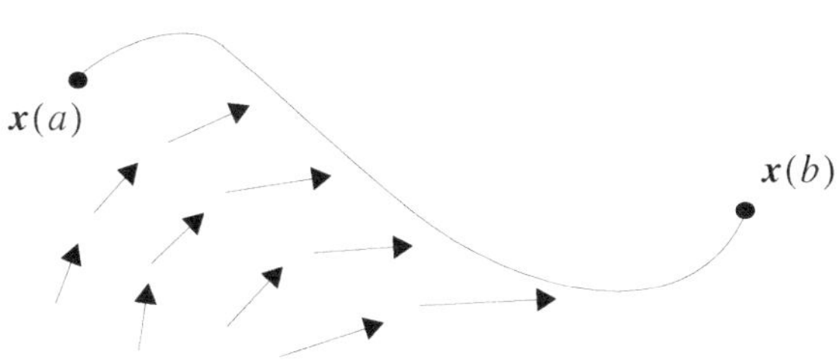

Figure 12-18

1. No looping back on itself,  and $\boldsymbol{x}$ having a continuous derivative.

positive when more fluid ends up *above* the curve. We'll soon see how this choice is determined.

In posing this question we are acting as if the amount of fluid over any parallelogram of fixed size entirely in the flow encloses a specified amount of fluid per unit height, independent of its location; we will simply refer to this as the *amount of fluid in the parallelogram.* So, with this usage, the amount of fluid in any parallelogram of area $\mathcal{A}$ is simply $\mathcal{A}$.

To return to the problem under consideration, pick any point $\boldsymbol{x}(t) = (x_1(t), x_2(t))$ on the curve $\boldsymbol{x}$ and draw a parallelogram, one of whose sides is the vector connecting $\boldsymbol{x}(t)$ to $\boldsymbol{x}(t + \Delta t)$, where $\Delta t > 0$ is small. Now draw the other side, consisting of the vector $\boldsymbol{v}(\boldsymbol{x}(t))\Delta\tau$. Note, $\boldsymbol{v}(\boldsymbol{x}(t))$ is the velocity associated with the point $\boldsymbol{x}(t)$, and $\Delta\tau > 0$ represents a small increment of time. Be careful to distinguish $\Delta\tau$, the small time increment, from Δt the increment in the curve parameter t. From its construction, with both

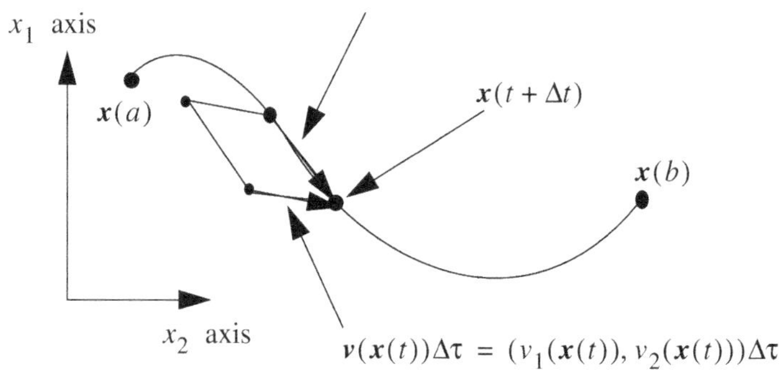

Figure 12-19

Δt and $\Delta\tau$ very small, and with continuity of $\boldsymbol{v}$, the velocity at all points in the shaded parallelogram is close to $\boldsymbol{v}(\boldsymbol{x}(t))$. Hence, the amount of fluid in the parallelogram crossing the curve in time period $\Delta\tau$ — signed in the sense that the amount crossing the curve $\boldsymbol{x}$ will change sign if the sign of $\boldsymbol{v}$ is reversed, as we should want. But using the fact that the determinant

$$det\begin{bmatrix} a & c \\ b & d \end{bmatrix} = \begin{pmatrix} -b \\ a \end{pmatrix} \cdot \begin{pmatrix} c \\ d \end{pmatrix} = ad - bc$$

represents the signed area of the parallelogram formed by the columns of $\begin{bmatrix} a & c \\ b & d \end{bmatrix}$

(see Definition 12.21 on page 466 and equation 12.15 on page 470) we see that the signed area of the parallelogram is

$$\begin{pmatrix} -x_2'(t) \\ x_1'(t) \end{pmatrix} \cdot \begin{pmatrix} v_1(\boldsymbol{x}(t)) \\ v_2(\boldsymbol{x}(t)) \end{pmatrix} \Delta t \Delta\tau = v_2(\boldsymbol{x}(t))x_1'(t) - v_1(\boldsymbol{x}(t))(x_2'(t))\Delta t \Delta\tau.$$

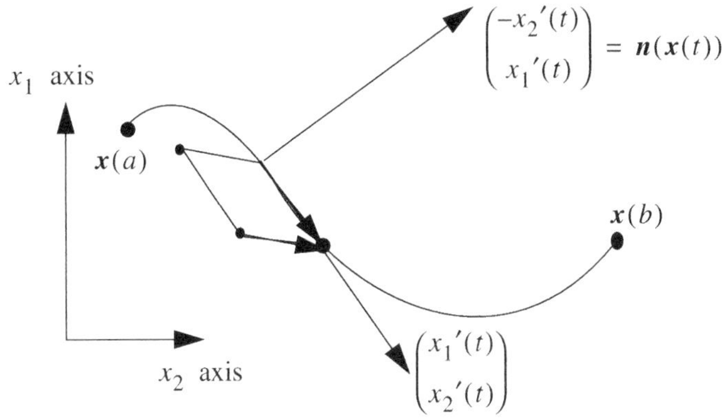

Figure 12-20

Note that as illustrated in Figure 12-20, we chose as the *positive* normal (perpendicular, see Theorem 8.15 on page 278) to (x_1, x_2) at t, the vector

$$\boldsymbol{n}(\boldsymbol{x}(t)) = \begin{pmatrix} -x_2'(t) \\ x_1'(t) \end{pmatrix}.$$

We could, just as properly, have chosen

$$\boldsymbol{n}(\boldsymbol{x}(t)) = \begin{pmatrix} x_2'(t) \\ -x_1'(t) \end{pmatrix}.$$

It's a matter of convention, and we must make the choice — whose effect is to determine which flows are positive and which are not.

The (signed) amount of fluid crossing the stretch of the curve $\boldsymbol{x}$ from $\boldsymbol{x}(t)$ to $x(t + \Delta t)$ per unit time is approximately $\boldsymbol{n}(\boldsymbol{x}(t)) \cdot \boldsymbol{v}(\boldsymbol{x}(t))\Delta t$.

With the given choice of positive normal to the curve $\boldsymbol{x}$, the signed net amount of fluid crossing the curve $\boldsymbol{x}$ per unit time is

$$\begin{aligned}\int_a^b \boldsymbol{n}(\boldsymbol{x}(t)) \cdot \boldsymbol{v}(\boldsymbol{x}(t))dt &= \int_a^b (-x_2'(t), x_1'(t)) \cdot (v_1(x_1(t), x_2(t)), v_2(x_1(t), x_2(t)))dt \\ &= \int_a^b -v_1(x_1(t), x_2(t))x_2'tdt + \int_a^b v_2(x_1(t), x_2(t))x_1'(t)dt \\ &= \int_{\boldsymbol{x}} (v_2(\boldsymbol{x})dx_1 - v_1(\boldsymbol{x})dx_2),\end{aligned}$$

where the curve $\boldsymbol{x}$ has domain $[a; b]$.

We summarize what has been established in Theorem 12.30.

Theorem 12.30: Fluid flux

Given a steady state two-dimensional velocity field (pair of continuous real valued functions) $\boldsymbol{v} = (v_1, v_2)$, where $\boldsymbol{v}(X_1, X_2)$ represents fluid velocity at the point (X_1, X_2) and a differentiable two-dimensional curve $\boldsymbol{x} = (x_1, x_2)$ with domain $[a; b]$ (see Definition A4.1 on page 773) we define the net rate of fluid flow across the curve $\boldsymbol{x}$ to be the integral

$$\int_a^b \boldsymbol{n}(\boldsymbol{x}(t)) \cdot \boldsymbol{v}(\boldsymbol{x}(t))dt = \int_a^b -v_1(x_1(t), x_2(t))x_2{}'tdt + \int_a^b v_2(x_1(t), x_2(t))x_1{}'(t)dt$$

$$= \int_{\boldsymbol{x}} (v_2(\boldsymbol{x})dx_1 - v_1(\boldsymbol{x})dx_2)$$

(a sum of two line integrals). The units are determined by the units in which velocity is measured; the sign of the *net rate of fluid flow* (*fluid flux*) reverses if the ***positive normal*** to $\boldsymbol{x}$ is chosen as $\begin{pmatrix} x_2{}' \\ -x_1{}' \end{pmatrix}$ rather than $\begin{pmatrix} -x_2{}' \\ x_1{}' \end{pmatrix}$; it also reverses if $\boldsymbol{x}$ is replaced by $\boldsymbol{y}$, where

$$\boldsymbol{y}(t) = x(-t), \quad -b \le t \le -a.$$

We leave these last assertions to Problem 12.16 to follow.

❑

Problem 12.16

Prove the final assertions of Theorem 12.30.

Hint: Compute $\int_a^b -v_1(x_1(t), x_2(t))x_2{}'tdt + \int_a^b v_2(x_1(t), x_2(t))x_1{}'(t)dt$ and

$$\int_{-b}^{-a} -v_1(y_1(t), y_2(t))y_2{}'tdt + \int_{-b}^{-a} v_2(y_1(t), y_2(t))y_1{}'(t)dt.$$

So far, we have seen two applications for line integrals — one concerned with the amount of work that must be supposed in moving an object along a curve through a force field, and the other with fluid flux (the net rate of fluid flow across a curve in a steady state velocity field). The computation of fluid flux can be extended to three-dimensional steady state fluid flow across a surface; but first we must provide a definition of *surface*, analogous to that of *curve. Again, our definition will be a restricted one that is still adequate to get the ideas across.*

Definition 12.31: Smooth surface (restricted version)

By a **smooth surface** is meant a triple, (X, Y, Z) of real valued functions of two real variables, having continuous gradients and a common two-dimensional interval domain (see Definition 1.1 on page 2, Definition 8.10 on page 272, Definition 8.7 on page 270, and Definition 12.1 on page 438) or a domain which is obtainable from an interval via a one-one function $\boldsymbol{M}$ (Definition 12.18 on page 460) such that $\boldsymbol{M}$ and its inverse (Definition 2.8 on page 37) have continuous gradients.

❑

To get a feel for this definition, note that as $(x,y,0)$ traces out a small rectangle in the base plane, $(X(x,y), Y(x,y), Z(x,y))$ approximately traces out a parallelogram in three dimensions — because, by the fundamental multivariate approximation, 8.17 on page 274,

$$\begin{pmatrix} X \\ Y \\ Z \end{pmatrix}(x_0 + h, y_0 + k) \cong 0\begin{pmatrix} X \\ Y \\ Z \end{pmatrix}(x_0, y_0) + D_1\begin{pmatrix} X \\ Y \\ Z \end{pmatrix}(x_0, y_0) + D_2\begin{pmatrix} X \\ Y \\ Z \end{pmatrix}(x_0, y_0) \qquad \textbf{\textit{12.27}}$$

$$= v_0 + hv_1 + kv_2,$$

as illustrated in Figure 12-21. The range of this surface is the way we visualize it, just as we visualize a curve. In neither case does the picture give the full information about how the curve or surface is *traversed.*

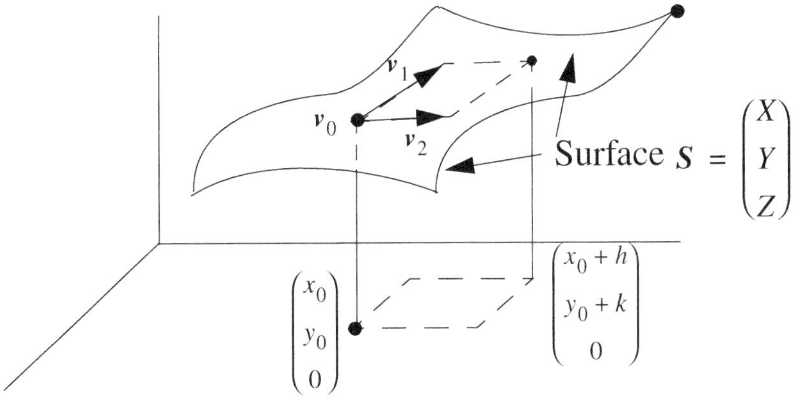

Figure 12-21

Example 12.17: Upper surface of unit sphere without north pole

For the most obvious representation of the upper hemisphere when you think of the usual X,Y,Z representation $X^2 + Y^2 + Z^2 = 1$, let

$$X(\xi, \eta) = \xi,\ Y(\xi, \eta) = \eta,\ Z(\xi, \eta) = \sqrt{1 - (\xi \ + \eta^2)}$$

where the domain will be taken to be $\{(\xi, \eta) \text{ in } \mathcal{E}_2 : 0 < \xi \ + \eta^2 \leq 1\}$. As (ξ, η) traverses the unit circle, we see that $(X, Y, Z)(\xi, \zeta)$ traverses the upper unit hemisphere, except for the north pole, in three dimensions.

A more natural representation of this surface uses polar coordinates, representing the coordinates of any point on this hemispheric surface as

$$(cos\,\vartheta\, cos\,\varphi, cos\,\vartheta\, sin\,\varphi, sin\,\vartheta),\ 0 \leq \vartheta < \pi/2,\ 0 \leq \varphi < 2\pi$$

(recall Figure 12-14 on page 470) i.e., via the triple $(X(M), Y(M), Z(M))$ where $(\xi, \eta) = M(\vartheta, \varphi) = (cos\,\vartheta\, cos\,\varphi, cos\,\vartheta\, sin\,\varphi) = cos\,\vartheta(cos\,\varphi, sin\,\varphi)$. M is a one-one mapping of the unit circle onto itself (missing the origin). Note that as ϑ, φ trace out the two-dimensional interval $0 \leq \theta \leq \pi/2, 0 \leq \varphi < 2\pi$, $M(\vartheta, \varphi)$ traces out the entire unit circle (perimeter and interior) in the base plane except for the point $(0,0)$. In the sense to be presented shortly, the triples (X, Y, Z) and $(X(M), Y(M), Z(M))$ are equivalent surfaces.

■

The concept of *equivalent surfaces* is also required here for any reasonably complete development, and we will soon provide a restricted definition which is adequate for our purposes.

Suppose we have a surface $\boldsymbol{S}$, embedded in a three-dimensional steady state (fluid) velocity field, as show in Figure 12-22.

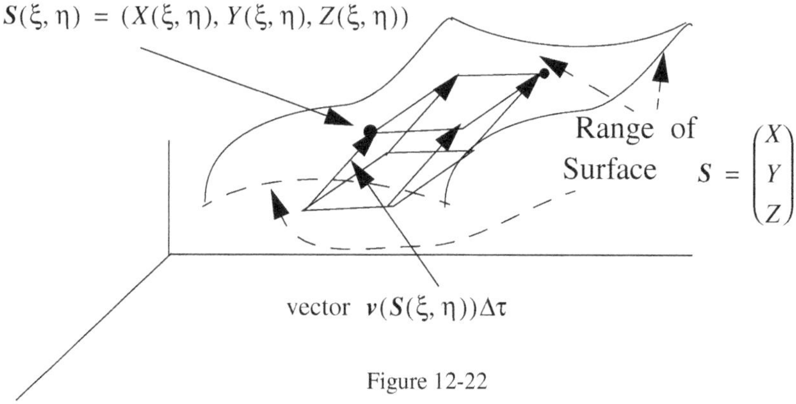

Figure 12-22

Because a smooth surface has a continuous gradient, let us choose a typical small approximate parallelogram on the range of this surface, generated by $\boldsymbol{S}(\xi, \eta), \boldsymbol{S}(\xi + \Delta\xi, \eta), \boldsymbol{S}(\xi, \eta + \Delta\eta)$, and draw, as shown, a vector $\boldsymbol{v}(\boldsymbol{S}(\xi, \eta))\Delta\tau$ (just as we did for two-dimensional fluid flow) where $\Delta\tau > 0$ represents a short increment of time. In a time period of duration $\Delta\tau$, at least approximately, the fluid in the parallelepiped formed by the parallelogram and the vector $\boldsymbol{v}(\boldsymbol{S}(\xi, \eta))\Delta\tau$ should cross the (range of the) surface $\boldsymbol{S}$, as should be evident from Figure 12-22. The net signed amount of fluid crossing should be the signed volume of this parallelepiped.

Using the multivariable fundamental approximation, 8.17 on page 274, recalling equation 12.27 on page 488, we see that the three vectors generating the approximate parallelepiped are

$$\boldsymbol{v}(\boldsymbol{S}(\xi, \eta))\Delta\tau, \quad D_1\boldsymbol{S}(\xi, \zeta)\Delta\xi, \quad D_2\boldsymbol{S}(\xi, \zeta)\Delta\eta.$$

Note that

$$D_i\boldsymbol{S}(\xi, \eta) = \begin{pmatrix} D_iX(\xi, \eta) \\ D_iY(\xi, \eta) \\ D_iZ(\xi, \eta) \end{pmatrix}.$$

From the results of Exercise 12.11.3 on page 473, the vector representing the signed area formed by the last two vectors is

$$D_1\boldsymbol{S}(\xi, \eta) \times D_2\boldsymbol{S}(\xi, \eta)\Delta\xi\Delta\eta \cong d\mathcal{A}(\boldsymbol{S}(\xi, \eta)) \qquad \textbf{\textit{12.28}}$$

and it is orthogonal to the parallelogram formed by these two vectors. The signed volume arising from this parallelogram crossing the surface of S per unit time is approximately the dot product

$$\boldsymbol{v}(\boldsymbol{S}(\xi, \eta)) \cdot d\mathcal{A}(\boldsymbol{S}(\xi, \eta))$$

and thus the net signed amount of fluid crossing the surface S per unit time (i.e., the *fluid flux*) is

$$\int_{\mathcal{D}} \boldsymbol{v}(\boldsymbol{S}(\xi, \eta)) \cdot (D_1\boldsymbol{S}(\xi, \eta) \times D_2\boldsymbol{S}(\xi, \eta))d\xi d\eta$$

where $\mathcal{D}$ is the domain of the surface S,

or

$$\int_{\mathcal{D}} \boldsymbol{v}(\boldsymbol{S}(\xi, \eta)) \cdot d\mathcal{A}(\boldsymbol{S}(\xi, \eta)) \quad \text{or} \quad \int_S \boldsymbol{v} \cdot d\mathcal{A}.$$

Note that once again we have made an arbitrary choice of the sense of positive crossing. Formally we give the following definition.

Definition 12.32: Surface integral

Let $\boldsymbol{S} = (X, Y, Z)$ represent a smooth surface with domain $\mathcal{D}$ (Definition 12.31 on page 487) and let $\boldsymbol{v}$ be a continuous three-dimensional vector valued function whose domain includes the range of $\boldsymbol{S}$ (see Definition 1.1 on page 2). The **surface integral of $\boldsymbol{v}$ over S**, denoted by

$$\int_S \boldsymbol{v} \cdot d\mathcal{A}$$

is defined as

$$\int_{\mathcal{D}} \boldsymbol{v}(\boldsymbol{S}(\xi, \eta)) \cdot (D_1\boldsymbol{S}(\xi, \eta) \times D_2\boldsymbol{S}(\xi, \eta))d\xi d\eta,$$

where the *cross product*, $\boldsymbol{m} \times \boldsymbol{n}$, of the three-dimensional vector $\boldsymbol{m}$ with the three-dimensional vector $\boldsymbol{n}$ is defined in Exercises 12.11.3 on page 473, and the dot product is given by Definition 8.10 on page 272.

❑

Note that although this definition was motivated by fluid flow considerations, it stands by itself, independent of its motivation.

Example 12.18: Flux of constant vertical fluid velocity field across upper unit hemisphere

To represent the surface of this hemisphere of unit radius, from Example 12.17 on page 488, we let

$$X(\xi, \eta) = \xi, Y(\xi, \eta) = \eta, Z(\xi, \eta) = \sqrt{1-(\xi \ \ + \eta^2)}.$$

A uniform vertical flow is represented by the constant vector field

$$(f, g, h) = (0, 0, c),$$

Then

$$v(\boldsymbol{S}(\xi, \eta)) = (0, 0, c)^T$$

and

$$D_1\boldsymbol{S}(\xi, \eta) = \begin{pmatrix} 1 \\ 0 \\ \dfrac{\xi}{\sqrt{1-(\xi^2+\eta^2)}} \end{pmatrix}, D_2\boldsymbol{S}(\xi, \eta) = \begin{pmatrix} 0 \\ 1 \\ \dfrac{\eta}{\sqrt{1-(\xi^2+\eta^2)}} \end{pmatrix}.$$

So

$$D_1\boldsymbol{S}(\xi, \eta) \times D_2\boldsymbol{S}(\xi, \eta) = \begin{pmatrix} \dfrac{-\xi}{\sqrt{1-(\xi^2+\eta^2)}} \\ -\dfrac{-\eta}{\sqrt{1-(\xi^2+\eta^2)}} \\ 1 \end{pmatrix}.$$

Thus the value we are seeking is given by

$$\int_{\mathcal{D}} c\,d\xi\,d\eta = \pi c,$$

where

$$\mathcal{D} = \{(\xi, \eta) \text{ in } \mathcal{E}_2 : 0 < \xi^2 + \eta^2 \le 1\}.$$

You should have been able to guess this result.

■

Exercises 12.19

1. Determine the surface area of the upper unit hemisphere.
   Hint: Let $\boldsymbol{v} = (f, g, h)$ be a vector field perpendicular to the surface of this hemisphere.
2. a. Determine the surface area of the ellipsoid $\frac{x^2}{2} + \frac{y^2}{3} + \frac{z^2}{4} = 1$.
   b. Determine the flux of the constant vector fields $(c_1, 0, 0)$ and $(0, c_2, 0)$ across the surface of the upper hemisphere.

To establish some of the main results concerning extensions of the fundamental evaluation theorem for integrals, we need to tell what is meant by equivalent surfaces. Recall Definition 12.28 on page 483 of equivalent curves. To avoid complications we'll restrict consideration to the functions z relating equivalent curves satisfying $z' > 0$, and insist that the curves $\boldsymbol{x}$ and $\boldsymbol{X}$ related by z themselves have nonzero derivatives. From Figure 12-23 we see that the normal

$$(-X_2'(\tau), X_1'(\tau)) = (-x_2'(z(\tau)), x_1'(z(\tau)))z'(\tau)$$

points in the same direction as the normal $(-x_2'(t), x_2'(t))$ where $t = z(\tau)$, because $z' > 0$. But z' is just the one-dimensional Jacobian determinant of the

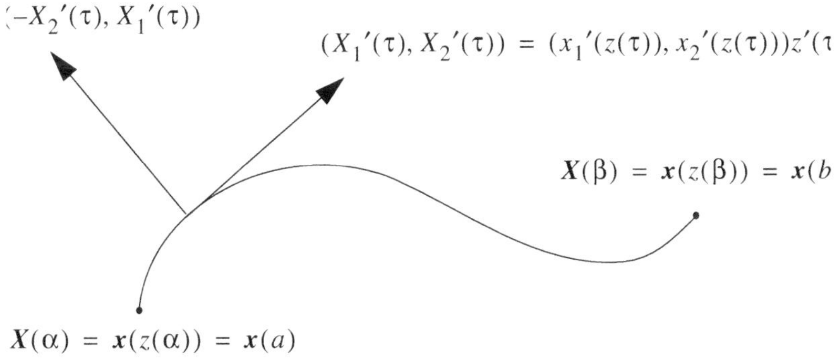

Figure 12-23

transformation z, *taking* $\boldsymbol{x}$ into $\boldsymbol{X}$ (see Theorem 12.23 on page 467). If two surfaces, $\boldsymbol{S}, \boldsymbol{U}$ yield the same range and their domains are in one-to-one correspondence with regard to their values — i.e.,

$$\boldsymbol{S} = \boldsymbol{U}(\mathbf{M}),$$

where $\mathbf{M}$ is a one-to-one function (see Definition 12.18 on page 460) whose range is the domain of $\boldsymbol{U}$ and whose domain is the domain of $\boldsymbol{S}$, and we want

$$\int_{\boldsymbol{S}} \boldsymbol{v} \cdot d\mathcal{A} = \int_{\boldsymbol{U}} \boldsymbol{v} \cdot d\mathcal{A},$$

the main matter that can go wrong is that the normal to S points in the opposite direction to that of U at corresponding domain values. This can be prevented by requiring the Jacobian determinant, J, of **M** (see Theorem 12.23 on page 467)

$$det(\mathbf{DM}^{\mathrm{T}}) = det\begin{bmatrix} D_1M_1 & D_1M_2 \\ D_2M_1 & D_2M_2 \end{bmatrix} = (D_1M_1)(D_2M_2) - (D_1M_2)(D_2M_1)$$

be always positive, just as with equivalent curves. In theory we can find out whether or not this is so, by finding its minimum value over the given domain to be positive, or showing that it is nonpositive somewhere. So we give the following definition.

Definition 12.33: Equivalent surfaces, simplest case

Two surfaces, $\boldsymbol{S}, \boldsymbol{U}$ (Definition 12.31 on page 487) with the same range (Definition 1.1 on page 2) related by the condition

$$\mathbf{S} = U(\mathrm{M}),$$

where **M** is a one-to-one function (Definition 12.18 on page 460) whose range is the domain of $\boldsymbol{U}$ and whose domain is the domain of $\boldsymbol{S}$, with **M** having a continuous invertible gradient matrix (see Theorem 10.8 on page 336 and Definition 8.5 on page 267)

$$\mathbf{DM}^{\mathrm{T}} = \begin{bmatrix} D_1M_1 & D_1M_2 \\ D_2M_1 & D_2M_2 \end{bmatrix}$$

are called *equivalent* if the following Jacobian determinant is always positive.

$$det(\mathbf{DM}^{\mathrm{T}}) = det\begin{bmatrix} D_1M_1 & D_1M_2 \\ D_2M_1 & D_2M_2 \end{bmatrix} = (D_1M_1)(D_2M_2) - (D_1M_2)(D_2M_1).$$

❑

The result we need is as follows.

Theorem 12.34: Integrals over equivalent surfaces

If $\boldsymbol{S}$ and $\boldsymbol{U}$ are equivalent surfaces (Definition 12.33 above) and $\boldsymbol{v}$ is a continuous three-dimensional vector valued function (Definition 8.7 on page 270) whose domain includes the range of $\boldsymbol{S}$ (Definition 1.1 on page 2) and whose gradients (Definition 8.10 on page 272) are continuous (see Theorem 8.8 on page 271) then using Definition 12.32 on page 490,

$$\int_{\boldsymbol{S}} \boldsymbol{v} \cdot d\mathcal{A} = \int_{\boldsymbol{U}} \boldsymbol{v} \cdot d\mathcal{A}.$$

❑

The proof, which we omit, is straightforward, but very tedious, requiring careful use of the multivariable chain rule (Theorem 12.25 on page 474) and the rule for change of variable in multiple integration (Theorem 12.23 on page 467).

Just as with curves, we must be able to piece together smooth surfaces in a continuous fashion, to form more general surfaces, and generalize the definition of equivalent surfaces, showing that integrals over equivalent surfaces are equal. This is needed just to deal properly with the simple case of the surface area of a cube. Unfortunately, carrying this development out properly would take more effort than we feel is warranted, since all we desire to provide here is an introduction to those who may encounter these tools in later work.

Both line and surface integrals have been related to fluid flux, and in many cases a curve will enclose some region, A, in the plane, or on the range of a surface while a surface will enclose some region B, in three-dimensional space. Intuitively these assertions aren't hard to understand, as illustrated in Figure 12-24.

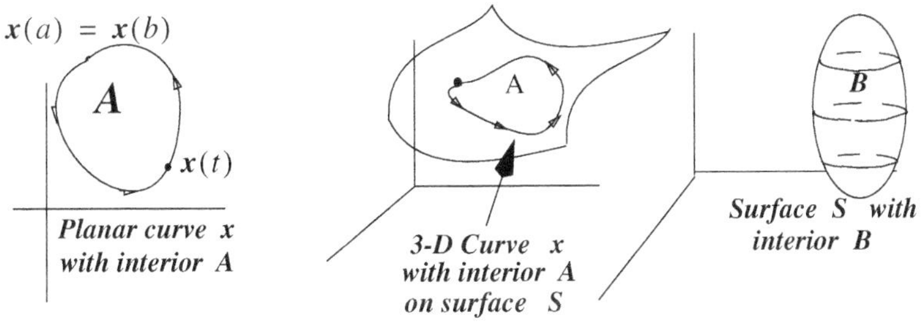

Figure 12-24

Giving these concepts precise meaning in the general case is, as already stated, beyond the scope of this work. So in what follows, we will restrict ourselves to examples which are intuitively understandable with the tools already introduced.

The first problem we'll look at is to determine a convenient representation for the *net signed rate of exit of fluid from a planar region* A enclosed by a curve $\boldsymbol{x}$ assuming a stationary fluid velocity field in the plane. We start with a simple case, namely, we look at the situation in which the planar region, A, has two representations:

$$A = \{(x, y) \text{ in } \mathcal{E}_2 : a \leq x \leq b \text{ and } \varphi(x) \leq y \leq \psi(x)\}$$
$$\text{and} \qquad\qquad \textbf{\textit{12.29}}$$
$$A = \{(x, y) \text{ in } \mathcal{E}_2 : \gamma \leq y \leq \delta \text{ and } \xi(y) \leq x \leq \zeta(y)\},$$

as illustrated in Figure 12-25, with

$$\varphi(a) = \psi(a), \varphi(b) = \psi(b), \xi(\gamma) = \zeta(\gamma), \xi(\delta) = \zeta(\delta),$$

and where $\varphi, \psi, \xi, \zeta$ are piecewise smooth (Definition 3.13 on page 89).

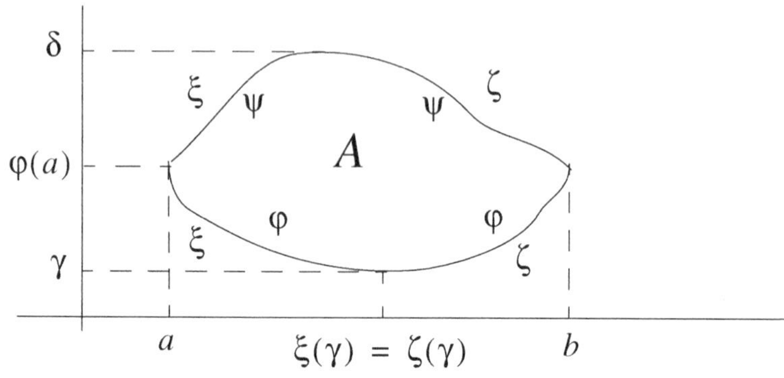

Figure 12-25

The region A is enclosed by the curve $\boldsymbol{x} = (x_1, x_2)$ defined by

$$x_1(t) = t, x_2(t) = \varphi(t) \quad \text{for} \quad a \le t \le b,$$
$$x_1(t) = -t + 2b, x_2(t) = \psi(-t + 2b) \quad \text{for} \quad b \le t \le 2b - a. \qquad \textbf{12.30}$$

The curve $\boldsymbol{x}(t)$ appears to traverse the boundary of A in a counterclockwise sense as t increases from a to $2b - a$. Now suppose the range of $\boldsymbol{x}$ is embedded in the two-dimensional steady state fluid velocity flow field, $\boldsymbol{v} = (v_1, v_2)$. Except maybe for a reversed algebraic sign, the flow rate we want is given by

$$\int_{\boldsymbol{x}} [v_2(\boldsymbol{x})dx_1 - v_1(\boldsymbol{x})dx_2]$$

(see Theorem 12.30 on page 487) where

$$\int_{\boldsymbol{x}} v_2(\boldsymbol{x})dx_1 = \int_a^{2b-a} v_2(x_1(t), x_2(t))x_1'(t)dt$$
$$= \int_a^b v_2(x_1(t), x_2(t))x_1'(t)dt + \int_b^{2b-a} v_2(x_1(t), x_2(t))x_1'(t)dt.$$

But

$$\int_a^b v_2(x_1(t), x_2(t))x_1'(t)dt = \int_a^b v_2(t, \varphi(t))dt$$

since $x_1'(t) = 1, x_2(t) = \varphi(t)$ for all t in $[a; b]$, while

$$\int_b^{2b-a} v_2(x_1(t), x_2(t))x_1'(t)dt. = -\int_b^{2b-a} v_2(-t + 2b, \psi(-t + 2b))dt$$
$$= \int_b^a v_2(\tau, \psi(\tau))d\tau,$$

the first sign change arising from $x_1'(t) = -1$ over this range and the final expression arising from the change of variable $\tau = -t + 2b$.

Hence

$$\int_{\boldsymbol{x}} v_2(\boldsymbol{x})dx_1 = \int_a^b v_2(t, \varphi(t))dt + \int_b^a v_2(\tau, \psi(\tau))d\tau, \\ = \int_a^b [v_2(t, \varphi(t)) - v_2(t, \psi(t))]dt. \qquad \textit{12.31}$$

(Note the sign change for the second term.)

We can similarly compute $\int_X v_1(X)dX_2$, where $\boldsymbol{X}$ is a curve whose range is the boundary of A, by using the second representation of A, in equations 12.29 on page 494. Suppose we take the natural approach of defining $\boldsymbol{X} = (X_1, X_2)$ by

$$X_1(t) = \xi(t), X_2(t) = t \quad \text{for} \quad \gamma \le t \le \delta \\ X_1(t) = \zeta(2\delta - t), X_2(t) = 2\delta - t \quad \text{for} \quad \delta \le t \le 2\delta - \gamma. \qquad \textit{12.32}$$

(Note that $\boldsymbol{X}$ goes clockwise, the reverse of $\boldsymbol{x}$. At the last stage we'll take this into account.) We find

$$\int_X v_1(\boldsymbol{X})dX_2 = \int_\gamma^\delta v_1(X_1(t), t)X_2'(t)dt + \int_\gamma^{2\delta - \gamma} v_1(X_1(t), t)X_2'(t)dt$$

$$= \int_\gamma^\delta v_1(\xi(t), t)dt - \int_\gamma^{2\delta - \gamma} v_1(\zeta(2\delta - t), 2\delta - t)dt$$

$$= \int_\gamma^\delta v_1(\xi(t), t)dt + \int_\delta^\gamma v_1(\zeta(t), t)dt$$

$$= \int_\gamma^\delta (v_1(\xi(t), t) - v_1(\zeta(t), t))dt.$$

Since $\boldsymbol{X}$ is traversed clockwise as t increases, we let $\boldsymbol{X}^*$ be $\boldsymbol{X}$ traversed in reverse (i.e., $X^*(-t) = X(t)$) so that

$$\int_{X^*} v_1(\boldsymbol{X}^*)d\boldsymbol{X}^* = -\int_\gamma^\delta (v_1(\xi(t), t) - v_1(\zeta(t), t))dt.$$

Now we note that $\boldsymbol{X}(\gamma) = \boldsymbol{X}(2\delta - \gamma)$, so that if we extend $\boldsymbol{X}^*$ periodically (for simplicity keeping the same name) it remains continuous. Now choose as $\boldsymbol{X}^{**}$ the portion of $\boldsymbol{X}^*$ that starts at the value t_0 for which $\boldsymbol{X}^*(t_0) = \boldsymbol{x}(a)$ and ends at $t + 2(\delta - \gamma)_0$. Geometrically it is evident that $\boldsymbol{X}^{**}$ is equivalent to $\boldsymbol{x}$ and (due to the periodicity of $\boldsymbol{X}^*$)

$$\int_{X^{**}} v_1(\boldsymbol{X}^{**})dX_2^{**} = \int_{\boldsymbol{x}} v_1(\boldsymbol{x})dx_2.$$

Hence

$$\int_{\boldsymbol{x}} v_1(\boldsymbol{x})dx_2 = -\int_\gamma^\delta [v_1(\xi(t), t) - v_1(\zeta(t), t)]dt. \qquad \textit{12.33}$$

Combining equations 12.31 and 12.33 yields

$$\int_{x} [v_2(\boldsymbol{x})dx_1 - v_1(\boldsymbol{x})dx_2] = \int_a^b [v_2(t, \varphi(t)) - v_2(t, \psi(t))]dt + \int_\gamma^\delta [v_1(\xi(t), t) - v_1(\zeta(t), t)]dt$$

But by the fundamental evaluation formula, Theorem 4.3 on page 103, we may write

$$v_2(t, \varphi(t)) - v_2(t, \psi(t)) = \int_{\psi(t)}^{\varphi(t)} D_2 v_2(t, y)dy,$$

$$v_1(t, \xi(t)) - v_1(t, \zeta(t)) = \int_{\zeta(t)}^{\xi(t)} D_1 v_1(t, y)dy.$$

Hence

$$\int_{x} [v_2(\boldsymbol{x})dx_1 - v_1(\boldsymbol{x})dx_2] = \int_a^b \left[\int_{\psi(t)}^{\varphi(t)} D_2 v_2(t, y)dy\right]dt + \int_\gamma^\delta \left[\int_{\zeta(t)}^{\xi(t)} D_1 v_1(t, y)dy\right]dt$$
$$= -\int_A \int (D_2 v_2 + D_1 v_1)dV_2,$$

this last step justified by Theorem 12.17 on page 455. We reverse signs to write this result as

$$\int_{x} [v_1(\boldsymbol{x})dx_2 - v_2(\boldsymbol{x})dx_1] = \int_A \int (D_1 v_1 + D_2 v_2)dV_2, \qquad \textbf{\textit{12.34}}$$

Now let's look at whether $\int_{x} [v_1(\boldsymbol{x})dx_2 - v_2(\boldsymbol{x})dx_1]$ represented flow *in* or *out* of A. As seen in Figure 12-26, since $\boldsymbol{x}$ is traversed counterclockwise, the normal, $\begin{pmatrix} -x_2' \\ x_1' \end{pmatrix}$, points into A. Hence, at any point, the contribution of the integrand of

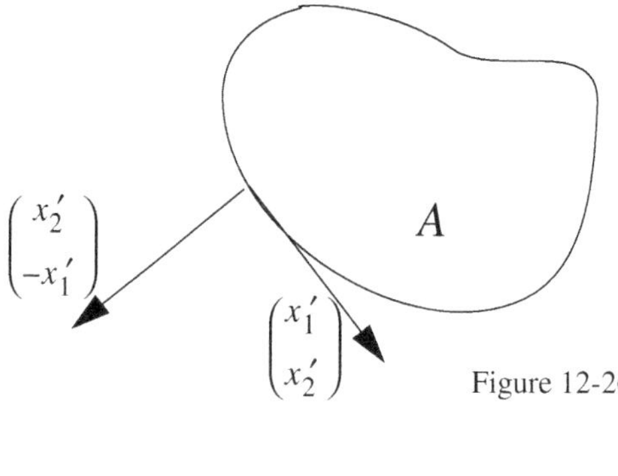

Figure 12-26

the left expression in equation 12.34 is to the net outward flow (a negative value if the flow at this point is inward). Reversing the direction in which the curve $\boldsymbol{x}$ traverses A would reverse the sign of the left integral in equation 12.34, making it yield the net outward flow, rather than the net inward flow.

We are now in a position to state Gauss's theorem — of which we have only established a simple case. This theorem does not depend on a fluid flow interpretation.

Theorem 12.35: Gauss's integral transformation theorem

Let $v = (v_1, v_2)$ be a two-dimensional continuous vector field (pair of real valued continuous functions of two real variables; see Definition 8.7 on page 270) and let A be a two-dimensional set defined by equation 12.29 on page 494. Let $\boldsymbol{x}$ be the curve defined by equations 12.30 on page 495. Then

$$\int_{\boldsymbol{x}} [v_1(\boldsymbol{x})dx_2 - v_2(\boldsymbol{x})dx_1] = \int\int_A (D_1 v_1 + D_2 v_2)dV_2. \qquad 12.35$$

❑

Both integrals in equation 12.35 represent the net outward flow rate of fluid from A when (v_1, v_2) is a stationary fluid velocity (vector) field.

Note that ***the fundamental summation theorem*** (Theorem 1.8 on page 12)

$$\sum_{j=m}^{k-1} \Delta f(j) = f(k) - f(m),$$

can be thought of as relating an integral over some region (the left-hand sum) to an integral over the boundary of this region (the right-hand sum). The same is true for the fundamental evaluation formula, equation 4.14 on page 103

$$\int_a^b g = \int_a^b f' = f(b) - f(a).$$

for one-dimensional integrals. Here, the boundary of $[a; b]$ consists of the two points, a, b (with an *orientation*). For Gauss's theorem, $\boldsymbol{x}$ is an oriented boundary of A. ***So we may think of Gauss's theorem as a multidimensional extension of the fundamental evaluation theorem.***

Definition 12.36: Divergence of a vector field

If $v = (v_1, v_2)$, $[v = (v_1, v_2, v_3)]$, is a two-dimensional [three-dimensional] vector field with continuous gradients (see Definition 8.10 on page 272 and Definition 8.7 on page 270) the vector

$$D_1 v_1 + D_2 v_2, \quad [D_1 v_1 + D_2 v_2 + D_3 v_3]$$

is called the divergence of v, and is denoted by $div\ v$. If $\mathbf{D}$ is the gradient operator, whose transpose $\mathbf{D}^T$ is the row matrix of partial derivative operators, then

$$div\ v = \mathbf{D}^T \mathbf{v},$$

where $\mathbf{v}$ is the column matrix form of the vector v.

❑

Gauss's theorem provides an interpretation of the divergence; because if A is a very small circle centered at a point (x_0, y_0), then

$$\int\int_A (D_1 v_1 + D_2 v_2) dV_2 = \int\int_A div\, \boldsymbol{v}\, dV_2 \cong div\, \boldsymbol{v}\,(x_0, y_0) V_2(A).$$

So, using the fluid flow interpretation of Gauss's theorem,

$$div\, \boldsymbol{v}\,(x_0, y_0) V_2(A) \cong \text{ net flow rate out of } A,$$

and hence $div\, \boldsymbol{v}\,(x_0, y_0)$ is the local flow rate *out* per unit two-dimensional volume at (x_0, y_0) (explaining why $D_1 v_1 + D_2 v_2$ is called the divergence).

Gauss's theorem in two dimensions can be extended to much more complicated curves which enclose a region, and its meaning stays the same. The difficulty with such extension is that great care must be taken to be sure we know what is meant by a curve enclosing a region. The region must be *simply connected*, implying that there are no *holes* in it; notice that the region A of Figure 12-25 on page 495 could not have any holes because of the simple way in which it was defined. In most of the commonly encountered applied problems, there is little difficulty in this regard. Another restriction is that the curve $\boldsymbol{x}$ cannot loop over itself. Again, the geometry of most of the common problems causes no trouble.

The extension of Gauss's integral transformation theorem to three dimensions relates the triple integral of the divergence of a three-dimensional vector field over some three-dimensional set, A, to the surface integral over the boundary of this set. The simplest extension occurs when the set A permits three sequential representations (Definition 12.16 on page 454)

$$A = \{(x_1, x_2, x_3) \text{ in } \mathcal{E}_3\colon a_1 \le x_1 \le b_1; a_2 \le x_2 \le b_2; \varphi(x_1, x_2) \le x_3 \le \psi(x_1, x_2)\}$$

$$A = \{(x_1, x_2, x_3) \text{ in } \mathcal{E}_3\colon a_1 \le x_1 \le b_1; a_3 \le x_3 \le b_3; \xi(x_1, x_3) \le x_2 \le \zeta(x_1, x_3)\}$$

and

$$A = \{(x_1, x_2, x_3) \text{ in } \mathcal{E}_3\colon a_2 \le x_2 \le b_2; a_3 \le x_3 \le b_3; \rho(x_2, x_3) \le x_1 \le \mu(x_2, x_3)\}.$$

If $\boldsymbol{v} = (v_1, v_2, v_3)$ is a three-dimensional vector field, and it as well as the functions defining A above are all well behaved, the three-dimensional version of Gauss's theorem concludes that

$$\int\int\int_A div\, \boldsymbol{v}\, dV_3 = \int\int_{\boldsymbol{S}} \boldsymbol{v}(\boldsymbol{S}) \cdot d\mathcal{A}(\boldsymbol{S}) \qquad \textbf{12.36}$$

(see Definition 12.32 on page 490) where $\boldsymbol{S}$ is a surface enclosing A, whose normal, $d\mathcal{A}(\boldsymbol{S})$, points outward from A (easy enough to determine in simple cases). In the case being looked at here, using the first of the sequential representations for A, we may take $\boldsymbol{S} = (X, Y, Z)$ with

$$X(\xi, \eta) = \xi,\ Y(\xi, \eta) = \eta,\ Z(\xi, \eta) = \varphi(\xi, \eta) \quad \text{for } a_1 \le \xi \le b_1,\ a_2 \le \eta \le b_2,$$

where

$$\varphi(\xi, \eta) = \psi(\xi, \eta)$$
$$\text{for all } \eta \text{ when } \xi = a_1 \text{ or } \xi = b_1$$
$$\text{for all } \xi \text{ when } \eta = a_2 \text{ or } \eta = b_2,$$

$$X(\xi, \eta) = 2a_1 - \xi,\ Y(\xi, \eta) = 2a_2 - \eta,\ Z(\xi, \eta) = \psi(2a_1 - \xi, 2a_2 - \eta)$$
$$\text{for } b_1 \le \xi \le 2a_1 - b_1,\ b_2 \le \eta \le 2a_2 + -b_2.$$

A sphere is representable in such a fashion, but a polar representation will allow better behaved gradients.

For the other important extension of the fundamental integral evaluation theorem, Stoke's theorem, we first rewrite the conclusion of the two-dimensional version of Gauss's theorem, equation 12.35 on page 498, with a seemingly trivial change — namely replace v_1 by $-v_1$, to obtain

$$\int_A\int (D_1 v_2 - D_2 v_1) dV_2 = \int_{\boldsymbol{x}} [v_1(\boldsymbol{x}) dx_1 + v_2(\boldsymbol{x}) dx_2].$$

Now if the vector field $\boldsymbol{v} = (v_1, v_2)$ is interpreted as a force field, the right-hand integral may be viewed as the work done by an object traversing the curve $\boldsymbol{x}$ which encloses A. Here A may be considered to be a **portion** of a surface in three dimensions. The left-hand double integral may be considered a surface integral, see Definition 12.32 on page 490 — i.e., the integral of a dot product

$$\int_S\int \boldsymbol{v}(\boldsymbol{S}) \cdot d\mathcal{A}(\boldsymbol{S}),$$

where $\boldsymbol{S}$ (the surface over which integration takes place) is A, *and* $\boldsymbol{v}(\boldsymbol{S})$ is the vector given by the symbolic determinant

$$det\begin{bmatrix} \boldsymbol{i} & D_1 & v_1 \\ \boldsymbol{j} & D_2 & v_2 \\ k & D_3 & 0 \end{bmatrix} = 0\boldsymbol{i} - 0\boldsymbol{j} + (D_1 v_2 - D_2 v_1)\boldsymbol{k}.$$

Here ***i, j, k*** are the standard unit axis vectors in three dimensions, as illustrated in Figure 12-27.

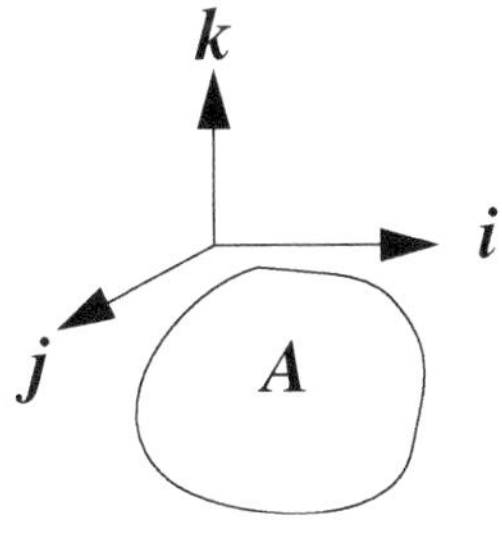

Figure 12-27

Definition 12.37: Curl of a vector field

If $\boldsymbol{v}$ is any well-behaved three-dimensional vector field, $\boldsymbol{v} = (v_1, v_2, v_3)$ (and with well-behaved components, $\boldsymbol{v} = (v_1, v_2, 0)$ is such a field), the vector

$$det\begin{bmatrix} \boldsymbol{i} & D_1 & v_1 \\ \boldsymbol{j} & D_2 & v_2 \\ k & D_3 & v_3 \end{bmatrix} = \begin{pmatrix} D_2v_3 - D_3v_2 \\ D_3v_1 - D_1v_3 \\ D_1v_2 - D_2v_1 \end{pmatrix}$$

is called the *curl* of **v**, and is denoted by ***curl* v**.

❑

Because of the formidable technical difficulties of stating Stoke's theorem properly, we are not providing a formal statement. Informally, what **Stoke's theorem** states is that if $\boldsymbol{S}$ is any reasonable (part of a) surface enclosed by the three-dimensional curve $\boldsymbol{x} = (x_1, x_2, x_3)$, then

$$\int\int_{\boldsymbol{S}} \boldsymbol{curl}\ \boldsymbol{v}(\boldsymbol{S}) \cdot d\mathcal{A}(\boldsymbol{S}) = \int_{\boldsymbol{x}} [v_1(x)dx_1 + v_2(x)dx_2 + v_3(x)dx_3] \qquad \textbf{12.37}$$

(see Definition 12.27 on page 481 and Definition 12.32 on page 490). We are avoiding considerations of the orientation of a surface corresponding to the orientation of its boundary curve, because our purpose is simply to provide some intuition as to the meaning of this theorem. The basic idea of surface and curve orientation is sort of as follows. The orientation of a curve can be specified by the order in which the points on its range are traversed. This doesn't seem to extend in any obvious manner to the orientation of a surface. But in two dimensions curve orientation can also be specified by the direction of the normal, $(-x_2', x_1')$, to the curve (which reverses sign when the curve $\boldsymbol{x}$ has its direction of traversal reversed). Similarly, the orientation of a surface can be specified by the direction of its normal, $d\mathcal{A}(\boldsymbol{S})$ (see Approximation 12.28 on page 490). To reverse the orientation of $\boldsymbol{S}$ we see that we need only reverse the sign of the normal being used — i.e., use $D_2\boldsymbol{S}(\xi, \eta) \times D_1\boldsymbol{S}(\xi, \eta)$ rather than $D_1\boldsymbol{S}(\xi, \eta) \times D_2\boldsymbol{S}(\xi, \eta)$. Suffice it to say here that the normal to the surface, $d\mathcal{A}(\boldsymbol{S})$, should point outward from the bounded set enclosed by this surface, and the normal to the curve $\boldsymbol{x}$ which lies on the surface should point outward from the surface set enclosed by $\boldsymbol{x}$. A more detailed discussion of this issue can be found in Courant [5].

Stoke's theorem is an extension of the fundamental evaluation theorem, in the sense that it relates a boundary integral to the integral over the region enclosed by this boundary. This result gives meaning to ***curl* v** as the local work per unit area near a given point, in traversing a small circle near this point. It explains why

***curl* v** is referred to as the *rotation of the field* **v**.

The integral evaluation theorems introduced here are of great importance in the study of partial differential equations, such as those governing three-dimensional vibrations, electrical fields and heat flow. One such equation, credited to **Poisson**[1] is

$$D_1^2 V + D_2^2 V + D_3^2 V = -\frac{\rho}{\varepsilon_0}, \qquad \textbf{\textit{12.38}}$$

which must be satisfied by any electrical potential function V in any region in which the (known) charge density function is denoted by ρ, and ε_0 is a known constant whose values depends on the units of measurement of length, electrical charge, etc. The electrical force field, $\boldsymbol{E} = (E_1, E_2, E_3)$ is a three-dimensional vector-valued function of position, where $\boldsymbol{E}(x, y, z)$ represents the force on a charged particle at position (x, y, z). The real valued *potential function, V,* is related to $\boldsymbol{E}$ by the equations

$$D_i V = E_i, \quad i = 1, 2, 3. \qquad \textbf{\textit{12.39}}$$

So it is evident why one would want to study the behavior of solutions of Poisson's equation, 12.38. Just as with ordinary linear differential equations, because the Laplacian operator Δ is linear (i.e., has the linear operator property, $\Delta(af + bg) = a\Delta f + b\Delta g$) ***each*** solution of Poisson's equation is a sum, $V_p + V_H$, where V_p is any specific solution of Poisson's equation, and V_H is some solution of **Laplace's equation**

$$D_1^2 V + D_2^2 V + D_3^2 V = 0. \qquad \textbf{\textit{12.40}}$$

For this reason a good deal of attention is given to investigating Laplace's equation. Here is an illustration of how the integral transformation theorems can be used in problems involving Laplace's equation.

Example 12.20: Laplace's equation in the unit sphere

Let A be the unit sphere in three dimensions; i.e.,

$$A = \{(x, y, z) \text{ in } \mathcal{E}_3 : x^2 + y^2 + z^2 \le 1\},$$

and suppose that

- f is a function of three real variables, which has all first and second partial derivatives each of which is continuous at every point in A (see Definition 8.5 on page 267 and Definition 8.7 on page 270);
- at all points on the boundary of the unit sphere (points for which $x^2 + y^2 + z^2 = 1$), we have $f(x, y, z) = 0$ and

  $$(D_1^2 + D_2^2 + D_3^2) f(x, y, z) = 0$$

  for all (x, y, z) in A.

---

1. This equation is often written as $\Delta V = -\rho/\varepsilon_0$ or $\nabla^2 V = -\rho/\varepsilon_0$. The symbol $\Delta = \nabla^2$ is called the *Laplacian operator.*

We will show that under these conditions

$$f(x, y, z) = 0 \quad \text{for all} \quad (x, y, z) \quad \text{in} \quad A\,.$$

Let S be the surface of A with normal pointing out. First we show that

$$\iiint_A \left((\boldsymbol{grad}\, f) \cdot (\boldsymbol{grad}\, f) + f \sum_{i=1}^{3} D_i^2 f \right) dV_3 = \iint_S (f\,\boldsymbol{grad}\, f) \cdot d\mathcal{A}. \qquad \mathbf{\textit{12.41}}$$

This follows from equation 12.36, the conclusion of the three-dimensional version of Gauss's theorem, by letting

$$\boldsymbol{v} = (fD_1 f, fD_2 f, fD_3 f) = f\,\boldsymbol{grad}\, f\,,$$

from which

$$div\ \boldsymbol{v} = \sum_{i=1}^{3} D_i v_i = \sum_{i=1}^{3} D_i(fD_i f) = \sum_{i=1}^{3} (D_i f)^2 + \sum_{i=1}^{3} f D_i^2 f$$

$$= (\boldsymbol{grad}\, f) \cdot (\boldsymbol{grad}\, f) + f \sum_{i=1}^{3} D_i^2 f\,.$$

In equation 12.41, note that since $f(x, y, z) = 0$ for points (x, y, z) on the sphere's surface, the surface integral must be 0. Because f satisfies Laplace's equation everywhere in A, the term

$$f \sum_{i=1}^{3} D_i^2 f = 0$$

everywhere in A. Hence equation 12.41 yields

$$\iiint_A (\boldsymbol{grad}\, f) \cdot (\boldsymbol{grad}\, f) dV_3 = 0,$$

i.e.,

$$\iiint_A \|\boldsymbol{grad}\, f\|^2 dV_3 = 0.$$

From this, it follows from the continuity of $\boldsymbol{grad}\, f$ that $\|\boldsymbol{grad}\, f\|$ is constant on A; but since f is 0 on the boundary of A, this constant must be 0, implying the claimed result.

We leave it as an exercise to show that this proves that there is only one function with prescribed values on the surface of the unit sphere which satisfies either Poisson's or Laplace's equation on and inside this sphere.

■

8 Complex Function Theory in Brief

The subject of differentiable complex valued functions of a complex variable is a vast one, with hosts of applications in physics, kinetics, and many other branches of mathematics. Fortunately there are many fine books at all levels on this subject. In this section we will only provide an introduction to four topics in this field:

- differentiation
- line integrals
- the Cauchy integral theorem
- power series, Laurent series, and the residue theorem,

because of their immediate importance in so many common applications. This list of topics seems to us the minimal set that you should have some familiarity within complex variable theory for applications.

Definition 12.38: Derivative of a complex function

Recalling Section 5 of Chapter 6 (starting page 199), let f be a ***complex valued function of a complex variable***, with $f(z)$ defined for all z close enough to some fixed complex number z_0. The **derivative**, $f'(z_0) = Df(z_0)$, is defined to be

$$\lim_{|z - z_0| \to 0} \frac{f(z) - f(z_0)}{z - z_0}.$$

❑

Formally this is just the familiar definition, and because of this formality, all the usual differentiation rules (see Theorem 2.13 on page 53) still hold. The reader should, however, keep in mind that the difference quotient

$$\frac{f(z) - f(z_0)}{z - z_0}$$

must be close to the fixed complex value $Df(z_0)$ ***for all complex*** z sufficiently close[1] to z_0; so differentiability of a complex valued function of a complex variable is a much stronger requirement than differentiability of a real valued function of a real variable. For instance, the complex function f given by

$$f(z) = \begin{cases} e^{-1/z^2} & \text{for } z \neq 0 \\ 0 & \text{for } z = 0 \end{cases} \qquad \textit{12.42}$$

1. At all points in some positive radius circle centered at z_0.

is not, as a function of a complex variable, differentiable at $z = 0$, even though, as a function of a real variable, $f'(0) = 0$. We leave the verification of these results to the following exercises.

Exercises 12.21

1. Show that as a function of a real variable, $f'(0) = 0$, for the function given by equations 12.42.
2. Show that as a function of a real variable, $f^{(n)}(0) = 0$, $n = 2,3,...$ for the function given by equations 12.42.
3. Show that as a function of a complex variable, f given by equations 12.42 is not differentiable at 0.
4. Compute $f'(z)$ for the following, or show it can't be done.
   a. $f(z) = z^n$, Hint: Make use of the result that the formal differentiation rules still hold.
   b. $f(z) = e^z$. Hint: Use Definition 6.9 on page 200 and the hint on part a of this exercise.
   c. $f(z) = sin(z) = \dfrac{e^{iz} - e^{-iz}}{2i}$.
   d. $f(z) = cos(z) = \dfrac{e^{iz} + e^{-iz}}{2}$.
   e. $f(z) = \bar{z} = \overline{x + iy} = x - iy$.

Remark: The result given in Exercise 12.21.2 displays a property of the function defined in equation 12.42 which is extremely useful in the theory of generalized functions, as we'll see in Chapter 15 (see Definition 15.18 on page 687). So this function is not merely an odd mathematical curiosity.

Definition 12.39: Regularity at a point

A complex valued function of a complex variable is said to be **regular at a point** z_0 if its derivative, $f'(z)$ (see Definition 12.38) exists at all points sufficiently close to z (including z_0 itself).

❑

Exercises 12.22

1. For f a function regular at $z = (x, y)$, let
$$f(z) = f_R(x, y) + i f_I(x, y)$$
where f_R and f_I are the real valued *real and imaginary* parts of f. *Show that*
$$D_1 f_R(x, y) = D_2 f_I(x, y) \text{ and } D_2 f_R(x, y) = -D_1 f_I(x, y).$$

Hint: $Df(z) = D_1 f_R(x, y) + iD_1 f_I(x, y).$

But also $Df(z) = \dfrac{D_2 f_R(x, y) + iD_2 f_I(x, y)}{i}.$

The equations

$$D_1 f_R = D_2 f_I,\ D_2 f_R = -D_1 f_I$$

are called the **Cauchy-Riemann** (partial) differential equations.

2. Show that under the conditions of exercise 1 above, if f' is continuous

$$(D_1^2 + D_2^2) f_R = 0 \quad \text{and} \quad (D_1^2 + D_2^2) f_I = 0.$$

The significance of this result is that the real and imaginary parts of a regular function of a complex variable satisfy Laplace's partial differential equations — an important result in fluid flow and electrical field theory.

We can now introduce the complex line integral over some curve C.

Definition 12.40: Piecewise smooth curve in the complex plane

A **piecewise smooth curve**, C, **in the complex plane** is a complex-valued function of a real variable, whose domain consists of a closed bounded interval, $[a;b] = \{real\ t : a \le t \le b\}$, whose components C_R, C_I are piecewise smooth (Definition 3.13 on page 89).

Definition 12.41: Complex integral $\int_C f(z)dz$

Let f be a complex valued function of a complex variable, defined and regular (Definition 12.39 on page 505) at every point of some bounded set (Definition 10.43 on page 406) A in the complex plane. Let $C = C_R + i\,C_I$ be a continuous, piecewise smooth curve in the complex plane, whose range (Definition 1.1 on page 2) lies in A. The **integral of f along** C (called a **contour integral**) ***denoted by*** $\int_C f(z)dz$ is given by

$$\int_C f(z)dz = \lim_{\|P\| \to 0} \sum_{i=0}^{n-1} f(C(\tau_i))(C(t_{i+1}) - C(t_i)),$$

where P is a partition $a = t_0 < t_1 < \cdots < t_n = b$ of $[a; b]$, with norm $\|P\|$ (Definition 4.1 on page 101) and τ_i is an arbitrary element of the interval $[t_i, t_{i+1}]$. As usual, the integral is the value the right-hand sum gets *close to* for small values of $\|P\|$.

❑

With all of the assumptions that we have made about the function f and the curve C, it turns out that this integral is well defined, and is just a special type of line integral treated previously in Section 7 of this chapter. The results here get interesting because of our consideration of functions f which are regular at all points of the set A.

Example 12.23: $\int_C f(z)dz$ in polar coordinates

To write this integral in polar coordinates, we represent C in rectangular coordinates as

$$C = C_R + iC_I \qquad \textbf{12.43}$$

or in polar coordinates as

$$C(t) = r(t)e^{i\vartheta(t)} = r(t)cos(\vartheta(t)) + ir(t)sin(\vartheta(t)), \quad a \le t \le b \qquad \textbf{12.44}$$

(see Definition 6.12 on page 200).

Substituting from equation 12.44 into the approximating sum, from Definition 12.41 of $\int_C f(z)dz$ and the mean value theorem, Theorem 3.1 on page 62, yields

$$\begin{aligned}\int_C f(z)dz &\cong \sum_i f(r(\tau_i))e^{i\vartheta(\tau_i)}\left(r(t_{i+1})e^{i\vartheta(t_{i+1})} - r(t_i)e^{i\vartheta(t_i)}\right)\\ &= \sum_i f(r(\tau_i))e^{i\vartheta(\tau_i)}\left(\{r(t_{i+1}) - r(t_i)\}e^{i\vartheta(t_{i+1})}\right.\\ &\quad \left. + r(t_i)\{e^{i\vartheta(t_{i+1})} - e^{i\vartheta(t_i)}\}\right)\\ &\cong \int_a^b f(r(t)e^{i\vartheta(t)})r'(t)e^{i\vartheta(t)}dt + \int_a^b f(r(t)e^{i\vartheta(t)})r(t)i\vartheta'(t)e^{i\vartheta(t)}dt.\end{aligned}$$

Similarly the rectangular coordinates form for C, equation 12.43, yields

$$\begin{aligned}\int_C f(z)dz &= \int_a^b f(C(t))C'(t)dt\\ &= \int_a^b [f_R(C(t)) + if_I(C(t))][C_R'(t) + iC_I'(t)]dt.\end{aligned}$$

So we have the two results

$$\int_C f(z)dz = \int_a^b f(r(t)e^{i\vartheta(t)})[r'(t)e^{i\vartheta(t)} + ir(t)\vartheta'(t)e^{i\vartheta(t)}]dt \qquad \textbf{12.45}$$

(polar) and

$$\int_C f(z)dz = \int_a^b [f_R(C(t)) + if_I(C(t))][C_R'(t) + iC_I'(t)]dt \qquad \textit{12.46}$$

(rectangular).

■

For integrals over circular paths, equation 12.45 is often the simplest to use, while for integrals over lines parallel to either axis, equation 12.46 is often preferable. Here is an example to illustrate the use of equation 12.45.

Example 12.24: $\int_C z^n dz$ n an integer, C a circle centered at 0

Here $r(t) = R > 0$ (radius) for $0 \le t < 2\pi$, $\vartheta(t) = t$ for $0 \le t < 2\pi$, so from equation 12.45

$$\begin{aligned}
\int_C z^n dz &= \int_0^{2\pi} f(Re^{it}) iRe^{it} dt \\
&= \int_0^{2\pi} (Re^{it})^n iRe^{it} dt \\
&= iR^{n+1} \int_0^{2\pi} e^{i(n+1)t} dt \\
&= \begin{cases} iR^{n+1} \dfrac{1}{i(n+1)} e^{i(n+1)t}\Big|_0^{t=2\pi} = 0 & \text{for} \quad n \ne -1 \\ iR^{n+1} 2\pi = 2\pi i & \text{for} \quad n = -1. \end{cases}
\end{aligned}$$

So we see that for C the counterclockwise traversed circle given by $C(t) = Re^{it}, 0 \le t < 2\pi$

$$\int_C z^n dz = \begin{cases} 0 & \text{for } n \ne -1 \\ 2\pi i & \text{for } n = -1. \end{cases}$$

■

Exercises 12.25

Compute $\int_C f(z)dz$ for the following.

1. $f(z) = z$ where C is the line segment from $\mathbf{0} = (0 + i0)$ to $\mathbf{i} = 0 + \mathbf{i}1$.

2. $f(z) = e^z$ over the same path as in question 1.
Hint: Write out the desired integral *longhand.*

3. $f(z) = z$ for $C = C_U$, the upper half of the unit circle traversed from $(1, 0)$ to $(-1, 0)$
Hint: Convert to polar coordinates as in Example 12.24.

4. $f(z) = z$ for $C = C_L$, the lower half of the unit circle traversed from $(1, 0)$ to $(-1, 0)$

5. Same as exercises 3 and 4, but with $f(z) = 1/z, \quad z \neq 0$.

Exercises 3, 4, and 5 above illustrate that sometimes the integral of a given complex function depends only on the (oriented) boundary of the path of integration i.e., (beginning point ; end point), and at other times it depends on other properties of the path. In order to investigate this question we introduce the concept of *deformation of a curve*[1] (our definition being a very restricted one) which is adequate for most standard applications.

Definition 12.42: Smooth curve deformation

Suppose C_0 and C_1 are continuous, piecewise smooth curves in the complex plane (Definition 12.40 on page 506) with common domain, $[a; b]$ having ranges (Definition 1.1 on page 2) in the bounded set A (Definition 6.22 on page 203), both curves being one-one[2] (Definition 12.18 on page 460), and for all τ with $a < \tau < b$ the condition $C_0(\tau) \neq C_1(\tau)$ is satisfied.

We will say that C_0 is **equicontinuously deformable into** C_1 *in* A, if for each t with $0 \le t \le 1$ there is a continuous piecewise smooth curve (Definition A4.1 on page 773 and Definition 3.13 on page 89) C_t whose range lies in A, with

- $C_t(a) = C_0(a), \quad C_t(b) = C_0(b)$,
- No value, $C_t(\tau)$ for $a < \tau < b$ is ever assumed by any

---

1. An alternate approach makes use of the concepts of *simple closed curves, interior and exterior of regions bounded by such curves.* Intuitively, a simple closed curve is just a curve whose range is a distorted circle, traversed once. Unfortunately, the proof that such curves divide the complex plane into two parts, the inside and outside, is usually taken for granted, since in many cases it seems obvious. The *Jordan curve theorem,* proving this claim is quite difficult, explaining the omission. We have chosen the *deformation* approach, because we feel that it takes less effort to come up with precise formulations. The big difficulty we have with the use of the Jordan curve theorem is that of deciding whether any given point is really inside such a curve, when the curve is given by some formula.
2. Which here means that for all τ, τ' with $\tau \neq \tau'$ we have $C_i(\tau) \neq C_i(\tau')$.

$C_{t'}(\tau')$ with $\tau' \neq \tau$ (i.e., for all t' with $0 < t' < 1$ and all $\tau' \neq \tau$, $C_{t'}(\tau') \neq C_t(\tau)$).

- For all t, t' sufficiently close to each other, the line segments connecting the points $C_t(\tau)$, $C_{t'}(\tau)$ all lie in A and the values $C_t(\tau)$, $C_{t'}(\tau)$ are all close — more precisely,

  for each positive number ε, there is a value

  $$\delta = q(C_0, C_1, A, \varepsilon), \text{ such that}$$

  if

  $$|t - t'| \leq \delta$$

  then

  for all τ in $[a; b]$ the line segment connecting $C_t(\tau)$ with $C_{t'}(\tau)$ lies in A

  and

  $$|C_t(\tau) - C_{t'}(\tau)| \leq \varepsilon.$$

❑

These conditions, although they may be more restrictive than are needed, are simple to visualize, and are satisfied in the cases we meet. Most important, they make the proofs relatively easy. Because of the restrictions built into this definition, we may have to resort to a small amount of extra work in situations where C_0 and C_1 have segments in common. We illustrate Definition 12.42 in Figure 12-28.

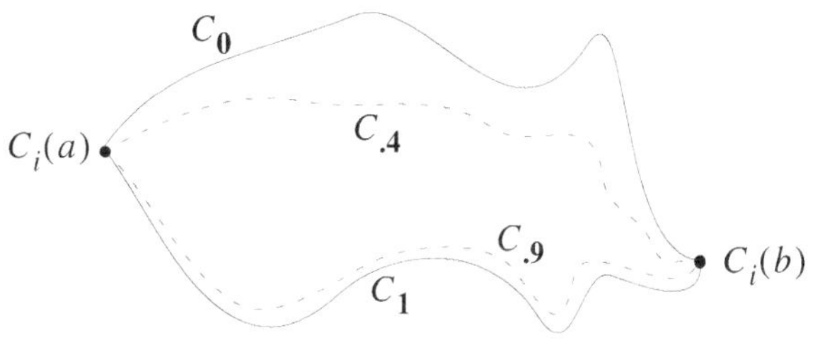

Figure 12-28

The important result ensuing from this definition is the following one.

Theorem 12.43: Cauchy's integral theorem

Suppose f is regular (Definition 12.39 on page 505) at every point of some bounded set A in the complex plane (Definition 6.22 on page 203) and C_0, C_1 are one-one curves (see Footnote 2. on page 509) lying entirely in A

(more precisely, whose ranges (Definition 1.1 on page 2) are subsets of A) with common domain the closed bounded interval $[a; b]$ (Definition 3.11 on page 88) satisfying $C_0(a) = C_1(a), C_0(b) = C_1(b)$.

If C_0 can be equicontinuously deformed to C_1 in A (Definition 12.42 on page 509) then

$$\int_{C_0} f(z)dz = \int_{C_1} f(z)dz$$

(see Definition 12.41 on page 506).

❑

In most common cases of practical interest, it will not be difficult to verify the hypotheses of this theorem.

This theorem sheds light on the results of Exercises 12.25 on page 508, because $f(z) = z$ and $f(z) = e^z$ are regular at all points of the complex plane, so the integrals of each of these functions over any non-self intersecting piecewise smooth curves beginning at z_0 and ending at z_1 must all be the same. On the other hand, since $f(z) = 1/z$ for $z \neq \mathbf{0}$ cannot be made regular at $z = \mathbf{0}$ no matter what value is chosen for $f(\mathbf{0})$, it isn't surprising that

$$\int_{C_U} \frac{dz}{z} \neq \int_{C_L} \frac{dz}{z},$$

since the upper unit circle curve ***cannot*** be equicontinuously deformed into the lower unit circle curve without crossing $z = \mathbf{0}$.

The essential reason for the validity of the Cauchy integral theorem appears to stem from the fact that if $\boldsymbol{T}$ is any curve doing a complete single traversal of the perimeter of any triangle in A, starting anywhere (without loss of generality at some vertex), and f is regular at every point in A, then[1]

$$\int_{\boldsymbol{T}} f(z)dz = 0. \qquad \textbf{\textit{12.47}}$$

Once equation 12.47 is established, the basic idea of the general proof (which will be presented in Appendix 9 starting page 811) is to show that

$$\int_{C_t} f(z)dz = \int_{C_{t'}} f(z)dz \quad \text{for } t' \text{ close to } t$$

1. This assertion will be verified shortly. It's not very difficult to define the interior of a triangle specified by three vertices — a point being there, if a line segment can be drawn from such a point to each vertex without intersecting any side except at the vertex.

in the following steps, which are illustrated graphically in Figure 12-29.

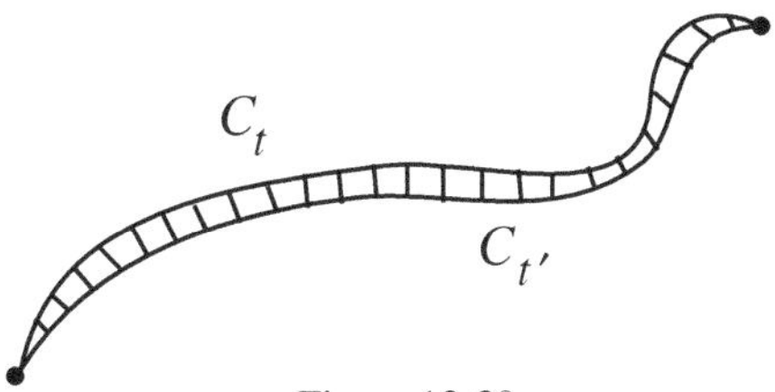

Figure 12-29

First, using a suitable partition of $[a; b]$, approximate C_t by P_t, a piecewise linear path (polygon), virtually indistinguishable from C_t, and $C_{t'}$ by $P_{t'}$ similarly. More precisely, we must establish that if the vertices of P_t, as well as those of $P_{t'}$, are sufficiently close together, we have the intuitively reasonable results that

$$\int_{C_t} f(z)dz \quad \text{is close to} \quad \int_{P_t} f(z)dz$$

and

$$\int_{C_{t'}} f(z)dz \quad \text{is close to} \quad \int_{P_{t'}} f(z)dz .$$

(This involves some topological concepts presented in Appendix 9 on page 811.) Then, using the result given in equation 12.47, we show that

$$\int_{P_t} f(z)dz = \int_{P_{t'}} f(z)dz; \qquad \textbf{12.48}$$

this shows that for t sufficiently close to t' we can choose P_t and $P_{t'}$ so that

$$\int_{C_t} f(z)dz \quad \text{and} \quad \int_{C_{t'}} f(z)dz$$

are arbitrarily close — hence, equal. Finally, just choose the sequence $t_i, 0 \le i \le k$ with $0 = t_0 < t_1 < \cdots < t_k = 1$ with successive values being sufficiently close so that

$$\int_{C_{t_i}} f(z)dz = \int_{C_{t_{i+1}}} f(z)dz$$

for all $i = 0,...,k - 1$, to conclude that

$$\int_{C_0} f(z)dz = \int_{C_1} f(z)dz .$$

The only part of these arguments that isn't intrinsically plausible, in our view, is the result culminating in equation 12.47 on page 511. For this reason, we include it in the main body of this discussion.

Lemma 12.44: $\int_T f(z)dz$ for f regular in A

Suppose f is a complex valued function of a complex variable, regular at every point of some set A (see Definition 12.39 on page 505). Let $\mathcal{T}$ be a triangle in A, determined by the points z_0, z_1, z_2, and let T be any piecewise smooth curve (see Definition 12.40 on page 506) traversing the perimeter of $\mathcal{T}$, say clockwise, with derivative $T'(t) \neq 0$ at all points where T' exists.

Under these conditions the complex integral (see Definition 12.41 on page 506)

$$\int_T f(z)dz = 0.$$

❑

To establish this lemma, bisect each side of $\mathcal{T}$ yielding four subtriangles $\mathcal{T}_1, \mathcal{T}_2, \mathcal{T}_3, \mathcal{T}_4$ as illustrated in Figure 12-30. Each of these four

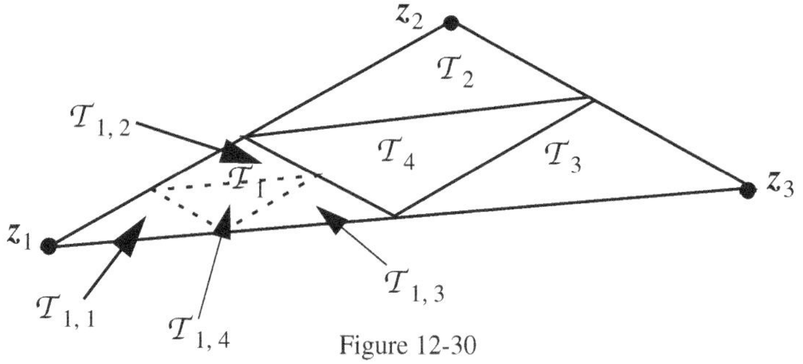

Figure 12-30

subtriangles have sides whose lengths are half that of the corresponding sides of $\mathcal{T}$ and thus each has perimeter half that of $\mathcal{T}$. It can be seen that

$$\int_T f(z)dz = \sum_{i=1}^{4} \int_{T_i} f(z)dz, \qquad \textbf{12.49}$$

where T_i is a curve whose range is the perimeter of $\mathcal{T}_i$, and which is traversed in the same orientation as $\mathcal{T}$ was by T (clockwise); because each side common to any pair of the triangles $\mathcal{T}_1, \mathcal{T}_2, \mathcal{T}_3, \mathcal{T}_4$ is, on the right side of equation 12.49, traversed twice, in reverse directions, canceling the contribution of these sides to the right-hand side. But this leads to the traversal of the perimeter of $\mathcal{T}$ as on the left-hand side, proving equality 12.49.

This shows that if the assertion $\int_T f(z)dz = 0$ were false, say,

$$\left|\int_T f(z)dz\right| = B > 0,$$

then for at least one value of i, say i_1,

$$\left|\int_{T_{i_1}} f(z)dz\right| \geq \frac{B}{4}.$$

Repeat this same bisection subdivision process on $\mathcal{T}_{i_1}$, obtaining $\mathcal{T}_{i_1, i_2}$ with

$$\left|\int_{T_{i_1, i_2}} f(z)dz\right| \geq \frac{B}{4^2},$$

and, in general, leading to triangles $\mathcal{T}_{i_1, i_2, \ldots, i_m}$ with side lengths $a/2^m, b/2^m, c/2^m$ when $a, b, c,$ were the side lengths of $\mathcal{T}$, and

$$\left|\int_{T_{i_1, i_2, \ldots, i_m}} f(z)dz\right| \geq \frac{B}{4^m}. \qquad \textbf{12.50}$$

Looking at Figure 12-30, with i_1 illustrated to be 1, we see that the sequence of triangles $\mathcal{T}_{i_1}, \mathcal{T}_{i_1, i_2}, \ldots$ each being contained in the previous one, and with side lengths approaching 0 defines a point z_0, which must be in A (since $\mathcal{T}$ is included in A, and z_0 must be in $\mathcal{T}$). Under the hypotheses of Lemma 12.44, because z_0 is in A, f must be regular at z_0, and we may thus write

$$f(z) = f(z_0) + (z - z_0)f'(z_0) + (z - z_0)\varepsilon(z - z_0), \qquad \textbf{12.51}$$

where $\varepsilon(z - z_0)$ is small for $0 < |z - z_0|$ sufficiently small.

But we can verify that

$$\int_{\Delta} (az+b)dz = 0$$

for Δ a piecewise smooth curve traversing any triangle in A and having a nonzero derivative, by choosing the following for Δ (a curve equivalent to any such curve, see Definition 12.28 on page 483) illustrated in Figure 12-31.

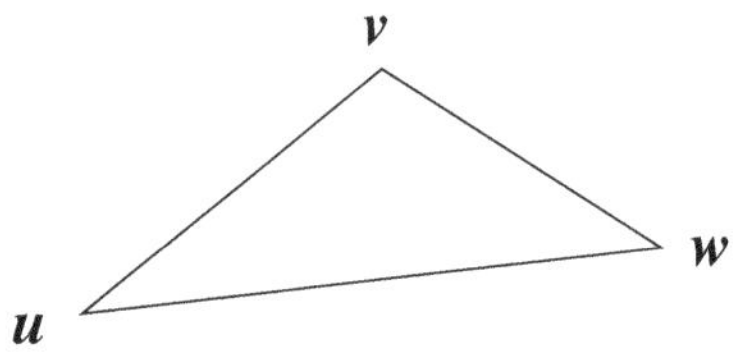

Figure 12-31

$$\Delta(t) = \begin{cases} (1-t)\boldsymbol{u} + t\boldsymbol{v} & \text{for } 0 \le t \le 1 \\ (2-t)\boldsymbol{v} + (t-1)\boldsymbol{w} & \text{for } 1 \le t \le 2 \\ (3-t)\boldsymbol{w} + (t-2)\boldsymbol{u} & \text{for } 2 \le t \le 3. \end{cases}$$

For this choice

$$\int_{\Delta} (az+b)dz = \int_0^1 [(1-t)\boldsymbol{u} + t\boldsymbol{v}](\boldsymbol{v}-\boldsymbol{u})dt$$
$$+ \int_1^2 [(2-t)\boldsymbol{v} + (t-1)\boldsymbol{w}](\boldsymbol{w}-\boldsymbol{v})dt$$
$$+ \int_2^3 [(3-t)\boldsymbol{w} + (t-2)\boldsymbol{u}](\boldsymbol{u}-\boldsymbol{w})dt.$$

This can be integrated directly, or the process simplified by changing variables, letting

$\tau = t$ in the first integral,
$\tau = t-1$ in the second integral,
$\tau = t-2$ in the third integral,

to find

$$\int_{\Delta}(az+b)dz$$

$$=\int_0^1\{[(1-\tau)\boldsymbol{u}+\tau\boldsymbol{v}](\boldsymbol{v}-\boldsymbol{u})+[(1-\tau)\boldsymbol{v}+\tau\boldsymbol{w}](\boldsymbol{w}-\boldsymbol{v})+[(1-\tau)\boldsymbol{w}+\tau\boldsymbol{u}](\boldsymbol{u}-\boldsymbol{w})\}d\tau$$

$$=\int_0^1\{\boldsymbol{u}(\boldsymbol{v}-\boldsymbol{u})+\boldsymbol{v}(\boldsymbol{w}-\boldsymbol{v})+\boldsymbol{w}(\boldsymbol{u}-\boldsymbol{w})+\tau[(\boldsymbol{v}-\boldsymbol{u})^2+(\boldsymbol{w}-\boldsymbol{v})^2+(\boldsymbol{u}-\boldsymbol{w})^2]\}d\tau$$

$$=\boldsymbol{u}(\boldsymbol{v}-\boldsymbol{u})+\boldsymbol{v}(\boldsymbol{w}-\boldsymbol{v})+\boldsymbol{w}(\boldsymbol{u}-\boldsymbol{w})+\frac{1}{2}[(\boldsymbol{v}-\boldsymbol{u})^2+(\boldsymbol{w}-\boldsymbol{v})^2+(\boldsymbol{u}-\boldsymbol{w})^2]$$

$$=0.$$

So

$$\left|\int_{\boldsymbol{T}_{i_1,i_2,\ldots,i_m}} f(z)dz\right| = \left|\int_{\boldsymbol{T}_{i_1,i_2,\ldots,i_m}}(z-z_0)\varepsilon(z-z_0)dz\right|$$

But

$$|z-z_0|\le\frac{a}{2^m}+\frac{b}{2^m}+\frac{c}{2^m}$$

because the distance between any two points in a triangle can't be greater than the sum of the side lengths. Hence

$$\left|\int_{\boldsymbol{T}_{i_1,i_2,\ldots,i_m}}(z-z_0)\varepsilon(z-z_0)dz\right|\le\frac{a+b+c}{2^m}\left(\frac{a}{2^m}+\frac{b}{2^m}+\frac{c}{2^m}\right)E_m$$

(the second factor being the perimeter length of $\mathcal{T}_{i_1,i_2,\ldots,i_m}$), where $|\varepsilon(z-z_0)|\le E_m$ for all z, z_0 in $\mathcal{T}_{i_1,i_2,\ldots,i_m}$; i.e.,

$$\left|\int_{\boldsymbol{T}_{i_1,i_2,\ldots,i_m}} f(z)dz\right|\le\frac{(a+b+c)^2}{4^m}E_m.$$

But for m large enough, E_m can be chosen arbitrarily small — i.e., small enough so that

$$(a+b+c)^2E_m<B,$$

from which it follows that for large enough m

$$\left| \int_{T_{i_1, i_2, \ldots, i_m}} f(z)dz \right| < \frac{B}{4^m}.$$

This inequality is incompatible with inequality 12.50 on page 514. The only way to remove this incompatibility, which arose from the possibility that $\int_T f(z)dz \neq 0,$ is to discard this possibility, proving Lemma 12.44.

Here is an example of application of the Cauchy integral theorem, Theorem 12.43 on page 510.

Example 12.26: Square in circle

Suppose C is a circle, inside of which is a square $\mathcal{S}$ and f is regular on the set consisting of the perimeters of C and $\mathcal{S}$ and those points inside C and outside of $\mathcal{S}$, as indicated by the shaded area below in Figure 12-32.

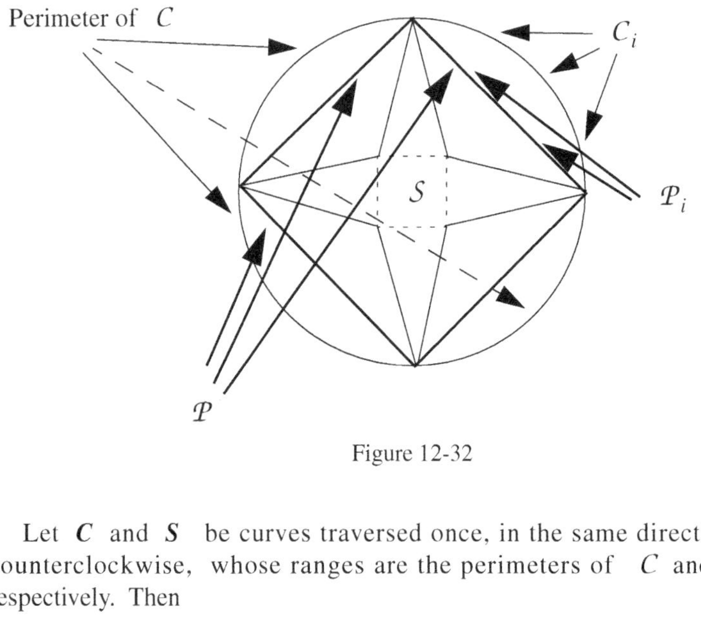

Figure 12-32

Let $\boldsymbol{C}$ and $\boldsymbol{S}$ be curves traversed once, in the same direction, say, counterclockwise, whose ranges are the perimeters of C and of $\mathcal{S}$, respectively. Then

$$\int_{\boldsymbol{C}} f(z)dz = \int_{\boldsymbol{S}} f(z)dz.$$

To see this first draw a polygon, $\mathcal{P}$, with vertices on the perimeter of C, and with $\mathcal{S}$ inside of the range of $\mathcal{P}$, as shown in Figure 12-32. (This can be done even if $\mathcal{S}$ isn't centered in the circle C.) Now subdivide the region inside of $\mathcal{P}$ and outside $\mathcal{S}$ into triangles, similar to the way illustrated in Figure 12-32. Associated with each side, $\mathcal{P}_i$, of the polygon $\mathcal{P}$ is a curve traversing $\mathcal{P}_i$ and the associated circular arc C_i counterclockwise. This curve is made up of two curves, reasonably denoted by $\boldsymbol{P}_i$ and $\boldsymbol{C}_i$, respectively. Let $\boldsymbol{P}_{-i}$ be the curve $\boldsymbol{P}_i$ traversed in reverse order. It's not difficult to deform $\boldsymbol{P}_{-i}$ equicontinuously into $\boldsymbol{C}_i$ (Definition 12.42 on page 509) simply by letting

$$\boldsymbol{C}_{ti} = (1-t)\boldsymbol{C}_i + t\boldsymbol{P}_{-i}$$

with $\boldsymbol{C}_{ti}$ playing the role of C_t, $\boldsymbol{C}_i$ playing the role of C_0 and $\boldsymbol{P}_{-i}$ playing the role of C_1 in Definition 12.42. If now $\boldsymbol{P}$ is a curve traversing the polygon $\mathcal{P}$ counterclockwise, it follows by successive applications of the Cauchy integral theorem that

$$\int_{\boldsymbol{C}} f(z)dz = \int_{\boldsymbol{P}} f(z)dz. \qquad \textbf{\textit{12.52}}$$

Now traversing each triangle once, counterclockwise yields

$$\int_{\boldsymbol{S}} f(z)dz = \int_{\boldsymbol{P}} f(z)dz, \qquad \textbf{\textit{12.53}}$$

since the *spoke* sides are traversed twice, once in each direction, yielding a net contribution of 0 from the spokes. But the total of these triangle integrals is 0 (since each triangle contributes 0 as shown in Lemma 12.44 on page 513).

So, combining equations 12.52 and 12.53 yields

$$\int_{\boldsymbol{C}} f(z)dz = \int_{\boldsymbol{S}} f(z)dz$$

as claimed.

■

Exercises 12.27

Evaluate

1. $\int_{C_U} e^z dz$ where C_U is the upper half of the unit circle traversed from $(1, 0)$ to $(-1, 0)$.

   Hint: It's very easy to compute $\int_C e^z dz$ where C is the straight line curve from $(1, 0)$ to $(-1, 0)$, since this is just a real integral.

2. $\int_C e^{1/z}\,dz$, where $C(t) = e^{it}, \quad 0 \le t \le 2\pi$.

   Hint: A valid proof takes a substantial amount of work. However, operating intuitively, if you expand the integrand in a Taylor series (using the Taylor series for e^x, Example 7.9 on page 231, which converges for all complex x) and use the fact that for z in the range of C, $|1/z| = 1$, it follows that this series converges uniformly (Chapter 7 Section 4 on page 232) and hence can be integrated term by term (Theorem 7.18 on page 236); you can now use the results of Example 12.24 on page 508 to get the desired value.

3. Same as exercise 2 for C *any rectangle containing 0 inside.*

4. Same as previous questions for the integrand e^{-1/z^2}

Problems 12.28

1. Extend the reasoning of Example 12.26 on page 517 to handle the case of a circle within a square.Hint: See Figure 12-33.

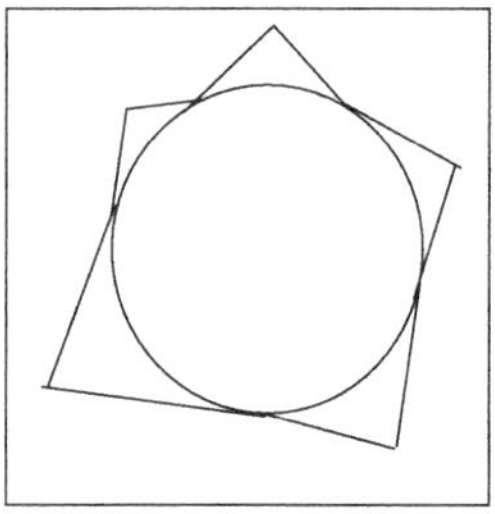

Figure 12-33

2. Let A be any annulus (ring) — i.e., the region between any two concentric circles, — as illustrated in Figure 12-34.

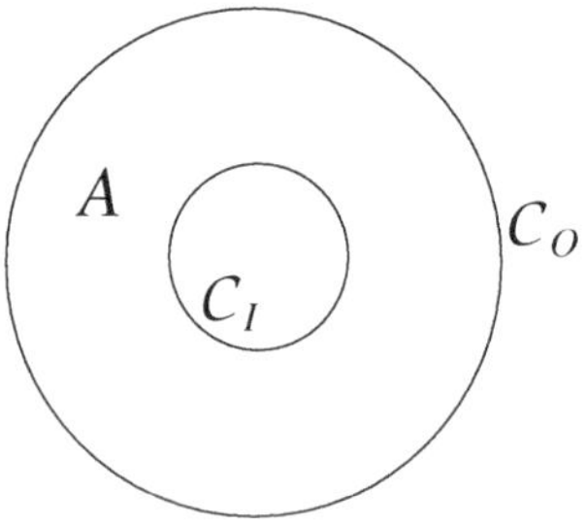

Figure 12-34

   Show that if $\boldsymbol{C_O}$ and $\boldsymbol{C_I}$ are curves traversing these circles in the same direction, and if F is regular at every point of A, then

$$\int_{C_O} f(z)dz = \int_{C_I} f(z)dz.$$

Hints: Consider the path P indicated in Figure 12-35, where the

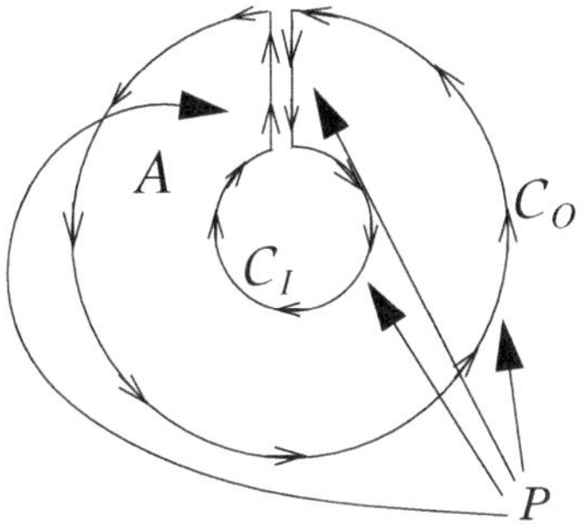

Figure 12-35

two indicated line segments are presumed to lie atop each other, so that the integrals over these segments cancel out. Use the Cauchy integral theorem (Theorem 12.43 on page 510) to show that

$$\int_{P} f(z)dz = 0,$$

as follows: you can easily show that the integral over each of the

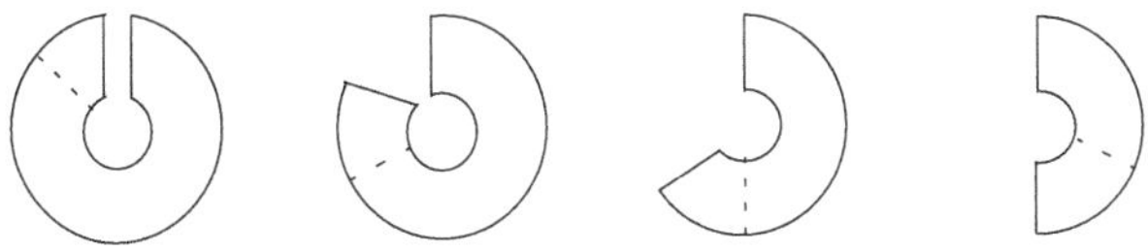

Figure 12-36

paths indicated in Figure 12-36 are the same, ultimately showing the original integral equal to that over a triangle in which f is regular. Hence by the Cauchy integral theorem, together with Lemma 12.44 on page 513, the original integral is zero, which establishes the desired result.

The next topic we take up is the *Cauchy integral formula.* It has both important theoretical and practical applications. On the theoretical side, it is used to prove that the values on the boundary of the domain of a regular function completely determine this function, to establish the Taylor and Laurent series for regular functions, and to prove the infinite differentiability of a regular function. It also leads to the residue theorem, which furnishes a practical means of evaluation of many complex integrals which arise in applied problems.

Theorem 12.45: Cauchy integral formula

If f is a complex valued function of a complex variable, regular (Definition 12.39 on page 505) on and inside a circle, C, of radius r centered at ζ, then using the definition of a complex integral (Definition 12.41 on page 506)

$$f(\zeta) = \frac{1}{2\pi i}\int_{\boldsymbol{C}} \frac{f(z)}{z-\zeta}\,dz, \qquad \textit{12.54}$$

where $\boldsymbol{C}$ is a curve whose range is the circle C, traversed counterclockwise.

❑

The proof is not difficult — the genius being thinking up this result. To establish the Cauchy integral formula, first note that

$$\frac{1}{2\pi i}\int_{\boldsymbol{C}} \frac{f(z)}{z-\zeta}\,dz = \frac{1}{2\pi i}\int_{\boldsymbol{C}_\delta} \frac{f(z)}{z-\zeta}\,dz,$$

where $\boldsymbol{C}_\delta$ is the curve corresponding to any circle centered at ζ of radius $\delta < r$; this follows from the result of Problem 12.28.2 on page 519. But from the differentiability of f, we may write

$$f(z) = f(\zeta) + (z-\zeta)f'(\zeta) + (z-\zeta)\varepsilon(z-\zeta),$$

where

$$\lim_{z \to \zeta} \varepsilon(z-\zeta) = 0.$$

So

$$\frac{1}{2\pi i}\int_{\boldsymbol{C}} \frac{f(z)}{z-\zeta}\,dz = \frac{1}{2\pi i}\left[\int_{\boldsymbol{C}_\delta} \frac{f(\zeta)}{z-\zeta}\,dz + \int_{\boldsymbol{C}_\delta} f'(\zeta)\,dz + \int_{\boldsymbol{C}_\delta} \varepsilon(z-\zeta)\,dz\right].$$

By Example 12.24 on page 508, the first of the integrals on the right side equals $2\pi i f(\zeta)$. The second integral is seen to be 0 because the integrand is a constant (special case of Example 12.24, the case $n = 0$) so that

$$\int_{\boldsymbol{C}_\delta} f'(\zeta)\,dz = f'(\zeta)\int_{\boldsymbol{C}_\delta} z^0\,dz = 0.$$

For the last integral on the right, choose δ small enough so that $|\varepsilon(z-\zeta)|$ is small (say $< \varepsilon'$, where ε' is some small positive number; this can be done as can be seen from a careful reading of the discussion following Definition 12.38 on page 504). Then

$$\left| \int_{C_\delta} \varepsilon(z-\zeta)\, dz \right| \leq 2\pi\delta\varepsilon',$$

where the product of the final two factors is small if δ is small. Since $\delta > 0$ can be chosen as small as we like, it now follows that

$$\frac{1}{2\pi i}\int_C \frac{f(z)}{z-\zeta}\, dz = f(\zeta),$$

as asserted.

An important consequence of the Cauchy integral formula is Theorem 12.46.

Theorem 12.46: Infinite differentiability of regular function

The derivative of a regular function (Definition 12.39 on page 505) is a regular function. Hence ***a regular function is infinitely differentiable*** (i.e., has an n-th derivative for all positive integers n).

❑

What must be shown first to establish Theorem 12.46 is that it is legitimate to differentiate the right side of equation 12.54 with respect to ζ under the integral sign — i.e., that

$$f'(\zeta) = D_\zeta \frac{1}{2\pi i}\int_C \frac{f(z)}{z-\zeta}\, dz = \frac{1}{2\pi i}\int_C D_\zeta \frac{f(z)}{z-\zeta}\, dz = \frac{1}{2\pi i}\int_C \frac{f(z)}{(z-\zeta)^2}\, dz. \qquad \textbf{12.55}$$

To do this, we first note that

$$\begin{aligned}\frac{f(\zeta+h)-f(\zeta)}{h} &= \frac{1}{2\pi i}\int_C f(z)\left[\frac{1}{z-(\zeta+h)} - \frac{1}{z-\zeta}\right] dz \\ &= \frac{1}{2\pi i}\int_C f(z)\left[\frac{1}{(z-(\zeta+h))(z-\zeta)}\right] dz.\end{aligned}$$

Hence

$$\begin{aligned}&\frac{f(\zeta+h)-f(\zeta)}{h} - \frac{1}{2\pi i}\int_C \frac{f(z)}{(z-\zeta)^2}\, dz \\ &= \frac{1}{2\pi i}\int_C f(z)\left[\frac{1}{(z-(\zeta+h))(z-\zeta)} - \frac{1}{(z-\zeta)^2}\right] dz \\ &= \frac{h}{2\pi i}\int_C f(z)\left[\frac{1}{(z-(\zeta+h))(z-\zeta)^2}\right] dz\end{aligned}$$

To establish equation 12.55 we need only show that the last expression above approaches 0 as h approaches 0. For this it suffices to show that the integrand is bounded for z on the circle C, traversed by $\boldsymbol{C}$. This claim follows from the inequality

$$\left|\int_{\boldsymbol{C}} g(z)\,dz\right| \le BL(\boldsymbol{C}), \qquad \textit{12.56}$$

where B is any upper bound for g on the circle $\boldsymbol{C}$ and $L(C)$ is the length of the curve $\boldsymbol{C}$ (here $2\pi r$). Inequality 12.56 is easily established from Definition 12.41 on page 506, of the complex integral.

By the reverse triangle inequality (Theorem 7.7 on page 218) for $|h| < r_0 < r$ the term in square brackets is seen to be bounded; we leave it to the reader to determine the bound. So it is only necessary to show that $f(z)$ is bounded for z on C. We establish this with the help of the following definition.

Definition 12.47: Continuous complex functions

For clarity, we extend the definition of continuity (Definition 8.7 on page 270) to include complex functions of complex variables as follows.

We say that **the complex valued function**, g, **of** n **complex variables is continuous at** $\boldsymbol{x}_0$ in the domain of g, if, for $\boldsymbol{x}$ in the domain of g, $g(\boldsymbol{x}) - g(\boldsymbol{x}_0)$ gets close to 0 as $\|\boldsymbol{x}_0 - \boldsymbol{x}\|$ gets small — i.e., small changes to the input, $\boldsymbol{x}_0$, yield small changes to the output $g(\boldsymbol{x}_0)$. More precisely,

> given any positive number, ε, there is a positive number, δ (determined by g, ε, and $\boldsymbol{x}_0$) such that
>
> whenever $\boldsymbol{x}$ is in the domain of g *and* $\|\boldsymbol{x}_0 - \boldsymbol{x}\| \le \delta$ *we have*
>
> $$|g(\boldsymbol{x}_0) - g(\boldsymbol{x})| \le \varepsilon.$$

The only changes that need be made are for the **length of a complex vector**, $(\boldsymbol{x}_1,...,\boldsymbol{x}_n)$, where the $\boldsymbol{x}_i$ are complex numbers; here we ***define***

$$\|(\boldsymbol{x}_1,...,\boldsymbol{x}_n)\| = \sum_{j=1}^{n} |\boldsymbol{x}_j|^2, \qquad \textit{12.57}$$

where the magnitude of the complex number $\boldsymbol{y} = y_R + \boldsymbol{i}y_I$ is given by $|\boldsymbol{y}| = \sqrt{y_R^2 + y_I^2}$ (see Definition 6.11 on page 200).

❑

It's not difficult to show that if a complex valued function of n complex variables is considered to be a pair of real valued functions of $2n$ real variables, then the complex valued function is continuous if and only if each of these real valued functions is a continuous function of its $2n$ variables, Definition 8.7 on page 270. We leave this to the reader.

Theorem 12.48: Boundedness of continuous function

Suppose f is a continuous real or complex valued function of n real or complex variables whose domain (Definition 1.1 on page 2) is a closed bounded set (Definition 10.43 on page 406, where the norm $\|\boldsymbol{x}\|$ is that of a complex vector, etc.[1]). Then f is bounded (Definition 10.45 on page 408).

❑

We only sketch this proof where f is a complex valued function of a single complex variable, because the extension to the general case poses no additional difficulties. The only real difference between this proof and the one already given for a real valued function of a real variable, in the section proving the mean value theorem (step c on page 64) where the set A is a closed bounded interval, is in the level of abstraction. The arguments are essentially the same.

In the case being considered, because of the boundedness of the domain of f, we may assume that the domain of f is included in some finite rectangle in the complex plane with sides parallel to the two coordinate axes. If the domain of f consists of only a finite number of points (complex values), the theorem's conclusion obviously holds, so we need only consider the case where the domain has an infinite number of points. As seen in Figure 12-37, if the function f were not bounded, then it would fail to be bounded on at least one of the four subrectangles generated by bisection of the original rectangle's two sides. Suppose this to be the shaded subrectangle in Figure 12-37. Repeat

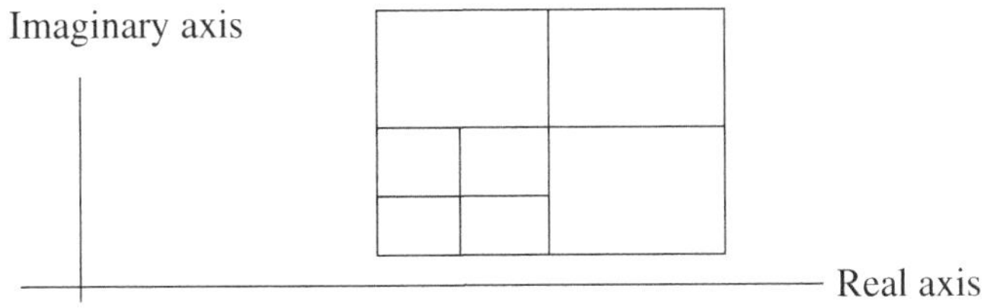

Figure 12-37

1. One slight complication, namely, the dimension of the space of complex vectors with n complex coordinates is still n — not $2n$, because, with $\boldsymbol{i}$ considered a scalar, the standard unit vectors, $\boldsymbol{e}_i$ (see Definition 8.2 on page 263) still form a basis (Definition 9.11 on page 299) for this space.

this process on the shaded rectangle, obtaining its darker subrectangle shown in Figure 12-38. Continuing this process, leads to a unique point, which must

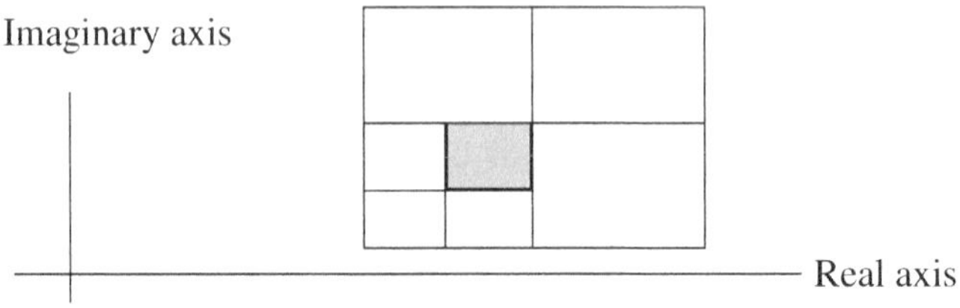

Figure 12-38

be in the domain of f, because the domain is a closed set. This point, call it $\boldsymbol{p}$, has the property that in every subrectangle defining it, the function f takes on values with an arbitrarily large magnitude. But from the continuity of f at every point in its domain, for all z close enough to $\boldsymbol{p}$, $f(z)$ is close to $f(\boldsymbol{p})$. So, f is bounded *close to* $\boldsymbol{p}$, but f would be unbounded in all nonzero area rectangles defining $\boldsymbol{p}$. This contradiction arose because of the assumptions that f is continuous, and f is not bounded. For continuous f the only way to remove this contradiction is to admit that f must be bounded, proving Theorem 12.48.

Thus we have established equation 12.55. We can now repeat the process to show that

$$\frac{f'(\zeta + h) - f'(\zeta)}{h} \quad \text{has limit} \quad \frac{1}{2\pi \boldsymbol{i}} \int_C \frac{2f(z)}{(z-\zeta)^3} dz.$$

But this does the job of proving Theorem 12.46, because we've shown that f' is regular. So wherever f is differentiable, so is f'. Inductively applying this result shows that a regular function is infinitely differentiable.

Note that from equation 12.55 on page 522 it now follows that for f a regular function on and inside a circle C of radius r centered at ζ

$$f^{(n)}(\zeta) = \frac{1}{2\pi \boldsymbol{i}} \int_C \frac{(n-1)!f(z)}{(z-\zeta)^n} dz. \qquad \textbf{\textit{12.58}}$$

Next we establish the power series and Laurent expansions of a regular function.

Theorem 12.49: Power series for regular f

If f is regular (Definition 12.39 on page 505) on and inside the circle of radius R centered at the complex number $\boldsymbol{a}$, then for ζ inside this set

$$f(\zeta) = \sum_{m=0}^{\infty} a_m(\zeta - \boldsymbol{a})^m \qquad \textbf{12.59}$$

(see Definition 7.4 on page 214), where

$$a_m = \frac{1}{2\pi \boldsymbol{i}} \int_{C_R} \frac{f(z)}{(z-a)^{m+1}} dz \qquad \textbf{12.60}$$

(see Definition 12.41 on page 506 for the definition of complex integral) where the curve $\boldsymbol{C}_R$ traverses the circle of radius R centered at $\boldsymbol{a}$ counterclockwise.

❑

We'll only establish this for the case $\boldsymbol{a} = \boldsymbol{0}$, to keep the notation more manageable. (Otherwise, we would have to rewrite $z - \boldsymbol{\zeta} = z - \boldsymbol{a} - (\boldsymbol{\zeta} - \boldsymbol{a})$.) By the Cauchy integral formula, Theorem 12.45 on page 521, for ζ inside $\boldsymbol{C}_R$

$$\begin{aligned} f(\zeta) &= \frac{1}{2\pi \boldsymbol{i}} \int_{C_R} \frac{f(z)}{z - \zeta} dz \\ &= \frac{1}{2\pi \boldsymbol{i}} \int_{C_R} \frac{1}{z} \frac{f(z)}{1 - \frac{\zeta}{z}} dz \\ &= \frac{1}{2\pi \boldsymbol{i}} \int_{C_R} \frac{1}{z} f(z) \sum_{m=0}^{\infty} \left(\frac{\zeta}{z}\right)^m dz, \end{aligned}$$

using the geometric series, Example 7.11 on page 233, clearly valid because $|\zeta/z| \le c < 1$ for all z on $\boldsymbol{C}_R$ (just draw a picture to see this). Hence the series

$$\sum_{m=0}^{\infty} \left(\frac{\zeta}{z}\right)^m$$

converges uniformly in z for ζ inside the circle associated with $\boldsymbol{C}_R$ (see Theorem 7.19 on page 237). Thus, using the boundedness of f on $\boldsymbol{C}_R$, the integrand series converges uniformly, and reusing inequality 12.56 on page 523 allows us to invoke Theorem 7.18 on page 236 to permit term-by-term integration, yielding

$$f(\zeta) = \frac{1}{2\pi \boldsymbol{i}} \sum_{m=0}^{\infty} \left(\int_{C_R} \frac{f(z)}{z^{m+1}} dz \right) \zeta^m$$

for ζ inside C_R, establishing the claimed power series expansion, equation 12.59 on page 526 for the case $a = 0$. The case of arbitrary a causes no extra difficulty.

Theorem 12.50: Laurent expansion

If f is regular (Definition 12.39 on page 505) in an annulus centered at the complex number a and determined by circles of radii $r,\ R,\ r < R,$ then for ζ inside this annulus (i.e., strictly between the two circles)

$$f(\zeta)=\frac{1}{2\pi i}\sum_{m=0}^{\infty}\left(\int_{C_R}\frac{f(z)}{(z-a)^{m+1}}dz\right)(\zeta-a)^m-\frac{1}{2\pi i}\sum_{j=0}^{\infty}\left(\int_{C_r}f(z)(z-a)^j dz\right)(\zeta-a)^{-j-1}.$$

This is conventionally written

$$f(\zeta) = \sum_{m=-\infty}^{\infty} a_m(\zeta-a)^m, \qquad \textbf{12.61}$$

where

$$a_m = \frac{1}{2\pi i}\int_{C_R}\frac{f(z)}{(z-a)^{m+1}}dz \quad m = 0, \pm 1, \pm 2, \ldots . \qquad \textbf{12.62}$$

❑

To establish this result, we first note, using the Cauchy integral formula, Theorem 12.45 on page 521, Cauchy integral theorem, Theorem 12.43 on page 510, and the methods developed in Problem 12.28.2 on page 519, that

$$f(\zeta) = \frac{1}{2\pi i}\int_{\mathcal{A}}\frac{f(z)}{z-\zeta}dz$$

where $\mathcal{A}$ is a curve tracing the indicated annulus path illustrated in Figure 12-39.

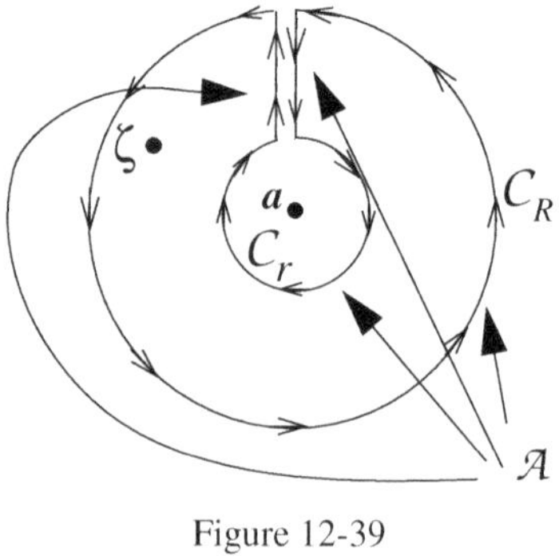

Figure 12-39

Again, with Figure 12-39 to guide us, we see that

$$f(\zeta) = \frac{1}{2\pi i}\int_{C_R} \frac{f(z)}{z-\zeta}dz - \frac{1}{2\pi i}\int_{C_r} \frac{f(z)}{z-\zeta}dz.$$

As before, for simplicity of presentation we treat only the case $\boldsymbol{a} = \mathbf{0}$. (To handle the situation generally, we would have to write $z - \boldsymbol{\zeta} = z - \boldsymbol{a} - (\boldsymbol{\zeta} - \boldsymbol{a})$.)

The first integral above is handled identically to the way we dealt with the power series expansion (Theorem 12.49 on page 525) for a function regular on and inside a circle, being equal to

$$\frac{1}{2\pi i}\sum_{m=0}^{\infty}\left(\int_{C_R} \frac{f(z)}{z^{m+1}}dz\right)\zeta^m. \qquad \textbf{\textit{12.63}}$$

To deal with the second integral, we rewrite

$$\frac{f(z)}{z-\zeta} = \frac{-f(z)}{\zeta\left(1-\frac{z}{\zeta}\right)} = -\sum_{j=0}^{\infty}\frac{f(z)}{\zeta}\frac{z^j}{\zeta^j},$$

which converges uniformly for $|z| \le r$ when ζ lies strictly inside the annulus. So

$$-\frac{1}{2\pi i}\int_{C_r} \frac{f(z)}{z-\zeta}dz = \frac{1}{2\pi i}\int_{C_r} \sum_{j=0}^{\infty}\frac{1}{\zeta^{j+1}}f(z)z^j dz$$

$$= \frac{1}{2\pi i}\sum_{j=0}^{\infty}\left(\int_{C_r} f(z)z^j dz\right)\zeta^{-j-1}$$

the term-by-term integration, justified as before by the uniform convergence. This establishes the Laurent expansion.

The final topic we take up is the ***residue theorem***, which permits simple evaluation of many integrals.

Suppose that the complex valued function f is regular in some annulus, $\mathcal{A}$ and has only a finite number of points, $\boldsymbol{a}_1,\ldots,\boldsymbol{a}_k$ inside the inner circle at which f is not regular, as indicated in Figure 12-40.

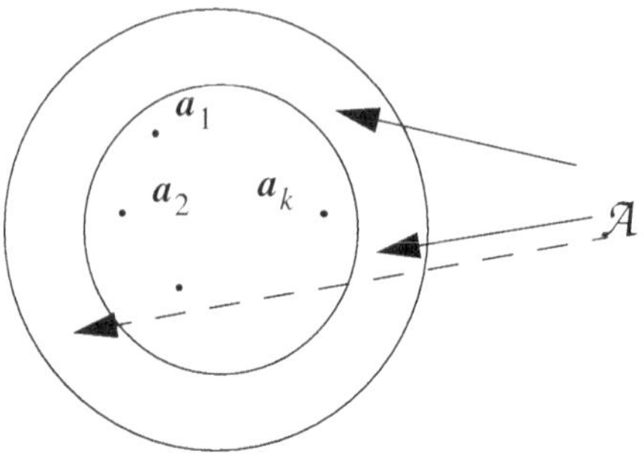

Figure 12-40

Also in this figure, around each of the $\boldsymbol{a}_q$, in some annulus, as indicated the function f possesses a uniformly convergent Laurent expansion

$$f(z) = \sum_{m=0}^{\infty} a_{qm}(z - \boldsymbol{a}_q)^m - \sum_{j=0}^{\infty} a_{-qj}(z - \boldsymbol{a}_q)^{-j-1}. \qquad \textbf{\textit{12.64}}$$

The values a_{-q1} are of particular importance, since we know from the results of Example 12.24 on page 508 that the integral

$$\int_{C_q} f(z)dz = 2\pi \boldsymbol{i}\, a_{-q1},$$

where C_q is any circular path traversed counterclockwise in the annulus associated with the point $\boldsymbol{a}_q$. Using the same techniques for application of the Cauchy integral theorem used in recent pages, we can prove the residue theorem.

Theorem 12.51: Residue theorem

Let the function f be regular (Definition 12.39 on page 505) in some annulus, $\mathcal{A}$ and assume that there are only a finite number of points, $\boldsymbol{a}_1,...,\boldsymbol{a}_k$, inside the inner circle defining this annulus at which f fails to be regular, there being a Laurent expansion (Theorem 12.50 on page 527) for f about each of those points, converging in some small annulus (as illustrated in Figure 12-40). Denote the coefficient a_{-q1}, of the -1 power, $(z - \boldsymbol{a}_q)^{-1}$, in the Laurent expansion about $\boldsymbol{a}_q$ by $R_{\boldsymbol{a}_q, f}$, calling this value the **residue of f associated with $\boldsymbol{a}_q$**.

Then for C any circular curve within the annulus $\mathcal{A}$ traversed counterclockwise

$$\int_C f(z)dz = \frac{1}{2\pi \boldsymbol{i}} \sum_{q=1}^{k} R_{\boldsymbol{a}_q}.$$

❑

Essentially this result has been proved. So we will provide an example and some exercises showing its applicability when the residues can be easily determined.

Example 12.29: $\int_C f(z)dz$ **for** $f(z) = \dfrac{1}{(z-2)^2(z-3)^3}$

We can try to write the Laurent expansion,

$$f(z) = \sum_{m=-\infty}^{\infty} a_m(z-\mathbf{a})^m$$

and use the fact that $(z-2)^2 f(z)$ is regular near $z = 2$. Thus, by Theorem 12.49 on page 525, $(z-2)^2 f(z)$ has a power series expansion about $(z-2)$, leading to

$$\sum_{m=-\infty}^{\infty} a_m(z-2)^m = \sum_{m=2}^{\infty} a_m(z-2)^m = f(z).$$

Thus letting

$$h(z) = (z-2)^2 f(z) = \sum_{m=2}^{\infty} a_{m-2}(z-2)^m,$$

we find

$$a_{-1} = (Dh)(2) = D\frac{1}{(z-3)^3}\bigg|_{z=2} = \frac{-3}{(z-3)^4}\bigg|_{z=2} = -3.$$

i.e., $R_{2,f} = -3$. Similarly, $R_{3,f} = 6$.

Hence, by the residue theorem, the integral $\int_C f(z)dz$, where C traverses counterclockwise a circle of radius 10 about $\mathbf{0}$, is

$$\int_C f(z)dz = \frac{1}{2\pi i}(6-3) = \frac{3}{2\pi i}.$$

■

Exercises 12.30

Let R_b be a circle of radius b about 0, traversed counterclockwise.

1. Find $\int_{R_5} sin(1/z)dz$.
2. Find $\int_{R_{10}} \dfrac{z}{(z+1)(z-10)^3}dz$.

Chapter 13

Preferred Coordinate Systems

Eigenvalues and Eigenvectors

1 Introduction

Many problems arise in applied areas such as queueing theory,[1] mechanics, and statistical multivariate analysis, where extensive knowledge of the properties of some $n \times n$ matrix is useful for developing and understanding the solution.

Example 13.1: m-Step transition matrix

In studying certain simple evolutionary processes, the system can be in any one of n states at times $t = 0,1,2,...$. For example, the state of a system of machines might simply be the number of machines undergoing repair. The statistical behavior of such a system is often described by a matrix **P**, whose *i-j* element is the probability that the system, known to be in state i at time t, will be in state j at time $t + 1$. We frequently want to determine the properties of $\mathbf{P}^m$ — because its *i-j* element represents the probability that the system, known to be in state i at time t, will be in state j ***at time*** $t + m$, and we want to find out how these probabilities behave for large m. So if the system describes a waiting line and state j represents the state that j people are waiting for service, there is great interest in knowing whether the waiting line does or does not tend to build up to levels which are irritatingly large.

■

Example 13.2: System of differential equations

Coupled mechanical or electrical systems are frequently described by a system

$$\mathbf{x}' = \mathbf{A}\mathbf{x} + \mathbf{f}, \qquad 13.1$$

where $\mathbf{x}$ is a vector of real valued functions of a real variable

---

1. The study of waiting lines, such as occur at traffic signals, or in a computer where jobs are waiting to be serviced, or at airports where planes are stacked up, etc.

$$\mathbf{x}(t) = \begin{bmatrix} x_1(t) \\ \cdot \\ \cdot \\ \cdot \\ x_n(t) \end{bmatrix}.$$

Here $x_1(t)$ might be the position of mass 1 at time t, $x_2(t)$ its velocity, $x_2(t)$ the position of mass 2 at time t, etc., and f a known *forcing function* vector. Any decent understanding of this system usually requires a great deal of knowledge about the matrix **A.**

Another situation leading to systems of the form given in equation 13.1 arises in dealing with the n-th order differential equation

$$(x^{(n)} + a_{n-1}x^{(n-1)}) + \cdots + a_1 x' + a_0 x = f. \qquad \textbf{13.2}$$

This equation can be reduced to the form of equation 13.1 by

first renaming x as x_1

and then (inductively) letting

$$\begin{aligned} x_2 &= x' = x_1' \\ x_3 &= x'' = x_2' \\ &\cdot \\ &\cdot \\ &\cdot \\ x_n &= x^{(n-1)} = x_{n-1}'. \end{aligned}$$

Substituting into equation 13.2 yields the system

$$\begin{aligned} x_n' &= -a_{n-1}x_n - a_{n-2}x_{n-1} - \cdots - a_0 x_1 = f \\ x_{n-1}' &= x_n \\ x_{n-2}' &= x_{n-1} \\ &\cdot \\ &\cdot \\ &\cdot \\ x_1' &= x_2 \end{aligned}$$

or

$$\mathbf{x'} = \mathbf{Ax} + \mathbf{f},$$

where

$$\mathbf{x'} = \begin{bmatrix} x_1' \\ x_2' \\ \cdot \\ \cdot \\ \cdot \\ x_n' \end{bmatrix} \qquad \mathbf{x} = \begin{bmatrix} x_1 \\ x_2 \\ \cdot \\ \cdot \\ \cdot \\ x_n \end{bmatrix} \qquad \mathbf{f} = \begin{bmatrix} 0 \\ 0 \\ \cdot \\ \cdot \\ 0 \\ f \end{bmatrix}$$

and

$$\mathbf{A} = \begin{bmatrix} 0 & 1 & 0 & \cdot & \cdot & 0 \\ 0 & 0 & 1 & 0 & \cdot & \cdot \\ \cdot & \cdot & \cdot & \cdot & \cdot & \cdot \\ \cdot & \cdot & \cdot & 0 & 1 & 0 \\ 0 & \cdot & \cdot & \cdot & 0 & 1 \\ -a_0 & -a_1 & \cdot & \cdot & -a_{n-2} & -a_{n-1} \end{bmatrix}.$$

When this system is solved, function x_1 is the desired x. For example, the second-order differential equation

$$x'' + a_1 x' + a_0 x = f \qquad \textbf{\textit{13.3}}$$

can be written as

$$\begin{bmatrix} x_1' \\ x_2' \end{bmatrix} = \begin{bmatrix} 0 & 1 \\ -a_0 & -a_1 \end{bmatrix} \begin{bmatrix} x_1 \\ x_2 \end{bmatrix} + \begin{bmatrix} \mathbf{0} \\ f \end{bmatrix}.$$

Notice that x_1 of this system satisfies equation 13.3, because written out longhand, this system is

$$x_1' = x_2 \qquad \textbf{\textit{13.4}}$$

$$x_2' = -a_0 x_1 - a_1 x_2 + f. \qquad \textbf{\textit{13.5}}$$

Differentiating equation 13.4 yields $x_1'' = x_2'$. Substituting this into the left of equation 13.5 and substituting equation 13.4 into the right of equation 13.5 yields

$$x_1'' = -a_0 x_1 - a_1 x_1' + f$$

or

$$x_1'' + a_1 x_1' + a_0 x_1 = f,$$

showing that $x = x_1$ satisfies equation 13.3.

■

2 Choice of Coordinate System to Study Matrix

If we think of an $n \times n$ matrix $\mathbf{A}$ as representing a function $\mathcal{A}$ whose value at $\mathbf{x}$ is $\mathbf{Ax}$, then we realize that the j-th column of $\mathbf{A}$ is just

$$\mathbf{Ae}_j = \mathcal{A}(\mathbf{e}_j),$$

the value of $\mathcal{A}$ at $\mathbf{e}_j$, ($\mathbf{e}_j$ being the unit vector directed along the positive j-th coordinate axis). So we may think specifically of $\mathbf{A}$ as representing the function $\mathcal{A}$ in the coordinate system specified by $\mathbf{e}_1,...,\mathbf{e}_n$.

This coordinate system is very likely one which was ***chosen to simplify setting up*** the description of the system being studied. Probably it was a coordinate system from which a local description was easy to obtain. Once this coordinate system has served its purpose, it may pay to look for better coordinates for *analysis* of the system.

In studying the properties of a matrix A, it has proved useful to choose as the new basis vectors, those vectors which $\mathbf{A}$ ***acts on in as simple a manner as possible. That is, the new basis should consist of vectors*** $\mathbf{v}_i$ ***which*** $\mathbf{A}$ **disturbs** ***as little as possible***. What this usually turns out to mean is the following:

> We try to choose as basis vectors, those vectors $\mathbf{v}_i$ for which
> $\mathbf{Av}_i$ is simply a scalar multiple of $\mathbf{v}_i$.

The theory requires that these scalar multiples and basis vectors be allowed to consist of complex numbers (see Section 5 of Chapter 6). For clarity and convenience of referencing, we repeat some previously given definitions and include some necessary extensions in the following.

Definition 13.1: Complex conjugate, complex dot product

If z is a complex number (Definition 6.8 on page 199) represented either as

$$z = z_R + \boldsymbol{i} z_I \qquad \text{(Cartesian form)}$$

or

$$z = re^{i\vartheta} \qquad \text{(Polar form)}$$

then its **complex conjugate**, denoted by $\bar{z}$ is defined by

$$\bar{z} = z_R - \boldsymbol{i} z_I$$

or equivalently

$$\bar{z} = re^{-i\vartheta}.$$

If

$$\boldsymbol{x} = \begin{pmatrix} x_1 \\ \cdot \\ \cdot \\ x_n \end{pmatrix}$$

is a complex valued vector with n coordinates (an n-dimensional complex valued vector) then its **complex conjugate**, $\bar{x}$ is given by

$$\bar{x} = \begin{pmatrix} \overline{x_1} \\ \cdot \\ \cdot \\ \overline{x_n} \end{pmatrix}.$$

The **complex conjugate of a complex matrix** is defined similarly, the operation of complex conjugation applying to each element of the matrix.

If **x** and **y** are complex n-dimensional vectors, their **dot product** (also called their **scalar product**) $\boldsymbol{x} \cdot \boldsymbol{y}$ is given by

$$\boldsymbol{x} \cdot \boldsymbol{y} = \sum_{j=1}^{n} x_i \bar{y}_i. \qquad \textit{13.6}$$

If x and y are considered as $n \times 1$ matrices (column matrices) $\mathbf{x}$ and $\mathbf{y}$, then this scalar product has the form

$$x \cdot y = \mathbf{x}^T \bar{\mathbf{y}}, \qquad 13.7$$

where T represents the transpose operator (Definition 10.13 on page 342).

Note that for ***complex*** dot products, it is generally ***not*** the case that $x \cdot y = y \cdot x$.

The above definitions were given in order that the *length*, $\|x\|$, of x, given by

$$\|x\| = \sqrt{x \cdot x}, \qquad 13.8$$

be a nonnegative real number.

***n*-dimensional Euclidean space** consists of all linear combinations

$$\sum_{j=1}^{n} a_i e_i$$

where the e_i are still the standard unit n-dimensional real vectors (Definition 8.2 on page 263) and the a_i may be any complex numbers. Subspaces, dimension, linear independence, orthogonality, basis, matrix products, etc, are still defined as before (see the relevant definitions in Chapter 9 and Chapter 10) with real numbers replaced by complex numbers. Just about all of the results in these chapters and Chapter 11 still hold.

❑

Definition 13.2: Eigenvalue, Eigenvector

Let $\mathbf{A}$ be an $n \times n$ real or complex matrix (Definition 10.1 on page 328). The *nonzero* complex valued vector x is said to be an **eigenvector of A with corresponding eigenvalue** λ if

$$\mathbf{A}x = \lambda x, \qquad 13.9$$

where λ is a complex number (Definition 6.8 on page 199).

❑

Directly from this definition, you can see that λ is an eigenvalue if and only if the following system of equations (considered as a complex matrix system) has a ***nontrivial solution*** (i.e., a solution, $\mathbf{x}$, other than the $\mathbf{0}$ column matrix).

$$(\mathbf{A}^{(n\times n)} - \lambda\mathbf{I}^{(n\times n)})\mathbf{x}^{(n\times 1)} = \mathbf{0}^{(n\times 1)}. \qquad \textit{13.10}$$

Here $\mathbf{I}$ is the $n \times n$ identity matrix (Definition 10.6 on page 334) and the superscripts $n \times n$ and $n \times 1$ are only written when needed for clarity, to display the number of rows and columns in the indicated matrices.

This is equivalent to saying that λ just be a value making the columns of the coefficient matrix $\mathbf{A} - \lambda\mathbf{I}$ in equation ***linearly dependent*** (see Definition 9.7 on page 297) for the following reasons:

> Let $\mathbf{C}$ denote the matrix $\mathbf{A} - \lambda\mathbf{I}$ (with the j-th column denoted by $\boldsymbol{c}_j$). From the definition of a matrix times a vector (equation 10.4 on page 328) equation may be written as
>
> $$\sum_{j=1}^{n} x_j \boldsymbol{c}_j = \mathbf{0}.$$
>
> Because the solution is assumed nontrivial, at least one of the x_j, say, x_{j_0}, must be nonzero. From this it follows that column $\boldsymbol{c}_{j_0}$ must be a linear combination of the other columns of the matrix $\mathbf{C}$, implying the claimed linear dependence.

Notice that once we've determined that some particular value of λ is an eigenvalue, it's theoretically and computationally quite easy to determine a corresponding eigenvector. Namely, solve equation using one of the methods presented in Section 3 of Chapter 9. ***For this reason we concentrate almost completely on the problem of finding the eigenvalues of*** $\mathbf{A}$, ***spending almost no effort on determining its eigenvectors***.

A good intuitive starting point is to realize that a square matrix (in this case $\mathbf{A} - \lambda\mathbf{I}$) has linearly dependent columns if and only if the n-dimensional volume of the parallelepiped formed by these columns is 0 — because the dependent column has no component orthogonal to the other columns to supply the necessary nonzero *height* in the direction perpendicular to these columns. Since the determinant of an $n \times n$ matrix is the (signed) volume of this parallelepiped (see Appendix 7 which starts on page 799, summed up in Theorem A7.1 on page 804) we would expect λ is an eigenvalue if and only if $\det(\mathbf{A} - \lambda\mathbf{I}) = 0$. This result, that the columns of a matrix are linearly dependent if and only if its determinant is 0, is also established in the same appendix, as Theorem A7.2 on page 804. But this linear dependence of the columns of $\mathbf{A} - \lambda\mathbf{I}$ when $\det(\mathbf{A} - \lambda\mathbf{I}) = 0$ guarantees a nontrivial solution to equation, as mentioned above. We state this result as it applies to eigenvalues, in Theorem 13.3.

Theorem 13.3: Characteristic equation for eigenvalues

The complex number λ is an eigenvalue of the $n \times n$ real or complex matrix $\mathbf{A}$ (Definition 13.2 on page 536) if and only if

$$det(\mathbf{A} - \boldsymbol{\lambda}\mathbf{I}) = 0, \qquad \textbf{\textit{13.11}}$$

where $\mathbf{I}$ is the $n \times n$ identity matrix (Definition 10.6 on page 334) and $det(\mathbf{C})$ is the determinant of the $n \times n$ matrix $\mathbf{C}$ (Definition 12.21 on page 466, suitably modified in light of Definition 13.1 on page 535). Equation 13.11 is called the **characteristic equation** (for the eigenvalues) of $\mathbf{A}$.

Note that the column expansion for determinants (Theorem 12.24 on page 472) together with induction, can be used to show that the left side of equation 13.11 is an n-th degree polynomial in the variable λ, called the **characteristic polynomial** of the matrix $\mathbf{A}$.

❑

This theorem is a ***good theoretical starting point*** for finding the eigenvalues of a matrix $\mathbf{A}$. However, ***don't get the idea that in order to solve the practical problem of finding eigenvalues, you should always try for a direct solution of equation 13.11***. And even if solving equation 13.11 is what you want to do, it's almost never efficient to determine the formula for $det(\mathbf{A} - \boldsymbol{\lambda}\mathbf{I})$ from the column expansion (Theorem 12.24 on page 472) for this determinant. ***With this caveat in mind, solely for illustrative purposes we give the following example***.

Example 13.3: Eigenvalues and Eigenvectors of $\begin{bmatrix} 1 & 2 \\ 3 & 5 \end{bmatrix}$

We find using the column expansion (Theorem 12.24 on page 472) for a determinant, that

$$det\left(\begin{bmatrix} 1 & 2 \\ 3 & 5 \end{bmatrix} - \lambda\begin{bmatrix} 1 & 0 \\ 0 & 1 \end{bmatrix}\right) = det\begin{bmatrix} 1-\lambda & 2 \\ 3 & 5-\lambda \end{bmatrix} = (1-\lambda)(5-\lambda) = \lambda^2 - 6\lambda - 1.$$

But $\lambda^2 - 6\lambda - 1 = 0$ if and only if $\lambda^2 - 6\lambda + 9 - 9 - 1 = 0$, i.e., if and only if

$$(\lambda - 3)^2 = 10.$$

Thus using Theorem 13.3, we find that the eigenvalues of the matrix $\begin{bmatrix} 1 & 2 \\ 3 & 5 \end{bmatrix}$ are $\lambda = 3 \pm \sqrt{10}$.

To find an eigenvector associated with the eigenvalue $\lambda_1 = 3 - \sqrt{10}$, we look for a nontrivial vector $v = \begin{pmatrix} x \\ y \end{pmatrix}$ satisfying $\mathbf{A}v = \lambda_1 v$; writing this equation in long form yields

$$\begin{bmatrix} 1 & 2 \\ 3 & 5 \end{bmatrix}\begin{pmatrix} x \\ y \end{pmatrix} = (3 - \sqrt{10})\begin{pmatrix} x \\ y \end{pmatrix}$$

or

$$\begin{aligned} [1 - (3 - \sqrt{10})]x + 2y &= 0 \\ 3x + [5 - (3 - \sqrt{10})]y &= 0 \end{aligned}$$

or

$$\begin{aligned} (-2 + \sqrt{10})x \qquad\qquad + 2y &= 0 \\ 3x + (2 + \sqrt{10})y &= 0 \end{aligned}$$

It's not hard to see that if we multiply the first equation by the constant $3/(-2 + \sqrt{10})$, we obtain the second equation. So we may discard one of these equations, say the second. Arbitrarily putting $x = 1$ in this equation yields $y = -\dfrac{-2 + \sqrt{10}}{2} = 1 - \sqrt{2.5}$. Thus we see that the eigenvalue $\lambda_1 = 3 - \sqrt{10}$ has eigenvector (any nonzero multiple of) $\begin{pmatrix} 1 \\ 1 - \sqrt{2.5} \end{pmatrix}$.

■

Exercises 13.4

Find the eigenvalues and corresponding eigenvectors for the following matrices. Your results can be verified using *Mathematica*, *MLab*, *Splus*, *Mathcad*, *Matlab*, or many other computer packages.

1. Find the other eigenvector in Example 13.3.

2. $\mathbf{A} = \begin{bmatrix} 1 & 0 \\ 0 & 1 \end{bmatrix}$.

3. $\mathbf{A} = \begin{bmatrix} 1 & 0 & 0 \\ 0 & 1 & 0 \\ 0 & 0 & 1 \end{bmatrix}$

4. $\mathbf{A} = \begin{bmatrix} 1 & 2 & 3 \\ 0 & 4 & 5 \\ 10 & 6 & 7 \end{bmatrix}$

   Hint: You can find at least one eigenvalue using bisection, page 76, since in this case $det(\mathbf{A} - \boldsymbol{\lambda}\mathbf{I})$ is a third-degree polynomial.

5. $\mathbf{A} = \begin{bmatrix} 1 & 0 \\ 1 & 1 \end{bmatrix}$

The first question that might come to mind in dealing with eigenvalues is whether we'd expect to find sufficiently many eigenvectors to form the basis for a coordinate system — i.e., n linearly independent vectors in n-dimensional space (see Definition 9.11 on page 299 and Theorem 9.15 on page 303). Because $det(\mathbf{A} - \lambda\mathbf{I})$ is an n-th degree polynomial when $\mathbf{A}$ is an $n \times n$ matrix (see Theorem 13.3 which starts on page 538) it can be factored[1] in the form

$$det(\mathbf{A} - \lambda\mathbf{I}) = c(\lambda - \lambda_1)(\lambda - \lambda_2)\cdots(\lambda - \lambda_n), \qquad \textit{13.12}$$

where $\lambda_1,...,\lambda_n$ are complex numbers, which aren't necessarily distinct. So there's hope that we can find n eigenvectors, because whenever $\lambda = \lambda_i$ $det(\mathbf{A} - \lambda\mathbf{I}) = 0$, and so there is an eigenvector corresponding to $\lambda = \lambda_i$ — but they have to be linearly independent in order to form a basis for a useful coordinate system. The results of Exercises 13.4.5 on page 540 show that some $n \times n$ matrices have fewer than n linearly independent eigenvectors. So the question is not going to be a trivially answered one.

There is one case where this question is rather easily answered, namely, when the matrix* A *has n distinct eigenvalues, the eigenvectors are linearly independent, and hence form a basis for a coordinate system for (real or complex) ***n-dimensional space***. We back up this assertion as follows.

Suppose $\mathbf{A}$ has n distinct eigenvalues, and one of its eigenvectors, denoted by $\boldsymbol{x}_k$, was linearly dependent on the remaining ones — i.e.,

$$\boldsymbol{x}_k = \sum_{j \neq k} a_j \boldsymbol{x}_j. \qquad \textit{13.13}$$

We can certainly restrict consideration to the case where the $\boldsymbol{x}_j$ on the right side of equation 13.13 are linearly independent (Definition 9.7 on page 297) because, if not, any single vector which is linearly dependent on the others could be replaced by a weighted sum of the others, and effectively eliminated from the sum on the right. Also, at least one of the a_j must be nonzero, since if they were all 0, then $\boldsymbol{x}_k$ would be $\mathbf{0}$, and hence could not be an

1. This result is referred to as *the fundamental theorem of algebra*. It can be found in Bell et al., *Modern University Calculus* [1]. Most people don't find this result too hard to accept without proof. The actual determination of eigenvalues $\lambda_1,...,\lambda_n$, even when they are all known to be distinct, can be quite difficult. See Ralston [14].

eigenvector (see Definition 13.2 on page 536). Now apply **A** to both sides of equation 13.13 (i.e., multiply both sides of this equation by **A**) and again use the definition of eigenvalues and eigenvectors (Definition 13.2) to obtain

$$\lambda_k \boldsymbol{x}_k = \sum_{j \neq k} a_j \lambda_j \boldsymbol{x}_j .$$

Now substitute for $\boldsymbol{x}_k$ using equation 13.13 into the left side of this last equation, to obtain

$$\lambda_k \sum_{j \neq k} a_j \boldsymbol{x}_j . = \sum_{j \neq k} a_j \lambda_j \boldsymbol{x}_j ,$$

which we can rewrite as

$$\sum_{j : j \neq k} a_j (\lambda_k - \lambda_j) \boldsymbol{x}_j = \mathbf{0}, \qquad \textit{13.14}$$

where this notation indicates that the sum, with k a fixed value, is over all terms with j different from k. In this equation, based on conclusions already drawn, there is some value of j, say j_0 for which the coefficient $a_j(\lambda_k - \lambda_j)$ is not zero. Now recall that we were able to restrict consideration to a sequence of linearly independent vectors, $\boldsymbol{x}_j$ in this sum, and equation 13.14 shows that $\boldsymbol{x}_{j_0}$ is linearly dependent on the remaining $\boldsymbol{x}_j$. This contradiction arose because of the assumption that equation 13.13 could hold. To eliminate this contradiction, we are forced to admit that equation 13.13 cannot hold, and hence the original set of eigenvectors must be linearly independent. We state this result formally in the following.

Theorem 13.4: Distinct eigenvalues yield linearly independent eigenvectors

If $\lambda_1,...,\lambda_q$ are distinct eigenvalues of the $n \times n$ real or complex matrix **A** (Definition 13.2 on page 536) then corresponding eigenvectors, $\boldsymbol{x}_1,...,\boldsymbol{x}_q$ are linearly independent (Definition 9.7 on page 297). If $q = n$, then these eigenvectors form a basis (Definition 9.11 on page 299) for a coordinate system for the space of n-dimensional vectors.

❑

If you want to see some immediate applications for eigenvalues and eigenvectors in studying matrix behavior, you can skip to Theorem 13.15 on page 556. To get a deeper understanding of the background theory, continue with what follows below.

To simplify the presentation of our next topic — *representing a linear operator in a new coordinate system* — we need to introduce vectors in a somewhat more general context. So far, all of our vectors have consisted of n-tuples of real or complex numbers. We could add such vectors and multiply them by real or complex numbers (scalars) to obtain other vectors. A subspace consisted of all vectors that could be generated from some finite set of n-tuples by these operations.

It's now useful to think of vectors simply as objects which can be added or multiplied by scalars to form other vectors. Such a collection is called a *vector space* (a concept we'll formalize in Definition 13.5 on page 543).

Example 13.5: Vector space of real polynomials of degree $\leq n$

The set of all real polynomial functions, given by $\sum_{j=1}^{n-1} a_j t^j$, of degree not exceeding the fixed value $n - 1$, with domain some finite nonzero length interval is a vector space, because the sum of two such polynomials and the product of a real scalar times such a polynomial is a polynomial of degree not exceeding $n - 1$.

Note that this vector space is not a set of n-tuples (although, as we'll soon see, it can be represented by such a set). This vector space is *generated by* the functions given by $f_k(t) = t^k$ $k = 0,...,n - 1$, via the operations of vector addition and multiplication of a vector by a scalar. We will see (Exercise 13.8.1 on page 553) that we can think of the f_k, $k = 0,...,n - 1$, as the original basis $\boldsymbol{E}_0,...,\boldsymbol{E}_{n-1}$, and hence this space can be represented by the set of all real n-tuples, where

$$\begin{pmatrix} a_0 \\ \cdot \\ \cdot \\ a_{n-1} \end{pmatrix}$$

represents the polynomial $\sum_{k=0}^{n-1} a_k t^k$. If the domain (Definition 1.1 on page 2) of these polynomials is the bounded real interval $[a; b]$, then if $\boldsymbol{v}$ and $\boldsymbol{w}$ are two such polynomials, a natural dot product $\boldsymbol{v} \cdot \boldsymbol{w}$ is given by

$$\boldsymbol{v} \cdot \boldsymbol{w} = \int_a^b v(t)w(t)dt \cdot$$

This dot product has all of the essential properties of the previous real dot products already encountered, namely,

- $\boldsymbol{v} \cdot \boldsymbol{w} = \boldsymbol{w} \cdot \boldsymbol{v}$,
- $v \cdot (aw_1 + bw_2) = a(v \cdot w_1) + b(v \cdot w_2)$
- $\boldsymbol{v} \cdot \boldsymbol{v} \geq 0$, equality holding if and only if $\boldsymbol{v} = \boldsymbol{0}$

  (here $\boldsymbol{0}$ is the zero *function*)

■

Another such non-n-tuple vector space is given in the following example.

Example 13.6: Imaginary exponential polynomials

Consider the vector space spanned by (Definition 9.4 on page 296) the imaginary exponential functions (Definition 6.9 on page 200) given by

$$f_k(t) = e^{ikt}, 0 \le t < 2, k = 0, \pm 1, ..., \pm m,$$

with the scalars being the complex numbers, and dot product

$$\boldsymbol{v} \cdot \boldsymbol{w} = \int_0^{2\pi} w(t)\overline{w(t)}dt.$$

This vector space can also be represented by the set of all complex n-tuples, where $n = 2m + 1$, with the n-tuple

$$\begin{pmatrix} a_{-m} \\ a_{-m+1} \\ \cdot \\ a_0 \\ \cdot \\ a_m \end{pmatrix}$$

representing the vector (function) whose formula is

$$\sum_{k=-m}^{m} a_k e^{ikt}.$$

■

We formalize the idea of a vector space in the following definition.

Definition 13.5: Abstract vector space, dimension, basis

A set of objects, denoted by ***v,w,x,...*** and referred to as **vectors**, is called a **vector space [over the real [complex] numbers]**, if there are operations of

- addition of vectors
- multiplication of a vector by a real [complex] number

both of which yield vectors, such that if a and b are real [complex] numbers,

- $(a+b)\boldsymbol{v} = a\boldsymbol{v} + b\boldsymbol{v}$
- $a(\boldsymbol{v}+\boldsymbol{w}) = a\boldsymbol{v} + a\boldsymbol{w}$
- $\boldsymbol{v}+\boldsymbol{w} = \boldsymbol{w}+\boldsymbol{v}$
- $(\boldsymbol{v}+\boldsymbol{w})+\boldsymbol{x} = \boldsymbol{v}+(\boldsymbol{w}+\boldsymbol{x})$
- $1\boldsymbol{v} = \boldsymbol{v}$

There exists a unique *zero vector*, $\mathbf{0}$, with the property that for all vectors $\boldsymbol{v}$

$$\mathbf{0} + \boldsymbol{v} = \boldsymbol{v}.$$

A vector space, $\mathcal{V}$, is said to be **finite dimensional with dimension n** if there is a finite sequence of vectors, $\boldsymbol{v}_1,...,\boldsymbol{v}_n$ such that every $\boldsymbol{v}$ in $\mathcal{V}$ can be written as a linear combination

$$\boldsymbol{v} = \sum_{j=1}^{n} a_{j,\boldsymbol{v}} \boldsymbol{v}_j, \qquad \textit{13.15}$$

where $a_{j,\boldsymbol{v}}$ are real [complex] and n is the smallest number allowing representation 13.15 for all $\boldsymbol{v}$ in $\mathcal{V}$. Any sequence $\boldsymbol{v}_1,...,\boldsymbol{v}_n$ with the above property is called a **basis** for $\mathcal{V}$, and is said to **span** $\mathcal{V}$.

A vector space which isn't finite dimensional is called **infinite dimensional**.

❑

In this chapter we will be dealing only with finite dimensional vector spaces. We'll meet infinite-dimensional vector spaces, and the extension of the concept of basis for these spaces in the succeeding chapters.

When the only kinds of vectors we were dealing with were n-tuples of numbers, it was evident that the $\mathbf{0}$ vector postulated in the definition of a vector space existed, and was the n-tuple of all zeros, with

$$\mathbf{0} = \boldsymbol{v} - \boldsymbol{v} = \begin{pmatrix} 0 \\ \cdot \\ \cdot \\ 0 \end{pmatrix}$$

for each n-tuple vector $\boldsymbol{v}$.

The result that $\mathbf{0} = \boldsymbol{v} - \boldsymbol{v}$ for each $\boldsymbol{v}$ in a finite-dimensional vector space is easy to establish in our more abstract setting; because if $\mathcal{V}$ is a finite dimensional vector space, there is some sequence of vectors, $\boldsymbol{v}_1,...,\boldsymbol{v}_n$, such that for each $\boldsymbol{v}$ we can write

$$\begin{aligned} \boldsymbol{v} &= \sum_j a_{j,\boldsymbol{v}} \boldsymbol{v}_j = \sum_j (a_{j,\boldsymbol{v}} + 0) \boldsymbol{v}_j \\ &= \sum_j a_{j,\boldsymbol{v}} \boldsymbol{v}_j + \sum_j 0 \, \boldsymbol{v}_j \\ &= \boldsymbol{v} + \sum_j 0 \, \boldsymbol{v}_j . \end{aligned}$$

Thus $\sum_j 0\,\boldsymbol{v}_j$ satisfies the postulated property of the $\mathbf{0}$ vector, and by the assumed uniqueness

$$\mathbf{0} = \sum_j 0\,\boldsymbol{v}_j$$

(where $0\boldsymbol{v}_j$ means the number 0 times the vector $\boldsymbol{v}_j$).

But since $\boldsymbol{v} - \boldsymbol{v} = \sum_j 0\,\boldsymbol{v}_j$, we have $\boldsymbol{v} - \boldsymbol{v} = \mathbf{0}$ for all vectors $\boldsymbol{v}$. Thus we have the following.

Theorem 13.6: 0

For all vectors $\boldsymbol{v}$ in a finite dimensional vector space, $\mathcal{V}$ (Definition 13.5 on page 543)

$$\boldsymbol{v} - \boldsymbol{v} = \mathbf{0}.$$

❑

It doesn't seem obvious how to establish this in the most general case.

Results which still hold for abstract vector spaces

Most of the important definitions and theorems preceding those dependent on orthogonality in Chapter 9 can be seen to hold for the more abstract vector spaces introduced here; in particular, Definition 9.4 on page 296 (linear combination, subspace), Definition 9.7 on page 297 (linear independence and linear dependence), Definition 9.8 on page 298 (dimension of a subspace), Definition 9.9 on page 298 (coordinates), Theorem 9.10 on page 299 (uniqueness of coordinates for linearly independent sequence, which depended on Theorem 13.6 above), Definition 9.11 on page 299 of *basis*, and Theorem 9.13 on page 302 that all bases have the same number of elements.

To make use of results on orthogonality, we need to introduce the general concept of dot products.

Definition 13.7: Dot product over a vector space

Let $\mathcal{V}$ be a given vector space (Definition 13.5 on page 543). A real [complex] valued function of pairs of vectors from $\mathcal{V}$ with value denoted by $\boldsymbol{v} \cdot \boldsymbol{w}$ is called a **dot product** (or **scalar product**) if for all vectors $\boldsymbol{v}, \boldsymbol{w}, \boldsymbol{x}, \ldots$ in $\mathcal{V}$

- $\boldsymbol{v} \cdot \boldsymbol{v} \geq 0$, equality holding only if $\boldsymbol{v} = \mathbf{0}$ (the *zero vector*),
- $\boldsymbol{v} \cdot \boldsymbol{w} = \overline{\boldsymbol{w} \cdot \boldsymbol{v}}$ (see Definition 13.1 on page 535),

- $(v + w) \cdot x = (v \cdot x) + (w \cdot x)$,
- $(a\, v) \cdot x = a(v \cdot x)$.

❑

Definition 13.8: Length, $\|v\|$ of vector, v, distance $\|v - w\|$

Given a vector space, $\mathcal{V}$ (Definition 13.5 on page 543) with dot product (Definition 13.7 on page 545), the **length**, $\|v\|$, **of** v is defined by

$$\|v\| = \sqrt{v \cdot v},$$

and the **distance between vectors** v **and** w is defined to be $\|v - w\|$. ❑

Results which still hold for abstract dot products

With the definition of a vector space (Definition 13.5 on page 543), dot product (Definition 13.7 on page 545) and length and distance between vectors (Definition 13.8 above) the definitions of projection (Definition 8.12 on page 276) and Theorem 8.13 on page 276 (projection in terms of dot products), Definition 8.14 on page 278 (orthogonality), Theorem 8.15 on page 278 (orthogonality in terms of dot products), the orthogonal breakup theorem Theorem 8.16 on page 278, and the Pythagorean theorem (Theorem 8.17 on page 279) remain unchanged.

The Schwartz inequality (Theorem 8.19 on page 282) must be changed with $(X \cdot Y)^2$ replaced by $|X \cdot Y|^2$. This is because equation 8.31 on page 282 must be replaced by the following:

$$\begin{aligned}\left\|\frac{X \cdot Y}{Y \cdot Y} Y\right\|^2 &= \left(\frac{X \cdot Y}{Y \cdot Y} Y\right) \cdot \left(\frac{X \cdot Y}{Y \cdot Y} Y\right) = \left(\frac{X \cdot Y}{Y \cdot Y}\right) Y \cdot \left(\frac{X \cdot Y}{Y \cdot Y} Y\right) \\ &= \frac{X \cdot Y}{Y \cdot Y}\overline{\left(\frac{X \cdot Y}{Y \cdot Y} Y\right) \cdot Y} \;=\; \frac{X \cdot Y}{Y \cdot Y}\,\frac{\overline{X \cdot Y}}{\overline{Y \cdot Y}}\,\overline{Y \cdot Y} \\ &= \frac{|X \cdot Y|^2}{Y \cdot Y}.\end{aligned}$$

Then Theorem 9.16 on page 303 (independence of nonzero orthogonal vectors), and the results of the definition of the Gram Schmidt procedure (Definition 9.18 on page 305 and Definition 9.19 on page 306), namely, Theorem 9.17 on page 304 (coordinates relative to orthogonal basis), Theorem 9.20 on page 307 (properties of the Gram Schmidt procedure), Theorem 9.21 on page 310 (orthogonality to *previous* subspaces), the definition of projection on a subspace (Definition 9.23 on page 316 of $P_S(Y)$), and Theorem 9.24 on page 316 on computation of $P_S(Y)$ ***all still hold***.

Having generalized the notion of vectors, we now find it useful to introduce the notion of a *linear operator* (also called a *linear function*) on a vector space into another (possibly different) vector space. This notion is just a slightly more abstract version of the function associated with a matrix. Recall that an $m \times n$ matrix **A** defines a function (see Chapter 10, Section 2 on page 329) whose domain (inputs) consists of n-tuples and whose range (outputs) consists of m-tuples; namely, the function $\mathcal{A}$ whose value at the n-tuple $\boldsymbol{x}$ is $\mathcal{A}(x) = \mathbf{A}\boldsymbol{x}$. Because of the nature of matrix multiplication (see Theorem 10.2 on page 330) it follows that

$$\mathcal{A}(a\boldsymbol{x} + b\boldsymbol{y}) = a\mathcal{A}(\boldsymbol{x}) + b\mathcal{A}(\boldsymbol{y}),$$

where a and b are scalars. This is the essence of a linear operator.

Definition 13.9: Linear operator

A function (Definition 1.1 on page 2) $\mathcal{A}$ whose domain is a vector space (Definition 13.5 on page 543) $\mathcal{V}$, and whose range is contained in a vector space $\mathcal{W}$ which possesses the **linear operator property**,

$$\mathcal{A}(a\boldsymbol{x} + b\boldsymbol{y}) = a\mathcal{A}(\boldsymbol{x}) + b\mathcal{A}(\boldsymbol{y}) \qquad \textit{13.16}$$

for all vectors $\boldsymbol{x}, \boldsymbol{y}$ in $\mathcal{V}$ and all scalars, a and b (real or complex as appropriate) is called **a linear operator from $\mathcal{V}$ into $\mathcal{W}$**.

❑

Because the only types of vectors that we dealt with earlier were n-tuples of numbers, there was little need to distinguish vectors from n-tuples. However, because, as we've already seen, vectors can be represented in terms of any basis, by n-tuples, it's necessary for clarity of presentation to make this distinction from here on. We formalize this change in the following two definitions.

Definition 13.10: *n*-Tuple representation of vector in basis

Let $\mathcal{V}$ be any finite dimensional vector space (Definition 13.5 on page 543) and $E = \boldsymbol{E}_1,...,\boldsymbol{E}_n$ a basis for $\mathcal{V}$. We say **the n-tuple**

$$\mathbf{v}_E = \begin{bmatrix} v_{1,E} \\ \cdot \\ \cdot \\ v_{n,E} \end{bmatrix}$$

of numbers represents the vector v in the basis E if

$$\boldsymbol{v} = \sum_{j=1}^{n} \mathrm{v}_{j,E} \boldsymbol{E}_j .$$

Note that the symbols representing such n-tuples will be printed in boldface, and they (and their coordinates) will be indexed by the associated basis, and will denote **column matrices**; usually the n-tuple associated with a vector and basis will be denoted by the same symbol as that denoting the vector, but in boldface and subscripted with the basis name, as in $\mathbf{v}_E$ associated with the vector $\boldsymbol{v}$ and the basis E above.

❑

To build the machinery necessary to see how eigenvectors furnish an appropriate basis in which to study matrices, we next need to see how to specify the matrix used to change coordinates from one basis to another. So suppose we have a vector space, $\mathcal{V}$, with two bases, $E = \boldsymbol{E}_1,...,\boldsymbol{E}_n$ and $F = \boldsymbol{F}_1,...,\boldsymbol{F}_n$, with the vector $\boldsymbol{w}$ represented by (see Definition 13.10 on page 547) the n-tuples, $\mathbf{w}_E$ in the basis E and $\mathbf{w}_F$ in the basis F. We want to determine the relationship between the n-tuples $\mathbf{w}_E$ and $\mathbf{w}_F$.

We know that

$$\boldsymbol{w} = \sum_{j=1}^{n} \mathrm{w}_{j,E} \boldsymbol{E}_j \quad \text{and} \quad \boldsymbol{w} = \sum_{j=1}^{n} \mathrm{w}_{j,F} \boldsymbol{F}_j . \qquad \textbf{\textit{13.17}}$$

To get what we need we must convert one of these equations, say the first, to look more like the other. Let's do this by writing the $\boldsymbol{E}_j$ in terms of the $\boldsymbol{F}_j$. All we need do is the same process used for the second equation above, but on each of the $\boldsymbol{E}_j$. That is write

$$\boldsymbol{E}_j = \sum_{k=1}^{n} \mathrm{E}_{j,k,F} \boldsymbol{F}_k , \qquad \textbf{\textit{13.18}}$$

where the n-tuple $\mathbf{E}_{j,F}$ whose k-th element is $\mathrm{E}_{j,k,F}$ represents the vector $\boldsymbol{E}_j$ in the basis $\boldsymbol{F}_1,...,\boldsymbol{F}_n$. Now substitute from equation 13.18 into the first of equations 13.17, to obtain

$$w = \sum_{j=1}^{n} \mathrm{w}_{j,E} \sum_{k=1}^{n} \mathrm{E}_{j,k,F} \boldsymbol{F}_k ,$$

which we rewrite as

$$\boldsymbol{w} = \sum_{k=1}^{n} \left(\sum_{j=1}^{n} \mathrm{w}_{j,E} \mathrm{E}_{j,k,F} \right) \boldsymbol{F}_k .$$

Since the coefficients in any basis representation of a vector are unique (see Theorem 9.10 on page 299 which we saw holds in the more abstract situation as discussed on page 545) it must be the case, comparing this equation with the second of equations 13.17, that

$$\mathrm{w}_{j,F} = \sum_{j=1}^{n} \mathrm{w}_{j,E} \mathrm{E}_{j,k,F}. \qquad \textit{13.19}$$

Note that this is a system of equations between numbers, which can be represented as a single matrix equation

$$\mathbf{w}_F = \mathbf{E}_F \mathbf{w}_E, \qquad \textit{13.20}$$

where $\mathbf{E}_F$ is the matrix whose j-th column is the n-tuple $\mathbf{E}_{j,F}$ representing the vector $\boldsymbol{E}_j$ in the basis $\boldsymbol{F}_1,...,\boldsymbol{F}_n$. In identical fashion we may write

$$\mathbf{w}_E = \mathbf{F}_E \mathbf{w}_F, \qquad \textit{13.21}$$

where $\mathbf{F}_E$ is the matrix whose j-th column is the n-tuple $\mathbf{F}_{j,E}$ representing the vector $\boldsymbol{F}_j$ in the basis $\boldsymbol{E}_1,...,\boldsymbol{E}_n$.

Substituting from equation 13.21 into equation 13.20 yields

$$\mathbf{w}_F = \mathbf{E}_F \mathbf{F}_E \mathbf{w}_F,$$

which, since it must hold for all n-tuples $\mathbf{w}_F$, shows that $\mathbf{E}_F\mathbf{F}_E$ is the $n \times n$ identity matrix (see Theorem 10.11 on page 337). Then it follows from Definition 10.7 on page 334, Definition 10.9 on page 336, and Theorem 10.8 on page 336, that the matrices $\mathbf{E}_F$ and $\mathbf{F}_E$ are invertible (have inverses) and are inverses of each other. Thus we may rewrite equation 13.20 as

$$\mathbf{w}_E = \mathbf{E}_F^{-1} \mathbf{w}_F. \qquad \textit{13.22}$$

Hence we have shown the following.

Theorem 13.11: Coordinate change via matrix multiplication

Suppose the vector $\boldsymbol{w}$ in the finite dimensional vector space $\mathcal{V}$ (see Definition 13.5 on page 543) is represented by the n-tuple $\mathbf{w}_E$ in the basis E (see Definition 13.10 on page 547) and by the n-tuple $\mathbf{w}_F$ in the basis F. The equations relating these n-tuples are

$$\mathbf{w}_F = \mathbf{E}_F \mathbf{w}_E \quad \text{and} \quad \mathbf{w}_E = \mathbf{E}_F^{-1} \mathbf{w}_F, \qquad \textit{13.23}$$

where $\mathbf{E}_F$ is the matrix whose j-th column is the n-tuple representing the j-th vector, E_j of the basis E, in the basis F, $j = 1,2,...,n$, and $\mathbf{E}_F^{-1}$ is the inverse of $\mathbf{E}_F$ (see Theorem 10.8 on page 336).

Note in equations 13.23 that for given bases E and F, $\mathbf{E}_F$ and $\mathbf{E}_F^{-1}$ are fixed matrices, and these equations hold for all n-tuples $\mathbf{w}_E$ and the corresponding $\mathbf{w}_F$ s. But these include all possible n-tuples, since each n-tuple $\mathbf{w}_E$ represents some vector in $\mathcal{V}$ (as does each $\mathbf{w}_F$).

The same proof that led to equations 13.23 also yields

$$\mathbf{w}_E = \mathbf{F}_E\mathbf{w}_F \quad \text{and} \quad \mathbf{w}_F = \mathbf{F}_E^{-1}\mathbf{w}_E. \qquad \textit{13.24}$$

Since equations 13.23 and 13.24 hold for all n-tuples $\mathbf{w}_E$ and $\mathbf{w}_F$ we see that

$$\mathbf{E}_F^{-1} = \mathbf{F}_E \quad \text{and} \quad \mathbf{F}_E^{-1} = \mathbf{E}_F, \qquad \textit{13.25}$$

(see Problem 13.9.1 on page 555 if you have difficulty establishing this yourself.)

Equations 13.23 show how to convert from the representation of a vector in one basis to its representation in another one.

❑

Our next tasks are to show how to represent a linear operator as a matrix associated with a pair of coordinate systems (one for the domain, one for the range), and how to change this representation when the coordinate systems are changed. We will then be ready to apply these results to a change of basis to the eigenvector coordinate system of a matrix, which will show why the eigenvector coordinate system is so often the system of choice. (It has other important uses as well, such as for developing the *singular value decomposition*, used for better understanding the behavior of $m \times n$ systems, see Golub and Van Loan [9]).

To investigate these problems, suppose $\mathcal{A}$ is a linear operator from the vector space $\mathcal{V}$ into the vector space $\mathcal{W}$ where $E = \boldsymbol{E}_1,...,\boldsymbol{E}_n$ is a basis for $\mathcal{V}$ and $G = \boldsymbol{G}_1,...,\boldsymbol{G}_n$ is a basis for $\mathcal{W}$. We want to produce a matrix $\mathbf{A}_{E,G}$ which represents $\mathcal{A}$ in these two bases; by which we mean that if

$$w = \mathcal{A}(v)$$

and $\mathbf{w}_G$ is the m-tuple representing w in the basis G (Definition 13.10 on page 547) with $\mathbf{v}_E$ representing v in the basis E, then[1]

$$\mathbf{w}_G = \mathbf{A}_{E,G}\mathbf{v}_E$$

1. Of course, we must show that there is such a relationship.

(see Definition 10.1 on page 328).

To find $\mathbf{A}_{E,G}$ notice that

$$\boldsymbol{v} = \sum_{j=1}^{n} \mathrm{v}_{j,E} \boldsymbol{E}_i$$

by Definition 13.10 on page 547. Since $\mathcal{A}$ is a linear operator (Definition 13.9 on page 547)

$$\mathcal{A}(\boldsymbol{v}) = \sum_{j=1}^{n} \mathrm{v}_{j,E}\, \mathcal{A}(\boldsymbol{E}_j). \qquad \mathbf{13.26}$$

But $\boldsymbol{w}_j = \mathcal{A}(\boldsymbol{E}_j)$ and $\boldsymbol{w} = \mathcal{A}(\boldsymbol{v})$ both belong to $\mathcal{W}$, so we can write

$$\boldsymbol{w}_j = \mathcal{A}(\boldsymbol{E}_j) = \sum_{k=1}^{m} \mathrm{w}_{j,k,G} \boldsymbol{G}_k,\ j = 1,...,n \ \text{ and } \ w = \mathcal{A}(\boldsymbol{v}) = \sum_{k=1}^{m} \mathrm{w}_{k,G} \boldsymbol{G}_k.$$

Hence substituting into equation 13.26 yields

$$\sum_{k=1}^{m} \mathrm{w}_{k,G} \boldsymbol{G}_k = \sum_{j=1}^{n} \mathrm{v}_{j,E} \sum_{k=1}^{m} \mathrm{w}_{j,k,G} \boldsymbol{G}_k,$$

which we rewrite as

$$\sum_{k=1}^{m} \mathrm{w}_{k,G} \boldsymbol{G}_k = \sum_{k=1}^{m} \left(\sum_{j=1}^{n} \mathrm{v}_{j,E}\, \mathrm{w}_{j,k,G} \right) \boldsymbol{G}_k .$$

Since the $\boldsymbol{G}_k$ form a basis for $\mathcal{W}$ and the coordinates in a basis are unique (Theorem 9.10 on page 299) it follows that

$$\mathrm{w}_{k,G} = \sum_{j=1}^{n} \mathrm{v}_{j,E}\, \mathrm{w}_{j,k,G} \ \text{ for } k = 1,...,n. \qquad \mathbf{13.27}$$

Equations 13.27 are easily seen to represent a matrix equation of the form

$$\mathbf{w}_G = \mathbf{A}_{E,G} \mathbf{v}_E ,$$

where $\mathbf{w}_G$ is the m-tuple representing $\boldsymbol{w} = \mathcal{A}(\boldsymbol{v})$ in the G basis for $\mathcal{W}$ and $\mathbf{v}_E$ is the n-tuple representing $\boldsymbol{v}$ in the E basis for $\mathcal{V}$ (again see Definition 13.10 on page 547). The matrix $\mathbf{A}_{E,G}$ is evidently the $m \times n$ matrix whose j-th column is

$$\mathbf{w}_{j,G} = \begin{pmatrix} \mathrm{w}_{j,1,G} \\ \cdot \\ \cdot \\ \mathrm{w}_{j,m,G} \end{pmatrix},$$

the m-tuple representing $\boldsymbol{w}_j = \mathcal{A}(\boldsymbol{E}_j)$ in the basis G for $\mathcal{W}$ (see equation 10.4 on page 328). We have thus shown the following result.

Theorem 13.12: Linear operator coordinate representation

Let $\mathcal{A}$ be a linear operator from the n-dimensional vector space $\mathcal{V}$ into the m-dimensional vector space $\mathcal{W}$ (see Definition 13.9 on page 547 and Definition 13.5 on page 543). Let v be a vector in $\mathcal{V}$ and $w = \mathcal{A}(v)$, with w represented by the m-tuple

$$\mathbf{w}_G = \begin{pmatrix} w_{1,G} \\ . \\ . \\ w_{m,G} \end{pmatrix}$$

in the basis G for $\mathcal{W}$ (see Definition 13.10 on page 547) and v represented by the n-tuple

$$\mathbf{v}_E = \begin{pmatrix} v_{1,E} \\ . \\ . \\ v_{n,E} \end{pmatrix}$$

in the basis E for $\mathcal{V}$. The linear operator $\mathcal{A}$ is represented by the matrix $\mathbf{A}_{E,G}$ in the coordinate systems related to these two bases — where the j-th column of $\mathbf{A}_{E,G}$ is the m-tuple $\mathbf{w}_{j,G}$ representing $w_j = \mathcal{A}(E_j)$ in the basis G for $\mathcal{W}$. By this we mean

$$\mathbf{w}_G = \mathbf{A}_{E,G}\mathbf{v}_E .$$

❑

Example 13.7: Matrix representing the derivative operator

In Example 13.5 on page 542 suppose we take the basis E to be the *vectors* given by the formulas $t^0, t, ..., t^{n-1}$. We leave the proof that these functions form a basis to Exercise 13.8.1 on page 553. Let D be the derivative operator on the (real) vector space of real polynomials of degree $\leq n - 1$, with domain $[a; b]$, $a < b$. What matrix represents D in the basis E?

Since $D_t^j t^j = j t^{j-1}$, $j > 0$, the j-th row of $\mathbf{A}$ is the n-tuple whose coordinates are all 0 except for coordinate $j + 1$ whose value is j, $j > 0$, and whose coordinates are all 0 when $j = n$, as indicated in equation 13.28.

$$\mathbf{A}^{n \times n} = \begin{bmatrix} 0 & 1 & 0 & 0 & . & . \\ 0 & 0 & 2 & 0 & . & . \\ . & . & 0 & 0 & . & . \\ . & . & . & . & . & . \\ 0 & 0 & . & . & 0 & n-1 \\ 0 & 0 & 0 & . & . & 0 \end{bmatrix} . \qquad \textit{13.28}$$

Try this for $n = 2$ to see that the bookkeeping is okay.

Here we are thinking of $\mathcal{V}$ and $\mathcal{W}$ as being n-dimensional, but then the image $D(\mathcal{V})$ (see Definition 12.20 on page 462) doesn't *take up* all of $\mathcal{W}$. So we could treat $\mathcal{W}$ as $n-1$ dimensional, and then the matrix $\mathbf{A}$ corresponding to D is $(n-1) \times (n-1)$, gotten by deleting the first column and last row of the matrix in equation 13.28, and reducing the dimension of $\mathcal{W}$ by eliminating the basis vector represented by t^{n-1}. It's simpler to keep $\mathcal{V}$ and $\mathcal{W}$ as we did originally.

■

Exercises 13.8

1. Referring to Example 13.5 on page 542, show that the vectors given by the formulas $t^0, t, ..., t^{n-1}$ form a basis for the vector space of real polynomials of degree $\leq n-1$.
   Hint: Suppose that it were the case that
   $$t^k = \sum_{\substack{j \neq k \\ 0 \leq j \leq n-1}} a_j t^j \quad \text{for all } t \text{ in } [a; b],\ a < b.$$ Try
   differentiating both sides of this equation a suitable number of times. You will see that each of the coefficients on the right have to be 0, which is quite impossible.
2. Find the matrix representing the derivative operator for the vector space of Example 13.6 on page 543.

We continue on by finding out how changes of bases affect the matrix representation of a linear operator. Thus, suppose that $\mathcal{A}$ is a linear operator (Definition 13.9 on page 547) from the finite dimensional space $\mathcal{V}$ into the finite dimensional space $\mathcal{W}$, represented by the matrix $\mathbf{A}_{E,G}$, where E is a basis for $\mathcal{V}$ and G is a basis for $\mathcal{W}$ (see Theorem 13.12 on page 552). How is $\mathbf{A}_{E,G}$ related to $\mathbf{A}_{F,H}$, where F and H are also bases for $\mathcal{V}$ and $\mathcal{W}$, respectively? By Theorem 13.11 on page 549 (see also Definition 13.10 on page 547)

$$\mathbf{G}_H \mathbf{w}_G = \mathbf{w}_H \quad \text{and} \quad \mathbf{E}_F \mathbf{v}_E = \mathbf{v}_F, \tag{13.29}$$

where $\mathbf{G}_H$ is the $m \times m$ whose j-th column represents the G basis vector $\mathbf{G}_j$ in the H basis, and $\mathbf{E}_F$ is the matrix whose k-th column represents the E basis vector $\mathbf{E}_k$ in the F basis. By Theorem 13.12 on page 552 we may write

$$\mathbf{w}_G = \mathbf{A}_{E,G}\mathbf{v}_E \qquad \mathbf{w}_H = \mathbf{A}_{F,H}\mathbf{v}_F. \tag{13.30}$$

From the second equations of 13.30 and 13.29 we find

$$\mathbf{w}_H = \mathbf{A}_{F,H}\mathbf{v}_F = \mathbf{A}_{F,H}\mathbf{E}_F\mathbf{v}_E. \qquad 13.31$$

From the first equation of 13.29 followed by the first equation of 13.30, we obtain

$$\mathbf{w}_H = \mathbf{G}_H\mathbf{w}_G = \mathbf{G}_H\mathbf{A}_{E,G}\mathbf{v}_E. \qquad 13.32$$

Hence from equations 13.31 and 13.32 we find

$$\mathbf{A}_{F,H}\mathbf{E}_F\mathbf{v}_E. = \mathbf{G}_H\mathbf{A}_{E,G}\mathbf{v}_E$$

for all vectors $\boldsymbol{v}$ — i.e. for all n-tuples $\mathbf{v}_E$ representing vectors $\boldsymbol{v}$ in the E basis, which means for all n-tuples $\mathbf{v}_E$.

But if $\mathbf{R}^{q\times n}\mathbf{v}_E = \mathbf{S}^{q\times n}\mathbf{v}_E$ for all n-tuples $\mathbf{v}_E$, then $\mathbf{R} = \mathbf{S}$, a result which we leave as Problems 13.9 on page 555 for the reader.

Hence, $\mathbf{A}_{F,H}\mathbf{E}_F = \mathbf{G}_H\mathbf{A}_{E,G}$, or

$$\mathbf{A}_{F,H} = \mathbf{G}_H\mathbf{A}_{E,G}\mathbf{E}_F^{-1}.$$

We have thus shown the following theorem.

Theorem 13.13: Representing matrix in changed coordinates

Let $\mathcal{A}$ be a linear operator (Definition 13.9 on page 547) from the finite dimensional vector space $\mathcal{V}$ into the finite dimensional space $\mathcal{W}$. Let $\mathbf{A}_{E,G}$ represent $\mathcal{A}$ in the bases E for $\mathcal{V}$ and G for $\mathcal{W}$, and $\mathbf{A}_{F,H}$ represent $\mathcal{A}$ in the bases F for $\mathcal{V}$ and H for $\mathcal{W}$ (see Theorem 13.12 on page 552).

Then

$$\mathbf{A}_{F,H} = \mathbf{G}_H\mathbf{A}_{E,G}\mathbf{E}_F^{-1},$$

where $\mathbf{E}_F$ is the matrix whose j-th column is the n-tuple representing the j-th vector, $\boldsymbol{E}_j$ of the basis E, in the basis F, $j = 1,2,...,n$, and $\mathbf{G}_H$ is the matrix whose k-th column is the m-tuple representing the k-th vector, $\boldsymbol{G}_k$ of the basis G, in the basis H, $k = 1,2,...,m$ (see Theorem 13.11 on page 549).

❑

As a corollary, we have the following result of great use in examining eigenvector bases.

Theorem 13.14: Corollary to Theorem 13.13

If $\mathcal{A}$ is a linear operator from the finite dimensional vector space $\mathcal{V}$ into itself (see Definition 13.9 on page 547) and $\mathbf{A}_{E,E}$ represents $\mathcal{A}$ in the bases E and E, then the representation $\mathbf{A}_{F,F}$ of $\mathcal{A}$ in the bases F,F is given by

$$\mathbf{A}_{F,F} = \mathbf{E}_F \mathbf{A}_{E,E} \mathbf{E}_F^{-1} \qquad 13.33$$

because here $G = E$ and $H = F$.

Alternatively we may write

$$\mathbf{A}_{F,F} = \mathbf{F}_E^{-1} \mathbf{A}_{E,E} \mathbf{F}_E \qquad 13.34$$

because $\mathbf{E}_F = \mathbf{F}_E^{-1}$ (as shown in Theorem 13.11 on page 549).

❑

Problems 13.9

1. Show that if $\mathbf{R}^{q \times n}\mathbf{v} = \mathbf{S}^{q \times n}\mathbf{v}$ for all n-tuples $\mathbf{v}$, then $\mathbf{R} = \mathbf{S}$.

   Hint: Letting

   $$\mathbf{v} = \mathbf{e}_1 = \begin{pmatrix} 1 \\ 0 \\ \cdot \\ 0 \end{pmatrix},$$

   show that the first column of $\mathbf{R}$ is the same as the first column of $\mathbf{S}$, etc.

2. For various values of n of your choice, find lots of really distinct $n \times n$ matrices which do not possess n distinct eigenvectors. Hint: use the results of exercise 13.4.5 on page 540 and Theorem 13.14 on page 555. The *Mathematica* program will be helpful.

Equation 13.34 sets the stage for showing why the basis of eigenvectors of a matrix $\mathbf{A}$ is so often the proper basis for studying this matrix; because, suppose we think of the matrix $\mathbf{A}$ as representing a linear operator, $\mathcal{A}$ in some basis, E (Theorem 13.12 on page 552) and suppose the eigenvectors of $\mathbf{A}$ (Definition 13.2 on page 536) form a basis for n-dimensional real [complex] space. Let F represent the eigenvector basis. From equation 13.34, in this basis $\mathcal{A}$ is represented by

$$\mathbf{A}_{F,F} = \mathbf{F}_E^{-1} \mathbf{A} \mathbf{F}_E. \qquad 13.35$$

The columns of $\mathbf{F}_E$ are the eigenvectors of $\mathbf{A}$ (in the original basis, E). But from the definition of the eigenvector, $\mathbf{x}_{(i)}$, and its corresponding eigenvalue, λ_i

$$\mathbf{A}\mathbf{x}_{(i)} = \lambda_i \mathbf{x}_{(i)} \quad i = 1,...,n. \qquad \textit{13.36}$$

(The eigenvalues need not be distinct, and the eigenvectors are only determined up to a nonzero multiplicative constant — but no matter.)

We can represent the n equations in 13.36 as one matrix equation,

$$\mathbf{A}^{n \times n}\mathbf{X}^{n \times n} = \mathbf{X}^{n \times n}\mathbf{\Lambda}^{n \times n}$$

where $\mathbf{\Lambda}$ is a diagonal matrix (0s for all off-diagonal elements) whose j-th element is the eigenvalue λ_i, and $\mathbf{X}$ is the matrix whose i-th column is $\mathbf{x}_{(i)}$, the i-th eigenvector. Since the eigenvectors are here assumed to form a basis, consisting of n linearly independent vectors, $\mathbf{X}$ is invertible (see Theorem 10.8 on page 336 ff.) and hence

$$\Lambda = \mathbf{X}^{-1}\mathbf{A}\mathbf{X}, \qquad \textit{13.37}$$

or, multiplying this equation on the left by $\mathbf{X}$ and on the right by $\mathbf{X}^{-1}$

$$\mathbf{A} = \mathbf{X}\Lambda\mathbf{X}^{-1}. \qquad \textit{13.38}$$

Equation 13.37 is exactly the same as equation 13.35 above, since $\mathbf{F}_E$ is the matrix of n-tuples representing the eigenvectors of $\mathbf{A}$ in the E basis. Hence $\mathbf{A}_{F,F} = \Lambda$, a matrix which is much simpler than $\mathbf{A}$ for performing various operations. We note that equation 13.38 does not require the *coordinate change* interpretation to see how useful it is as a way of representing $\mathbf{A}$, as we shall see shortly; but this interpretation is very useful for understanding what's going on.

We state the result just derived formally as Theorem 13.15.

Theorem 13.15: Eigenvector representation of square matrix

Let $\mathbf{A}$ be an $n \times n$ matrix with n linearly independent eigenvectors (see Definition 13.2 on page 536, Theorem 13.4 on page 541). Let $\mathbf{X}$ be the matrix whose columns are the eigenvectors of $\mathbf{A}$. Then

$$\mathbf{A} = \mathbf{X}\Lambda\mathbf{X}^{-1},$$

where Λ is the $n \times n$ matrix whose diagonal consists of the eigenvalues corresponding to the columns of $\mathbf{X}$, with all *off-diagonal* elements equal to 0.

❑

In the section to follow we will examine some interesting applications of Theorem 13.15.

Remarks

Notice that matrices seem to have several distinct uses. They are used in

- representing systems of linear equations
- changing coordinates to obtain more convenient descriptions of systems
- representing linear operators.

3 Some Immediate Eigenvector Applications

As we'll see shortly, there are many situations in which we want to find the behavior of high powers of a matrix. For example, in trying to determine the general behavior of compartmental models (Example 4.5 on page 96) we will be led to examining the infinite matrix series

$$\sum_{m=0}^{\infty} \frac{\mathbf{A}^m}{m!};$$

also studies of the long run behavior of waiting lines will lead in the same direction. For this reason we will start of with a general look at this problem in

Example 13.10: Behavior of $\mathbf{A}^m$

When the eigenvectors of the matrix $\mathbf{A}$ form a basis, we can make use of Theorem 13.15 above. In such a case, because we can write $\mathbf{A} = \mathbf{X}\Lambda\mathbf{X}^{-1}$, it follows that

$$\mathbf{A}^2 = (\mathbf{X}\Lambda\mathbf{X}^{-1})(\mathbf{X}\Lambda\mathbf{X}^{-1}) = \mathbf{X}\Lambda^2\mathbf{X}^{-1}$$

$$\mathbf{A}^3 = (\mathbf{X}\Lambda\mathbf{X}^{-1})(\mathbf{X}\Lambda\mathbf{X}^{-1})^2 = \mathbf{X}\Lambda^3\mathbf{X}^{-1}$$

and by induction, generally

$$\mathbf{A}^m = \mathbf{X}\Lambda^m\mathbf{X}^{-1}. \qquad \textit{13.39}$$

What makes equation 13.39 so useful is that

$$\Lambda^m = \begin{bmatrix} \lambda_1^m & 0 & . & . & 0 \\ 0 & \lambda_2^m & 0 & . & 0 \\ 0 & . & . & . & 0 \\ 0 & . & 0 & \lambda_{n-1}^m & 0 \\ 0 & . & . & 0 & \lambda_n^m \end{bmatrix}$$

so that computation of $\mathbf{A}^m$ requires only two matrix multiplications and use of the symbols λ_j^m, the latter often requiring no computation to determine the behavior of the product on the right of equation 13.39. We'll see some illustrations of this assertion in some of the examples to follow.

■

Example 13.11: Three state Markov process

The matrix

$$\mathbf{P} = \begin{bmatrix} 0.8 & 0.2 & 0.0 \\ 0.1 & 0.85 & 0.05 \\ 0.0 & 0.3 & 0.7 \end{bmatrix} \qquad \textit{13.40}$$

might be the transition probability matrix of some three-state system; for example, your telephone with *call-waiting* service could be in one of three states

1. not in use, 2. in use — no call is waiting, 3. in use — a call is waiting.

The above matrix indicates that the probability that the system changes from state 2 to state 3 in one step (one time unit, say one minute) is 0.05. Under the Markov process assumption[1] it can be shown that the matrix

$$\mathbf{P}^2 = \begin{bmatrix} 0.66 & 0.33 & 0.01 \\ 0.165 & 0.7575 & 0.0775 \\ 0.03 & 0.465 & 0.505 \end{bmatrix}$$

consists of the two-step transition probabilities — which indicates that the probability of going from state 2 to state 3 in two steps is 0.0775.

---

1. The meaning of this assumption is that in this system, knowledge of all of the parameters describing the present state yields the same prediction of the probabilities of future states as does additional information about the states at times preceding the present.

One of the questions of greatest interest is to determine the probability of going from state i to state j in m steps, for large m (which also turns out to be the long-term probability of being in state j). Manually we can compute, using any one of many available packages (*Mathcad* was used in the case) the values

$$\mathbf{P}^4 = \begin{bmatrix} 0.49 & 0.472 & 0.037 \\ 0.236 & 0.664 & 0.099 \\ 0.112 & 0.597 & 0.291 \end{bmatrix} \qquad \mathbf{P}^8 = \begin{bmatrix} 0.356 & 0.568 & 0.076 \\ 0.284 & 0.612 & 0.104 \\ 0.228 & 0.623 & 0.148 \end{bmatrix}$$

$$\mathbf{P}^{16} = \begin{bmatrix} 0.305 & 0.597 & 0.097 \\ 0.209 & 0.601 & 0.101 \\ 0.292 & 0.604 & 0.104 \end{bmatrix} \qquad \mathbf{P}^{32} = \begin{bmatrix} 0.3 & 0.6 & 0.1 \\ 0.3 & 0.6 & 0.1 \\ 0.3 & 0.6 & 0.1 \end{bmatrix}.$$

So it looks as if $\mathbf{P}^m \cong \mathbf{P}^{32}$ for large m.

To get a better theoretical handle on $\mathbf{P}^m$ we find the eigenvalues of $\mathbf{P}$ using *Mathcad* 5.0 to be

$$\lambda_1 = 1, \lambda_2 = 0.75, \lambda_3 = 0.6$$

with eigenvectors

$$\mathbf{x}_{(1)} = \begin{bmatrix} 0.577 \\ 0.577 \\ 0.577 \end{bmatrix}, \quad \mathbf{x}_{(2)} = \begin{bmatrix} -0.549 \\ 0.137 \\ 0.824 \end{bmatrix}, \quad \mathbf{x}_{(3)} = \begin{bmatrix} 0.302 \\ -0.302 \\ 0.905 \end{bmatrix},$$

which you can certainly verify yourself. We also find

$$\mathbf{X}^{-1} = [\mathbf{x}_{(1)}, \ \mathbf{x}_{(2)}, \ \mathbf{x}_{(3)}]^{-1} = \begin{bmatrix} 0.52 & 1.04 & 0.173 \\ -0.971 & 0.485 & 0.486 \\ 0.553 & -1.105 & 0.552 \end{bmatrix}$$

from which Theorem 13.15 on page 556 yields that

$$\mathbf{P} = \mathbf{X}\Lambda\mathbf{X}^{-1} = \begin{bmatrix} 0.577 & -0.549 & 0.302 \\ 0.577 & 0.137 & -0.302 \\ 0.577 & 0.824 & 0.905 \end{bmatrix} \begin{bmatrix} 1 & 0 & 0 \\ 0 & 0.75 & 0 \\ 0 & 0 & 0.6 \end{bmatrix} \begin{bmatrix} 0.52 & 1.04 & 0.173 \\ -0.971 & 0.485 & 0.486 \\ 0.553 & -1.105 & 0.552 \end{bmatrix}$$

and thus, as with Example 13.10 on page 557, we find

$$\mathbf{P}^m = \mathbf{X}\Lambda^m\mathbf{X}^{-1} \cong \begin{bmatrix} 0.577 & -0.549 & 0.302 \\ 0.577 & 0.137 & -0.302 \\ 0.577 & 0.824 & 0.905 \end{bmatrix} \begin{bmatrix} 1 & 0 & 0 \\ 0 & 0 & 0 \\ 0 & 0 & 0 \end{bmatrix} \begin{bmatrix} 0.52 & 1.04 & 0.173 \\ -0.971 & 0.485 & 0.486 \\ 0.553 & -1.105 & 0.552 \end{bmatrix}$$

for large m. Carrying out these computations, we find

$$\mathbf{P}^m \cong \begin{bmatrix} 0.577 & -0.549 & 0.302 \\ 0.577 & 0.137 & -0.302 \\ 0.577 & 0.824 & 0.905 \end{bmatrix} \begin{bmatrix} 0.52 & 1.04 & 0.173 \\ 0 & 0 & 0 \\ 0 & 0 & 0 \end{bmatrix} = \begin{bmatrix} 0.3 & 0.6 & 0.1 \\ 0.3 & 0.6 & 0.1 \\ 0.3 & 0.6 & 0.1 \end{bmatrix}$$

as we found on page 546.

So with the transition probability matrix given by equation 13.40 on page 558, we find that in the long run your phone stands

- a 30% chance of not being in use,
- a 60% chance of being in use with a call waiting,

and

- a 10% chance of being in use with no call waiting.

■

Exercises 13.12

1. Using eigenvalue—eigenvector methods, find $\mathbf{A}^m$ or explain why it can't be done.

a. $\mathbf{A} = \begin{bmatrix} 1 & 1 \\ 4 & 1 \end{bmatrix}$

b. $\mathbf{A} = \begin{bmatrix} 1 & 9 \\ 1 & 1 \end{bmatrix}$

c. $\mathbf{A} = \begin{bmatrix} 1 & 0 \\ 1 & 1 \end{bmatrix}$

d. $\mathbf{A} = \begin{bmatrix} 1 & 0 & 0 \\ 0 & 1 & 1 \\ 0 & 4 & 1 \end{bmatrix}$

2. What is $\mathbf{A}^m$ for exercise 1c above?
3. For the following three-state phone system, find $\mathbf{P}^m$,

$$\mathbf{P} = \begin{bmatrix} 0.9 & 0.1 & 0 \\ 0.3 & 0.65 & 0.05 \\ 0 & 0.15 & 0.85 \end{bmatrix}.$$

4. Find the form of $\mathbf{P}^m$ from Example 13.11 which will enable you to estimate easily how large m should be taken so that $\mathbf{P}^m$ is within a specified distance from its limiting value (see Definition 10.18 on page 350).

Another important use for eigenvalues arises in solving systems of constant coefficient differential equations. We illustrate this in

Example 13.13: $\mathbf{y}' = \mathbf{A}\mathbf{y}$ $\quad \mathbf{A} = \begin{bmatrix} 1 & 1 \\ 4 & 1 \end{bmatrix}$

The eigenvalues of $\mathbf{A}$ are easily obtained from the characteristic equation (see Theorem 13.3 on page 538) which yields

$$\lambda_1 = 3, \lambda_2 = -1 \quad \text{with corresponding eigenvectors } \begin{bmatrix} 1 \\ 2 \end{bmatrix} \text{ and } \begin{bmatrix} 1 \\ -2 \end{bmatrix}$$

(which can be verified directly). Hence, from Theorem 13.15 on page 556, we may write

$$\mathbf{A} = \begin{bmatrix} 1 & 1 \\ 2 & -2 \end{bmatrix}\begin{bmatrix} 3 & 0 \\ 0 & -1 \end{bmatrix}\begin{bmatrix} 1 & 1 \\ 2 & -2 \end{bmatrix}^{-1} = \begin{bmatrix} 1 & 1 \\ 2 & -2 \end{bmatrix}\begin{bmatrix} 3 & 0 \\ 0 & -1 \end{bmatrix}\begin{bmatrix} 1/2 & 1/4 \\ 1/2 & -1/4 \end{bmatrix}.$$

The system $\mathbf{y}' = \mathbf{A}\mathbf{y}$ becomes

$$\mathbf{y}' = \begin{bmatrix} 1 & 1 \\ 2 & -2 \end{bmatrix}\begin{bmatrix} 3 & 0 \\ 0 & -1 \end{bmatrix}\begin{bmatrix} 1/2 & 1/4 \\ 1/2 & -1/4 \end{bmatrix}\mathbf{y}.$$

Now multiplying on the left by

$$\begin{bmatrix} 1 & 1 \\ 2 & -2 \end{bmatrix}^{-1} = \begin{bmatrix} 1/2 & 1/4 \\ 1/2 & -1/4 \end{bmatrix}, \qquad \textit{13.41}$$

we find

$$\left(\begin{bmatrix} 1/2 & 1/4 \\ 1/2 & -1/4 \end{bmatrix}\mathbf{y}'\right) = \begin{bmatrix} 3 & 0 \\ 0 & -1 \end{bmatrix}\left(\begin{bmatrix} 1/2 & 1/4 \\ 1/2 & -1/4 \end{bmatrix}\mathbf{y}\right),$$

which suggests letting

$$\mathbf{z} = \begin{bmatrix} 1/2 & 1/4 \\ 1/2 & -1/4 \end{bmatrix} \mathbf{y} \qquad \textit{13.42}$$

to obtain the equation

$$\mathbf{z}' = \begin{bmatrix} 3 & 0 \\ 0 & -1 \end{bmatrix} \mathbf{z}.$$

Written out in expanded form, this matrix equation becomes the easily solved *decoupled* system

$$z_1' = 3z_1$$
$$z_2' = -z_2.$$

The solution to this system, using the shift rule (Theorem 5.11 on page 146) is found to be

$$z_1(t) = c_1 e^{3t} \qquad z_2(t) = c_2 e^{-t}. \qquad \textit{13.43}$$

From equations 13.41 and 13.42 we obtain

$$\mathbf{y} = \begin{bmatrix} 1/2 & 1/4 \\ 1/2 & -1/4 \end{bmatrix}^{-1} \mathbf{z} = \begin{bmatrix} 1 & 1 \\ 2 & -2 \end{bmatrix} \mathbf{z},$$

i.e.,

$$y_1 = z_1 + z_2 \qquad y_2 = 2z_1 - 2z_2.$$

The constants c_1 and c_2 can be chosen to satisfy initial conditions on z_1 and z_2.

■

In Chapter 6, Exercise 6.12.3 on page 208, with some effort we found the formula solution of the glycerol kinetics problem. We now illustrate the more elegant approach to this solution using eigenvalues.

Example 13.14: Matrix solution of glycerol kinetics system

The system of differential equations (see equation 6.22 on page 188) for the glycerol kinetics problem of Example 6.7 on page 188 can be written in matrix form as

$$\mathbf{x}' = \mathbf{A}\mathbf{x} + \mathbf{f} \qquad \mathbf{x}(0) = \begin{bmatrix} 1.41 \\ 0 \end{bmatrix} \qquad \mathbf{f} = \begin{bmatrix} f_1 \\ 0 \end{bmatrix},$$

where

$$f_1(t) = \begin{cases} 0.441 & \text{for } t > 0 \\ 0 & \text{for } t \le 0 \end{cases}$$

and

$$\mathbf{A} = \begin{bmatrix} -0.571 & 0.085 \\ 0.26 & -0.085 \end{bmatrix}.$$

Because

$$\begin{aligned} det(\mathbf{A} - \boldsymbol{\lambda}\mathbf{I}) &= det\begin{bmatrix} -0.571 - \lambda & 0.085 \\ 0.26 & -0.085 - \lambda \end{bmatrix} \\ &= (0.571 + \lambda)(0.085 + \lambda) + 0.085 \times 0.26 \\ &= \lambda^2 + 0.656\lambda + (0.328)^2 - (0.328)^2 + 0.026435 \\ &= (\lambda + 0.328)^2 - 0.081149, \end{aligned}$$

we find that $det(\mathbf{A} - \boldsymbol{\lambda}\mathbf{I}) = 0$ for

$$\lambda_1 = -0.328 + \sqrt{0.081149} \qquad \lambda_2 = -0.328 - \sqrt{0.081149}.$$

The eigenvectors corresponding to these eigenvalues (see Definition 13.2 on page 536) are seen to be

$$\mathbf{v}_1 = \begin{bmatrix} -0.159 \\ -0.987 \end{bmatrix} \qquad \mathbf{v}_2 = \begin{bmatrix} -0.897 \\ 0.442 \end{bmatrix}.$$

Hence the matrix $\mathbf{V}$ of eigenvectors and its inverse, $\mathbf{V}^{-1}$ are given by

$$\mathbf{V} = \begin{bmatrix} -0.159 & -0.897 \\ -0.987 & 0.442 \end{bmatrix} \qquad \mathbf{V}^{-1} = \begin{bmatrix} 0.463 & -0.939 \\ -1.033 & 0.166 \end{bmatrix} \qquad \textbf{\textit{13.44}}$$

(the inversion of the matrix[1] $\mathbf{V}$ being done by *Mathematica*, or any of the methods presented in Chapter 10). The eigenvalue matrix

$$\Lambda = \begin{bmatrix} -0.043 & 0 \\ 0 & -0.613 \end{bmatrix}. \qquad \textbf{\textit{13.45}}$$

---

1. The notation change for the eigenvector matrix from $\mathbf{X}$ to $\mathbf{V}$ is necessary since the symbol $\mathbf{X}$ was preempted by the differential equation.

Hence, from Theorem 13.15 on page 556 (with $\mathbf{V}$ in place of $\mathbf{X}$)

$$\mathbf{A} = \mathbf{V\Lambda V}^{-1}.$$

This allows us to write the equation $\mathbf{x}' = \mathbf{Ax} + \mathbf{f}$ as

$$\mathbf{x}' = \mathbf{V\Lambda V}^{-1}\mathbf{x} + \mathbf{f}$$

or

$$(\mathbf{V}^{-1}\mathbf{x})' = \Lambda(\mathbf{V}^{-1}\mathbf{x}) + \mathbf{V}^{-1}\mathbf{f}.$$

Letting $\mathbf{z} = \mathbf{V}^{-1}\mathbf{x}$, this equation becomes

$$\mathbf{z}' = \mathbf{\Lambda z} + \mathbf{V}^{-1}\mathbf{f}. \qquad \textit{13.46}$$

Since

$$\mathbf{V}^{-1}\mathbf{f} = \begin{bmatrix} -0.463 & -0.939 \\ -1.033 & 0.166 \end{bmatrix}\begin{bmatrix} 0.441 \\ 0 \end{bmatrix} = \begin{bmatrix} -0.2042 \\ -0.4556 \end{bmatrix}$$

we see using equations 13.44 and 13.45 that equation 13.46 becomes

$$\begin{aligned} z_1' &= -0.43z_1 - 0.2042 \\ z_2' &= -0.613z_2 - 0.4556. \end{aligned} \qquad \textit{13.47}$$

To solve this we could use the shift rule, Theorem 5.11 on page 146; instead, we first note that a particular solution of these equations is

$$z_{p1} = -0.2042/0.043 = -4.749, z_{p2} = -0.4556/0.613 = -0.7432.$$

By the remark following Theorem 11.1 on page 410 and Theorem 5.12 on page 147, we find the general solution (Definition 11.3 on page 412) of equations 13.47 to be

$$z_1(t) = -4.749 + c_1e^{-0.043t} \qquad z_2(t) = -0.7432 + c_2e^{-0.613t}.$$

Since $\mathbf{z} = \mathbf{V}^{-1}\mathbf{x}$ we have $\mathbf{x} = \mathbf{Vz}$, and thus $x_1 = -0.159z_1 - 0.897z_2$ $x_2 = -0.987z_1 + 0.442z_2$,

or

$$x_1(t) = -0.159(-4.749 + c_1e^{-0.043t}) - 0.897(-0.7432 + c_2e^{-0.613t})$$

$$x_2(t) = -0.987(-4.749 + c_1e^{-0.043t}) + 0.442(-0.7432 + c_2e^{-0.613t}).$$

To evaluate the constants c_1 and c_2 corresponding to the initial conditions $x_2(0) = 0, x_1(0) = 1.41$, we rewrite the above equations as

$$\begin{aligned} x_1(t) &= 1.4218 - 0.159c_1e^{-0.043t} - 0.897c_2e^{-0.613t} \\ x_2(t) &= 4.3588 - 0.987c_1e^{-0.043t} + 0.442c_2e^{-0.613t}. \end{aligned} \qquad \textbf{\textit{13.48}}$$

Substituting in for the initial conditions on the left, and using $e^0 = 1$, we find

$$\begin{aligned} 0.159c_1 + 0.897c_2 &= 1.4218 - 1.41 = 0.0118 \\ 0.987c_1 - 0.442c_2 &= 4.3588, \end{aligned}$$

from which it easily follows (by solving the first equation for c_1 and substituting into the second one, etc.) that $c_1 = 4.0976$ and $c_2 = -0.713$. Hence, substituting into equations 13.48 yields

$$\begin{aligned} x_1(t) &= 1.4218 - 0.6514e^{-0.043t} + 0.6396e^{-0.613t} \\ x_2(t) &= 4.3588 - 4.0436e^{-0.043t} - 0.3152e^{-0.613t}. \end{aligned}$$

The discrepancy between these answers and those in equations 6.27 on page 189 arises from the small number of significant digits being carried in the computations done here.

Remark: Plateau values

It should be noted that the limiting values

$$\lim_{t \to \infty} x_1(t) \cong 1.4218 \quad \text{and} \quad \lim_{t \to \infty} x_2(t) \cong 4.3588$$

can be computed without solving differential equations 6.22 on page 188, because, as one would guess, for large t in such a compartmental system, we would expect $x_1'(t)$ and $x_2'(t)$ to approach 0, once the system has *settled down*, which we expect it to. Then we need only solve the algebraic equations

$$\mathbf{Ax} + \mathbf{f} = \mathbf{0} \qquad \textbf{\textit{13.49}}$$

for these limiting values — i.e.,

$$\begin{aligned} -0.571x_1 + 0.085x_2 &= -0.441 \\ 0.26x_1 - 0.085x_2 &= 0. \end{aligned}$$

Adding these two equations yields $0.311x_1 = 0.441$, from which we find

$$x_1(\infty) = 1.418006, x_2(\infty) = \frac{0.26}{0.085}x_1(\infty) = 4.3374.$$

■

Exercises 13.15

1. Redo Example 13.14 on page 562 with greater accuracy than was done in text, to reproduce the results of equations 6.27 on page 189. Note that the exact formulas give you information about the rate of approach to the plateau values.

2. Solve the differential equation $x''' + 3x'' - 21x' + x = sin$, $x(0) = x'(0) = x''(0) = 0$, by converting to a system of first order differential equations and using the eigenvalue method of solution. Check your results by using the shift rule, Theorem 5.11 on page 146, as well as by use of some trustworthy computer package, such as *Mathematica*, *Maple*, *Matlab*, *Mlab*, etc.
Hint for use of shift rule: $D^3 + 3D^2 - 21D + 1 = (D-2)(\cdots)$.

3. Show that if the $n \times n$ matrix $\mathbf{A}$ has distinct eigenvalues $\lambda_1,...,\lambda_n$, hence n linearly independent eigenvectors, $\mathbf{v}_1,...,\mathbf{v}_n$, corresponding to them (see Theorem 13.4 on page 541) then the solution of

$$\mathbf{x}' = \mathbf{Ax} \quad \text{subject to} \quad \mathbf{x}(0) = \mathbf{x}_0$$

is

$$\mathbf{x}(t) = \mathbf{V}\begin{bmatrix} e^{\lambda_1 t} & 0 & . & . & 0 \\ 0 & e^{\lambda_2 t} & 0 & . & . \\ . & . & . & . & . \\ 0 & . & 0 & e^{\lambda_{n-1} t} & 0 \\ 0 & . & . & 0 & e^{\lambda_n t} \end{bmatrix}\mathbf{V}^{-1}\mathbf{x}_0$$

Hints: As in Example 13.14 on page 562, the given differential equation may be written $(\mathbf{V}^{-1}\mathbf{x})' = \Lambda(\mathbf{V}^{-1}\mathbf{x})$, so that this differential equation may be rewritten by letting $\mathbf{z} = \mathbf{V}^{-1}\mathbf{x}$, as $\mathbf{z}' = \Lambda\mathbf{z}$, where $\mathbf{V}$ is the matrix of eigenvectors of $\boldsymbol{A}$. The solution of this equation is of the form

$$\mathbf{z}(t) = \begin{bmatrix} e^{\lambda_1 t} & 0 & . & . & 0 \\ 0 & e^{\lambda_2 t} & 0 & . & . \\ . & . & . & . & . \\ 0 & . & 0 & e^{\lambda_{n-1} t} & 0 \\ 0 & . & . & 0 & e^{\lambda_n t} \end{bmatrix}\begin{bmatrix} c_1 \\ . \\ . \\ . \\ c_n \end{bmatrix}$$

so that

$$\mathbf{x}(t) = \mathbf{V}\begin{bmatrix} e^{\lambda_1 t} & 0 & . & . & 0 \\ 0 & e^{\lambda_2 t} & 0 & . & . \\ . & . & . & . & . \\ 0 & . & 0 & e^{\lambda_{n-1} t} & 0 \\ 0 & . & . & 0 & e^{\lambda_n t} \end{bmatrix}\begin{bmatrix} c_1 \\ . \\ . \\ . \\ c_n \end{bmatrix}$$

But from this last equation we see that

$$\mathbf{x}(0) = \mathbf{V}\begin{bmatrix} c_1 \\ . \\ . \\ c_n \end{bmatrix}$$

so that

$$\begin{bmatrix} c_1 \\ . \\ . \\ c_n \end{bmatrix} = \mathbf{V}^{-1}\mathbf{x}(0) = \mathbf{V}^{-1}\mathbf{x}_0.$$

4. Generalize the result of exercise 3 above to solving $\mathbf{x}' = \mathbf{A}\mathbf{x} + \mathbf{f} \qquad \mathbf{x}(0) = \mathbf{x}_0$.

5. Define $e^{\mathbf{B}} = \mathbf{I}^{n \times n} + \sum_{k=1}^{\infty} \frac{\mathbf{B}^k}{k!}$, where $\mathbf{B}$ is any $n \times n$ matrix (recall Definition 10.22 on page 351, of a matrix infinite series).

   a. Show that the solution of the system $\mathbf{x}' = \mathbf{A}\mathbf{x} \qquad \mathbf{x}(0) = \mathbf{x}_0$ is $\mathbf{x}(t) = e^{\mathbf{A}t}\mathbf{x}_0$, where $\mathbf{A}$ is an $n \times n$ matrix.

   b. For the case where the $n \times n$ matrix $\mathbf{A}$ has n linearly independent eigenvectors, determine $e^{\mathbf{A}}$ in terms of the matrix $\mathbf{V}$ of eigenvectors and the matrix Λ of corresponding eigenvalues of $\mathbf{A}$.

6. It's important to avoid being too cavalier in manipulating the matrices $e^{\mathbf{A}}$ and $e^{\mathbf{B}}$. The law of exponents **does not hold** when $\mathbf{A}$ and $\mathbf{B}$ are $n \times n$ matrices with $n > 1$. To see this, show that the equality $e^{\mathbf{A}}e^{\mathbf{B}} = e^{\mathbf{B}}e^{\mathbf{A}}$ does not hold generally — so that $e^{\mathbf{A}+\mathbf{B}}$ **cannot be written** as $e^{\mathbf{A}}e^{\mathbf{B}}$.
   Hint: Try

$$\mathbf{A} = \begin{bmatrix} 0 & 1 \\ 0 & 0 \end{bmatrix}, \mathbf{B} = \begin{bmatrix} 0 & 0 \\ 1 & 0 \end{bmatrix}.$$

4 Numerical Determination of Eigenvalues

Because determination of eigenvalues is so vital to many applications, we'll now devote some space to introducing a few of the principal methods that have been developed for this purpose. That likelihood that you will ever be required to write your own eigenvalue program is small, due to the ready availability of powerful computer packages having this capability on PCs (*Mathematica*, *Mlab*, *Mathcad*, *Matlab*, *Splus*, to name a few). Still, it's worth knowing something about the available methods, to provide insight into what is involved in this task.

The first method that we'll discuss is the *power method*, which is directly useful when there is a unique eigenvalue with largest magnitude. We are allowing the possibility that the subspace of eigenvectors associated with this eigenvalue has dimension higher than 1; however, for the derivation being presented, we are assuming that the $n \times n$ matrix $\mathbf{A}$ does possess n linearly independent eigenvectors (Definition 13.2 on page 536 and Definition 9.7 on page 297) $\mathbf{x}_1,...,\mathbf{x}_n$. Then, if $\mathbf{v}$ is an arbitrary n-dimensional real or complex vector, it follows (see Theorem 9.13 on page 302, Definition 9.11 on page 299 and Theorem 9.15 on page 303 if needed) that we may write

$$\mathbf{v} = \sum_{j=1}^{n} c_i \mathbf{x}_i .$$

Then from the linear operator property of matrices (Theorem 10.2 on page 330) it follows that

$$\mathbf{A}^m \mathbf{v} = \sum_{j=1}^{n} c_j \lambda_j^m \mathbf{x}_j. \qquad \textit{13.50}$$

Suppose now that

$$\lambda_1 = \lambda_2 = \cdots = \lambda_k$$

$$\text{and} \qquad \textit{13.51}$$

$$|\lambda_{k+j}| < |\lambda_1| \quad \text{all} \quad j \geq 1.$$

Then, provided that $c_1^2 + ... + c_k^2 \neq 0$, for large enough m, we would expect $\mathbf{A}^m \mathbf{v}$ to consist principally of

$$\sum_{j=1}^{k} c_j \lambda_j^m \mathbf{x}_j.$$

This prompts us to write

$$\mathbf{A}^m \mathbf{v} = \sum_{j=1}^{n} c_j \lambda_j^m \mathbf{x}_j = \lambda_1^m \left[\sum_{j=1}^{k} c_j \mathbf{x}_j + \sum_{j=k+1}^{n} c_j \left(\frac{\lambda_j}{\lambda_1} \right)^m \mathbf{x}_j \right]. \qquad \textit{13.52}$$

Note that under the assumptions that we have made, the magnitude of the second sum in the square brackets is approaching 0 for large m. Hence we can see that so long as, say the q-th coordinate value of $\sum_{j \le k} c_j \mathbf{x}_j$ is nonzero, then the computable ratio

$$\frac{(\mathbf{A}^{m+1}\mathbf{v})_q}{(\mathbf{A}^m\mathbf{v})_q} \qquad \textit{13.53}$$

of the q-th coordinate values of these matrices, will be close to λ_1 for large enough m. It's reasonable, when using this *power* method, to choose q as the coordinate maximizing the magnitude of the denominator in equation 13.53, to control the effect of roundoff error. We summarize this discussion in Definition 13.16

Definition 13.16: Power method for eigenvalue computation

The **power method for estimating the assumed unique largest magnitude eigenvalue** (Definition 13.2 on page 536) of the $n \times n$ matrix $\mathbf{A}$ (where $\mathbf{A}$ is assumed to have n linearly independent eigenvectors, see Definition 9.7 on page 297) is to use the quantity

$$\frac{(\mathbf{A}^{m+1}\mathbf{v})_q}{(\mathbf{A}^m\mathbf{v})_q}$$

(see Definition 10.1 on page 328) where the subscript q represents the q-th coordinate value of the given vector (assumed nonzero) and m chosen large enough for the above ratio to have stabilized. This method tends to yield good answers provided that $\mathbf{v}$ has a nonzero *component* in the subspace generated by the vectors $\mathbf{x}_1,...,\mathbf{x}_k$, having this eigenvalue — i.e., has a nonzero projection on this subspace (see Definition 9.22 on page 313).

❑

Example 13.16: Power method application

Let $\mathbf{A} = \begin{bmatrix} 1 & 1 & 2 \\ 1 & 2 & 3 \\ 0 & 3 & 4 \end{bmatrix}$. We examine the ratio of the 1,1 elements of the matrices $\mathbf{A}^{m+1}$ and $\mathbf{A}^m$ for $m = 25$ and $m = 36$. We find that

$$\frac{(\mathbf{A}^{26})_{1,1}}{(\mathbf{A}^{25})_{1,1}} = \frac{(\mathbf{A}^{36})_{1,1}}{(\mathbf{A}^{35})_{1,1}} = 6.39936.$$

Note that since $\mathbf{A}^m = \mathbf{A}^m\mathbf{I} = \mathbf{A}^m[\mathbf{e}_1,...,\mathbf{e}_n]$, where $\mathbf{I}$ is the $n \times n$ identity matrix (Definition 10.6 on page 334) and the $\mathbf{e}_j$ are the standard n-dimensional unit vectors (Definition 8.2 on page 263) by using the ratios above, we are in fact looking at

$$\frac{(\mathbf{A}^{m+1}\mathbf{v})_q}{(\mathbf{A}^m\mathbf{v})_q}$$

where $\mathbf{v} = \mathbf{e}_1$ **and** $q = 1$**.**

The value $\lambda = 6.39936$ obtained from this method does, in fact, make $det(\mathbf{A} - \boldsymbol{\lambda}\mathbf{I}) = -\lambda^3 + 7\lambda^2 - 4\lambda + 1 = .000416$ which is very close to 0 (see Theorem 13.3 on page 538). By examining the derivative of this polynomial, it's not difficult to see that this eigenvalue is accurate to the number of digits displayed. (You can use one of the methods in Section 4 of Chapter 3 to verify this claim.)

■

Problems 13.17

1. What would you expect the behavior of the power method to be in a case where there is more than one eigenvalue with largest magnitude? How would you use the answer to this question to see whether or not to trust the results of the power method in any given case?
   Hint: Start with two real eigenvalues of largest magnitude and opposite sign, and write out an expression for $\mathbf{A}^m\mathbf{v}$ separating out the first two terms (similar to what was done in the discussion prior to Definition 13.16 on page 569).

2. Suppose that the real $n \times n$ matrix $\mathbf{A}$ has n linearly independent eigenvectors, and a unique pair of complex conjugate eigenvalues of largest magnitude. Modify the power method to handle this situation. This result is important because it arises at some stage of most processes for eigenvalue determination when not all eigenvalues are real.
   Note that if there are two eigenvalues of largest magnitude, then either they are real, in which case they must be negatives of each other (as assumed in the previous problem) or they **must** be complex conjugates (Definition 13.1 on page 535) of each other; this follows because, since $\mathbf{A}$ is assumed real, if λ is one of the two largest magnitude eigenvalues assumed nonreal, then its corresponding eigenvector, $\mathbf{x}$, cannot be real. But
   $$\mathbf{A}\mathbf{x} = \boldsymbol{\lambda}\mathbf{x} \text{ implies } \mathbf{A}\bar{\mathbf{x}} = \overline{\mathbf{A}\mathbf{x}} = \overline{\boldsymbol{\lambda}\mathbf{x}} = \bar{\lambda}\bar{\mathbf{x}}$$
   so that $\bar{\lambda}(\neq \lambda)$ is an eigenvalue of $\mathbf{A}$ or the same (largest) magnitude as λ. Since distinct eigenvalues yield linearly independendent eigenvectors (Theorem 13.4 on page 541) $\bar{\mathbf{x}}$ is

not a scalar multiple of $\mathbf{x}$. Thus $\bar{\lambda}$ is the other largest magnitude eigenvalue, with corresponding eigenvector $\bar{x}$.

Hints: Denote the pair of largest magnitude eigenvalues by $re^{i\vartheta}$ and $re^{-i\vartheta}$ (see Definition 6.12 on page 200, where we may assume from the geometry of complex numbers, that $0 < \vartheta < \pi$) and the corresponding eigenvectors by $\mathbf{x}_+$ and $\mathbf{x}_-$. We'll assume that we have scaled the eigenvectors so that $\|\mathbf{x}_+\| = \|\mathbf{x}_-\|$, with

$$\mathbf{x}_+ = \overline{\mathbf{x}_-}.$$

Now let $\mathbf{v}$ be any real vector which we will assume is not totally missing both $\mathbf{x}_+$ and $\mathbf{x}_-$ in its eigenvector expansion. (This will certainly hold for at least one of the standard unit vectors (Definition 8.2 on page 263) i.e., for $\mathbf{v}$ being some member of the set $\{\mathbf{e}_1,...,\mathbf{e}_n\}$.) Then

$$\mathbf{v} = c_+\mathbf{x}_+ + c_-\mathbf{x}_- + \sum_{j=3}^{n} c_i\mathbf{x}_i$$

where $\mathbf{x}_3,...,\mathbf{x}_n$ are the remaining eigenvectors. We may legitimately write

$$\mathbf{A}^m\mathbf{v} = c_+(re^{i\vartheta})^m\mathbf{x}_+ + c_-(re^{-i\vartheta})^m\mathbf{x}_- + \sum_{j=3}^{n} c_i\lambda_i^m\mathbf{x}_i.$$

But because $\mathbf{A}^m\mathbf{v}$ is real, and λ_i^m/r^m approaches 0 for large m, and $\mathbf{x}_+ = \overline{\mathbf{x}_-}$, it is seen that we must have $c_- = \overline{c_+}$. Hence for large m, with small relative error, we must have

$$\begin{aligned}\mathbf{A}^m\mathbf{v} &\cong c_+(re^{i\vartheta})^m\mathbf{x}_+ + c_-(re^{-i\vartheta})^m\mathbf{x}_-\\ &= r^m[c_+e^{im\vartheta}\mathbf{x}_+ + c_-e^{-im\vartheta}\mathbf{x}_-]\\ &= r^m[c_+e^{im\vartheta}\mathbf{x}_+ + \overline{c_+e^{im\vartheta}\mathbf{x}_+}]\ .\end{aligned}$$

provided (as we'll see below) that m is such that $cos(m\vartheta + \varphi)$ is not near 0; now we denote the q-th coordinate of $c_+\mathbf{x}_+$ by $Re^{i\varphi}$ and recognize from the properties of the imaginary exponential (Definition 6.9 on page 200) that under these circumstances

$$\begin{aligned}\mathbf{A}^m\mathbf{v} &\cong r^m R[e^{i(m\vartheta+\varphi)} + e^{-i(m\vartheta+\varphi)}]\\ &= 2r^m R\,cos(m\vartheta + \varphi).\end{aligned}$$

So whenever $cos(m\vartheta + \varphi)$ isn't near 0, for large enough m, the approximation

$$\mathbf{A}^m \mathbf{v} \cong 2r^m R\,cos(m\vartheta + \varphi)$$

is a good one. Now let

$$c_j = 2r^{m+j} R\,cos([m+j]\vartheta + \varphi)$$

for $j = 0,1,2,...$.
Using the equality

$$cos(x)cos(y) = \frac{cos(x+y) + cos(x-y)}{2}$$

(easily established using Definition 6.8 on page 199 and Theorem 6.10 on page 200) we find that

$$q_j = c_{j+1}^2 - c_j c_{j+2} = (2r^{m+j+1}R)^2\,\frac{1 - cos(2\vartheta)}{2}.$$

Since $0 < \vartheta < \pi$ the right-hand side is never 0. Hence it is easy to verify that

$$r^2 = \frac{q_1}{q_0} \qquad 2r\,cos(\vartheta) = \frac{c_1 c_2 - c_0 c_3}{q_0}.$$

Thus, so long as

$$\mathbf{A}^{m+j}\mathbf{v} = C_j \cong 2r^{m+j} R\,cos([m+j]\vartheta + \varphi)$$

is a good approximation for $j = 0,1,2,3$, so, too, is the approximation

$$Q_j = C_{j+1}^2 - C_j C_{j+2} \cong (2r^{m+j+1}R)^2\,\frac{1 - cos(2\vartheta)}{2}.$$

Assuming this to be the case, we find that the approximations

$$r^2 \cong \frac{Q_1}{Q_0} \qquad 2r\,cos(\vartheta) \cong \frac{C_1 C_2 - C_0 C_3}{Q_0}$$

are good ones. From these we can easily find good approximations to r and ϑ, because $0 < \vartheta < \pi$. This yields the conplex conjugate pair of eigenvalues we were seeking. There is the matter of whether the approximation

$$\mathbf{A}^{m+j}\mathbf{v} = C_j \cong 2r^{m+j} R\,cos([m+j]\vartheta + \varphi)$$

is a good one for $j = 0,1,2,3$. The details here get messy, but probably the simplest approach is to see whether the C_j are behaving as expected with the given estimates of r and ϑ. For this you need estimates of R and φ. These are not hard to find from the approximation

$$\mathbf{A}^{m+j}\mathbf{v} = C_j \cong 2r^{m+j} R\,cos([m+j]\vartheta + \varphi)$$

once r and ϑ have been estimated. If results turn out badly, experiment with a larger value of m, and other Q_j.

3. Handle the case where the largest magnitude eigenvalues are $\lambda, -\lambda$.

4. For those with access to a digital computer, write programs to implement the methods presented here on matrices of your own choice. (This can be done with *Mathematica* or *Splus*, and checked with their eigenvalue programs.)

Matrix deflation

The last technique we'll look at to help in the numerical determination of eigenvalues is called matrix deflation. ***Deflation replaces the matrix* A *with a matrix* B *where* precisely one of the known nonzero eigenvalues of A is changed to the value** 0, ***and all other eigenvalues of* A *are the same as the eigenvalues of* B**. Properly applied, this method frequently permits repeated application of the methods already discussed, to find all eigenvalues of a matrix; in particular when the magnitudes of all real and all complex conjugate pairs of eigenvalues are distinct.

Here's the basis of matrix deflation. Suppose $\mathbf{x}$ is an eigenvector of the $n \times n$ matrix $\mathbf{A}$ ($\mathbf{x}$ represented as a column matrix) associated with the known eigenvalue $\lambda \neq 0$. Let $\mathbf{v}$ be a (column) vector such that the dot product

$$\mathbf{v}^{\mathrm{T}}\mathbf{x} = 1 \qquad \textit{13.54}$$

(T is the transpose operator, see Definition 10.13 on page 342), so $\mathbf{v}$ has a nonzero component in the direction of $\mathbf{x}$ (see Definition 8.12 on page 276). Note that once the eigenvalue λ is known, it is not difficult to compute the corresponding eigenvectors (see the discussion between equation on page 537 and Theorem 13.3). Now let

$$\mathbf{B} = \mathbf{A} - \lambda\mathbf{x}\mathbf{v}^{\mathrm{T}}. \qquad \textit{13.55}$$

Note that $\mathbf{x}\mathbf{v}^{\mathrm{T}}$ is an $n \times n$ matrix. Then the eigenvector $\mathbf{x}$, of $\mathbf{A}$, associated with the eigenvalue λ of $\mathbf{A}$, is an eigenvector of $\mathbf{B}$ associated with the eigenvalue 0 of $\mathbf{B}$, since

$$\mathbf{Bx} = \mathbf{Ax} - \lambda\mathbf{xv}^T\mathbf{x} = \lambda\mathbf{x} - \lambda\mathbf{x} = 0\mathbf{x}. \qquad 13.56$$

Before proceeding on to show that we can choose $\mathbf{v}$ so that this method really works, let's see what's going on a little better. The matrix $\mathbf{xv}^T$ is an $n \times n$ matrix, each of whose columns is a multiple of (the column matrix) $\mathbf{x}$. Thus when $\mathbf{xv}^T$ acts on (multiplies) some vector, the result is always a multiple of $\mathbf{x}$ (see Definition 10.1 on page 328 and equation 10.4 following). Scaling $\mathbf{v}^T$ so that $\mathbf{xv}^T = 1$ just ensures that $\lambda\mathbf{xv}^T$ acting on $\mathbf{x}$ yields precisely $\lambda\mathbf{x}$ so that $\mathbf{Bx} = 0\mathbf{x}$. There is a great deal of flexibility in choosing $\mathbf{v}$ since we can satisfy equation 13.54 by adding any element of the orthogonal complement of $\mathbf{x}$ to the appropriate multiple of $\mathbf{x}$. This flexibility will be what allows us to choose $\mathbf{v}$ in such a way that no other eigenvalues except the specified one are affected.

Next we'll show that any *nonzero eigenvalues* associated originally with eigenvectors of $\mathbf{A}$ which are linearly independent of $\mathbf{x}$ are still eigenvalues of $\mathbf{B}$, associated with the same number of linearly independent eigenvectors possessed by $\mathbf{A}$, so long as equation 13.54 holds.

So let $\mathbf{x}'$ be any eigenvector of $\mathbf{A}$, which

- is linearly independent of the eigenvector $\mathbf{x}$
- has eigenvalue $\lambda' \neq 0$ (possibly $\lambda' = \lambda$).

Then for any number c, from equation 13.56 and equation 13.55 (the definition of $\mathbf{B}$)

$$\begin{aligned} \mathbf{B}(\mathbf{x}' - c\mathbf{x}) &= \mathbf{Bx}' - \mathbf{B}c\mathbf{x} = \mathbf{Bx}' \\ &= (\mathbf{A} - \lambda\mathbf{xv}^T)\mathbf{x}' = \mathbf{Ax}' - (\lambda\mathbf{xv}^T)\mathbf{x}', \end{aligned}$$

i.e.,

$$\mathbf{B}(\mathbf{x}' - c\mathbf{x}) = \lambda'\mathbf{x}' - d\mathbf{x}, \qquad 13.57$$

(this latter because $\mathbf{xv}^T$ takes every vector into a multiple of $\mathbf{x}$ [or alternatively, using the associativity $(\mathbf{AB})\mathbf{C} = \mathbf{A}(\mathbf{BC})$ of matrix multiplication, and the fact that $\mathbf{v}^T\mathbf{x}'$ is a scalar]). We note that d depends only on $\mathbf{v}, \lambda$ and $\mathbf{x}'$, but not on c. Since we are treating the case $\lambda' \neq 0$, we may rewrite equation 13.57 as

$$\mathbf{B}(\mathbf{x}' - c\mathbf{x}) = \lambda'\left(\mathbf{x}' - \frac{d}{\lambda'}\mathbf{x}\right)$$

for all choices of d. Choosing $c = d/\lambda'$ yields

$$\mathbf{B}(\mathbf{x}' - c\mathbf{x}) = \lambda'(\mathbf{x}' - c\mathbf{x}) \ ,$$

which shows (because the linear independence of $\mathbf{x}$ and $\mathbf{x}'$ implies $\mathbf{x}' - c\mathbf{x} \neq \mathbf{0}$) that λ' is an eigenvalue of $\mathbf{B}$.

We still have to worry about eigenvectors of $\mathbf{A}$ which had eigenvalue 0, being eigenvectors of $\mathbf{B}$ with eigenvalue 0. To ensure this, recall the flexibility we still have in choosing $\mathbf{v}$. The situation we are examining is that in which

$$\begin{array}{ll} \mathbf{A} \text{ has linearly independent eigenvectors} & \\ & \mathbf{x}_1,...,\mathbf{x}_k, \mathbf{x}_{k+1},...,\mathbf{x}_F \\ \text{with eigenvalues} & 0, ... ,0, \lambda_{k+1},...,\lambda_F \end{array} \qquad \textbf{\textit{13.58}}$$

with $\lambda_{k+j} \neq 0$ for $j \geq 1$, and we are *deflating* $\mathbf{x}_{k+1}, \lambda_{k+1}$.

Note that it's easy enough for us to determine a basis for the entire sequence of eigenvectors having eigenvalue 0, by obtaining the general solution of $\mathbf{Ax} = \mathbf{0}$ (see Theorem 11.4 on page 415); i.e., we can determine an acceptable sequence, $\mathbf{x}_1,...,\mathbf{x}_k$, which for all practical purposes is equivalent to the original sequence of linearly independent eigenvectors with eigenvalue 0. Let's assume that we have determined $\mathbf{x}_{k+1}, \lambda_{k+1}$, by, say, the power method, or one of its variations. Choose $\mathbf{v}$ so that

$$\mathbf{v}^T\mathbf{x}_{k+1} = 1 \qquad \textbf{\textit{13.59}}$$

(so that the eigenvalue λ_{k+1} has been converted to the eigenvalue 0 of $\mathbf{B}$, associated with the eigenvector $\mathbf{x}_{k+1}$ of both $\mathbf{A}$ and $\mathbf{B}$). Now for $j \leq k$, $\mathbf{x}_j$ is an eigenvector of $\mathbf{A}$ with eigenvalue 0, and hence

$$\begin{aligned} \mathbf{Bx}_j &= (\mathbf{A} - \lambda_{k+1}\mathbf{x}_{k+1}\mathbf{v}^T)\mathbf{x}_j \\ &= -\lambda_{k+1}\mathbf{x}_{k+1}\mathbf{v}^T\mathbf{x}_j. \end{aligned}$$

We see from this, that if $\mathbf{v}$ can also be chosen so that $\mathbf{v}^T\mathbf{x}_j = 0$ for all $j \leq k$, then $\mathbf{x}_j$ will still be an eigenvector of $\mathbf{B}$, with eigenvalue 0. So, to accomplish the deflation of $\mathbf{x}_{k+1}, \lambda_{k+1}$ and still leave the subspace of vectors with eigenvalue 0 unchanged, we want to choose $\mathbf{v}$ to satisfy

$$\mathbf{v} \text{ must belong to } [[\mathbf{x}_1,...,\mathbf{x}_k]]^{\perp} \qquad \textbf{\textit{13.60}}$$

(see Definition 11.2 on page 412), and

$$\mathbf{v}^T\mathbf{x}_{k+1} = 1. \qquad \textbf{\textit{13.61}}$$

Now ***if we could not satisfy condition*** 13.60 and equation 13.61, it would be the case that whenever $\mathbf{v}$ is a member of $[[\mathbf{x}_1,...,\mathbf{x}_k]]^{\perp}$ then $\mathbf{v}^T\mathbf{x}_{k+1} = 0$.

But this means that $[\![\mathbf{x}_1,...,\mathbf{x}_k]\!]^{\perp}$ is a subset of $[\![\mathbf{x}_{k+1}]\!]^{\perp}$, which is equivalent to $[\![\mathbf{x}_{k+1}]\!]$ being a subset of $[\![\mathbf{x}_1,...,\mathbf{x}_k]\!]$. But then $\mathbf{x}_{k+1}$ would be a linear combination of the vectors $\mathbf{x}_1,...,\mathbf{x}_k$, which is impossible under the assumed linear independence of all of the eigenvectors $[\![\mathbf{x}_1,...,\mathbf{x}_k, \mathbf{x}_{k+1}]\!]$. Since the impossibility of satisfying condition 13.60 and equation 13.61 leads to an impossible situation, it must be possible to satisfy these two conditions.

In fact, because $\mathbf{A}\mathbf{x}_j = \mathbf{0}$ for $j \le k$, choosing $\mathbf{v}$ to satisfy condition 13.60 and equation 13.61 is equivalent to choosing $\mathbf{v}$ so that

$$\mathbf{v} \text{ belongs to } \mathbf{A}_{row} \text{ and } \mathbf{v}^{\mathrm{T}}\mathbf{x}_{k+1} = 1, \qquad \textbf{\textit{13.62}}$$

where $\mathbf{A}_{row}$ is the subspace generated by the rows of $\mathbf{A}$ (see Definition 9.4 on page 296). Determining $\mathbf{v}$ can be accomplished by solving the system of linear equations associated with condition 13.62.

We have shown that if $\mathbf{A}$ satisfies condition 13.58 on page 575 and $\mathbf{v}$ satisfies conditions 13.62 above, then $c_{k+2},...,c_F$ can be chosen so that $\mathbf{B}$ has eigenvectors

$$\mathbf{x}_1,...,\mathbf{x}_k, \mathbf{x}_{k+1}, x_{k+2} - c_{k+2}\mathbf{x}_{k+1},..., \mathbf{x}_F - c_F x_{k+1}$$

and corresponding eigenvalues

$$0,......,0, \quad 0, \qquad \lambda_{k+2}, \; . \; . \; . \; . \; . \; . \; , \lambda_F.$$

Note that given linearly independent eigenvectors $\mathbf{x}_1,...,\mathbf{x}_k,\mathbf{x}_{k+1},...,\mathbf{x}_F$ of $\mathbf{A}$, then so is the sequence of eigenvectors of $\mathbf{B}$.

We summarize these results in the following theorem.

Theorem 13.17: Matrix deflation for finding eigenvalues

Suppose that the $n \times n$ matrix $\mathbf{A}$ has linearly independent eigenvectors

$$\mathbf{x}_1,...,\mathbf{x}_k, \mathbf{x}_{k+1},...,\mathbf{x}_F$$

$$\text{with eigenvalues} \quad 0, \;....\; ,0, \lambda_{k+1},...,\lambda_F,$$

with λ_{k+1} known and $\lambda_{k+j} \ne 0$ for $j \ge 1$. Let $\mathbf{A}_{row}$ be the subspace generated by the rows of $\mathbf{A}$ (see Definition 9.4 on page 296). Choose $\mathbf{v}$ so that

$$\mathbf{v} \text{ belongs to } \mathbf{A}_{row} \text{ and } \mathbf{v}^{\mathrm{T}}\mathbf{x}_{k+1} = 1,$$

(which can be accomplished by finding any nontrivial solution of the system of of linear equations just specified), and let

$$\mathbf{B} = \mathbf{A} - \lambda_{k+1}\mathbf{x}_{k+1}\mathbf{v}^{\mathrm{T}}.$$

Then for

$$c_{k+j} = \lambda_{k+j}\mathbf{v}^{\mathrm{T}}\mathbf{x}_{k+j} \qquad j = 2,...,F$$

(where we need not determine the c_{k+j}, but we do need $\mathbf{v}$ and λ_{k+1}) $\mathbf{B}$ has linearly independent eigenvectors

$$\mathbf{x}_1,...,\mathbf{x}_k, \mathbf{x}_{k+1}, x_{k+2} - c_{k+2}\mathbf{x}_{k+1},...,\mathbf{x}_F - c_F x_{k+1}$$

with corresponding eigenvalues

$$0,......,0, \quad 0, \qquad \lambda_{k+2}, \; . \; . \; . \; . \; . \; . \; ,\lambda_F.$$

❑

The practical use of this theorem is that we can now repeat our use of such techniques as the power method to determine the remaining λ_{k+j} just as we determined λ_{k+1}.

Computational observation

If we choose a value j, such that the j-th row of $\mathbf{A}$ doesn't consists solely of zeros, and choose $\mathbf{v}^{\mathrm{T}}$ to be the multiple of the j-th row of $\mathbf{A}$, satisfying the condition $\mathbf{v}^{\mathrm{T}}\mathbf{x}_{k+1} = 1$, then the j-th row of the deflated matrix $\mathbf{B}$ will be $\mathbf{0}$. We leave the details of this to the next problem.

Problem 13.18

Suppose that $\mathbf{A}_j = (A_{j,1}, A_{j,2},...,A_{j,n})$, the j-th row of $\mathbf{A}$ (treated as a $1 \times n$ matrix) is not the $\mathbf{0}$ vector, and the hypotheses of Theorem 13.17 are satisfied. Also choose $\mathbf{v}^{\mathrm{T}}$ to be proportional to $\mathbf{A}_j$ (i.e., equal to $c\mathbf{A}_j$) with c chosen so that $\mathbf{v}^{\mathrm{T}}\mathbf{x}_{k+1} = 1$. Under these circumstances, show that the j-th row of the deflated matrix $\mathbf{B}$ consists of all zeros. (This choice is called *Weilandt's deflation.* We'll see its significance shortly.)
Hints: From its definition the j-th row of $\mathbf{B}$ is $\mathbf{A}_j - (\lambda_{k+1}\mathbf{x}_{k+1}\mathbf{v}^{\mathrm{T}})_j$, where $(\lambda_{k+1}\mathbf{x}_{k+1}\mathbf{v}^{\mathrm{T}})_j$ is the j-th row of the $n \times n$ matrix $\lambda_{k+1}\mathbf{x}_{k+1}\mathbf{v}^{\mathrm{T}}$, i.e., the j-th row of the matrix $\lambda_{k+1}\mathbf{x}_{k+1}c\mathbf{A}_j$. From Definition 10.3 on page 331 of matrix multiplication, this is simply $\lambda_{k+1}\mathbf{x}_{k+1,j}c\mathbf{A}_j$, where $\mathbf{x}_{k+1,j}$ is the j-th coordinate of $\mathbf{x}_{k+1}$. So all we need show is that $\mathbf{A}_j - \lambda_{k+1}\mathbf{x}_{k+1,j}c\mathbf{A}_j$ is the $\mathbf{0}$ vector, which will surely follow if we can prove that

$$\lambda_{k+1}\mathbf{x}_{k+1,j}c = 1. \qquad \textit{13.63}$$

To do this recall that c was defined so that $\mathbf{v}^{\mathrm{T}}\mathbf{x}_{k+1} = 1$, i.e., so that

$$c\mathbf{A}_j\mathbf{x}_{k+1} = 1. \qquad \textit{13.64}$$

But because $\mathbf{A}\mathbf{x}_{k+1} = \lambda_{k+1}\mathbf{x}_{k+1}$ it follows from the rules of matrix multiplication that $\mathbf{A}_j\mathbf{x}_{k+1} = \lambda_{k+1}x_{k+1,\,j}$. Substituting this into equation 13.64 yields

$$c\lambda_{k+1}x_{k+1,\,j} = 1$$

establishing equation 13.63, and proving the asserted result.

We shall now see that the result of Problem 13.18 implies that choosing $\mathbf{v}^{\mathrm{T}}$ to be proportional to $\mathbf{A}_j$ (i.e., equal to $c\mathbf{A}_j$) with c chosen so that $\mathbf{v}^{\mathrm{T}}\mathbf{x}_{k+1} = 1$ effectively leaves $\mathbf{B}$ with one less row and one fewer column than $\mathbf{A}$, substantially reducing the labor. Because, although $\mathbf{B}$ is an $n \times n$ matrix, in the equation

$$\mathbf{B}\mathbf{x} = \lambda\mathbf{x} \qquad \textit{13.65}$$

the j-th row of $\mathbf{B}$ is $\mathbf{0}$. Thus the j-th coordinate of $\mathbf{B}\mathbf{x}$ is 0. Therefore, since deflation is only concerned with solutions of this equation for nonzero λ (since we already know the eigenvectors of $\mathbf{B}$ associated with $\lambda = 0$) the j-th coordinate of any such eigenvector, $\mathbf{x}$, must be 0. This means that in equation 13.65 we can set the j-th coordinate of $\mathbf{x}$ equal to 0 on both sides.

The j-th of the scalar equations in this system $(\mathbf{0}^{\mathrm{T}}\mathbf{x} = \lambda x_j = \lambda 0)$ is useless, as far as finding λ, and so can be eliminated leaving us with the equation

$$\mathbf{B}^{(j)}\mathbf{x} = \lambda\mathbf{x}^{(j)}, \qquad \textit{13.66}$$

where $\mathbf{B}^{(j)}$ is the matrix gotten from $\mathbf{B}$ by deleting its j-th row, and $\mathbf{x}^{(j)}$ the vector gotten from $\mathbf{x}$ by eliminating its j-th coordinate (row). But now the j-th coordinate of $\mathbf{x}$ on the left side makes no contribution whatsoever, so it can be eliminated as long as the associated column of $\mathbf{B}^{(j)}$ is deleted as well. So, as far as determining the remaining eigenvalues of $\mathbf{A}$, we can now replace equation 13.66 with

$$\mathbf{B}^{(j,\,j)}\mathbf{x}^{(j)} = \lambda\mathbf{x}^{(j)} \qquad \textit{13.67}$$

where $\mathbf{B}^{(j,\,j)}$ is the matrix obtained from $\mathbf{B}$ by deleting its j-th row and column and $\mathbf{x}^{(j)}$ is a vector with $j - 1$ coordinates. Thus we reduce the dimension of our problem by 1 at each step. Of course, if we want to determine the eigenvectors, all we need do is solve the systems $\mathbf{A}\mathbf{x}_i = \lambda_i\mathbf{x}_i$ for each of the eigenvalues that have been determined.

We provide a simple example to illustrate the above technique.

Example 13.19: Deflation of a 3×3 matrix

Recall the matrix

$$A = \begin{bmatrix} 1 & 1 & 2 \\ 1 & 2 & 3 \\ 0 & 3 & 4 \end{bmatrix}$$

of Example 13.16 on page 569. We found its largest magnitude eigenvalue, $\lambda_1 = 6.39336$ by the power method. In determining the vector $\mathbf{v}$ to be used in the deflation, we have to scale $\mathbf{v}$ so that $\mathbf{v}^T\mathbf{x}_1 = 1$, where $\mathbf{x}_1$ is the eigenvector we associated with λ_1, which means that we have to determine $\mathbf{x}_1$. We thus need a nontrivial solution $\mathbf{x} = \mathbf{x}_1$ of

$$(\mathbf{A} - \lambda_1\mathbf{I})\mathbf{x} = \mathbf{0}.$$

The simplest approach to solution is to arbitrarily choose the third coordinate value, $\mathbf{x}_{1,3}$, of $\mathbf{x}$ to be 1. Then use the third equation of this system to determine $\mathbf{x}_{1,2}$ and the second equation of the system to then determine $\mathbf{x}_{1,1}$. Doing this we find

$$\mathbf{x}_1 = \begin{bmatrix} 0.51854 \\ 0.799787 \\ 1 \end{bmatrix}.$$

Now setting

$$1 = \mathbf{v}^T\mathbf{x}_1 = c\mathbf{A}_3\mathbf{x}_1 = c\begin{bmatrix} 0 & 3 & 4 \end{bmatrix}\begin{bmatrix} 0.51854 \\ 0.799787 \\ 1 \end{bmatrix} = c\,6.39937$$

yields $c = .156265$. So

$$\mathbf{v}^T = c\mathbf{A}_3 = 0.156265[0, 3, 4] = [0, 0.468796, 0.625063]$$

and thus

$$\mathbf{B} = \mathbf{A} - \lambda_1\mathbf{x}_1\mathbf{v}^T = \begin{bmatrix} 1 & -0.55426 & -0.07222 \\ 1 & -0.39711 & -0.19615 \\ 0 & 0.00282 & 0.00375 \end{bmatrix}.$$

The last row differs from 0 due to the limited number of places we were carrying. Had we computed λ_1 to greater accuracy, $\lambda_1 = 6.3399356$, we would have obtained

$$\mathbf{B} = \begin{bmatrix} 1 & -0.55562 & -0.77416 \\ 1 & -0.39936 & -0.19914 \\ 0 & 0 & 0 \end{bmatrix}.$$

The eigenvalues we are looking for are those of the matrix

$$\mathbf{B}^{(3,\,3)} = \begin{bmatrix} 1 & -0.55562 \\ 1 & -0.39936 \end{bmatrix}$$

obtained from the matrix $\mathbf{B}$ by deleting its third row and column. From its characteristic equation (Theorem 13.3 on page 538) we know that the eigenvalues of $\mathbf{B}^{(3,\,3)}$ must satisfy the condition

$$(1 - \lambda)(-0.39936 - \lambda) + 0.55562 = 0,$$

i.e.,

$$\lambda^2 - 0.60064\lambda + 0.15626 = 0$$

(which can be found by completion of the square, see page 276). We can check the validity of this answer by noting that this equation is satisfied by the same λ values as satisfy

$$\frac{det[\mathbf{A} - \lambda\mathbf{I}]}{\lambda - \lambda_1} = 0.$$

■

Before moving on to our final topics in this chapter, we have a few concluding remarks about the numerical determination of eigenvalues.

One set of methods that we have not investigated is where the matrix is transformed to another one with the same set of eigenvalues, but in which the characteristic equation is relatively simple to solve. When this technique is combined with rather reliable methods for finding all of the roots of a polynomial, it can be very useful. Sometimes the *power method* for finding the unique largest magnitude eigenvalue will behave poorly, because of another eigenvalue with close to the same magnitude. There are a variety of methods which can improve conditions, in particular, the very simple δ^2 method.

The subject of calculation of eigenvalues is a large one, and a variety of methods, some good, some poor have been developed. The books of Ralston [14], Wilkinson [17] and Golub and Van Loan [9], contain large bibliographies. Journals of applied math also have many articles in this area. It is unlikely that any one book will be thoroughly up-to-date if your particular application is a tough one. For the standard applications the above-mentioned books should provide just about anything regarding the theory needed in this area at present. Computer packages such as *Mlab* and *Mathematica* provide excellent numerical determination of eigenvalues for most matrices you're likely to meet.

5 Eigenvalues of Symmetric Matrices

Now we turn to the study of eigenvalues of symmetric matrices. One very important application of the study of these eigenvalues is the theory of sufficient conditions for a relative maximum or minimum of a function of several variables.[1] To investigate how the eigenvector coordinate system can help in this area, note that the second-order terms of the multidimensional Taylor series (Theorem 10.33 on page 371 ff.) are of the form

$$\left(\frac{1}{2}(\boldsymbol{h} \cdot \boldsymbol{D})^2 f\right)(\boldsymbol{x}).$$

If we write out this expression in expanded form, we find

$$\left(\frac{1}{2}(\boldsymbol{h} \cdot \boldsymbol{D})^2 f\right)(\boldsymbol{x}) = \frac{1}{2} \sum_{i,\, j = 1}^{n} a_{ij} h_i h_j \qquad \textbf{\textit{13.68}}$$

where $a_{ij} = (D_i D_j f)(\boldsymbol{x})$.

Recall (Definition 10.36 on page 377) that the matrix $\mathbf{A}$ whose i-j element is $a_{ij} = (D_i D_j f)(\boldsymbol{x})$ is called the *Hessian* matrix (associated with f at $\boldsymbol{x}$). In Section 8 of Chapter 10, starting page 372, we saw the role of the Hessian in developing conditions to help decide whether a maximum or a minimum is achieved if $\boldsymbol{grad} f(\boldsymbol{x}) = 0$. But the general theory of the local behavior of f at $\boldsymbol{x}$ when $\boldsymbol{grad} f(\boldsymbol{x}) = 0$ is probably better understood by changing to the natural coordinate system of the Hessian matrix. In fact, when we search for maxima and minima it may often pay to go ***directly*** to this coordinate system. To start our investigation, we first show that under usually satisfied conditions, the eigenvectors of a Hessian matrix are orthogonal. This arises because under the conditions of Theorem 10.32 on page 369

$$D_i D_j f = D_j D_i f$$

so that

$$a_{ij} = a_{ji}.$$

That is,

$$\mathbf{A}^{\mathrm{T}} = \mathbf{A}, \qquad \textbf{\textit{13.69}}$$

where T is the transpose operator (Definition 10.13 on page 342).

Since satisfaction of equation 13.69 occurs so often, we formalize it in the following definition.

---

1. Another important application arises in statistical principal component analysis, which is concerned with *data reduction* with a minimal loss of information.

Definition 13.18: Symmetric matrix

If $\mathbf{A}$ is a real $n \times n$ matrix whose transpose (Definition 10.13 on page 342), $\mathbf{A}^T$, satisfies the condition $\mathbf{A}^T = \mathbf{A}$, then $\mathbf{A}$ is called **symmetric**.

❑

We can now establish the following theorem, whose content was probably discovered by everyone computing eigenvalues and eigenvectors of symmetric matrices.

Theorem 13.19: Eigenvalues and eigenvectors of symmetric matrices

If $\mathbf{A}$ is a real, symmetric matrix (Definition 13.18 above) then all of its eigenvalues (Definition 13.2 on page 536) are real, and eigenvectors corresponding to distinct eigenvalues are orthogonal (Definition 8.14 on page 278).

❑

For every $n \times n$ (possibly complex) matrix $\mathbf{A}$, using the complex definition of dot product (Definition 13.1 on page 535) for all complex n-dimensional vectors $\boldsymbol{x}, \boldsymbol{y}$

$$(\mathbf{A}\boldsymbol{x}) \cdot \boldsymbol{y} = (\mathbf{A}\mathbf{x})^T \bar{\mathbf{y}} = \mathbf{x}^T \mathbf{A}^T \bar{\mathbf{y}} = \mathbf{x}^T \overline{\bar{\mathbf{A}}^T \mathbf{y}} = \boldsymbol{x} \cdot (\bar{\mathbf{A}}\boldsymbol{y}). \qquad \textit{13.70}$$

(Note that in the above equation we write $\boldsymbol{x},\boldsymbol{y}$ when $\boldsymbol{x}$ and $\boldsymbol{y}$ are thought of as vectors, and $\mathbf{x},\mathbf{y}$ when they are treated as $n \times 1$ matrices.)

For a real symmetric matrix, $\mathbf{A}$, the result of equation 13.70 becomes

$$(\mathbf{A}\boldsymbol{x}) \cdot \boldsymbol{y} = \boldsymbol{x} \cdot (\mathbf{A}\boldsymbol{y}). \qquad \textit{13.71}$$

It's convenient to formalize the property specified by equation 13.71 in the following definition.

Definition 13.20: Self-adjointness of linear operator, matrix

An arbitrary $n \times n$ complex matrix satisfying equation 13.71 is called **self-adjoint**, and the function $\mathcal{A}$ associated with the matrix $\mathbf{A}$ (see Section 2 of Chapter 10, starting page 329) is called a **self-adjoint operator**.

❑

It is this self-adjoint property of the function $\mathcal{A}$ that makes many of our results go through.

To establish Theorem 13.19 suppose λ is a (possibly complex) eigenvalue of $\mathbf{A}$ with eigenvector $\mathbf{x}$. From equation 13.71 we find, choosing $\boldsymbol{y} = \boldsymbol{x}$, that

$$\begin{aligned}\lambda(\boldsymbol{x} \cdot \boldsymbol{x}) &= (\lambda \boldsymbol{x}) \cdot \boldsymbol{x} = (\lambda \boldsymbol{x}) \cdot \boldsymbol{y} \\ &= (\mathbf{A}\boldsymbol{x}) \cdot \boldsymbol{y} = \boldsymbol{x} \cdot (\mathbf{A}\boldsymbol{y}) \\ &= \boldsymbol{x} \cdot (\mathbf{A}\boldsymbol{x}) = \boldsymbol{x} \cdot (\lambda x) \\ &= \bar{\lambda}(\boldsymbol{x} \cdot \boldsymbol{x}).\end{aligned}$$

But since $\boldsymbol{x}$ is an eigenvector, $\boldsymbol{x} \cdot \boldsymbol{x} = \sum_{j=1}^{n} x_j^2 \neq 0$, and thus $\lambda = \bar{\lambda}$, proving that λ is real.

Now if λ and μ are distinct eigenvalues of $\mathbf{A}$ with eigenvectors $\boldsymbol{x}_\lambda$ and $\boldsymbol{x}_\mu$, then from equation 13.71 with $\boldsymbol{x} = \boldsymbol{x}_\mu$ and $\boldsymbol{y} = \boldsymbol{x}_\lambda$, we find

$$\mu \boldsymbol{x}_\mu \cdot \boldsymbol{x}_\lambda = \lambda \boldsymbol{x}_\mu \cdot \boldsymbol{x}_\lambda$$

and since $\mu \neq \lambda$ we must have $\boldsymbol{x}_\mu \cdot \boldsymbol{x}_\lambda = 0$, proving that $\boldsymbol{x}_\mu \perp \boldsymbol{x}_\lambda$ (Definition 8.14 on page 278) as claimed, establishing Theorem 13.19.

At this stage we present one of the main results about symmetric matrices.

Theorem 13.21: Eigenvectors of symmetric matrix as a basis

Every symmetric $n \times n$ matrix (Definition 13.18) possesses n mutually orthogonal (Definition 8.14 on page 278) eigenvectors, which are suitable as a basis for n-dimensional space (see Theorem 9.13 on page 302 and Theorem 9.16 on page 303).

❑

For the purposes of this proof it's convenient to recall that in Chapter 10, Section 2 on page 329 we identified any $k \times k$ matrix as a function (linear operator, Definition 13.9 on page 547) whose domain is a k-dimensional vector space (Definition 13.5 on page 543); and to realize that a linear operator whose domain is a k-dimensional vector space, $\mathcal{V}_k$, can be represented as a $k \times k$ matrix tied to a particular basis for $\mathcal{V}_k$ (Theorem 13.12 on page 552).

We need the following definition for our investigation.

Definition 13.22: Restriction of a function

Let f be a function (Definition 1.1 on page 2) and let S be a subset of *dmn f*. The function f_S with domain S is called the **restriction of *f* to *S***, if

$$f_S(x) = f(x) \quad \text{for all } x \text{ in } S.$$

❑

That is, f_S may have the same formula as f, but its domain is usually smaller. To see that we are not just nitpicking, and that f_S and f can be quite different, just let

$$f(x) = x \quad \text{for all real } x$$

and let S be the set of positive reals. The graphs of f and f_S are nowhere near the same, as can be see in Figure 13-1.

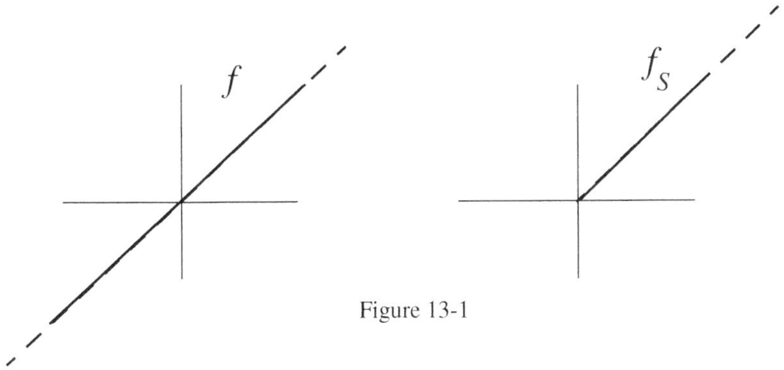

Figure 13-1

We also need the following.

Definition 13.23: Invariant subspace

Let $\mathcal{L}$ be a linear operator (Definition 13.9 on page 547) and S a subspace of the domain of $\mathcal{L}$ (Definition 9.4 on page 296). S is called an **invariant subspace of** $\mathcal{L}$ if

$$\text{whenever } \boldsymbol{x} \text{ is in } S, \ \mathcal{L}(\boldsymbol{x}) \text{ is in } S.$$

❑

Theorem 13.24: Invariance of $S^{\perp}$ for self-adjoint $\mathcal{L}$

Let S be an invariant subspace (Definition 13.23 above) of the self-adjoint linear operator $\mathcal{L}$ (Definition 13.20 on page 582) associated with the symmetric matrix L (see Chapter 10, Section 2 on page 329 and Definition 13.18 on page 582). Then the orthogonal complement $S^{\perp}$ (Definition 11.2 on page 412) is also an invariant subspace of $\mathcal{L}$.

❑

Proof: Suppose $\boldsymbol{y}$ is in $S^{\perp}$. Then from Theorem 8.15 on page 278 it follows that for all $\boldsymbol{x}$ in S, $\boldsymbol{x} \cdot \boldsymbol{y} = 0$. By the self-adjointness of $\mathcal{L}$ it is then seen that

$$x \cdot \mathcal{L}(y) = \mathcal{L}(x) \cdot y.$$

But

$$\mathcal{L}(x) \cdot y = 0 \quad \text{because} \quad \mathcal{L}(x) \text{ is in } S$$

(this latter from the invariance of S.) Hence

$$x \cdot \mathcal{L}(y) = 0 \text{ for all } x \text{ in } S.$$

Thus $\mathcal{L}(y)$ is in $S^{\perp}$. Since this holds whenever y is in $S^{\perp}$, we see that $S^{\perp}$ is also an invariant subspace of $\mathcal{L}$, as claimed.

Now to finish up the proof of Theorem 13.21.

Given the symmetric $n \times n$ matrix $\mathbf{A}$ we know from the fundamental theorem of algebra (Theorem 6.4 on page 197) that equation 13.11 on page 538 has at least one solution — eigenvalue λ_0, and hence from Theorem 13.3 on page 538, that $\mathbf{A}$ has eigenvalue λ_0 and associated nonzero *eigen-n-tuple* which we will call $\mathbf{x}_0$. The function $\mathcal{A}$, associated with $\mathbf{A}$ also has eigenvalue λ_0, and eigenvector x_0 (which is represented by the n-tuple $\mathbf{x}_0$ relative to some basis in *dmn* $\mathcal{A}$).

It is evident that the subspace $S_0 = [[x_0]]$ generated by x_0 is an invariant subspace of $\mathcal{A}$. By Theorem 13.24 on page 584 so is $C_1 = S_0^{\perp}$. Now the restriction (Definition 13.22 on page 583) $\mathcal{A}_{C_1}$ of the linear operator $\mathcal{A}$ to C_1 is evidently self-adjoint, and is thus associated with an $(n-1) \times (n-1)$ symmetric matrix, we'll call $\mathbf{A}_1$. The basis with which this matrix is associated can be the vectors associated with a set of orthonormal n-tuples coming from the columns of the $n \times n$ identity matrix, $\mathbf{I}^{n \times n}$, which are generated by the Gram-Schmidt procedure (see Theorem 10.31 on page 366) on the columns of $\mathbf{I}^{n \times n}$ preceded by the n-tuple $\mathbf{x}_1$. Note that it doesn't really matter whether we're working with vectors or with the n-tuples representing them. Most importantly, although $\mathbf{A}_1$ is an $(n-1) \times (n-1)$ matrix, the $n-1$-tuples it operates on represent vectors in C_1. In fact, if $\mathbf{x}^{n \times 1}$ represents a given vector from C_1 in the original coordinate system and $\mathbf{w}^{(n-1) \times 1}$ represents this same vector in the coordinate system associated with $\mathbf{A}_1$, then $\mathbf{Ax}$ and $\mathbf{A}_1\mathbf{w}$ represent the same vector, albeit in two different coordinate systems.

But it's clear from our earlier reasoning that $\mathbf{A}_1$ has an eigenvalue, λ_1 (possibly equal to λ_0, a fact of no significance) and an *eigen-n-tuple* representing an eigenvector, x_1, of $\mathcal{A}_{C_1}$, which also is an eigenvector of $\mathcal{A}$ distinct[1] from its first eigenvector x_0. Now $S_2 = [[x_1, x_2]]$ is an invariant subspace of $\mathcal{A}$, because

1. Because neither x_0 nor x_1 is the 0 vector, and $x_0 \cdot x_1 = 0$.

$$\begin{aligned} \mathcal{a}(\alpha x_0 + \beta x_1) &= \alpha \mathcal{a}(x_0) + \beta \mathcal{a}(x_1) \\ &= \alpha\lambda_0 x_0 + \beta\lambda_1 x_1. \end{aligned}$$

Continue on with this process, next forming the restriction $\mathcal{a}_{C_2}$ of $\mathcal{a}$ to C_2, where $C_2 = S_2^{\perp}$, to find an eigenvector of x_2 of $\mathcal{a}_{C_2}$,... .

The process continues until we have found n orthogonal eigenvectors — which establishes the result asserted by Theorem 13.21.

Although it has been implied earlier in our discussion of the results which still hold for abstract vector spaces, page 545, and for abstract dot products, page 546, it's worth repeating that ***if a sequence*** $v_1,...,v_n$ ***of real n-tuples form a basis for real n-dimensional space, these same vectors form a basis for complex n-dimensional space***. This follows because each complex n-tuple can be written as a complex weighted sum of the standard real unit vectors $e_1,...,e_n$, where e_j has a 1 in the j-th coordinate, and 0 in all others. For this reason, we don't make much of a distinction between a real basis for real n-dimensional space and a real basis for complex n-dimensional space.

With the results concerning symmetric matrices now available, we can take a better look at how to study the local behavior of functions of several real variables.

First let's see what a changeover to the proper coordinate system tells us about the nature of a function f near a point x such that [1] $\textbf{\textit{grad}}\, f(x) = \mathbf{0}$.

From Taylor's theorem in several variables (Theorem 10.33 on page 371) and equation 13.68 on page 581 we can see that in the situation being considered, for small h (using matrix notation selectively for convenience

$$f(x+h) \cong f(x) + \frac{1}{2}\mathbf{h}^{\mathrm{T}}\mathbf{A}\mathbf{h},$$

where $\mathbf{A}$ is the *Hessian* matrix $(a_{ij} = D_i D_j f(x))$ at x, and T is the transpose operator (Definition 10.13 on page 342).

Let Γ be the matrix of eigenvectors of the Hessian matrix, which we are assuming is symmetric (conditions for this being given in Theorem 10.32 on page 369). From Theorem 13.19 on page 582 we know that these eigenvectors are orthogonal, and from Theorem 13.21 on page 583 they form a basis (Definition 9.11 on page 299). We can therefore assume that these eigenvectors have been scaled to have unit length — i.e., any eigenvector $\mathbf{x}$

1. Recall from Section 8 of Chapter 10 that in most cases we already know how to determine whether $f(x)$ is a maximum, minimum, or saddle point. The approach taken here should provide more insight as to what's actually going on — insight which can be critical when the problem is a tough one.

replaced by $\mathbf{x}/\|\mathbf{x}\|$, where the denominator is the length of $\mathbf{x}$. From these facts it's not difficult to see that

$$\Gamma^T\Gamma = \mathbf{I}, \qquad \text{i.e.,} \qquad \Gamma^T = \Gamma^{-1},$$

(see Definition 10.6 on page 334 and Theorem 10.8 on page 336).

Now express $\mathbf{h}$ of approximation in terms of the eigenvectors of $\mathbf{A}$, via

$$\Gamma\mathbf{q} = \mathbf{h} \tag{13.72}$$

($\mathbf{q}$ is the vector of coordinates of $\mathbf{h}$ in the eigenvector coordinate system, see Theorem 13.11 on page 549). Substituting this equation into approximation shows it to be equivalent to

$$f(\boldsymbol{x}+\boldsymbol{h}) \cong f(\boldsymbol{x}) + \frac{1}{2}\mathbf{q}^T\Gamma^T\mathbf{A}\Gamma\mathbf{q}. \tag{13.73}$$

But from Theorem 13.15 on page 556 (see also the immediately preceding equations 13.37 and 13.38) and the fact that $\Gamma^T = \Gamma^{-1}$, we know

$$\Gamma^{-1}\mathbf{A}\Gamma = \Gamma^T\mathbf{A}\Gamma = \Lambda, \tag{13.74}$$

where Λ is the matrix with zeros off diagonal, whose diagonal elements are the eigenvalues corresponding to the eigenvectors in Γ. Since

$$\mathbf{q}^T\Lambda\mathbf{q} = \sum_{i=1}^{n} \lambda_i q_i^2$$

(from the definition of matrix multiplication, Definition 10.3 on page 331) we see that approximation 13.73 may be written as

$$f(\boldsymbol{x}+\boldsymbol{h}) \cong f(\boldsymbol{x}) + \frac{1}{2}\sum_{i=1}^{n} \lambda_i q_i^2, \tag{13.75}$$

where q_i are the coordinate values of the vector $\boldsymbol{h}$ expressed in terms of the eigenvectors of the Hessian matrix.

It's evident that a matrix $\mathbf{A}$ has eigenvalue 0, if and only if $\mathbf{A}$ does not have an inverse; because in precisely such a case, from the definition of eigenvalue, there would be a nonzero vector $\mathbf{x}$ such that $\mathbf{Ax} = \mathbf{0}$. But from the definition of multiplication of a matrix times a vector (10.4 on page 328) and the definition of linear independence of a sequence of vectors (Definition 9.7 on page 297), it would follow that the columns of $\mathbf{A}$ are linearly dependent; so that the columns of $\mathbf{A}$ are linearly dependent if and only if $\mathbf{A}$ has an eigenvalue of 0. But our claim then follows from Theorem 10.8 on page 336. We state this result formally as follows.

Theorem 13.25: Eigenvalues and invertibility of a matrix

An $n \times n$ matrix is invertible (has an inverse, Theorem 10.8 on page 336) if and only if 0 is not one of its eigenvalues (Definition 13.2 on page 536).

❑

With this result we can see the following.

Theorem 13.26: Eigenvalues and relative extrema

Suppose f is a real valued function of n real variables (Definition 8.1 on page 261) satisfying $\boldsymbol{grad} f(x) = 0$ (Definition 8.10 on page 272) and suppose it has continuous first and second partial derivatives, within some circle of positive radius centered at $\boldsymbol{x}$. Then the Hessian matrix, $\mathbf{A}$, of f at $\boldsymbol{x}$ (Definition 10.36 on page 377) will exist and be symmetric (from Theorem 10.32 on page 369) and hence from Theorem 13.21 on page 583 the eigenvectors of the Hessian form a basis for n-dimensional space. If the Hessian has an inverse (Theorem 10.8 on page 336) then none of its eigenvalues are 0. From approximation 13.75 it follows that if all of the eigenvalues are positive, then f has a relative minimum at $\boldsymbol{x}$ (see Theorem 10.35 on page 376). If all of the eigenvalues are negative, then f has a relative maximum at $\boldsymbol{x}$.

Translating into the terminology used in establishing Theorem 10.37 on page 381, focusing on equation 10.76 on page 383, we see that ***the quadratic form*** $\sum_{i,j} h_i h_j H_{i,j}$ ***associated with the Hessian is positive definite if and only if all of the eigenvalues of the Hessian are positive (negative definite if all of them are negative) and neither if at least one of these conditions is not fulfilled***.

No matter whether or not the Hessian is invertible, ***if some of the eigenvalues are positive and some negative, then*** f ***has a saddle-point at*** $\boldsymbol{x}$, **(maximum in the directions of the eigenvectors with negative eigenvalues, a minimum in the directions of the eigenvectors with positive eigenvalues)** ***and hence, neither a maximum nor a minimum at*** $\boldsymbol{x}$.

If the Hessian has a 0 eigenvalue, then no conclusions concerning maxima or minima can be drawn solely knowing that all of the other eigenvalues are of one algebraic sign. In this situation further information, such as knowledge of the behavior of higher derivatives, is needed.

❑

To conclude, let's start from scratch, not even assuming knowledge that $\boldsymbol{grad} f(\boldsymbol{x}) = \mathbf{0}$ and see how the proper choice of coordinate system (that of the eigenvectors of the Hessian) aids our understanding in the search for maxima and minima of a function of several variables.

So, suppose we are looking for the extrema of $f(\boldsymbol{x})$, where $\boldsymbol{x} = (x_1,...x_n)$. One reasonable approach is to write the second-order Taylor approximation (see Theorem 10.33 on page 371)

$$f(\boldsymbol{x}_0 + \boldsymbol{h}) \cong \sum_{k=0}^{2} \frac{(\boldsymbol{h} \cdot \boldsymbol{D})^k f}{k!}(\boldsymbol{x}_0)$$

$$= f(\mathbf{x}) + \mathbf{h}^{\mathrm{T}}\mathbf{D}f(\mathbf{x}) + \frac{1}{2}\mathbf{h}^{\mathrm{T}}\mathbf{A}\mathbf{h}$$

where $\mathbf{A}$ is the Hessian matrix whose *i-j* element is $D_i D_j f(\boldsymbol{x})$, and make a direct try at finding $\mathbf{h}$ minimizing or maximizing the last expression on the right. Let us start with an initial guess, $\mathbf{x}_0$, and make a sequence of *improvements,* $\mathbf{h}_k$, designed to maximize (or minimize)

$$f(\mathbf{x}_k) + \mathbf{h}_k^{\mathrm{T}}\mathbf{D}f(\mathbf{x}_k) + \frac{1}{2}\mathbf{h}_k^{\mathrm{T}}\mathbf{A}_k\mathbf{h},$$

where $\mathbf{x}_{k+1} = \mathbf{x}_k + \mathbf{h}_k$, in hopes that the sequence $\mathbf{x}_k$ will converge to the value we seek that maximizes or minimizes $f(\boldsymbol{x})$.

The only apparent obstacle is finding an easy way of maximizing or minimizing

$$f(\mathbf{x}) + \mathbf{h}^{\mathrm{T}}\mathbf{D}f(\mathbf{x}) + \frac{1}{2}\mathbf{h}^{\mathrm{T}}\mathbf{A}\mathbf{h}. \qquad \textit{13.76}$$

To overcome this obstacle, let us convert to the coordinate system based on the eigenvectors of the Hessian, $\mathbf{A}$, which, as before, we will assume are orthogonal and of unit length (look at the first part of the statement of Theorem 13.26 for conditions under which all of this works). Letting Γ be the matrix whose columns are the eigenvectors of $\mathbf{A}$, the coordinates for $\mathbf{h}$ in this eigenvector coordinate system are given by the n-tuple $\mathbf{q}$, which satisfies

$$\mathbf{\Gamma q} = \mathbf{h} \quad \text{or} \quad \mathbf{q} = \Gamma^{-1}\mathbf{h} = \Gamma^{\mathrm{T}}\mathbf{h} \qquad \textit{13.77}$$

(see Theorem 13.11 on page 549 and the discussion in the paragraph preceding equation 13.72 on page 587). Substituting the first of equations 13.77 into expression 13.76 converts that expression to

$$f(\boldsymbol{x}) + \mathbf{q}^{\mathrm{T}}\Gamma^{\mathrm{T}}\mathbf{D}f(\mathbf{x}) + \frac{1}{2}\mathbf{q}^{\mathrm{T}}\Gamma^{\mathrm{T}}\mathbf{A}\Gamma\mathbf{q},$$

which we recognize from equation 13.74 on page 587 as

$$f(\boldsymbol{x}) + \mathbf{q}^{\mathrm{T}}\Gamma^{\mathrm{T}}\mathbf{D}f(\mathbf{x}) + \frac{1}{2}\mathbf{q}^{\mathrm{T}}\Lambda\mathbf{q}.$$

Writing this out in somewhat expanded form yields

$$f(\mathbf{x}) + \sum_{i=1}^{n} q_i(\Gamma^T \mathbf{D}f(\mathbf{x}))_i + \frac{1}{2}\sum_{i=1}^{n} \lambda_i q_i^2 .$$

Choosing q_i to maximize or minimize this can be done via differentiation, which yields for each *i*

$$(\Gamma^T \mathbf{D}f(\mathbf{x}))_i + \lambda_i q_i = 0$$

or

$$q_i = -\frac{(\Gamma^T \mathbf{D}f(\mathbf{x}))_i}{\lambda_i} .$$

Hence

$$\mathbf{q} = -\Lambda^{-1}\Gamma^T \mathbf{D}f(\mathbf{x})$$

and thus from the latter equations of 13.77

$$\mathbf{h} = \Gamma\mathbf{q} = -\Gamma(\Lambda^{-1}\Gamma^T \mathbf{D}f(\mathbf{x})),$$

i.e., again making use of equation 13.74

$$\mathbf{h} = -\mathbf{A}^{-1}\mathbf{D}f(\mathbf{x}). \qquad \textit{13.78}$$

A careful comparison will show that this approach is precisely the same as using the multidimensional Newton's method (Definition 10.39 on page 387) on $\boldsymbol{grad} f(x) = \mathbf{D}f(\mathbf{x})$ to determine where $\boldsymbol{grad} f(x) = \mathbf{0}$. Thus we see that conversion to the eigenvector coordinate system yields a considerable simplification for investigating maxima and minima for functions of several variables.

Exercises 13.20

1. Show that a symmetric positive definite matrix is invertible.
2. Handle the problems treated in Example 10.18 on page 379 and Example 10.19 on page 380 via eigenvalues.
3. Redo Problems 10.20.3 and 10.20.4 (page 382) using eigenvalues.
4. Construct some symmetric matrices and determine which of them are positive definite and which are not from their eigenvalues. (Use of eigenvalue computer packages is encouraged.)

Chapter 14

Fourier and Other Transforms

1 Introduction

Fourier series, Fourier integrals, generating functions, and Laplace transforms are usually presented as merely those tricks of the trade which just happen to be among the better performers for solving certain types of equations. Now it's true that you can look at these transforms from this narrow view, and learn how to make use of them for the solution of standard problems. If this is your aim, then any well written standard text on engineering mathematics, e.g., Churchill [3], [4], will serve this purpose. Our goals are

- to show how these methods are simply another aspect of choosing appropriate coordinate systems in which to study problems of interest (in these cases, eigenvector coordinate systems of constant coefficient differential or difference operators);
- to point out the nature of the solutions obtained in systems driven by discontinuous driving forces — an aspect often disregarded in many standard treatments.

2 Fourier Series

In electronic systems, such as stereo units, and in the study of mechanical vibrations, there is a great deal of interest in finding the long-run response to periodic inputs. In the stereo hi-fi case this is fundamental for describing the system's behavior. In the mechanical system, such as an airplane wing and engine, or a bridge driven by wind, this response provides vital information, such as how far the wing or bridge will be periodically bent. This background is needed to understand the more realistic statistical theory of such time series.

Definition 14.1: Periodic function, period

A function, f, of a real variable is said to be **periodic with period p**, if there is a real number p such that for all real t in the domain of f (Definition 1.1 on page 2) $t + p$ is in the domain of f and

$$f(t+p) = f(t). \qquad \textbf{\textit{14.1}}$$

If there is a smallest $p > 0$, say $p = p_F$, for which equation 14.1 holds, then p_F is called the **fundamental period of** f.

❑

The *sin, cos,* and imaginary exponential function (whose value at t is e^{it}) are periodic, all with fundamental period 2π.

For the first part of this section we will therefore investigate the problem of finding a periodic solution, f, of the equation

$$P(D)f = g, \qquad \textbf{\textit{14.2}}$$

where g is a given well-behaved (i.e., continuous) periodic function with period p_0 and $P(D)$ is a q-th-degree polynomial in D for which each solution of the homogeneous equation $P(D)f = 0$ approaches 0 for large t. It will turn out that there is only one such solution of equation 14.2, and under the specified conditions this well characterizes the long run-response of the electrical or mechanical system to the periodic input g.

The *transient* part of the solution of $P(D)f = g$ (i.e., the solution of $P(D)f = 0$) is only guaranteed to approach 0 for large t if the real part of each root of the equation

$$P(\lambda) = 0 \qquad \textbf{\textit{14.3}}$$

is negative. (This claim can be established using the results of Exercise 6.12.11 on page 209 and Problem 6.13.3 on page 209.) If this condition is satisfied, we say that the system governed by equation 14.2 is **stable**. One way to determine whether the system is stable is to determine all of the roots of equation 14.3. In many cases today, this is easily done using computer packages, such as *Mathematica* or *Mlab.* For example, to find all roots of the equation

$$P(\lambda) = \lambda^4 + \lambda^3 + 3\lambda^2 + \lambda + 1 = 0$$

using *Mathematica*, we need only type the following, exactly as printed.

```
NSolve[x^4 + x^3 + 3 x^2 + x + 1 == 0,x]
```

(Note the switch to English letters in using computer packages.) In *Mlab* use the function Proot.

Exercises 14.1

Determine which of the following systems are stable.

1. $(D^2 + 2D + 1)f = g$
2. $(D^2 + 1)f = g$
3. $(D^2 - 2D + 1)f = g$
4. $(D^4 + D^3 + 3.5D^2 + D + 2)f = g$

5. $(D^2 - 1)f = g$
6. $(D^3 - D + 1)f = g$
7. $(D^2 + 0.2D + 1)f = g$

To find the periodic solution we're looking for, it's reasonable here to try the same type *eigenvalue-eigenvector* approach that was used to investigate matrix systems **Ax** = **y**, because the derivative operator D and polynomials in D can meaningfully be approximated by matrices, as we know from the advanced discussion starting on page 433. Since these operators are limits of such matrices, we shouldn't be surprised if we can find sufficiently many *eigenfunctions* for them. We should also not be surprised that new obstacles will have to be overcome as well.

So we start by looking for the eigenfunctions of the operator $P(D)$.

Definition 14.2: Eigenfunctions and eigenvalues of $P(D)$

If $P(D)$ is a q-th-degree polynomial in the derivative operator D (Definition 2.5 on page 30 ff.) then the nonzero, q-times differentiable function $\boldsymbol{x}$ (with continuous q-th derivative) is an **eigenfunction of** $\boldsymbol{P(D)}$ **corresponding to the eigenvalue** λ if

$$P(D)\boldsymbol{x} = \lambda\boldsymbol{x}.$$

❑

For the problem under consideration we will search for periodic eigenfunctions having a specified period (Definition 14.1 on page 591) which, with no loss of generality, we can assume equal to 1. We will then try to determine whether these eigenfunctions will serve as a basis of a coordinate system for a reasonably large set of periodic functions with the specified period.

So we are searching for periodic nontrivial solutions, λ, x of $P(D)\boldsymbol{x} = \lambda\boldsymbol{x}$, or letting $P(D) - \lambda = Q(D)$, for periodic solutions, x of

$$Q(D)x = 0, \qquad \textit{14.4}$$

where $Q(D)$ is a q-th-degree polynomial in D.

Justified by the fundamental theorem of algebra (Theorem 6.4 on page 197) we may write equation 14.4 as

$$(D - a_1)[(D - a_2)[\dots (D - a_q)]]x = 0. \qquad \textit{14.5}$$

The values a_j need not be distinct. Equation 14.5 can be viewed as a sequence of first-order constant coefficient differential equations

$$\begin{aligned}
&(1) \quad && (D-a_1)g_1 = 0 \ \text{ where } \ g_1 = (D-a_2)\cdots(D-a_q)x \\
&(2) \quad && (D-a_2)g_2 = 0 \ \text{ where } \ g_2 = (D-a_3)\cdots(D-a_q)x \\
& && \vdots \\
&(q-1) \quad && (D-a_{q-1})g_{q-1} = 0 \ \text{ where } \ g_{q-1} = (D-a_q)x.
\end{aligned}$$

Successively solve equations $(1),(2),...,(q-1)$ for $g_1, g_2,...,g_{q-1}$ using the shift rule (Theorem 5.11 on page 146 and Exercise 6.9.4 on page 199)

$$(D-\alpha)g = e^{\alpha t}D(e^{-\alpha t}g)$$

and, finally, solve the last equation

$$(D-a_q)x = g_{q-1}$$

in the same way.

For distinct a_j we obtain solutions of the form

$$\sum_j c_j e^{a_j t}.$$

A multiple root $(D-\alpha)^k$ yields a solution $cR_{k-1}(t)e^{\alpha t}$ where $R_{k-1}(t)$ is a polynomial in t of degree $k-1$ with nonzero leading coefficient, the coefficient of t^{k-1} (see Problem 6.13.3 on page 209 and Exercise 6.9.4 on page 199).

As far as the problem of finding period* 1 *solutions of* $\boldsymbol{P(D)x = \lambda x}$*, the only* $\boldsymbol{x}$ *that qualify are of the form

$$\boldsymbol{x(t) = e^{i2\pi nt}},$$

where n is an integer.

With this set of eigenfunctions of $P(D)$, the next question to be tackled concerns the search for a reasonably *rich* class of functions which can be expressed as linear combinations (or limits of linear combinations) of these eigenfunctions. We'll see that the span of these eigenfunctions (i.e., the set of linear combinations, or limits of linear combinations of these functions) includes the set of piecewise smooth functions (see Definition 3.13 on page 89) with period 1, which is surely a reasonable size class of functions. There is a complication that we have not mentioned — namely, because we have limits of linear combinations available, a feature not arising in finite dimensional spaces, more than one form of convergence (i.e., more than one type of limit) can be used.[1] To keep matters simple in this chapter, here we will work with *pointwise convergence* — i.e., $g_n \rightarrow g$ for piecewise continuous g (see Definition 3.13 on page 89) meaning that $\lim_{n \to \infty} g_n(t) = g(t)$ at each continuity point of g, and convergence to the average of the left- and right-hand limits of g at its simple jump discontinuities.

To investigate the problem of our ability to express a piecewise smooth period 1 function g as a linear combination of the eigenfunctions given by $e^{i2\pi nt}$ of $P(D)$ (or limits of these), we start by asking the question

If

$$g(t) = \sum_n a_n(g) e^{i2\pi nt} \qquad \textbf{14.6}$$

how would we find the coefficients[2] $a_n(g)$?

To help answer this question recall that in three dimensions, if $\boldsymbol{w}_1, \boldsymbol{w}_2, \boldsymbol{w}_3$ are orthonormal vectors (Definition 9.19 on page 306 and the discussion preceding it) and $\boldsymbol{v}$ is some three-dimensional vector, then by Theorem 9.17 on page 304 and the fact that $\boldsymbol{w}_j \cdot \boldsymbol{w}_j = 1$, we have

$$\boldsymbol{v} = P_{\boldsymbol{w}_1}(\boldsymbol{v}) + P_{\boldsymbol{w}_2}(\boldsymbol{v}) + P_{\boldsymbol{w}_3}(\boldsymbol{v})$$

where

$$P_{\boldsymbol{w}_j}(\boldsymbol{v}) = (\boldsymbol{v} \cdot \boldsymbol{w}_j)\boldsymbol{w}_j .$$

Remembering that finite dimensional vectors are functions (sequences, Definition 1.2 on page 3) which can be used to approximate real or complex valued functions of a real variable, it's natural to extend the definition of dot

---

1. A most natural definition of convergence seems to be *mean square convergence,* where we take $g_n \underset{ms}{\rightarrow} g$ to mean that $\lim_{n \to \infty} \|g_n - g\|^2 = \lim_{n \to \infty} \int_0^1 |g_n(t) - g(t)|^2 dt = 0$. This will be discussed in Chapter 15.
2. Usually, equation 14.6 is written $g(t) = \sum_n a_n e^{i2\pi nt}$. We use the symbol $a_n(g)$ rather than a_n here to stress the dependence of these coefficients on the function g.

product (Definition 13.7 on page 545) to the period 1 functions now being considered, via the following definition.

Definition 14.3: $F \cdot G$ $\|F\|$ for periodic functions

If F and G are periodic piecewise continuous functions of period 1 (Definition 3.13 on page 89 and Definition 14.1 on page 591) we let their **dot product**, $F \cdot G$, be given by

$$F \cdot G = \int_0^1 F(t)\overline{G(t)}dt,$$

where $\overline{G(t)}$ is the complex conjugate (Definition 13.1 on page 535) of $G(t)$.

The **norm**, $\|F\|$, of F is defined as previously, namely, by

$$\|F\| = \sqrt{F \cdot F}.$$

The Schwartz inequality (Theorem 8.21 on page 284) and the triangle inequality (Theorem 10.20 on page 350) extend to these *vectors*.

❑

With this definition, we see that the eigenfunctions given by $e^{i2\pi nt}$ are orthonormal; i.e., using the formula to denote the function, where m and n are integers,

$$e^{i2\pi nt} \cdot e^{i2\pi mt} = \begin{cases} 0 \text{ if } & m \neq n \\ 1 \text{ if } & m = n. \end{cases}$$

Hence, were the right side of equation 14.6 a finite sum, by taking the dot product of both sides with the function given by $e^{i2\pi mt}$, we would find that

$$g(t) \cdot e^{i2\pi mt} = \sum_n a_n(g) e^{i2\pi nt} \cdot e^{i2\pi mt} = a_m(g).$$

That is,

$$a_m(g) = \int_0^1 g(t)e^{-i2\pi mt}dt. \qquad \textbf{\textit{14.7}}$$

We know (Theorem 7.18 on page 236) that if equation 14.6 were valid, with the right side converging uniformly, then the term-by-term integration involved in the dot product would be valid. But since a uniformly convergent series of continuous functions is continuous (Theorem 7.17 on page 235) and we have only been assuming g piecewise smooth (Definition 3.13 on page 89) our ***previous theory* does not provide *an instantaneous* justification *of***

the term-by-term integration which would yield the validity of equations **14.7**.

Before examining how to modify our approach to this obstacle, we introduce a definition and some notation.

Definition 14.4: Fourier coefficients, Fourier series

If a function g is piecewise smooth (Definition 3.13 on page 89) and of period 1 (Definition 14.1 on page 591) the numbers

$$a_m(g) = \int_0^1 g(t)e^{-i2\pi mt}\,dt$$

are called the **Fourier coefficients of** g.

If the series (Definition 7.4 on page 214) denoted by

$$\sum_{n=-\infty\to\infty} a_n(g)\,e^{i2\pi nt},$$

stands for

$$a_0(g) + \sum_{n=1\to\infty} [a_n(g)e^{i2\pi nt} + a_{-n}(g)e^{-i2\pi nt}]$$

(this representing the sequence $a_0(g), a_0(g) + a_1(g)e^{i2\pi t} + a_{-1}(g)e^{-i2\pi t}, \ldots,$ convergent or not), this series is called the **Fourier series for** g.

❑

You might expect that if g is piecewise continuous (Definition 3.13 on page 89) then its Fourier series would converge to g at all continuity points of g. This is ***not*** the case and in fact, the area of convergence properties of Fourier series appears somewhat more like a mine field, loaded with unexpected results. To name but two:

- even if g is periodic of period 1 and continuous, the Fourier series for $g(t)$ need not converge to $g(t)$. (See Knopp [10] pages 355 ff.)

On the other hand, under these conditions

- the series obtained from $\sum_{n=-\infty\to\infty} a_n(g)\,e^{i2\pi nt}$ by integrating term by term from 0 x converges to $\int_0^x g(t)dt$ even if the series $\sum_{n=-\infty\to\infty} a_n(g)\,e^{i2\pi nt}$ fails to converge to $g(t)$ at various t (See Zygmund [18] page 27).

These difficulties explain our restriction to piecewise smooth, rather than the more general piecewise continuous periodic functions. Therefore you might feel that investigating Fourier series is a hopeless task. This is really not so, as we will see in the brief optional discussion to follow. However, the effort required for implementation of this more satisfying approach is greater than we feel is warranted here. With the tools we have readily available we can construct a much simpler, yet adequate theory for our current purposes.

In this optional portion we will try to give some idea of what is required to develop the more satisfying theory of Fourier series that we referred to above. The basic obstacle to this development is that it requires familiarity with the Lebesgue integral (see Section 5 of Chapter 4 beginning on page 114). It can be shown that if g is real valued and *Lebesgue measurable* (a requirement that is not very stringent from apractical viewpoint), and the Lebesgue integral $\int_0^1 g^2(t)dt$ is finite, then

$$\lim_{N \to \infty} \int_0^1 \left[g(t) - \left\{ a_0(g) + \sum_{n=1}^{N} [a_n(g)e^{i2\pi nt} + a_{-n}(g)e^{-i2\pi nt}] \right\} \right]^2 dt = 0. \quad \textbf{14.8}$$

That is, the Fourier series $\sum_{n = -\infty \to \infty} a_n(g)\, e^{i2\pi nt}$ ***converges in mean square to g***.

Convergence in mean square doesn't imply pointwise convergence — an assertion whose proof we leave to Problem 14.2 below, but is a reasonable approach to the concept of convergence, in the sense that *most values* of

$$a_0(g) + \sum_{n=1}^{N} [a_n(g)e^{i2\pi nt} + a_{-n}(g)e^{-i2\pi nt}]$$

will be close to $g(t)$ for large N.

Problem 14.2

Show that convergence in mean square, as defined in equation 14.8 above, even for a sequence of Riemann integrable functions, doesn't imply pointwise convergence anywhere in their domain, say the interval $[0;1]$.

Hint: Geometrically define $g_n(t)$ to equal $\sqrt[4]{n}$ for t in an interval of length $1/n$, and 0 elsewhere. We let these intervals repeatedly *sweep out* $[0;1]$, relying on the fact that $\sum_{k=1 \to \infty} 1/k$ diverges to ∞ (Example 7.6 on page 224). So

$$g_1(t) = 1 \quad \text{on} \quad [0;1] \quad 0 \text{ elsewhere}$$

$$g_2(t) = \sqrt[4]{2} \quad \text{on} \quad \left[0;\frac{1}{2}\right] \quad 0 \text{ elsewhere}$$

$$g_3(t) = \sqrt[4]{3} \quad \text{on} \quad \left[\frac{1}{2};\frac{1}{2}+\frac{1}{3}\right] \quad 0 \text{ elsewhere}$$

$$g_4(t) = \sqrt[4]{4} \quad \text{on} \quad \left[\frac{1}{2}+\frac{1}{3};1\right] \cup \left[0;\frac{1}{2}+\frac{1}{3}+\frac{1}{4}-1\right] \quad 0 \text{ elsewhere}$$

$$\vdots$$

(each time the right hand end point would exceed 1, *wrap it around* to continue over again from 0).

Then

$$\int_0^1 g_n^2(t)dt = \frac{1}{n}\sqrt{n} = \frac{1}{\sqrt{n}} \to 0,$$

but $g_n(t)$ does not converge for any t in [0;1]; for *most* t values $g_n(t)$ is 0; however, for any chosen value of t in [0;1], there are an infinity of values of n (getting sparser as n grows) at which $g_n(t)$ becomes arbitrarily large. A readable reference on this topic is Zygmund [18], pages 74-75.

Although we don't feel that there is much to be gained by supplying proofs that the Fourier series of a periodic piecewise smooth function converges to the function,[1] it is very much worth illustrating this situation geometrically in the following example.

Example 14.3: Convergence of square wave Fourier series

Let

$$S(t) = \begin{cases} 1 \text{ for } n \le t \le n + 0.5 & n = 0, \pm 1, \pm 2,... \\ 0 \text{ otherwise.} \end{cases}$$

Then

$$a_n(S) = \int_0^1 S(t)e^{-i2\pi nt}dt = \begin{cases} 1/2 & \text{for } n = 0 \\ \dfrac{1}{i\pi n} & \text{for } n \text{ odd} \\ 0 & \text{otherwise.} \end{cases}$$

1. A comprehensible proof is given in Courant [5], Vol. 1 page 447.

We find

$$\sum_{n=-\infty\to\infty} a_n(S)e^{i2\pi nt} = \frac{1}{2} + \sum_{\substack{n=1\to\infty \\ n \text{ odd}}} \frac{2}{\pi n} sin(2\pi nt).$$

The plot, given in Figure 14-1, consisting of $1+8=9$ terms of this series, is generated by the *Mathematica* command, exactly typed in as follows.

```
x=Plot[.5 + Sum[(2/Pi) 1/(2 n + 1) Sin[ 2 Pi (2 n + 1) t],{n,0,8}],{t,-1,3}]
```

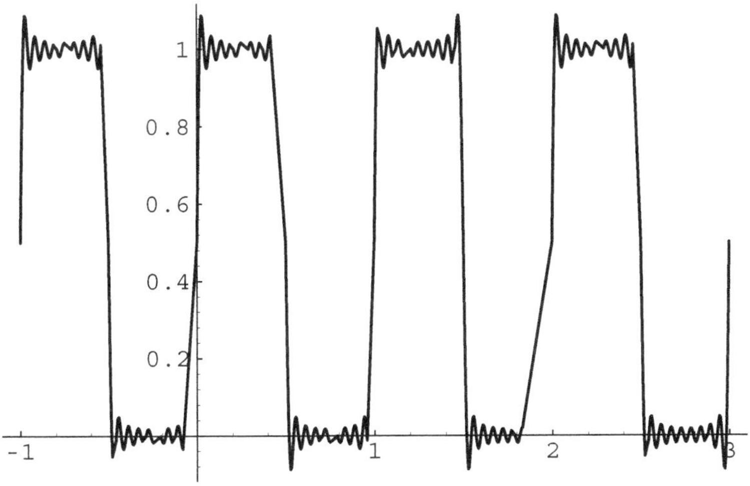

Figure 14-1

As can be seen, convergence seems uniform inside the smooth (here flat) parts of S, necessarily being nonuniform near the jumps.

■

Exercises 14.4

Plot several partial sums for Fourier series for the indicated functions.

1. $f(t) = 1-t$ for $0 \le t < 1$ periodically extended, (*the sawtooth*)
2. $f(t) = 0.5 + -|t-0.5|$ for $0 \le t < 1$ periodically extended,
3. $f(t) = 0.5 + -|t|$ for $-0.5 \le t < 0.5$ periodically extended,
4. $f(t) = 1-|t|$ for $-1 \le t < 1$ periodically extended. (For this, notice that the period is of length 2, so you need to modify the theory a trifle.)

The key to the minimal theory to be presented is the concept of completeness of a sequence of orthonormal functions.

Definition 14.5: Complete orthonormal sequence

Let φ_j, $j = 0, \pm 1, \pm 2, ...$ be a doubly infinite sequence of complex valued functions of a real variable, periodic with period 1 (Definition 14.1 on page 591) and orthonormal relative to the dot product (Definition 14.3 on page 596)

$$\varphi_j \cdot \varphi_k = \int_0^1 \varphi_j(t)\overline{\varphi_k(t)}dt,$$

where orthonormality means

$$\varphi_j \cdot \varphi_k = \begin{cases} 1 & \text{if } j = k \\ 0 & \text{otherwise.} \end{cases}$$

This sequence is said to be **complete with respect to the continuous period 1 functions** if for each continuous function ψ of period 1 with $\psi \cdot \psi = 1$, the sequence

$$\psi, \varphi_0, \varphi_{\pm 1}, \varphi_{\pm 2}, ...$$

fails to be a mutually orthogonal sequence. (So, if behavior here is like that in finite dimensional vector spaces, it *looks* as if each continuous period 1 function must be a linear combination (or limit of linear combinations) of the φ_j, because such a function cannot have a *nontrivial* component orthogonal to all of them.)

❑

Before justifying the concept of completeness we prove the following result.

Theorem 14.6: Completeness of $e^{\pm i2\pi nt}$

If $\varphi_j(t) = e^{i2\pi jt}$, then the sequence $\varphi_0, \varphi_{\pm 1}, \varphi_{\pm 2}, ...$ is complete with respect to the continuous period 1 functions (Definition 14.5 above).

❑

For this proof it's simpler to work purely in the real domain, with sines and cosines. So we will show that this theorem holds if ψ is real and continuous, with $\psi \cdot \psi = 1$. (The extension of this results to complex exponentials is not difficult, and is left to Exercise 14.5.1 on page 604). From the assumed normality ($\psi \cdot \psi = 1$) of ψ,

a. there is a value t_0 in the open interval $(-1;1)$ with $\psi(t_0) \neq 0$; without loss of generality, we may assume $\psi(t_0) = 2\varepsilon > 0$, since if $\psi(t_0) < 0$, just use $-\psi$ in place of ψ.

and from the assumed continuity of ψ (Definition 7.2 on page 213),

b. there is a $\delta > 0$ such that $\psi(t) > \varepsilon$ for t in the open interval (Definition 3.11 on page 88) $(t_0 - \delta; t_0 + \delta)$, where $(t_0 - \delta; t_0 + \delta)$ is contained in the open interval $(0;1)$, as illustrated in Figure 14-2.

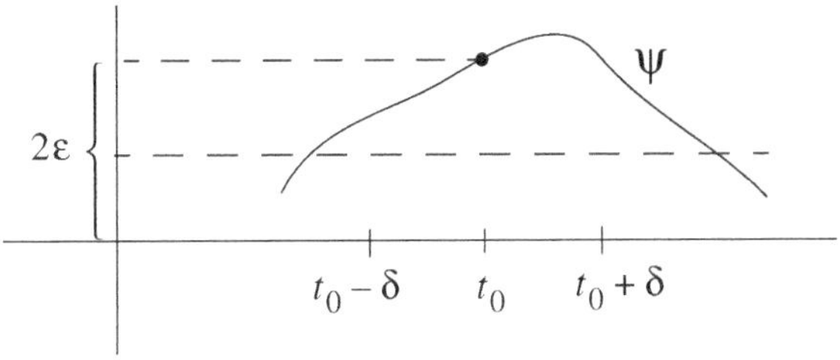

Figure 14-2

If this theorem were false, then there would be some ψ satisfying the above conditions, orthogonal to all functions represented by a series formed from the sequence $1 = cos(0\pi t), sin(n2\pi t), cos(n2\pi t), n = 1, 2, 3, ...$ and hence ψ would be orthogonal to ***all*** so-called *trigonometric polynomials*

$$T_n(t) = \frac{1}{2}a_0 + \sum_{k=1}^{n} \{a_k cos(k2\pi t) + b_k sin(k2\pi t)\}. \qquad \textbf{\textit{14.9}}$$

i.e.,

$$\int_0^1 \psi(t) T_n(t) dt = 0 \quad \text{for all such } T_n. \qquad \textbf{\textit{14.10}}$$

But we shall now show that under these circumstances there would be a sequence of such trigonometric polynomials such that

c. $T_n(t)$ tends uniformly to ∞ for all t in each closed interval I contained in $(t_0 - \delta; t_0 + \delta)$ (recall Definition 3.11 on page 88 of closed and open intervals).

d. $T_n(t) > 1$ for all t in $(t_0 - \delta; t_0 + \delta)$.

e. The T_n are uniformly bounded for t outside $(t_0 - \delta; t_0 + \delta)$.

Provided we can show the existence of such a sequence, T_n, since ψ is continuous, and thus bounded (Theorem 3.12 on page 88) for large enough n we must have

$$\int_0^1 \psi(t) T_n(t) dt > 0. \qquad \textbf{\textit{14.11}}$$

Since equations 14.11 and 14.10 are incompatible, this would show that conditions a on page 601 and b on page 602 (which led to them) cannot hold — which means that there cannot exist a function ψ such that the sequence $\psi, 1 = cos(0\pi t), sin(n2\pi t), cos(n2\pi t), n = 1, 2, 3, ...$ is orthonormal.

To show why equation inequality 14.11 on page 602 follows from conditions c, d, e above, you're best off drawing a picture illustrating the three assertions above. Basically, the behavior of the T_n on any fixed closed interval I lying inside $(t_0 - \delta; t_0 + \delta)$ *overpowers* its behavior anywhere else, for large *n*.

To clinch the argument, we choose

$$T_n(t) = [1 + cos(2\pi\{t - t_0\}) - cos(2\pi\delta)]^n$$

and need only show that

- this expression satisfies the three conditions, c, d, e presented above;
- it is of the form given in equation 14.9 on page 602.

The second of these assertions follows easily from the substitution

$$cos(2\pi\{t - t_0\}) = \frac{e^{i2\pi(t - t_0)} + e^{-i2\pi(t - t_0)}}{2}$$

(see Definition 6.9 on page 200).

The first of these assertions follows easily from the fact that for $t - t_0$ strictly between $-\delta$ and δ

$$cos(2\pi\{t - t_0\}) - cos(2\pi\delta) > 0,$$

while for other t in $[0;1]$ which are outside $(t_0 - \delta; t_0 + \delta)$

$$cos(2\pi\{t - t_0\}) - cos(2\pi\delta) \le 0,$$

as illustrated in Figure 14-3.

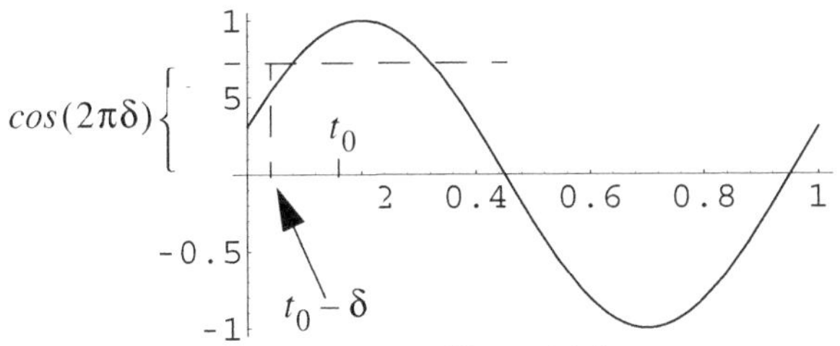

Figure 14-3

Thus we have shown that the orthonormal system $1 = cos(0\pi t), sin(n2\pi t), cos(n2\pi t), n = 1, 2, 3,...$ cannot be enlarged, via appending a continuous function of period 1, and still remain orthonormal. Hence this system is complete with respect to the continuous period 1 functions, proving Theorem 14.6.

Exercises 14.5

1. Verify the assertion that proving Theorem 14.6 for real valued period 1 continuous functions, ψ, and the trigonometric functions $1 = cos(0\pi t), sin(n2\pi t), cos(n2\pi t), n = 1, 2, 3,...$ and real dot product is equivalent to proving it for complex period 1 functions ψ, the system $e^{i2\pi nt}$, $n = 0, \pm 1, \pm 2,...,$ with complex dot product

The operational significance of the completeness of the orthonormal system $1 = cos(0\pi t), sin(n2\pi t), cos(n2\pi t), n = 1, 2, 3,...$ is shown in Theorem 14.7

Theorem 14.7: Equality of functions with identical Fourier coefficients

If f_1 and f_2 are continuous period 1 functions (Definition 7.2 on page 213 and Definition 14.1 on page 591) with the same Fourier coefficients (Definition 14.4 on page 597) i.e., $a_n(f_1) = a_n(f_2)$ for all n, then

$$f_1 = f_2.$$

❑

So ***even if the Fourier series fails to converge to the function, these coefficients determine the function***; hence they represent a coordinate system which includes the period 1 functions.

To establish Theorem 14.7 just note that if $f_1 \neq f_2$, then letting $g = f_1 - f_2$, it would be the case that

$$\sqrt{g \cdot g} = \|g\| = \sqrt{\int_0^1 |g^2(t)| dt} > 0.$$

But because $g = f_1 - f_2$ has all Fourier coefficients

$$g \cdot e^{i2\pi nt} = a_n(g) = a_n(f_1) - a_n(f_2) = 0,$$

it would follow, since we can normalize g so that $\|g\| = 1$ without (in this case) altering its Fourier coefficients, that

$$\|g\| = 1 \text{ and } g \text{ is orthogonal to all } e^{i2\pi nt} \quad n = 1,2,.... \qquad \textbf{\textit{14.12}}$$

Since g, being the difference between two continuous function, is itself continuous (Theorem 2.10 on page 38) this would violate the completeness of the $e^{i2\pi nt}$ with respect to the continuous period 1 functions, established in Theorem 14.6 on page 601. Because it leads to a recognizably false conclusion, we must give up on the possibility that $f_1 \neq f_2$, establishing Theorem 14.7.

Exercises 14.6

1. Extend the theory just developed to continuous functions with period p.
2. In Theorem 7.31 on page 260, we saw that two power series expanded about some value t_0 and which are equal on some positive length interval including t_0, are everywhere equal. If two Fourier series converge to the same values for all t in a given interval, are they equal everywhere?

We now return to the problem of finding a period 1 solution of $P(D)f = g$, where g has period 1. Here we will restrict consideration to functions g which are smooth enough — say, the derivative, Dg, exists everywhere. This may seem like a strong restriction, but from a physical viewpoint it is not; no input function, g, occurring in systems with inertia can have true jump discontinuities — for instance, an actual switch cannot be turned *on* or *off* instantaneously. Hence, it seems reasonable to believe that for practical modeling purposes, we may approximate any actual input with a rather smooth function, and the corresponding actual output should be reasonably close to the output that the model produces from this smooth input.[1]

Using integration by parts (Theorem 4.9 on page 121) if $D^4 g$ exists and is continuous, we see from Definition 14.4 on page 597, that for g of period 1

---

1. This question will be investigated in more detail in Chapter 15 on generalized functions.

$$a_n(g') = \int_0^1 g'(t)e^{-i2\pi nt}dt = \left.\frac{g(t)e^{-i2\pi nt}}{-i2\pi n}\right|_0^{t=1} + \int_0^1 g(t)i2\pi ne^{-i2\pi nt}dt$$

$$= i2\pi na_n(g).$$

That is,

$$a_n(g') = i2\pi na_n(g).$$

Hence,

$$a_n(g) = \frac{1}{i2\pi n}a_n(g') \quad n = \pm 1, \pm 2, \ldots .$$

Since $a_n(D^4 g)$ exists and is bounded, continuing in the same manner, we find

$$a_n(g) = \frac{1}{(2\pi n)^4}a_n(D^4 g).$$

So, from Theorem 7.19 on page 237, the Fourier series for $g(t)$,

$$\sum_{n = -\infty \to \infty} a_n(g)e^{i2\pi nt} \quad \text{where} \quad a_0(g) = \int_0^1 g(t)dt$$

certainly converges uniformly (as do the derived series for Dg and $D^2 g$). The Fourier series sum can therefore be integrated term by term (Theorem 7.18 on page 236) and thus, since the function representing the sum of the Fourier series for g, and g itself, have the same Fourier coefficients, by Theorem 14.7 on page 604 they are equal; i.e., for g smooth enough

$$g(t) = \sum_{n=-\infty}^{\infty} a_n(g)e^{i2\pi nt} = a_0(g) + \sum_{n=1}^{\infty} [a_n(g)e^{i2\pi nt} + a_{-n}(g)e^{-i2\pi nt}].$$

Now recall that the problem we are investigating is that of a solution of $P(D)f = g$, where we are assuming that the system being described is stable (see the discussion surrounding equation 14.3 on page 592). So the roots of $P(\lambda)$ must all have a negative real part, and thus $P(\pm i2\pi n)$ is not only never 0, but the set of these values is bounded away from 0; i.e., there is a positive number b such that $|P(\pm i2\pi n)| \ge b > 0$. This guarantees that the series

$$\frac{a_0(g)}{P(0)} + \sum_{n = 1 \to \infty}\left[\frac{a_n(g)e^{i2\pi nt}}{P(i2\pi n)} + \frac{a_{-n}(g)e^{-i2\pi nt}}{P(-i2\pi n)}\right]$$

converges uniformly to some period 1 function *f*, and the derivatives $D^j f$, where $j \le k$ = the degree of the polynomial *P*, are equal to the sums of the uniformly convergent series

$$\sum_{n=1 \to \infty} D^j \left[\frac{a_n(g) e^{i2\pi n t}}{P(i2\pi n)} + \frac{a_{-n}(g) e^{-i2\pi n t}}{P(-i2\pi n)} \right].$$

Hence the sum

$$f(t) = \frac{a_0(g)}{P(0)} + \sum_{n=1}^{\infty} \left[\frac{a_n(g) e^{i2\pi n t}}{P(i2\pi n)} + \frac{a_{-n}(g) e^{-i2\pi n t}}{P(-i2\pi n)} \right]$$

satisfies $P(D)f = g$ and is evidently periodic with period 1. Because of the stability of the system, no nontrivial solution of $P(D)f = 0$ can be periodic. (Remember that every solution of this homogeneous equation must approach 0 for large *t*, and this can't happen for nontrivial periodic functions.) This shows that if f_1 and f_2 are period 1 solutions of $P(D)f = g$, then $f_1 - f_2$ is a solution of $P(D)f = 0$, so we must have $f_1 - f_2 = 0$, showing that $f_1 = f_2$, establishing the claimed uniqueness of the solution we just found. We summarize what we have shown so far.

Theorem 14.8: Unique periodic solution of $P(D)f = g$

If g is a period 1 function (Definition 14.1 on page 591) which is sufficiently smooth (e.g., $D^4 g$ exists and is continuous) and $P(D)f = g$ is stable (all solutions, *f*, of $P(D)f = 0$ tend to 0 for large t — which is equivalent to the condition that the real parts of all roots of $P(\lambda) = 0$ are negative — see discussion surrounding equation 14.3 on page 592) then the unique periodic period 1 solution of $P(D)f = g$ is given by

$$f(t) = \frac{a_0(g)}{P(0)} + \sum_{n=1 \to \infty} \left[\frac{a_n(g) e^{i2\pi n t}}{P(i2\pi n)} + \frac{a_{-n}(g) e^{-i2\pi n t}}{P(-i2\pi n)} \right],$$

and for large t every solution tends to this one (though some solutions take a good deal longer than others to exhibit this behavior).

❑

Now for g real and $P(D)$ having real coefficients, we see from the definition of the Fourier coefficients (Definition 14.4 on page 597) that

$$a_n(g) = \overline{a_{-n}(g)} \quad \text{and} \quad P(i2\pi n) = \overline{P(-i2\pi n)},$$

where the *overbar* denotes complex conjugation (Definition 13.1 on page 535). Thus if we write

$$a_n(g) = r_n e^{i\vartheta_n} \quad \text{and} \quad P(i2\pi n) = R_n e^{i\varphi_n},$$

then the output resulting from the *input* terms of *frequency* $2\pi n$,

$$a_n(g)e^{i2\pi nt} + a_{-n}(g)e^{-i2\pi nt} = 2r\left(\frac{a_n(g)e^{i2\pi nt}}{P(i2\pi n)} + \frac{a_{-n}(g)e^{-i2\pi nt}}{P(-i2\pi n)}\right)$$

is

$$\frac{a_n(g)e^{i2\pi nt}}{P(i2\pi n)} + \frac{a_{-n}(g)e^{-i2\pi nt}}{P(-i2\pi n)} = \frac{r_n}{R_n}\left(e^{i(2\pi nt + \vartheta_n - \varphi_n)} + e^{-i(2\pi nt + \vartheta_n - \varphi_n)}\right)$$

$$= 2\frac{r_n}{R_n}cos(2\pi nt + \vartheta_n - \varphi_n).$$

So the effect of the inverse of the operator $P(D)$ is to shift the input sinusoidal wave, $cos(2\pi nt + \vartheta_n)$, by a *phase* of $-\varphi_n$ and amplify its magnitude by the factor $1/R_n = 1/|P(i2\pi n)|$. This leads us to give our next definition.

Definition 14.9: Transfer function (period 1 input)

The function with value $1/|P(i2\pi n)|$ is called the **transfer function** of the system governed by the equation $P(D)f = g$ when g has period 1 (Definition 14.1 on page 591). Its values represent the amplification of the input component with frequency $2\pi n$.

❑

In applications to vibrating mechanical or electronic systems, if the ratio r_n/R_n (representing the magnitude of the output to an input of frequency $2\pi n$ and magnitude r_n) is too large, this may indicate serious system problems.

Example 14.7: Spring/mass with friction

A system of a spring with Hooke's law spring coefficient k, mass m, and friction proportional to velocity, subjected to a period 1 driving force g, is usually described by the differential equation

$$\left(D^2 + \frac{c}{m}D + \frac{k}{m}\right)x = \frac{g}{m},$$

(see Chapter 5, Section 6). Suppose $m = 1$, $c = 1$, and $k = 4000\pi^2$, representing a stiff spring with relatively little friction. Below we present a list of pairs $(n, |P(i2\pi n)|)$ for $n = 0,1,...,15$.

(0, 3947.84), (1, 3908.36), (2, 3789.93), (3, 3592.54), (4, 3316.19)

(5, 2960.88), (6, 2526.62), (7, 2013.40), (8, 1421.23), (9, 750.11),

(10, 6.28), (11, 829.08), (12, 1737.07), (13, 2724.02), (14, 3789.94),

(15, 4934.81)

We see that the magnitude of the transfer function is relatively large ($|P(i2\pi n)|$ is very small) at $n = 10$ (a frequency of 20π, about 60 Hz). If $a_{10}(g)$ is also relatively large, say $a_{10}(g) = 40$ and $|a_n(g)| \ll 240$ for $|n| \neq 10$, then

$$\left|\frac{a_{10}(g)}{P(i20\pi)}\right| = 40$$

with all the other

$$\left|\frac{a_n(g)}{P(i2\pi n)}\right| << 40.$$

Assuming that no other frequencies contribute much, if $x(t)$ is measured in feet we expect a maximum value of about 40 feet for $x(t)$. If this is beyond the breaking point extension of the spring, we'd expect the spring to break (e.g., the Tacoma narrows bridge to snap).

■

This model for response to vibration is an extremely simplified one — and in any realistic situation, a good description will be more complicated. In particular, in the airplane engine case, the driving force is only approximately periodic; a more realistic approach would involve a statistical description of the driving force. This is treated in courses on time series analysis, for which the concepts introduced here are essential background.

Exercises 14.8

From the viewpoint of dangerous frequencies to examine, what are they for the following spring/mass friction systems?

1. $(D^2 + 0.2D + 4000)f = g$
2. $(D^2 + 0.3D + 7000)f = g$
3. $(D^2 + 0.3D + 2000)f = g$

It's worth noting that in many computer packages, there are Fast Fourier Transform (FFT) programs specifically written to approximate the Fourier coefficients, $a_n(g)$, from a list representing sampled data — i.e., discrete values corresponding to $g(kh), k = 0, \pm 1, \pm 2, \ldots$, where h is fixed. They are useful for determining if observed data seems to have a large component at any specified frequency.

For reference purposes we quote the following result, whose proof, as already mentioned, can be found in Courant [5], Vol. 1, page 437.

Theorem 14.10: Convergence of Fourier series

If the real valued periodic function f (Definition 14.1 on page 591) of a real variable is piecewise smooth (Definition 3.13 on page 89) then

- Its Fourier series (Definition 14.4 on page 597) converges at every point x (Definition 7.4 on page 214) to

  $$\frac{f(x+) + f(x-)}{2}$$

  where $f(x+) = \lim_{\substack{t \to x \\ t > x}} f(t)$ and $f(x-) = \lim_{\substack{t \to x \\ t < x}} f(t)$.
- In every closed interval (Definition 3.11 on page 88) in which f is continuous (Definition 7.2 on page 213) its Fourier series converges uniformly (Definition 7.16 on page 234).
- If, in addition f has no discontinuities (i.e., is continuous at every point in its domain) then its Fourier series converges absolutely (Definition 7.25 on page 253) and hence, unconditionally (Definition 7.24 on page 252).

❑

3 Fourier Integrals and Laplace Transforms

By means of suitable limit arguments, the results on Fourier series can be extended to provide expansions in the eigenfunctions of polynomials $P(D)$ in the derivative operator D for functions f which are piecewise smooth (Definition 3.13 on page 89) and whose absolute values are integrable (a property not possessed by nontrivial periodic functions) and which satisfy the condition

$$f(x) = \frac{f(x+) + f(x-)}{2}, \qquad \textit{14.13}$$

where

$$f(x+) = \lim_{\substack{t \to x \\ t > x}} f(t) \text{ and } f(x-) = \lim_{\substack{t \to x \\ t < x}} f(t) \qquad \textit{14.14}$$

for all real t.

The precise statement of this extension is given in the following theorem.

Theorem 14.11: Fourier integral theorem

Let f be a real or complex valued piecewise smooth function (Definition 3.13 on page 89) satisfying condition 14.13 for all x, and for which

$$\lim_{x \to \infty} \int_{-x}^{x} |f(t)| dt \quad \text{exists} \qquad \textbf{14.15}$$

(assumed finite). Then the function F (also written $\mathcal{F}(f)$) given by

$$\mathcal{F}(f)(s) = F(s) = \int_{-\infty}^{\infty} f(t) e^{-i2\pi ts} dt \qquad \textbf{14.16}$$

(see Definition 12.11 on page 449) is defined for all real s, and

$$f(t) = \int_{-\infty}^{\infty} F(s) e^{i2\pi ts} ds \qquad \textbf{14.17}$$

for all real t.

Notice that equation 14.17 is an expansion for f in the eigenfunctions of the derivative operator D. This expansion, being an integral, is still essentially a sum (i.e., a *linear combination*, see Definition 9.4 on page 296). The ***coordinates*** in this eigenfunction basis are the values $F(s)$.

The function $F = \mathcal{F}(f)$ is called the **Fourier integral of** f.

❑

Condition 14.15, which relates to integrals the same way *absolute convergence* (Definition 7.25 on page 253) relates to infinite series, is sufficiently important to warrant a formal definition of its own.

Definition 14.12: Absolute integrability

When condition 14.15 holds, namely, that

$$\lim_{x \to \infty} \int_{-x}^{x} |f(t)| dt \quad \text{exists,}$$

f is said to be **absolutely integrable**. If the condition that

$$\int_{-x}^{x} |f(t)| dt \quad \text{exists}$$

holds, then f is said to be **absolutely integrable on $[-x;x]$**.

❑

The Fourier integral theorem has been used extensively in probability theory, the major area being the investigation of the behavior of sums of random variables (see Lukacs [12]). Our main use here will be for application to Laplace transforms. To start we will provide a *bare-bones* outline furnishing only the essential ideas for derivation of the Fourier integral theorem from the Fourier series results.

Initially, we suppose f is real valued, and will define a sequence $f_m \quad m = 1, 2,...$ of functions with period m such that

$$\lim_{m \to \infty} f_m = f.$$

and see how the Fourier series representations of the f_m behave. For convenience, the basic period (Definition 14.1 on page 591) of duration m, will extend from $-m/2$ to $m/2$.

We let

$$f_m(t) = f(t) \quad \text{for} \quad -m/2 \le t < m/2 \qquad \textbf{\textit{14.18}}$$

and extend f_m to be periodic, i.e., for t in $[-m/2; m/2]$

$$f_m(t + km) = f_m(t) \qquad k = \pm 1, \pm 2, \ldots .$$

Now the function φ_m given by

$$\varphi_m(t) = f_m(mt) \qquad \textbf{\textit{14.19}}$$

is periodic with period 1, so from Theorem 14.10 on page 610 it follows that for

$$a_n(\varphi_m) = \int_{-1/2}^{1/2} \varphi_m(t) e^{-i\pi 2nt} dt \qquad \textbf{\textit{14.20}}$$

we have

$$\varphi_m(t) = \sum_{n=-\infty}^{\infty} a_n(\varphi_m) e^{i\pi 2nt}. \qquad \textbf{\textit{14.21}}$$

Then from equations 14.19 and 14.21

$$f_m(t) = \varphi_m\left(\frac{t}{m}\right) = \sum_{n=-\infty}^{\infty} a_n(\varphi_m) e^{i\pi 2 \frac{n}{m} t},$$

where from equations 14.20, 14.19, and 14.18, respectively,

$$a_n(\varphi_m) = \int_{-1/2}^{1/2} \varphi_m(t) e^{-i\pi 2nt} dt = \int_{-1/2}^{1/2} f(mt) e^{-i\pi 2nt} dt$$

$$= \frac{1}{m} \int_{-1/2}^{1/2} f(\tau) e^{-i\pi 2 \frac{n}{m} \tau} d\tau = \frac{1}{m} \int_{-1/2}^{1/2} f(t) e^{-i\pi 2 \frac{n}{m} t} dt$$

(Note that the last step just changed the dummy variable of integration to t from τ.)

Letting

$$A_n(f_m) = \int_{-1/2}^{1/2} f(t) e^{-i\pi 2 \frac{n}{m} t} dt,$$

then

$$f_m(t) = \sum_{n=-\infty}^{\infty} A_n(f_m) e^{i\pi 2 \frac{n}{m} t} \frac{1}{m}.$$

We find, denoting n/m by s and setting

$$a_{s,m}(f) = \int_{-m/2}^{m/2} f(t) e^{-i\pi 2 \frac{n}{m} t} dt, \qquad \textbf{\textit{14.22}}$$

that

$$f_m(t) = \sum_{n=-\infty}^{\infty} a_{s,m}(f) e^{-i\pi 2 s t} \frac{1}{m}. \qquad \textbf{\textit{14.23}}$$

As m grows large, we see from 14.22 above that $a_{s,m}(f)$ approaches a limit as m grows large, which we denote by $F(s)$, given by

$$F(s) = \int_{-\infty}^{\infty} f(t) e^{-i2\pi s t} dt .$$

The series sum in equation 14.23 for $f_m(t)$ looks like an approximating sum for

$$\int_{-\infty}^{\infty} F(s) e^{i2\pi s t} ds$$

(since $a_{s,m}(f) \underset{m \to \infty}{\to} F(s)$).

Because $f_m(t) \underset{m \to \infty}{\to} f(t)$, it thus seems that

$$f(t) = \int_{-\infty}^{\infty} F(s) e^{i2\pi s t} ds .$$

i.e., ***it seems plausible that the Fourier integral theorem holds for real valued functions f***.

If f is complex valued, say $f = f_R + i f_I$, defining

$$F_R(s) = \int_{-\infty}^{\infty} f_R(t)e^{-i2\pi st}dt \quad \text{and} \quad F_I(s) = \int_{-\infty}^{\infty} f_I(t)e^{-i2\pi st}dt,$$

we then have

$$f_R(t) = \int_{-\infty}^{\infty} F_R(s)e^{i2\pi st}ds \quad \text{and} \quad f_I(t) = \int_{-\infty}^{\infty} F_I(s)e^{i2\pi st}ds$$

and hence, setting $F(s) = F_R(s) + \boldsymbol{i}F_I(s)$,

$$F(s) = \int_{-\infty}^{\infty} [f_R(t) + \boldsymbol{i}f_I(t)]e^{-i2\pi st}dt = \int_{-\infty}^{\infty} f(t)e^{-i2\pi st}dt$$

and

$$\begin{aligned} f(t) = f_R(t) + \boldsymbol{i}f_I(t) &= \int_{-\infty}^{\infty} F_R(s)e^{i2\pi st}ds + \int_{-\infty}^{\infty} \boldsymbol{i}F_I(s)e^{i2\pi st}ds \\ &= \int_{-\infty}^{\infty} F(s)e^{i2\pi st}ds, \end{aligned}$$

establishing the Fourier integral theorem for complex valued functions f from its validity for real valued functions.

Both Fourier series and Fourier integrals have serious limitations for solving many engineering type problems — the former because it restricts analysis to periodic inputs extending from $-\infty$ to ∞, and the latter because of its absolute integrability requirement (Definition 14.12 on page 611). These limitations can be overcome rather easily with the Laplace transform, which is obtained from a simple modification of the Fourier integral. The Laplace transform applies to the class of functions which *start out* at time 0, and are allowed to grow fairly fast. The degree of growth is defined as follows.

Definition 14.13: Functions of exponential order

A real or complex valued function, f, of a real variable, satisfying the condition

$$f(t) = 0 \quad \text{for} \quad t < 0$$

is said to be **of exponential order** if there are positive numbers, K, c such that for all real t

$$|f(t)| \le Ke^{ct}.$$

We will refer to such a c as an (**exponential**) **growth constant**. So, the growth of a function of exponential order is not greatly restricted as its argument grows large.

❑

Example 14.9: $f_1(t) = t^2, \quad f_2(t) = exp(t^2)$

Since

$$e^{\sqrt{2}t} = 1 + \sqrt{2}t + \frac{2t^2}{2} + \cdots \geq t^2 \text{ for } t \geq 0$$

(Example 6.2 on page 181), if

$$f_1(t) = \begin{cases} t^2 & \text{for} \quad t \geq 0 \\ 0 & \text{for} \quad t < 0, \end{cases}$$

we see that $|f_1(t)| \leq e^{\sqrt{2}t}$ for all real t, so that f_1 is of exponential order with growth constant $c = \sqrt{2}$.

For f_2, if we write

$$e^{(t^2)} = e^{t^2/2 + t^2/2} = e^{t^2/2}e^{t^2/2} = e^{t^2/2}e^{\left(\frac{t}{2}\right)t}$$

and take t large enough so that

$$e^{t^2/2} > K \quad \text{and} \quad t/2 > c,$$

no matter what K and c we can see that for large enough t, $e^{t^2/2} > Ke^{ct}$.

Hence, f_2 is not of exponential order — ***it grows too fast***.

■

Exercises 14.10

Determine which of the following functions, whose formula for $t \geq 0$ is given, and which are 0 for $t < 0$, are of exponential order.

1. $f(t) = (e^t)^2$
2. $f(t) = cos(t)e^{3t}$
3. $f(t) = e^{(e^t)}$
4. $f(t) = e^{(e^{-t})}$

If g is a function of exponential order, it needn't have a Fourier integral, but the function H given by

$$H(t) = e^{-ut}g(t)$$

does have a Fourier integral if u exceeds the growth constant, c, of g. The Laplace transform of g that we're about to introduce is essentially the Fourier integral of a rescaled H; its important properties will stem from those of Fourier integrals.

Definition 14.14: Laplace transform

Let g be a piecewise smooth real valued function of a real variable, of exponential order with exponential growth constant c (Definition 3.13 on page 89 and Definition 14.13 on page 614). The **Laplace transform G of g** is a function defined for all complex $z = u + iv$ with the real part $u > c$, by the formula

$$G(z) = \int_0^{\infty} g(t)e^{-zt}dt.$$

This *improper integral* (Definition 12.11 on page 449) converges absolutely (Definition 14.12 on page 611) and **uniformly**[1] for all $z = u + iv$ such that $u \geq c_0 > c$, where c_0 is any fixed value exceeding c.

For each z, the value of the Laplace transform of g at z is essentially a weighted sum of g values.

Note that the lower limit of integration can be taken as any negative number, or $-\infty$, when it is assumed that $g(t) = 0$ for $t < 0$.

Sometimes the Laplace transform G is denoted by $\mathcal{L}(g)$, so that

$$G(z) = \mathcal{L}(g)(z).$$

❑

Making the substitution $t = 2\pi\tau$, we find

1. By this *uniform convergence,* we mean that so long as the *real part,* u of z satisfies the condition $u \geq c_0$, then for each $\varepsilon > 0$ there is a value N_ε (dependent, of course, on the function g, and on the value c_0 but otherwise independent of z) such that $\left|G(z) - \int_0^w g(t)e^{-zt}dt\right| \leq \varepsilon$ provided only that $w \geq N_\varepsilon$. This concept is a natural extension of uniform convergence of a series, Definition 7.16 on page 234, the *set* of the series definition corresponding to those z with $u \geq c_0$.

$$G(z) = \int_0^{\infty} g(t)e^{-ut}e^{-ivt}dt = \int_0^{\infty} g(2\pi\tau)e^{-u2\pi\tau}e^{-iv2\pi\tau}2\pi d\tau$$

$$= \int_0^{\infty} [2\pi e^{-u2\pi\tau}g(2\pi\tau)]e^{-iv2\pi\tau}d\tau.$$

Thus, the Laplace transform of g evaluated at $z = u + iv$ is the Fourier integral evaluated at v, of the function H_u whose value is given by

$$H_u(\tau) = 2\pi e^{-u2\pi\tau}g(2\pi\tau). \qquad \textbf{14.24}$$

Symbolically (for clarity) for $z = u + iv$

$$G(z) = \mathcal{L}(g)(z) = \int_{-\infty}^{\infty} H_u(\tau)e^{-i2\pi\tau v}d\tau = \mathcal{F}(H_u)(v) \qquad \textbf{14.25}$$

(where $\mathcal{F}(H_u)$ denotes the Fourier integral of H_u and $\mathcal{F}(H_u)(v)$ the value of this Fourier integral at v).

From the Fourier integral theorem (Theorem 14.11 on page 611) it follows that

$$H_u(\tau) = \int_{-\infty}^{\infty} \mathcal{F}(H_u)(v)\, e^{i2\pi\tau v}dv$$

So, substituting from equation 14.24 into the left side of this equation, and replacing $G(z)$ by $G(u + iv)$ for $\mathcal{F}(H_u)(v)$ for the first factor in the integrand, we have

$$2\pi e^{-u2\pi\tau}g(2\pi\tau) = \int_{-\infty}^{\infty} G(u + iv)e^{i2\pi\tau v}dv, \qquad \textbf{14.26}$$

(so long as $u > c$, the growth constant for g). Note that in order for this to be valid, at any jump discontinuity, t_d of g, we require

$$g(t_d) = \frac{1}{2}[g(t_d+) + g(t_d-)],$$

(see equation 14.14 on page 610 and the reliance of Theorem 14.11 on page 611 on equation 14.13 on page 610).

We can rewrite equation 14.26, by substituting $t = 2\pi\tau$, as

$$2\pi e^{-ut}g(t) = \int_{-\infty}^{\infty} G(u + iv)e^{ivt}dv$$

or

$$g(t) = \frac{1}{2\pi}\int_{-\infty}^{\infty} G(u+iv)e^{(u+iv)t}dv,$$
$$= \frac{1}{2\pi i}\int_{u-i\infty}^{u+i\infty} G(z)e^{zt}dz.$$

(The mysterious i appears because $z = u+iv, \quad dz = idv$.)

This latter integral represents an improper (i.e., limiting) contour integral over a directed line parallel to the imaginary axis in the right half plane (see Definition 12.41 on page 506).

Thus we see that the Laplace transform of g represents the coordinates of g in the eigenfunctions (Definition 14.2 on page 593) of the derivative operator D, where g is a function of exponential order (Definition 14.13 on page 614). As we'll see, these coordinates can greatly simplify solution of differential equations $P(D)f = g$ related to *driving functions, g,* of exponential order.

We summarize the results just derived in the following theorem.

Theorem 14.15: Laplace transform, existence and uniqueness

If g is a piecewise smooth function of exponential order with growth constant c (see Definition 3.13 on page 89 and Definition 14.13 on page 614) then its Laplace transform

$$\mathcal{L}(g)(z) = G(z) = \int_0^{\infty} g(t)e^{-zt}dt$$

is defined for all complex $z = u+iv$ for which $u>c$. Furthermore, provided that for all x it's the case that g satisfies the condition

$$g(x) = \frac{1}{2}[g(x+)+g(x-)] \qquad \textit{14.27}$$

(see equation 14.14 on page 610) the *inversion formula*,

$$g(t) = \frac{1}{2\pi i}\int_{u-i\infty}^{u+i\infty} G(z)e^{zt}dz \qquad \textit{14.28}$$

holds for all real t (this being an [*improper*] contour integral in the complex plane, see Chapter 12 Section 8); this implies that if f and g are such that

$$f(x) = \frac{1}{2}[f(x+)+f(x-)] \quad \text{and} \quad g(x) = \frac{1}{2}[g(x+)+g(x-)]$$

for all x with $\mathcal{L}(f)(z) = \mathcal{L}(g)(z)$ for $u>c$, then $f = g$.

❑

We point out that we will very rarely make direct use of the inversion formula, equation 14.28, for determining a function from its Laplace transform. Its importance arises because it assures us that the transform does determine the function. Hence, ***if we recognize an expression as the Laplace transform of some known function, we can conclude that this function is the only one with this transform***.

To apply Laplace transforms to the solution of differential equations, we need the following result.

Theorem 14.16: $\mathcal{L}(f')$

If for $t > 0$, f is a continuous piecewise smooth function of exponential order with growth constant c (Definition 3.13 on page 89 and Definition 14.13 on page 614) with the possible exception of a jump discontinuity at $t = 0$, and f' is extended to have value

$$\frac{f'(t_d+) + f'(t_d-)}{2}$$

(see equation 14.14 on page 610) at each of its jump discontinuities,[1] t_d, then its Laplace transform (Definition 14.14 on page 616) $\mathcal{L}(f')$ exists for u (the real part of $z = u + \boldsymbol{i}\, v$) satisfying $u > c$ and

$$\mathcal{L}(f')(z) = z\, \mathcal{L}(f)(z) - f(0+).$$

❑

To establish this result we examine

$$\int_0^T f'(t)\, e^{-zt} dt = \sum_{j=1}^{N(T)-1} \int_{t_{d_j}}^{t_{d_{j+1}}} f'(t)\, e^{-zt} dt + \int_0^{t_{d_1}} f'(t)\, e^{-zt} dt + \int_{t_{d_{N(T)}}}^{T} f'(t)\, e^{-zt} dt, \qquad \textbf{\textit{14.29}}$$

where $t_{d_1}, \ldots, t_{d_{N(T)}}$ are the jump discontinuities of f' in the interval $[0;T]$.

For each of the integrals, integration by parts (Theorem 4.9 on page 121) is allowed, yielding for those in the summation,

1. We'll continue using the symbol f' for this extension, even though the extended function cannot be the derivative of f at such a jump discontinuity.

$$\int_{t_{d_j}}^{t_{d_{j+1}}} f'(t)\, e^{-zt} dt = f(t)\, e^{-zt}\Big|_{t_{d_j}}^{t = t_{d_{j+1}}} + \int_{t_{d_j}}^{t_{d_{j+1}}} z f(t)\, e^{-zt} dt;$$

if $t_{d_1} \neq 0$, for the next to last term in equation 14.29, integration by parts yields

$$\int_0^{t_{d_1}} f'(t)\, e^{-zt} dt = f(t)\, e^{-zt}\Big|_0^{t = t_{d_1}} + \int_0^{t_{d_1}} z f(t)\, e^{-zt} dt,$$

and finally, assuming T is not a jump discontinuity of f', we obtain

$$\int_{t_{d_{N(T)}}}^{T} f'(t)\, e^{-zt} dt = f(t)\, e^{-zt}\Big|_{t_{d_{N(T)}}}^{t = T} + \int_{t_{d_{N(T)}}}^{T} z f(t)\, e^{-zt} dt.$$

For $u > c$ (where $z = u + iv$)

$$\lim_{t \to \infty} f(t)\, e^{-zt} = 0 \qquad \textbf{\textit{14.30}}$$

(same arguments as in Example 14.9 on page 615).

But the sum of the integrals

$$\int_0^{t_{d_1}} z f(t)\, e^{-zt} dt + \sum_{j=1}^{N(T)-1} \int_{t_{d_j}}^{t_{d_{j+1}}} z f(t)\, e^{-zt} dt + \int_{t_{d_{N(T)}}}^{T} z f(t)\, e^{-zt} dt = \int_0^T z f(t)\, e^{-zt} dt,$$

while due to the assumed continuity of f, the sum

$$f(t)\, e^{-zt}\Big|_0^{t = t_{d_1}} + \sum_{j=1}^{N(T)-1} f(t)\, e^{-zt}\Big|_{t_{d_j}}^{t = t_{d_{j+1}}} + f(t)\, e^{-zt}\Big|_{t_{d_{N(T)}}}^{t = T}$$

is *telescoping*, with only the terms

$$f(t)\, e^{-zt}\Big|_0^{t = t_{d_1}} \quad \text{and} \quad f(T) e^{-zt}$$

remaining. Due to the possibility of a jump discontinuity in f at 0, the first of these terms must be taken as

$$-f(0+) = -\lim_{\substack{t \to 0 \\ t < 0}} f(t)$$

(the value of f at 0 itself being irrelevant).

Hence we find

$$\int_0^T f'(t)e^{-zt}dt = z\int_0^T f(t)e^{-zt}dt + f(T)e^{-zT} - f(0+).$$

From the equation 14.30 we know that the middle term on the right has limit 0 as T becomes large, while the limit of the integral in the first term on the right is the Laplace transform of f at z. Hence the right side has a limit as T becomes large, and hence so does the left, yielding

$$\mathcal{L}(f')(z) = \int_0^\infty f'(t)e^{-zt}dt = z\int_0^\infty f(t)e^{-zt}dt - f(0+) = z\,\mathcal{L}(f)(z) - f(0+),$$

establishing Theorem 14.16.

One more needed result is easily established.

Theorem 14.17: Laplace transform of shifted function

If the function g has a Laplace transform (Definition 14.14 on page 616) then for fixed $a > 0$ the *shifted function* f given by $f(t) = g(t-a)$ has Laplace transform given by

$$\mathcal{L}(f)(z) = e^{-az}\mathcal{L}(g)(z)$$

whenever the right side exists. Similarly, for fixed b, by writing

$$\int_0^\infty e^{-zt}f(t)dt = \int_0^\infty e^{-(z-b)t}[e^{-bt}f(t)]dt,$$

we find that if φ is the Laplace transform of f, then the function whose value at z is $\varphi(z-b)$ is the Laplace transform of the *function* given by $e^{-bt}f(t)$.

❑

At this stage we're ready to furnish a neat method for solution of

$P(D)f = g$ subject to initial conditions $f^{(j)}(0) = 0,\ j = 0,...,k-1,$

where k is the degree of $P(D)$ and g is continuous and of exponential order.

Now we already know how to solve this equation using the shift rule (see Section 6 of Chapter 6) so you might wonder what is the merit of another method of solution. The justification is that we can frequently take advantage of being able to recognize the expressions we encounter as Laplace transforms of known functions. For several reasons we don't have to establish conditions justifying our operations separately for each case being considered:

- First, we can always check on the validity of the generated solution after the fact.

- Second, and more importantly (even though we leave it as Problem 14.11 below) it is not hard to show that if g is continuous and of exponential order, so is f (the known solution developed earlier) and its k derivatives.

Problem 14.11

Show that if g is continuous and of exponential order (Definition 14.13 on page 614) and $P(D)f = g$, then f is of exponential order.

> Hint: First, write the polynomial $P(D)$ as $(D-\alpha)Q(D)$, where $Q(D)$ is a polynomial one degree lower than $P(D)$. The equation $P(D)f = g$ becomes $(D-\alpha)Q(D)f = g$. Denote $Q(D)f$ by H, show that the solution H of $(D-\alpha)H = g$ must be of exponential order. This is easy because you can obtain an explicit formula for H, using the shift rule (Theorem 5.11 on page 146 extended in Theorem 6.18 on page 202). This process can be repeated, you show that you obtain exponential order solutions at each stage — from which the assertion about f is evident.

In order to carry through our demonstration, we need to develop the formulas for several Laplace transforms.

Definition 14.18: *Re*(z)

If $z = u + iv$ is a complex number with u and v real, then we denote **its real part**, u, by $Re(z)$.

❑

Theorem 14.19: Some specific Laplace transforms

The given functions g are assumed to have value 0 for $t < 0$.

If $g(t) = e^{ikt}$ then $\mathcal{L}(g)(z) = \dfrac{1}{z - ik}$ for $Re(z) > 0$,

if $g(t) = e^{\lambda t}$, λ real then $\mathcal{L}(g)(z) = \dfrac{1}{z - \lambda}$ for $Re(z) > \lambda$,

if $g(t) = sin(kt)$ then $\mathcal{L}(g)(z) = \dfrac{k}{z^2 + k^2}$ for $Re(z) > 0$,

if $g(t) = cos(kt)$ then $\mathcal{L}(g)(z) = \dfrac{z}{z^2 + k^2}$ for $Re(z) > 0$, if $g(t) = H(t) = \begin{cases} 0 & \text{for } t < 0 \\ 1 & \text{for } t > 0 \end{cases}$ then $\mathcal{L}(g)(z) = \frac{1}{z}$ for $Re(z) > 0$,

(H is the *Heaviside unit step*),

if $g_k(t) = t^k \quad k = 1, 2, ...$ then $\mathcal{L}(g_k)(z) = \frac{k!}{z^{k+1}}$ for $Re(z) > 0$.

For the *ramp* function, R_n

$$R_n(t) = \begin{cases} nt & \text{for } 0 \le t \le \frac{1}{n} \\ 1 & \text{for } t > \frac{1}{n} \end{cases} \quad \text{then} \quad \mathcal{L}(R_n)(z) = \frac{n}{z^2}(1 - e^{-z/n})$$

for $Re(z) > 0$.

❑

The result for e^{ikt} is evident by simple direct antidifferentiation; the next two results are obtained from the imaginary and real parts of the e^{ikt} result. The result for H, the Heaviside unit step, is by direct integration. The $\mathcal{L}(t^k)$ result is most easily seen by letting

$$\varphi(r) = \int_0^\infty e^{rt} dt = -\frac{1}{r} \quad \text{for} \quad Re(r) < 0. \qquad \textbf{\textit{14.31}}$$

Then, using Theorem 12.26 on page 477 (differentiation under the integral sign) we find

$$(D^k \varphi)(r) = \int_0^\infty t^k e^{rt} dt \quad \text{for } Re(r) < 0. \qquad \textbf{\textit{14.32}}$$

Therefore

$$(D^k \varphi)(r) = D^k\left(-\frac{1}{r}\right) = (-1)^{k+1} \frac{k!}{r^{k+1}}.$$

Looking at equation 14.32, you can see that we want

$$(D^k \varphi)(-z) = \frac{k!}{z^{k+1}}.$$

For the final result of Theorem 14.19, if $I(t) = t$ represents the identity function defined for t in the nonnegative reals, this transform is simply the Laplace transform of the function whose value at t is

$$n[I(t) - I(t - 1/n)],$$

which we recognize as the formula for the *ramp* function, R_n (the last function defined in Theorem 14.19) as you can see from Figure 14-4.

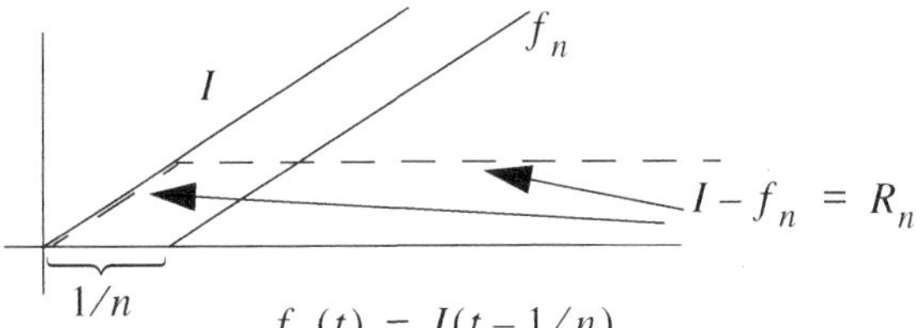

$f_n(t) = I(t - 1/n)$

I shifted by *1/n, value at t* is *I(t-1/n)*

Figure 14-4

Here are some illustrations, in which we let the formula represent the function whenever it's convenient.

Example 14.12: $(D^2 + 1)f = sin(2t),\ f(0) = f'(0) = 0$

Take Laplace transforms of both sides, and use the results of Theorem 14.16 on page 619, Theorem 14.19 just proved, Problem 14.11 on page 622, and the given initial conditions above, to see that

$$\mathcal{L}(Df)(z) = z\mathcal{L}(f)(z) - f(0) = z\mathcal{L}(f)(z)$$

$$\mathcal{L}(D^2 f)(z) = z\mathcal{L}(f')(z) - f'(0) = z\mathcal{L}(f')(z) = z^2\mathcal{L}(f)(z)$$

$$\mathcal{L}(sin(2t))(z) = \frac{2}{z^2 + 4}.$$

Hence

$$z^2\mathcal{L}(f)(z) + \mathcal{L}(f)(z) = \frac{2}{z^2 + 4},$$

or

$$\mathcal{L}(f)(z) = \frac{2}{(z^2 + 1)(z^2 + 4)}. \qquad \textit{14.33}$$

We can write the expression to the right of the *equal sign* as

$$\left(\frac{a}{z^2 + 1} + \frac{b}{z^2 + 4} = \frac{a(z^2 + 4) + b(z^2 + 1)}{(z^2 + 1)(z^2 + 4)}\right), \qquad \textit{14.34}$$

and since the numerators on the right side of equations 14.33 and 14.34 must be the same for all z with $Re(z) > 0$, it follows that

$$(a+b)z^2 + 4a + b = 2 .$$

From this, it follows from Theorem 7.31 on page 260 on the uniqueness of power series coefficients, that

$$a + b = 0 \quad \text{and} \quad 4a + b = 2 .$$

Hence $a = 2/3$ and $b = -2/3$. The desired solution therefore has Laplace transform

$$\mathcal{L}(f)(z) = \frac{2}{3}\frac{1}{z^2+1} - \frac{2}{3}\frac{1}{z^2+4} .$$

Using the easily verified fact that the Laplace transform represents a linear operator (Definition 13.9 on page 547) and the results of Theorem 14.19 above, we see that

$$f(t) = \frac{2}{3}sin(t) - \frac{1}{3}sin(2t) \quad \text{for} \quad t > 0.$$

■

We can formally apply Laplace transforms to solve $P(D)f = g$ where g has jump discontinuities, ***but we know that there can be no real valued function of a real variable which can satisfy this equation at the discontinuities of g*** — for then $D^n f$ would have to exist at all points, including these discontinuities (where n is the degree of the polynomial $P(D)$). Thus $D^k f$ would have to be differentiable, and hence continuous (see paragraph a on page 63) for all $k \le n - 1$. So the jump discontinuity in g would have to arise from a jump discontinuity in $D^n f$. In Figure 14-5 we illustrate what $D^n f$ and $D^{n-1} f$ would have to look like at such a discontinuity. From the figure it is clear that $D^{n-1} f$ could not have a derivative at the discontinuity, since the limit defining the derivative has to be unique (the unique value approached by the difference quotient, $(r(x+h) - r(x))/h$, when the nonzero value h is close enough to 0. Here $r = D^{n-1} f$). But here two different values are being approached — one from the left, and a different value from the right. So $D^n f$ cannot exist at the value where there is a jump discontinuity, and hence whatever we got from applying Laplace transforms cannot be a solution of the differential equation $P(D)f = g$ at this point.

This situation is similar to that in which we try to solve the matrix equation $\mathbf{Ax} = \mathbf{y}$, by least squares — i.e., choose $\mathbf{x}$ to minimize the distance $\|\mathbf{Ax} - \mathbf{y}\|$. If there is a solution to $\mathbf{Ax} = \mathbf{y}$ this gets it. If there is no solution to $\mathbf{Ax} = \mathbf{y}$, we nonetheless obtain something useful, namely, the *least squares projection* (see the discussion just preceding Definition 9.25 on page 320). Similarly here, while Laplace transforms cannot generate a solution when one doesn't exist, the function f we obtain is very close to what we want. It turns out to be the so-called *weak*, or *generalized solution* of the differential equation $P(D)f = g$, at least in the simple cases we will examine.

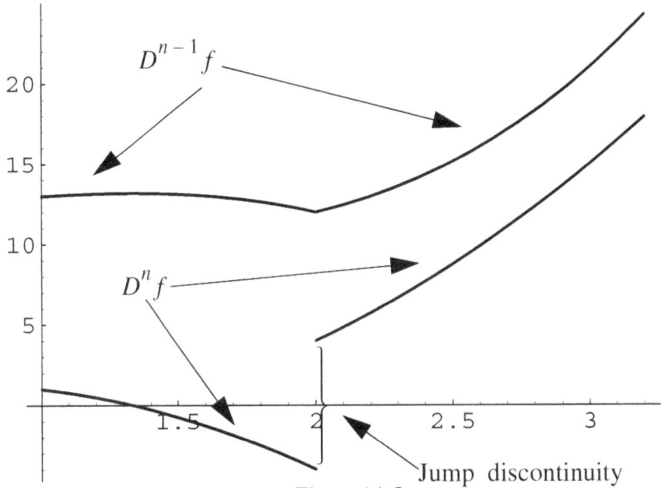

Figure 14-5

The **generalized solution** (which we will treat in much greater detail in Chapter 15) is the function $\lim_{j \to \infty} f_j$, where f_j is the solution of the differential equation

$$P(D)f_j = g_j, \quad f^{(k)}(0) = 0 \quad \text{for} \quad k = 0, 1, ..., n-1$$

and

$$\lim_{j \to \infty} g_j = g \quad \text{at all continuity points of } g.$$

Before we can proceed with our illustrative example (in which the Laplace transform fails to solve the differential equation $P(D)f = g$, but nonetheless provides a useful generalized solution) it will make life easier if we first establish some results about convolutions.

Definition 14.20: Convolution, $f*G$, of f and G

Let f and G be two real or complex valued piecewise continuous functions (Definition 3.13 on page 89) defined for all real t, with value 0 for $t < 0$. The **convolution**, $f*G$, is defined by

$$(f*G)(t) = \int_0^\infty f(\tau)G(t-\tau)d\tau = \int_0^t f(\tau)G(t-\tau)d\tau .$$

Note that the convolution exists at each t because the function Q_t, given by $Q_t(\tau) = f(\tau)G(t-\tau)$, is piecewise continuous for each t.

❑

As we'll see, convolutions arise naturally in the use of Laplace transforms to solve differential equations. The will also come up in the discussion of *impulse response* in Chapter 15.

Theorem 14.21: Laplace transform of convolution

If f and G are piecewise smooth functions (Definition 3.13 on page 89) whose values are 0 for negative arguments, both of exponential order (Definition 14.13 on page 614) then the Laplace transform (Definition 14.14 on page 616) of their convolution is given by

$$\mathcal{L}(f*G)(z) = \mathcal{L}(f)(z)\mathcal{L}(G)(z)$$

for $Re(z) > max(c_f, c_G)$ (where c_f and c_G are the growth constants of f and $G,$ respectively (Definition 14.13 on page 614) $Re(z)$ is the *real part of z* (Definition 14.18 on page 622) and $max(c_f, c_G)$ is the larger of the two numbers c_f, c_G).

❑

We establish this result in the following straightforward way:

First we see immediately that $f*G$ is of exponential order, because for all t, $f(t) \le k_f e^{c_f t}$ and for all $t - \tau$, $G(t-\tau) \le k_G e^{c_G(t-\tau)}$. Thus for $q = max(c_f, c_G)$

$$\left|\int_0^\infty f(t)G(\tau-t)dt\right| = \left|\int_0^\tau f(t)G(\tau-t)dt\right| \le \left|\int_0^\tau k_f e^{qt} k_G e^{q(t-\tau)}dt\right|$$

$$= k_f k_G e^{-q\tau}\int_0^\tau e^{2qt}dt = k_f k_G e^{-q\tau}\,\frac{e^{2q\tau}-1}{2q}$$

$$\le k_f k_G\,\frac{e^{q\tau}}{2q},$$

establishing the exponential order of $f*G$. Thus $f*G$ does have a Laplace transform.

Now examine

$$\int_0^T e^{-zt}\left[\int_0^\infty f(\tau)G(t-\tau)d\tau\right]dt = \int_0^T e^{-zt}\left[\int_0^T f(\tau)G(t-\tau)d\tau\right]dt$$

$$= \int_0^T\int_0^T e^{-zt} f(\tau)G(t-\tau)d\tau dt,$$

where we recognize that this last expression is equal to the double integral over the square determined by the points (0,0) and (T,T) (see Theorem 12.17 on page 455). Hence, applying this same theorem again, we are allowed to reverse the order of integration, obtaining

$$\int_0^T e^{-zt}\left[\int_0^\infty f(\tau)G(t-\tau)d\tau\right]dt = \int_0^T\left[\int_0^T e^{-zt}f(\tau)G(t-\tau)dt\right]d\tau,$$
$$= \int_0^T e^{-z\tau}f(\tau)\left[\int_0^T e^{-z(t-\tau)}G(t-\tau)dt\right]d\tau.$$

Now make the substitution $u = t-\tau$, $du = dt$ in the bracketed integral on the previous line (see equation 4.29 on page 123 and the surrounding discussion) to obtain that the last iterated integral is

$$\int_0^T e^{-z\tau}f(\tau)\left[\int_{-\tau}^{T-\tau} e^{-zu}G(u)du\right]d\tau = \int_0^T e^{-z\tau}f(\tau)\left[\int_0^{T-\tau} e^{-zu}G(u)du\right]d\tau.$$

But for the bracketed integral on the right, we may write

$$\int_0^{T-\tau} e^{-zu}G(u)du = \int_0^T e^{-zu}G(u)du - \int_{T-\tau}^T e^{-zu}G(u)du,$$

so that

$$\int_0^T e^{-zt}\left[\int_0^\infty f(\tau)G(t-\tau)d\tau\right]dt = \int_0^T e^{-z\tau}f(\tau)\left[\int_0^T e^{-zu}G(u)du\right]d\tau - \int_0^T e^{-z\tau}f(\tau)\left[\int_{T-\tau}^T e^{-zu}G(u)du\right]d\tau. \qquad \textbf{14.35}$$

To just about complete our argument, we show that for

$$Re(z) > q_0 > q = max(c_f, c_G)$$

the last term in equation 14.35 gets small in magnitude as T gets large; because

$$\left|\int_0^T e^{-z\tau}f(\tau)\left[\int_{T-\tau}^T e^{-zu}G(u)du\right]d\tau\right| \le \left|\int_0^T e^{-q_0\tau}k_f e^{q\tau}\left[\int_{T-\tau}^T e^{-q_0u}k_G e^{qu}du\right]d\tau\right|$$
$$= \int_0^T k_f k_G e^{-(q_0-q)\tau}\left[\int_{T-\tau}^T e^{-(q_0-q)u}du\right]d\tau$$
$$= \int_0^T k_f k_G e^{-(q_0-q)\tau}\frac{-1}{q_0-q}e^{-(q_0-q)u}\Bigg|_{T-\tau}^{u=T}d\tau$$

$$= \int_0^T k_f k_G e^{-(q_0 - q)\tau} \frac{1}{q_0 - q}[e^{-(q_0 - q)(T - \tau)} - e^{-(q_0 - q)T}]d\tau$$

$$\leq k_f k_G e^{-(q_0 - q)T} \int_0^T d\tau = k_f k_G e^{-(q_0 - q)T} T,$$

this latter known to get small for large T, since $q_0 - q > 0$ (using arguments similar to those in Example 14.9 on page 615).

Since the left side of equation 14.35 approaches $\mathcal{L}(f*G)(z)$ and the rightmost term in equation 14.35 approaches 0 as T becomes large, the first term on the right side of equation 14.35 must also approach $\mathcal{L}(f*G)(z)$. But this term clearly approaches $\mathcal{L}(f)(z)\mathcal{L}(G)(z)$ so long as $Re(z) > q_0 > q = max(c_f, c_G)$, establishing Theorem 14.21.

Notice that Theorem 14.21 proves Theorem 14.22.

Theorem 14.22: Associativity and commutativity of convolution

Convolution (Definition 14.20 on page 626) is associative and commutative for functions of exponential order. That is, if f, G, H are of exponential order (Definition 14.13 on page 614) then

$$\mathcal{L}(f*G) = \mathcal{L}(G*f) \quad \text{(commutativity)},$$

$$\mathcal{L}((f*G)*H) = \mathcal{L}(f*(G*H)) \quad \text{(associativity)}.$$

❑

We now illustrate what happens when there is no solution to the differential equation $P(D)f = g$, ***but the Laplace transform method gets a result anyway***.

Example 14.13: "*Solution*" of $(D^2+1)f=H(t)+H(t-1), f(0)=f'(0)=0$

Here H is the Heaviside unit step (Theorem 14.19 on page 622). Taking Laplace transforms of both sides (Theorem 14.15 on page 618) yields, with the aid of Theorem 14.16 on page 619, Theorem 14.19 on page 622, and Theorem 14.17 on page 621, that

$$(z^2 + 1)\mathcal{L}(f)(z) = \frac{1}{z} + \frac{e^{-z}}{z}$$

and thus

$$\mathcal{L}(f)(z) = (1 + e^{-z})\frac{1}{z(z^2 + 1)}. \qquad \textbf{\textit{14.36}}$$

We write the right-hand side of this equation in the form

$$(1+e^{-z})\left(\frac{a}{z}+\frac{b+cz}{z^2+1}\right) = (1+e^{-z})\left(\frac{a(z^2+1)+(b+cz)z}{z(z^2+1)}\right).$$

In order that this expression match that on the right of equation 14.36, we require

$$a(z^2+1)+(b+cz)z = 1 \quad \text{for all } z,$$

i.e.,

$$(a+c)z^2+bz+a = 1 \quad \text{for all } z.$$

With any knowledge of calculus (and more formally from Theorem 7.31 on page 260) we see that this requires

$$a+c=0 \quad b=0 \quad a=1$$

or

$$a=1, c=-1, b=0.$$

Thus

$$\mathcal{L}(f)(z) = (1+e^{-z})\left(\frac{1}{z}-\frac{z}{z^2+1}\right) = \frac{1}{z}-\frac{z}{z^2+1}+\frac{e^{-z}}{z}-\frac{ze^{-z}}{z^2+1}.$$

Thus from Theorem 14.19 on page 622 and Theorem 14.17 on page 621, we find that if there is a function f satisfying the differential equation of this example, then

$$\begin{aligned} f(t) &= H(t)-cos(t)H(t)+H(t-1)-cos(t-1)H(t-1) \\ &= H(t)(1-cos(t))+H(t-1)(1-cos(t-1)). \end{aligned}$$

Now the graph of $H(t)(1-cos(t))$ looks as shown in Figure 14-6 while that of $H(t-1)(1-cos(t-1))$ is this curve shifted one unit to the right. But $H(t-1)(1-cos(t-1))$, while it is differentiable at $t=1$, has no second derivative there (the derivative from the right being 1, and the derivative from the left being 0). So the function f that we have obtained is not a solution of the given differential equation. It is in fact the *generalized solution*, as we show in the example to follow this one.

”

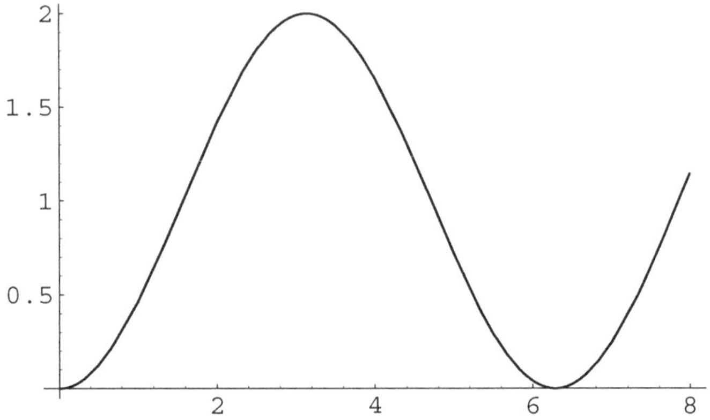

Figure 14-6

■

It will now be shown that the Laplace transform approach that we have just seen yields the *generalized* or *weak solution* of the given differential equation. In Problem 14.17 we will indicate how to extend the method used in the following example to convince us that this is the case in general for the solution of $P(D)f = g$ where g is of exponential order with a finite number of jump discontinuities on any finite interval.

Example 14.14: Modification of Example 14.13 to determine the meaning of its *solution*

We will use continuous approximations to the Heaviside unit step H, namely, the ramp functions, R_n (see Theorem 14.19 on page 622) to determine a generalized solution to the differential equation of Example 14.13. That is, we will solve the equation

$$(D^2+1)f_n(t) = R_n(t) + R_{n-1}(t),\ f_n(0) = f_n{}'(0) = 0.$$

As in Example 14.13 (page 629) we find

$$\begin{aligned} \mathcal{L}(f_n)(z) &= \frac{1}{z^2+1}\left[\frac{n}{z^2}(1-e^{-z/n}) + e^{-z}\ \frac{n}{z^2}(1-e^{-z/n})\right] \\ &= \frac{1}{(z^2+1)z}\left[(1+e^{-z})\frac{n}{z}(1-e^{-z/n})\right] \\ &= \left[\left(\frac{1}{z} - \frac{z}{z^2+1}\right)(1+e^{-z})\right]\frac{n}{z}(1-e^{-z/n}) \end{aligned}$$

The expression in the square brackets is the Laplace transform of the solution obtained in Example 14.13 on page 629. The remaining factor (from the Heaviside unit step results of Theorem 14.19 on page 622 and the result of Theorem 14.17 on page 621 on the Laplace transform of a shifted function) is the Laplace transform of

$$G_n(t) = n[H(t) - H(t-1/n)] = \begin{cases} 0 & \text{for } t \le 0 \\ n & \text{for } 0 < t \le 1/n \\ 0 & \text{for } t > 1/n; \end{cases}$$

see Figure 14-7.

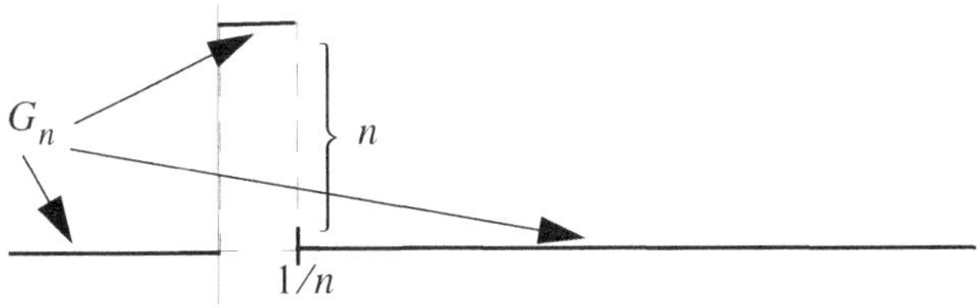

Figure 14-7

Using the results of Theorem 14.21 on page 627 on the Laplace transform of a convolution, we see that f_n is the convolution of G_n with the continuous function given by the formula

$$H(t)(1 - cos(t)) + H(t-1)(1 - cos(t-1)).$$

As we saw at the end of Example 14.13 on page 629, this function fails to have a second derivative at $t = 1$. But the convolution of G_n with this function must have a second derivative at $t = 1$, because $R_n(t) + R_n(t-1)$ is continuous, and, hence, the equation

$$(D^2 + 1)f_n(t) = R_n(t) + R_{n-1}(t), f_n(0) = f_n'(0) = 0$$

has a solution everywhere (as can be verified by the shift rule method discussed in the Remark on page 207) so that

$$f_n(t) = (G_n * f)(t) = \int_0^{\infty} G_n(\tau)f(t-\tau)d\tau = \int_0^{1/n} G_n(\tau)f(t-\tau)d\tau \cong f(t)$$

for n large enough, which shows f to be a *generalized* (or *weak*) *solution*, as claimed. (Note that we used $f*G = G*f$, Theorem 14.22 on page 629.) In Chapter 15 we'll show that all generalized solutions are the same, no matter what continuous sequence, g_n, is used to approximate a discontinuous driving force g, of the equation $P(D)f = g$.

■

Exercises 14.15

Use Laplace transforms to rapidly solve $P(D)f = g$, where $g(t) = 0$ for $t < 0$, subject to $D^j f(0) = 0, j = 0,...,k$, where k is the degree of $P(D)$.

1. $(D^2 + 2D + 1)f = e^t - 1$,
2. $(D^2 + 1)f = 1 - cos(2t)$,
3. $(D^3 + D^2 + D + 1)f = sin(t)$. Hint: $P(-1) = 0$.

Problems 14.16

Use Laplace transforms to get a formal solution to the following, where the right side, g, has jump discontinuities, and then show that this formal solution is really a generalized solution (i.e., for a chosen sequence, g_n, of continuous right-hand sides with a limit, the solution sequence, f_n, approaches the limit f).

1. $(D^2 + D + 1)f = \begin{cases} 0 & \text{for } t < 1 \\ cos(t) & \text{for } t \geq 1 \end{cases}$

   $f(1) = f'(1) = 0$

2. $(D^2 + D + 1)f = \begin{cases} 0 & \text{for } t < 1 \\ exp(t) & \text{for } t \geq 1 \end{cases}$

   $1) = f\xi\zeta cd'(1) =$

Let S be the following *square wave* defined for positive real t:

$$S(t) = \begin{cases} 1 & \text{for } 0 \leq t \leq 1 \\ 0 & \text{for } 1 < t < 2 \end{cases}$$

S periodic with period 2 (Definition 14.1 on page 591).

3. Find $\mathcal{L}(S)(z)$. Hint: Examine $S(t) + S(t-1)$.
4. Referring to the result of the previous problem, find the Laplace transform solution of $(D^2 + 1)f = S, f(0) = f'(0) = 0$, and show that it is a weak solution.
5. For g continuous, and 0 for $t \leq 0$, show that the solution of $(D^2 + 1)f = g$ subject to $f(0) = f'(0) = 0$ is the convolution of g with the sine function. This convolution formula is a special case of the impulse response form of the solution to this

differential equation, and will be investigated further in Chapter 15.

Problem 14.17

Come up with a reasonably convincing argument for the assertion that the Laplace transform provides a generalized solution for most interesting cases of exponential order driving forces whose discontinuities consist solely of a finite number of jump discontinuities in any finite length interval, with the discontinuity jumps growing at most exponentially with time, and the same holding true for the continuous portion of g (i.e., the function g_c that you get when removing the jump discontinuities).

Hints: First examine the differential equation $P(D)f = H(t-a)$ and its relation to the equations $P(D)f = R_n(t-a)$, where H and R_n are defined in Theorem 14.19 on page 622 (see Example 14.14 on page 631).

To do this write $P(D) = (D-\alpha)Q(D)$ and look first at the *Laplace transform solutions* of

$$(D-\alpha)F = H(t-a) \quad \text{and} \quad (D-\alpha)F_n = R_n(t-a),$$

showing that as n gets large F_n approaches F when their initial conditions are the same. Since F_n and F are both continuous, the *Laplace transform solutions* of $Q(D)f = F$ and $Q(D)f_n = F_n$ are the conventional solutions of these differential equations. From this, using say the shift rule and properties of integrals, show that as n gets large, f_n gets close to f. Finally, for the equation $P(D)f = g$, write $g = g_d + g_c$, where g_d consists of the jump discontinuities of g — i.e., g_d is constant except at the jump discontinuities of g, at which it jumps the same amount as g and $g_c = g - g_d$. Show then that the Laplace transform method applied to the equation $P(D)f_d = g_d$ gets the generalized solution of this equation over any finite interval by writing g_d as a finite sum

$$g_d = \sum a_k H(t-t_k).$$

Now patch things together to obtain the generalized solution to the driving force g, knowing that the Laplace transform method gives the legitimate solution to the equation $P(D)f_c = g_c$.

While the details have been glossed over, the essential idea is that the most important aspect consisted of determining that the *Laplace transform solution* to the equation $P(D)f(t) = H(t-a)$ is the generalized solution to this equation.

Example 14.18: *Deconvolution* and related problems

In the cases that we've examined so far it's assumed that we know the differential equation, $P(D)f = g$, describing the system, and the driving function g, and want to determine the solution f. Quite frequently we have a system in which the *input* g and the *output* f can be measured and we want to use these measurements to determine something about the system — for instance, if $P(D)$ is of the form $D^2 + a$, we might want to determine a. In more complex cases we might assume we know the degree, k, of $P(D)$, and want to determine its coefficients from these measurements. If we knew that $f(0) = f'(0) = \cdots = f^{(k)}(0) = 0$, and g was continuous and of exponential order (Definition 14.13 on page 614) then taking Laplace transforms of both sides of $P(D)f = g$, we find, making multiple use of Theorem 14.16 on page 619, that

$$\mathcal{L}(f)(z) = \frac{\mathcal{L}(g)(z)}{P(z)}, \qquad \textit{14.37}$$

from which, using Definition 14.20 on page 626 of convolution and Theorem 14.21 on page 627 on the Laplace transform of a convolution, it follows that

$$f(t) = (G*g)(t), \qquad \textit{14.38}$$

where G is the function whose Laplace transform is $1/P$.

From measurements on f and g, we might be able to approximate $\mathcal{L}(f)$ and $\mathcal{L}(g)$, and then, using equation 14.37, estimate P, yielding the type of information about the system that we might need. Since equations 14.37 and 14.38 are equivalent, it is evident why this process is referred to as *deconvolution*. Notice that we might plug the measured values of f and g into a set of linear equations approximating equation 14.38 to attempt to estimate G (from which we might be able to approximate P).

Alternatively, we might have knowledge of the solution f of $P(D)f = g$, and of the system (i.e., of $P(D)$) and want to determine what the driving force, g, is. This is also deconvolution, and equations 14.37 or 14.38 could be used in a similar fashion for this purpose.

■

4 Generating Functions and Extensions

The ***generating functions*** (or ***z-transforms***) that we are introducing in this section ***are appropriate coordinates for many problems involving*** **difference equations**.

The first difference equation encountered here was in Example 1.2 on page 6, describing a simple model of population growth. In Chapter 5, the approximation $x(t+h) \cong (1+h)x(t)$ to the differential equation $x' = x$ is a difference equation, as is equation 5.46 on page 155 approximating the trigonometric functions. And difference equations appeared throughout Chapter 6, Section 3 in applying Taylor's theorem to numerical solution of differential equations. Difference equations arise in counting problems fundamental to probability. Here are a few *combinatorial* (counting) examples.

Example 14.19: Number of *head-tail* sequences with k heads

In running a sequence of experiments (or trials) each of which can result in either Success or Failure,[1] we often want to know how many of these sequences have exactly k Successes in n trials.

For one trial $(n = 1)$ the answer is easy

There's one sequence with $k = 0$ Successes (F)
There's one sequence with $k = 1$ Success (S)

and no (0) such sequences with k Successes for any other k.

For $n = 2$ trials the answer is also easy:

one sequence with $k = 0$ Successes (F,F)
two sequences with $k = 1$ Success (F,S), (S,F)
one sequence with $k = 2$ Successes (S,S)

This process begins to get tedious for large values of n, but we can develop a difference equation for the quantity $N_{k,n}$, the number of S-F sequences of length n having exactly k Ss and $n - k$ Fs, as follows.

Any such sequence must end in either S or F — i.e.,

(, , ... ,F) or (, , ... ,S).

Those Success-Failure sequences of length n having exactly k Successes and ending in F must have exactly k Ss in the first $n - 1$ trials (sequence positions); there are $N_{k,n-1}$ such sequences — because we are essentially looking at the set of S-F sequences of length $n - 1$ having exactly k Ss.

Similarly, those Success-Failure sequences of length n having exactly k Successes and ending in S must have exactly $k - 1$ Ss in the first $n - 1$ trials. There are exactly $N_{k-1,n-1}$ such sequences. Hence we find

$$N_{k,n} = N_{k,n-1} + N_{k-1,n-1}. \qquad \textbf{\textit{14.39}}$$

Our examination of the case $n = 2$ showed that

---

1. Such experiments are often thought of as coin tosses.

$$N_{0,2} = 1, N_{1,2} = 2, N_{2,2} = 1, \quad \text{with } N_{k,2} = 0 \text{ otherwise}. \qquad \textbf{14.40}$$

Equations 14.39 and 14.40 allow determination of $N_{k,n}$ for all integers k,n. For example, putting $n = 3$, $k = 0,1,2,3$, respectively, yields

$$N_{0,3} = N_{-1,2} + N_{0,2} = 0 + 1 = 1$$
$$N_{1,3} = N_{0,2} + N_{1,2} = 1 + 2 = 3$$
$$N_{2,3} = N_{1,2} + N_{2,2} = 2 + 1 = 3$$
$$N_{3,3} = N_{2,2} + N_{3,2} = 1 + 0 = 1,$$

($N_{k,3} = 0$ for any other values of k.)

This process can be continued indefinitely for successive values of n.

We'll soon see that proper choice of a coordinate system will allow easy derivation of a formula for $N_{k,n}$.

■

All sorts of counting problems can be solved via difference equations. Here are a few problems to help you learn to set up the difference equations for combinatorial situations.

Problems 14.20

1. Derive a difference equation for the number of possible subsets of size k that can be produced from the elements of a set of size n (i.e. a set having n distinct elements).
   Hint: Let E denote some particular element of the set from which the size k subsets are to be formed. Every such subset either includes E or fails to include E.

2. How come the answer to problem 1 is the same difference equation that was obtained from Example 14.19?
   Hint: Put the collection of size k subsets into one-to-one correspondence with the set of sequences of n Ss and Fs having exactly k Ss by putting a 1 in each sequence position corresponding to a chosen individual. Choosing a subset of k sequence positions is, in fact, choosing a subset of size k from the set of n available sequence positions.

3. A die is tossed repeatedly. Derive a difference equation for the number of die-toss sequences with n tosses, whose **sum** is k.
   Hint: if the sum on n tosses is k, then the sum on the previous $n - 1$ tosses cannot be other than $k - 1, k - 2, ..., k - 6$.

4. This problem is a real challenge. How many sequences of 9 S-F trials have a run of at least three successive Ss? Note that if you made a list of all sequences of 9 trials of this kind, for each of these sequences you could easily determine whether or not it had such a run. The problem is that there are $2^9 = 512$ such sequences, which means that to list them you would have to write $9 \times 512 = 4584$ symbols.
   Hints: You may assume that the total number of S-F sequences corresponding to n trials is 2^n. Each sequence of the desired type either achieved such a run prior to the last trial or completed the first such run at the last trial.

Example 14.21: Queueing line

In the study of random processes we encounter many *differential-difference* equations. One such equation determines the probability that a waiting line will have n customers after the line has been open for t hours. The *difference* part of this equation comes from the integer number of customers. The *differential* part comes from the *time* variable.

The equation is set up by observing that if at time t there are n customers in line, then a very short time later there are only three significant possibilities:

$n - 1$ customers in line (1 customer lost)
n customers in line (nothing happened)
$n + 1$ customers in line (a new arrival to the line)

If $P(n, t)$ denotes the probability that n customers are in line at time t, then assuming a uniform random average rate of arrivals of a people per unit time, and a uniform random average rate of departures of d people per unit time, we might very well have over a period of very small duration h,

$$\begin{aligned} P(n, t+h) \cong\ & P(n, t)(1 - [a+d]h) \quad \text{(no arrival or departure)} \\ & + P(n+1, t)dh \quad \text{(one departure, no arrival)} \\ & + P(n-1, t)ah \quad \text{(one arrival, no departure)} \end{aligned} \qquad \textbf{\textit{14.41}}$$

for $n \geq 1$. Rewrite this as

$$\frac{P(n, t+h) - P(n, t)}{h} \cong -P(n, t)[a+d] + P(n+1, t)d + P(n-1, t)a$$

and, letting the nonzero value h approach 0, obtain

$$D_2 P(n, t) = -P(n, t)(a+d) + P(n+1, t)d + P(n-1, t)a \qquad \textbf{\textit{14.42}}$$

for $n \geq 1$ (D_2 is the partial derivative operator with respect to the second coordinate; see Definition 8.5 on page 267).

A distinct additional equation is needed for the case $n = 0$, namely,

$$P(0, t+h) \cong \underset{\text{no change}}{P(0, t)(1 - ah)} + \underset{\text{departure}}{P(1, t)dh} \qquad \textbf{14.43}$$

because there is no possibility of an arrival resulting in an empty line.

Proceeding as in the derivation of equation 14.42, we are led to

$$D_2 P(0, t) = -P(0, t)a + P(1, t)d. \qquad \textbf{14.44}$$

Differential-difference equations 14.42 and 14.44 may be solved numerically using approximations 14.41 and 14.43 repeatedly. Initial conditions specifying $P(n, 0)$ would have to be supplied for all n. Often, but not always, we assume

$$P(n, 0) = \begin{cases} 1 & \text{if} \quad n = 0 \\ 0 & \text{otherwise.} \end{cases}$$

These conditions might be suitable for certain situations, such as when a computer gets turned on, there being no jobs to run immediately. On the other hand, the size of the initial waiting line for Super Bowl tickets, or the size of the waiting line for airport landings might rarely be 0.

■

In order to develop some techniques for useful examination of difference equations, we introduce the following.

Definition 14.23: The shift operator, T

Let f be a given sequence (Definition 1.2 on page 3) where it is common to use the symbol f_n to stand for $f(n)$. So the sequence f is often written $(f_0, f_1, ...)$.

The **(right) shift operator**, T, is a function whose inputs and outputs (domain and range) are sequences, Tf, with value at n given by

$$Tf(n) = f(n+1) \qquad n = 0, 1, ... \,. \qquad \textbf{14.45}$$

i.e.,[1] Tf is the sequence $(f_1, f_2, ...)$.

If f is a sequence with more than one argument, T will be subscripted to indicate which argument is affected — e.g.,

1. This is the usual abuse of notation. Strictly speaking, equation 14.45 should be written $(T(f))(n) = f(n+1)$, but this complicates the reading too much. In most cases no confusion should arise.

$$T_2 f(m, n) = f(m, n+1).$$

$D_1 T_2 f(x, n)$ will involve first ***shifting*** the second argument of the function f, and then ***differentiating*** with respect to the first argument of the function $T_2 f$.

■

Example 14.22: Use of the shift operator, *T*

Difference equation 14.39 on page 636 can be written using shift operators as follows:

first realize that $N_{k,n}$ is the value of the function N at (k,n), and could be written $N(k, n)$. Realizing this, we see that difference equation 14.39 can be written

$$N = T_2 N + T_1 T_2 N$$

or

$$(I - T_2 - T_1 T_2) N = \mathbf{0},$$

where I is the identity operator, $IN = N$, and $\mathbf{0}$ is the sequence of all zeros.

Similarly, by rewriting equation 14.42 on page 638 as

$$D_2 P(n+1, t) = -P(n+1, t)(a+d) + P(n+2, t)d + P(n, t)a$$

for $n \geq 0, t \geq 0$, and just repeating equation 14.43 on page 639,

$$D_2 P(0, t) = -P(0, t)a + P(1, t)d,$$

waiting line behavior can be described using the shift and partial derivative operators via the equations

$$(D_2 T_1 + (a+d)T_1 - dT_1^2 - aI)P = \mathbf{0}$$

$$[(D_2 + aI - dT_1)P](0, t) = 0 \text{ for all } t \geq 0.$$

■

Our reason for introducing the shift operator is to analyze difference equations in the coordinate system consisting of eigenvectors of this operator (see Definition 13.2 on page 536). It's remarkably easy to determine these eigenvectors, since if

$$Tf = \lambda f, \quad \lambda \neq 0,$$

we must have

$$f(1) = \lambda f(0)$$

$$f(2) = \lambda f(1) = \lambda^2 f(0)$$

$$f(3) = \lambda f(2) = \lambda^3 f(0)$$

$$\vdots$$

and thus for nonzero (possibly complex) λ the *eigenvectors* of T consist of the sequences $(1, \lambda, \lambda^2, \lambda^3, ...)$. That is, if E is an eigenvector of the shift operator, T, then

$$E_k = E(k) = \lambda^k, \quad k = 0, 1, 2, ...$$

We state this result formally as Theorem 14.24.

Theorem 14.24: Eigenvalues and eigenvectors of the shift operator

Each nonzero complex number is an eigenvalue of the shift operator, T (Definition 14.23 on page 639), on sequences $(f_0, f_1, ...)$. The eigenvector corresponding to the eigenvalue λ is the sequence

$$\boldsymbol{E}_\lambda = (1, \lambda, \lambda^2, \lambda^3, ...), \text{ i.e., } \boldsymbol{E}_{\lambda, k} = \boldsymbol{E}_\lambda(k) = \lambda^k, \quad k = 0, 1, 2, ...$$

These eigenvectors are also eigenvectors of any polynomial $P(T)$ in T.

❑

The ***next issue*** is that of ***finding the eigenvector expansion for a given sequence whose elements do not grow too fast*** — i.e., finding out how to express such sequences as weighted sums of the eigenvectors. More precisely, given a sequence g, where $g(n) = g_n$, how can we write g as some sort of integral

$$g_n = g(n) = \int G(\lambda)\lambda^n d\lambda \qquad \textit{14.46}$$

or, using functional notation, to write

$$g = \int G(\lambda)\boldsymbol{E}_\lambda d\lambda, \qquad \textit{14.47}$$

where $\boldsymbol{E}_\lambda$ is the eigenvector $(\lambda^0, \lambda, \lambda^2, \lambda^3, ...)$

That is, how can we choose G to make equation 14.47 valid? The answer, from the residue theorem (Theorem 12.51 on page 529) is that if $G(\lambda)$ is chosen as a Laurent expansion (Theorem 12.50 on page 527) convergent on and beyond some circle, C centered at 0,

$$G(\lambda) = \sum_{m=0}^{\infty} g(m)\lambda^{-m-1} \qquad \textit{14.48}$$

the only term in the expression

$$G(\lambda)\lambda^n = \sum_{m=0}^{\infty} g(m)\lambda^{-m-1+n}$$

in equation 14.46 contributing to its right-hand side is the coefficient of λ^{-1}; this coefficient is $g(n)$. So we find from the residue theorem that if $G(\lambda)$ is chosen according to equation 14.48, then

$$g(n) = \frac{1}{2\pi i}\int_C \sum_{m=0}^{\infty} g(m)\lambda^{-m-1}\lambda^n d\lambda. \qquad \textit{14.49}$$

So, remembering that equation 14.46 (or 14.47) represents the eigenvector representation of the sequence g, and comparing with the result in equation 14.49, we conclude that the coordinate of the eigenvector $\boldsymbol{E}_\lambda = (1, \lambda, \lambda^2, \lambda^3, \ldots)$ in the eigenvector expansion for the fairly arbitrary sequence g is

$$\sum_{m=0}^{\infty} g(m)\lambda^{-m-1}, \qquad \textit{14.50}$$

provided that $g(m)$ doesn't grow so fast with m that this series diverges for too many λ. (This is like the exponential order assumption, Definition 14.13 on page 614, for Laplace transforms.)

The eigenvector expansion argument shows why we expect the expression in 14.50 above to furnish useful coordinates.

For greater simplicity, we use $\sum_{m=0}^{\infty} g(m)z^m$ instead of the expression in 14.50 as our coordinates, since this just multiplies the above series by $2\pi i\lambda$ and then substitutes $1/z$ for λ. This leads to the following definition.

Definition 14.25: Generating function (or z transform)

Given a sequence of real or complex numbers, $g = (g_0, g_1, \ldots)$, we define the **generating function** G_g (also called the **z-transform**) of g by

$$G_g(z) = \sum_{m=0}^{\infty} g_m z^m$$

provided that this sum exists (Definition 7.4 on page 214) for some nonzero value, z_0, of z.

❑

The following result, easily established using the comparison test, Theorem 7.13 on page 228, is needed in what follows.

Theorem 14.26: Circle of convergence

If the series $\sum_{m = 0 \to \infty} g_m z^m$ converges for some nonzero value, z_0, of z, then it converges for all z such that $|z| < |z_0|$, and converges uniformly (Definition 7.16 on page 234) for all z such that $|z| \le |z_1| < |z_0|$. From this it is not difficult to show that the sum of this series represents a regular function (Definition 12.39 on page 505) in this set. From the Cauchy integral formula (Theorem 12.45 on page 521) the values of this sum on the circle consisting of those z satisfying $|z| = |z_1|$ determine this sum for all z inside this circle.

❑

In most of the standard practical problems involving difference equations, we will develop some relatively simple formula for the generating function G_g, in which case it follows from the differentiability of G_g that

$$g_n = \frac{1}{n!}(D^n G_g)(0) = \frac{1}{n!} G_g^{(n)}(0). \qquad \textbf{14.51}$$

We illustrate this with the following example.

Example 14.23: Solution of $f(n+1) + f(n) - 2^n = 0, \; f(0) = 1$

We use this equation to determine

$$G_f(z) = \sum_{m=1}^{\infty} f(m) z^m$$

by multiplying the given equation by z^n and adding over all $n \ge 0$, obtaining

$$\sum_{m=0}^{\infty} f(n+1)z^n + \sum_{m=0}^{\infty} f(n)z^n - 2^n z^n = 0.$$

Since

$$\sum_{m=0}^{\infty} f(n+1)z^n = \frac{1}{z}\sum_{n=0}^{\infty} f(n+1)z^{n+1} = \frac{1}{z}(G_f(z) - f(0))$$

(note the similarity of this result to Theorem 14.16 on page 619 for Laplace transforms), and using the geometric series result (Example 7.11 on page 233)

$$\sum_{n=0}^{\infty} 2^n z^n = \frac{1}{1-2z}$$

for $|z| \le 1/2$, we find

$$\frac{1}{z}(G_f(z) - f(0)) + G_f(z) - \frac{1}{1-2z} = 0$$

or

$$G_f(z)\left(\frac{1}{z}+1\right) = \frac{f(0)}{z} + \frac{1}{1-2z}.$$

Hence

$$G_f(z) = \frac{z}{z+1}\left(\frac{f(0)}{z} + \frac{1}{1-2z}\right) = \frac{1}{z+1} + \frac{z}{(z+1)(1-2z)}$$

$$= \frac{1}{z+1} + \frac{a}{z+1} + \frac{b}{1-2z},$$

where

$$\frac{a}{z+1} + \frac{b}{1-2z} = \frac{a+b+(b-2a)z}{(z+1)(1-2z)}.$$

Since we must have $a+b+(b-2a)z = z$, it follows that $a = -b$ and $3b = 1$. Hence $b = 1/3$, $a = -1/3$. Thus

$$G_f(z) = \frac{2}{3}\frac{1}{z+1} + \frac{1}{3}\frac{1}{1-2z}.$$

In order to determine the sequence generating G_f, rather than using formula 14.51 on page 643, it's easier to expand each of the expressions on the right in a geometric series, Example 7.11 on page 233, getting

$$G_f(z) = \sum_{k=0}^{\infty}\left[\frac{2}{3}(-1)^k z^k + \frac{1}{3}2^k z^k\right],$$

so that

$$f_k = \frac{2}{3}(-1)^k + \frac{1}{3}2^k.$$

We see easily that $f_0 = 1$ and that

$$f_{k+1} + f_k = \frac{2}{3}(-1)^{k+1} + \frac{1}{3}2^{k+1} + \frac{2}{3}(-1)^k + \frac{1}{3}2^k = \frac{2}{3}2^k + \frac{1}{3}2^k = 2^k,$$

showing that we got the right solution.

■

Exercises 14.24

1. Solve $f_{n+2} + f_n = 3^n$, $f(0) = f(1) = 0$.
2. Roughly speaking, how large is the solution of the previous problem for large n?
3. Find the generating function for the number of dice toss sequences with sum k in n tosses.
4. Make up some of your own problems of this type.

In handling difference equations involving more than one variable (e.g., the Success/Failure example giving rise to equation 14.39 on page 636) we could determine the eigenvectors for T_1T_2, and for differential difference equations (e.g., the waiting time problem, Example 14.21 on page 638) we could determine the eigenvectors for D_1T_1. We will leave these as problems, and show instead how such equations are ***usually*** handled.

Problems 14.25

1. For *double sequences* $f_{n,\,m}$
   a. Verify that the sequences $f_{m,\,n} = \lambda^m\mu^n$ are eigenvectors of D_1T_1 for each nonzero λ, μ.
   b. Derive this result from basic principles.
   Hint: If $T_1f = \lambda f$, then for each n, $f_{m,\,n} = c_n\lambda^m$. But then $T_2(T_1f) = \mu c_{n+1}\lambda^m$. Then by induction on n,

   $$T_1f = c_0\mu^{n+1}\lambda^m$$

   and hence $$f_{m,\,n} = c_0\mu^n\lambda^m.$$

   Since c_0 is nonzero, we may take

   $$f_{m,\,n} = \mu^n\lambda^m.$$

2. Derive the result that the eigenvectors f for the product, D_1T_2, of the derivative with respect to the first coordinate and the shift with respect to the second has as its eigenvectors

   $$f(t, n) = e^{\lambda t}\mu^n, \quad \mu \neq 0.$$

3. Show that if G_g is the generating function for the sequence $g = (g_0, g_1, ...)$ (Definition 14.25 on page 642) then for T the shift operator (Definition 14.23 on page 639)

   $$G_{Tg}(z) = \begin{cases} \frac{1}{z}G_g(z) - g_0 & \text{for } z \neq 0 \\ g_1 & \text{for } z = 0. \end{cases}$$

   Note the similarity with the corresponding result for Laplace transforms, Theorem 14.16 on page 619.

4. Show that we may take as the generating function for double sequences $f_{m,\,n}$ (i.e., the coordinates in the eigenvector expansion of f)

   $$G_f(\lambda, \mu) = \sum_{m=0}^{\infty}\sum_{n=0}^{\infty} f_{m,\,n}\lambda^m\mu^n,$$

   which is essentially the m-generating function of the n-generating function.

5. Given a two-variable difference equation of the form

$$P(T_1, T_2)f = g$$

with appropriate initial conditions to guarantee a unique solution — such as given in Example 14.19 on page 636, where $N_{k,1}$ are assigned for all integers k, show how to obtain the generating function of f from that of g.

Hints: A typical term of $P(T_1, T_2)$ is $a_{j,k}T_1^jT_2^k$, where $a_{j,k}$ is some constant. Next note that if

$$P(T_1, T_2) = \sum_{j,k=0}^{n} a_{j,k}T_1^jT_2^k$$

then (see Problem 4 above)

$$G_{P(T_1,T_2)}(\lambda, \mu) = \sum_{j,k=0}^{n} G_{a_{j,k}T_1^jT_2^kf}(\lambda, \mu).$$

But

$$G_{a_{j,k}T_1^jT_2^kf}(\lambda, \mu) = \sum_{m=0}^{\infty}\sum_{n=0}^{\infty} f_{m+j,n+k}\lambda^{m+j}\mu^{n+k}$$

(again from Problem 4 above), which can be rewritten essentially as was suggested in Problem 3 above — i.e., in terms of $G_f(\lambda, \mu)$, λ, μ, and initial values of f.

This procedure is what is essentially done in the example to follow. As we'll see, this can be carried out in several steps, one for each of the summations in the double sum above.

We now continue on with the solution of the two-variable difference equation of Example 14.19 on page 636.

Example 14.26: Solution of $N_{k,n} = N_{k-1,n-1} + N_{k,n-1}$

Instead of determining the two-variable generating function (see Problem 4 above) we will first determine the generating function φ_k given by

$$\varphi_k(\lambda) = \sum_{n=1}^{\infty} N_{k,n}\lambda^b.$$

(This is half way toward determining the two-variable generating function $G_N(\lambda, \mu)$.) To do this, multiply both sides of

$$N_{k,n} = N_{k-1,n-1} + N_{k,n-1} \qquad \textbf{\textit{14.52}}$$

by λ^n, and sum from[1] $n = 2$ to ∞, obtaining

1. Quite soon we'll see why the lower limit is chosen as $n = 2$. If you prefer $n = 1$, rewrite equation 14.52 as $N_{k,n+1} = N_{k-1,n} + N_{k,n}$, which has the identical meaning.

$$\sum_{n=2}^{\infty} N_{k,n}\lambda^n = \sum_{n=2}^{\infty} N_{k-1,n-1}\lambda^n + \sum_{n=2}^{\infty} N_{k,n-1}\lambda^n. \qquad 14.53$$

Now recognize that

$$\sum_{n=2}^{\infty} N_{k-1,n-1}\lambda^n = \lambda \sum_{n=2}^{\infty} N_{k-1,n-1}\lambda^{n-1} = \lambda\varphi_{k-1}(\lambda),$$

$$\sum_{n=2}^{\infty} N_{k,n-1}\lambda^n = \lambda\varphi_k(\lambda),$$

$$\sum_{n=2}^{\infty} N_{k,n}\lambda^n = \varphi_k(\lambda) - N_{k,1}\lambda.$$

Hence, substituting into equation 14.53, we find

$$\varphi_k(\lambda) - N_{k,1}\lambda = \lambda(\varphi_{k-1}(\lambda) + \varphi_{k-1}(\lambda)),$$

or, solving for $\varphi_k(\lambda)$,

$$\varphi_k(\lambda) = \frac{\lambda}{1-\lambda}[\varphi_{k-1}(\lambda) + N_{k,1}], \quad k = 2, 3, \ldots \,. \qquad 14.54$$

We know

$$N_{k,1} = \begin{cases} 1 & \text{for} \quad k = 0 \\ 1 & \text{for} \quad k = 1 \\ 0 & \text{otherwise.} \end{cases} \qquad 14.55$$

Using equations 14.55, we may rewrite equations 14.54 as

$$\varphi_k(\lambda) = \frac{\lambda}{1-\lambda}\varphi_{k-1}(\lambda), \quad k = 2, 3, \ldots \,,$$

which are easily solved, since each time the subscript is reduced by one unit, a factor of $\lambda/(1-\lambda)$ is introduced. This leads to

$$\varphi_k(\lambda) = \left(\frac{\lambda}{1-\lambda}\right)^{k-1}\varphi_1(\lambda), \quad k = 2, 3, \ldots \,. \qquad 14.56$$

By definition

$$\varphi_1(\lambda) = \sum_{n=1}^{\infty} N_{1,n}\lambda^n.$$

Furthermore, $N_{1,n} = n$ for $n = 1,2,...$ (number of S-F sequences of length n with exactly one S — one sequence for each possible position of the single S). Hence

$$\varphi_1(\lambda) = \sum_{n=1}^{\infty} n\lambda^n = \lambda \sum_{n=1}^{\infty} n\lambda^{n-1} = \lambda \sum_{n=1}^{\infty} D(\lambda^n)$$

$$= \lambda D \sum_{n=1}^{\infty} \lambda^n \quad \text{for} \quad |\lambda| < 1,$$

where these last steps are justified by the known convergence of the geometric series for $|\lambda| < 1$, Example 7.11 on page 233, and the ability to interchange summation and differentiation, justified here by Theorem 7.20 on page 238. Again invoking the formula for the sum of the geometric series, we find therefore that

$$\varphi_1(\lambda) = \lambda D\left(\frac{1}{1-\lambda} - 1\right) = \lambda \frac{1}{(1-\lambda)^2} \quad \text{for} \quad |\lambda| < 1.$$

Substituting this into equation 14.56 yields

$$\varphi_k(\lambda) = \frac{\lambda^k}{(1-\lambda)^{k+1}} \quad \text{for} \quad k = 1, 2,.... \qquad \textbf{\textit{14.57}}$$

To see what power series this represents, using the allowable interchange of differentiation and summation, note that

$$D^k \frac{1}{1-\lambda} = \frac{k!}{(1-\lambda)^{k+1}}$$

(the proof being inductive — you can see the pattern by trying a few cases). Hence

$$\frac{\lambda^k}{(1-\lambda)^{k+1}} = \frac{\lambda^k}{k!} D^k \frac{1}{1-\lambda} = \frac{\lambda^k}{k!} D^k \sum_{j=1}^{\infty} \lambda^j = \frac{\lambda^k}{k!} \sum_{j=1}^{\infty} D^k \lambda^j$$

$$= \frac{\lambda^k}{k!} \sum_{j=k}^{\infty} j(j-1)\cdots(j-[k-1])\lambda^{j-k}.$$

Simplifying just a little bit more, introducing a different variable of summation, leads to

$$\varphi_k(\lambda) = \sum_{n=k}^{\infty} \frac{n(n-1)\cdots(n-[k-1])}{k!} \lambda^n.$$

But by definition, and the fact that $N_{k,n} = 0$ for $k < n$,

$$\varphi_k(\lambda) = \sum_{n=1}^{\infty} N_{k,n}\lambda^n = \sum_{n=k}^{\infty} N_{k,n}\lambda^n.$$

Comparing these last two expressions for $\varphi_k(\lambda)$ yields the final result that

$$N_{k,n} = \frac{n(n-1)\cdots(n-[k-1])}{k!}.$$

We did the harder version of this problem. The easier one is left as Problem 14.27.1.

■

Problems 14.27

1. Find $N_{k,n}$ of Example 14.26 on page 646 using the generating function
$$\psi_n(\mu) = \sum_{k=0}^{n} N_{k,n}\mu^k.$$

2. Find the coordinates equivalent to those in the eigenfunction expansion for the operator $D_1 T_2$, extending the results of Problem 14.25.2 on page 645.
Hint: You should be able to guess that it's a combination Laplace transform - generating function.

3. Using a combination of generating functions and Laplace transforms, try solving for $P(n, t)$, the probability of a waiting line of length n at time t (of Example 14.21 on page 638) if the waiting line is empty at time 0, as a function of the average arrival rate, a, and average departure rate, d.

4. For the important one-dimensional heat equation
$$D_1^2 u = D_2 u$$
where $u(x, t)$ is the temperature in an insulated rod at time t subject to a specified temperature distribution at time 0, and fixed end point temperatures, our previous results, Problems 14.25.1 on page 645 and Problems 14.25.2 on page 645 suggest that maybe some of the eigenfunctions for the operator $D_1^2 - D_2$ are functions f of the form
$$f(x, t) = r(x)s(t).$$

Investigate this question to see if this is so, and if so, what conditions the functions r and s must satisfy. (This approach to finding eigenfunctions is referred to as *separation of variables* — a method which is usually presented as a sort of magic trick with no prior motivation.)

Chapter 15

Generalized Functions

1 Introduction

In a sense this chapter represents an even more radical departure in its approach than the previous ones, for several reasons. First, it covers material that has usually been omitted from the mathematics most scientists are exposed to. Second, it illustrates the types of blind alleys that are usually encountered in trying to solve problems not well handled by known and well-understood methods. Also, some of the material is very advanced, with extremely involved technical details.

For these reasons we have some advice to keep you from getting lost in the technical forest, and to make it easier to get comfortable with the subject.

More than anything else, first understand the *motivation* for the definitions. Next, understand the *meaning* of the definitions and statements of theorems. The nature of the difficulties to be encountered is next in importance, followed by the general idea of how to overcome these difficulties. Last in order of importance, though necessary for technical competence, are the specific proofs. In brief: *more important than the specific details, are the meaning and the reasons for these details.*

The material which we consider vital is set in standard size font, with the most important landmarks emphasized in some fashion. As in previous chapters, detailed derivations involving a lot of algebraic detail are set in smaller type, set off between the following matching symbols: ⌈ and ⌊. These sections should probably be skipped or scanned lightly on first reading.

For variety, many of the theorems (those which we do not prove, due to the broad sweep of an introduction) are stated as exercises for those who want to master the more technical aspects of this kind of analysis.

We hope that the reader who omits only the proofs and verification of the exercises will nevertheless be able to imbibe the spirit and essence of one of the most exciting and useful developments in modern analysis.

The motivating force behind our development is the posing of a very simple problem, which seems to have a physical solution, but nevertheless has no convincing and intellectually satisfying mathematical solution within the previously developed framework. The attempt to develop a suitable description for this problem leads to an extension of the concept of a real valued function of a real variable which permits a revitalization and simplification of much of analysis. It presents us with new improved tools to handle both old and new problems.

2 Overview

Instead of introducing the subject of generalized functions as a full-blown, formal theory, we have chosen to let the subject develop as an outgrowth of our attempts to find a *generalized* or *weak* solution to a simple R-C (resistance-capacitance) electrical circuit problem. This problem, with the help of Kirchoff's laws describing electrical circuit behavior, leads to a first- order, linear, constant coefficient, ordinary differential equation, and it is for this equation that we seek a solution. We eventually define generalized functions in terms of so-called *regular* sequences of test functions — test functions being *infinitely differentiable functions* (functions which have an n-th derivative for each positive integer n) with *bounded support* (functions whose values are always 0 outside of some finite interval).

The actual development in this chapter proceeds as follows: in Section 3 the problem is posed, the differential equation is derived, and the appropriate differential operator, L, is defined. Because the forcing function is discontinuous, the previous *classical* theory does not apply. The notion of *solution* is extended and a generalized solution is obtained. So that experience may help to point the way to the most widely usable definition of a generalized function, we postpone this definition until Section 5, and in Section 4 we extend the notion of an impulse response function (see Theorem 11.15 on page 435 as well as Problem 14.16.5 on page 633) — this extension also known as a *Green's function.*[1]

We become more formal in Section 5 and define the notions of *support, test function, regular sequence*, the set $\mathcal{G}$, of *generalized functions*, and various operations on elements of $\mathcal{G}$.

In Section 6 we consider simple existence and uniqueness theorems for differential equations involving generalized functions. These theorems are intended to indicate how the theory might be developed, were it pursued in fuller detail.

The original R-C circuit problem is reconsidered in Section 7. At this point, the notion of generalized function is well defined, and the existence and

1. There is a distinction between the concepts of Green's functions and impulse response function, even though they turn out to be the same function. A *Green's function* is a function used to generate a particular solution of an equation in response to a driving force, while the *impulse response function* is the solution to a particular type of driving force.

uniqueness theorems of Section 6 can be brought to bear. A particular regular sequence, which defines a generalized solution, is presented, graphed, and discussed from a physical point of view.

In Section 8 we return to the impulse response function for L, and discuss it as the response to an impulse (Dirac's *delta*) δ, which is now a legitimate generalized function.

Section 9 alters notation a bit, moving from the precise notation we built up for pedagogical purposes to notation more like that the reader will probably encounter in the literature. Section 10 contains a very brief and heuristic discussion of generalized eigenfunction expansions, extending some of the concepts introduced in Chapter 14. In Section 11 we discuss the set $\mathcal{T}''$ of *continuous linear functionals on our space of test functions*, $\mathcal{T}$. It is pointed out that $\mathcal{T}''$ is essentially the same as $\mathcal{T}$, but is considered by those with experience in this area, a more appropriate vehicle for further theoretical study than $\mathcal{T}$.

We conclude this chapter with Section 12, which contains some remarks on directions for further generalizations.

3 A Circuit Problem and Its Differential Operator L

Consider the circuit in Figure 15-1 and imagine the five elements wired together as depicted, with the switch open, the capacitor uncharged, and the voltmeter of such a character that its effect on the remainder of the circuit is virtually negligible. Think next of an experiment which will begin, by

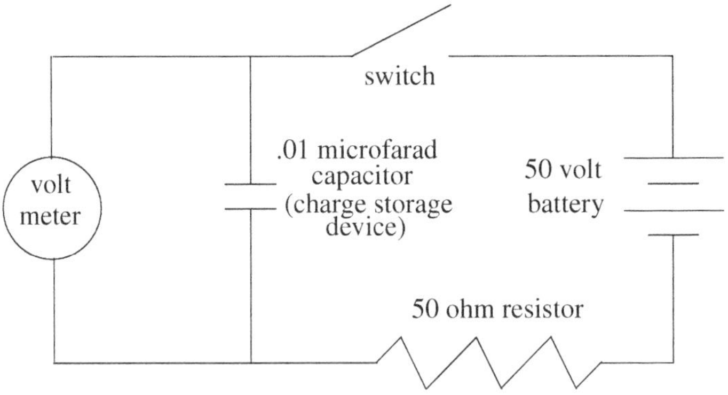

Figure 15-1

definition, with the start of a timer. The switch is to be closed precisely 1 microsecond after the timer starts, and the ***problem*** is ***to predict the voltmeter reading*** as a function of time.

Let t denote the time elapsed, in microseconds, from the start of the experiment, and let $Q(t)$ denote the *charge on the capacitor*, in micro-coulombs, at time t. To mirror the dormancy of the system for earlier times, we assume that $Q(t)$ is 0 for all $t \le 0$. From the definition of capacitance, the voltage drop (electrical pressure drop) across the capacitor at time t will be given by $Q(t)/0.01$, or $100Q(t)$. Since electrical current is rate of change of charge, the voltage drop across the resistor is $50Q'(t)$ (Ohm's empirical law; more properly, Ohm's law should be thought of as the definition of a resistor — namely, a *resistor* is any device whose voltage drop is proportional to the current through it). Now introduce the *switching function*, E, defined by

$$E(t) = \begin{cases} 0 & \text{if } t < 1 \\ 1 & \text{if } t \ge 1 \end{cases}$$

15.1

which is the *Heaviside unit step* at 1 (see Theorem 14.19 on page 622) illustrated in Figure 15-2. Then ***Kirchoff's empirical law***, which asserts that

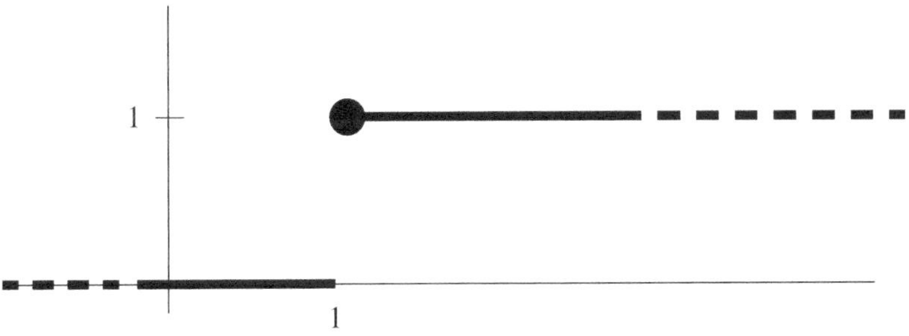

Figure 15-2

the total voltage drop around a closed loop is zero, yields

$$50Q'(t) + 100Q(t) = 50E(t)$$

or

$$Q'(t) + 2Q(t) = E(t). \qquad \textbf{\textit{15.2}}$$

Let us define the operator

$$L = D + 2, \qquad \textbf{\textit{15.3}}$$

where L is assumed to act on functions which have a derivative at all real values. Then equation 15.2 may be written

$$LQ = E, \qquad \textbf{15.4}$$

and our problem would be solved if we could invert L $(Q = L^{-1}E)$.

In order to carry on the discussion in a more economical manner, let us introduce the following notation.

Notation 15.1: C, C^n, C^∞

Let C denote the set of continuous real valued functions whose domain (Definition 1.1 on page 2) is the set $\mathcal{R}$, the set of real numbers (the *real line*). For each positive integer n let C^n denote the set of functions whose nth derivatives are members of C. (This, of course, implies the existence of the nth derivatives at all real values.) Let C^∞ denote the set of functions whose derivatives of all orders exist at all real values.

Note that $C^\infty \subseteq C^n \subseteq C$ (where the symbol $\subseteq$ means *is included in*, i.e., every element of C^∞ is an element of C^n, etc.).

For the problem under discussion, the domain of L will be restricted to the set of real valued functions X in C^1 with $X(0) = 0$. For X in the domain of L, DX is in C, $2X$ is in C, and thus LX is in C. This shows that the range of L, $rngL \subseteq C$. By application of the extended shift rule (Theorem 6.18 on page 202) it can be seen that for every function F in C, a solution to the differential equation

$$LX = F \qquad \textbf{15.5}$$

does exist. Consequently $rngL = C$. (From the uniqueness theorem, it follows that L is one-to-one (Definition 12.18 on page 460) and that $invL = L^{-1}$ (Definition 2.8 on page 37) exists. Actual application of the shift rule shows that a formula for $L^{-1}F$ is given by

$$X(t) = (L^{-1}F)(t) = e^{-2t}\int_0^t e^{2u}F(u)du \qquad \textbf{15.6}$$

for F in $rngL$, a fact we shall soon make use of.)

Now since E has a jump discontinuity at $t = 1$, there is no way that equation 15.4 has a solution function, X, using the current definition of derivative (Definition 6.16 on page 202; see the discussion on page 625 starting ***but we know ...***) and ***hence E is not in the range of L***. (We might be tempted to believe that $Q = E/2$ is *almost* a solution, since $E(0) = 0$, and $LE(t) = 2E(t)$ for all t except $t = 1$. But, because of the resistance in the circuit, we would expect it to take time for the voltage across the capacitor to build up; therefore, ***on physical grounds***, ***50E(t) seems an unreasonable***

prediction of the voltmeter reading, and "almost," in the above sense, is not very useful.)

If we were ignorant of the theory, we might be tempted to search for a solution in the following *reasonable* manner: Assume Q exists, and note that $Q(0) = 0$ and $LQ(t) = 0$ for $t < 1$. Since we are assuming that Q possess a derivative everywhere, Q must be continuous, and $Q(1) = 0$. We now solve the equation

$$LQ(t) = 1 \quad \text{for} \quad t \geq 1$$

subject to $Q(1) = 0$, and obtain the unique solution

$$Q(t) = 0.5[1 - e^{-2(t-1)}] \quad \text{for} \quad t \geq 1.$$

Thus, if a solution, Q, exists, if must be of the form

$$Q(t) = \begin{cases} 0 & \text{if} \quad t < 1 \\ 0.5[1 - e^{-2(t-1)}] & \text{if} \quad t \geq 1. \end{cases} \qquad \textbf{\textit{15.7}}$$

Now Q is continuous at $t = 1$, but for $t < 1$, $Q'(t) = 0$, while for $t \geq 1$, $Q'(t) = e^{-2(t-1)}$. Thus it is evident that Q fails to have a derivative at $t = 1$, and is not in the domain of L. This shouldn't be surprising, since we already know that equation 15.4 does not possess a conventional solution.

But although equation 15.4 does not possess a solution within the framework of our present mathematical theory, $100Q(t)$ **does almost perfectly, in fact, as a predictor of the voltmeter reading. Consequently, we may feel that the difficulty is in the mathematical description; so we should be able to enlarge the theory to encompass the above result.**

Since the discontinuity in E is the reason that a solution to equation 15.4 fails to exist, **maybe E is the wrong mathematical object to associate with the physical observation of the switch closing.** Besides, does a switch *really* close instantaneously? Perhaps it just closes very fast, in a manner which can be better described by a continuous function. That is, rather than thinking of the jump in E as a *break*, we might be better off thinking of this jump as representing a very steep continuous rise, so that E represents a *limiting* case of continuous functions, rather than a discontinuous one. To this end let us define E_N for $N = 1, 2,...$ by

$$E_N(t) = \begin{cases} 0 \text{ if } & t < 1, \\ (t-1)N \text{ if } & 1 \le t \le 1 + 1/N, \\ 1 \text{ if } & t > 1 + 1/N, \end{cases} \qquad \textbf{15.8}$$

as illustrated in Figure 15-3, for $1 \le N \le 6$.

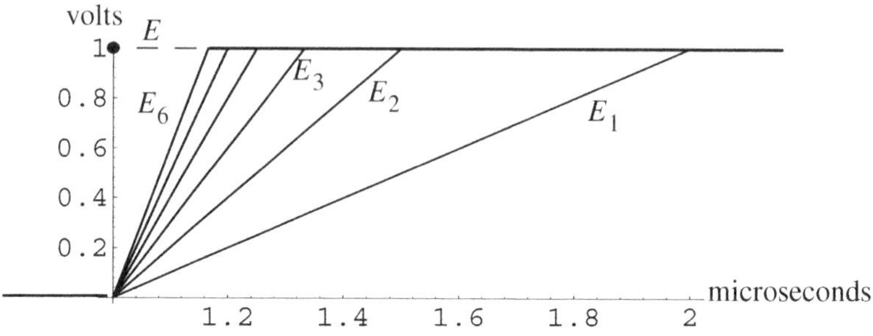

Figure 15-3

Now E_N is a function which is continuous and which, for very large N, changes from 0 to 1 quite rapidly. Maybe, for some very large integer N, the equation $LQ = E_N$ is a better description of the given circuit than $LQ = E$. (That is, maybe we should amend our definition of a switch.) Let us find the solution of $LQ = E_N$ for N an arbitrary integer. Probably the simplest way to do this is via Laplace transforms. Recall the following results, all of which appear in Section 3 of Chapter 14, starting on page 610.

If I_+ stands for the function satisfying

$$I_+(t) = \begin{cases} 0 & \text{for} \quad t \le 0, \\ t & \text{for} \quad t > 0, \end{cases} \qquad \textbf{15.9}$$

then its Laplace transform

$$\mathcal{L}(I_+)(s) = \frac{1}{s^2} \quad \text{for} \quad s > 0 \qquad \textbf{15.10}$$

(Theorem 14.19 on page 622). The Heaviside unit-step function, H, also defined in Theorem 14.19, has Laplace transform

$$\mathcal{L}(H)(s) = \frac{1}{s} \quad \text{for} \quad s > 0. \qquad \textbf{15.11}$$

Now let f be a function of exponential order (Definition 14.13 on page 614) hence having a Laplace transform, and which satisfies the condition $f(\tau) = 0$ for $\tau < 0$. Let $a > 0$ and

$$g(t) = f(t-a)$$

for all t. Then, from Theorem 14.17 on page 621, g has a Laplace transform and

$$\mathcal{L}(g)(s) = e^{-as}\mathcal{L}(f)(s) \qquad \textbf{15.12}$$

at all s at which f is guaranteed to have a Laplace transform by the exponential order condition (see Theorem 14.15 on page 618). Also for fixed b, if φ is the Laplace transform of f, then the function (whose value at z is)

$$\varphi(z-b) \qquad \textbf{15.13}$$

is the Laplace transform of the function (whose value at t is)

$$e^{-bt}f(t). \qquad \textbf{15.14}$$

Finally, from Theorem 14.16 on page 619

$$\mathcal{L}(f')(s) = s\mathcal{L}(f)(s) - f(0+) \qquad \textbf{15.15}$$

We know from application of the extended shift rule (see Theorem 6.18 on page 202) that the equation

$$(D+2)Q_N = E_N, \quad Q_N(0) = 0 \qquad \textbf{15.16}$$

has a unique solution. From the results of Problem 14.11 on page 622, this solution is of exponential order, and hence from Theorem 14.15 on page 618, this solution has a Laplace transform. It is thus legitimate to write

$$\mathcal{L}((D+2)Q_N)(s) = (s+2)\mathcal{L}(Q_N)(s) - Q_N(0). \qquad \textbf{15.17}$$

Using the definition of E_N, equation 15.8 on page 657 and equation 15.9 on page 657 (the definition of I_+) we find that

$$E_N(t) = N\{I_+(t-1) - I_+(t-[1+1/N])\}$$

(drawing a picture can be a quick convincer). Hence, using equations 15.10 and 15.12, we find

$$\mathcal{L}(E_N)(s) = N\left(e^{-s}\frac{1}{s^2} - e^{-s(1+1/N)}\frac{1}{s^2}\right)$$

or

$$\mathcal{L}(E_N)(s) = N\frac{1}{s^2}e^{-s}(1-e^{-s/N}). \qquad \textbf{15.18}$$

Now using the condition $Q_N(0) = 0$ from equations 15.16 above, and taking Laplace transforms of both sides of the left equation in 15.16, using equations 15.17 and 15.18, we find, after a minor amount of algebra, that

$$\mathcal{L}(Q_N)(s) = \frac{N}{s^2(s+2)}e^{-s}(1-e^{-s/N}).$$

By writing

$$\frac{1}{s^2(s+2)} = \frac{a+bs}{s^2} + \frac{c}{s+2} = \frac{(a+bs)(s+2)+cs^2}{s^2(s+2)}$$

$$= \frac{(b+c)s^2+(a+2b)s+2a}{s^2(s+2)}$$

and remembering that since we want

$$(b+c)s^2+(a+2b)s+2a = 1$$

for all sufficiently large s, we must have[1] $b+c = 0, a+2b = 0, 2a = 1$, we find $a = 0.5$, $b = -0.25$, $c = 0.25$. Hence

$$\mathcal{L}(Q_N)(s) = \left[\frac{0.5N}{s^2} - \frac{0.25N}{s} + \frac{0.25N}{s+2}\right]e^{-s}(1-e^{-s/N}). \qquad \textbf{15.19}$$

From equations 15.10 and 15.11, and the result leading to expression 15.14, the square-bracketed factor in equation 15.19 is the (value of the) Laplace transform of the function G_N given by

$$G_N(t) = 0.5NI_+(t) - 0.25N\{H(t) - e^{-2t}H(t)\} \qquad \textbf{15.20}$$

(whose value is $0.5Nt - 0.25N(1-e^{-2t})$ for $t \geq 0$, and 0 for $t < 0$).

From equation 15.12 we see that the factor e^{-s} shifts the function G_N one unit to the right; for simplicity, refer to this function as $G_N(t-1)$; the last factor subtracts from $G_N(t-1)$ the *function* $G_N(t-1-1/N)$.

---

1. This is the method of *partial fractions.* We need only verify the final result, $\frac{1}{s^2(s+2)} = \frac{0.5-0.25s}{s^2} + \frac{0.25}{s+2}$.

Thus

$$Q_N(t) = G_N(t-1) - G_N(t-1-1/N). \qquad \textbf{\textit{15.21}}$$

From this we conclude the following.

For $t < 1$

$$Q_N(t) = 0.$$

For $1 \le t \le 1 + 1/N$

$$Q_N(t) = 0.5N(t-1) - 0.25N(1-e^{-2(t-1)}) = N[0.25e^{-2(t-1)} - 0.25 + 0.5(t-1)].$$

For $t > 1 + 1/N$

$$Q_N(t) = N\left[0.25e^{-2(t-1)} - 0.25 + 0.5(t-1) - \left\{0.25e^{-2(t-1-1/N)} - 0.25 + 0.5\left(t-1-\frac{1}{N}\right)\right\}\right]$$

$$= N[0.25e^{-2(t-1)}\{1-e^{2/N}\} + 0.5/N].$$

Summarizing, we have

$$Q_N(t) = \begin{cases} 0 \quad \text{for} \quad t < 1 \\ N[0.25e^{-2(t-1)} - 0.25 + 0.5(t-1)] \ \text{for}\ 1 \le t \le 1 + 1/N \\ N[0.25e^{-2(t-1)}\{1-e^{2/N}\} + 0.5/N] \ \text{for}\ t > 1 + 1/N. \end{cases} \qquad \textbf{\textit{15.22}}$$

It is seen that Q_N satisfies equations 15.16 on page 658.

We graph Q and Q_N for $N =$ 1,2,3,6 in Figure 15-4.

From these graphs, it appears that the Q_N are approaching a limit as N grows large; in fact, it looks as if $N = 6$ is *reasonably large* in this case. We leave the remaining properties of Q_N to Exercises 15.1.

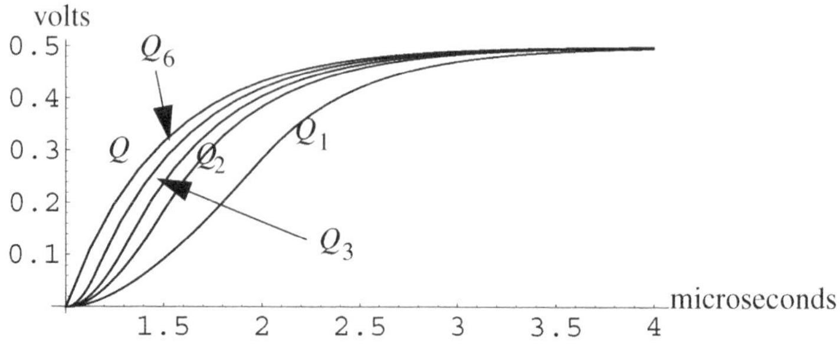

Figure 15-4

Exercises 15.1

1. Show directly that Q_N defined above in equation 15.22 is everywhere continuous.
2. Show that $\lim_{N \to \infty} Q_N(t) = Q(t)$, where $Q(t)$ is defined by equation 15.7 on page 656.
   Hint: Make appropriate use of Taylor's theorem (Theorem 6.2 on page 179).
3. Show **directly** that DQ_N (defined in equation15.22 on page 660) is everywhere continuous; also show that this result follows indirectly from the continuity of Q_N, E_N and $(D+2)Q_N = E_N$. We graph DQ_N in Figure 15-5, for $N = 1,2,3,6$ to give you a feel for these functions.

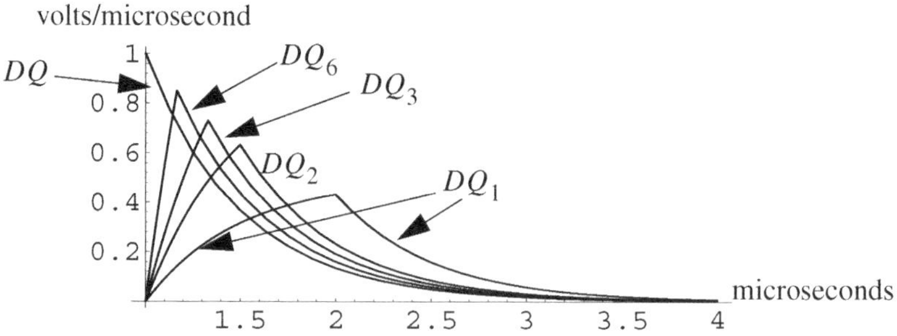

Figure 15-5

From the above graphs you should have little difficulty determining the behavior of D^2Q_N, which we nevertheless plot in Figure 15-6.

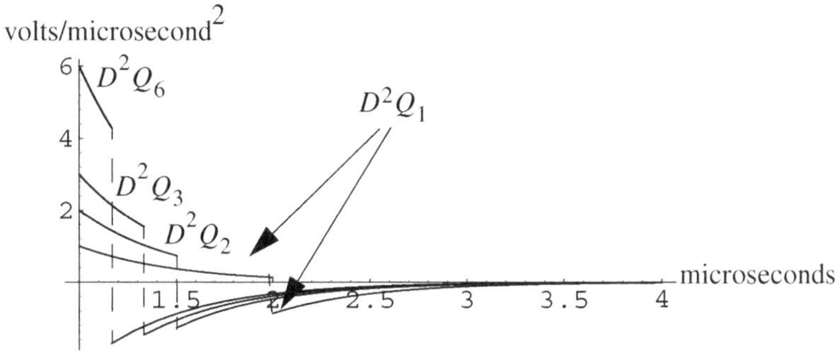

Figure 15-6

It's worth looking a little closer at the relation between the E_N and E. We see from their definitions, equations 15.8 on page 657 and 15.1 on page 654 that

$$\lim_{N \to \infty} E_N(t) = \begin{cases} 0 & \text{if } t < 0 \\ 1 & \text{if } t > 0. \end{cases}$$

15.23

Thus, $\lim_{N \to \infty} E_N(t) = E(t)$, except at $t = 1$ (at which the limit fails to exist). Now physically, if $\lim_{N \to \infty} E_N$ and E were voltages, we would have a difficult job measuring a difference between them, due to the *impossibility* of pinpointing the time $t = 1$. Generally speaking, the best we can hope to do in practice is to make measurements which are close to weighted averages; because any actual measuring instrument, at the very least, is influenced not just by a single point in space or time, but by values at *close-by* points. In fact, the concept of velocity *at some time point* (recall Example 4.1 on page 91) is defined by the position values at times close to this time point. Even though E_N does not approach E pointwise, it does approach E in *mean square* (mentioned in Footnote 1 on page 595). That is,

$$\lim_{N \to \infty} \int_{-\infty}^{\infty} [E_N(t) - E(t)]^2 dt = 0.$$

We stress, that although it may seem more natural to you, due to its familiarity, pointwise convergence is more of a *mathematical* notion, while mean square convergence is more *physical*. In this situation, however, Q_N approaches Q pointwise as well as in the mean square sense.

Definition 15.2: Mean square convergence

If f and f_j, $j = 1,2,...$ are real valued functions of a real variable (Definition 2.1 on page 21) for which the equation

$$\lim_{j \to \infty} \int_{-\infty}^{\infty} [f_j(t) - f(t)]^2 dt = 0$$

is both meaningful[1] and true, then we say that the sequence f_j, $j = 1,2,...$ **approaches (or converges to) the function f in mean square**. To indicate this type of convergence, we write

$$f_j \overset{ms}{\to} f \qquad \text{or} \qquad ms \lim_{j \to \infty} f_j = f.$$

❑

The reader may note that we can have both $f_j \overset{ms}{\to} f$ and $f_j \overset{ms}{\to} g$ with f different from g, e.g.,

> f_j and f being functions that are 0 for all values of t, and g being 0 at all values of t, except for $t = 0$, with $g(0) = 1$.

As already mentioned, from a physical viewpoint this shouldn't cause much concern.

More important than the technical details, is that we have learned the following. In looking for the *impossible*, i.e., for a solution of $LQ = E$, we have found

$$(LQ_N)(t) = E_N(t), \ E_N \overset{ms}{\to} E, \ Q_N \overset{ms}{\to} Q. \qquad \textbf{\textit{15.24}}$$

In light of this discussion, it's probably worthwhile to look again at the graphs of E, E_N and Q, Q_N (Figure 15-3 on page 657 and Figure 15-4 on page 661).

Exercises 15.2

For the following, recall the meaning of mean square convergence (Definition 15.2 above).

1. Verify that E_N approaches E in mean square, and Q_N approaches Q in mean square (for their definitions see equations 15.8 on page 657, 15.1 on page 654, 15.22 on page 660, and 15.7 on page 656).

---

1. The integral in this definition is an improper one, see Definition 12.11 on page 449; for *lim* see Definition 7.1 on page 213.

2. Provide an example of a sequence of functions which converges pointwise, but does not converge in the mean square sense.
Hint: Look at the sequence given by $f_N(t) = 1$ for $N \le t < N+1$, and $f_N(t) = 0$ otherwise.

3. Define

$$P_N(t) = \begin{cases} 0 \text{ if } t < 1-1/N \\ Ne^{-2/N}e^{-2(t-1)} + 2N(t-1) + 2 - N \text{ if } 1-1/N \le t \le 1 \\ N(e^{-2/N}-1)e^{-2(t-1)} + 2 \text{ if } t > 1. \end{cases}$$

Show that P_N is continuously differentiable (has a continuous derivative) for all t. Determine where P_N has a second derivative.

4. Show that P_N of the previous exercise, approaches Q (equation 15.7 on page 656) pointwise and in the mean square sense.

5. Define $R_N(t) = E_N(t + 1/N)$ and show that $LP_N(t) = R_N(t)$, where $L = D + 2$.

6. Show that R_N of the previous exercise, approaches E (equation 15.1 on page 654) in the mean square sense, and pointwise for all t. Compare this latter result with the result following equation 15.23 on page 662.

If we take the mean square difference[1] as a measure of *closeness*, or distance, between functions, we can make the following statement.

Given any $\varepsilon > 0$, *we can choose an integer, N, such that* LQ_N *is no further than* ε *from E.*

Now we are in a position to generalize the notion of the solution of $LQ = E$. We would like to say that Q is a solution, if there are functions $\mathbf{Q}$ in the domain of L, close to Q, for which $L\mathbf{Q}$ is close to E. But *how close?* **Arbitrarily close!** We need to use sequences to make this idea more precise. *We might say that* Q *is a solution of* $LQ = E$ *if there is a sequence of functions*[2] $\{Q_N\}\big|_{N=1}^{\infty}$ *such that each* Q_N *is in the domain of L, and*

$$Q_N \overset{ms}{\to} Q \quad \text{and} \quad LQ_N \overset{ms}{\to} E.$$

1. Actually, to match the Pythagorean theorem definition of distance (Definition A4.3 on page 774 we should use the quantity $\sqrt{\int_{-\infty}^{\infty} |f(t) - g(t)|^2 dt}$ as the distance between functions f and g.
2. The notation to follow is a common way of denoting sequences. Often we use the less precise notation Q_N, $N = 1, 2, \ldots$.

This seems to be one reasonable way, both mathematically and physically, to handle the problem we have posed. However, we should realize that by handling the problem in this fashion, we are defining a new operator — an *extension* of L, since this operator is now acting on a wider class of functions. We have enlarged the domain of L to include functions close to those in its former domain. The distinction between L and this, the first of its extensions, will turn out to be extremely important for a clear understanding of the theory we are developing. We therefore formalize these ideas in the following definition.

Definition 15.3: Extension L_1 of the operator L

Let L be the operator[1] $D+2$ whose domain consists of those functions f in C^1 (see Notation 15.1 on page 655) with $f(0) = 0$. Let L_1 be an operator whose domain consists of all real valued functions F whose domain is the set of real numbers, and for which, for some sequence of functions $\{F_N\}\big|_{N=1}^{\infty}$ with F_N in the domain of L

$$F_N \overset{ms}{\rightarrow} F,$$

i.e., the domain of L_1 consists of mean square limits of sequences of functions F_N in C^1.

The value L_1F is then defined[2] to be the **mean square limit of the sequence**

$$\{LF_N\}\big|_{N=1}^{\infty}$$

i.e.,

$$LF_N \overset{ms}{\rightarrow} L_1F$$

We think of L_1 as an ***extension*** of L by identifying any function F in the domain of L with the ***sequence*** $F,F,...$ (i.e., with $F_N = F$ for all N). This sequence is in the domain of L_1. Then if $LF = Y$, the equation

$$L_1X = Y$$

has as its solution, the sequence $F,F,...$ which is in the domain of L_1.

---

1. An operator is just a function whose domain and range consist of functions, see Definition 1.1 on page 2.
2. We won't spend much effort *here* on the significance of this definition. The real question concerns just how large and useful is the class of functions in the domain of L_1. Already in trying to solve $LQ = E$, we saw the use of L_1. We will follow up on its significance in greater detail in the next section.

Notice here that ***solutions are* sequences *of real valued functions;* not *a single real valued function of a real variable.*** The solutions obtained in earlier chapters to differential equations of the form $LX = Y$ appear to be embedded in this newer structure, which handles certain limiting (ideal) cases as well.

❑

As yet, however, we do not know *how far* L_1 can *take us*, or how *good*, i.e., generally useful it is. In fact, we do not yet have much depth of understanding of what we have done, nor have we explored any significant implications. ***If this theory is to be worthwhile, it must do more than just solve one simple circuit problem***. We therefore have to investigate the structure of L_1 a little further, in order to see how to proceed. For this it is very important to notice that F must have the property that for each positive integer N,

$$\int_{-\infty}^{\infty} [F_N(t) - F(t)]^2 dt$$

makes sense. Which mathematical objects are in the domain of L_1 depends critically on our definition of convergence, and the notion of convergence depends, in turn, on our definition of integral. Hence we must put some thought into these concepts in the important process of becoming aware of the overall structure of the theory as it is developed, and the role it plays in determining what it means to speak of a solution to, say,

$$L_1 X = Y, \qquad \textbf{\textit{15.25}}$$

where Y is some given mathematical object. In one context (or theory) equation 15.25 might have no (nonempty[1]) solution, in another context, a unique solution, and in still a third, a multiplicity of functions X satisfying equation 15.25. Again, observe that although $LX = E$ has no nonempty solution, we found that $L_1 X = E$ does have ***a*** (nonempty) solution; namely, equation 15.24 on page 663 shows that

$$L_1 Q = E, \qquad \textbf{\textit{15.26}}$$

1. We are being careful, because, according to Definition 11.3 on page 412, by the *general solution* we mean the set of all functions satisfying this equation. If this general solution is empty, it is common practice to say that the equation *has no solution*. If the general solution consists of more than one function, we often say that it has *multiple solutions*. If the general solution consists of precisely one function, the solution is said to be *unique*.

where Q is given by equation 15.7 on page 656. We see that the theory has been generalized in such a way that the function Q, which seemed fairly reasonable physically, and which ***was not*** a solution to $LX = E$, has been admitted to the rank of *solution*, in the extension from L to L_1. But the work has just begun! Within this new generalized theory one would also hope to derive, among other things, existence and uniqueness theorems. It turns out (and here our guide is our own experience and the judgement of others who have had the time to investigate) that the theory we have generated is but one step in the right direction. In the interest of economy, we turn to one approach which has proved of wide use.

4 Green's Function for L

In the previous section we focused our attention on a very specific problem in order to *get our feet wet* and to determine the nature of the difficulties likely to be encountered. In this section we set our sights on a higher goal and attempt to build a fairly general and useful theory which will solve a wide class of problems. The general problem of inverting linear differential operators (to solve differential equations) is the higher goal, and though it is sure that we can never solve this problem in complete generality, we can give an introduction to some methods which have yielded substantial progress.

We want to investigate the determination of an object x which satisfies (in some sense) a set of initial conditions and also (in some sense) the equation

$$\mathsf{L}x = y, \qquad \textbf{\textit{15.27}}$$

where y is a given object and L is some linear differential (or maybe an integral) operator, such as L_1 or L.

Let us continue to think in terms of the equation

$$L_1 x = y, \qquad \textbf{\textit{15.28}}$$

but with the more general goal in mind. The work which we have done on Green's functions (impulse response functions) for a system of linear algebraic equations in Section 6 of Chapter 11, starting on page 432, which allows us to write the solution of

$$\sum_{j=1}^{k+m} A(i,j)x(j) = y(i) \quad ,i = 1, 2,...,k+m \qquad \textbf{\textit{15.29}}$$

subject to

$$x(1) = x(2) = \cdots = x(k) = 0, y(1) = y(2) = \cdots = y(k) = 0 \qquad \textbf{\textit{15.30}}$$

in the form

$$x(j) = \sum_{i=1}^{k+m} G(j, i)y(i) \qquad \textbf{15.31}$$

is our point of departure.

The notation here is somewhat changed from that in Section 6 of Chapter 11, for its suggestive value. The vector $(x_1,...,x_{k+m})$ of that section has been rewritten in functional notation as $(x(1),...,x(k+m))$ and similarly for $(y_1,...,y_{k+m})$ and G.

Here

$$\boldsymbol{G}_i = \begin{pmatrix} G(1, i) \\ . \\ . \\ G(k+m, i) \end{pmatrix}$$

may be thought of as the *response* of a system described by A to the *impulse forcing function*

$$\boldsymbol{e}_i = \boldsymbol{y} = \begin{pmatrix} y(1) \\ . \\ . \\ y(k+m) \end{pmatrix}$$

which has a 1 in position i, $i = k+1,...,k+m$, and 0s in all other positions. That is, we interpret y in equation 15.29 as an object similar to a forcing function of a mechanical system. In attempting to solve equation 15.25, it is natural to try to generalize equation 15.31 to

$$x(t) = \int G(t, s)y(s)ds \qquad \textbf{15.32}$$

and ask for the corresponding *impulse forcing function* corresponding to the vector $\boldsymbol{y} = \boldsymbol{e}_i$. ***Here*** $y(s)$ would have to be 0 except at one point, $s = t$, at which it would concentrate all of its *energy* (hammer blow). But if y is an ordinary function, this leads us nowhere, because the integral would always be 0. So we have to back up a bit; we'll try a more careful *takeoff* from a finite number of dimensions by rewriting equation 15.32 as a limit of (upper) sums

$$x(t) = \lim_{N \to \infty} x_N(t), \qquad \textbf{15.33}$$

where

$$x_N(t) = \sum_{i=1}^{N} G(t, s_i)y(s_i)\Delta s. \qquad \textbf{15.34}$$

In this discrete approximation to the integral, the forcing *vector* is $(y(s_1)\Delta s, y(s_2)\Delta s,...,y(s_N)\Delta s)$. We assume that the integration interval in equation 15.32 is finite, and we can then think of $y(t)$ as being approximated by a step function (see Figure 15-7) where, for simplicity in our development,

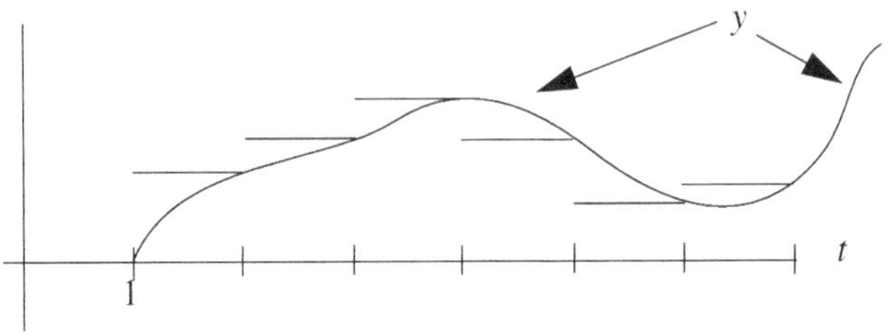

Figure 15-7

we assumed $y(s) = 0$ for $s < 1$. For fixed N, $G(t, s_i)$ is the response of the approximated system to $(0, 0,...,0, 1_i, 0,...,0)$, the 1 in the i-th coordinate corresponding geometrically to a rectangular pulse with unit area, centered at position s_i. The desired $G(t, s)$ (the one yielding equation 15.32, **not** the one from equations 15.29 through 15.31) ***is beginning to look as if it might be the following limit. Take a rectangle of unit area, centered about s, with base length 1/N. Associate with this rectangle a function $I_N(\ , s)$, whose value at t is $I_N(t, s)$, where***

$$I_N(t, s) = \begin{cases} N & \text{if } s - \frac{1}{2N} \le t \le s + \frac{1}{2N} \\ 0 & \text{otherwise.} \end{cases} \qquad \textbf{15.35}$$

Find the response, X_s, to the function $I_N(\ , s)$, i.e., the solution

$$X_s = G_N(\ , s) \qquad \textbf{15.36}$$

to the equation

$$L_1 X_s = I_N(\ , s) \qquad \textbf{15.37}$$

and then take the limit of $G_N(t, s)$ as $N \to \infty$ to obtain G(t, s).

We are now forced to wade through some tedious, but not especially difficult analysis in order to find the response $G_N(\ , s)$ to a *rectangle* (rectangular-looking function) $I_N(\ , s)$, of unit area, centered about s, with base length $1/N$ — i.e., to solve equation 15.37, where $I_N(\ , s)$ is given by equation 15.35 and we are assuming $s \ge 1$. We see that

$$I_N(t, s) = N\left[H\left(t - \left[s - \frac{1}{2N}\right]\right) - H\left(t - \left[s + \frac{1}{2N}\right]\right)\right], \qquad \textbf{15.38}$$

where H is the Heaviside unit step (see Theorem 14.19 on page 622) since $I_N(t, s)$ describes turning on a switch at time $s - 1/(2N)$ (the first term) and turning it off at time $s + 1/(2N)$ (the second term).

In order to find the solution of equation 15.37, first write it as

$$(L_1 X_s)(t) = I_N(t, s) = N\left[H\left(t - \left[s - \frac{1}{2N}\right]\right) - H\left(t - \left[s + \frac{1}{2N}\right]\right)\right]. \qquad \textbf{15.39}$$

Next notice that for all τ, $E(\tau) = H(\tau - 1)$. Then make use of the linear operator property[1] of L_1, to see that the solution we're seeking is just the difference $X_{s-} - X_{s+}$, where

$$(L_1 X_{s-})(t) = NE\left(t + 1 - \left[s - \frac{1}{2N}\right]\right), \quad (L_1 X_{s+})(t) = NE\left(t + 1 - \left[s + \frac{1}{2N}\right]\right).$$

But we showed that the solution of the equation $L_1 Q = E$ is given by equation 15.7 on page 656 (see 15.24 on page 663 and the meaning of L_1, Definition 15.3 on page 665).

From all of this it follows that

$$G_N(t, s) = \begin{cases} 0 \text{ if } t \le s - \dfrac{1}{2N} \\ \dfrac{N}{2}\left[1 - exp\left(-2\left\{t - s + \dfrac{1}{2N}\right\}\right)\right] \text{ if } s - \dfrac{1}{2N} < t \le s + \dfrac{1}{2N} \\ \dfrac{N}{2}[e^{1/(2N)} - e^{-1/(2N)}]exp(-2\{t - s\}) \text{ if } s + \dfrac{1}{2N} < t \end{cases} \qquad \textbf{15.40}$$

and

$$L_1 G_N(t, s) = I_N(t, s) \quad \text{for} \quad s \ge 1. \qquad \textbf{15.41}$$

Thus we have found the solution $X_s = G_N(\ , s)$ to equation 15.37. Recall that our motivation in finding G_N was the hope that we could write the solution of $L_1 x = y$ in the form

$$x(t) = \int_{\infty}^{\infty} G(t, s)y(s)ds. \qquad \textbf{15.42}$$

Our previous reasoning led us to believe that it might be the case that

$$G(t, s) = \lim_{N \to \infty} G_N(t, s).$$

---

1. Namely, $L_1(aX + bY) = aL_1(X) + bL_1(Y)$, see Definition 13.9 on page 547.

In fact, from equation 15.40 we find using Taylor's theorem (Theorem 6.2 on page 179) just as we did in finding $\lim_{N \to \infty} Q_N$ in Exercises 15.1.2 on page 661, that

$$\lim_{N \to \infty} G_N(t, s) = \begin{cases} 0 & \text{if } t < s \\ 0.5e^{-2(t-s)} & \text{if } t \geq s, \end{cases} \qquad \textbf{15.43}$$

for $s \geq 1$. For the moment, let us define

$$G^*(t, s) = \lim_{N \to \infty} G_N(t, s) \qquad \textbf{15.44}$$

(since we do not know that $\lim_{N \to \infty} G_N(t, s)$ will satisfy equation 15.42).

What we would like to know is: *to what equation (related to our problem) is $G^*(,s)$ a solution?*

From our development we might conjecture that

$$L_1 G^*(\ , s) = I^*(\ , s),$$

where

$$I^*(t, s) = \lim_{N \to \infty} I_N(t, s) \quad \text{or} \quad I^*(t, s) = ms \lim_{N \to \infty} I_N(t, s).$$

However, from equation 15.35 on page 669 we see that

$$\lim_{N \to \infty} I_N(t, s) = 0 \text{ for } t \neq s \text{ and } \lim_{N \to \infty} I_N(s, s) \text{ does not exist,}$$

so we can hardly use the notion of pointwise (ordinary) limit to define $I^*(\ , s)$. Intuitively it can also be seen that we aren't likely to have any success finding a function $I^*(\ , s)$ for which

$$I_N(\ , s) \xrightarrow{ms} I^*(\ , s),$$

since such a function would (with respect to integration) effectively have to be 0 on any interval not containing the point $t = s$. But the value at $t = s$ is then the only flexibility we have, and we know that the value of a function at any one point cannot affect its integral.

Thus it appears that we cannot meaningfully assert that $L_1 G^*(\ , s) = I^*(\ , s)$. But our previous searches suggest a solution to our dilemma. We arrived at L_1 by a *limiting* process, and we were led to investigate $L_1 G_N(\ , s) = I_N(\ , s)$ by limiting considerations. ***This suggests***

that we might be able to better utilize I_N and G_N by creating a further generalization, which extends L_1 to $\mathcal{L}_1$ and in which

$$\mathcal{L}_1 G = \mathcal{G},$$

where $\mathcal{G}$ is some sort of generalized objected related to the sequence $\{I_N\}\big|_{N=1}^{\infty}$.

That is, it looks as if we may yet have to *pile on* another limiting abstraction. This will make the going a bit tougher, but seems unavoidable if real progress is to be made.

To reach such a stage of generalization, we need to define the unfamiliar in terms of the familiar. We want somehow to describe $\mathcal{G}$ in terms of $\{I_N\}\big|_{N=1}^{\infty}$. But it appears that we cannot even say *ms* $I_N \rightarrow \mathcal{G}$ because there seems to be no function $\mathcal{G}$ for which

$$\int [I_N - \mathcal{G}]^2 dt \rightarrow 0 \text{ as } N \rightarrow \infty .$$

The above discussion should remind us of the problem we encounter when the number system in current use has only the rationals; the Pythagorean theorem (see the *proof* provided in the preface) indicates that there should be a square root of 2, but there is no rational number whose square is 2. To make available objects such as $\sqrt{2}$, we needed to extend the number system, using only the objects already available, to define this extension; in this case we require a *coordinate system* using *only the rational numbers*, to describe the extension we want. We end up describing this extension (the *real numbers*) using infinite decimals, which are just special types of sequences of pairs of rational numbers. Each element of such an infinite decimal may be thought of as a single coordinate for the real number being specified by the infinite decimal (see Appendix 1 on page 757).

Our development **here** is analogous to the development of the reals. But in this case, since we are describing a system which extends the set of real valued functions of a real variable, we expect the coordinates of the new objects to resemble coordinates that we have used to describe functions. Because of our experience with Fourier series, Fourier integrals, and Laplace transforms, we should not be surprised if the coordinates we end up using are also integrals[1] which are similar to those used in our previous transform development. This

---

1. Recall from Definition 14.4 on page 597 that the coordinates we used to describe periodic functions, *f*, were the Fourier coefficients, $a_m(g)$, which are integrals, given by $a_m(g) = \int_0^1 g(t)e^{-i2\pi mt}dt$.

makes it reasonable to introduce the notation to follow, whose purpose is to make the details of this development more readable.

Notation 15.4: $\int F, \int F(x)dx$

For any function F for which $\int_{-\infty}^{\infty} F(x)dx$ is defined, in some sense, let $\int F$ and $\int F(x)dx$ denote

$$\int_{-\infty}^{\infty} F(x)dx.$$

❑

There are many quantities which are determined by the sequence $\{I_N(\ ,s)\}\Big|_{N=1}^{\infty}$ which might be used as coordinates to characterize $\mathcal{A}(\ ,s)$. For example, one *coordinate* for $\mathcal{A}(\ ,s)$ might be $\lim_{N \to \infty} \int I_N(t,s)dt$. If we defined $\int \mathcal{A}(\ ,s) = \int \mathcal{A}(t,s)dt$ to be this limit (extending the integral operator) we might ask whether $\int \mathcal{A}(\ ,s)$ gives us enough information to completely characterize $\mathcal{A}(\ ,s)$. The answer to this question is almost certainly that *by itself,* $\int \mathcal{A}(\ ,s)$ *would not be sufficient to completely characterize the quantity which* $\mathcal{A}(\ ,s)$ *represents.* This is because $\mathcal{A}(\ ,s)$ physically represents an impulse (hammer blow) which takes place at time s, and clearly, therefore we want to distinguish $\mathcal{A}(\ ,1)$ from $\mathcal{A}(\ ,2)$. But from equation 15.35 on page 669 we see that

$$\int \mathcal{A}(\ ,s) = \lim_{N \to \infty} \int I_N(t,s)dt = 1$$

for all s, so that this *coordinate* by itself would **not** distinguish $\mathcal{A}(\ ,1)$ from $\mathcal{A}(\ ,2)$.

So here's the position we are in at this stage. We want to use sequences, say $\{J_N\}\Big|_{N=1}^{\infty}$, to define some generalized object $\mathcal{J}$ but we don't yet know how to do so, because, as we'll see, sometimes we want two different sequences to determine the same object,[1] $\mathcal{J}$, and sometimes we want them to represent different objects. We need some criteria to *help determine when two sequences* $\{J_N\}\Big|_{N=1}^{\infty}$ *and* $\{K_N\}\Big|_{N=1}^{\infty}$ *should represent the same object, and when they should represent different objects.* These criteria will be used to determine our crucially needed definitions of equality and inequality, just as in the definition of real numbers in the Appendix on real numbers, see page 758.

To help us in finding the necessary criteria, we look at any sequence I_N, representing an impulse, and note that we probably want

1. Just like 1.00000... and .999999... determine the same real number.

$$\{I_1(t,1), I_2(t,1), I_3(t,1),...\} \quad \text{and} \quad \{I_2(t,1), I_4(t,1), I_6(t,1),... \}$$

(although they are different sequences of functions) to determine or *represent* (be representations of ... be names of) the same generalized object.

On the other hand, we want

$$\{I_1(t,1), I_2(t,1), I_3(t,1),...\} \quad \text{and} \quad \{I_1(t,2), I_1(t,2), I_1(t,2),... \}$$

to be representations of different generalized objects, in spite of the fact that

$$\lim_{N \to \infty} \int I_N(t,1)dt = \lim_{N \to \infty} \int I_N(t,2)dt = 1, \qquad \textbf{15.45}$$

since, as we know, these sequences $\{I_N(\ ,1)\}\Big|_{N=1}^{\infty}$ and $\{I_N(\ ,2)\}\Big|_{N=1}^{\infty}$ stand for impulses happening at different times.

Now for the particular sequence $\{I_N\}\Big|_{N=1}^{\infty}$ given by equation 15.35 on page 669 there are functions f for which

$$\lim_{N \to \infty} \int I_N(t,1)f(t)dt \neq \lim_{N \to \infty} \int I_N(t,2)f(t)dt\,;$$

for example,

$$f(t) = \begin{cases} 1 & \text{for} \quad 0.5 \le t \le 1.5 \\ 0 & \text{otherwise.} \end{cases}$$

So maybe it is possible to distinguish generally between sequences $\{F_N\}\Big|_{N=1}^{\infty}$ and $\{G_N\}\Big|_{N=1}^{\infty}$ representing different generalized objects by finding a so-called *test function*, f, for which

$$\lim_{N \to \infty} \int F_N f \neq \lim_{N \to \infty} \int G_N f. \qquad \textbf{15.46}$$

Note that we would have to choose the test function f so that the integrals make sense.

Now we shall be forced to put certain restrictions on the sequences used to define generalized objects (for example, so that both sides of inequality 15.46 be defined) and within this framework, we shall also see that this restricts the class of test functions, f, which can be used (again so that both sides of inequality 15.46 would be meaningful).

As a matter of fact, inequality 15.46 comes pretty close to a good definition of inequality of generalized objects. To put this another way, we shall later call two sequences $\{F_N\}\Big|_{N=1}^{\infty}$ and $\{G_N\}\Big|_{N=1}^{\infty}$ names of the same generalized object, if for all f from some prescribed set of functions (called *test functions*)

$$\lim_{N \to \infty} \int F_N f = \lim_{N \to \infty} \int G_N f. \qquad 15.47$$

There must be sufficiently many test functions to discriminate between sequences which we would like to call different, and yet not so many as to distinguish between sequences which we feel represent the same object.

We can think of a test function f as yielding a coordinate, with $\lim_{N \to \infty} \int F_N f$ as the "*f coordinate value*" of the sequence $\{F_N\}\big|_{N=1}^{\infty}$, just as in Fourier series, the function given by e^{imt} yields the coordinate value (Fourier coefficient, see Definition 14.4 on page 597)

$$a_m(g) = \int_0^1 g(t) e^{-i2\pi mt} dt.$$

It would be rather satisfying, if from these coordinate values, $\lim_{N \to \infty} \int F_N f$, we could construct a sequence of exactly the same type as $\{F_N\}\big|_{N=1}^{\infty}$. In the case where the sequence $\{F_N\}\big|_{N=1}^{\infty}$ represents a *reasonable* real valued function of a real variable this can often be done. But just as with Fourier series, this may not be so easy (see the discussion on page 597 following Definition 14.4).

Test functions may be thought of as measuring instruments, in that, just like measuring instruments, they provide information about the object being measured. This is an especially reasonable interpretation, in light of the fact that we may think of many measuring instruments as calculating some weighted average of the *variable* being measured over some small region of space or time — just think of a thermometer, or a metal detector. With a *rich enough* class of measuring instruments, you should be able to get just about all of the information you need about the object being measured.

Choosing the class of test functions, f, is a task to which we could devote a great deal of thought. (A similar assertion holds for choosing the class of functions of the defining sequences for generalized objects.) Others have carried out such studies, and we will take advantage of the conclusions they reached.

There are several ways of choosing the classes just referred to, and any such choice determines which particular theory of generalized functions we end up with. Each of the several theories has its advantages and disadvantages. From the applied point of view, the thing to do is choose the theory that is most useful for the problem at hand. An analogy in physics would be the choice of whether to use Newtonian or relativistic mechanics; Newtonian mechanics *may* be simpler if it is applicable, but on the other hand, there are times when it is not, and relativistic mechanics is the only feasible tool.

In the following section we will concentrate on a single set of formal definitions which have already proved their merit. After developing the necessary machinery for solving our original problem, we shall return to the Green's function (impulse response) in Section 8.

5 Generalized Functions: Definition, Some Properties

To allow performance of the integration $\int_{-\infty}^{\infty} F_N(t) f(t) dt$ for almost all reasonable functions, F_N, it's reasonable to restrict test functions, f, to vanish (be 0) outside some bounded interval. This also simplifies integration by parts (Theorem 4.9 on page 121) so that for $H_N' = F_N$, the term $H_N(t) f'(t)\Big|_{-\infty}^{t=\infty}$ conveniently vanishes. We also know that for the stated purpose, test functions which are continuous can have quite sharp discriminating power. With some rather artificial looking analysis, it can be shown that it is possible to achieve the same discrimination available from the class of continuous test function using a much more manageable, and smaller class of functions, each of which vanishes outside of some bounded interval. The class of test functions that will be used here is the set of infinitely differentiable functions,[1] each of which vanishes outside of *some* interval determined by the function. This class of test functions has the advantage of being easy to operate with. The requirement that elements of the sequences defining generalized functions also must be infinitely differentiable will make it quite simple to define differentiation of generalized functions, and in addition yields the property that *all generalized functions are infinitely differentiable.* This property seems amazing, when you realize that all continuous functions may be thought of as generalized functions.

Our definitions will yield a mathematical system with certain *closure properties* — i.e., properties which help ensure that many problems posed within the system can be solved within the system; recall that the problem of solving $LX = E$ was unsolvable where E and X were arbitrary real valued functions of a real variable; the notion of *solution* had to be extended. We saw that the extended system we developed for this problem itself needed extending if we wanted to find a Green's (impulse response) function.

With this as an introduction, let us now proceed to introduce some theory. As might be expected, in the space we have available, it is, at best, only possible to *outline* a small portion of this subject, and to indicate the type of development that would be attempted if space were available.

---

1. Functions possessing an n-th derivative for all positive integers, n.

Definition 15.5: Point of accumulation

The point x is said to be a **point of accumulation of the set** A if there are point(s) in A arbitrarily close to x; i.e., if for each $\varepsilon > 0$ there is at least one point p in A, with $p \neq x$, whose distance[1] from x is less than ε.

❑

Definition 15.6: Closure of a set

The **closure of a set**, A, of real numbers is the *union*[2] of A and the set of points of accumulation of A.

❑

Definition 15.7: Support (or carrier) of a function

The **carrier** (or **support**) of a real valued function, f, of a real variable (Definition 2.1 on page 21) is the closure (Definition 15.6 above) of the set of real numbers, x, for which $f(x) \neq 0$.

❑

Definition 15.8: Bounded support

A real valued function of a real variable (Definition 2.1 on page 21) is said to have **bounded support** if its support is contained in a closed bounded interval (Definition 3.11 on page 88).

❑

Thus if a real valued function, f, of a real variable is of bounded support, there is a positive number, say B (dependent on f) such that $f(x) = 0$, if x is outside the interval $[-B;B]$ — i.e., if either $x > B$ or $x < -B$.

Definition 15.9: Set $\mathcal{T}$ of test functions

The set $\mathcal{T}$ of **test functions** is chosen to be the set of all real valued functions, f, of a real variable for which

- f is in the collection C^{∞} (Notation 15.1 on page 655)

and

- f has bounded support (Definition 15.8).

❑

---

1. To keep matters simple, we use as the distance between points x and y the quantity $\|x - y\|$ presented in Definition 8.6 on page 270..
2. The *union* of sets consists of the collection of those points in at least one of these sets.

We shall see that $\mathcal{T}$ is not vacuous in Theorem 15.20 on page 689.

While every function in C^{∞} has bounded support, there is no bounded interval which supports every member of $\mathcal{T}$.

Note that $\int f(t)dt$ exists for each test function f in $\mathcal{T}$. (You should verify this assertion.) *Thus $\mathcal{T}$ has very pleasant mathematical properties, while at the same time, we shall see that just about every real valued physical phenomenon of a real variable should be describable by one element, or a sequence of elements, of $\mathcal{T}$.*

Theorem 15.10: Test functions form vector space

$\mathcal{T}$ is a vector space over the real numbers (Definition 13.5 on page 543).

❑

We leave this routine proof as an informal exercise.

As mentioned already, analysis to come makes it appear than any continuous phenomenon can be approximated well by some element of $\mathcal{T}$; hence in the limit, it can be perfectly described by a sequence of elements of $\mathcal{T}$. In fact, just about every physically meaningful real valued function of a real variable should be describable by means of a sequence of elements of $\mathcal{T}$. Even discontinuous phenomena (such as E defined in equation 15.1 on page 654) appear to be *almost* mean square limits of sequences of elements of $\mathcal{T}$.

It's worth repeating that most physical measurements are usefully thought of as weighted averages. For example, the speed shown on an automobile speedometer is not instantaneous speed, but more likely is closer to being some sort of weighted average of speeds in a short time interval preceding the time considered. This inability to give a *true* reading comes about as follows: due to the finite speed of transmission of vibrations, as well as reflections of waves and multiple paths involved in transmitting motion, the system is unable to respond instantaneously, and responds to a pulse in a multitude of ways. Furthermore, due to inertia, the system does not stop responding instantly. Thus the system of speed measuring appears to have a memory which seems describable in terms of a weighted average, i.e., as a integral of the form $\int gT$, where g is the instantaneous speed and T characterizes the particular speedometer.

Another example would be furnished by an ammeter used to measure voltage, in which a switch permits current to flow for a short time. Assume the meter averages what it sees, but that the imperfect switch permits only a small amount of current to pass until *good contact* is made. The net result is a weighted average, which emphasizes voltage at times in the middle of the period that the switch is on.

So generally, all of our physically measured knowledge of a phenomenon represented by g in $\mathcal{T}$ may be thought of as being given by integrals

$$\int g(t)T_k(t)dt, \qquad \textbf{\textit{15.48}}$$

where T_k in $\mathcal{T}$, $k = 1,2,...,n$ represent our n measuring instruments.

For the voltmeter mentioned above, if g stands for the *true* voltage, the function T describing this measuring device may look as shown in Figure 15-8.

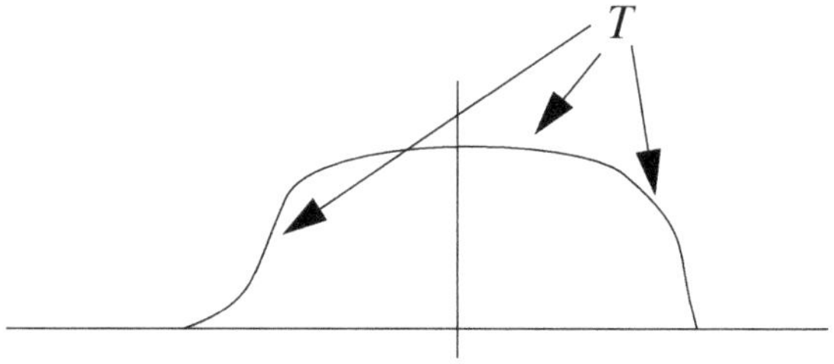

Figure 15-8

For mathematical simplicity, we assume that our knowledge of g in $\mathcal{T}$ is provided by the set of values

$$\int g(t)T(t)dt \qquad \textbf{\textit{15.49}}$$

generated by all T in $\mathcal{T}$.

Similarly, our knowledge of the objects which are described by sequences, $\{g_N\}\big|_{N=1}^{\infty}$ of elements of $\mathcal{T}$ is assumed given by

$$\lim_{N\to\infty} \int g_N(t)T(t)dt \qquad \textbf{\textit{15.50}}$$

generated[1] by all T in $\mathcal{T}$. These considerations motivate Definition 15.11.

---

1. Our justification for assuming that we do not require the entire sequence, $\{\int g_N(t)T(t)dt\}\big|_{N=1}^{\infty}$, but can make do with the limit $\lim_{N\to\infty} \int g_N(t)T(t)dt$, to get the information we want about the object described by the sequence $\{g_N\}\big|_{N=1}^{\infty}$, of course, comes from the results we will derive. However, it shouldn't be a surprise, because we already know (see the development of the real numbers in Appendix 1 on page 757) that for describing a real number by a sequence of intervals, each contained in the previous one, with lengths tending to 0, the same information is obtained from any arbitrary infinite subsequence of the given sequence of intervals (in this case, because such a subsequence squeezes down on the same point of the real line).

Definition 15.11: Regular sequences of test functions

A sequence of test functions $\{g_N\}\Big|_{N=1}^{\infty}$ (Definition 15.9 on page 677) is called a **regular sequence**, if for every test function T,

$$\lim_{N \to \infty} \int g_N(t)T(t)dt$$

exists.

❑

Note that the union of the supports of the g_N need not be a bounded set (Definition 3.10 on page 88).

In line with the previous discussion of equality, we give Definition 15.12.

Definition 15.12: Equivalent regular sequences

Two regular sequences, $\{f_N\}\Big|_{N=1}^{\infty}$ and $\{g_N\}\Big|_{N=1}^{\infty}$ are called **equivalent**, if for every test function T (Definition 15.9 on page 677)

$$\lim_{N \to \infty} \int f_N(t)T(t)dt = \lim_{N \to \infty} \left[\int g_N(t)T(t)dt \right].$$

The class of regular sequences equivalent to the sequence $\{f_N\}\Big|_{N=1}^{\infty}$ is called the **equivalence class determined by** $\{f_N\}\Big|_{N=1}^{\infty}$.

❑

Definition 15.13: Generalized function

The set $\mathcal{G}$ of generalized functions is the set of equivalence classes of regular sequences (see Definition 15.12 above). As we will see, ***a given generalized function can be represented by any one of the equivalent regular sequences defining it***.

If $\{f_N\}\Big|_{N=1}^{\infty}$ represents the generalized function F and $\{g_N\}\Big|_{N=1}^{\infty}$ represents the generalized function G, then $F = G$ if and only if, for every test function T, we have

$$\lim_{N \to \infty} \int f_N(t)T(t)dt = \lim_{N \to \infty} \int g_N(t)T(t)dt;$$

i.e., $F = G$ if and only if $\{f_N\}\Big|_{N=1}^{\infty}$ is in the equivalence class determined by $\{g_N\}\Big|_{N=1}^{\infty}$.

❑

Having said what a generalized function *is* (the verb *to be* can cause quite a bit of trouble), we should pause and recall the development of the notion of real numbers (Appendix 1 starting on page 757). For *practical purposes* of computation and measurement, we could make do with those rationals between -10^{95} and 10^{95} which carry 100 digits to the right of the decimal point (similar to the finite number of measuring instruments that might be all we need to measure some phenomena as indicated by the T_k in equation 15.48 on page 679). But for theoretical purposes, we need not only ***all of the rationals between*** -10^{95} and 10^{95}, but their limits, the reals (analogous to what is given in equation 15.50 on page 679). As the theory progresses, it is instructive to be recalling this analogy between *generalized rational numbers* (the reals, as summarized in the appendix referred to above) and generalized functions.

In order to be sure that Definition 15.12 is consistent, we should check that Definition 15.11 actually yields an **equivalence relation** in the set of regular sequences — in the sense that

- f is equivalent to itself
  (*reflexivity*)
- if f is equivalent to g then g is equivalent to f
  (*symmetry*)
- if f is equivalent to g and g is equivalent to r, then f is equivalent to r.
  (*transitivity*).

We see immediately that the set $\mathcal{T}$ (Definition 15.9 on page 677) of test functions can be thought of as a set of generalized functions with T in $\mathcal{T}$ associated with the regular sequence $T,T,T,...$.

Problem 15.3

Show that if S and T are in $\mathcal{T}$ with $S \neq T$, then S and T give rise to different generalized functions
Hint: If $S = kT$ with $k = 0$, then since T can't be 0, from its continuity it follows that $\int ST = 0 < \int TT$. If $S = kT$ with $k \neq 0, 1$, then

$$\int ST = \int kTT = k\int TT \neq \int TT.$$

Thus, whenever $S = kT$ with $k \neq 1$ it follows that

$$\int ST \neq \int TT.$$

If S is not a multiple of T, without loss of generality we may suppose that $\int S^2 \leq \int T^2$. By the Schwartz inequality (Theorem 8.21 on page 284) in this case we know that $\int ST < \sqrt{\int S^2}\sqrt{\int T^2}$. Combining these last inequalities, we find

$$\int ST < \int TT.$$

So in all of the cases

$$\int ST \neq \int TT.$$

Considering the second factor as the test function, we see that the generalized function $S,S,...$ is a different generalized function than $T,T,...$. So we have the following theorem.

Theorem 15.14: Distinct generalized functions

If S and T are distinct test functions (Definition 15.9 on page 677) their corresponding generalized functions $(S,S,...)$ $(T,T,...)$ are distinct (not equal — Definition 15.13 on page 680).

❑

Now that we know what we mean by a generalized function (Definition 15.13) let us define some operations on $\mathcal{G}$, ***the set of generalized functions***, which at least show that the algebraic operations already familiar to us on the elements of the ***set of test functions***, $\mathcal{T}$ (Definition 15.9 on page 677) are preserved among those generalized functions we associated with elements of $\mathcal{T}$. (Actually, operations other than algebraic ones will also be preserved.)

Definition 15.15: Operations on generalized functions

Given any two generalized functions (elements of $\mathcal{G}$) F and G, choose regular sequences (Definition 15.11 on page 680) $\{F_N\}\Big|_{N=1}^{\infty}$ and $\{G_N\}\Big|_{N=1}^{\infty}$, representing F and G, respectively.

We define

- **Addition**: $F + G$ is that element, H, of $\mathcal{G}$ represented by

$$\{F_N + G_N\}\Big|_{N=1}^{\infty}.$$

- **Multiplication by a real number**: For a real, aF is that element H, of $\mathcal{G}$, represented by $\{aF_N\}\Big|_{N=1}^{\infty}$.

- **Multiplication by a test function**: For T in $\mathcal{T}$ (Definition 15.9 on page 677) TF is that element H, of $\mathcal{G}$, represented by $\{TF_N\}\Big|_{N=1}^{\infty}$.

- **Subtraction**: $F - G = F + [(-1)G]$.

- **Differentiation**: With D representing the differentiation operator, DF is that element H, of $\mathcal{G}$, represented by $\{DF_N\}\Big|_{N=1}^{\infty}$ (so that every generalized function has a derivative). We sometimes use the symbol F'.

- **Integration**: For each test function T (see Definition 15.9 on page 677)

$$\int FT = \lim_{N \to \infty} \int F_N(u)T(u)du.$$

❑

You may note that *antidifferentiation* was ***not*** defined here. We shall see that all generalized functions have antiderivatives. However, we cannot do the *obvious*, i.e., use the sequence $\{\int_{-\infty}^{x} F_N(t)dt\}\Big|_{N=1}^{\infty}$ to represent the desired antiderivative, since this sequence does not have the required bounded support (see Definition 15.8 on page 677). In the proof of Theorem 15.41 on page 722 we will go through the *natural* patch to handle this problem. Examples of application of these definitions are best postponed until after Theorem 15.20 on page 689, where we show that the set $\mathcal{T}$ of test functions is not vacuous.

There are several points which must be checked in order to ensure that these definitions make sense.

First, when $\{F_N\}\Big|_{N=1}^{\infty}$ and $\{G_N\}\Big|_{N=1}^{\infty}$ are regular sequences (Definition 15.11 on page 680) what about $\{F_N + G_N\}\Big|_{N=1}^{\infty}$? It too is a regular sequence, because

$$\begin{aligned} &\lim_{N\to\infty} \int [F_N(t)+ G_N(t)]T(t)dt \\ &= \lim_{N\to\infty} \int F_N(t)T(t)dt + \lim_{N\to\infty} \int G_N(t)T(t)dt \end{aligned} \qquad \textbf{15.51}$$

and the limits on the right exist, because $\{F_N\}\Big|_{N=1}^{\infty}$ and $\{G_N\}\Big|_{N=1}^{\infty}$ are regular sequences.

Second, is $F + G$ uniquely defined? Yes, because the limits on the right of equation 15.51 are independent of the particular representatives $\{F_N\}\Big|_{N=1}^{\infty}$ and $\{G_N\}\Big|_{N=1}^{\infty}$.

Third, is $\{aF_N\}\Big|_{N=1}^{\infty}$ in $\mathcal{T}$? Yes, because for each T in $\mathcal{T}$ (Definition 15.9 on page 677),

$$\lim_{N \to \infty} \int aF_N T = a \lim_{N \to \infty} \int F_N T.$$

Fourth, for T in $\mathcal{T}$, is TF_N in $\mathcal{T}$? Again, yes, because for T_1 in $\mathcal{T}$ $\lim_{N \to \infty} \int TF_N T_1$ exists, because the product TT_1 is seen to belong to $\mathcal{T}$.

Fifth, we need to show that $\{DF_N\}\big|_{N=1}^{\infty}$ is a regular sequence. To do this we use integration by parts (Theorem 4.9 on page 121) and utilize the fact that test functions (Definition 15.9 on page 677) have bounded support. Since each DF_N is in $\mathcal{T}$ (because each F_N is in $\mathcal{T}$) $\{DF_N\}\big|_{N=1}^{\infty}$ is a sequence of test functions; for each T in $\mathcal{T}$

$$\lim_{N \to \infty} \int (DF_N)(t)T(t)dt = -\lim_{N \to \infty} \int F_N(t)(DT)(t)dt \qquad \textbf{15.52}$$

(the constant $F_N(t)T(t)\big|_{-\infty}^{t=\infty}$ disappearing due to the bounded support), and since $\{F_N\}\big|_{N=1}^{\infty}$ is a regular sequence, and DT is in $\mathcal{T}$, the limit on the right exists, establishing that $\{DF_N\}\big|_{N=1}^{\infty}$ is a regular sequence.

Finally, is DF uniquely defined? Yes, it is, because the limit on the right side of equation 15.52 is independent of the choice of the representative, $\{F_N\}\big|_{N=1}^{\infty}$ of F.

Exercises 15.4

1. Show that the operator D defined in Definition 15.15 on page 682 is a linear operator.
2. Show that the integral operator, $\int$, defined in Definition 15.15 on page 682 is a linear operator.

We remark that with respect to the operations so far defined in the set of generalized function, $\mathcal{G}$, in Definition 15.15, whether we work with $\mathcal{T}$ or with the subset of $\mathcal{G}$ which corresponds to $\mathcal{T}$ (sequences $(T,T,T,...)$ with T in $\mathcal{T}$) does not matter, since sum, product, and derivative were defined in $\mathcal{G}$ in a way corresponding to that in $\mathcal{T}$, i.e., the derivative DT corresponds to the *generalized derivative* $(DT,DT,...)$.

This is again similar to the embedding of the rationals in the reals. We shall see that we get most of the previous theory preserved in the one we are developing.

One of the reasons for introducing the theory of generalized functions is so that we can examine the equation $LX = Y$ in a much more general way. For example, if Y is an arbitrary generalized function, *does there exist a generalized function* X (and is it unique? — recall our definition of sameness, Definition 15.12 on page 680), *such that* $LX = Y$? As we'll see, the answer does turn out to be *yes*.

But before we can make use of this theory, even on our motivating problem (solving $(D+2)Q = E$, equation 15.4 on page 655) it will be convenient to *embed* the class of ordinary functions in the larger class of generalized functions. (Informally, we already embedded the class $\mathcal{D}$ of test functions in $\mathcal{G}$, but the test functions have bounded support, and the ordinary functions we are referring to will not generally have this property.) What we want is a mapping (function) say, $\mathcal{M}$, that associates a unique generalized function F with each *ordinary function* f; i.e., $\mathcal{M}f = F$. (Note that in the interest of simpler notation, we are writing $\mathcal{M}f$ instead of the more proper $\mathcal{M}(f)$.) ***We should say more carefully exactly what we mean by an ordinary function.*** We are not really interested in all real valued functions of a real variable. It's just too big a set. We want to choose a set that's big enough to include just about all of those that would interest us, but not so big that the definition of the mapping $\mathcal{M}$ is unduly complicated. For this compromise, we will choose the class of real valued functions whose domain (Definition 1.1 on page 2) is the set of real numbers, and which are *piecewise continuous* (Definition 3.13 on page 89) *on each closed bounded interval.*

Notation 15.16: Piecewise continuous functions, $\mathcal{PC}$

We denote by $\mathcal{PC}$ the class of functions which are piecewise continuous (in the sense of Definition 3.13 on page 89) on each closed bounded interval (Definition 3.11 on page 88). It is shown in Theorem A3.4 on page 769, that each function f in $\mathcal{PC}$ has the property any bounded interval can be expressed as a partition (Definition 4.1 on page 101) consisting of a finite number of intervals, on each of which the restriction (Definition 13.22 on page 583) of f to this interval is *uniformly continuous* (Definition A3.3 on page 769).

❑

Recall that the notion of interval includes the degenerate case, where an interval consists of a single point; any function restricted to such an interval is *trivially* uniformly continuous; because establishing a lack of uniform continuity requires that the function in question be defined at an infinite number of points (see the proof of Theorem A3.4 which starts on page 769). So if a function is defined at only one point, it is impossible to prove that it fails to be uniformly continuous. But in the logic we are using, if you show the impossibility of the denial of some result, then this result holds — even (and especially) if there are no situations to which it applies.

Example 15.5: Illustrating piecewise continuity or its absence

- $$f(x) = \begin{cases} x^n & \text{for } x \neq 0 \\ 0 & \text{for } x = 0 \end{cases}$$

  is not piecewise continuous for any $n < 0$.

- $$f(x) = \begin{cases} 0 & \text{for } x < 0 \\ x+1 & \text{for } x \geq 0 \end{cases}$$

  is piecewise continuous.

- $$f(x) = \begin{cases} 1 & \text{if } x \text{ is not an integer} \\ e^x & \text{if } x \text{ is an integer} \end{cases}$$

  is piecewise continuous.

■

For the purposes of the previous chapters it was sufficient for us to show that monotone functions and continuous functions were Riemann integrable. It now becomes vital to our theory to show that piecewise continuous functions are integrable over any bounded interval. This should not surprise us, since step functions, such as illustrated below in Figure 15-9, are nonetheless known to be integrable.

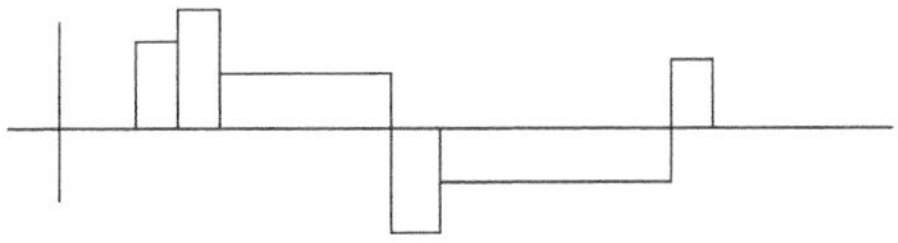

Figure 15-9

Theorem 15.17: Integrability of functions in $\mathcal{PC}$

If the real valued function, f, of a real variable is in $\mathcal{PC}$ (Notation 15.16 on page 685) then $|f|$ is in $\mathcal{PC}$ and f and $|f|$ are Riemann integrable (Definition 4.2 on page 102) on any bounded interval (Definition 3.11 on page 88). Integrability of $|f|$ is referred to as the **absolute integrability of f**.

❑

Problem 15.6

Prove Theorem 15.17. For this it may be worthwhile to first review the relevant parts of Chapter 4 and Appendix 3. Then show that the contribution to the integral *near* a jump discontinuity is *small.*

There are several problems which must be handled to embed $\mathcal{PC}$ in $\mathcal{G}$. They are handled by a two-stage procedure called *regularization,* which does the following. For each *ordinary function* f, we determine a sequence of *very nice smooth functions* approaching f in some reasonable sense. The actual process of regularization is somewhat tricky and delicate (as you'll see if you go through the proof of Theorem 15.22 on page 691). The powerful, yet simple results which follow make it worthwhile (very much like the payoff from the heavy investment that can be made in computer programming).

Definition 15.18: Regularization of elements of $\mathcal{PC}$

Given any element of $\mathcal{PC}$ (Notation 15.16 on page 685) **regularization consists of** finding a sequence of C^{∞} functions (Notation 15.1 on page 655) which

- *approximates* this element (smoothing out the *bumps* and everything else too);
- converts **this** C^{∞} sequence to a C^{∞} sequence, each of whose elements has bounded support (Definition 15.7 on page 677).

❑

As we shall see in examining the three sequences to follow, the function given by $\varphi_N(t)$ in Notation 15.19, is a major tool for *patching things together very smoothly.*[1]

The quantity $\rho_N(t)$ in the notation below is a multiplicative *Finagle factor* to achieve bounded support, and yet leave *most things* unaltered as N gets large. With hard work it is shown that $\mathcal{M}(f, N)(t)$ is close to $f(t)$ for large N.

---

1. Probabilists will recognize that our use of φ_N in defining $\mathcal{M}(f, N)(t)$, which associates a generalized function with the ordinary function f, is similar to the conversion of discrete to continuous data by adding a random quantity which can take values continuously, to the discrete data, thus *smearing* its distribution.

Notation 15.19: $\varphi_N(t)$, $\rho_N(t)$, $\mathcal{M}(f, N)(t)$

For $N = 1,2,3,...$ and f in $\mathcal{PC}$ (Notation 15.16 on page 685) we let

1.
$$\varphi_N(t) = \begin{cases} exp\left(-\dfrac{1}{1-N^2t^2}\right) & \text{if } N^2t^2 < 1 \\ 0 & \text{if } N^2t^2 \geq 1, \end{cases}$$

2.
$$\rho_N(t) = KN\int_{-N}^{+N} \varphi_N(t-u)du, \quad \text{where} \quad \frac{1}{K} = \int_{-1}^{1} \varphi_1(u)du,$$

and

3.
$$\mathcal{M}(f, N)(t) = \rho_N(t)KN\int f(u)\varphi_N(t-u)du.$$

This last sequence, more properly denoted by $(\mathcal{M}(f, N))(t)$, will be used to define the mapping $\mathcal{M}$, which embeds ordinary functions in the class of generalized functions.

❑

We illustrate the functions ρ_2 and φ_2 in Figure 15-10.

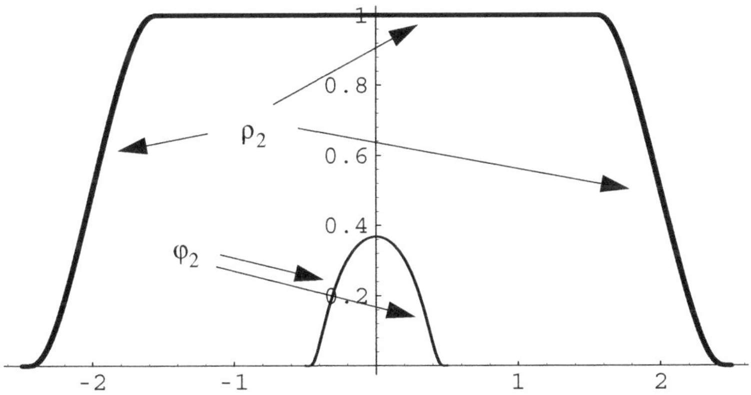

Figure 15-10

The behavior of the function φ defined in equation 15.19.1 is quite striking, in that it is *very very flat* near $t = \pm 1/N$, allowing it to be used for very smooth patching. Its behavior is related to the strange properties of the function g given by

$$g(x) = \begin{cases} e^{-1/x^2} & \text{for} \quad x \neq 0 \\ 0 & \text{for} \quad x = 0. \end{cases} \qquad \textit{15.53}$$

As a function of a real variable, $D^k g(0) = 0$ for all nonnegative integers, k, but g is not identically 0. Hence, it cannot have a Taylor series about 0 which converges to g in some positive radius circle centered at 0; because, if it had such a series, from Taylor's theorem (Theorem 6.2 on page 179) g would have to be 0 at all points in this circle, which it clearly isn't. The strange behavior of g is best understood by letting x be complex; it is seen that $g(x)$ is *very badly behaved* for x near 0 — just look at it along the imaginary axis. But it is precisely this bad behavior which allows the properties so useful for patching.

Theorem 15.20: φ_N is a test function

For each N, the function φ_N of Notation 15.19 on page 688 is in $\mathcal{T}$ (Definition 15.9 on page 677).

❑

Problem 15.7

Prove Theorem 15.20.

Hints: The only worry is differentiability at $t = \pm 1/N$. You must show that the left-handed derivative,

$$\lim_{\substack{h \to 0 \\ h > 0}} \varphi_N^{(k)}(1 - h) = 0$$

for all nonnegative integers, k (since the right-hand derivative is evidently 0 at 1). No generality is lost by taking $N = 1$. Ultimately you will see that the problem reduces to that of showing that

$$\lim_{\substack{h \to 0 \\ h > 0}} \frac{\varphi_1(1 - h)}{h^k} = 0,$$

or equivalently that

$$\lim_{\substack{h \to 0 \\ h > 0}} \frac{exp\left(\frac{-1}{2h}\right)}{h^k} = 0 \cdot$$

This last is equivalent to showing that

$$\lim_{s \to \infty} s^k e^{-s} = 0, \qquad \textbf{15.54}$$

where by $\lim_{s \to \infty} s^k e^{-s}$, we mean the value $s^k e^{-s}$ approaches as s becomes algebraically quite large. Equation 15.54 follows from the Taylor series for e^s (see Problem 7.15.3 on page 241).

Note that $KN\int \varphi_N(t)dt = 1$ (see Notation 15.19 on page 688) and $\varphi_N(t)$ tends to *peak* more and more near $t = 0$ for large N. Thus for large N, $\varphi_N(t-u)$ peaks for u near t, *so that* $KN\int f(u)\varphi_N(t-u)dt$ will be close to $f(t)$ for large N, if f is continuous at t.

Now seems a convenient time to establish the properties of $\rho_N(t)$ indicated in Figure 15-10.

Theorem 15.21: Behavior of ρ_N

The function ρ_N of Notation 15.19.2 on page 688 satisfies the conditions that

for all real t, and $N = 1,2,...$	$0 \le \rho_N(t) \le 1,$
if $t \ge N + \frac{1}{N}$ or $t \le -N - \frac{1}{N}$	$\rho_N(t) = 0,$
if $-N + \frac{1}{N} \le t \le N - \frac{1}{N}$	$\rho_N(t) = 1,$
for $-N - \frac{1}{N} \le t < -N + \frac{1}{N}$	$\rho_N(t)$ increases as t increases
for $N - \frac{1}{N} \le t < N + \frac{1}{N}$	$\rho_N(t)$ decreases as t increases.

❑

A convincing argument for these results is obtained by carefully examining Figure 15-11 and using the result, obtainable from Notation 15.19 on page 688, that

$$KN\int \varphi_N(z)dz = KN\int_{-1/N}^{1/N} \varphi_N(z)dz = K\int_{-1}^{1} \varphi_1(z)dz = 1. \qquad \textbf{15.55}$$

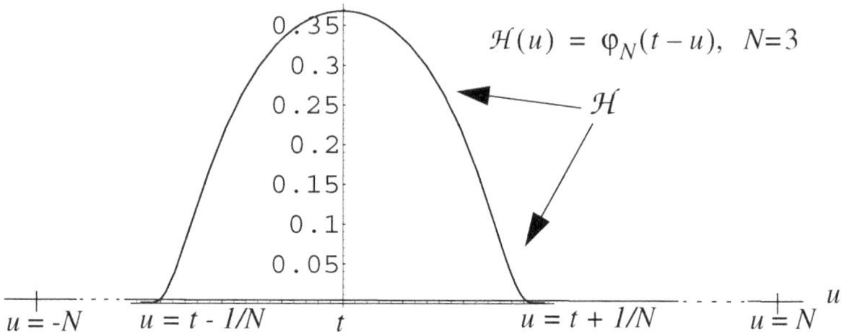

Figure 15-11

The important technical details nailing down the trickier parts of regularization are to be found in the proof of Theorem 15.22. Those who desire to see more or less precisely the difficulties which must be overcome and how they are handled can get a good idea from this proof. The following problems are handled in the proof just mentioned, but it's worth trying them in advance, to get a feel for the difficulties to be encountered.

Problems 15.8

1. Show that $\{\varphi_N(t)\}\big|_{N=1}^{\infty}$ is a regular sequence (Definition 15.11 on page 680).
2. Show that $\{\rho_N(t)\}\big|_{N=1}^{\infty}$ is a regular sequence (Definition 15.11 on page 680).
3. Compute $\rho_2(t)$ for $t < -2.5$, and for $t = 2$.
4. Compute $\lim_{N \to \infty} \int \rho_N(t)\varphi_2(t)dt$.

Theorem 15.22: Embedding $\mathcal{PC}$ in $\mathcal{G}$

If the function f belongs to $\mathcal{PC}$ (Notation 15.16 on page 685) then the sequence $\{\mathcal{M}(f,N)\}\big|_{N=1}^{\infty}$ given in Notation 15.19 on page 688 is well defined and is a regular sequence (Definition 15.11 on page 680).

❑

First note that the integral

$$\int f(u)\varphi_N(t-u)du \qquad \textbf{\textit{15.56}}$$

is defined for each $t, N = 1,2,3,...$ because

$$\varphi_N(t-u) = 0 \text{ for } t - \frac{1}{N} \leq u \leq t + \frac{1}{N}$$

which allows use of Theorem 15.20 on page 689, which (with the piecewise continuity of f) shows the integrand to be piecewise continuous and with bounded support for each t. Hence by Theorem 15.17 on page 686 the expression in 15.56 is meaningful.

The factor $\rho_N(t)$ is well defined, via even simpler reasoning than proved the existence of the integral in 15.56 above. Furthermore, $\rho_N(t)$ has bounded support (Definition 15.8 on page 677) because for $|t| > N + 1/N$ and $-N \leq u \leq N$, we have $|t-u| > 1/N$, and hence $\varphi_N(t-u) = 0$; this assertion then follows from Notation 15.19.2 on page 688.

Combining the two results just established, using Notation 15.19.3 on page 688 shows that $\mathcal{M}(f, N)$ is well defined and has bounded support.

We next want to show that the functions given by $\rho_N(t)$ and $\mathcal{M}(f, N)$ are in C^∞ (Notation 15.1 on page 655). In order to do this efficiently, it's convenient to examine the expression

$$z(t) = \int g(u)\varphi_N(t-u)du, \qquad \textbf{\textit{15.57}}$$

where g represents some piecewise continuous function. First, as with 15.56, this integral is meaningful.

It should be fairly evident that we'll need to show that z is differentiable.[1] To be precise, we would like to show for any fixed t we choose, that $z'(t)$ exists, and determine a formula for this derivative. In order to do this we first choose some fixed value, t_0, such that t is ***strictly inside*** the interval $[t_0 - 1; t_0 + 1]$. Then, for u outside the interval $[t_0 - 2, t_0 + 2]$, and $N = 1,2,3,...$ we see that

$$\varphi_N(t-u) = 0.$$

---

1. This will be needed both to show that the function ρ_N is differentiable, and that the same holds for the function given by $\mathcal{M}(f, N)(t)$.

Hence

$$z(\tau) = \int_{t_0-2}^{t_0+2} g(u)\varphi_N(\tau - u)du \qquad \textbf{15.58}$$

for all τ sufficiently close to t. To establish the differentiability of z at t, we turn to Theorem 12.26 on page 477, which shows that to prove $z'(t)$ exists, and

$$z'(t) = \int_{t_0-2}^{t_0+2} g(u)\varphi_N'(t - u)du \qquad \textbf{15.59}$$

we need only verify that

- $\varphi_N'(t-u)$ exists for t and the relevant values of u,N

(which follows from the results of Theorem 15.20 on page 689) and that

$$g(u)\frac{\varphi_N(t+h-u) - \varphi_N(t-u)}{h} \to g(u)\varphi_N'(t-u) \qquad \textbf{15.60}$$

as h approaches 0, uniformly for u in $[t_0 - 2, t_0 + 2]$. For this last we use the mean value theorem, Theorem 3.1 on page 62, to write

$$g(u)\frac{\varphi_N(t+h-u) - \varphi_N(t-u)}{h} = g(u)\varphi_N'(t-u+\vartheta h),$$

where $|\vartheta| \le 1$ (ϑ depending on t,h,u,N). To establish the result in 15.60, we therefore need only show that

$$g(u)\varphi_N'(t-u+\vartheta h) \to g(u)\varphi_N'(t-u, N)$$

as h approaches 0, uniformly for u in $[t_0 - 2, t_0 + 2]$, or that

$$g(u)[\varphi_N'(t-u+\vartheta h) - \varphi_N'(t-u)] \to 0$$

as h approaches 0, uniformly for u in $[t_0 - 2, t_0 + 2]$.

But since g is assumed piecewise continuous (Definition 3.13 on page 89) and a continuous function whose domain is a closed interval is bounded (Theorem 3.12 on page 88) it's not hard to see that a piecewise continuous function is bounded on each closed bounded interval. Thus, we need only show that

$$[\varphi_N'(t-u+\vartheta h) - \varphi_N'(t-u)] \to 0$$

as h approaches 0, uniformly for u in $[t_0 - 2, t_0 + 2]$. But this now follows from the uniform continuity (see Theorem A3.4 on page 769) of a continuous function whose domain is a closed bounded interval, applied to the function given by $\varphi_N'(t-u)$. Thus we have established equation 15.59. Since for each N, $\varphi_N(t)$ is in C^∞ (Theorem 15.20 on page 689) this argument may be repeated inductively, to show that for all positive integers n, $D^n z(t)$ exists with

$$D^n z(t) = \int_{t_0-2}^{t_0+2} g(u)D^n\varphi_N(t - u)du. \qquad \textbf{15.61}$$

Since t_0 can always be found, it follows that z given by 15.57 on page 692 is in C^∞. $\mathcal{M}(f, N)$ is in C^∞, and so $\mathcal{M}(f, N)$ is in $\mathcal{T}$ for each N.

Now $\rho_N(t)$ is of the form of z given by equation 15.57 on page 692 (just examine Notation 15.19 on page 688 if you need to) so it is in C^∞, and it has been shown to have bounded support, hence $\{\rho_N(t)\}\big|_{N=1}^{\infty}$ is in $\mathcal{T}$ (Definition 15.9 on page 677). The function $\mathcal{M}(f, N)$, being the product of $\rho_N(t)\,KN$ with a function of the form of z given by 15.57 is in C^∞ and (due to the first factor) also has bounded support, and therefore is also in $\mathcal{T}$.

For the final step in proving that $\{\mathcal{M}(f, N)\}\big|_{N=1}^{\infty}$ is a regular sequence (Definition 15.11 on page 680) we must show that

$$\lim_{N\to\infty} \int \mathcal{M}(f, N)(t)T(t)dt = \lim_{N\to\infty} \int \rho_N(t)KN\{\int f(u)\varphi_N(t-u)du\}T(t)dt \qquad \textbf{15.62}$$

exists for each T in $\mathcal{T}$.

To carry this out, let us choose an arbitrary T in $\mathcal{T}$. Then

$$\begin{aligned} N\int\int f(u)\varphi_N(t-u)T(t)dudt &= N\int f(u)\{\int \varphi_N(t-u)T(t)dt\}du \\ &= \int f(u)\left\{\int \varphi_N\left(\frac{v}{N}\right)T\left(u+\frac{v}{N}\right)dv\right\}du \\ &= \int_{-1}^{+1} \varphi_N\left(\frac{v}{N}\right)\left\{\int_{-B}^{B} f(u)T\left(u+\frac{v}{N}\right)du\right\}dv, \end{aligned} \qquad \textbf{15.63}$$

where $v = N(t-u)$ and $B > 1$ is chosen so that the support of T (Definition 15.7 on page 677) is contained in $[-B+1; B-1]$. Also, using the definition of K (Notation 15.19 on page 688) which shows that

$$K\int \varphi_N\left(\frac{v}{N}\right)dv = 1,$$

we have

$$\begin{aligned} \int f(t)T(t)dt &= K\int \varphi_N\left(\frac{v}{N}\right)dv\int f(t)T(t)dt \quad \left(\text{since } \varphi_N\left(\frac{v}{N}\right) = \varphi_1(v)\right) \\ &= K\int \varphi_N\left(\frac{v}{N}\right)\{\int f(u)T(u)du\}dv \quad (\text{using the bounded support of } T). \end{aligned} \qquad \textbf{15.64}$$

Using the positivity of φ, equations 15.63 and 15.64 we find that

$$\begin{aligned} &\left|KN\int\int f(u)\varphi_N(t-u)T(t)dudt - \int f(t)T(t)dt\right| \\ &\le K\int_{-1}^{1} \varphi_N\left(\frac{v}{N}\right)\left\{\int_{-B}^{B} |f(u)|\left|T\left(u+\frac{v}{N}\right) - T(u)\right|du\right\}dv. \end{aligned} \qquad \textbf{15.65}$$

Since T is in C^{∞}, DT is continuous (see paragraph a on page 63, and its proof on page 65); since $[-B+1; B-1]$ supports both DT and T, we can choose a number M_1 such that

$$|DT(u)| \le M_1$$

for all real u (Theorem 3.12 on page 88). Then, by the mean value theorem (Theorem 3.1 on page 62),

$$\left|T\left(u+\frac{v}{N}\right)-T(u)\right| \le |v|\frac{M_1}{N}. \qquad \textbf{\textit{15.66}}$$

We let

$$M_2 = \int_{-B}^{B} |f(u)|du \qquad \textbf{\textit{15.67}}$$

(we assumed that f is in $\mathcal{PC}$, so it is absolutely integrable, see Theorem 15.17 on page 686, so that equation 15.67 is meaningful). It follows also from Theorem 15.17, that because $|f|$ is in $\mathcal{PC}$, the right side of inequality 15.65 is dominated by (*less than or equal to*)

$$K\int_{-1}^{+1} \varphi_N\left(\frac{v}{N}\right)\frac{M_1M_2v}{N}dv \le \frac{M_1M_2}{N}. \qquad \textbf{\textit{15.68}}$$

Since M_1 and M_2 are independent of N, we see that the right side of inequality 15.65 approaches 0 as N approaches ∞. Hence

$$\lim_{N\to\infty} KN\int\int f(u)\varphi_N(t-u)T(t)dudt = \int f(t)T(t)dt. \qquad \textbf{\textit{15.69}}$$

(While we didn't explicitly verify that all of the expressions occurring above do exist, this is the case, and the reader should verify that this is so.)

From equation 15.69 we see that in order to establish the existence of the limit in equation 15.62, we must show that $\rho_N(t)$ plays no essential role there. But we know, because T is a test function, that T has bounded support and from Theorem 15.21 on page 690, that $\rho_N(t) = 1$ for $-N+1/N \le t \le N-1/(N.)$ So, for N sufficiently large (say $N \ge N_T$) we have

$$\rho_N(t)T(t) = T(t) \quad \text{for all real } t.$$

So, for $N \ge N_T$ we see that

$$\int\rho_N(t)KN\{\int f(u)\varphi_N(t-u)du\}T(t)dt = KN\int\int f(u)\varphi_N(t-u)T(t)dudt.$$

Since equation 15.69 shows that the right side has a limit as N grows large, so does the left side — proving the existence of the limit in equation 15.62.

So we have finally shown that for each test function T,

$$\lim_{N\to\infty}\int \mathcal{M}(f,N)(t)T(t)dt \text{ exists, and in fact equals } \int f(t)T(t)dt.$$

This establishes that $\{\mathcal{M}(f, N)\}\big|_{N=1}^{\infty}$ is a regular sequence, completing the proof of Theorem 15.22 on page 691.

Theorem 15.23: $\mathcal{M}(\mathcal{K}, N)$

Let $\mathcal{K}$ stand for the real valued constant function of a real variable whose value is k. Then $\{\mathcal{M}(\mathcal{K}, N)\}\big|_{N=1}^{\infty} = \{k\rho_N\}\big|_{N=1}^{\infty}$, both expressions being defined in Notation 15.19 on page 688.

❑

Notation 15.19 on page 688 and equation 15.55 on page 691 yield

$$\mathcal{M}(\mathcal{K}, N)(t) = \rho_N(t)KN\int k\varphi_N(t-u)du = k\rho_N(t).$$

With Theorem 15.22 at our disposal, we know that the following definition makes sense.

Definition 15.24: Embedding of $\mathcal{PC}$ in $\mathcal{G}$ via the operator $\mathcal{M}$

For each function f in $\mathcal{PC}$, Notation 15.16 on page 685, define $\mathcal{M}f$ in[1] $\mathcal{G}$ (Definition 15.13 on page 680) by the representation $\{\mathcal{M}(f, N)\}\big|_{N=1}^{\infty}$ of Notation 15.19 on page 688.

❑

Now we notice that, in a sense, every piecewise continuous function is infinitely differentiable (see Definition 15.15 on page 682) a result which may appear quite surprising.

Theorem 15.25: Linearity of embedding in $\mathcal{G}$

$\mathcal{M}$ is a linear operator (Definition 13.9 on page 547).

❑

Theorem 15.26: Commutativity of D and $\mathcal{M}$

If f' is in $\mathcal{PC}$ (Notation 15.16 on page 685) then

$$\mathcal{M}(Df) = D(\mathcal{M}f) = D\mathcal{M}f,$$

(see Definition 15.24 above), there being no danger of confusion from dropping the parentheses, because $\mathcal{M}$ is not in the domain (Definition 1.1 on page 2) of the derivative operator D (see Definition 2.5 on page 30 for Df,

1. More properly we should write $\mathcal{M}(f)$ and $(\mathcal{M}(f, N))(t)$ instead of $\mathcal{M}f$ and $\mathcal{M}(f, N)(t)$.

with f a real valued function of a real variable, and Definition 15.15 on page 682 for the definition of the generalized function derivative operator, D).

❑

Problem 15.9

Prove Theorems 15.25 and 15.26. The proof of Theorem 15.26 involves showing that the sequences given by $(f')_N$ and by f_N' yield the same generalized function, where

$$(f')_N(t) = \int f'(u)\varphi_N(t-u)du \text{ and } f_N(t) = \int f(u)\varphi_N(t-u)du.$$

The ability to interchange order of differentiation and integration and an integration by parts are needed on the second function f_N.

These theorems, together with Problem 15.11.1 on page 699 show that, as far as differentiation of elements f with f' in $\mathcal{PC}$ (Notation 15.16 on page 685) we can just as well work with the set of generalized functions, $\mathcal{G}$.

We shall soon be able to show that if f and g are in $\mathcal{PC}$ (i.e., piecewise continuous, Definition 3.13 on page 89) and differ at least at one point which is a continuity point of both f and g, then $\mathcal{M}f \neq \mathcal{M}g$, so that this is a *real* embedding and does preserve the inequalities between functions which we wanted unequal.

The sequence $\{I_N(\ ,s)\}\Big|_{N=1}^{\infty}$ given by equation 15.35 on page 669, used to introduce the development of Green's (impulse response) functions in Section 4 of this chapter, represented a hammer blow or impulse of unit strength (due to the unit area) occurring at time s. Because $I_N(\ ,s)$ are *rectangles*, they are in $\mathcal{PC}$; but obviously $\{I_N(\ ,s)\}\Big|_{N=1}^{\infty}$ is not a regular sequence (Definition 15.11 on page 680) because its elements have jump discontinuities. Since we so obviously want an impulse to be represented by some generalized function, it is clear that we would like to find some regular sequence which *behaves like* $\{I_N(\ ,s)\}\Big|_{N=1}^{\infty}$, relative to the criterion of giving the same limiting behavior as

$$\lim_{N\to\infty} \int I_N(t,s)T(t)dt.$$

Examination of the sequence of functions

$$\{NK\varphi_N\}\Big|_{N=1}^{\infty}$$

illustrated in Figure 15-12, shows that it might be a suitable *regular* replacement for $\{I_N(\ ,s)\}\Big|_{N=1}^{\infty}$.

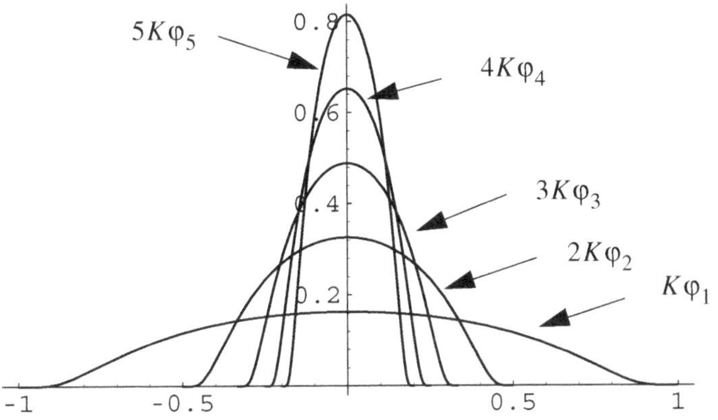

Figure 15-12

Note that we obtain this impulse by squeezing one axis (replacing t by NT in going from $\varphi_1(t)$ to $\varphi_N(t)$ which has the effect of shrinking the support of φ from $[-1;1]$ to $[-1/N;1/N]$) and expanding the other axis (i.e., multiplying by N). *This expands the range* by a factor of N, and keeps the area equal to 1.

Problems 15.10

Referring to Definition 15.11 on page 680,

1. Show that $\{KN\varphi_N\}\Big|_{N=1}^{\infty}$ is a regular sequence.

2. Show that $\{KN\varphi_N(at+b)\}\Big|_{N=1}^{\infty}$ is a regular sequence, where t denotes the identity function, $t(\tau) = \tau$ for τ real, and a,b are real numbers.

Definition 15.27: The Dirac delta (δ)

The Dirac δ is that generalized function defined by the regular sequence

$$\{KN\varphi_N\}\Big|_{N=1}^{\infty}.$$

❑

δ represents an impulse occurring at time 0. It will be significant in the general definition of Green's functions (impulse response functions) as well as in defining the concept of generalized function initial conditions — similar to the ordinary initial conditions of the form $x(0) = a, Dx(0) = b,...$ etc.

Problems 15.11

1. Use the Dirac δ to show that if f and g are in $\mathcal{PC}$ (Notation 15.16 on page 685) and s is a continuity point of both f and g (Definition 7.2 on page 213) such that $f(s) \neq g(s)$, then $\mathcal{M}f \neq \mathcal{M}g$.

2. Show that if f is a function in C^{∞} (Notation 15.1 on page 655) whose derivative is bounded away from 0, and if $\{g_N\}\Big|_{1=N}^{\infty}$ is a regular sequence (Definition 15.11 on page 680) then the sequence given by $\{c_N\}\Big|_{N=1}^{\infty}$, where $c_N(t) = \rho_N(t) g_N(f(t))$ is a regular sequence.
   Hints: The bounded support is taken care of by ρ_N. The infinitely differentiable property follows from the product and chain rules for differentiation. All that needs to be shown is that
   $$\int c_N(t) T(t) dt$$
   has a limit as N grows large, for each test function T. Because of the bounded support of T, we need only show that
   $$\int g_N(f(t)) T(t) dt$$
   has a limit for each such T. Make the substitution $u = f(t)$, with $t = f^{-1}(u)$, and show that
   $$\mathcal{T}(u) = \frac{T(f^{-1}(u))}{f'(f^{-1}(u))}$$
   represents a test function, $\mathcal{T}$ — the bounded support not being difficult because of the condition on the derivative of f. The infinitely differentiable property takes some work.

In attempting to extend the notion of function, we have tried to preserve as many of the familiar operations as was conveniently possible. The idea of composition of functions *seems*, to the authors, to generalize best to the next definition.

Definition 15.28: Composition

The *composition*, $g(m)$ of a generalized function g, represented by the sequence $\{g_N\}\Big|_{N=1}^{\infty}$ with an infinitely differentiable monotone function m whose derivative is bounded away from 0 (i.e., $|m'(t)| > c > 0$ for all t) is the generalized function represented by the sequence

$$\{c_N\}\Big|_{N=1}^{\infty}, \text{where} \quad c_N(t) = \rho_N(t) g_N(f(t))$$

(see Notation 15.19 on page 688).

❑

This definition is justified by the results of Problem 15.11.2 just preceding. It is aimed, in particular, at giving meaning to $\delta(t-x)$, where δ is the Dirac delta function (Definition 15.27 on page 698) t is the identity function on the reals ($t(\tau) = \tau$) and x is a real constant (standing for the constant function with this value); because the generalized function $\delta(t-x)$ corresponds to an impulse occurring at time x. These *delta functions* furnish a convenient way of describing *sampled data systems.* If the input to some system consists of a function, say g, sampled at times spaced $\mathcal{T}$ time units apart, then it is useful to replace g by the generalized function $\boldsymbol{g}$, where

$$\boldsymbol{g}(t) = g(0)\delta(t) + g(\mathcal{T})\delta(t-\mathcal{T}) + \cdots = \sum_k g(k\mathcal{T})\delta(t-k\mathcal{T}) .$$

While we have not defined this infinite series, it is intuitively clear what is meant in this case, namely, $\boldsymbol{g}$ has the property that for each test function T,

$$\int \boldsymbol{g}T = \sum_{k=q}^{m} g(k\mathcal{T})T(k\mathcal{T})$$

where the support of T lies in the interval $[q-1;m+1]$. We can establish that there is such a generalized function by letting it be represented by the regular sequence

$$\left\{\sum_{k=0}^{N-1} g(k\mathcal{T})KN\varphi_N(t-k\mathcal{T})\right\}\Bigg|_{N=1}^{\infty} .$$

We shall speak a bit more about infinite series of generalized functions in Section 10 of this chapter.

Theorem 15.29: Properties of Dirac's δ

If t is the real identity function, $t(\tau) = \tau$, T is in $\mathcal{T}$ (Definition 15.9 on page 677) a and b are real with $a \neq 0$ and δ is Dirac's delta (see Definition 15.27 on page 698) then

- δ is in $\mathcal{G}$ (Definition 15.13 on page 680) but δ is not one of the embedded generalized $\mathcal{PC}$ functions (see Notation 15.16 on page 685); i.e., δ is not in $\mathcal{M}(\mathcal{PC})$, the image (Definition 12.20 on page 462) of $\mathcal{PC}$.
- $\{KN\varphi_N(at+b)\}\Big|_{N=1}^{\infty}$ is a sequence representing $\delta(at+b)$, where K and φ are defined in Notation 15.19 on page 688. This definition is a bit more direct than using Definition 15.28 on page 699 to define $\delta(at+b)$ (since we needn't use ρ). Note that this is ***composition*** that we are dealing with. We are ***not*** talking about the value of δ at a real number $at+b$, a concept which will not be defined.

- $\int \delta(at+b)T(t) = \frac{1}{|a|}T\left(-\frac{b}{a}\right)$.

- $\delta(t) = \delta(-t)$.

- $\int D^n\delta(t)T(t) = (-1)^n D^n T(0)$. Recall the parts of Definition 15.15 on page 682 concerning differentiation and integration of generalized functions.

- For the Heaviside unit step, H (see Theorem 14.19 on page 622)
$$D[\mathcal{M}(H)] = \delta.$$
(See Theorem 15.22 on page 691 for $\mathcal{M}$.)

❑

Problem 15.12

Prove Theorem 15.29. Hint for the first part: Note from the conclusion of the proof of Theorem 15.22 on page 695 that $\lim_{N \to \infty} \int T\mathcal{M}(f, N) = \int fT$.

Let us pause here and recall our goal. We eventually want to interpret the equation $LX = Y$ in a generalized (or *weak*) sense; however, an initial condition of the form $X(0) = 0$ *does not make sense for X a generalized function.* **In an attempt to arrive at the equivalent of an initial condition for a generalized function** X, consider a continuous function, f, and the corresponding generalized function $F = \mathcal{M}f$. Now the function f has *values*; i.e., to each real x there corresponds a real value $f(x)$. We know more details about the elements of C (continuous functions) than we do about the elements of the larger class $\mathcal{G}$ (generalized functions). How much detail do we want about each element g of $\mathcal{G}$? Recalling our discussion considering physical measurement as a weighted average (near equation 15.48 on page 679) *we would expect that all we need to know about the generalized function g is how it integrates near 0, in order to obtain something equivalent to an initial condition. If we only need to know how each g integrates near zero,* we could consider

$$\lim_{N \to \infty} \int g(t)KN\varphi_N(t)dt.$$

This last expression looks like $\int g\delta$, but we do not yet have a definition of the product $g\delta$ (which is a product of two generalized functions) or of $\int g\delta$. We *might* try to define $\int g\delta$ by having it equal $\lim_{N \to \infty} \int g\Delta_N$, where $\{\Delta_N\}\Big|_{N=1}^{\infty}$ is any representative of δ.

To understand why we don't do this, consider the following problems.

Problems 15.13

Let $w_N(t) = KN\varphi_N(t - 1/N)$ (Notation 15.19 on page 688). Show that

1. $\lim_{N \to \infty} w_N(t) = 0$ for all real t.

2. $\int w_N(t)dt = 1$ for $N = 1, 2, 3, \ldots$.

3. $\{w_N\}\Big|_{N=1}^{\infty}$ represents Dirac's δ (Definition 15.27 on page 698).

4. $\lim_{N \to \infty} \lim_{M \to \infty} \int w_M(t)KN\varphi_N(t)dt$ doesn't exist (is infinite).

5. $\lim_{M \to \infty} \lim_{N \to \infty} \int w_M(t)KN\varphi_N(t)dt = 0$.

The last two results warn us of trouble ahead when we try to define the product of two generalized functions.

Having seen from Problems 15.13 above, and the discussion following, that we must proceed with care, we now define the product of two generalized functions under certain restraints on the factors — in order to arrive at an acceptable generalization of the concept of *initial conditions*.

Since we want the product rule (see Theorem 2.13 on page 53) to hold wherever possible, we *back into* the following definition, with the aid of *Leibnitz's rule* for real valued functions, q, r, which have n derivatives, which states that

$$D^n(qr) = \sum_{j=0}^{n} \binom{n}{j} D^j q\, D^{n-j} r, \qquad \textit{15.70}$$

with the binomial coefficient, $\binom{n}{j}$, defined for all nonnegative integers, n, and all integers j, being given by the formula

$$\binom{n}{j} = \begin{cases} \dfrac{n!}{j!(n-j)!} & \text{for } 0 \le j \le n, \\ 0 & \text{otherwise,} \end{cases} \qquad \textit{15.71}$$

where

$$0! = 1, \quad (k+1)! = (k+1)k!, \qquad \textit{15.72}$$

for all nonnegative integers k,

For our purposes, a somewhat more useful version of formula 15.70 is the following:

$$qD^n r = \sum_{j=0}^{n} (-1)^j \binom{n}{j} D^{n-j}[(D^j q)r]. \qquad \textbf{\textit{15.73}}$$

To convince yourself of this, try comparing formulas 15.70 and 15.73 for $n = 1,2,3$.

Equations 15.70 and 15.73 can be proved directly by induction on n and are left as exercises in the following set.

Exercises 15.14

1. Verify that for each integer j and nonnegative integer n

$$\binom{n}{j-1} + \binom{n}{j} = \binom{n+1}{j}$$

   where the above are defined via equations 15.71 and 15.72. Hint: Use induction; i.e., show that the above holds for $n = 0$, and that whenever it holds for $n = k$, it also holds for $n=k+1$.

2. Verify that if $D^n q$ and $D^n r$ exist, then equation 15.70 holds. Hints: The proof is inductive, see previous exercise. By the product rule, Theorem 2.13 on page 53, the result is seen to hold for $n = 1$. Examining any k for which the result holds, we have

$$\begin{aligned} D^{k+1}(qr) &= D^k[D(qr)] \\ &= D^k[(Dq)r + q\,Dr] \\ &= \sum_{j=0}^{k} \binom{k}{j} [(D^{j+1}q)(D^{k-j}r)] \\ &\quad + \sum_{j=0}^{k} \binom{k}{j} [(D^j q)(D^{k-j+1}r)]. \end{aligned}$$

   Now in the first sum, substitute $J = j + 1$ and then replace J by j — so that the terms in square brackets are the same. Then, separating out the $j = 0$ and $j = k + 1$ terms, combine the sums and use the results of the previous exercise.

3. Verify that if q and r have n-th derivatives then equation 15.73 holds.
   Hints: Again, the proof is inductive — starting the same way as that of the previous exercise. Then for any k at which the desired result is known to hold, we have

$$
\begin{aligned}
q(D^{k+1}r) &= q(D^k[Dr]) \\
&= \sum_{j=0}^{k} (-1)^j \binom{k}{j} D^{k-j}[(D^j q)(Dr)] \\
&= \sum_{j=0}^{k} (-1)^j \binom{k}{j} D^{k-j}[D\{(D^j q)r\} - (D^{j+1} q)r] \\
&= \sum_{j=0}^{k} (-1)^j \binom{k}{j} D^{k-j+1}\{(D^j q)r\} \\
&\quad - \sum_{j=0}^{k} (-1)^j \binom{k}{j} D^{k-j}[(D^{j+1} q)r].
\end{aligned}
$$

Now proceed as in the previous exercise, beginning by substituting $J = j + 1$ in the second sum, then replacing J by j, etc.

Definition 15.30: Product of generalized functions

If

- n denotes a nonnegative integer,
- $D^n q$ is in $\mathcal{PC}$ (Notation 15.16 on page 685),
- $Q = \mathcal{M}q$ (Definition 15.24 on page 696),
- r is in $\mathcal{PC}$,
- $R = \mathcal{M}r$,
- t denotes the real identity function, $t(\tau) = \tau$,
- a and b are real numbers,

then we define the product QD^nR of the generalized functions Q and D^nR by the formula

$$
QD^nR = \sum_{j=0}^{n} (-1)^j \binom{n}{j} D^{n-j} \mathcal{M}[(D^j q)r], \qquad \textbf{15.74}
$$

where $\binom{n}{j}$ is the binomial coefficient, defined in equation 15.71 on page 702.

❑

Note that formula 15.74 is in complete analogy with equation 15.73, the corresponding formula for real valued functions of a real variable. Note also that we are defining the product of a very nice generalized function Q with a *true* generalized function D^nR (one which need not itself be an embedded real valued function of a real variable, even though it is the n-th generalized derivative of such an embedded function). Finally, we can see that formula 15.74 really does define the quantity QD^nR, because the quantities on the right side are all well-defined — $(D^jq)r$ being an ordinary piecewise continuous real valued function of a real variable, hence having a well-defined associated generalized function, $\mathcal{M}[(D^jq)r]$, for which all derivatives, in particular $D^{n-j}\mathcal{M}[(D^jq)r]$, are well defined.

Another problem that we run into is illustrated by the Heaviside unit step function H (see Theorem 14.19 on page 622). It can be seen that $H^k = H$ for each positive integer k, and it's not hard to show that if $\mathcal{H} = \mathcal{M}H$, then it's also true that $\mathcal{H}^k = \mathcal{H}$ (take $n = 0$, $q = H$ and $r = H$ in Definition 15.30). Using the fact that $D\mathcal{H} = \delta$ (see the last result from Theorem 15.29, appearing on page 701) **if the product rule were to hold unrestrictedly**, we would have

$$\delta = D\mathcal{H} = D\mathcal{H}^2 = 2\mathcal{H}D\mathcal{H} \quad \text{and} \quad \delta = D\mathcal{H} = D\mathcal{H}^3 = 3\mathcal{H}^2D\mathcal{H} = 3\mathcal{H}D\mathcal{H},$$

which surely cannot be.

As we will see, Definition 15.30 just ensures, for example, that when FD^2G exists, then so do FG, $(DF)(DG)$, $(D^2F)G$, and Leibniz's rule for $n = 2$,

$$D^2(FG) = (D^2F)G + 2(DF)(DG) + G(D^2F),$$

still holds.

Definition 15.30 implies that the product of any function in C^∞ (Notation 15.1 on page 655) with any generalized function which is the n-th derivative of a piecewise continuous function is defined. It is not difficult to verify that the C^∞ functions embedded in the set $\mathcal{G}$ of generalized functions obey the same old formal rules. For example,

$$\mathcal{M}(e^{at}) \cdot \mathcal{M}(e^{bt}) = \mathcal{M}(e^{(a+b)t},)$$

where t is the identity function, $t(\tau) = \tau$.

Problems 15.15

1. Establish the assertion of the paragraph just preceding.
2. Establish that multiplication of generalized functions, as defined in Definition 15.30 on page 704 is
   commutative, $[fg = gf]$,
   associative, $[f(gh) = (fg)h]$,
   and
   distributive, $[f(g+h) = (fg) + (fh)]$.

If f is a function whose n-th derivative is piecewise continuous, and $F = \mathcal{M}f$ is the corresponding generalized function, Definition 15.30 is not applicable for defining the product $F\delta^{(n)}$. Since this product will be useful in our later work, we need to establish it separately.

Definition 15.31: $F\delta^{(n)}$

Let $\delta^{(n)} = D^n\delta$, where δ is Dirac's generalized *delta* function (Definition 15.27 on page 698). If f is a real valued function of a real variable whose n-th derivative, $D^n f$ is in $\mathcal{PC}$ (Notation 15.16 on page 685) and $F = \mathcal{M}f$ (Definition 15.24 on page 696) then we define

$$F\delta^{(n)}$$

to be the generalized function represented by the sequence

$$\{s_{N,n}\}\Big|_{N=1}^{\infty} = \{\mathcal{M}(f,N)KND^n\varphi_N\}\Big|_{N=1}^{\infty} \qquad \textbf{15.75}$$

(see Notation 15.19 on page 688).

❑

It remains to show that for n a nonnegative integer, the sequence $\{s_{N,n}\}\big|_{N=1}^{\infty}$ given in equation 15.75 is a regular sequence (Definition 15.11 on page 680) thus properly defining a generalized function. We leave this to the following problem.

Problem 15.16

Show that the sequence $\{s_{N,n}\}\big|_{N=1}^{\infty}$ given in equation 15.75 is a regular sequence (Definition 15.11 on page 680) thus properly defining a generalized function.

Hints: We'll establish the regularity for the cases $n = 0,1$ — from which the general inductive proof is not hard to divine.

For the case $n = 0$ we want to show that if f is piecewise continuous and T is an arbitrary test function (Definition 15.9 on page 677) then

$$\lim_{N\to\infty} \int s_{N,0}T \quad \text{exists.}$$

That is, we want to show that

$$\int [\rho_N(t)KN\int f(u)\varphi_N(t-u)du]KN\varphi_N(t)T(t)dt$$

has a limit as N gets large, or, more specifically, that

$$\int_{-1/N}^{1/N}\left[\rho_N(t)KN\int_{t-1/N}^{t+1/N} f(u)\varphi_N(t-u)du\right]KN\varphi_N(t)T(t)dt \qquad \textbf{15.76}$$

has a limit as N gets large.

In this regard, there are only two cases to consider, namely

- f is continuous at 0,
- f has a jump discontinuity at 0.

When f is continuous at 0, first choose N sufficiently large so that for all u between $-2/N$ and $2/N$, $f(u)$ is close to $f(0)$. Also make N sufficiently large so that also the product

$$KN\int_{t-1/N}^{t+1/N} f(u)\varphi_N(t-u)du$$

is close to $f(0)$, simultaneously for all t between $-1/N$ and $1/N$. (We can do this because for each fixed t, as a function of u, $KN\varphi_N(t-u)$ is concentrating all of its unit area very near t, and for all u in this interval, $f(u)$ is close to $f(0)$.)

Since T is a test function, it is surely continuous at 0. Thus we can easily see, examining expression 15.76, and noting that for all t between $-1/N$ and $1/N$, $\rho_N(t) = 1$ (see Theorem 15.21 on page 690) repeating the same argument yields a limit of $f(0)T(0)$.

The case where f has a jump discontinuity at 0 requires a more delicate argument. The algebra (which we will leave to the reader to supply) becomes much more comprehensible with the aid of a suitable picture. Namely, near enough to, but still to the left of 0, we can think of the function f as the sum of a constant function (whose value is the left-hand limit,[1] $f(0-)$) and a very small function ε_L, while on the right and near enough to this discontinuity, the function is the sum of the constant function whose value is the right-hand limit, $f(0+)$, and a very small function ε_R. Thus we may think of f very near the jump discontinuity at 0 as the sum of the displaced multiple, f_H, of the Heaviside unit step, pictured in Figure 15-13, and a very small function, ε, gotten by pasting ε_L and ε_R together.

The contribution of f to expression 15.76 is the sum of the contributions of f_H and ε. The contribution of ε is small (for large N), while the contribution of f_H is, due to the symmetry of φ, just the factor

1. Where $f(0-) = \lim_{\substack{h \to 0 \\ h < 0}} f(h)$ and $f(0+) = \lim_{\substack{h \to 0 \\ h > 0}} f(h)$.

$[f(0-) + f(0+)]/2$. Hence, taking into account the additional factor $T(t)$, we find the limit in the case of a jump discontinuity at 0 to be

$$T(0)\frac{[f(0-) + f(0+)]}{2}.$$

Note that this formula also holds when 0 is a continuity point of f.

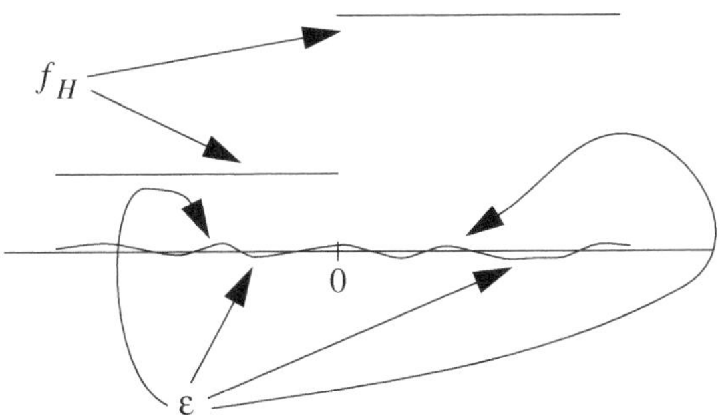

Figure 15-13

We now treat the case $n = 1$. Here we want to show that for each test function T, the quantity

$$\int_{-1/N}^{1/N} \left[\rho_N(t) KN \int_{t-1/N}^{t+1/N} f(u)\varphi_N(t-u)du\right] KN\varphi_N'(t)T(t)dt \qquad \textbf{15.77}$$

has a limit as N grows large.

As previously, we can forget the term $\rho_N(t)$, since it equals 1 for all t between $-1/N$ and $1/N$ (Theorem 15.21 on page 690).

We integrate the outer integral above by parts (Theorem 4.9 on page 121) taking

$$v'(t) = KN\varphi_N'(t), \quad u(t) = KNT(t)\int_{t-1/N}^{t+1/N} f(u)\varphi_N(t-u)du. \qquad \textbf{15.78}$$

Now the function u is infinitely differentiable (the limits of integration for its integrand can be taken as any interval containing $[-2.1/N;2.1/N]$). Furthermore, independent of whether f' does or does not have a discontinuity at 0, we find

$$u'(t) = KNT'(t)\int_{t-1/N}^{t+1/N} f(u)\varphi_N(t-u)du + KNT(t)\int_{t-1/N}^{t+1/N} f(u)\varphi_N'(t-u)du, \qquad \textbf{15.79}$$

the differentiation under the integral sign, which is established using Theorem 12.26 on page 477, the mean value theorem, Theorem 3.1 on page 62, and the uniform continuity of φ' (see Definition A3.3 on page 769 in Appendix 3) which follows from the fact that φ is infinitely differentiable, hence continuous (see paragraph a on page 63 and its establishment, on page 65) hence uniformly continuous (see Theorem A3.4 on page 769). In this integration by parts, the term

$$u(t)v(t)\Big|_{t=-1/N}^{t=1/N}$$

disappears due to the presence of the factors $\varphi(\pm 1/N, N)$. Thus we need only be concerned with

$$-\int_{-1/N}^{1/N} u'(t)v(t)dt\,,$$

which equals the integral in expression 15.77.

Before substituting for $v(t)$ and $u'(t)$ in this integral, we integrate by parts in the second integral of equation 15.79 for $u'(t)$, obtaining that

$$KNT(t)\int_{t-1/N}^{t+1/N} f(u)\varphi_N'(t-u)\,du \;=\; KNT(t)\int_{t-1/N}^{t+1/N} f'(u)\varphi_N(t-u)\,du\,.$$

(Note the absence of the " – " sign, since in φ' the variable u is preceded with a " – " sign.)

After all of this manipulation, we find the integral in expression 15.77 to equal

$$-\int_{-1/N}^{1/N} u'(t)v(t)dt \qquad \mathbf{\mathit{15.80}}$$

$$=\int_{-1/N}^{1/N} -\left\{T'(t)KN\int_{t-1/N}^{t+1/N} f(u)\varphi_N(t-u)\,du+T(t)KN\int_{t-1/N}^{t+1/N} f'(u)\varphi_N(t-u)\,du\right\}KN\varphi_N(t)dt$$

This last expression is the sum of two quantities,

$$\int_{-1/N}^{1/N} -KN\left\{\int_{t-1/N}^{t+1/N} f(u)\varphi_N(t-u)\,du\right\}KN\varphi_N(t)T'(t)dt \qquad \mathbf{\mathit{15.81}}$$

and

$$\int_{-1/N}^{1/N} -KN\left\{\int_{t-1/N}^{t+1/N} f'(u)\varphi_N(t-u)\,du\right\}KN\varphi_N(t)T(t)dt, \qquad \mathbf{\mathit{15.82}}$$

each of which has the same form as the integral in expression 15.76 (since $\rho_N(t) = 1$ in the region of integration) and hence each approaches a limit as N grows large. Thus we have shown that for $n = 0$ and $n = 1$, the sequence given by $s_{N,n}$ in equation 15.75 on page 706 is a regular sequence, and thus, $F\delta^{(n)}$ defined in Definition 15.31 on page 706 yields a legitimate generalized function for $n = 0,1$. The justification for larger n is similar, and is left to the reader.

When G is a generalized function represented by a regular sequence, $\{G_N\}\big|_{N=1}^{\infty}$ (Definition 15.11 on page 680 and Definition 15.13 on page 680) and T is a test function (Definition 15.9 on page 677) we defined

$$\int GT = \lim_{N \to \infty} \int G_N T$$

(see the last part of Definition 15.15 on page 683, with G replacing F).

Now suppose F is a generalized function representing the real valued function f, where $D^n f$ exists and is piecewise continuous. In Definition 15.31 on page 706 we just established a reasonable definition for $F\delta^{(n)}$ as a generalized function. The functions $\rho_k,\ k = 1, 2,...$ (see Notation 15.19 on page 688 and Theorem 15.21 on page 690) are all test functions, with $\rho_k(t) = 1$ for $-k \le t \le k$ (and, in fact, the sequence $\{\rho_N\}\big|_{N=1}^{\infty}$ represents the generalized function corresponding to the real valued constant function, **1**, whose value is 1 (Theorem 15.23 on page 696)). Hence the integrals

$$\int F\delta^{(n)}\rho_k$$

are defined for all positive integers, k, and are all equal. Also, the support of all the ρ_k always includes that of $\delta^{(n)}$.

In light of these facts, it seems eminently reasonable to define

$$\int F\delta^{(n)} = \int F\delta^{(n)}\rho_k. \qquad \textit{15.83}$$

For $n = 0$, from the results of Problem 15.16 on page 706 we therefore see that for f piecewise continuous

$$\begin{aligned}\int F\delta &= \lim_{N \to \infty} \int_{-1/N}^{1/N} KN\left\{\int_{t-1/N}^{t+1/N} f(u)\varphi_N(t-u)du\right\}KN\varphi_N(t)dt \qquad \textit{15.84}\\ &= \frac{f(0-)+f(0+)}{2}\rho(0,k) = \frac{f(0-)+f(0+)}{2}.\end{aligned}$$

For the case $n = 1$, with f' piecewise continuous, since $T'(t) = D\rho_k(t) = 0$ (because $T(t) = \rho_k(t) = 1$ for the t values of interest) we find using expression 15.77 (substituting there for $T(t)$ the value 1) that

$$\int F\delta' = \lim_{N \to \infty} \int_{-1/N}^{1/N} KN\left[\int_{t-1/N}^{t+1/N} f(u)\varphi_N(t-u)du\right] KN\varphi_N'(t)\,dt.$$

Utilizing the derivation following expression 15.77, we find the desired limit to be the sum of the limits as N grows large, of expressions 15.81 and 15.82 with $T(t) = 1$ and $T'(t) = 0$. (This leaves only expression 15.82). That is,

$$\int F\delta' = -\lim_{N \to \infty} \int_{-1/N}^{1/N} KN\left\{\int_{t-1/N}^{t+1/N} f'(u)\varphi_N(t-u)\right\} KN\varphi_N(t)dt$$

$$= -\frac{1}{2}[f'(0-) + f'(0+)] = -\int F'\delta,$$

with the last two steps both justified by equation 15.84.

By inductively continuing this same process, we will have shown the following, which we state formally, for future reference.

Definition 15.32: $\int FD^n\delta$

Suppose f is a real valued function of a real variable whose n-th derivative is piecewise continuous (Definition 3.13 on page 89) with corresponding generalized function $F = \mathcal{M}f$ (Definition 15.24 on page 696) and δ is Dirac's generalized delta function (Definition 15.27 on page 698). We define

$$\int FD^n\delta = \int F(D^n\delta)\rho_k$$

where k can be any positive integer, ρ is defined in Notation 15.19 on page 688, $FD^n\delta$ is defined in Definition 15.31 on page 706, and integrals of the form on the right are defined in the last part of Definition 15.15 on page 683.

❑

Theorem 15.33: $\int FD^n\delta$

Under Definition 15.32 we have

$$\int FD^n\delta = (-1)^n\int(D^nF)\delta = \frac{f^{(n)}(0-) + f^{(n)}(0+)}{2},$$

where

$$f^{(n)}(0-) = \lim_{\substack{h \to 0 \\ h < 0}} f^{(n)}(h) \quad \text{and} \quad f^{(n)}(0+) = \lim_{\substack{h \to 0 \\ h > 0}} f^{(n)}(h) \cdot$$

❑

The result of Theorem 15.33 is just the familiar integration by parts formula applied n times to the left integral, to yield the right-hand average. Note that it applies to continuity points, in such cases the right-hand side being just $f^{(n)}(0)$.

We need to extend these results to the more general case in which δ above is replaced by the composition (Definition 15.28 on page 699)

$$\boldsymbol{d}_{a,\,b} = \delta(at + b), \ a \neq 0, \ \text{and } t \text{ the real identity, } t(\tau) = \tau.$$

To do this we introduce the extension, Definition 15.34, of Definition 15.32.

Definition 15.34: $F\boldsymbol{d}^{(n)}_{a,\,b}$

Let a and b be real numbers, δ be the Dirac delta (Definition 15.27 on page 698) $\boldsymbol{d}_{a,\,b}$ the composition (Definition 15.28 on page 699) given by

$$\boldsymbol{d}_{a,\,b} = \delta(at + b), \ a \neq 0, \ \text{and } t \text{ the real identity, } t(\tau) = \tau\,.$$

Suppose that f is a real valued function of a real variable with piecewise continuous n-th derivative (Definition 3.13 on page 89) and

$$F = \mathcal{M}f = \{F_N\}\Big|_{N=1}^{\infty}$$

(see Definition 15.24 on page 696). The product $F\boldsymbol{d}^{(n)}_{a,\,b}$ is defined as the generalized function (Definition 15.13 on page 680) represented by the regular sequence

$$\{F_N D^n \psi_{a,\,b,\,N}\}\Big|_{N=1}^{\infty},$$

where

$$\psi_{a,\,b,\,N}(t) = KN\varphi(at + b, N)$$

Notation 15.19 on page 688

❑

Establishing that $F\boldsymbol{d}^{(n)}_{a,\,b}$ is well defined is a simple extension of the proof for $F\delta^{(n)}$, and defining the integral, $\int F\boldsymbol{d}^{(n)}_{a,\,b}$, is similarly easy, provided you exercise adequate care.

To guess correctly the result that this integral evaluates to, first look at

$$\int F\boldsymbol{d}_{a,\,b} = \lim_{N\to\infty}\int F_N(\tau)\psi_{a,\,b,\,N}(\tau)d\tau = \lim_{N\to\infty}\int F_N(\tau)KN\varphi_N(a\tau+b)d\tau.$$

With the substitution $w = a\tau + b$, considering separately the cases $a > 0$ and $a < 0$, we find the last expression above to be

$$\lim_{N\to\infty}\int F_N\left(\frac{w-b}{a}\right)KN\varphi_N(w)\frac{1}{|a|}dw = \frac{1}{|a|}\frac{f\left(\frac{-b}{a}-\right)+f\left(\frac{-b}{a}+\right)}{2}.$$

When $n = 1$,
$$\begin{aligned}\int F\boldsymbol{d}'_{a,\,b} &= \lim_{N\to\infty}\int F_N(\tau)\psi'_{a,\,b,\,N}(\tau)d\tau\\ &= -\lim_{N\to\infty}\int F_N'(\tau)\psi_{a,\,b,\,N}(\tau)d\tau\\ &= -\frac{1}{|a|}\frac{f'\left(\frac{-b}{a}-\right)+f'\left(\frac{-b}{a}+\right)}{2}\end{aligned}$$

and in general,

$$\int F\boldsymbol{d}^{(n)}_{a,\,b} = (-1)^n\frac{1}{|a|}\frac{f^{(n)}\left(\frac{-b}{a}-\right)+f^{(n)}\left(\frac{-b}{a}+\right)}{2}.$$

We state this result formally in the following definition.

Definition 15.35: $\int F\boldsymbol{d}^{(n)}_{a,\,b}$

For $F\boldsymbol{d}^{(n)}_{a,\,b}$ as given in Definition 15.34 on page 712, define $\int F\boldsymbol{d}^{(n)}_{a,\,b}$ to be

$$\int F\boldsymbol{d}^{(n)}_{a,\,b} = (-1)^n\frac{1}{|a|}\frac{f^{(n)}\left(\frac{-b}{a}-\right)+f^{(n)}\left(\frac{-b}{a}+\right)}{2}$$

where
$$g(x-) = \lim_{\substack{h\to 0\\ h<0}} g(x+h), \quad g(x+) = \lim_{\substack{h\to 0\\ h>0}} g(x+h)$$

❑

Notice that in the process of embedding the piecewise continuous function, $\mathcal{PC}$ in the class $\mathcal{G}$, of generalized functions, in a sense we make the natural simplification of letting the value of f at a jump be the average of its left- and right-hand limits. That is, for f piecewise continuous and each real t

$$\lim_{N \to \infty} \mathcal{M}(f, N)(t) = \frac{f(t-) + f(t+)}{2}, \qquad \textbf{15.85}$$

the essence of which was established in the hints to Problem 15.16 on page 706. This is reflected in Theorem 15.33 and in Definition 15.35

We now establish the generlized function product rule.

Theorem 15.36: The product rule for generalized functions

Let F and G be generalized functions, for which $F'G$ exists in conformity with Definition 15.30 on page 704. Then the generalized functions FG and FG' exist, and

$$(FG)' = F'G + FG'. \qquad \textbf{15.86}$$

❑

To establish the product rule, we have to be fairly systematic in our approach in order to keep from descending into utter confusion.

From the original hypotheses that f' is piecewise continuous, and g is piecewise continuous, recall Definition 15.30, and apply it with $n = 1$, $q = f$, $Q = F$, $R = G$, and $r = g$, to conclude that

$$FG' \text{ exists and } FG' = D\mathcal{M}(fg) - \mathcal{M}(f'g). \qquad \textbf{15.87}$$

Because when f' is piecewise continuous (in $\mathcal{PC}$) f is in $\mathcal{PC}$, under the original hypotheses both f and g are in $\mathcal{PC}$. Hence, applying Definition 15.30 with $q = f$, $Q = F$, $R = G$, and $r = g$, but now with $n = 0$; we conclude that

$$FG \text{ exists and } FG = \mathcal{M}(fg).$$

Hence

$$(FG)' = D\mathcal{M}(fg). \qquad \textbf{15.88}$$

Finally applying Definition 15.30 with $q = f'$, $Q = F'$, $R = G$, $r = g$, with $n = 0$; we conclude that

$$F'G \text{ exists and } F'G = \mathcal{M}(f'g). \qquad \textbf{15.89}$$

Combining equations 15.87, 15.88, and 15.89 yields

$$(FG)' = F'G + FG',$$

establishing the product rule.

Note that we cannot operate with quite as free a hand in using the product rule here, as we could in the class of differentiable functions. This is not too surprising, when you consider how many more objects have *generalized derivatives* than have derivatives.

Proper use of the previous definition and theorem is a fairly delicate matter, as can be seen from the following example.

Example 15.17: $\mathcal{M}(e^{2t})\delta$ and its consequences

Let t be the real identity function ($t(\tau) = \tau$). Recalling Definition 15.24 on page 696 of the operator $\mathcal{M}$, the definition of $F\delta$ (from Definition 15.31 on page 706) and the relevant part of Definition 15.15 on page 682, we have the equations

$$\int \mathcal{M}(e^{2t})\delta T = \lim_{N \to \infty} \int [KN\varphi_N(t-u)e^{2u}du]KN\varphi_N(t)T(t)dt = T(0) \quad \textbf{\textit{15.90}}$$

because the term in the square brackets gets close to e^{2t} which are uniformly close to 1 when t is near 0.

$$\int \delta T = T(0).$$

From the definition of equality of generalized functions (see Definitions 15.12 and 15.13, page 680) we therefore have

$$\mathcal{M}(e^{2t})\delta = \delta. \quad \textbf{\textit{15.91}}$$

Now from the product rule, Theorem 15.36 on page 714, we find that

$$2\mathcal{M}(e^{2t})\delta + \mathcal{M}(e^{2t})\delta' = \delta''$$

or, using equation 15.91,

$$\delta' = 2\delta + \mathcal{M}(e^{2t})\delta'.$$

You shouldn't make the mistake of believing, just because of equation 15.91, that when multiplying δ by $\mathcal{M}(e^{2t})$ (where t is the real identity function) where $\mathcal{M}(e^{2t})$ behaves like the number 1, that when $\mathcal{M}(e^{2t})$ multiplies δ', it still behaves like the number 1 (which would imply that $\mathcal{M}(e^{2t})\delta' = \delta'$). An explanation for the behavior we see is that δ' ***looks at the behavior of the*** **derivative** ***of what it multiplies***, and though $e^{2\tau}$ looks like 1 for τ near 0, its derivative doesn't.

■

Problems 15.18

1. Establish Leibniz's rule (15.70 on page 702) for generalized functions, when the product QD^nR is defined in accordance with Definition 15.30 on page 704.
   Hints: What we'd like to show is that when $D^n f$ and g are piecewise continuous, with $F = \mathcal{M}f,\ G = \mathcal{M}g$, then the following equation is both meaningful and true:

$$D^n(FG) = \sum_{k=0}^{n} \binom{n}{k}(D^k F)(D^{n-k}G).$$

It's not hard to show that Definition 15.30 applies to FG with $D^n(FG) = D^n\mathcal{M}(fg)$. Also using this definition we find all products on the right side defined with

$$(D^k F)(D^{n-k}G) = \sum_{j=0}^{n-k} (-1)^j \binom{n-k}{j} D^{n-(k+j)}\mathcal{M}[(D^{k+j}f)\,g].$$

So, our object is to show that

$$D^n\mathcal{M}(fg) = \sum_{k=0}^{n} \binom{n}{k} \sum_{j=0}^{n-k} (-1)^j \binom{n-k}{j} D^{n-(k+j)}\mathcal{M}[(D^{k+j}f)\,g]$$

or

$$D^n\mathcal{M}(fg) = \sum_{k=0}^{n} \sum_{j=0}^{n-k} \binom{n}{k}(-1)^j \binom{n-k}{j} D^{n-(k+j)}\mathcal{M}[(D^{k+j}f)\,g].$$

The important points to realize are that we can't do anything to manipulate *inside* the expression $\mathcal{M}[(D^{k+j}f)g]$ and that we are dealing with a double sum over a set written in sequential form (see Definition 12.16 on page 454). Just as with integration, for determination of the sum,

$$\sum_{\{(k,\,j):0 \le k \le n,\, 0 \le j \le n-k\}} v(k,j,n)$$

a change of variables is suggested — in this case, $u = k + j,\ v = j$. Then the substitution is made, and the set $\{(k,j):0 \le k \le n, 0 \le j \le n-k\}$ is rewritten in sequential form but in the variables u,v. When this is done, the set of interest is $\{(u,v):0 \le u \le n, 0 \le v \le u\}$. With the aid of the binomial expansion[1] of $(1-1)^u$, which is 0 for u a positive integer, and

---

1. We prove the binomial expansion, $(a+b)^n = \sum_{j=0}^{m} \binom{n}{j} a^j b^{n-j}$, by induction.

the value of the first term in the double sum for $u = 0$, the desired result falls out.

2. Show that if f is in the class C^{∞} (see Notation 15.1 on page 655) and g is in $\mathcal{PC}$ (Notation 15.16 on page 685) with $F = \mathcal{M}f$, $G = \mathcal{M}g$ (Definition 15.24 on page 696) and that the sequence $\{\gamma_N\}\Big|_{N=1}^{\infty}$ represents $\mathcal{M}g$, but ***not assuming that***

$$\gamma_N = \mathcal{M}(f, N),$$

show that the product FD^nG exists for each nonnegative integer n, and is represented by the sequence $\{fD^n\gamma_N\}\Big|_{N=1}^{\infty}$.

Hints: The existence of FD^nG follows from Definition 15.30 on page 704. We'll only show how the product rule establishes the desired result for $n = 1$, the general case being similar, but using Leibniz's rule, see Problem 1.
The heart of the solution is showing that the sequence $\{f\gamma_N\}\Big|_{N=1}^{\infty}$ represents the generalized function FG, because remember from Definition 15.30 on page 704, FG ***is defined*** as $\mathcal{M}(fg)$ (the generalized function represented by the embedding sequence $\{(fg)_N\}\Big|_{N=1}^{\infty}$ given by

$$(fg)_N(t) = \mathcal{M}(fg, N)(t) = KN\int f(u)g(u)\varphi_N(t-u)du,$$

which is not the same sequence as $\{f\gamma_N\}\Big|_{N=1}^{\infty}$).
But we know that G is represented by both $\{\gamma_N\}\Big|_{N=1}^{\infty}$ and (from the embedding of g) by $\{g_N\}\Big|_{N=1}^{\infty}$, where

$$g_N(t) = \mathcal{M}(g, N)(t) = KN\int g(u)\varphi_N(t-u)du.$$

From this, we know that for all test functions $\boldsymbol{T}$

$$\lim_{N\to\infty}\int g_N T = \lim_{N\to\infty}\int \gamma_N T = L_T.$$

Thus to show that $FG = \mathcal{M}(fg)$ we need to show that

$$\lim_{N\to\infty}\int (fg)_N T = \lim_{N\to\infty}\int f\gamma_N T .$$

This takes a little bit of work. Since f is infinitely differentiable, we know that fT is a test function, and thus the right hand integral is L_{fT}. Because f has a continuous derivative, we know that f' is bounded in the support of T (Theorem 3.12 on page 88). This shows that f is uniformly continuous on each closed bounded real interval (see Appendix 3, Definition A3.3 on page 769 for the definition of uniform continuity; the proof of uniform continuity follows from the mean value theorem, Theorem 3.1 on page 62). From the uniform continuity of f we can conclude that for all t in the support of each test function T (Definition 15.7 on page 677) for any given positive number ε, we can choose a number $\delta > 0$, dependent on ε, and T, but not on t, such that if

$$|t - u| < \delta, \quad \text{then} \quad |f(t) - f(u)| < \varepsilon.$$

This means that for all t in the support of T we can write

$$f(u) = f(t) + E(u) \quad \text{where} \quad |E(u)| < \varepsilon$$

provided $|t - u| < \delta$. In the equation

$$(fg)_N(t) = KN\int f(u)g(u)\varphi_N(t-u)\,du = KN\int_{t-1/N}^{t+1/N} f(t)g(u)\varphi_N(t-u)du$$

choose $\delta = 1/N$ and replace $f(u)$ by $f(t) + E(u)$, to see that we can make this integral arbitrarily close to

$$KN\int_{t-1/N}^{t+1/N} f(t)g(u)\varphi_N(t-u)du,$$

in which case $\int (fg)_N T$ is arbitrarily close to

$$\int f(t)g_N(t)T(t)dt = L_{fT}.$$

Thus we have finally shown that

$$\lim_{N \to \infty} \int (fg)_N T = L_{fT},$$

and we already knew that

$$\lim_{N \to \infty} \int f\gamma_N T = L_{fT}.$$

Therefore the sequence $\{f\gamma_N\}\Big|_{N=1}^{\infty}$ represents the generalized function $FG = \mathcal{M}(fg)$. In fact our proof shows that the generalized function represented by the sequence $\{(D^n f)\gamma_N\}\Big|_{N=1}^{\infty}$ is $(D^n F)G$. Now invoke the product rule,

$(FG)' = F'G + FG'$ (Theorem 15.36 on page 714) which we rewrite as

$$FG' = (FG)' - F'G.$$

From the product rule for differentiable real valued functions of a real variable (see Theorem 2.13 on page 53))

$$f\gamma_N' = (f\gamma_N)' - f'\gamma_N.$$

Since the sequence $\{(f\gamma_N)' - f'\gamma_N\}|_{N=1}^{\infty}$ represents the generalized function $(FG)' - F'G = FG'$, it follows that the sequence $\{f\gamma_N'\}|_{N=1}^{\infty}$ represents FG', which is the desired result for $n = 1$.

We now come to an important result which we state without proof, because we don't have the proper *measure-theoretic* tools available. However, in discussing it, we will indicate what would be needed for its establishment.

Theorem 15.37: Dominated convergence theorem

If for all $N = 1, 2, 3, \ldots$

- g_N belongs to the class $\mathcal{T}$ of test functions (Definition 15.9 on page 677)
- the g_N are *uniformly bounded* (i.e., there is a constant, B such that for all real t and all N, $|g_N(t)| \le B$)

and

- $\lim_{N \to \infty} g_N = f$ exists with f in the class $\mathcal{PC}$ of piecewise continuous functions (Notation 15.16 on page 685)

then the sequence $\{g_N\}|_{N=1}^{\infty}$ represents the generalized function $\mathcal{M}f$ (Definition 15.24 on page 696) corresponding to f.

❑

From the definition of $\mathcal{M}f$ and Definition 15.13 on page 680, of generalized functions, we know that in order to establish this result, we must show that for all test functions T

$$\lim_{N \to \infty} \int g_N T = \lim_{N \to \infty} KN \int \left[\int f(u)\varphi_N(t-u)du \right] T(t)dt.$$

We already know from equation 15.69 on page 695 that

$$\lim_{N \to \infty} KN \int \left[\int f(u)\varphi_N(t-u)du \right] T(t)dt = \int f(t)T(t)dt.$$

So all we we would have to show is that under our hypotheses above,

$$\lim_{N \to \infty} \int g_N T = \int f(t)T(t)dt. \qquad \mathbf{15.92}$$

The condition of uniform boundedness of the g_N is to avoid a situation like $g_N(t) = f(t) + 1/(Nt)$. When the uniform boundedness is satisfied, then it turns out that with the convergence of the g_N to f, we can show that the *size* of the set of t values for which $g_N(t)$ isn't near $f(t)$, is small. That is, for each $\varepsilon > 0$, the so-called *measure*, m, of the set of t such that $|g_N(t) - f(t)| > \varepsilon$ gets small. The contribution of this set to the integral $\int g_N T$ can't exceed mB, which is therefore small. Off of this set, $g_N(t)$ is close to $f(t)$, so that equation 15.92 holds. The trouble is, that although it seems intuitively reasonable, proving the assertion about the measure m becoming small requires more development than we can afford.

A rather easy-to-prove corollary of Theorem 15.37 is the following.

Theorem 15.38

If the uniformly bounded sequence $\{g_N\}\Big|_{N=1}^{\infty}$ represents the generalized function G and $\lim_{N \to \infty} g_N = f$ exists with f in the class $\mathcal{PC}$, of piecewise continuous functions (Notation 15.16 on page 685) then

$$\int G\,\delta(t - x) = \frac{f(x-) + f(x+)}{2}.$$

Here δ is Dirac's *delta* (Definition 15.27 on page 698) and, for any g,

$$g(x-) = \lim_{\substack{h \to 0 \\ h < 0}} g(x + h), \quad g(x+) = \lim_{\substack{h \to 0 \\ h > 0}} g(x + h)$$

❑

These last results get us even closer to the handle we need on generalized function initial conditions. They will be of use in recognizing when a given generalized function specified by some regular sequence represents a piecewise continuous function; for example, when the generalized function solution of a given differential equation corresponds to a piecewise continuous function, and which particular piecewise continuous function.

Problem 15.19

Let f and g be in the class $\mathcal{PC}$ (Notation 15.16 on page 685) and $f \sim g$ be an abbreviation for

$$\int (\mathcal{M}f)\delta(t - b) = \int (\mathcal{M}g)\delta(t - b) \text{ for all real } b$$

where t is the real identity function ($t(\tau) = \tau$) δ is Dirac's *delta*, Definition 15.27 on page 698, and $\mathcal{M}$ is defined in Definition 15.24 on page 696. Show that $f \sim g$ defines an equivalence relation over the class $\mathcal{PC}$ (see Definition 15.12 on page 680, with the understanding that an equivalence relation is a partition of a set into mutually exclusive equivalence classes). What, if any, is the significance of this result?

We have finally reached a point where we can generalize more completely the notion of a solution to the equation $LX = Y$. The generalization from L to L_1 (see Definition 15.3 on page 665) was not nearly enough. We need to be able to handle the case $Y = \delta$, Dirac's impulse as a forcing function. And we must adapt the theory to deal with initial conditions. We begin the next section with a formal definition which extends the operator L.

6 $\mathcal{L}X = Y$, Existence and Uniqueness Theorems

Definition 15.39: The operator $\mathcal{L}$

The domain (Definition 1.1 on page 2) of $\mathcal{L}$ consists of those elements g of $\mathcal{G}$ (Definition 15.13 on page 680) for which $g\delta$ is defined (see Definition 15.31 on page 706) and

$$\int g\delta = 0$$

(see Definition 15.32 on page 711 and Theorem 15.33 on page 711). For g in the domain of $\mathcal{L}$, $\mathcal{L}g$ is defined by

$$\mathcal{L}g = (D+2)g.$$

Note that relying on Theorem 15.33, the condition $\int g\delta = 0$ corresponds to the previous initial condition $X(0) = 0$, associated with equation 15.5 on page 655. We abbreviate the problem

$$\text{Solve } (D+2)X = Y \text{ subject to } \int X\delta = 0,$$

where Y is a generalized function and D is the derivative operator on generalized functions, simply as

$$\text{Solve } \mathcal{L}X = Y.$$

❑

Now the range of $\mathcal{L}$ (see Definition 1.1 on page 2) is a subset of the class $\mathcal{G}$ of generalized functions. But does this range include all generalized functions? In other words, given Y in $\mathcal{G}$, does the equation

$$\mathcal{L}X = Y \qquad \textbf{\textit{15.93}}$$

necessarily possess a *nonempty solution*, i.e., at least one generalized function X for which equation 15.93 is satisfied. ***The answer to this is NO***, but to see this, we must develop some machinery. (Recall that in dealing with ordinary real valued functions of a real variable, if Y were continuous, then there is a solution, X, to $(D+2)X = Y$ subject to $X(0) = 0$.)

Definition 15.40: Antiderivative of a generalized function

The generalized function (Definition 15.13 on page 680) f is said to be an **antiderivative** or **primitive** of the generalized function g if

$$Df = g$$

(see Definition 15.15 on page 682).

❑

Referring to the definition just given, we can immediately, however, establish the existence of such primitives.

Theorem 15.41: Existence of generalized function primitive

Every generalized function has a primitive — or, in other words,

> if g is a generalized function, then the equation $Df = g$ has a nonempty solution (at least one generalized function f satisfying this equation).

❑

To prove this result, in light of the definition of generalized function (Definition 15.13 on page 680) we must construct a regular sequence $\{f_N\}\big|_{N=1}^{\infty}$, such that $\{f'_N\}\big|_{N=1}^{\infty}$ is equivalent to $\{g_N\}\big|_{N=1}^{\infty}$ (see Definition 15.12 on page 680). The obstacle to this construction is that while its easy enough to construct f_N such that $f'_N = g_N$, such f_N cannot, in general, have bounded support. Just look at what happens if $g_1 = \rho_1$. Then f_1 is 0 to the left of -2, and on the support of $g_1 = \rho_1$ must be increasing, beyond which it must be constant. So its support is not bounded. However, this example should point the way to what we have to do. It's worth trying to do this yourself before going on. Here's the proof.

┌

Let the sequence $\{g_N\}\big|_{N=1}^{\infty}$ represent the generalized function g, and define Υ and F_N by

$$\Upsilon(t) = K\varphi_1(t),\ F_N(t) = \int_{-\infty}^{t} g_N(z)dz - \int_{-\infty}^{\infty}\left[\Upsilon(s)\int_{-\infty}^{s} g_N(z)dz\right], \tag{15.94}$$

and

$$f_N(t) = \rho_N(t)F_N(t). \tag{15.95}$$

(For φ, K, ρ, recall Notation 15.19 on page 688.) It seems reasonable that the f_N will define the generalized function f that we're seeking.

Since the g_N are infinitely differentiable (being a regular sequence, Definition 15.11 on page 680, because they represent a generalized function) from the fundamental theorem of calculus (Theorem 4.10 on page 127) F'_N exists, and is infinitely differentiable, with ,

$$F'_N = g_N, \tag{15.96}$$

and hence from equation 15.95, the f'_N belong to the class $\mathcal{T}$ of test functions (Definition 15.9 on page 677). From equation 15.95 we know that the f_N have bounded support, because of the factor ρ_N, and are differentiable, so that the f_N belong to the class $\mathcal{T}$ of test functions.

We next want to establish that the f_N sequence is regular (Definition 15.11 on page 680). To accomplish this we go through several steps. First, for all $N =$ 1,2,3,..., we can see directly from equation 15.94, that

$$\int \Upsilon = 1 \tag{15.97}$$

using Notation 15.19 on page 688, and therefore also that

$$\int F_N \Upsilon = 0. \tag{15.98}$$

Now choose an arbitrary test function, T (Definition 15.9 on page 677) and define

$$\chi(t) = \int_{-\infty}^{t} T(z)dz - \left(\int_{-\infty}^{t} \Upsilon(z)dz\right)\int T. \tag{15.99}$$

One can readily see that χ is a test function; because Υ is a test function (from equation 15.94 and Notation 15.19 on page 688) so for t to the left of the support of T and Υ, we have $\chi(t) = 0$, while from equation 15.97 it's also true that $\chi(t) = 0$ for t to the right of the support of T and Υ; so χ has bounded support; furthermore, from the fundamental theorem of calculus, Theorem 4.10 on page 127,

$$\chi'(t) = T(t) - \Upsilon(t)\int T, \qquad \textbf{15.100}$$

so that χ is infinitely differentiable, thus establishing our claim that it is a test function.

It follows from this equation that

$$\begin{aligned}\int F_N T &= \int F_N[\chi' + \Upsilon\int T] \\ &= \int F_N\chi' + (\int F_N\Upsilon)\int T \qquad \textbf{15.101} \\ &= \int F_N\chi' \quad \text{this last step justified by equation 15.98.}\end{aligned}$$

Now integrating by parts we find

$$\begin{aligned}\int F_N\chi' &= -\int F'_N\chi \qquad \textbf{15.102} \\ &= -\int g_N\chi \quad \text{this last step justified by equation 15.96.}\end{aligned}$$

The limit of this last integral exists, since χ is a test function. Hence, combining equations 15.101 and 15.102, we find that

$$\lim_{N\to\infty}\int F_N T = \lim_{N\to\infty} -\int g_N T \quad \text{exists.} \qquad \textbf{15.103}$$

But for large enough N, for all t in the support of T we know from equation 15.95 and the definition of ρ_N (Notation 15.19 on page 688) that $f_N(t) = F_N(t)$. Hence

$$\lim_{N\to\infty}\int f_N T$$

exists. So f_N is now seen to generate a regular sequence of test functions, and thus defines a generalized function that we will call f. All we need finally show is that $f' = g$ to prove Theorem 15.41.

To carry out this last step, we see that for each test function T, for large enough N, for all t in the support of T $\rho_N(t) = 1$ and $\rho'_N(t) = 0$. Hence, using equation 15.95 (the definition of f_N) we find

$$\begin{aligned}\lim_{N\to\infty}\int f'_N T &= \lim_{N\to\infty}\int[\rho'_N(t)\, F_N(t) + \rho_N(t)F'_N(t)]T(t)dt \\ &= \lim_{N\to\infty}\int F'_N(t)T(t)dt \\ &= \lim_{N\to\infty}\int g_N(t)T(t)dt \quad \text{this last step using equation 15.96.}\end{aligned}$$

Thus the sequence $\{f'_N\}\Big|_{N=1}^{\infty}$ represents the generalized function g, so that $f' = g$.

This shows that indeed, each generalized function g has a primitive, f, i.e., a generalized function f with $f' = g$.

❑

To help determine existence and uniqueness of solutions of $\mathcal{L}X = g$, we need the following.

Theorem 15.42: Generalized functions with same derivative

If two generalized functions have the same derivative, then they differ by a constant. In other words, if f and g are in the class $\mathcal{G}$ (Definition 15.13 on page 680) and $f' = g'$ (see Definition 15.15 on page 682 and Definition 15.13 on page 680) then there is a constant real valued function, $\mathcal{K}$, of a real variable (whose value is k) such that

$$f - g = \mathcal{M}\mathcal{K}$$

(see Definition 15.24 on page 696).

❑

From the condition $f' = g'$ we know from the definitions referred to above that if f and g are represented by the sequences $\{f_N\}\big|_{N=1}^{\infty}$ and $\{g_N\}\big|_{N=1}^{\infty}$, respectively, that for each test function T

$$\lim_{N \to \infty} \int f_N' T = \lim_{N \to \infty} \int g_N' T.$$

Hence, for all test functions T

$$\lim_{N \to \infty} \int [f_N - g_N]' T = 0.$$

From the definition of subtraction of generalized functions (see Definition 15.15 on page 682) because the sequence $\{0\rho_N\}\big|_{N=1}^{\infty}$ represents the *zero constant generalized function,* $\mathbf{0}$ (see Theorem 15.23 on page 696 and Definition 15.24 on page 696) that

$$(f - g)' = \mathbf{0}.$$

So, letting $r = f - g$, we need only prove the following.

> If $r' = \mathbf{0}$, then $r = \mathcal{M}\mathcal{K}$ for some constant real valued function $\mathcal{K}$ (again see Theorem 15.23 and Definition 15.24).

Since for each test function, T,

$$\lim_{N \to \infty} \int k\rho_N T = k\int T,$$

we want to show for each test function T, that

since

$$\lim_{N \to \infty} \int r_N' T = 0 \qquad \textbf{15.104}$$

it follows that

$$\lim_{N \to \infty} \int r_N T = k\int T \qquad \textbf{15.105}$$

for some real constant k.

For test functions T with $\int T = 0$, we therefore need only show that

$$\lim_{N \to \infty} \int r_N T = 0. \qquad \textbf{15.106}$$

But for such a test function, it's not at all hard to see that the function $\mathcal{T}$ defined by

$$\mathcal{T}(t) = \int_{-\infty}^{t} T(u)du$$

is a test function, because it is clearly infinitely differentiable (from the fundamental theorem of calculus (Theorem 4.10 on page 127) and the infinite differentiability of the [test function] T) and if it did not have bounded support, since T **does** have bounded support, there is no way we could have $\int T = 0$. Suppose that support to lie in the interval $[a; b]$. Then we can use integration by parts in this case, to see that

$$\int r_N T = \int_a^b r_N T = r_N(t)\,\mathcal{T}(t)\Big|_{t=a}^{b} - \int_a^b r_N' \,\mathcal{T} = -\int_a^b r_N' \,\mathcal{T}$$

from which we can conclude from equation 15.104 that

$$\lim_{N \to \infty} \int r_N T = 0.$$

This establishes Theorem 15.42 for the case $\int T = 0$.

For the case $\int T \neq 0$, pick any test function T_1 for which $\int T_1 \neq 0$. Here all we need show is that if A is defined by

$$\lim_{N \to \infty} \int r_N T_1 = A\int T_1;$$

then for all other test functions T with $\int T \neq 0$, we have

$$\lim_{N \to \infty} \int r_N T = A \int T. \qquad \textit{15.107}$$

That is, we must show that

$$\lim_{N \to \infty} \int r_N T = \frac{\lim_{N \to \infty} \int r_N T_1}{\int T_1} \int T$$

which we may write as

$$\lim_{N \to \infty} \int r_N \frac{T}{\int T} = \lim_{N \to \infty} \int r_N \frac{T_1}{\int T_1}. \qquad \textit{15.108}$$

Now let

$$T_1^* = \frac{T_1}{\int T_1} \quad \text{and} \quad T^* = \frac{T}{\int T}.$$

Note that T_1^* and T^* are both test functions, and that

$$\int T_1^* = \int T^* = 1.$$

Establishing equation 15.108 is thus seen to be equivalent to establishing the equation

$$\lim_{N \to \infty} \int r_N T^* = \lim_{N \to \infty} \int r_N T_1^*$$

or

$$\lim_{N \to \infty} \int r_N (T^* - T_1^*) = 0.$$

But since $T^* - T_1^*$ is itself a test function, and $\int (T^* - T_1^*) = 0$, this follows from the result of the first part of our proof, yielding the result that if two generalized functions have the same derivative, they differ by a (generalized) constant function.

Problems 15.20

1. Let t be the real identity function ($t(\tau) = \tau$ for all real τ). Show that for X a generalized function corresponding to a piecewise continuous function,

$$X' + 2X = (\mathcal{M}e^{-2t})[(\mathcal{M}e^{2t})X]'.$$

(This is the generalized version of the shift rule (Theorem 5.11 on page 146).)
Hint: You may want to recall the product rule (Theorem 15.36 on page 714).

2. Using the results of Problem 1, and Theorem 15.41 on page 722 to show that the differential equation $X' + 2X = g$ possesses a nonempty[1] generalized function solution for each generalized function g.
Hints: Besides Theorem 15.41 you may want to recall Definition 15.30 on page 704 (product) and the discussion following it.

Because of the initial condition restriction $\int X\delta = 0$ we know that the domain (Definition 1.1 on page 2) of $\mathcal{L}$ (Definition 15.39 on page 721) does not consists of the entire set, $\mathcal{G}$, of generalized functions. It should therefore not surprise us too much that the range of $\mathcal{L}$ is smaller than $\mathcal{G}$. We have in fact the following theorem.

Theorem 15.43: Nonexistence result

Recall the Dirac δ, Definition 15.27 on page 698, and the definition of the derivative of a generalized function, Definition 15.15 on page 682.

The equation $\mathcal{L}X = \delta'$ does not have a nonempty solution (see Footnote 1 on page 728).

❑

To see this we start by noticing that from the constructions resulting from Problems 15.20, **all** solutions of $\mathcal{L}X = \delta'$ are of the form

$$X = \mathcal{M}(e^{-2t})Y, \qquad \textbf{\textit{15.109}}$$

where Y is a *primitive* of $\mathcal{M}(e^{2t})\delta'$ (see Definition 15.40 on page 722) i.e.,

$$Y' = \mathcal{M}(e^{2t})\delta',$$

where we know from Theorem 15.41 on page 722 that Y exists.

The initial condition $\int X\delta = 0$ (see Definition 15.32 on page 711) which because of equation 15.109 may be written as

$$\int \mathcal{M}(e^{-2t})Y\delta = 0, \qquad \textbf{\textit{15.110}}$$

is ***meaningless*** — in the sense that it does not fall under Definition 15.32, as we now show; because from the product rule (Theorem 15.36 on page 714)

1. That is, there is at least one generalized function X for which $X' + 2X = g$.

$$[M(e^{2t})\delta]' = \mathcal{M}(e^{2t})\delta' + \mathcal{M}(2e^{2t}\delta)$$

and hence, from Example 15.17 on page 715, we know that

$$Y' = \mathcal{M}(e^{2t})\delta' = \delta' - 2\delta. \qquad \textbf{\textit{15.111}}$$

Now, therefore, using Theorem 15.42 on page 725 and the last result in Theorem 15.29 on page 700, we find that every primitive, Y, of $\mathcal{M}(e^{2t})\delta'$ must be of the form

$$Y = \delta - 2\mathcal{M}(H) + \mathcal{M}(\mathcal{K}), \qquad \textbf{\textit{15.112}}$$

where H is the Heaviside unit step function (again, see Theorem 15.29) and $\mathcal{K}$ is a real valued constant function with value k, of a real variable.

Thus the initial conditions may be written

$$\int \mathcal{M}(e^{-2t})[\delta - 2\mathcal{M}(H) + \mathcal{M}(\mathcal{K})]\,\delta = 0,$$

or, again using the results of Example 15.17, and the commutativity of generalized function multiplication (see Problem 15.15.2 on page 705)

$$\int [\delta - 2\mathcal{M}(H) + \mathcal{M}(\mathcal{K})]\,\delta$$

or, using Theorem 15.33 on page 711

$$\int \delta^2 - 1 + k = 0. \qquad \textbf{\textit{15.113}}$$

Since δ is not a generalized function corresponding to an embedded piecewise continuous function, we cannot apply Definition 15.32 on page 711 (definition of the product of generalized functions) so that equation 15.110 makes no sense under our current system. This establishes Theorem 15.43.

We remark further that **we would not want $\mathcal{L}X = \delta'$ to have a solution** for several reasons. Mathematically, any reasonable attempt at defining $\int \delta^2$ would lead to *the value of* δ *at* 0, which could not be a finite number. Physically, if $\int \delta^2$ were to mean anything, it would have to be the limiting energy dissipated across a one-ohm resistor by a voltage I_N which is a rectangular pulse of unit area whose base length is $1/N$. The energy from I_N (considered as a voltage) across an R-ohm resistor, from consideration of physics, is

$$\int \frac{I_N^2(t)}{R}\,dt,$$

and hence the energy from such a pulse is $\int I_N^2(t)dt = \int_{-1/(2N)}^{1/(2N)} N^2 dt = N,$ and the limiting energy is infinite.

We now present fundamental theorems for more general derivative operator polynomials than $D + 2$.

Theorem 15.44: General solution of $P(D)X = g$

Let $P(D) = \sum_{j=1}^{n} a_j D^j$ be an n-th degree polynomial in the derivative operator D, acting on elements of the set $\mathcal{G}$ of generalized functions (Definition 15.13 on page 680). Then for each g in $\mathcal{G}$, there are solutions of the equation $P(D)X = g$, and every generalized function X satisfying this equation can be written in a form involving n *arbitrary* constants; by which we mean that for given P and g there is a set of generalized functions, each of which can be denoted by $R(k_1,...,k_n)$ (one generalized function $R(k_1,...,k_n)$ for each choice of the sequence $(k_1,...,k_n)$ of complex numbers) which constitutes the set of all solutions of $P(D)X = g$.

❑

We establish this result formally by the methods illustrated in Section 6 of Chapter 6, which starts on page 203, being justified in all of the analogous operations by the product rule for differentiation of generalized functions (Theorem 15.36 on page 714) the existence of primitives of generalized functions (Theorem 15.41 on page 722) and Theorem 15.42 on page 725 which extends to generalized functions the result that two functions whose derivatives are the same differ by a constant.

Remark: If g is the generalized function corresponding to the real valued function $\mathcal{g}$ of a real variable having a continuous n-th derivative, then our solution is *the same* as the solution of $P(D)\mathcal{X} = \mathcal{g}$, where $P(D)$ is assumed to be acting on functions having a continuous n-th derivative. That is, the solution, X, of the generalized function equation $P(D)X = g$, and the solution, $\mathcal{X}$, of the ordinary differential equation $P(D)\mathcal{X} = \mathcal{g}$ are related by the equation $X = \mathcal{M}(\mathcal{X})$ (see Definition 15.24 on page 696). This follows from Theorem 15.25 on page 696 (linearity of the operator $\mathcal{M}$) and Theorem 15.26 on page 696 ($D\mathcal{M} = \mathcal{M}D$).

Theorem 15.45: Uniqueness result for $\mathcal{P}(D)X = g$

Let $\mathcal{P}(D)$ be an n-th degree polynomial in the derivative operator D acting on those generalized functions X such that

$$\int X\delta = \int X\delta' = \cdots = \int X\delta^{(n-1)} = 0. \qquad \textit{15.114}$$

Note that by Definition 15.30 on page 704, this then restricts X to be of the form $X = \mathcal{M}(f)$, where $f^{(n-1)} = D^{n-1}f$ is in $\mathcal{PC}$ (Notation 15.19 on page 688). Then if g is in $\mathcal{PC}$, the equation $\mathcal{P}(D)X = g$ has a unique solution.

Actually, with respect to the solution X, $D^n X$ is the embedding of a piecewise continuous function (see Theorem 15.22 on page 691).

❑

We give only the highlights of the proof, with more vigor than rigor.

In finding the most general solution of $P(D)X = g$ (where $P(D)$ stands for the same polynomial as specified by $\mathcal{P}(D)$, but without the restriction specified by equations 15.114) we can discern a particular solution consisting of that part of the formula we have obtained (using the shift rule, etc., see Section[1] 6 of Chapter 6), which does not involve terms containing arbitrary constants arising from antidifferentiation. The terms involving the arbitrary constants actually constitute the general solution of the homogeneous equation — i.e., the equation with $g = \mathbf{0}$. These terms are of the form $\mathcal{M}(q_j(t)e^{c_j t})$ (see Definition 15.24 on page 696) where the c_j are distinct complex numbers, and q_j is a polynomial of degree $m(j)$, with

$$\sum_j [m(j)+1] = n$$

(q_j having $m(j)+1$ arbitrary constants). It is not *too* difficult to show[2] that the resulting system of algebraic equations

$$\int X\delta = \int X\delta' = \cdots = \int X\delta^{(n-1)} = 0$$

is uniquely solvable. We again note that the reason for the trouble with the equation $\mathcal{P}(D)X = \delta'$ is that in this case the condition $\int X\delta = 0$ is meaningless.

7 Solution of the Original Circuit Equations

Using our new tools, we descend from the rather sketchy treatment of the previous section and solve $\mathcal{L}X = \mathcal{M}(E)$, where $\mathcal{L} = D+2$, restricted to acting on those generalized functions (elements of $\mathcal{G}$, see Definition 15.13 on

1. The complex notation used there does not create any problems, since each complex equation is just a pair of real equations.
2. It's easy to see that this is equivalent to showing that the homogeneous solution can satisfy any n initial conditions. For the specific case in which there are n distinct c_j, this result was established in Exercise 13.15.3 on page 566. We will not prove the general case here.

page 680) satisfying the condition $\int X\delta = 0$. We'll see how simple our new tools are to operate with. For economy of notation we will use the symbol $\Leftrightarrow$ preceding an equation, to indicate the equivalence of this equation to the one on the previous line.

First we find the general solution of

$X' + 2X = \mathcal{M}(E)$ (recall the definition of the embedding operator, $\mathcal{M}$, Definition 15.24 on page 696) using essentially the methods of Section 6 of Chapter 6

$\Leftrightarrow \quad (D+2)X = \mathcal{M}(E)$

$\Leftrightarrow \quad \mathcal{M}(e^{-2t})[\mathcal{M}(e^{2t})X]' = \mathcal{M}(E)$ using Theorem 15.36 on page 714 (product rule), the commutativity, $\mathcal{M}D = D\mathcal{M}$ (Theorem 15.26 on page 696) and the derivative of the exponential, which justified the shift rule

$\Leftrightarrow \quad [\mathcal{M}(e^{2t})X]' = \mathcal{M}(e^{2t})\,\mathcal{M}(E)$ see discussion immediately preceding Problems 15.15 on page 705

$\Leftrightarrow \quad \mathcal{M}(e^{2t})X = R$ where R is a primitive (generalized antiderivative) of $\mathcal{M}(e^{2t})\,\mathcal{M}(E)$ (see Theorem 15.41 on page 722)

$\Leftrightarrow \quad X = \mathcal{M}(e^{-2t})R$ (note that $R' = \mathcal{M}(e^{2t})\,\mathcal{M}(E) = \mathcal{M}(e^{2t}E)$.) This last step justified by the definition of the product of functions in $\mathcal{G}$, Definition 15.30 on page 704.

Note that together with Theorem 15.41 (which proves the existence of R) Theorem 15.42 on page 725 shows how to obtain all such primitives.

Recalling that E is a Heaviside unit step starting at $t = 1$, the graph of $e^{2t}E$ has roughly the shape shown (darkened) in Figure 15-14.

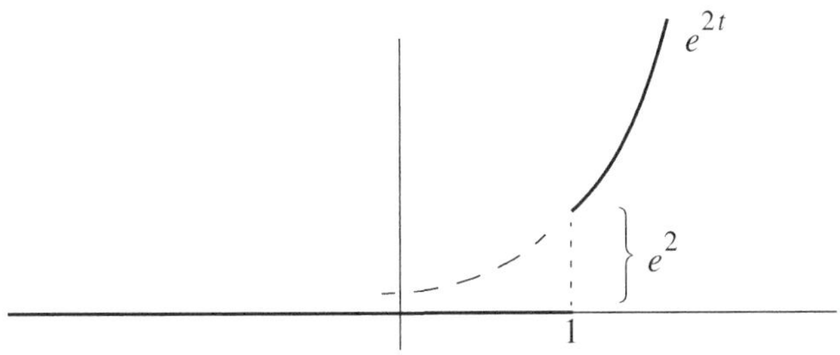

Figure 15-14

So we can guess that $R = \mathcal{M}(S) + \mathcal{M}(\mathcal{K})$, where

$$S(t) = \begin{cases} 0 & \text{for} \quad t < 1 \\ \dfrac{e^{2t}}{2} - \dfrac{e^2}{2} & \text{for} \quad t \geq 1 \end{cases}$$

and $\mathcal{K}$ is a constant function with value k (by Theorem 15.42 on page 725). So the general solution[1] of $(D+2)X = E$ is

$$X = \mathcal{M}(e^{-2t})[\mathcal{M}(S) + \mathcal{M}(\mathcal{K})]$$
$$= \mathcal{M}(e^{-2t}S) + \mathcal{M}(\mathcal{K}e^{-2t}),$$

i.e.,

$$X = \mathcal{M}(Z) + \mathcal{M}(\mathcal{K}e^{-2t}), \qquad \textbf{\textit{15.115}}$$

where

$$Z(t) = \begin{cases} 0 & \text{for} \quad t < 1 \\ \dfrac{1}{2} - \dfrac{e^{2-2t}}{2} & \text{for} \quad t \geq 1 \end{cases}$$

is the general solution of $(D+2)X = \mathcal{M}(E)$. The condition $\int X\delta = 0$ is

$$\int \mathcal{M}(Z)\delta + \int \mathcal{M}(\mathcal{K}e^{-2t})\delta = 0,$$

which, using Definition 15.32 on page 711, yields the value k, of $\mathcal{K}$ to be 0. Thus the solution of $\mathcal{L}X = \mathcal{M}(E)$ is

$$X = \mathcal{M}(Z), \qquad \textbf{\textit{15.116}}$$

where

$$Z(t) = \begin{cases} 0 & \text{for} \quad t < 1 \\ \dfrac{1}{2} - \dfrac{e^{-2(t-1)}}{2} & \text{for} \quad t \geq 1. \end{cases} \qquad \textbf{\textit{15.117}}$$

---

1. Set of all (generalized) functions satisfying the given differential equation.

(Same answer as all the previous ones.) That is, Z is equal to the previous Q (see equation 15.7 on page 656).

It's worthwhile to look at the graphs of some regular sequences,

$$\{e_N\}\Big|_{N=1}^{\infty} \text{ representing } \mathcal{M}(E),$$

and

$$\{q_N\}\Big|_{N=1}^{\infty}, \quad \{q_N'\}\Big|_{N=1}^{\infty}, \text{ and } \{q_N''\}\Big|_{N=1}^{\infty}$$

representing

$$\mathcal{M}(Z), \mathcal{M}(Z'), \text{ and } \mathcal{M}(Z''), \text{ respectively,}$$

(in Figures 15-15 through 15-18) where $(D+2)q_N = e_N$, and compare them with their counterparts in Figure 15-3 on page 657 through Figure 15-6 on page 662.

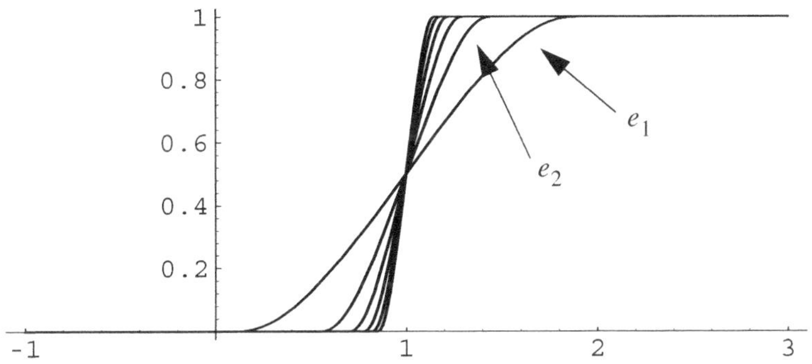

Figure 15-15

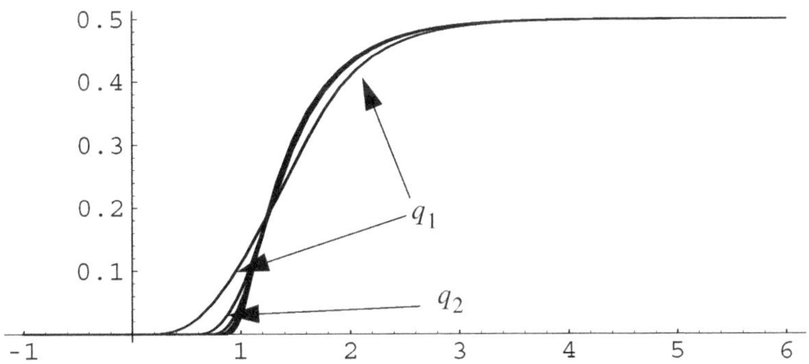

Figure 15-16

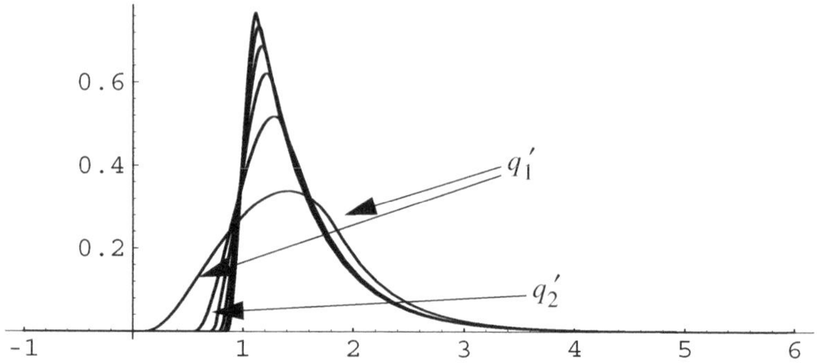

Figure 15-17

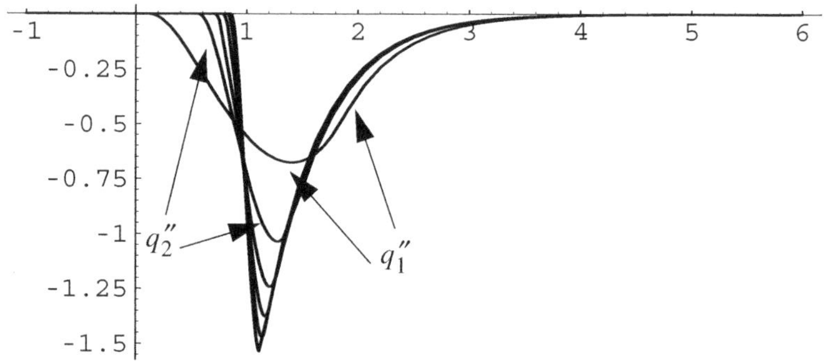

Figure 15-18

We restate the reason that E is a function of interest — namely, because it is a limiting case, i.e.,

$$e_N \xrightarrow{ms} E.$$

As with other objects in analysis which are limits (integrals, derivatives) E is an idealization which is easier to manipulate. *In a similar way,* Q is an idealization, and if we expect the voltmeter reading, as a function of time, to be fairly smooth, we would expect some q_N, or maybe Q_N rather than Q, to *best describe* (or at least *reasonably well describe*) the voltage reading.

Exercises 15.21

Recalling that $E(t)$ is 0 for $t < 1$ and 1 for $t > 1$, solve

1. $(D+1)(D+2)X = E, \int X\delta = 0 = \int X\delta'$

2. $(D+1)(D-2)X = \delta(t-s)$ t the real identity function, s real

3. $(D+1)(D-2)X = E$

4. $(D^2 + D + 1)X = E$

8 Green's Function for *P*(*D*); Solution to *P*(*D*)*X* = δ

One of our motives for developing generalized functions is the desire to write the solution of the ordinary differential equation

$$P(D)X = Y, \quad X(0) = X'(0) = \cdots = X^{(n-1)}(0) = 0 \qquad \textbf{\textit{15.118}}$$

(where $P(D)$ is an n-th degree polynomial in the derivative operator D, and Y is a given reasonably behaved function), in the form[1]

$$X(t) = \int G(t, s)Y(s)ds, \qquad \textbf{\textit{15.119}}$$

where t is the real identity function, $t(\tau) = \tau$.

We conjectured that generalized functions ***had to be introduced***, from our belief that for each constant s, $X_s = \mathcal{M}(G[t, s])$ (see Definition 15.24 on page 696) was the impulse response function, i.e., the solution of

$$\mathcal{P}(D)X = \delta(t-s), \qquad \textbf{\textit{15.120}}$$

an equation which had no analog in real valued functions of a real variable.

Before going on with our investigation, we remark that Green's functions were known before the advent of generalized functions, but were quite awkward and mysterious prior to this theory. While we cannot go any more deeply into the theory, it should be clear from what has been said, that linear

---

1. This type of result is important because it furnishes us with a very convenient way to represent a solution, a representation which can be very useful for establishing important results. G is called the *Green's function for the operator* $P(D)$.

operators other than polynomials in D, such as $\mathcal{L} = D^n + A_{n-1}D^{n-1} + \cdots + A_1 D + A_0$, where the A_j are reasonably well behaved real valued functions of a real variable, possess Green's functions, which turn out to be the functions G for which $X_s = \mathcal{M}(G[t, s])$ solves

$$\mathcal{L}X_s = \delta(t - s) \qquad \textbf{15.121}$$

(t being the real identity function), subject to some initial conditions. We should also expect that every solution of $\mathcal{L}X = Y$ should be of the form $X_P + X_H$, where X_p is given by equation 15.119, and X_H is a solution of $\mathcal{L}X = 0$. The reason we should expect $\mathcal{L}X_P = Y$ is that we expect[1]

$$\begin{aligned} \mathcal{L}X_p &= \mathcal{L}\int G(t, s)Y(s)ds = \int \mathcal{L}G(t, s)Y(s)ds \\ &= \int \delta(t - s)Y(s)ds = Y(t) = Y, \end{aligned} \qquad \textbf{15.122}$$

provided that all of the integrals above make sense, and that we can interchange the order of integration and differentiation.

The reasons for the difficulties in proving such general results are twofold. First, establishing existence theorems for, and the properties of, solutions of such differential equations, are nontrivial tasks — especially finding just the right conditions that make the results valid. Second, we haven't defined the first integral of equations 15.122 when the integrand is a generalized function (which it must be in order to apply $\mathcal{L}$ under the integral sign). This problem can be handled provided care to embed all the relevant objects in equations 15.122 as generalized functions — noting that because the unembedded $G(\tau, s)$ is 0 for $s > \tau$ (because you wouldn't expect a system to respond in the present to a future input) and the unembedded function Y is 0 for arguments less than 0, the sequence of integrals arising from the generalized function integrand can include the factor $\rho_N(s)$ (see Theorem 15.23 on page 696) for large enough N, and this factor can serve as the test function in the definition of integral of a generalized function given in Definition 15.15 on page 682. Third, with regard to justifying the interchange of limit operations ($\mathcal{L}\int = \int \mathcal{L}$) as already mentioned in the discussion surrounding Theorem 15.37 on page 719, we have not developed the appropriate measure theoretic tools — i.e., tools which deal with the measure (size) of a set. But at least intuitively, since over very short intervals, we expect operators like $\mathcal{L}$ to behave very much like polynomials $P(D)$, if the functions A_j are locally linear, we shouldn't be surprised that under many circumstances,

---

1. Keeping in mind that t stands for the identity function, and being somewhat cavalier about distinguishing between real valued functions and their associated generalized functions.

the results we have developed hold up under much more general conditions[1].

$\lfloor$

We will turn our attention to the specific case $\mathcal{L} = D + 2$ acting on those X satisfying $\int X\delta = 0$, and we find that, at least in this specific case, our original conjecture, as expressed in the material surrounding equations 15.118, 15.119 and 15.120, is correct.

We note that the methods used in Chapter 6, Section 6 can also be used easily on $\mathcal{P}(D)X = \delta(t-s)$, where $\mathcal{P}(D)$ is the operator acting on those generalized functions satisfying

$$\int X\delta = \int X\delta' = \cdots = \int X\delta^{(n-1)} = 0$$

(analogous to initial conditions $X(0) = X'(0) = \cdots = X^{(n-1)}(0) = 0$).

Example 15.22: Green's function for $\mathcal{L} = D + 2$

We start out by solving

$$\mathcal{L}X = \delta(t-s), \qquad \textbf{\textit{15.123}}$$

i.e., finding the *impulse response function*, X_s, and then seeing if $X_s = \mathcal{M}(G[t, s])$ with G actually living up to our hopes that it is a Green's function of $\mathcal{L} = D + 2$ acting on X for which $X'(0) = 0$. That is, satisfying the condition that the solution to $\mathcal{L}X = Y$ really is given by

$$X(t) = \int G(t, s)Y(s)ds. \qquad \textbf{\textit{15.124}}$$

With the symbol $\Leftrightarrow$ introduced at the beginning of Section 7 of this chapter, and using the shift rule (see Theorem 6.18 on page 202) we find

$$(D+2)X = \delta(t-s)$$

$$\Leftrightarrow \quad \mathcal{M}(e^{-2t})D[\mathcal{M}(e^{2t})X] = \delta(t-s)$$

$$\Leftrightarrow \quad D[\mathcal{M}(e^{2t})X] = \mathcal{M}(e^{2t})\delta(t-s)$$

$$\Leftrightarrow \quad D[\mathcal{M}(e^{2t})X] = \mathcal{M}(e^{2s})\delta(t-s)$$

(since $\delta(t-s)$ concentrates all of its mass at $t = s$. *Note* that $\mathcal{M}(e^{2s})$ is the constant generalized function which corresponds to the constant function e^{2s}.)

---

1. And we shouldn't be surprised that they are much tougher to establish.

$\Leftrightarrow \quad \mathcal{M}(e^{2t})X = \mathcal{M}(e^{2s})\mathcal{M}(H(t-s)) + \mathcal{M}(\mathcal{K})$, having made use of Theorem 15.29 on page 700, and Theorem 15.42 on page 725, where H is the Heaviside unit step (see Theorem 14.19 on page 622) $\mathcal{K}$ is a constant function,

$\Leftrightarrow \quad X = \mathcal{M}(e^{2(s-t)}H(t-s)) + \mathcal{M}(\mathcal{K}e^{-2t})$.

For simplicity, we restrict to $s \geq 1$ (i.e., the forcing function acts only after time $s = 1$) and note that the condition $\int X\delta = 0$ yields that $k = 0$ (where k is the value of the constant function $\mathcal{K}$). Thus ***we conjecture*** that the Green's function for $\mathcal{L}$ is given by

$$G(t, s) = \begin{cases} 0 & \text{for} \quad t < s \\ e^{2(s-t)} & \text{for } t \geq s \end{cases}$$

for $s \geq 1$; i.e., dealing with real valued functions of a real variable, if Y is a given continuous function, with $Y(s) = 0$, and

$$(D+2)X = Y \quad \text{with} \quad X(0) = 0,$$

then we conjecture that

$$X(t) = \int G(t, s)Y(s)ds = \int_0^t e^{2(s-t)}Y(s)ds.$$

In fact, we know this to be the case from equation 15.6 on page 655. We can go even further, pointing out that if Y is in the class $\mathcal{PC}$ (Notation 15.16 on page 685) of piecewise continuous functions, then

$$X = \mathcal{M}\left[\int_0^t e^{2(s-t)}Y(s)ds\right]$$

solves

$$\mathcal{L}X = \mathcal{M}(Y).$$

We leave this to the reader to verify.

■

Problem 15.23

Handle the case of general Y in Example 15.22, i.e., eliminate the restriction $Y(s) = 0$ for $s \leq 1$.

We give one more illustration.

Example 15.24: Two-point boundary problem

In many problems involving ordinary differential equations arising from spatial problems governed by partial differential equations, the specified constraints occur as boundary conditions that are not initial conditions (conditions on the solution at some initial time). For instance, in problems of heat flow, the endpoints of a rod are sometimes held at a fixed temperature (such as when the two ends are in boiling water). This example illustrates such a situation.

Find the Green's function (see Footnote 1. on page 736) for $L = D^2$ acting on X for which $X(0) = 0$, $X(1) = 1$.

Recall the Heaviside unit step, defined in Theorem 14.19 on page 622, the symbol $\Leftrightarrow$ introduced at the beginning of Section 7, the shift rule (Theorem 6.18 on page 202) the embedding operator, $\mathcal{M}$ (Definition 15.24 on page 696) and the Dirac delta (Definition 15.27 on page 698). For $j = 1,2$ and each real s, we let $\mathcal{K}_{j,s}$ denote a constant function (whose value may depend on j, s).

First we solve the generalized equation $D^2X = \delta(t-s)$,

$$
\begin{aligned}
& D^2X = \delta(t-s) \\
\Leftrightarrow\quad & DX = \mathcal{M}(H(t-s)) + \mathcal{M}(\mathcal{K}_{1,s}) \quad \text{using Theorem 15.42 on page 725} \\
\Leftrightarrow\quad & X = \mathcal{M}[S(t-s)] + \mathcal{M}[t\mathcal{K}_{1,s}] + \mathcal{M}(\mathcal{K}_{2,s}) \quad (\text{where } S(y) = yH(y)) \\
\Leftrightarrow\quad & X = \mathcal{M}[S(t-s) + t\mathcal{K}_{1,s} + \mathcal{K}_{2,s}].
\end{aligned}
$$

The condition $X(0) = 0$ becomes $\int X\delta = 0$, or $S(-s) + \mathcal{K}_{2,s} = 0$, while $X(1) = 1$ becomes $\int X\delta(t-1) = 1$, or $S(1-s) + \mathcal{K}_{1,s} + \mathcal{K}_{2,s} = 1$.

We obtain

$$
\begin{aligned}
\mathcal{K}_{2,s} &= -S(-s) = sH(-s) \\
\mathcal{K}_{1,s} &= 1 - \mathcal{K}_{2,s} - S(1-s) \\
&= 1 - sH(-s) - (1-s)H(1-s).
\end{aligned}
$$

Thus the Green's function is given by

$$
\begin{aligned}
G(t,s) &= (t-s)H(t-s) + t - tsH(-s) - t(1-s)H(1-s) + sH(-s) \\
&= t + (t-s)H(t-s) + s(1-t)H(-s) - t(1-s)H(1-s).
\end{aligned}
$$

We leave it as an exercise to check that G is in fact a Green's function (as defined in Example 15.22 on page 738).

■

Exercises 15.25

1. Find the Green's function (as defined in Example 15.22 on page 738) for $L = D^2 + 4$ acting on those X for which $X(0) = 0 = X(1)$.
2. Find the Green's function (as defined in Example 15.22 on page 738) for $L = D^2 + k^2$ acting on those X for which $X(0) = 0 = X(1)$.

Theorem 15.46: Main result on Green's functions

Let $P(D)$ be an n-th degree polynomial in the derivative operator, D, and let $\mathcal{P}(D)$ be this same polynomial restricted to act on those elements X of the class $\mathcal{G}$ of generalized functions (Definition 15.13 on page 680, also see Definition 15.15 on page 682) such that

$$\int X\delta = \int X\delta' = \dots = \int X\delta^{(n-1)} = 0$$

(see Definition 15.32 on page 711 and Theorem 15.33 on page 711). Then for each real number s, there is a unique solution, X_s, to the differential equation

$$\mathcal{P}(D)X = \delta(t-s),$$

where t is the real identity function and $\delta(t-s)$ is defined by specialization of Theorem 15.29 on page 700.

Further, this solution, X_s, which is the *impulse response* (generalized) *function*, $X_s = \mathcal{M}[G(t,s)]$ (for the meaning of $\mathcal{M}$, see Definition 15.24 on page 696) where G is in the class $\mathcal{PC}$ of piecewise continuous functions (see Notation 15.16 on page 685) as a function of either of its arguments.[1] For each τ, $\int G(\tau, s)ds$ exists, and for Y in the class $\mathcal{PC}$,

$$X = \mathcal{M}\left[\int G(t,s)Y(s)ds\right]$$

exists and is the unique solution of

$$\mathcal{P}(D)X = \mathcal{M}(Y).$$

If Y is continuous, then $X = \int G(t,s)Y(s)ds$ is the unique solution to

$$P(D)X = Y, \quad X(0) = X'(0) = \dots = X^{(n-1)}(0) = 0.$$

❑

1. Coordinates of elements of its domain; conventionally we would say that $G(t,s)$ is continuous both in t for fixed s, and in s for fixed t.

The basic idea of the proof is to first find the most general solution of

$$P(D)X = \delta(t-s),$$

using the techniques of Section 6 of Chapter 6, verifying that the solution has the asserted piecewise continuity properties. Then apply the initial conditions

$$\int X\delta = \int X\delta' = \dots = \int X\delta^{(n-1)} = 0.$$

Finally, verify that $\mathcal{M}[\int G(t, s)Y(s)ds]$ solves $\mathcal{P}(D)X = \mathcal{M}(Y)$, using the commutativity of $\mathcal{M}$ with $\int$ and with $P(D)$ (the latter following from Theorem 15.26 on page 696) and the dominated convergence theorem, Theorem 15.37 on page 719 to show that $P(D)\int = \int P(D)$.

Exercises 15.26

Solve the following generalized function differential equations by first finding a Green's function and then invoking Theorem 15.46 on page 741. For relevant examples see Example 15.22 on page 738. We will make use of the displaced Heaviside unit step, E, defined by $E(t) = 0$ for $t < 1$, $E(t) = 1$ for $t \geq 1$.

1. $(D+1)X = \mathcal{M}(E\ cos),\ \int X\delta = 0$
2. $(D+1)^2X = \mathcal{M}(E\ cos),\ \int X\delta = 0 = \int X\delta'$
3. $(D^2 - D - 2)X = \mathcal{M}(E\ sin),\ \int X\delta = 0 = \int X\delta'$

Problem 15.27

For g a generalized function with $\int g\delta = 0$, does the equation

$$(D+2)X = g,\ \int X\delta = 0$$

have a solution?

9 Notational Change

Up to this point we have put extra stress on the distinction between a real valued piecewise continuous function of a real variable, f, and the corresponding generalized function $\mathcal{M}(f)$ (also sometimes denoted by $\mathcal{M}f$).

Since we were often dealing simultaneously with f and $\mathcal{M}f$, failing to distinguish would have muddied the waters too much. Also the mapping between piecewise continuous functions and their corresponding generalized functions is not one to one — two functions differing at exactly one domain value will yield the same generalized function.

However, in most of the literature, this careful distinction is abandoned. To prepare the reader for the cruel outside world, we will deemphasize the distinction, as follows. If f is a piecewise continuous function, we will denote the corresponding generalized function by the symbol $\boldsymbol{f}$. Usual functions will be denoted in Times Roman italic font. Their corresponding generalized functions will be denoted by the same letter in bold italic Helvetica font.

If x is a real number, by $\boldsymbol{f}(x)$ we shall mean

$$\int \boldsymbol{f}\,\delta(t-x) = \frac{f(x-)+f(x+)}{2}, \quad \text{see} \quad \text{Definition 15.35 on page 713;}$$

i.e., $\boldsymbol{f}(x) = \dfrac{f(x-)+f(x+)}{2}$.

If t is the real identity function, then $\mathcal{M}(e^t)$ will be denoted by $\boldsymbol{e}^t$.

10 Generalized Eigenfunction Expansions, Series

The following heuristic discussion is intended only to give a preview of some of the results which one might anticipate. Let us assume that we are given a differential operator L and a corresponding complete orthonormal set of real eigenfunctions, φ_n (see Definition 14.5 on page 601) and nonzero eigenvalues, λ_n —

i.e.,

$$L\varphi_n = \lambda_n \varphi_n. \qquad \textit{15.125}$$

Assume also that we are given a function f which may be written

$$f = \sum_{n=1}^{\infty} a_n \varphi_n. \qquad \textit{15.126}$$

Since the φ_n are orthonormal, taking dot products of both sides of this equation with φ_j (not worrying about questions of convergence) and assuming that $f \cdot g = \int fg$), we find

$$a_j = \int f(x)\varphi_j(x)dx. \qquad \textit{15.127}$$

Proceeding heuristically, operating in $\mathcal{G}$ (the set of generalized functions, Definition 15.13 on page 680) where t is the real identity function ($t(\tau) = \tau$) taking into account the notation of Section 9 of this chapter, and Theorem 15.29 on page 700 concerning $\delta(at + b)$,

$$\begin{aligned}\int \boldsymbol{f}\,\delta(t-y)\,dt &= \boldsymbol{f}(y) = \sum_{n=1}^{\infty} a_n \varphi_n(y)\\ &= \sum_{n=1}^{\infty}\{\int f(t)\varphi_n(t)\,dt\}\varphi_n(y) = \sum_{n=1}^{\infty}\int \boldsymbol{f}(t)\,\phi_n(t)\,\phi_n(y)\,dt \qquad \textit{15.128}\\ &= \int \boldsymbol{f}(t)[\sum_{n=1}^{\infty}\phi_n(t)\phi_n(y)]dt,\end{aligned}$$

where ϕ_n is the generalized function corresponding to φ_n.

Formally, we have, equating the two extremes of equations 15.128, that

$$\delta(t-y) = \sum_{n=1}^{\infty}\phi_n(t)\phi_n(y) \qquad \textit{15.129}$$

The Green's function (see Section 8 starting page 736) satisfies

$$L\boldsymbol{G}(t, y) = \delta(t-y), \qquad \textit{15.130}$$

and because of this and equation 15.125 on page 743, we obtain **formally**

$$\boldsymbol{G}(t, y) = \sum_{n=1}^{\infty}\frac{1}{\lambda_n}\phi_n(t)\phi_n(y). \qquad \textit{15.131}$$

Before getting too carried away with the formalism, notice that in equation 15.129, δ is a generalized function, and the ϕ_n on the right side of this equation are the generalized function counterparts of the ordinary functions φ_n. But then, *what do we mean by an infinite series of generalized functions?* We answer this with the following .

Definition 15.47: *Convergence* for generalized functions

A sequence of generalized functions, $\{G_j\}\big|_{j=1}^{\infty}$ is said to **approach** (or **converge to**) the generalized function G, if for all test functions T (Definition 15.9 on page 677)

$$\lim_{j\to\infty}\int G_j T = \int GT,$$

(see Definition 15.15 on page 682).

If this generalized limit exists, we write $\lim_{j \to \infty} G_j = G$.

❑

Note that even if $\lim_{j \to \infty} \int G_j T$ exists for every test function, T, we do not know that there is a generalized function G such that $\lim_{j \to \infty} G_j = G$.. (This is a question of completeness similar to the question raised by the Cauchy criterion, Theorem 7.8 on page 219. It is actually answered in the next section, and shows the need for the tools of linear topological spaces.)

Definition 15.48: Sum of generalized series

Keeping in mind Definition 15.47, a series of generalized functions, $\{g_j\}\Big|_{j=1}^{\infty}$ is said to **sum to the generalized function G**, if

$$\lim_{j \to \infty} \sum_{k=1}^{j} g_k = G.$$

If this sum exists, we denote it by

$$\sum_{k=1}^{\infty} g_k.$$

Theorem 15.49: Generalized sum rule

Referring to Definition 15.48, if $g_1, g_2, ...$ are generalized functions and

$$G = \sum_{k=1}^{\infty} g_k,$$

then

$$G' = \sum_{k=1}^{\infty} g_k'.$$

❑

Problem 15.28

Prove Theorem 15.49. This is not as tough as it may look.

Hints: Use the definitions of generalized functions (Definition 15.13 on page 680) of test functions (Definition 15.9 on page 677, etc.) and work with sequences. If

$$G = \lim_{n \to \infty} G_n = \lim_{n \to \infty} \sum_{j=1}^{n} g_j, \text{ and if } \{G_{n,k}\}\Big|_{k=1}^{\infty} \text{ represents } G_n,$$

we know that

$$\lim_{n \to \infty} \lim_{k \to \infty} \int G_{n,k} T$$

exists for each test function T. We want to show that this implies that $\lim_{n \to \infty} \lim_{k \to \infty} \int G_{n,k}' T$ also exists for every test function, T, and that $\lim_{n \to \infty} \int G_n' T$ exists for each test function, and that

$$\int G'T = \lim_{n \to \infty} \int G_n' T = \lim_{n \to \infty} \lim_{k \to \infty} \int G_{n,k}' T.$$

The basic tools used are integration by parts and the bounded support of each test function. There may, however, be some hidden technical difficulties.

Thus any convergent infinite series of generalized functions can be differentiated term by term any number of times, and still be meaningful and convergent in the generalized sense. That is, series of generalized functions behave like convergent power series inside their circles of convergence (see Theorem 7.21 on page 242).

Example 15.29

Consider the Heaviside unit step, defined in Theorem 14.19 on page 622, and let

$$F(x) = \frac{\pi}{4}(1 - H(-x) - H(x - \pi)).$$

The graph of F looks as shown in Figure 15-19.

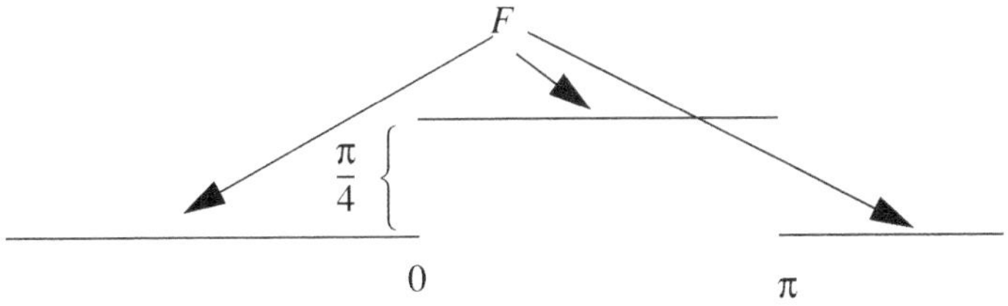

Figure 15-19

It can be shown that on the open interval $(0; \pi)$,

$$\sum_{n=1}^{\infty} \frac{\sin[(2n-1)x]}{2n-1} = 1.$$

Hence

$$F(x) = \sum_{n=1}^{\infty} \frac{sin[(2n-1)x]}{2n-1} F(x), \qquad \textit{15.132}$$

where the convergence is pointwise. Also, F is piecewise continuous (Notation 15.16 on page 685) and with x denoting the real identity function, $sin[(2n-1)x]$ is infinitely differentiable. Consider now the *generalized* counterpart of equation 15.132. With x as above, via Theorem 15.49 on page 745 and the last result from Theorem 15.29 on page 700, we obtain

$$\frac{\pi}{4}[\delta(x) - \delta(x-\pi)] = \boldsymbol{F}'(x)$$

$$= \sum_{n=1}^{\infty} \left\{ [\boldsymbol{cos}((2n-1)x)]\boldsymbol{F}(x) + \frac{\boldsymbol{sin}((2n-1)x)}{2n-1}\boldsymbol{F}'(x) \right\}$$

$$= \sum_{n=1}^{\infty} \left\{ [\boldsymbol{cos}((2n-1)x)]\boldsymbol{F}(x) + \frac{\boldsymbol{sin}((2n-1)x)}{2n-1}[\delta(x) - \delta(x-\pi)] \right\}.$$

By Definition 15.30 on page 704 it follows that

$$\delta \boldsymbol{sin} = \mathcal{M}[(H sin)'] - \mathcal{M}[H cos] = 0$$

(recall the meaning of $\mathcal{M}$ from Definition 15.24 on page 696), since

$$H(t) sin(t) = \begin{cases} 0 & \text{for} \quad t < 0 \\ sin(t) & \text{for} \quad t > 0 \end{cases}$$

and

$$cos(t) = H(t) cos(t) \quad \text{for} \quad t > 0 .$$

Similarly, we find for each n

$$\frac{\boldsymbol{sin}[(2n-1)x]}{2n-1}\delta(x) = 0 = \frac{\boldsymbol{sin}[(2n-1)x]}{2n-1}\delta(x-\pi).$$

So

$$\delta(x) - \delta(x-\pi) = \frac{4}{\pi}\sum_{n=1}^{\infty} \boldsymbol{cos}[(2n-1)x]\boldsymbol{F}(\boldsymbol{x}). \qquad \textit{15.133}$$

This series, which makes no sense pointwise for $0 < x < \pi, x \neq \pi/2$ is meaningful in the generalized sense. By a similar argument

$$\delta'(x) - \delta'(x-\pi) = \frac{4}{\pi}\sum_{n=1}^{\infty}(2n-1)\,\boldsymbol{cos}[(2n-1)x]\boldsymbol{F}(x),$$

and this series is even more patently divergent in the ordinary sense.

Exercises 15.30

1. Define $T(x) = x\rho_{10}(x)$ (recall Notation 15.19 on page 688) and note that T is in the class $\mathcal{T}$ of test functions (Definition 15.9 on page 677). Then from the third result of Theorem 15.29 on page 700, we find

$$\int T(x)[\delta(x) - \delta(x-\pi)] = -\pi.$$

Check out equation 15.133 via Definition 15.47 on page 744 and Definition 15.48 on page 745, with

$$g_k(x) = \frac{4}{\pi}\,\boldsymbol{cos}[(2n-1)x]\boldsymbol{F}(x).$$

2. It is well known that for $0 < x < 2\pi,\quad x = \pi - 2\sum_{n=1}^{\infty}\frac{sin(nx)}{n}$.

This may be rewritten, for all real x as

$$\left(\frac{\pi-x}{2}\right)(1-H(-x)-H(x-2\pi))=\sum_{n=1}^{\infty}\left[\frac{sin(nx)}{n}(H(x)-H(x-2\pi))\right].$$

Using Theorem 15.49 on page 745 and Theorem 15.29 on page 700 show that

$$-[1-\boldsymbol{H}(-x)-\boldsymbol{H}(x-2\pi)] + \pi\delta(x-2\pi)$$
$$= 2\sum_{n=1}^{\infty}\left[\frac{\boldsymbol{sin}(nx)}{n}\{\boldsymbol{H}(x)-\boldsymbol{H}(x-2\pi)\}\right]$$

Does it make any sense to multiply by $\delta(x-\pi/4)$ and integrate over the entire real line?

3. Show that $\sum_{n=0}^{\infty}\delta'(t-n)$ converges in the generalized sense, by finding a series for a piecewise continuous function and using Theorem 15.49.

Problem 15.31

Prove or disprove $\lim_{n \to \infty} g_n = \delta$ where $g_n(x) = \frac{sin(nx)}{\pi x}$.

Hint: Use

$$\int_0^{\infty} \frac{sin(x)}{x} dx = \frac{\pi}{2}.$$

11 Continuous Linear Functionals

Standing alone, this section has no immediate practical application for the student. Its purpose is to point out the connections between the theory we have developed, and the more sophisticated approach to generalized functions in common use by mathematicians. In a sense, this section is a dictionary, so that a reader who encounters generalized functions described as continuous linear functionals can relate results given in this latter formulation to his/her own needs.

An alternative way of regarding generalized functions is as *functionals* acting on the set $\mathcal{T}$ of test functions (Definition 15.9 on page 677). Suppose that g is in the class $\mathcal{G}$ of generalized functions (Definition 15.13 on page 680). Then, choosing a representative of g, say $\{g_N\}\Big|_{N=1}^{\infty}$, we have for each T in $\mathcal{T}$, that

$$\lim_{N \to \infty} \int g_N T$$

exists, and is, by definition, independent of the particular representative $\{g_N\}\Big|_{N=1}^{\infty}$. Thus, for each T in $\mathcal{T}$, g determines a real number $\int gT$ — i.e., associated with each g we have a function G, defined by

$$G(T) = \int gT. \qquad \textbf{\textit{15.134}}$$

A function whose domain (see Definition 1.1 on page 2) is a vector space (Definition 13.5 on page 543) and whose range is contained in the field of scalars of the vector space is called a *functional*. Since for all real a,b and all test functions, T_1, T_2,

$$G(aT_1 + bT_2) = aG(T_1) + bG(T_2), \qquad \textbf{\textit{15.135}}$$

each g in $\mathcal{G}$ defines a *linear* functional, G.

Is G a continuous function? That is, if the sequence $T_1, T_2,...$ converges to T, does the sequence $G(T_1), G(T_2),...$ converge to $G(T)$?

We know what is meant by convergence of $G(T_1), G(T_2),...$ to $G(T)$; namely, $\lim_{n \to \infty} |G(T_n) - G(T)| = 0$, or, using the linearity of G

$$\lim_{n \to \infty} |G(T_n - T)| = 0.$$

However, we have not yet decided on what would be an appropriate, useful definition of convergence of a sequence $T_1, T_2,...$ of test functions, or, equivalently, an appropriate definition of convergence of $T_1 - T, T_2 - T,...$ to **0**, since the linearity of G has reduced all continuity questions to questions of continuity at **0**. (Note that due to the nature of the vector space $\mathcal{T}$, **0** must be the function whose value is 0 at every real.) We will soon introduce a notion of convergence in $\mathcal{T}$, such that each generalized function g makes G a continuous linear function. That is, we shall define what is meant by $\lim_{n \to \infty} T_n = \mathbf{0}$ in such a way that

$$\text{whenever} \quad \lim_{n \to \infty} T_n = \mathbf{0}, \text{ we have } \lim_{n \to \infty} G(T_n) = \lim_{n \to \infty} \int gT_n = 0.$$

This will prove to be very useful in finding what many mathematicians consider to be a more desirable coordinate system in which to describe generalized functions (see Gelfand and Shilov [6]).

In order to motivate the definition for convergence of a sequence of test functions that we will choose, we give two counterexamples. To introduce these examples, recall the definitions of **least upper bound** (**sup**) and **greatest lower bound** (**inf**) Definition 10.44 on page 407.

> If A is a nonempty set of real numbers which is bounded from above (i.e., there is a real number, B, such that for all x in A, $x \le B$) then the **least upper bound** (or **supremum**) **of** $\boldsymbol{A}$ (denoted by *sup* A) is that value b, such that
>
> $$\text{for all } x \text{ in } A, \ x \le b$$
>
> and if $C < b$
>
> ***it is false that*** for all x in A, $x \le C$.
>
> The **greatest lower bound** (or **infimum**) **of a set** A (denoted by *inf* A) which is bounded from below is defined by replacing the < symbol in all of its occurrences above by the > symbol.

It can be shown that every nonempty set of real numbers which is bounded from above [below] has a real least upper bound [greatest lower bound]. (This is one of the important differences between the set of rationals and the set of reals.)

Example 15.32

Recalling Notation 15.19 on page 688, define $T_n(x) = \varphi_n(x)/n^2$. Then for each real x

$$\lim_{n\to\infty} \sup_x |T_n(x)| = 0 \text{ and even } \lim_{n\to\infty} |T_n'(x)| = 0.$$

Now let $g = \delta''$ (see Definition 15.27 on page 698) which we know is in $\mathcal{G}$. Then using Theorem 15.33 on page 711 we find

$$\lim_{n\to\infty} G(T_n) = \lim_{n\to\infty} \int \delta'' T_n = \lim_{n\to\infty} T_n''(0) = -\frac{2}{e}.$$

So we see that if this sequence $T_1, T_2,...$ converged to $\mathbf{0}$, G would not be a continuous linear functional. But the reason for this bad behavior is that for this sequence $\lim_{n\to\infty} \sup_x |T_n^{(m)}(x)| \neq 0$ for $m \geq 2$. To make G continuous, convergence to $\mathbf{0}$ must exclude such behavior. So, it seems reasonable to require that the convergence of $T_1, T_2,...$ implies that

$$\lim_{n\to\infty} \sup_x |T_n^{(m)}(x)| = 0 \text{ for all } m = 0,1,2,... .$$

■

Example 15.33

Recall Notation 15.19 on page 688, and define

$$T_n(x) = \frac{1}{n}\varphi_1(x/n).$$

This sequence $T_1, T_2,...$ converges to zero uniformly (Definition 7.16 on page 234) over the entire set of real numbers, as do each of its higher derivatives. But letting

$$U(t) = \begin{cases} 0 & \text{if} \quad t < 0 \\ 1/2 & \text{if} \quad t = 0 \\ 1 & \text{if} \quad t > 0, \end{cases}$$

and

$$g = U, G(T) = \int UT,$$

then

$$\lim_{n \to \infty} G(T_n) = \lim_{n \to \infty} \int U\, T_n = \lim_{n \to \infty} \frac{1}{n} \int_0^n exp\left[\frac{-1}{1-(x/n)^2}\right] dx$$

$$= \lim_{n \to \infty} \int_0^1 exp\left[\frac{-1}{1-z^2}\right] dz > 0.$$

If this sequence $T_1, T_2, ...$ converged to 0, then G would fail to be a continuous linear functional. Since we want it to be continuous, we must prohibit the behavior exhibited by this sequence, which in this case is that there is no bounded interval which contains the supports of all of the T_n.

■

We've now seen enough examples to clue us in on what will prove to be a useful definition of convergence in $\mathcal{T}$ and which will make continuous all Gs with

$$G(T) = \int gT, \quad g \text{ in } \mathcal{G}, \text{ and } T \text{ in } \mathcal{T} \text{ (the class of test functions).}$$

Definition 15.50: Convergence criterion for test functions

Let N_{T_n} be the product of the following two numbers

- the supremum of $|x|$ (Definition 10.44 on page 407) for x in the support of T_n (Definition 15.7 on page 677)
- $\sup_{p \text{ a nonnegative integer}} \sup_x \left|T_n^{(p)}(x)\right|$ — if this is finite (otherwise set $N_{T_n} = 2$).

We say $\boldsymbol{T_1, T_2, ...}$ **converges to 0** if $\lim_{n \to \infty} N_{T_n} = 0$.

❑

Under this definition of convergence, it turns out that this makes the set $\mathcal{T}$ of test functions into what is called a *linear topological space.*

Definition 15.51: Dual space

The set of continuous linear functionals acting on the set $\mathcal{T}$ of test functions is called the **dual of** $\mathcal{T}$ and is denoted by $\mathcal{T}'$.

❑

Thus if F is in $\mathcal{T}'$, then $dmnF = \mathcal{T}, rngF = \mathcal{R}$ (the real numbers) and F is linear and continuous with respect to the definition of convergence given above in Definition 15.50.

Before stating the next theorem we introduce the notion of isomorphism, as the natural formalism to describe a change of description in which operations can be equivalently carried out under what seem to be quite different descriptions.

Definition 15.52: Algebraic and analytic isomorphism

Suppose $\mathcal{V}$ and $\mathcal{W}$ are two vector spaces (Definition 13.5 on page 543) over the same scalars.[1] Then $\mathcal{V}$ and $\mathcal{W}$ are said to be **algebraically isomorphic** if there is a function $\mathcal{J}$, called an **isomorphism**, with the following properties

- $dmn\ \mathcal{J} = \mathcal{V}$, $rng\ \mathcal{J} = \mathcal{W}$ (see Definition 1.1 on page 2) (*dmn f* is an abbreviation for the domain of *f*, and *rng f* the corresponding abbreviation for the range of *f*.)
- $\mathcal{J}$ has an inverse function, $\mathcal{J}^{-1}$ (Definition 10.5 on page 333)
- $\mathcal{J}$ is a linear operator, i.e.,
$$\mathcal{J}(a_1 v_1 + a_2 v_2) = a_1 \mathcal{J}(v_1) + a_2 \mathcal{J}(v_2)$$
Here + on the left side denotes addition in $\mathcal{V}$ while on the right it denotes addition in $\mathcal{W}$.

Note that if $\mathcal{V}$ and $\mathcal{W}$ are algebraically isomorphic, we can, for example, perform the operation of adding $(v_1 + v_2)$ in either $\mathcal{V}$ or $\mathcal{W}$ as follows:

since

$$\mathcal{J}(v_1 + v_2) = \mathcal{J}(v_1) + \mathcal{J}(v_2)$$

we have

$$v_1 + v_2 = \mathcal{J}^{-1}[\mathcal{J}(v_1) + \mathcal{J}(v_2)] \cdot$$

On the left the addition is performed in $\mathcal{V}$, while on the right it is performed in $\mathcal{W}$. Conceivably, the computation of the right side, though it looks more difficult, is actually easier than the direct computation of $v_1 + v_2$ (as is the case of multiplication by means of logarithms).

If $\mathcal{V}$ and $\mathcal{W}$ have a notion of convergence, for either space we will write simply $\lim_{n \to \infty} T_n = T$ if the sequence $T_1 - T, T_2 - T, \ldots$ converges to $\mathbf{0}$ in that space.

---

1. We have only defined vector spaces over the real or complex scalars. There are other extensions, though we needn't be concerned here with them.

We say that $\mathcal{V}$ and $\mathcal{W}$ are **analytically isomorphic** if

- they are algebraically isomorphic (with, say, isomorphism $\mathcal{J}$) and further
- whenever $\lim_{n \to \infty} T_n = T$ in $\mathcal{V}$, then

$$\lim_{n \to \infty} \mathcal{J}(T_n) = \mathcal{J}(\lim_{n \to \infty} T_n) = \mathcal{J}(T) \text{ in } \mathcal{W},$$

and

- whenever $\lim_{n \to \infty} W_n = W$ in $\mathcal{W}$, then

$$\lim_{n \to \infty} \mathcal{J}^{-1}(W_n) = \mathcal{J}^{-1}(\lim_{n \to \infty} W_n) = \mathcal{J}^{-1}(W).$$

❑

Note that if two spaces are analytically isomorphic, we can do our analysis (algebra, plus limiting operations) equally well in either space.

It is now possible to state the following result.

Theorem 15.53: Equivalence of dual to original space

The set $\mathcal{G}$ of generalized functions (Definition 15.13 on page 680) and its dual (Definition 15.51 on page 752) are analytically isomorphic (Definition 15.52 on page 753) if we define the notion of convergence of the sequence $F_1, F_2, \ldots$ to $\mathbf{0}$ in $\mathcal{T}'$ (i.e., $\lim_{n \to \infty} F_n = \mathbf{0}$) by

$$\lim_{n \to \infty} \sup_{\substack{N_T \le 1 \\ T \text{ in } \mathcal{T}}} |F_n(T)| = 0$$

using Definition 15.50 on page 752 for convergence to $\mathbf{0}$ in $\mathcal{T}$.

❑

Since much is known about linear topological spaces and their duals, it is easier in the sense of available powerful results, to do our work in $\mathcal{T}'$ rather than in $\mathcal{G}$. (We may consider $\mathcal{T}'$ as a better coordinate system for generalized functions. For example, see Gelfand and Shilov [6].)

$\mathcal{T}'$ (and thus $\mathcal{G}$) has another important property which we state without proof.

Theorem 15.54: Completeness of $\mathcal{T}'$

$\mathcal{T}'$ is *complete* — meaning that if $F_1, F_2, \ldots$ are in $\mathcal{T}'$, and if for every T in $\mathcal{T}$, $\lim_{N \to \infty} F_N(T)$ exists, then there exists a unique F in $\mathcal{T}'$ such that

$$\lim_{N \to \infty} F_N = F$$

or, in other words, for each T in $\mathcal{T}$

$$\lim_{N \to \infty} F_N(T) = F(T).$$

❑

Because of the isomorphism between $\mathcal{T}'$ and $\mathcal{G}$, Theorem 15.54 tells us the following about $\mathcal{G}$.

Theorem 15.55: Completeness of $\mathcal{G}$

If $g_1, g_2, \ldots$ are in $\mathcal{G}$ (Definition 15.13 on page 680) and $\lim_{n \to \infty} \int g_n T$ exists for each test function T, then there is a unique element g in $\mathcal{G}$ such that

$$\lim_{n \to \infty} g_n = g.$$

❑

This result assures us that in $\mathcal{G}$ any limiting process (in the sense of the left side of Definition 15.47 on page 744) if it converges, will not *take us outside*[1] *of* $\mathcal{G}$.

12 Further Extensions

In this chapter we have only provided an introduction to what is now a vast subject. There are other useful ways to define the class of test functions — one possibility being to choose semi-infinite intervals instead of bounded ones as the support. Generalized Fourier and Laplace transforms may also be of interest. And certainly, generalized functions of several variables are needed, for application to the subject of partial differential equations.

Probably the easiest way to learn more about this area is to start with the monographs on this subject which are listed in the bibliography.

---

1. The real number system has this completeness property (see the discussion surround Figure A1-2 on page 759) i.e., that the reals also were constructed so that the limiting operations did not lead to numbers outside the system. The rational numbers do not possess this property.

Epilogue

As with any book, the amount of material that can be covered is limited. In this epilogue our aim is to indicate the most important aspects of the analysis for modeling which were hardly touched on, but are necessary for a reasonable command of this area. They are the subjects of

- Integral equations — used as an alternative to local description in many cases. In fact, existence theorems for most differential equations are usually proved by converting them to integral equations.
- Partial differential equations, used as local descriptions of phenomena involving space and time. Especially important here are the heat, wave and electrical and magnetic field equations.
- Nonlinear differential equations, which often yield more accurate, if more difficult to treat, descriptions of many phenomena, and are sometimes unavoidable.
- Probability, including stochastic processes, which is used to describe phenomena where there is variability not explicitly accounted for in the model.
- Statistics, which is concerned with determining appropriate probability models based on observed data.
- Measure theory, which provides a more satisfactory basis than the one provided here for the theory of integration.

The bibliography presents some references to look into these areas.

One final point — we made no attempt to present the most efficient algorithms for carrying out numerical computations. Our aim was to provide understanding of the basic tools used in such algorithms. Books such as Golub and Van Loan [9] provide that type of information for those who need it.

Appendix 1

The Real Numbers

For the practical purposes of computation, finite decimals, which are only a small part of the set of rational numbers, are more than adequate; since computers get along with only a small part of the set of finite decimals. But in trying to discover useful results, restriction to even the rationals would prove extremely clumsy because there would be no numbers representing most square roots, and no number representing the ratio perimeter/diameter of a circle. With only the rationals we would be hobbled in our search for new results. The basic idea behind the extension of our system of numbers to include these *reals*, is to use **pairs of rationals** to *zero in* on any point on a line. For example, to narrow down on $\sqrt{2}$, we use the infinite decimal

$$1.414213562373095....$$

by which we mean the sequence of intervals

$I_1 = (1, 2)$ which *pins down* $\sqrt{2}$ to the numbers between 1 and 2 (because $1^2 < 2$ and $2 < 2^2$)

$I_2 = (1.4, 1.5)$ which pins down $\sqrt{2}$ to the numbers between 1.4 and 1.5 (because $1.4^2 < 2$ and $2 < 1.5^2$)

$I_3 = (1.41, 1.42)$, etc.

The basic properties of these intervals are

- Each interval is part of the previous one.
- As n grows large, the length of the n-th intervals gets close to 0. (In the decimal case, the length of the n-th interval above is $1/10^{n-1}$; in the binary case used in computers, the length would be $1/2^{n-1}$.)

A sequence of intervals with the properties above is called a **nested sequence of intervals**. Each nested sequence determines a single point on a line, and for us, a **real number**.

One problem: a single point can be specified by lots of different nested sequences — like 1.999... and 2.000... both determine the same point (the real number 2). So, we need a means of deciding when two nested sequences (which you can think of as infinite decimals) stand for the same number. (We had this same type of problem for rationals, where two different fractions could represent the same rational number, e.g., 1/2 and 3/6.) It's simple.

Two nested sequences represent the same number if each interval from one of the sequences overlaps with all intervals from the other sequence.

Denote two such sequences by R and S. If one of the intervals in R lies completely to the left of any of the intervals of S, then $R < S$, as illustrated in Figure A1-1.

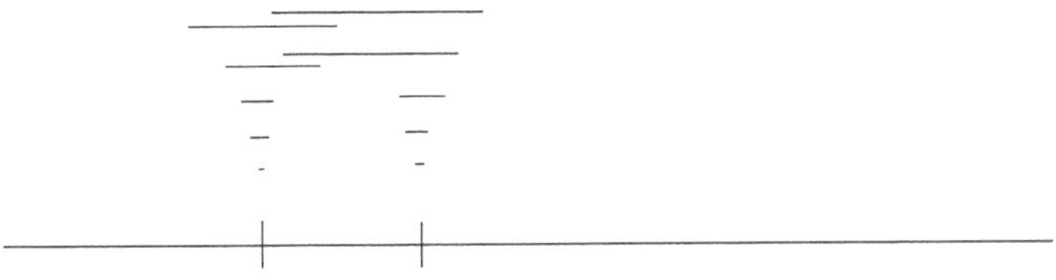

Figure A1-1

If the intervals from one of the sequences always overlap with the intervals from the other sequence, then the two nested sequences must zero down on the same point, and therefore stand for the same real number.

The old rules for inequality still hold — i.e.,

$$\text{if } R < S \text{ and } S < T, \text{ then } R < T, \text{ etc.}$$

We need to know how to add and multiply real numbers. For the positive reals (those representing points to the right of 0, if the n-th interval S_n of S is written $S_n = (S_{nL}, S_{nR})$, that of T as $T_n = (T_{nL}, T_{nR})$, then the n-th interval of the sum

$$S + T \text{ is } (S_{nL} + T_{nL}, S_{nR} + T_{nR})$$

and of the product

$$ST \text{ is } (S_{nL}T_{nL}, S_{nR}T_{nR}).$$

A real R represents a rational r if r is in every interval of R.

It's known that nested sequences can represent numbers such as $\sqrt{2}$, which aren't rational.[1] If we now form ***nested sequences of reals***, it's easy to

1. See *Modern University Calculus*, by Bell, Blum, Lewis and Rosenblatt, [1]

see that we don't create any additional numbers — because given any nested sequence of reals, we can create a nested sequence of rational intervals, equal to it. Just make the corresponding rational intervals a trifle bigger, as illustrated in Figure A1-2 (the intervals with rational end-points indicated as darker).

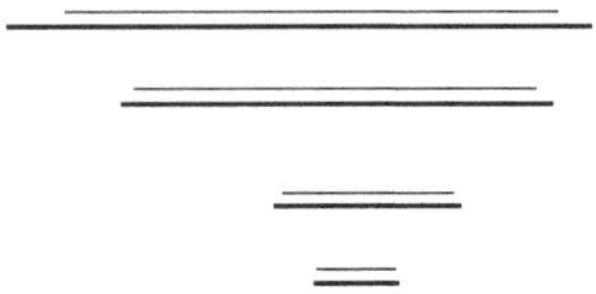

Figure A1-2

You might want to think about just how you would accomplish this. (Hint: Each nested sequence of reals has each of its end-points determined by a nested sequence of rational intervals. As the rational end-point associated with one of the *left* real end-points, use one of the *left* rationals used to define this *left* real end-point — going far down enough in this sequence, so that this rational is getting very close to the real left end when you get far down the real sequence.)

Because repeating this *zeroing in* process which created the reals from the rationals doesn't enlarge the reals, we say that the reals are **complete**. Completeness will be an important property of other mathematical structures which arise from a similar extension (e.g., the set of Lebesgue integrals, and the set of generalized functions).

In many situations we will solve a problem using a process of bisection, to settle down on a number with some property. This could have been done to find $\sqrt{2}$, as follows:

$$1 < \sqrt{2} < 2$$

$$1 < \sqrt{2} < 1.5$$

$$1.25 < \sqrt{2} < 1.5$$

$$\vdots$$

and then showing that the real number we constructed has the property we wanted — in this illustration, calling this number R, that $R^2 = 2$.

The real numbers have other properties which are important for deriving essential results, which we introduce in the main body of the book as needed.

Appendix 2

Inverse Functions

This appendix is devoted to furnishing some precise proofs of a few intuitively reasonable results which seemed a bit harder to nail down than we expected. Recall Definition 2.8 on page 37 or inverse functions. Let f be a function (Definition 1.1 on page 2). The function g is said to be the **inverse function of f** if

- $dmn\ g = rng\ f$ ($dmn\ g$ stands for the domain of g, $rng\ f$ stands for the range of f)
- for each x in $dmn\, f$, $\quad g(f(x)) = x$.

It seems almost obvious that f is also the inverse function of g — i.e., that for each y in $dmn\ g$

$$f(g(y)) = y.$$

And this isn't hard to establish, as follows. Choose any y in $dmn\ g$. Because $dmn\ g = rng\ f$, $y = f(x)$ for at least one x in $dmn\, f$.

Then

$$f(g(y)) = f(g(f(x))) = f(x) = y,$$

establishing the asserted result.

Now lots of functions have inverse functions, but one class of real-valued functions of a real variable is particularly important in this regard — the strictly monotone functions (Definition 2.9 on page 37). We want to show that every member of the class of strictly monotone functions has an inverse function. We'll only treat the monotone increasing case (the decreasing case being handled in the same way). If f is monotone increasing, suppose y is in $rng\, f$. Then for at least one x in $dmn\, f$,

$$y = f(x). \qquad \textbf{\textit{A2.1}}$$

But there can only be one such x satisfying equation A2.1, because if $x \neq x^*$, either

$$x < x^* \text{ implying } f(x) < f(x^*)$$

or

$$x > x^* \text{ implying } f(x) > f(x^*).$$

Define g with $dmn\ g = rng\, f$ by the equation

$$g(y) = x.$$

Then since $g(y) = g(f(x))$, we have $g(f(x)) = x$, proving our assertion.

Notice that the inverse function g of a strictly increasing real-valued function of a real variable is itself strictly increasing.

It's worth noting that the heart of a function having an inverse function is the property of being *one-to-one* (see Definition 12.18 on page 460). Strictly monotone functions are one-to-one, and it is this property which makes them invertible (having an inverse). If f is a one-to-one function, then (by definition) for all pairs x, x^* in *dmn f*,

$$\text{if } x \neq x^* \text{ then } f(x) \neq f(x^*).$$

To define the inverse function, g of f, let y be in *rng f*. Then $y = f(x)$ for at least one x in *dmn f*, and this x is unique for any given y (since if $x^* \neq x$, then $f(x^*) \neq f(x)$. Hence define g by the equation

$$g(y) = x \quad \text{where } y = f(x).$$

Then

$$g(f(x)) = g(y) = x.$$

Also, as before in the monotone case, f is also the inverse of g. (Same arguments as before; if y is in *dmn g* then $y = f(x)$ for some unique x in *dmn f*, and

$$f(g(y)) = f(g(f(x))) = f(x) = y.$$

Clearly if f isn't one-to-one, it cannot have an inverse function.

We summarize the results established here in the following theorem.

Theorem A2.1: Inverses and one to one functions

The function f (Definition 1.1 on page 2) has an inverse function g (i.e., a function satisfying $g(f(x)) = x$ for all x in *dmn* f) if and only if f is *one to one* (i.e., for all x, x^* in *dmn f*, if $x \neq x^*$ then $f(x) \neq f(x^*)$).

If f is one to one with inverse function g, then g is one to one with inverse function f.

If f is a strictly monotone real-valued function of a real variable (Definition 2.9 on page 37) it has an inverse function which is of the same type as f (increasing if f is increasing, decreasing if f is decreasing).

❑

Appendix 3

Riemann Integration

For completeness, we now present an *outline* of a more consistent theoretical development of Riemann integration,[1] one more suited to establishing mathematical results than that given in Definition 4.2 on page 102; after which, we will show that this integral is the same as given by Definition 4.2. What we are presenting here is mainly for those few who are unsatisfied when there are almost no clues to what is necessary for proper establishment of important, but intuitively reasonable results.

We are given a real valued function, f, whose domain (inputs) includes the closed interval $[a, b]$, where $a < b$, and want to determine

- when it is reasonable to define $\int_a^b f$,

and if so,

- how to do it in a manner simplifying theoretical development.

Thinking of an integral as an area, we subdivide the interval $[a; b]$ into sub-intervals

$$I_1 = [a_0;a_1], I_2 = [a_1;a_2],...,I_J = [a_{J-1};a_J],$$

where $a_0 = a$ and $a_J = b$; i.e., we partition the interval $[a; b]$ (recall Definition 4.1 on page 101). For reasons that will soon become evident, we allow these intervals to have different lengths.

We won't even try to define $\int_a^b f$ if f is not bounded on $[a;b]$ (see paragraph b on page 63 for what is meant by a *bounded* function).

For each $j = 1,...,J$ let

- m_j be the *greatest lower bound* of f on I_j,
  i.e., $f(x) \geq m_j$ for all x in I_j
  and if $m > m_j$ then for at least one x in I_j, we have $f(x) < m$.
- M_j be the *least upper bound*[2] of f on I_j,
  i.e., $f(x) \leq M_j$ for all x in I_j
  and if $M < M_j$ then for at least one x in I_j, we have $f(x) > M$.
- l_j be the length, $a_j - a_{j-1}$, of I_j.

Notice that if f is not bounded, at least one of the above bounds will not be

---

1. A detailed development is provided in *Modern University Calculus* [1].
2. We can construct least upper bounds and greatest lower bounds via bisection, using essentially the same technique as in the discussion starting with equation 3.16 on page 76.

defined, and then we would be unable to proceed with the definition we are developing.

Now, if f has anything resembling a signed area over $[a, b]$, it should lie between the *lower* and *upper sums,*

$$L = \sum_{j=1}^{J} l_j m_j \quad \text{and} \quad U = \sum_{j=1}^{J} l_j M_j. \qquad \textbf{\textit{A3.1}}$$

Lower and upper sums are illustrated in Figure A3-1 and Figure A3-2. Area below the x-axis counts as negative. In these figures it would appear that the signed area between the graph of f and the x-axis lies between the lower sum, L and the upper sum, U.

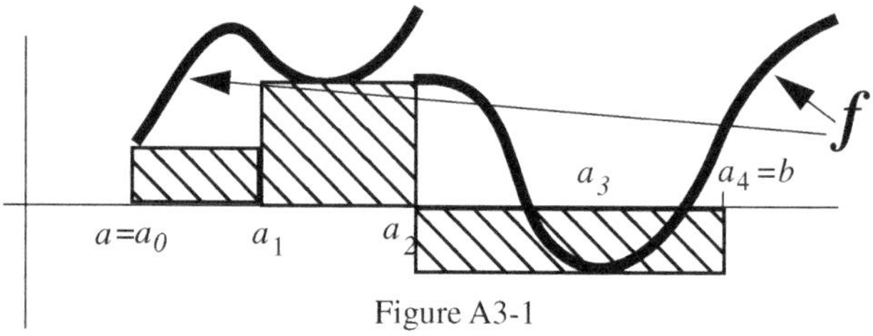

Figure A3-1

Lower Sum

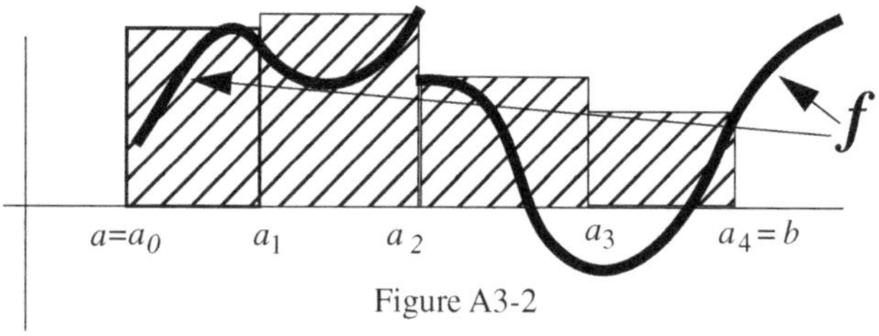

Figure A3-2

Upper Sum

In general, it seems reasonable to insist that the integral only be defined under the circumstances that when $[a;b]$ is *cut up finely enough,* the upper and lower sums for f should get *close — sandwiching* the integral. To make this easier to verify (when it happens), we establish the result that ***no lower sum can ever exceed any upper sum.*** Specifically, we can show that if L^* and U^{**} are ***any*** lower and upper sums for f corresponding to ***any*** two partitions of $[a;b]$, then

$$L^* \le U^{**}. \qquad \textbf{\textit{A3.2}}$$

This is established in three steps

- Prove that $L \leq U$ for lower and upper sums based on the same partition. This is easy, following essentially from the basic rules governing inequalities — namely,

  $$\text{if} \quad m < M \quad \text{and} \quad l > 0, \quad \text{then} \quad l\,m < l\,M$$

  and

  $$\text{if} \quad A < B \;\;\&\;\; C < D, \quad \text{then} \quad A + C < B + D.$$

- Show that if a partition is *refined* by adding in more subdivision points, then the *lower sum cannot decrease*, and the *upper sum cannot increase*. In fact, you need only refine by appending a single additional point, because any refinement can be obtained by adding points one at a time. If you prove that the assertion holds for such a simple refinement, then you're home free — because any refinement can be obtained by repeatedly adding a single point; for each such addition the lower sums can't decrease, and the upper ones can't increase. Investigating what happens when a single point is added to the partition is quite easy, as illustrated for upper sums in Figure A3-3. The old upper sum contribution is the area of the largest rectangle, while the new contribution is the crosshatched area in Figure A3-3. Here, as asserted, the upper sum has not increased (in fact it decreased) under refinement.

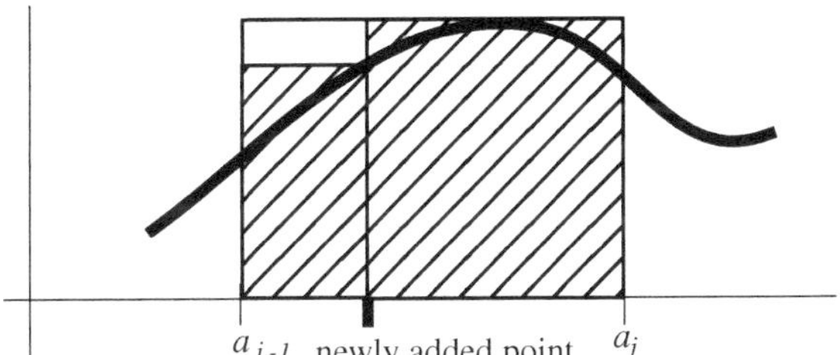

Figure A3-3

- Finally, refine the partitions used to define L^* and U^{**} by combining them — designating the lower and upper sums based on this combined partition by L and U.

Then, using all of these results, we have

$$L^* \leq L \leq U \leq U^{**}$$

from which inequality 3.2 follows.

Now we're close to being able to show that if, by suitably choosing the partition, we can make $U - L$ at least as close a nonzero distance to 0 as we desire, then a reasonable definition of $\int_a^b f$ can be made; choose a sequence, $P_1, P_2, \ldots$

of partitions of $[a, b]$ with corresponding lower and upper sums $(L_1, U_1), (L_2, U_2), \ldots$ with $U_n - L_n$ approaching 0 as n becomes large. To better see what is going on, we may toss out any partitions P_n for which $L_n < L_{n-1}$ or $U_n > U_{n-1}$. (We will still be left with a sequence of partitions having the originally specified property that the differences between upper and lower sums are approaching 0.) This situation is illustrated in Figure A3-4.

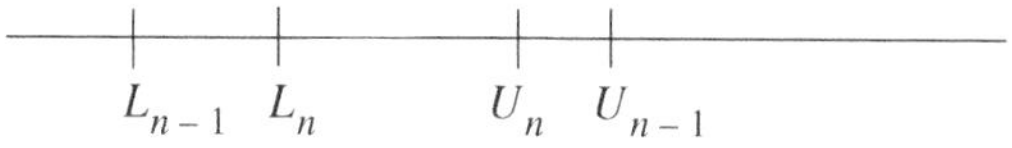

Figure A3-4

We see that there is precisely one point in all of the intervals $[L_n, U_n]$. We state this result formally as Theorem A3.1.

Theorem A3.1: Definition and existence of the Riemann integral $\int_a^b f$

Suppose $[a. b]$ can be partitioned so that the nonnegative difference $U - L$ defined by some partition can be made close to 0 — i.e., given any positive value, d, we can make $U - L < d$ by suitably choosing the partition of $[a; b]$. Then there is precisely one value, called the *Riemann integral* of f on $[a; b]$ and denoted by

$$\int_a^b f \qquad \text{or}^1 \qquad \int_a^b f(x)dx,$$

satisfying

$$L \leq \int_a^b f \leq U \qquad \textbf{A3.3}$$

for all L, U.

Each estimate

$$\frac{1}{2}(U + L) \quad \text{of} \quad \int_a^b f \qquad \textbf{A3.4}$$

is within

$$\frac{1}{2}(U - L) \quad \text{of} \quad \int_a^b f. \qquad \textbf{A3.5}$$

❑

We use the upper-sum lower-sum approach because of the simplicity of dealing with only two numbers for each partition. The price paid for this simplicity is that we had to introduce least upper bounds and greatest lower bounds; in practical work we usually don't compute the upper or lower sums, because finding them can be a formidable task. Almost all of the numerical estimates of integrals are

1. In the alternate integral symbol, the variable x is called a *dummy variable of integration.* Any symbol not conflicting with previously used ones may be used as the dummy variable of integration; this alternate notation is useful for functions specified by their formulas.

Riemann sums of one kind or another. We examine some of them in Section 6 of Chapter 4.

In order to connect the two approaches that we have been pursuing for dealing with integrals, we need to establish the following.

Theorem A3.2: Equivalence of *Riemann sum* and *upper/lower sum* approaches

- If the integral in the Riemann sum sense exists, then it exists in the upper/lower sum sense, and they are equal.
- If the integral in the upper/lower sum sense exists, so does the Riemann sum defined integral, and they are equal.

❑

The first part is easy, since (in theory) we can always find Riemann sums close to the upper and lower sums for any partition. But for small norm partitions, all Riemann sums are close together when the Riemann sum integral exists, so that in this situation for small norm partitions, the upper and lower sums must be close together; from which we see that the integral must be defined in the upper/lower sum sense.

Now to prove the second part, ***we need to show*** that so long as we can make $U - L$ small by suitably partitioning $[a; b]$, then if $P_1, P_2....$ is any sequence of partitions whose norms are getting small, any associated sequence of Riemann sums must be approaching a limit, namely, the *upper/lower sum integral.*

To do this, first choose a partition P so that $U - L$ is small — thus L and U are both close to the upper/lower sum defined integral (which must lie between them). Now we use the boundedness of f — i.e., $|f(x)| \le B$ for all x in $[a; b]$. Choose n so large that the total length l_n of the intervals from P_n which *sloppily* overlap the partitioning points of P is small, as illustrated in Figure A3-5.

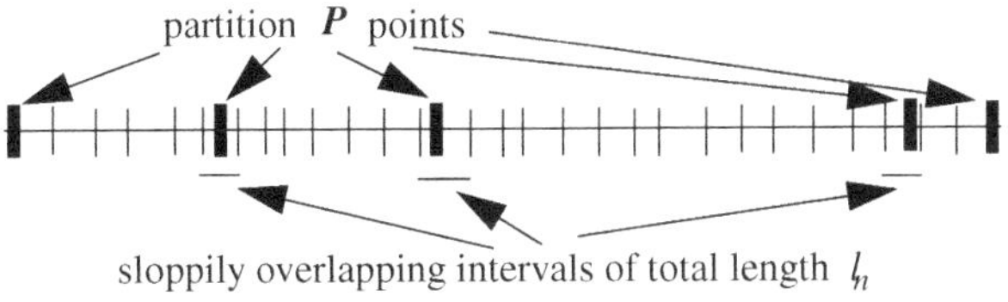

Figure A3-5

Specifically, make l_n ***small*** enough so that $2Bl_n$ is ***small***.

Now combine the partitions P and P_n into a partition we will call Q, whose upper and lower sums are denoted by U_Q and L_Q. Because Q is a refinement of P, we have $L \le L_Q \le U_Q \le U$. Any Riemann sum associated with partition Q lies between L_Q and U_Q. Any original Riemann sum, R_n, associated with partition P_n can't differ by more than $2Bl_n$ from a Riemann sum, R_Q, associated with Q, which is obtained by modifying R_n only on the *sloppily overlapping intervals.*

So R_n is within $2Bl_n$ of being between L and U. Hence it is close to L and U, and thus to the upper/lower sum integral. This establishes the connection we needed, proving Theorem A3.2.

With the results just established, we can now justify Theorem 4.4 on page 104. We need only note that because all Riemann sums for partition P_n are within d_n of each other, with d_n getting small for large n, for P_n the lower and upper sums, L_n, U_n are within d_n of each other. Hence every Riemann sum, R_n, associated with partition P_n lies between L_n and U_n. Applying Theorem A3.1 (on page 766) we know that $\int_a^b f$ must exist, with inequalities 3.3 through 3.5 holding for $L = L_n$ and $U = U_n$. This situation is illustrated in Figure A3-6, from

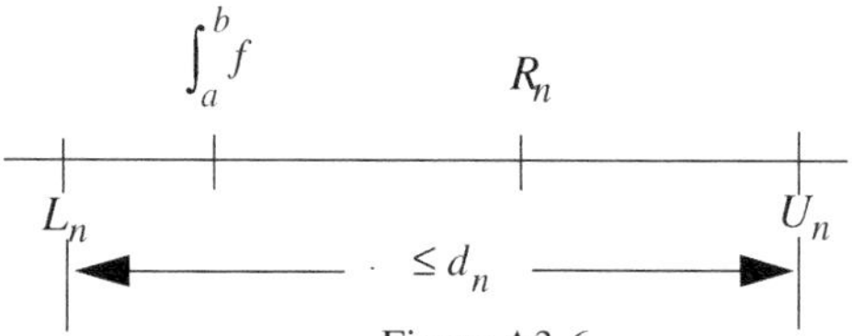

Figure A3-6

which it is evident that

$$\left|\int_a^b f - R_n\right| \leq d_n,$$

proving Theorem 4.4 on page 104.

We'll postpone further illustration of these results to nonoptional sections of Chapter 4 and conclude this appendix with a brief look at integration (and certain other important properties) of continuous functions

A function can be continuous at each point of its domain, but not even be bounded — for example, $f(x) = 1/x$ for $x > 0$, assumes arbitrarily large values as x gets close to 0. However, if the domain of the continuous function f is a **closed bounded** interval (one which includes its endpoints), then f is bounded, (see paragraph b on page 63). The boundedness of such continuous functions is not adequate for easy establishment of their integrability. For this, another property of a continuous function with closed bounded interval domain is appropriate; this is the property of ***uniform continuity***. To see how this property arises, let us re-examine the continuous function $f(x) = 1/x$, $x > 0$. The continuity of f follows, because for each $x > 0$, by restricting the nonzero value of h to be *close enough* to 0, we can keep $1/(x+h)$ as near to $1/x$ (but not equal to $1/x$) as desired. ***But***, the h values needed to do this are required to be considerably smaller for x values near 0 than for large x's; e.g.,

$$\text{for } x = .01, \quad \frac{1}{x} - \frac{1}{x - 0.005} = 100 - 200 = -100,$$

$$\text{while for } x = 1, \quad \frac{1}{x} - \frac{1}{x - 0.005} = 1 - \frac{1}{1 - 0.005} \cong 1 - (1 + 0.005) = -0.005$$

That is, the restrictions on h to restrict the output changes are not ***uniform*** (the same) for all xs. On this account we say that $f(x) = 1/x$ for $x > 0$ ***fails to be uniformly continuous.*** To be more positive in our approach, we give the following definition.

Definition A3.3: Uniform continuity

The real valued function, f, of a real variable, is said to be *uniformly continuous* if each positive **output change restriction**, $\varepsilon > 0$, ***can be achieved by*** an **input change restriction**, $\delta > 0$, dependent on ε, but ***not dependent on x.*** More quantitatively, we can ensure

$$|f(x+h) - f(x)| \le \varepsilon$$

by restricting $$|h| \le \delta,$$

where δ may depend on ε, but not on x.

Even more explicitly, we may write that f is *uniformly continuous* if there is a positive ***function*** δ of a single positive real variable, such that for each $\varepsilon > 0$,

$$\text{whenever } |h| \le \delta(\varepsilon) \text{ is satisfied, we have } |f(x+h) - f(x)| \le \varepsilon. \qquad \textbf{\textit{A3.6}}$$

If f has a bounded derivative, f' say, $|f'(x)| \le M$ for all x, then from the *mean value theorem*, Theorem 3.1 on page 62, we have

$$|f(x+h) - f(x)| \le |hf'(x^*)| \quad (\text{where } x^* \text{ is between } x \text{ and } x+h)$$

$$\le |hM|, \text{ for all } x.$$

This shows that such a function, f, is uniformly continuous; since by choosing $\delta(\varepsilon) = \varepsilon/M$ (which, as can be seen, is independent of x), it follows that:

$$\text{when } |h| \le \delta(\varepsilon) \text{ then } |f(x+h) - f(x)| \le \left|\frac{\varepsilon}{M} M\right| = \varepsilon.$$

The asserted uniform continuity now follows from condition A3.6 above.

For advanced theoretical purposes, a stronger result, which also extends to higher dimensions, is often needed. This is given in the following theorem.

Theorem A3.4: Uniform continuity of continuous function

Let f be a real valued function which is continuous and whose domain includes the closed interval, $[a; b]$, $a < b$. Then f is uniformly continuous on $[a; b]$ — i.e., there is a positive real valued function, δ, of a single positive real variable, such that for each $x; x + h$ in $[a; b]$ and each $\varepsilon > 0$,

$$\text{if } |h| \le \delta(\varepsilon), \text{ then } |f(x+h) - f(x)| \le \varepsilon.$$

❑

The proof is by contradiction — i.e., to determine that if f failed to be uniformly continuous on $[a;b]$, then there would be at least one x in $[a;b]$ at which f would fail to be continuous; since the theorem is concerned solely with functions f which are continuous at each x in $[a;b]$, it would follow that it could not be *dealing* with any functions which are not uniformly continuous on $[a;b]$, establishing the desired result, providing we can establish the above-specified contradiction.The essence of establishing the desired contradiction lies in properly phrasing the denial of uniform continuity, a somewhat delicate process,[1] which we now present.

We want to see how to deny[2] (assert the falsity of) the statement

For each $\varepsilon > 0$ *there is a* value[3] $\delta > 0$ *such that for all* $x, x+h$ *in* $[a;b]$,

$$\text{if } |h| \le \delta \text{ then } |f(x+h) - f(x)| \le \varepsilon. \qquad \textbf{A3.7}$$

This statement is of the form:

For each $\varepsilon > 0$ there is a value $\delta > 0$ such that statement A3.7 is true.

Its denial is a statement of the form:

There is at least one $\varepsilon > 0$ for which ***it is false that***

there is a value $\delta > 0$ such that statement A3.7 is true.

i.e., There is at least one $\varepsilon > 0$ such that for all $\delta > 0$, statement A3.7 is false.

Applying this to statement A3.7 yields that the denial of A3.7 is the statement:

There is at least one $\varepsilon > 0$ *such that for all* $\delta > 0$ *it is false that whenever* $|h| \le \delta$ *and* x *and* $x + h$ *are in* $[a;b]$, *we have* $|f(x+h) - f(x)| \le \varepsilon$. **A3.8**

It is evident that we can rewrite statement A3.8 as

There is at least one $\varepsilon > 0$ *such that for all* $\delta > 0$
there are some x, $x + h$ *in* $[a;b]$ *with* $|h| \le \delta$ *such that*

$$|f(x+h) - f(x)| > \varepsilon. \qquad \textbf{A3.9}$$

The validity of this statement would imply that letting $\delta_i = 1/i$ for $i = 1,2,..$ there would be a sequence of pairs, (x_i, h_i) , $i = 1,2,...$ with

$$x_i \text{ and } h_i \text{ in } [a, b] \text{ and } |h_i| \le 1/i \text{ for which}$$
$$|f(x_i + h_i) - f(x_i)| > \varepsilon \text{ for all } i, \qquad \textbf{A3.10}$$

where ε is the positive value specified in condition A3.9.

We are now getting close to the end of the proof, because, using bisection (or *decisection*) on $[a;b]$, we could then always choose one of the two (or one of the ten) subintervals for which there are infinitely many i for which statement A3.10 is satisfied. This process would define a number, X, in $[a;b]$. We could then choose a subsequence $x_{i_1}, x_{i_2}, \ldots$ of the original sequence, $x_1, x_2, \ldots$ such that x_{i_j} is inside the *j-th* interval defining X *(just choose as* x_{i_j} any element of the sequence $x_1, x_2, \ldots$ in the *j-th* interval defining X, whose subscript is beyond any of the previously chosen subscripts; that is, make sure $i_j > i_{j-1}$ for $j = 2,3,\ldots$).

---

1. Which I got wrong the first time I tried to write this paragraph; following this failure, I broke the process into a sequence of pieces, each of which, I hope, is completely natural and understandable. It is not unusual to be forced into such an approach.
2. Some familiarity with formal mathematical logic makes this process easier.
3. Determined by ε, a, b, and the function f.

It is evident that then

$$\lim_{j \to \infty} x_{i_j} = X \text{ and, of course } \lim_{j \to \infty} (x_{i_j} + h_{i_j}) = X. \qquad \textbf{A3.11}$$

But if statement A3.10 holds, then for all i

$$|f(x_i + h_i) - f(x_i)| > \varepsilon \qquad \textbf{A3.12}$$

and because of the continuity of f at X (since X is in $[a; b]$) both $f(x_{i_j})$ and $f(x_{i_j} + h_{i_j})$ would approach $f(X)$ for j sufficiently large, which certainly would prohibit A3.12.

Since the assumption of continuity of f on $[a; b]$ but lack of uniform continuity have led to a contradiction (i.e., a false statement, namely, the assertion that inequality A3.12 holds for all i, and the impossibility that it holds for sufficiently large i), we must give up on the possibility of continuity of f on $[a; b]$ in the absence of its uniform continuity on $[a; b]$, proving, at last, Theorem A3.5.

We can now establish the integrability of functions which are continuous on a closed bounded interval, by first showing that uniformly continuous functions are integrable on closed bounded intervals.

Theorem A3.5: Integrability of uniformly continuous functions

Let f be a real valued function of a real variable which is uniformly continuous, (Definition A3.3 on page 769), on the closed bounded interval $[a; b]$, (Definition 3.11 on page 88). Then f is Riemann integrable on $[a; b]$.

❑

Proof: Let ε^* be an arbitrary positive number, and let $\varepsilon = \varepsilon^*/(b-a)$. By the uniform continuity of f (Definition A3.3 on page 769) δ can be chosen so that $|f(x+h) - f(x)| \le \varepsilon$ for $|h| \le \delta$ and x, $x+h$ in $[a; b]$. Then with only moderate effort, we can see that for any partition of $[a; b]$ whose norm (maximum interval length) doesn't exceed δ, it follows (recalling 3.1 on page 764) that

$$0 \le U - L = \sum_{j=1}^{J} (M_j - m_j) l_j \le \varepsilon \sum_{j=1}^{J} l_j = \varepsilon^*.$$

So $U - L$ can be made as small as desired by suitable choice of partitions for a function uniformly continuous on $[a; b]$. By Theorem A3.1 on page 766, f is integrable on $[a; b]$ as asserted.

Corollary to Theorem A3.5: Integrability of continuous functions

If $[a; b]$ is a closed bounded interval and f is continuous on $[a; b]$, then f is Riemann integrable on $[a; b]$.

❑

This result follows by applying Theorem A3.5 to the result of Theorem A3.4 on page 769.

Appendix 4

Curves and Arc Length

From the discussion of Section 7 of Chapter 5, it doesn't seem at all unreasonable to conjecture that *arcsin*(t) represents the radian measure of the *angle* (signed length of the arc of the unit circle) going from (1,0) to the point (*COS* (*arc sin*(t)), *SIN* (*arc sin* (t))). In order to verify this conjecture, we have to better pin down the concept of angle. To do this we will introduce the concepts of *curve* and *length of arc* (curve length).

Intuitively, a curve in two or three dimensions is just a set of points, $(x_1(t)\ x_2(t))$ or $(x_1(t), x_2(t), x_3(t))$ indexed by some variable, t, which could (but need not) represent time. If t does stand for time, then $(x_1(t), x_2(t))$ might represent the position in plane of some object at time t. Similarly, $(x_1(t), x_2(t), x_3(t))$ would represent the position of an object in three-dimensional space at time t. We illustrate this concept in Figure A4-1.

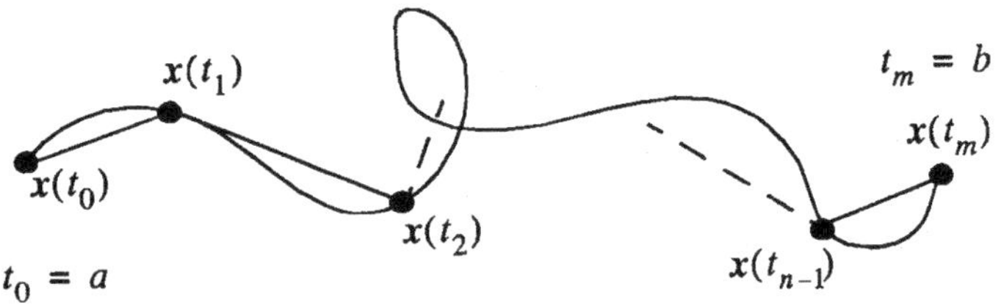

Figure A4-1

No understanding should be lost by giving the following general definition of curve.

Definition A4.1: Curve

A curve in n dimensions is a sequence $\boldsymbol{x} = (x_1, x_2, ..., x_n)$ of real valued functions of a real variable (Definition 2.1 on page 21) having a common interval domain (Definition 1.1 on page 2).

❑

As already mentioned, a curve may be viewed as an ordered set of points being traced out in n-dimensional space over time. Although our development for our particular application only requires two dimensions ($n = 2$) it should be evident that this restriction is not necessary for the algebraic manipulations we will soon perform.

It is convenient to introduce the concept of tangent line to a curve at this point.

Definition A4.2: Tangent to a curve

Suppose $\boldsymbol{x} = (x_1, x_2, ..., x_n)$ is a curve (Definition A4.1) whose derivative (Definition 2.5 on page 30)

$$\boldsymbol{x'} = (x_1', x_2', \ldots, x_n')$$

exists at $t = t_0$. The tangent to $\boldsymbol{x}$ at $t = t_0$ is the vector valued function whose value, $T_{t_0}(t)$ at t is given by

$$T_{t_0}(t) = \boldsymbol{x}(t_0) + (t - t_0)\boldsymbol{x}'(t_0)$$

$$= (x_1(t_0) + (t - t_0)x_1'(t_0), x_2(t_0) + (t - t_0)x_2'(t_0), \ldots, x_1(t_0) + (t - t_0)x_1'(t_0))$$

❑

We note that under our definition, the tangent to a two-dimensional curve is not restricted to being nonvertical, and if the range (see Definition 1.1 on page 2) of a curve intersects itself, there are likely to be two different tangent lines at the point of intersection, as can be seen in Figure A4-1.

We can approximate the length of the curve illustrated in Figure A4-1 from $t = a$ to $t = b$ by the sum of the lengths of the line segments connecting the indicated points in this figure.

Recall that the Pythagorean definition of length (Definition 8.6 on page 270) as motivated in this book's preface where we establish the Pythagorean theorem, is given by the following.

Definition A4.3: Pythagorean length, $\| \ \|$

The length, $\|\boldsymbol{a} - \boldsymbol{b}\|$, of the line segment connecting the points $\boldsymbol{a} = (a_1, a_2, ..., a_n)$ and $\boldsymbol{b} = (b_1, b_2,...,b_n)$ in n dimensions is given by

$$\|\boldsymbol{a} - \boldsymbol{b}\| = \sqrt{\sum_{i=0}^{n} (a_i - b_i)^2}.$$

The quantity $\|\boldsymbol{a} - \boldsymbol{b}\|$ is also referred to as the *distance* between the vectors or points $\boldsymbol{a}$ and $\boldsymbol{b}$.

❑

Applying Definition A4.3 to the hypotenuse line segment in Figure A4-2, we see that its length is

$$\sqrt{[x_2(t_{i+1}) - x_2(t_i)]^2 - [x_1(t_{i+1}) - x_1(t_i)]^2}.$$

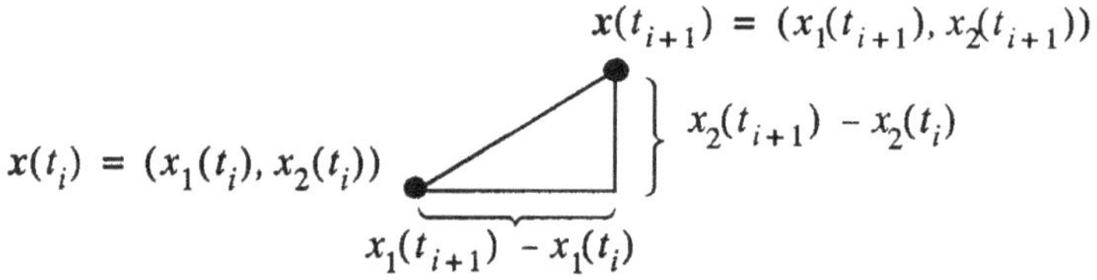

Figure A4-2

Hence the approximate length of the planar curve $\boldsymbol{x}$ from $t = a$ to $t = b$ is given by

$$\sum_{i=0}^{n-1} \sqrt{[x_2(t_{i+1}) - x_2(t_i)]^2 - [x_1(t_{i+1}) - x_1(t_i)]^2}.$$

By the mean value theorem (Theorem 3.1 on page 62) this expression may be written in the form

$$\sum_{i=0}^{n-1} \sqrt{[x_2'(\tau_i)(t_{i+1} - t_i)]^2 - [x_1'(\tau_i^*)(t_{i+1} - t_i)]^2},$$

where τ_i and τ_i^* are (unspecified) points between t_i and t_{i+1}. This last sum may be rewritten as

$$\sum_{i=0}^{n-1} \sqrt{[x_2'(\tau_i)]^2 - [x_1'(\tau_i^*)]^2}(t_{i+1} - t_i).$$

This last form of the approximation to the length of the planar curve $\boldsymbol{x}$ from $t = a$ to $t = b$ is also seen to be essentially (if these derivatives are continuous, Definition 2.6 on page 34) a Riemann sum approximation to the integral

$$\int_a^b \sqrt{[x_1'(\tau)]^2 + [x_2'(\tau)]^2}\, d\tau.$$

(see Definition 4.2 on page 102). This then leads to the definitions for curve length in two and in n dimensions.

Definition A4.4: Length of a curve (arc length)

If $\boldsymbol{x} = (x_1, x_2)$ is a two-dimensional curve whose domain is the interval $[a;b]$, then the length of $\boldsymbol{x}$ is defined to be

$$\int_a^b \sqrt{[x_1'(\tau)]^2 - [x_2'(\tau)]^2}\, d\tau$$

provided this integral exists. Similarly, if $\boldsymbol{x} = (x_1, x_2, \ldots, x_n)$ is an n-dimensional curve, whose domain is the interval [a;b], the following integral (provided it exists) is the length of $\boldsymbol{x}$:

$$\int_a^b \sqrt{\sum_{i=1}^{n} [x_1'(\tau)]^2}\, d\tau.$$

The extension of these definitions of a portion of a curve (i.e., to the part of $\boldsymbol{x}$ in which t is restricted to a subinterval of its domain) is simply to carry out the above integration over this subinterval.

❑

Example A4.1: Arc length for $x = (t^2/2, (2t + 4)^{3/2}/3)$

We want to find the length of this curve from $t = 2$ to $t = 7$. Using Definition A4.4, we need $x_1'(t) = t, x_2'(t) = \sqrt{2t+4}$, so that

$$\begin{aligned}\int_a^b \sqrt{[x_1'(\tau)]^2 + [x_2'(\tau)]^2}\, d\tau &= \int_2^7 \sqrt{\tau^2 + 2\tau + 4}\, d\tau \\ &= \int_2^7 \sqrt{(\tau+2)^2}\, d\tau \\ &= \int_2^7 (\tau + 2)\, d\tau \\ &= \left.\frac{(\tau+2)^2}{2}\right|_2^{\tau=7},\end{aligned}$$

is the desired arc length.

■

At this point I should point out that almost any real-life arc length problems that you will run into are not likely to have simple answers like the one above; most will require numerical integration. But if you feel compelled to go through some exercises which yield such simple answers, here is the way this example was constructed. Starting off with some integrand which is non-negative and whose integral allows a formula evaluation, such as $\tau + 2$ for τ going from 2 to 7 above, we write this integrand in the form

$$\sqrt{(\tau + 2)^2}$$

which we expand out to

$$\sqrt{\tau^2 + 2\tau + 4}.$$

Then choose some of the terms under the square root sign to represent $[x_1'(\tau)]^2$ and the remaining terms to represent $[x_2'(\tau)]^2$, these choices being made so that it is possible to determine formulas for $x_1(\tau)$ and $x_2(\tau)$ easily. In this case we chose $[x_1'(\tau)]^2 = \tau^2, [x_2'(\tau)]^2 = 2\tau + 4$ so that we were led to the original formulas for $x_1(\tau)$ and $x_2(\tau)$.

Exercises A4.2

For those with access to the *Mathematica* program, make up your own arc length exercises following the suggestions of Example A4.1. You can check the various stages of your computation as well as your answers using the *Mathematica* tools we introduced earlier (see the discussion just preceding Exercises 2.15 on page 54, as well as Exercise 4.20.5 on page 125).

Returning from the arc length definition diversion, if we look specifically at the curve x given for $-1 < t < 1$ by

$$x_1(t) = \sqrt{1 - t^2}, \quad x_2(t) = t,$$

we see that

- This range of this curve is the part of the unit circle centered at the origin, which is at (1,0) for $t = 0$, and goes 1/4 revolution counterclockwise as t goes from 0 to 1, and 1/4 revolution clockwise as t goes from 0 to -1, as illustrated in Figure A4-3.

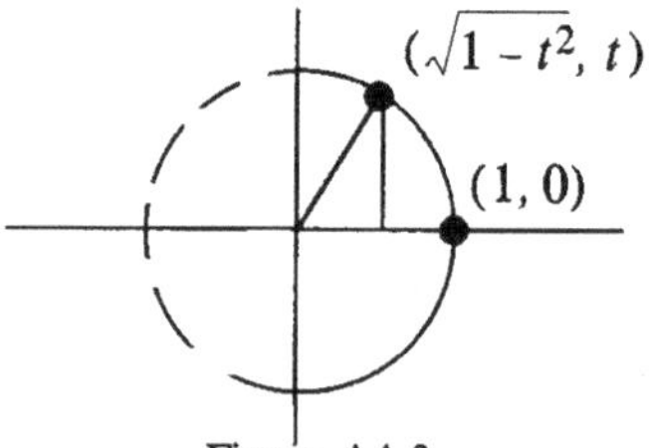

Figure A4-3

- It is not difficult to verify that

$$\sqrt{[x_1'(\tau)]^2 + [x_2'(\tau)]^2} = \sqrt{\frac{\tau^2}{1 - \tau^2} + 1} = \frac{1}{\sqrt{1 - \tau^2}}.$$

This shows that the value, *arcsin*(t), of the *arcsine* function given in Definition 5.13 on page 160, for $-1 < t < 1$, represents the arc length along the unit circle centered at the origin starting from the point (1,0) ending at the point $(\sqrt{1-t^2}, t)$. So $\vartheta = arc\ sin(t)$ is the signed *radian measure of the angle* (wedge) determined by the origin, (0,0), and the ordered pair of points $(1,0)$, $(\sqrt{1-t^2}, t)$. The quantity $t = SIN(\vartheta) = SIN(arc\ sin(t))$, being the *signed length* of the vertical leg of the *right triangle* in Figure A4-3 is the familiar ratio *opposite/hypotenuse* since the hypotenuse is of unit length.

The quantity $COS(\vartheta) = SIN(\vartheta) = \sqrt{1 - SIN^2(\vartheta)} = \sqrt{1-t^2}$ is the familiar *adjacent/hypotenuse*.

So we see that for any given angle (wedge) as illustrated in Figure A4-4.,

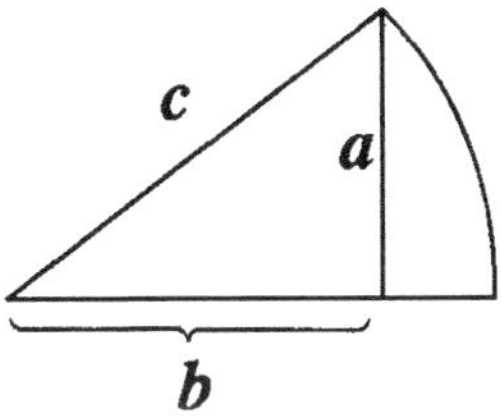

Figure A4-4

whose radian measure ϑ is defined by the equation

$$\vartheta = \frac{\text{signed length of indicated circular arc}}{c}$$

we find that

$$SIN(\vartheta) = \frac{opposite}{hypotenuse} = \frac{a}{c}, \quad COS(\vartheta) = \frac{adjacent}{hypotenuse} = \frac{b}{c},$$

where a is the *signed length*[1] of the vertical leg of the right triangle in Figure A4-4.

The introduction of the usual geometric interpretation of *SIN* and *COS* furnishes a reasonable way to extend the definitions of these functions beyond their original domains. First of all we notice that from examination of Figure A4-3, we expect

$$arc\ sin(t) = \int_0^t \frac{1}{\sqrt{1-\tau^2}} d\tau$$

1. In referring to the signed length, a is taken to be positive if the vertical leg is above the horizontal one. The circular arc is assigned a positive length if it arises from a counterclockwise rotation.

to have a finite limit as t approaches 1 (from below, of course), since this limit corresponds to the length of 1/4-th of the unit circle. We leave the proof of this assertion of Problem A4.3 on page 779. The length of the entire unit circle is denoted by 2π. (π itself is the arc length of any circle measured in diameters of this circle.) From a practical viewpoint, if we want to compute π with the knowledge we currently have, it's better to compute $\pi/4 = arc\ sin(1/\sqrt{2})$ (see the discussion preceding equation 5.59 on page 162).

Referring to our copy of Figure A4-3 given below as Figure A4-3, if we continue moving counterclockwise on the unit circle, having started at (1,0),

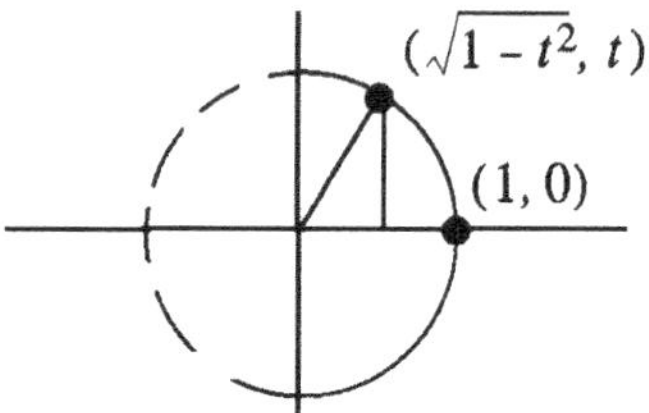

Figure A4-5

the vertical line of the right triangle continues to exist, remaining positive until a half circle has been traversed (covering an arc length of π), and then becoming negative until an arc length of 2π has been traversed. The symmetry of this process, together with the analogous reasoning if we had chosen to move clockwise starting from (1,0) provides the geometric motivation for the definition of the *sine* function that we are about to introduce. We will do this formally for reference purposes.

Definition A4.5: π

The number π represents the arc length of a half-cricle of radius 1 unit.[1]One formula for π which is amenable to digital computer evaluation is

$$\pi = 4\int_0^{1/\sqrt{2}} \frac{1}{\sqrt{1-\tau^2}} d\tau.$$

❑

Problem A4.3

Show that $\lim_{\substack{t \to 1 \\ t<1}} \int_0^t \frac{1}{\sqrt{1-\tau^2}} d\tau$ exists.

1. Arc length measured in the same units being used to measure the radius length.

Hints: $\dfrac{1}{\sqrt{1-\tau^2}} = \dfrac{1}{\sqrt{(1-\tau)(1+\tau)}} \le \dfrac{1}{\sqrt{1-\tau}}$ for $0 \le \tau < 1$. But using standard techniques the right side is easy to integrate. You'll also need equation 4.18 on page 107 of Theorem 4.5, and Theorem 3.8 on page 86, which you may need to modify a little to apply to the given situation.

Appendix 5

MLAB Dofiles for Numerical Solution of Differential Equations

MLAB Code in GLYCEROL1.do

```
TYPE "Please enter step length (minutes)"; SL = KREAD();
    TYPE "Please Enter bolus amount (mmol sugg 1.41 same as tracee in
    cmpt 1)"; BOL = KREAD();
    TYPE "Please enter k12 (cmpt2->cmpt1,sugg.085/min)";K12=KREAD();
    TYPE "Please enter k21 (cmpt1->cmpt2, sugg .26/min)";K21=KREAD();
    Type "Please enter k01 (cmpt1->outside,sugg .311/min)";K01=KREAD();
    TYPE "Please enter Ra (sugg .441 mmol C/min, same as tracee)";
    RA=KREAD();X1 = BOL; X2 = 0; MINUTES = 0;
    FOR J = 1:20000 { TYPE "Enter # to carry out the next iteration # times,
    0 terminates.";
    INP = KREAD(); If INP = 0 THEN DONE;
    FOR I = 1:INP {X1TEMP = X1 - SL*((K21 + K01)*X1 - K12*X2 - RA);
    X2=X2+SL*(K21*X1-K12*X2);X1=X1TEMP};
    MINUTES=MINUTES + INP*SL; TYPE MINUTES,X1,X2 };
```

MLAB Code changes to GLYCEROL1.do, in GLYCEROL2.do

The only change made in the code above is in the FOR I=1:INP loop, which is changed to:

```
FOR I = 1:INP {X1TEMP = X1 - SL*((K21 + K01)*X1 - K12*X2 - RA) +
(SL*SL/2)*((K21+K01)*((K21+ K01)*X1 - K12*X2 - RA)+ K12*(K21*X1
- K12*X2));
X2 = X2 + SL*(K21*X1 - K12*X2) + (SL*SL/2)*(K21*((K21 + K01)*X1 -
K12*X2 - RA) - K12*K21*X1 + K12*K12*X2);
 X1 = X1TEMP;};
```

In this code a semicolon (not a newline character) is used for termination.

MLAB Code in GLYCEROLMLI.do

```
METHOD=GEAR;JACSW=1;ERRFAC=.0000001;
    NAMESW=FALSE;
    Type "Please enter the distance between computed outputs (minutes)";
    SL = KREAD();
    Type "Please enter the final time";
```

```
FTIME = KREAD();
TYPE "Enter bolus amount (mmol sugg 1.41 same as tracee in cmpt 1)";
BOL = KREAD();
TYPE "Please enter k12 (cmpt 2 -> cmpt 1, sugg .085/min)";
K12 = KREAD();
TYPE "Please enter k21 (cmpt 1 -> cmpt 2, sugg .26/min)";
K21 = KREAD();
Type "Please enter k01 (cmpt 1 -> outside, sugg .311/min)";
K01 = KREAD();
TYPE "Please enter k02 (cmpt 2 -> outside, sugg 0/min)";
K02 = KREAD();
TYPE "Please enter Ra (sugg .441 mmol C/min, same as tracee)";
RA = KREAD();
FCT X1'T(T) = -(K21+K01)*X1(T)+K12*X2(T)+RA;INIT X1(0)=BOL;
FCT X2'T(T) = K21*X1(T) -(K12 + K02)*X2(T);INIT X2(0) = 0;
M = INTEGRATE(X1'T,X2'T,0:FTIME:SL);
S = "  X1          X1'          X2          X2'";
PRINT S,M IN Out;
```

Appendix 6

Newton's Method Computations

The implementation of the computations below, for minimization of the expression in equation 10.85 on page 389 will be in Splus, as follows.

The argument, (a, d, k) will be abbreviated in Splus as the vector variable

$$aa, \quad \text{where} \quad aa[1] = a, aa[2] = d, aa[3] = k,$$

The Splus function FF corresponds to the formula

$$a + de^{-kt_i}$$

while the Splus function g corresponds to the function g of equation 10.85 on page 389.

The Splus functions gi represent the partial derivatives $D_i g$, and the Splus functions gij correspond to the second partial derivatives $D_i D_j g$. The code for these functions was written in two steps — the first step being the computation of a typical term in the sum (using the notation giterms or gijterms) and the second one, adding up these terms by means of the Splus *sum* function. To accomplish the minimization we want to solve the system of equations

$$grad\ g(a, d, k) = \mathbf{0}. \qquad \textbf{\textit{A6.1}}$$

The vector function ***grad*** g is represented by the Splus matrix valued function DgT (T being the transpose). The matrix of gradients of the elements, gi, of ***grad*** g, is represented by the Splus matrix valued function DDgT.

The Splus function newtiter (for Newton Iterator) is the function whose input is the current Newton iterate (a column matrix) and whose output is the *next* Newton iterate.

The symbol tt represents the sequence of time points and y the sequence of corresponding data values in defining the various functions just introduced.

Before presenting the Splus functions that are used for this example, a few comments on some of the Splus notation are worth making.

- To combine a sequence of numbers, say q,r,s into a vector denoted by x, we type x _ c(q,r,s). The underscore, _, is the assignment operator.
- The Splus function, *matrix*, is the method used here to convert a vector into a matrix.
- Typing the name of a function, e.g., g1 <return>, yields the definition of the function; typing the function with appropriate arguments, e.g., g1(aa) <return> results in the evaluation of this function at these arguments.
- The value in the j-th coordinate of the vector aa is denoted by aa[j].
- The actual definition of a function, say g1, in Splus is of the form

```
g1_function(<arguments>){
    splus statements
                    }
```

What follows are the definitions that are required.

```
FF
function(aa) { return(aa[1] + aa[2] * exp( - aa[3] * tt)) }

g
function(aa) { return(sum(y - (FF(aa))^2)) }

g1terms
function(aa){return((-2 * ((y - (aa[1] + aa[2] * exp( - aa[3] * tt))))))}

g2terms
function(aa){ return((-2*((y - (aa[1] + aa[2]*exp( -aa[3]*tt)))*exp( -aa[3]*tt))))}

g3terms
function(aa){ return((2*((y - (aa[1] + aa[2]*exp( - aa[3]*tt)))*aa[2]*tt*exp(-aa[3]*tt))))}
g1
function(aa) { return(sum(g1term(aa))) }
```

```
g2
function(aa) { return(sum(g2term(aa))) }

g3
function(aa) { return(sum(g3term(aa))) }

DgT
function(aa){ return(matrix(c(g1(aa), g2(aa), g3(aa)), byrow = T, ncol = 1)) }

g11terms
function(aa) { return(2*exp(0*tt)) }

g12terms
 function(aa) { return(2*exp( -aa[3]*tt)) }

g13terms
function(aa) { return(-2*aa[2]*tt*exp( -aa[3]*tt)) }

g21terms
function(aa) { return(2*exp( -aa[3]*tt)) }

g22terms
function(aa) { return(2*exp(-2*aa[3]*tt)) }

g23terms
function(aa){return(2*(y-(aa[1] + 2*aa[2]*exp( -aa[3]*tt)))*tt*exp( -aa[3]* tt)) }

g31terms
function(aa) { return(-2*aa[2]*tt*exp( - aa[3]*tt)) }

g32terms
function(aa){ return(2*(y - (aa[1] + 2*aa[2]*exp( - aa[3]*tt)))*tt*exp( - aa[3]*tt))}

g33terms
function(aa){return(-2*(y-(aa[1]+2*aa[2]*exp(-aa[3]*tt)))*aa[2]*tt^2 *exp(-aa[3]*tt))}

g11
function(aa) { return(sum(g11terms(aa))) }

g12
function(aa) { return(sum(g12terms(aa))) }
g13
function(aa) { return(sum(g13terms(aa))) }
```

```
g21
function(aa) { return(sum(g21terms(aa))) }
g22
function(aa) { return(sum(g22terms(aa))) }

g23
function(aa) { return(sum(g23terms(aa))) }

g31
function(aa) { return(sum(g31terms(aa))) }

g32
function(aa) { return(sum(g32terms(aa))) }

g33
function(aa) { return(sum(g33terms(aa))) }

DDgT
function(aa){return(matrix(c(g11(aa), g12(aa), g13(aa), g21(aa), g22(aa), g23(aa),
                g31(aa), g32(aa), g33(aa)), byrow = T, ncol = 3))}

newtiter

function(aa){return(aa - solve(DDgT(aa)) %*% DgT(aa))}
```

The function solve, above, when called with a single argument consisting of an invertible square matrix, computes and returns the inverse of this matrix. Matrix multiplication is denoted by the symbol %*%.

```
sigma
function(aa){return(sqrt((1/(length(tt) - 3)) * sum((FF(aa) - y)^2)))}
```

The length function above returns the number of coordinates in its argument.

Several situations are illustrated in what follows. The first assumes noise-free data. The “true” parameters were put in aa, and are obtained by typing

aa <return> obtaining

```
> aa
[1] 2.0 -2.0 0.5
```

(The bracketed numbers just give the starting index of the coordinate values on the given line.)

The chosen time points are in tt, given by

> tt

[1] 0.0 0.5 1.0 1.5 2.0 2.5 3.0 3.5 4.0 4.5 5.0

For these choices, the value of y is FF(aa), which is called yperf

> y

[1] 0.0000000 0.4423984 0.7869387 1.0552669 1.2642411 1.4269904 1.5537397

[8] 1.6524521 1.7293294 1.7892016 1.8358300

> As a starting value for the Newton iterate to estimate aa, in this case we choose

> x0

[1] 1.80 -1.84 0.60

A few of the defined functions are evaluated at x0:

> g1terms(x0)

[1] -0.080000000 -0.011007920 0.006495818 -0.006710137 -0.036876935

[6] -0.075099802 -0.115779268 -0.155543882 -0.192500935 -0.225719389 -0.254876417

> g1(x0)

[1] -1.147619

> g12terms(x0)

[1] 2.00000000 1.48163644 1.09762327 0.81313932 0.60238842 0.44626032

[7] 0.33059778 0.24491286 0.18143591 0.13441103 0.09957414

> g12(x0)

[1] 7.431979

> DgT(x0)

[,1]

```
[1,] -1.1476189
[2,] -0.1986901
[3,] -0.7256147
> DDgT(x0)
         [,1]     [,2]      [,3]
[1,] 22.000000 7.431979 16.663096
[2,]  7.431979 4.426708  5.287081
[3,] 16.663096 5.287081 17.582633
> solve(DDgT(x0))
           [,1]        [,2]        [,3]
[1,]  0.2414535 -0.20609066 -0.16685472
[2,] -0.2060907  0.52840593  0.03642135
[3,] -0.1668547  0.03642135  0.20405098
> solve(DDgT(x0))%*%DgT(x0)
            [,1]
[1,] -0.11507622
[2,]  0.10509662
[3,]  0.03618667
> x0 - solve(DDgT(x0))%*%DgT(x0)
           [,1]
[1,]  1.9150762
[2,] -1.9450966
[3,]  0.5638133

> Now we run through a few Newton iterations starting with x0

> x1_newtiter(x0)
> x1
           [,1]
```

```
[1,]  1.9150762
[2,] -1.9450966
[3,]  0.5638133
```

> resid(x1) This is the current sum of squares, g(x1) , called the *residual*

```
[1] 0.005149181
> x2_newtiter(x1)
> x2
         [,1]
[1,]  1.9877148
[2,] -1.9902242
[3,]  0.5025259
> resid(x2)
[1] 0.000439786
> x3_newtiter(x2)
> x3
         [,1]
[1,]  1.9990944
[2,] -1.9994106
[3,]  0.5006244
> resid(x3)
[1] 4.998028e-07
> x4_newtiter(x3)
> x4

         [,1]
[1,]  1.9999987
[2,] -1.9999989
[3,]  0.5000001
> resid(x4)
```

```
[1] 7.076606e-12
> x5_newtiter(x4)
> x5
    [,1]
[1,]  2.0
[2,] -2.0
[3,]  0.5
> resid(x5)
[1] 6.853945e-23
> x6_newtiter(x5)
> x6
    [,1]
[1,]  2.0
[2,] -2.0
[3,]  0.5
> resid(x6)
[1] 4.930381e-31
>
```

> So the process has effectively converged to aa at the 5th iteration. We next try a new value of x0, print out iterations until reaching x35. To facilitate matters, we abbreviate the function newtiter simply as n.

```
> n_newtiter
> x0_c(.1,.1,-2)
> x1_n(x0)

> x1
            [,1]
[1,] -180.55819470
[2,]    0.09967731
[3,]   -1.91019697
```

```
> x2_n(x1)
> x2
            [,1]
[1,] -109.65856911
[2,]    0.09966835
[3,]   -1.81113007
> x3_n(x2)
> x3
           [,1]
[1,] -69.10604967
[2,]   0.09963496
[3,]  -1.71246641
> x10_n(n(n(n(n(n(n(x3)))))))
> x10
           [,1]
[1,] -2.12468849
[2,]  0.09817212
[3,] -1.03008121
> x15_n(n(n(n(n(x10)))))
> x15
           [,1]
[1,]  0.59179256
[2,]  0.08547458
[3,] -0.63767406
> x20_n(n(n(n(n(x15)))))
> x20
           [,1]
[1,] -0.09008429
[2,]  0.32017812
```

```
[3,] 1.04156306
> x25_newtiter(n(n(n(n(x20)))))
> x25
        [,1]
[1,] 1.9804461
[2,] -2.0371821
[3,] 0.6116159
> x30) _n(n(n(n(n(x20)))))
> x30
        [,1]
[1,] 1.9804461
[2,] -2.0371821
[3,] 0.6116159

> x35_n(n(n(n(n(x30)))))
> x35
    [,1]
[1,] 2.0
[2,] -2.0
[3,] 0.5
> resid(x35)
[1] 8.164709e-16
```

> So this takes quite a bit longer but still yields effective convergence to the correct vector value. But don't be fooled, because if instead we try

```
> x0_c( -.1, -.1,50)
> x0
[1] -0.1 -0.1 50.0
```

we obtain

```
> x1_newtiter(x0)
```

```
> x1
          [,1]
[1,]   1.353639
[2,]  -1.353639
[3,]  71.712736
> x5_n(n(n(n(x1))))
> x5
          [,1]
[1,]   1.353639
[2,]  -1.353639
[3,]  79.712736
> x15_n(n(n(n(n(n(n(n(n(n(x5))))))))))
> x15
          [,1]
[1,]   1.353639
[2,]  -1.353639
[3,]  99.712736
> FF(x15)
 [1] 0.000000 1.353639 1.353639 1.353639 1.353639 1.353639 1.353639
1.353639
 [9] 1.353639 1.353639 1.353639
> resid(x15)
[1] 1.946603
> x20_n(n(n(n(n(x15)))))

> x20
          [,1]
[1,]    1.353639
[2,]   -1.353639
[3,]  109.712736
```

```
> FF(x20)
 [1] 0.000000 1.353639 1.353639 1.353639 1.353639 1.353639 1.353639
1.353639
 [9] 1.353639 1.353639 1.353639
> resid(x20)
[1] 1.946603
```

It's pretty evident that k will increase to arbitrarily high values. Furthermore, we find

```
> DgT(x20)
          [,1]
[1,] 4.440892e-16
[2,] 2.734293e-24
[3,] 1.850623e-24
> DDgT(x20)
          [,1]         [,2]         [,3]
[1,] 2.200000e+01  2.000000e+00  2.030883e-24
[2,] 2.000000e+00  2.000000e+00 -1.367146e-24
[3,] 2.030883e-24 -1.367146e-24 -9.253113e-25
```

So we have effectively found x5 or x20 as a solution of equation 6.1 on page 783, but they clearly are not solutions we want — illustrating the danger of using Newton's method mindlessly. The matrix DDgT(x20) is reasonable — the second partial derivatives not involving the third coordinate, k, are nowhere near 0, but the values in these coordinates are not changing, so it doesn't matter. For third coordinate of x20, namely, k, this value is changing, but the effect on FF(x20) is negligible.

Next we make the problem being treated a bit more realistic by using *noisy data* — i.e., the perfect data to which random noise is added. The noise is chosen to have a Gaussian distribution with expectation 0 and standard deviation .22.

```
> noise
 [1]  0.195370 -0.283420  0.164895  0.045127  0.099031 -0.155640
-0.033570
 [8] -0.153650 -0.480590  0.362271  0.066895
> yperf
 [1] 0.0000000 0.4423984 0.7869387 1.0552669 1.2642411 1.4269904
1.5537397 1.6524521 1.7293294 1.7892016 1.8358300

> ynoisy_yperf+noise
> ynoisy
 [1] 0.1953700 0.1589784 0.9518337 1.1003939 1.3632721 1.2713504
1.5201697
 [8] 1.4988021 1.2487394 2.1514726 1.9027250
> y_ynoisy

> x0_c(1.8,-1.84,.6)
> x0
[1]  1.80  -1.84  0.60

> resid(x0)
[1] 0.6064414

> x1n_newtiter(x0)
> x1n
          [,1]
[1,]  1.9415963
[2,] -1.8679858
[3,]  0.4736086
> resid(x1n)
[1] 0.5599425

> x2n_newtiter(x1n)
> x2n
```

```
        [,1]
[1,]  2.0381248
[2,] -1.9471924
[3,]  0.4307136

> resid(x2n)
[1] 0.5484198

> x3n_newtiter(x2n)
> x3n
        [,1]
[1,]  2.1091776
[2,] -2.0010992
[3,]  0.3961791

> resid(x3n)
[1] 0.5460741

> x4n_newtiter(x3n)
> x4n
        [,1]
[1,]  2.1331390
[2,] -2.0204633
[3,]  0.3878273

> resid(x4n)
[1] 0.5457878
> x5n_newtiter(x4n)
> x5n
        [,1]
[1,]  2.1386948
[2,] -2.0248311
```

```
[3,] 0.3855702

> resid(x5n)
[1] 0.5457807

> x6n_newtiter(x5n)
> x6n
          [,1]
[1,]  2.1388367
[2,] -2.0249463
[3,]  0.3855219

> resid(x6n)
[1] 0.5457807

> x8n_newtiter(x7n)
> x8n
          [,1]
[1,]  2.1388369
[2,] -2.0249465
[3,]  0.3855218
```

So we seem to have effective convergence at the 8-th iteration with the noisy data. Notice that the noise error induced an error in the estimated constants, which should have been 2, -2, .5. In Figure A6-1 we plot the noisy data (diamonds) the *true* regression function (solid curve) and the estimated least squares regression function fitted to the noisy data (dashed curve). In the given range we see that the fitted curve approximates the *true* regression curve pretty well.

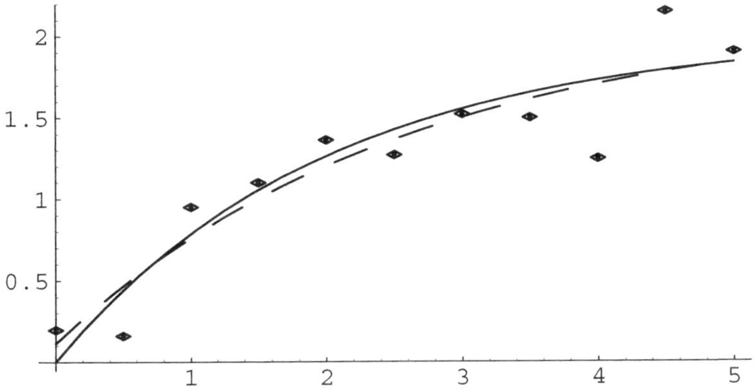

Figure A6-1

Appendix 7

Evaluation of volume of $A^{n \times n} I_{u_n}$

Definition 12.21 on page 466 of the determinant of an $n \times n$ matrix, A, is not the kind that permits immediate evaluation of $det(A)$. In order to accomplish this task, we have to introduce some ways to transform the matrix A, which

- don't affect its determinant, or affect it in a particularly simple way;
- after sufficiently many such transformations, yield a simple form whose determinant is immediately recognizable.

The ideas behind the type of transformations we are seeking are based on

a. the same type of matrix row operations used for matrix inversion (see Theorem 10.12 on page 339 and Example 10.7 on page 339) namely, adding a multiple of one row to another, and interchanging rows;

b. showing that the volume (Jordan content)

$$V(\mathbf{M}I_{u_n}) = V(I_{u_n}) = 1$$

where I_u is the n-dimensional unit interval (see Definition 12.1 on page 438, where $\boldsymbol{a}$ and $\boldsymbol{b}$ are n-dimensional vectors, $\boldsymbol{a}$ consisting of all 0s and $\boldsymbol{b}$ of all 1s) $\mathbf{M}$ is the $n \times n$ matrix obtained by carrying out the matrix operation of adding a multiple of one row to another, or obtained by interchanging two rows, on the $n \times n$ identity matrix (Definition 10.6 on page 334) and MI_u is the *image* of I_u under $\mathbf{M}$ (see Definition 12.20 on page 462);

c. extending this result to show that $V(\mathbf{M}A) = V(A)$ whenever A is an n-dimensional set with well-defined Jordan content (Definition 12.7 on page 442).

Before embarking on this proof, you should be curious about why the result is believable. That is, *why would you expect the volume of the image of the unit cube, $V(\mathbf{M}I_u)$ to be unchanged when the matrix* $\mathbf{M}$ *is of the above form?* Well, recall that this image in two dimensions is just the area of the indicated parallelogram, whose sides are the columns of **M.**

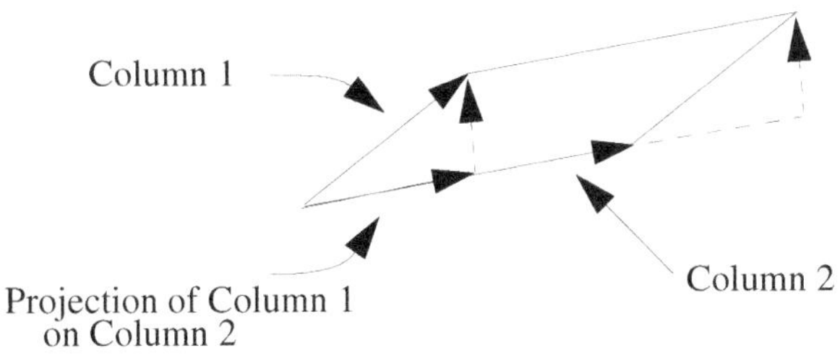

Figure A7-1

Subtracting this multiple of Column 2 from Column 1 changes the shape of the parallelogram — converting it to a rectangle, but, as you can see, doesn't change the area; it just shifts the triangular piece to the other side. Algebraically, it's because the total area is the sum of the areas formed by Column 2 with each of the legs of the triangle whose legs add up, vectorially, to Column 2. But since the parallelogram formed by the Projection and Column 2 has 0 area (due to the linear dependence of this projection and Column 2) only the component of Column 1 orthogonal to Column 2 contributes to the area. Although this may seem to represent only a special case, a little thought should convince you that the idea extends not only to all two-dimensional cases, but to any number of dimensions.

We'll work backwards, first showing that if step b above has been proved, then step c follows. To do this, we approximate A closely by a finite union (see Footnote 1 on page 453) of intervals

$$\left(\bigcup_{j=1}^{m} I_j\right) \cup \left(\bigcup_{k=1}^{q} B_k\right),$$

where the (interior intervals) I_j lie completely within the set A, and the (boundary intervals) B_k, include both points in A and points not in A (see Figure 12-3 on page 443 and the surrounding discussion). Because A is assumed to have a well-defined Jordan content (Definition 12.7 on page 442) the total volume of the boundary intervals can be made small, by keeping the partition norm (Definition 12.2 on page 439) which generated these intervals, sufficiently small. Now letting the symbol $\subseteq$ denote *set inclusion*,[1] because

$$\bigcup_{j=1}^{m} I_j \subseteq A \subseteq \left(\bigcup_{j=1}^{m} I_j\right) \cup \left(\bigcup_{k=1}^{q} B_k\right) \qquad \textbf{\textit{A7.1}}$$

it follows that

1. That is, $A \subseteq B$ means that each element of the set A is also an element of the set B.

$$\mathbf{M}\left(\bigcup_{j=1}^{m} I_j\right) \subseteq \mathbf{M}(A) \subseteq \boldsymbol{M}\left(\left(\bigcup_{j=1}^{m} I_j\right) \cup \left(\bigcup_{k=1}^{q} B_k\right)\right). \qquad \textbf{\textit{A7.2}}$$

By the invertibility of $\mathbf{M}$, which is easy to justify,[1] $\mathbf{M}$, considered as a function is one-one (Definition 12.18 on page 460). So, because the intervals I_j and B_k are non-overlapping, the images, $\mathbf{M}(I_j)$ and $\mathbf{M}(B_k)$, also are nonoverlapping. Hence, using the fact that $V(\ )$ denotes an integral, from Theorem 12.15 on page 453, we see that

$$V\left(\bigcup_{j=1}^{m} I_j\right) = \sum_{j=1}^{m} V(I_j) \text{ and } V\left(\left(\bigcup_{j=1}^{m} I_j\right) \cup \left(\bigcup_{k=1}^{q} B_k\right)\right) = \sum_{j=1}^{m} V(I_j) + \sum_{k=1}^{q} V(B_k) \qquad \textbf{\textit{A7.3}}$$

and

$$V\left(\mathbf{M}\left(\bigcup_{j=1}^{m} I_j\right)\right) = V\left(\bigcup_{j=1}^{m} \mathbf{M}(I_j)\right) = \sum_{j=1}^{m} V(\mathbf{M}(I_j)) = \sum_{j=1}^{m} V(I_j). \qquad \textbf{\textit{A7.4}}$$

Similarly

$$V\left(\boldsymbol{M}\left(\left(\bigcup_{j=1}^{m} I_j\right) \cup \left(\bigcup_{k=1}^{q} B_k\right)\right)\right) = \sum_{j=1}^{m} V(I_j) + \sum_{k=1}^{q} V(B_k). \qquad \textbf{\textit{A7.5}}$$

Applying the volume operator, V, to the elements expression A7.1 above yields

$$V\left(\bigcup_{j=1}^{m} I_j\right) \le V(A) \le V\left(\left(\bigcup_{j=1}^{m} I_j\right) \cup \left(\bigcup_{k=1}^{q} B_k\right)\right) \qquad \textbf{\textit{A7.6}}$$

while applying V to the elements of expressions A7.2 yields

$$V\left(\mathbf{M}\left(\bigcup_{j=1}^{m} I_j\right)\right) \le V(\mathbf{M}(A)) \le V\left(\mathbf{M}\left(\bigcup_{j=1}^{m} I_j\right) \cup \left(\bigcup_{k=1}^{q} B_k\right)\right). \qquad \textit{A7.7}$$

Now substituting from equations A7.4 and A7.5 into inequalities yields

$$\sum_{j=1}^{m} V(I_j) \le V(A) \le \sum_{j=1}^{m} V(I_j) + \sum_{k=1}^{q} V(B_k) \qquad \textbf{\textit{A7.8}}$$

and

$$\sum_{j=1}^{m} V(I_j) \le V(\mathbf{M}(A)) \le \sum_{j=1}^{m} V(I_j) + \sum_{k=1}^{q} V(B_k). \qquad \textbf{\textit{A7.9}}$$

---

1. Since the operation is reversed by ***subtracting*** the same multiple of the first row being considered from the second one.

Now, comparing inequalities A7.8 and A7.9, and using the fact that by choosing sufficiently small the norm of the partition giving rise to these sets makes $\sum_k V(B_k)$ arbitrarily small, shows that

$$V(\mathbf{M}(A)) = V(A), \qquad \textbf{\textit{A7.10}}$$

provided that we know $V(\mathbf{M}I_{u_n}) = V(I_{u_n}) = 1$. We'll save this step for last.

If we have shown that $V(\mathbf{M}(A)) = V(A)$, for $\mathbf{M}$ an elementary row ***addition*** operation matrix, and A any n-dimensional set with well-defined Jordan content, we can now go to work and reduce the matrix $\mathbf{C} = \mathbf{DH}(\boldsymbol{w})$ to diagonal form by a series of successive row operations of the type mentioned at the start of this appendix (you might want to look at Example 10.7 on page 339, but omitting multiplication of a row by a scalar, which may be needed to generate the identity from the original matrix, but is not needed to convert to diagonal form). In the diagonal form, only the diagonal elements are nonzero. Now note that each such operation can be carried out by multiplying on the left by the corresponding elementary row operation matrix — e.g., to add $2 \times$ row 3 to row 1 in the matrix

$$\mathbf{B} = \begin{bmatrix} a & b & c \\ d & e & f \\ g & h & i \end{bmatrix} \qquad \textbf{\textit{A7.11}}$$

multiply $\mathbf{B}$ on the left by

$$\mathbf{M} = \begin{bmatrix} 1 & 0 & 2 \\ 0 & 1 & 0 \\ 0 & 0 & 1 \end{bmatrix}. \qquad \textbf{\textit{A7.12}}$$

Exercises

1. Verify that multiplying $\mathbf{MB}$ from equations A7.11 and A7.12 does have the effect of adding $2 \times$ row 3 to row 1.
2. Generalize the result of the previous problem to handle all such situations.

So we have shown that we can find elementary row operation matrices, say $\mathbf{M}_1,...,\mathbf{M}_r$, such that

$$\mathbf{M}_1\mathbf{M}_2\cdots\mathbf{M}_r\mathbf{C} = \mathcal{D}, \qquad \textbf{\textit{A7.13}}$$

where $\mathcal{D}$ is a diagonal matrix, whose i-th diagonal element is d_{ii}. The volume, $V(\mathcal{D}(I_{u_n}))$, is easily seen to be given by

$$V(\mathcal{D}(I_{u_n})) = \prod_{i=1}^{n} |d_{ii}|. \qquad \textbf{\textit{A7.14}}$$

So there remains only the first step, to show that if **M** is a matrix obtained from the identity matrix via an elementary row operation, then

$$V(\mathbf{M}I_{u_n}) = V(I_{u_n}) = 1.$$

We will leave it to the reader to verify this if **M** represents a row interchange. As for showing this result when **M** corresponds to adding a multiple of one row to another, this is where we must perform an integration. For simplicity, we will only consider the situation of adding a multiple of row 2, $c \times$ row 2, to row 1. This really loses no generality, since row interchange can handle any other cases. So we want to prove that

$$V\left(\begin{bmatrix} 1 & c & 0 & 0 & . & 0 \\ 0 & 1 & 0 & . & . & 0 \\ 0 & 0 & 1 & 0 & . & 0 \\ . & . & . & . & . & . \\ 0 & . & . & . & . & . \\ 0 & 0 & . & . & 0 & 1 \end{bmatrix} I_{u_n}\right) = 1. \qquad \textbf{\textit{A7.15}}$$

This is where we need to represent volume (Jordan content) as an integral. Letting $\boldsymbol{m}_j$ denote the j-th column of **M**, we have

$$\mathbf{M}\boldsymbol{x} = \sum_{j=}^{n} \boldsymbol{m}_j x_j,$$

where

$$\boldsymbol{m}_2 = c\boldsymbol{e}_1 + \boldsymbol{e}_2 \quad \text{and} \quad \boldsymbol{m}_k = \boldsymbol{e}_k \quad \text{for} \quad k \neq 2,$$

with $\boldsymbol{e}_j$ being the standard n-dimensional j-th unit vector (Definition 8.2 on page 263) and a matrix times a vector given in Definition 10.1 on page 328. Hence, using the set notation introduced in Definition 12.16 on page 454, together with the definition of the image of a set under a function (Definition 12.20 on page 462) we find

$$\mathbf{M}(I_{u_n}) = \{y \text{ in } \mathcal{E}_n : \boldsymbol{y} = (x_1 + cx_2)\boldsymbol{e}_1 + \sum_{j=2}^{n} x_j \boldsymbol{e}_j \text{ for } 0 \leq x_i \leq 1, \text{ all } i = 1,\dots,n\}.$$

To determine $V(\mathbf{M}(I_{u_n}))$ we first write $\mathbf{M}(I_{u_n})$ in sequential form (again see Definition 12.16); here we let y_1 be the last coordinate determine. In this sequential form we find

$$\mathbf{M}(I_{u_n})=\{(y_1,y_2,...,y_n) \text{ in } \mathcal{E}_n : 0\le y_2\le 1; 0\le y_3\le 1;...;0\le y_n\le 1; cy_2\le y_1\le 1\},$$

because the limits for $y_2,...,y_n$, set first, are evident. But for each choice of y_2 and x_1, $y_1 = cy_2 + x_1$ — so as x_1 varies from 0 to 1 in the non-sequential form of $\mathbf{M}(I_{u_n})$ just preceding the sequential form above, for chosen y_2, y_1 varies from cy_2 to $cy_2 + 1$. Then since

$$V(\mathbf{M}(I_{u_n})) = \int_{\mathbf{M}(I_{u_n})} \mathbf{1}$$

(the right side being a multiple integral (see Definition 12.3 on page 440) where $\mathbf{1}$ is the constant function of n variables whose value is 1) from Theorem 12.17 on page 455, on evaluation of a multiple integral as an iterated integral, we have

$$V(\mathbf{M}(I_{u_n})) = \int_0^1 \cdots \int_0^1 \left[\int_{cy_2}^{cy_2+1} 1\, dy_1\right] dy_2 \cdots dy_n = 1,$$

establishing the asserted result.

Thus we have shown the following result.

Theorem A7.1: Volume of image of unit interval in $\mathcal{E}_n$

The volume of the image (Definition 12.20 on page 462) of the unit n-dimensional interval (Definition 12.1 on page 438) $\mathbf{A}(I_{u_n})$ under an $n \times n$ matrix, $\mathbf{A}$ is the magnitude (absolute value) of the determinant of $\mathbf{A}$, as defined in Definition 12.21 on page 466. Symbolically

$$V(\mathbf{A}(I_{u_n})) = \int_{\mathbf{A}(I_{u_n})} \mathbf{1} \qquad = |det\mathbf{A}|.$$

❑

The next result is needed for connection of the geometric meaning of determinants with the suitability of the associated matrix for coordinate changes.

Theorem A7.2: Determinants and linear dependence

If $\mathbf{A}$ is a real $n \times n$ matrix (Definition 10.1 on page 328) with determinant $det(\mathbf{A})$ (Definition 12.21 on page 466) then the columns of $\mathbf{A}$ are linearly dependent (Definition 9.7 on page 297) if and only if

$$det(\mathbf{A}) = 0.$$

❑

Proof: It isn't hard to see that the row operations applied to a matrix to convert it to the diagonal form needed for simple evaluation of its determinant do not alter the linear dependence/independence status of the set of its column vectors, (or its row vectors). But in the diagonal form, it is easy to see that the columns of this final matrix are linearly dependent if and only if at least one of the diagonal elements, d_{ii}, is 0, and this occurs if and only if the determinant,

$$\prod_{i=1}^{n} d_{ii} = 0,$$

proving the desired result.

Appendix 8

Determinant Column and Row Expansions

Most people meet determinants for the first time from the row or column expansion which will be presented here. This makes determinants somewhat more mysterious than they need be. Here determinants arose from the need to determine the volume of the image of the unit n-dimensional interval, in order to study change of variable in integration. If we then write down the form of the determinant for the 2×2 case, we find

$$det\begin{bmatrix} a & b \\ c & d \end{bmatrix} = ad - bc = (a, c) \cdot (d, -b).$$

With a bit more effort we also find for the 3×3 case that

$$det\begin{bmatrix} a & b & c \\ d & e & f \\ g & h & i \end{bmatrix} = (a, d, h) \cdot \left(\begin{bmatrix} e & f \\ h & i \end{bmatrix}, -\begin{bmatrix} b & c \\ h & i \end{bmatrix}, \begin{bmatrix} b & c \\ e & f \end{bmatrix} \right)$$

In both cases the first of the two vectors in the right-hand dot product is the first column of the matrix being examined. Since the determinant represents a volume (a *signed* volume, it turns out) it's sort of evident that the second vector must stand for a vector whose magnitude is the $n - 1$ dimensional volume of the figure generated by the remaining columns, and this vector must be orthogonal to the subspace generated by these columns, if the dot product is to generate the volume of the figure generated by all of the columns of the matrix. From this viewpoint, and examining the objects in the second vector of the dot product, you might very well guess the form of the column expansion.

The column and row expansions are not especially useful for actually evaluating determinants (the definition, Definition 12.21 on page 466, is really better for that purpose in the few cases where it is needed for change of variable in integration). However these expansions are essential for developing the theory of eigenvalues and eigenvectors of Chapter 13.

Theorem A8.1: Determinant row and column expansions

Let $\mathbf{A}$ be a given real $n \times n$ matrix (Definition 10.1 on page 328) with i, j element $a_{i,j}$. Let $\mathcal{A}_{i,j}$ denote the **submatrix** obtained from $\mathbf{A}$ by deletion of row i and column j, and let $\mathfrak{a}_{i,j} = det(\mathcal{A}_{i,j})$ — this determinant being called the ***i,j*-minor of A**.

Then

$$det(\mathbf{A}) = \sum_{i=1}^{n} (-1)^{i+1} a_{i,1} \mathcal{a}_{i,1} \quad \textit{1st column expansion}$$

$$det(\mathbf{A}) = \sum_{j=1}^{n} (-1)^{j+1} a_{1,j} \mathcal{a}_{1,j} \quad \textit{1st row expansion.}$$

❑

The proof of this theorem is inductive;[1] the inductive hypothesis being examined is that ***both row and column expansions are valid expression for the determinant when*** $n = k$. We then show that if the hypothesis holds for $n = k$

a. The interchange of rows changes the sign of the column expansion (but not its magnitude) when $n = k + 1$. This should come as no surprise, if you just look at

$$det\begin{bmatrix} a & b \\ c & d \end{bmatrix} \text{ and } det\begin{bmatrix} c & d \\ a & b \end{bmatrix}.$$

b. Adding a multiple of one row to another doesn't alter the value of the column expansion when $n = k + 1$.

(Similar statements hold for row expansions and column operations.)

With this shown, we can go through the elementary row operations on the column expansion of a $(k+1) \times (k+1)$ matrix, to obtain a diagonal matrix $\mathcal{D}$. The value of the column expansion of $\mathcal{D}$ is easily seen to be the product

$$\prod_{i=1}^{n} d_{i,i}$$

of the diagonal elements of $\mathcal{D}$. If proper track is kept of the number of row interchanges, you can determine whether the original column expansion of **A** is

$$\prod_{i=1}^{n} d_{i,i} \quad \text{or} \quad -\prod_{i=1}^{n} d_{i,i}.$$

From this the desired result follows. Thus if we establish steps a and b above, knowing that the determinant of every $k \times k$ matrix is given by the row and column expansion, we then know that this holds for every $(k+1) \times (k+1)$ matrix. (In proving a and b, we will observe the strange situation that the proof turns out much easier if we use row operations to

1. Considering any value $n = k$ for which the result holds ($n = 2,3$ are already known to be such k) and showing that from this *inductive hypothesis*, the result then holds for $n = k + 1$.

establish column expansion results and column operations to establish row expansion results.)

To establish a we will just consider the case of interchanging rows 1 and 2, (any other interchange treated almost identically). Write the column expansion as

$$a_{1,1}\mathcal{a}_{1,1} - a_{2,1}\mathcal{a}_{2,1} + \sum_{i=3}^{k+1} (-1)^{i+1} a_{i,1}\mathcal{a}_{i,1}. \qquad \textbf{\textit{A8.1}}$$

Under interchange of rows 1 and 2 the first two terms become $a_{2,1}\mathcal{a}_{2,1} - a_{1,1}\mathcal{a}_{1,1}$, while in the summation, since rows 1 and 2 are present in all the minors, we see from the inductive hypothesis that all minors change sign. Thus under interchange of rows 1 and 2 the column expansion becomes

$$a_{2,1}\mathcal{a}_{2,1} - a_{1,1}\mathcal{a}_{1,1} - \sum_{i=3}^{k+1} (-1)^{i+1} a_{i,1}\mathcal{a}_{i,1},$$

i.e., the column expansion changes sign under interchange of rows 1 and 2 in **A**.

To prove part b again we'll only consider the typical case of adding a multiple, $c \times$ row 2, to row 1. We again consider the column expansion as displayed in expression A8.1. The first term changes to $(a_{1,1} + ca_{2,1})\mathcal{a}_{1,1}$, because $\mathcal{a}_{1,1}$ isn't affected. In the second term, the factor $a_{2,1}$ is unaffected, but the first row of the submatrix $\mathcal{A}_{2,1}$ giving rise to the minor $\mathcal{a}_{2,1}$ is now $a_{1,2} + ca_{2,2} \quad a_{1,3} + ca_{2,3} \quad \cdots\cdots \quad a_{1,n} + ca_{2,n}$

so that this submatrix looks like

$$\begin{bmatrix} a_{1,2} + ca_{2,2} & a_{1,3} + ca_{2,3} & \cdots\cdots & a_{1,n} + ca_{2,n} \\ a_{3,2} & a_{3,3} & \cdots\cdots & a_{3,n} \\ a_{4,2} & a_{4,3} & \cdots\cdots & a_{4,n} \\ . & . & \cdots\cdots & \\ . & . & \cdots\cdots & \\ a_{n,2} & a_{n,3} & \cdots\cdots & a_{n,n} \end{bmatrix}.$$

Expressing the corresponding minor using the row expansion shows it to be

$$\mathcal{a}_{2,1} + c\mathcal{a}_{1,1}.$$

Thus the first two terms of the column expansion of $det(\mathbf{A})$ become

$$(a_{1,1} + c a_{2,1})\boldsymbol{a}_{1,1} - a_{2,1}(\boldsymbol{a}_{2,1} + c\boldsymbol{a}_{1,1}) = a_{1,1}\boldsymbol{a}_{1,1} - a_{2,1}\boldsymbol{a}_{2,1}.$$

Since in all other $\boldsymbol{a}_{i,1}$, $i > 2$, $c \times \text{row } 2$ is added to the first row, by the inductive hypothesis, none of these minors are changed, and it follows that the column expansion is unchanged, proving part b and establishing Theorem A8.1.

Appendix 9

Cauchy Integral Theorem Details

For the proof of the Cauchy integral theorem, several not-so-obvious results are needed. One particular such result, of a topological[1] nature, is the following.

Theorem A9.1: Regularity in a band

Let f be a function of a complex variable which is ***regular*** on a set A, i.e., ***regular at every point of the set*** A (Definition 12.39 on page 505) and let $\mathcal{C}$ be a subset of A which is the range (Definition 1.1 on page 2) of a continuous, piecewise smooth curve, $\mathcal{C}$ (Definition 12.40 on page 506) whose domain is the closed bounded real interval $[a;b]$ (Definition 3.11 on page 88). Then there is a positive number p, such that f is regular at each point z satisfying the condition $|z - z_{\mathcal{C}}| \le p$, for all $z_{\mathcal{C}}$ in $\mathcal{C}$; that is, f is regular in a *band of width p* around $\mathcal{C}$, as illustrated in Figure A9-1.

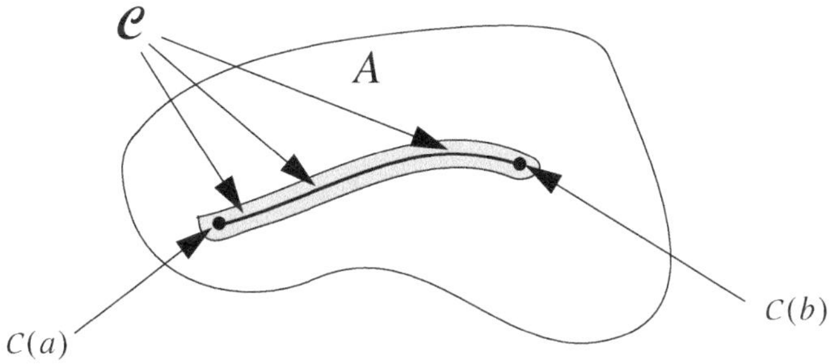

Figure A9-1

❑

We break the proof into several parts for pedagogical reasons.

Lemma A9.2: Closed range

Under the hypotheses of Theorem A9.1, the range, $\mathcal{C}$, of the curve C is a closed bounded set (Definition 10.43 on page 406) in the complex plane.[2]

❑

1. The word *topological* refers to the behavior of continuous functions.
2. Note that for our current purposes we are considering complex numbers to be vectors in two dimensions.

To establish this result, we note that it follows from Theorem A3.4 on page 769 that the curve C is uniformly continuous; hence, the range, $\mathcal{C}$, of C is bounded. To show that $\mathcal{C}$ is closed, consider any sequence $z_1, z_2, \ldots$ in $\mathcal{C}$ for which $\lim_{j \to \infty} z_j = L$, where we may assume $z_j = C(t_j)$, $a \le t_j \le b$.

We can choose a subsequence, $t_{j_1}, t_{j_2}, \ldots$ of the sequence $t_1, t_2, \ldots$, which converges, as follows: using successive bisections of $[a;b]$, each of which contains infinitely many elements from the original sequence $t_1, t_2, \ldots$ defines a value t^*. Because $[a;b]$ is closed, this value, t^*, belongs to $[a;b]$. From each of the intervals defining t^* choose one of the t_j, calling t_{j_k} the value from the k-th interval defining t^*, always making certain that the indices are increasing; that is, obtain $t_{j_1}, t_{j_2}, \ldots$ with $j_1 < j_2 < \cdots$. It is evident that

$$\lim_{k \to \infty} t_{j_k} = t^* \quad \text{and} \quad \lim_{k \to \infty} C(t_{j_k}) = L.$$

But since t^* belongs to $[a;b]$ and C is assumed continuous at all points of its domain, $[a;b]$, we have

$$C(t^*) = \lim_{k \to \infty} C(t_{j_k}) = L,$$

showing that L is in the range $\mathcal{C}$ of C. Thus the range $\mathcal{C}$ of the curve C is closed, proving Lemma A9.2.

Next

Lemma A9.3

If $\mathcal{C}$ is a closed bounded subset of the set A on which the complex valued function, f, of a complex variable is regular (i.e., f is regular at each point of $\mathcal{C}$) then there is a positive number p, such that f is regular at all z satisfying the condition that for at least one $z_{\mathcal{C}}$ in $\mathcal{C}$,

$$|z - z_{\mathcal{C}}| \le p. \qquad \textbf{\textit{A9.1}}$$

That is, f is regular at all z *not further than* p from the set $\mathcal{C}$.

❑

Proof: For each $z_{\mathcal{C}}$ in $\mathcal{C}$ there is a circle of positive radius centered at $z_{\mathcal{C}}$ at each of whose points f is regular (because, there is a positive radius circle centered at $z_{\mathcal{C}}$ at every point of which f is differentiable, from the definition of *regularity*, Definition 12.39 on page 505. The circle with half this radius will serve as the claimed one). Next we examine the corresponding circles with radii half the length of those just chosen. ***We claim that the union of a***

finite number of these circles will include every point of $\mathcal{C}$; the proof is, as it must be, by contradiction. For if an infinite number of these circles were required, then bisecting the two-dimensional interval containing $\mathcal{C}$ on each axis, would yield that at least one of the four subintervals just produced would require (the union of) an infinite number of these circles for coverage of the part of $\mathcal{C}$ in this subinterval. Because $\mathcal{C}$ is closed, continuing in this manner generates a point z^* in $\mathcal{C}$, in all of whose circular neighborhoods, an infinite number of the original (1/4 radius length) circles are required to cover the part of $\mathcal{C}$ in these neighborhoods. But after sufficiently many bisections, all of these circular neighborhoods lie within the one circle centered at z^* (from the original infinite set of 1/4 radius circles). So the claimed infinity of such circles that we are led to, as being still required for coverage, in fact aren't required. The assumption leading to this false statement was that an infinite number of the 1/4 radius circles we started with were needed to cover $\mathcal{C}$. This assumption must be abandoned. Only a finite number of these circles are needed to cover $\mathcal{C}$. Because of our conservatism, we can now double the length of the radii of the finite number of circles covering $\mathcal{C}$, and the function f is still regular at each point of the union of these enlarged circles. Since the union of the unenlarged circles still covers $\mathcal{C}$, we have a *buffer* around $\mathcal{C}$ whose size is the radius, p, of the smallest of the finite number of unenlarged circles. By including in this union the boundary of each of the finite number of enlarged circles, we see that this covering union is closed.

So we have constructed a closed bounded set containing $\mathcal{C}$, with the buffer band illustrated in Figure A9-1, completing the proof of Lemma A9.3, and hence of Theorem A9.1.

The next result needed to establish the Cauchy integral theorem is to show that the integral of f along a *close polygonal approximation to* $\mathcal{C}$, is *close to the integral of* f *over* $\mathcal{C}$ *itself.*

Theorem A9.4: Closeness of $\int_C f(z)dz$ and $\int_{\mathcal{P}} f(z)dz$

Let C be a continuous piecewise smooth curve (Definition 12.40 on page 506) whose domain is the closed bounded real interval $[a;b]$ (Definition 3.11 on page 88) and let f be a complex valued function of a complex variable which is regular (Definition 12.39 on page 505) at each point of the set A, which includes the range, $\mathcal{C}$, of the curve C. For simplicity assume that $a = 0$, and the length of the curve C from 0 to $t \le b$ (Definition A4.4 on page 775) is t (*so that* C *is parametrized by arc length*). Because C is piecewise smooth, it has a finite arc length. So we can always replace C by the equivalent curve parametrized by arc length, without affecting the integral (see Theorem 12.29 on page 483). Thus this assumption loses no generality.

Let P represent a partition of $[a;\, b]$ (Definition 4.1 on page 101) generated by t_j, where

$$0 = a = t_0 < t_1 < \cdots < t_m = b\ ,$$

then for norm P (the *maximum* value of $t_{j+1} - t_j$) sufficiently small

$$\int_{\mathcal{P}} f(z)dz \quad \text{exists and is } \textit{close} \text{ to} \quad \int_C f(z)dz,$$

where $\mathcal{P}$ represents the polygonal curve generated by the points $z_0 = C(t_0)\ , \ldots,\ z_n = C(t_m)$.

❑

To establish this result, first using Lemma A9.2 on page 811 and Lemma A9.3 on page 812 to choose $p > 0$ for which inequality A9.1 on page 812 is satisfied, and denote the *buffer* (the corresponding closed bounded set of z values) by $\mathcal{B}$. Note that $\mathcal{B}$ includes $\mathcal{C}$.

Since $\mathcal{B}$ is closed and bounded, and is included in the set A on which f is regular, f is uniformly continuous on $\mathcal{B}$ (see Theorem 3.14 on page 89 which shows that differentiability implies continuity, and Theorem A3.4 on page 769, both of which easily extend to complex valued functions of a complex variable).

Now let norm P be sufficiently small so that

- all approximating sums $\sum_j f(\zeta_j)\Delta z_j$ for $\int_C f(z)dz$ are *close* to $\int_C f(z)dz$, where $\zeta_j = C(\tau_j)$ for $t_{j-1} \le \tau_j < t_j$,

- norm $P \le p$,

- for all z, z' in $\mathcal{B}$, if $|z - z'| \le$ norm P then $|f(z) - f(z')|$ is uniformly small.[1]

---

1. Meaning that given any $\varepsilon > 0$ by choosing norm P small enough (how small, determined by ε) for all z, z' in $\mathcal{B}$ with $|z - z'| \le$ norm P, we have $|f(z) - f(z')| \le \varepsilon$. That this can be done follows from the uniform continuity of f in the buffer, $\mathcal{B}$.

To conclude the establishment of Theorem A9.4 we will show that under the above conditions, the approximating sums for the two integrals are close to each other and close to their respective integrals.

We can take as the approximating sum for $I = \int_C f(z)dz$ the sum $S = \sum_j f(C(t_{j-1}))\Delta z_j$, and we know that under the above conditions,

$$|S - I| \text{ is small.}$$

By our construction, all approximating sums $\sum_{j^*} f(\zeta'_{j^*})\Delta z'_{j^*}$ for $\int_{\mathcal{P}} f(z)dz$ satisfy

$$\left|\sum_{j^*} f(\zeta'_{j^*})\Delta z'_{j^*} - \sum_j f(C(t_{j-1}))\Delta z_j\right|$$

$$= \left|\sum_{j^*} [f(C(t_{j-1})) + \varepsilon_{j^*}]\Delta z_{j^*} - \sum_j f(C(t_{j-1}))\Delta z_j\right|,$$

where $|\varepsilon_{j^*}| \le \varepsilon$, and the value of $j - 1$ in the first sum corresponds to the segment of the polygon determined by j^*. But from this last fact we see that

$$\sum_{j^*} f(C(t_{j-1}))\Delta z_{j^*} = \sum_j f(C(t_{j-1}))\Delta z_j .$$

Hence

$$\mathcal{D} = \left|\sum_{j^*} f(\zeta'_{j^*})\Delta z'_{j^*} - \sum_j f(C(t_{j-1}))\Delta z_j\right|$$

$$\le \left|\sum_{j^*} \varepsilon_{j^*}\Delta z_{j^*}\right| \le \varepsilon\left|\sum_{j^*} \Delta z_{j^*}\right| \le = \varepsilon L_{\mathcal{P}},$$

where $L_{\mathcal{P}}$ is the length of the approximating polygon. But $L_{\mathcal{P}} \le L_C$, where L_C is the length of the curve C, which we know to be finite. Thus $\mathcal{D}$ is small when norm P is small enough. With their approximating sums seen to be close for small norm P, and the approximating sums easily made close to the integrals being approximated, we see that the integrals themselves must be close for norm P small enough, proving Theorem A9.4.

The last piece needed to nail down the Cauchy integral theorem is to show that the integrals over the two polygonal paths, P_t and $P_{t'}$ are equal when $|t - t'|$ is small enough (see equation 12.48 on page 512). We reproduce Figure 12-29, shown here as a slightly altered Figure A9-2, to recall the issue being discussed.

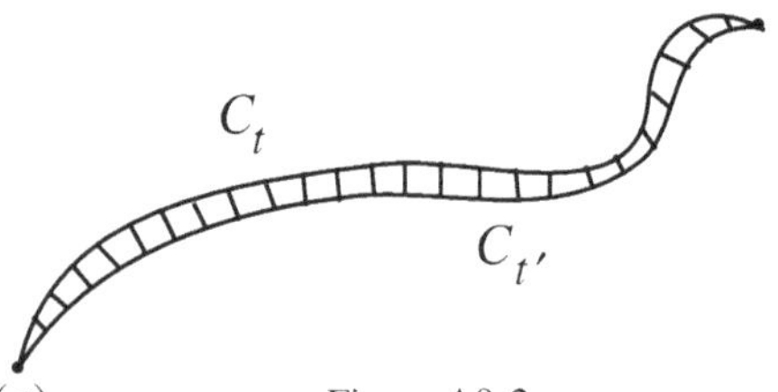

Figure A9-2

This is easy; using the condition $|t-t'| \le q(C_0, C_1, A, \varepsilon)$ from Definition 12.42 on page 509 together with the buffer around C_t and $C_{t'}$, we find that for norm $P_{t'}$ and norm P_t small enough, all of the displayed small rectangles and triangles (of which there are two of the latter) lie in A. Since a rectangle can be decomposed into two triangles, and from Lemma 12.44 on page 513, the integral over any triangle in A is 0, it follows that the integrals over the two indicated polygonal paths connecting $C(a)$ with $C(b)$ are the same, finally concluding our labors.

Glossary

$f, G, F \ldots$ functions 2

f_k k-th element of sequence f 3

Δf forward difference sequence 5

Δ forward difference operator 5

$\Delta^2 f$ second forward difference sequence 8

Δ^2 second forward difference operator 8

$\sum_{k=m}^{n}$ summation operator 11-12

$\cong$ approximately equal 17

$(x_0, f(x_0))$ point on graph of function f 22

$\frac{f(x_0+h)-f(x_0)}{h}$ difference quotient 29

$\lim_{x \to x_0} F(x)$ limit of $F(x)$ at the point x_0 30, 213

$f'(x_0)$ derivative of f at the point x_0 30

$Df(x_0)$ derivative of f at the point x_0 30

D derivative operator 30

$f^{(n)}(x_0)$ n-th derivative of f at the point x_0 30

$D^n f(x_0)$ n-th derivative of f at the point x_0 30

D derivative operator

D^n n-th derivative operator 30

δh small change in h 34

$H(f)$ composition of function H with function f 35

f^{-1} inverse function of function f 37

$invf$ inverse function of function f 37

$|x|$ magnitude of x 40, 200

I identity function 42

$b_k \ldots b_1 . a_1 a_2 \ldots$ infinite decimal 757

$<, \leq, >, \geq$ inequality signs 3-817

⌈,⌊ beginning and end of advanced section

❑ end of theorem statement

■ end of example

$\lim_{n \to \infty} x_n$ limit of sequence 85,213

$[a; b]$ closed interval 88

$(a; b)$ open interval 88

$\int$ integral sign 103

$\|x\|$ length of vector x 270

$\|x - y\|$ distance between vectors x and y 270

Df or $\textbf{\textit{grad}}\, f$ gradient vector of function f 272

D or $\textbf{\textit{grad}}$ gradient operator 273

$x \cdot y$ dot product of vectors x,y 272

$P_Y(X)$ projection of vector X on vector Y 276

sgn the signum function 283

$cos(X,Y)$ cosine between vectors X and Y 283

$[[v_1, v_2,...,v_j]]$ subspace spanned by vectors $v_1, v_2,...,v_j$ 296

$dim\ S$ dimension of subspace S 298

$\mathcal{R}_n$ real n-dimensional space 303

$\mathcal{E}_n$ Euclidean n-dimensional space 303

$\in$ membership symbol 303

$P_S(w)$ projection of vector w on subspace S 316

$$\mathbf{V} = [v_1, v_2,...,v_k] = \begin{bmatrix} v_{11} & v_{12} & \cdots & v_{1k} \\ v_{21} & v_{22} & \cdots & v_{2k} \\ . & . & \cdots & . \\ . & . & \cdots & . \\ v_{n1} & v_{n2} & \cdots & v_{nk} \end{bmatrix}$$

notation for matrices 327

$\mathbf{V}a$ matrix times vector 328

$\mathbf{WV}$ matrix product

$\mathbf{I}^{n \times n}$ the n by n identity matrix 334

$\mathbf{V}^{-1}$ inverse of matrix $\mathbf{V}$ 334

$\mathbf{W}^{\mathrm{T}}$ transpose of matrix $\mathbf{W}$ 342

$\|\mathbf{A}\|$ norm of matrix $\mathbf{A}$ 350

$\sum_{j = r \to \infty} \mathbf{M}_j$ matrix infinite series 352

$\sum_{j=r}^{\infty} \mathbf{M}_j$, sum of matrix infinite series 352

$\dfrac{(h \cdot D)^k f}{k!}$ 372

inf infimum 407

sup supremum 407

glb greatest lower bound = *inf* 407

lub least upper bound = *sup* 07

$S^{\perp}$ orthogonal complement of subspace S 412

$S \cap T$ intersection of subspaces S and T 419

$rnk\,\mathbf{A}$ rank of matrix $\mathbf{A}$ 430

$R(\boldsymbol{X}, P)$ multivariate Riemann sum 440

$\int_a^b f$, $\int_{[\boldsymbol{a};\boldsymbol{b}]} f(\boldsymbol{x})d\boldsymbol{x}$, $\int_{[\boldsymbol{a};\boldsymbol{b}]} f dV_n$ and

$\int_{[\boldsymbol{a};\boldsymbol{b}]} f(x_1,...,x_n)\, dx_1 \cdots dx_n$

multiple integral of f over an interval 440

$\int_A f dV_n$ multiple integral over set A 443

$\int_{a_n}^{b_n}\left[\int_{a_{n-1}}^{b_{n-1}} \cdots \left[\int_{a_1}^{b_1} f(x_1,...,x_n) dx_1\right] dx_2 \cdots dx_{n-1}\right] dx_n$ iterated integral 446

$\int_{-\infty}^{b} f(\tau)d\tau$, $\int_a^{\infty} f(\tau)d\tau$, $\int_{-\infty}^{\infty} f(\tau)d\tau$ improper Riemann integrals 449

$\int_{\mathcal{E}_n} f dV_n$ improper Riemann integral over Euclidean n-dimensional space 451

$\int_A f dV_n$ improper multiple integral over set A 452

$A \cup B$ union of sets A and B 453

$\bigcup_{t \text{ in } T} A_t$ union of sets A_t 453

$G(A)$ image of set A under function G 462

$\mathbf{M}A$ *image of set* A under matrix $\mathbf{M}$ 462

$det\,\mathbf{B}$ determinant of matrix $\mathbf{B}$ 467

$\prod_{i=1}^{n} d_i$ product of d_1 through d_n 467

J Jacobian determinant 467

$\boldsymbol{v} \times \boldsymbol{w}$ cross-product of the three-dimensional vectors $\boldsymbol{v}$ and $\boldsymbol{w}$ 473

$\int_{\boldsymbol{x}} \boldsymbol{F}(\boldsymbol{x})\cdot d\boldsymbol{x}$ and $\int_{\boldsymbol{x}} F_k(\boldsymbol{x})dx_k$ $k=1,...,n$ line integrals 481

$\int_C f(x, y)dx$, $\int_C g(x, y)dy$ line integrals 482

$\int_S \boldsymbol{v} \cdot d\mathcal{A}$ surface integral 490

$div\,\boldsymbol{v}$ divergence of vector field 498

$\Delta = \nabla^2$ Laplacian operator 502

$\int_C f(z)dz$ complex integral 506

$F \cdot G$ dot product for periodic complex valued functions 596

$F\delta^{(n)}$ product of generalized functions F and $\delta^{(n)}$

$\int FD^n\delta$ integral of the product of generalized functions D^nF and δ 711

$F\boldsymbol{d}_{a,\,b}^{(n)}$ product of the generalized function F and the scaled, shifted n-th derivative of Dirac's delta 712

$\int F\boldsymbol{d}_{a,\,b}^{(n)}$ integral of the above product 713

$\mathcal{L}$ generalized function differential operator related to $D + 2$, subject to homogeneous initial conditions 721

$\mathcal{P}(D)$ generalized function n-th degree polynomial differential operator, subject to homogeneous initial conditions 730

$\boldsymbol{f}$ notation for generalized function associated with continuous function f 743

$\mathcal{T}'$ space dual to the set of test functions $\mathcal{T}$ 752

$\mathcal{J}$ isomorphism 753

$a_{i,\,j}$ i-j minor of matrix $\mathbf{A}$ 807

Bibliography

[1] Bell, S., Blum, J.R., Lewis, J.V. and Rosenblatt, J., *Modern University Calculus*, Holden Day, San Francisco, 1996.

[2] Carson, E.R., Cobelli, C. and Finkelstein, L., *The Mathematical Modeling of Metabolic and Endocrine Systems*, John Wiley & Sons, New York, 1983.

[3] Churchill, R.V., *Fourier Series and Boundary Value Problems*, McGraw Hill, New York.

[4] Churchill, R.V., *Operational Mathematics*, 3rd Edition, McGraw Hill Book Company, New York, 1972.

[5] Courant, R. *Differential and Integral Calculus*, Interscience Publishers, New York, 1949.

[6] Gelfand, I.M. and Shilov, G.E., *Generalized Functions*, Volumes 1–5, Academic Press, London, 1964-1968.

[7] Godfrey, *Compartmental Models and Their Application*, Academic Press, London, 1983.

[8] Golomb, M. and Shanks, M., *Elements of Ordinary Differential Equations*, McGraw Hill Book Company, New York, 1965.

[9] Golub, G. and Van Loan, C.F., *Matrix Computations*, Johns Hopkins University Press, Baltimore, 1996.

[10] Knopp, K., *Theory and Application of Infinite Series*, Blackie & Son Limited, London, 1928.

[11] Luenberger, David G., *Introduction to Linear and Nonlinear Programming*, Addison-Wesley Publishing Company, Reading, 1973.

[12] Lukacs, E., *Characteristic Functions*, Hafner Publishing Company, New York, 1970.

[13] Nehari, Zeev, *Introduction to Complex Analysis*, Allyn and Bacon Incorporated, Boston, 1968.

[14] Ralston, A. *A First Course in Numerical Analysis*, McGraw-Hill Book Company, New York, 1965.

[15] Rosenblatt, J., "A more direct approach to compartmental modeling". Progress in Food and Nutrition Science, Vol. 12. pp 315–324, 1988.

[16] Smirnov, V.I., *Linear Algebra and Group Theory*, McGraw Hill Book Company, New York, 1961.

[17] Wilkinson, J.H., *The Algebraic Eigenvalue Problem*, Oxford University Press, Fair Lawn, 1965.

[18] Zygmund, A., *Trigonometrical Series*, Chelsea Publishing Company, New York, 1952.

9780367400101
T - #0289 - 071024 - C836 - 234/156/37 - PB - 9780367400101 - Gloss Lamination